Math Study Skills

Your overall success in mastering the material in your course depends on you. You must be **committed** to doing your best in this course. This commitment means dedicating the time needed to study math and to do your homework.

In order to succeed in math, you must know how to study it. The goal is to study math so that you understand and not just memorize it. The following tips and strategies will help you develop good study habits.

General Tips

ATTEND EVERY CLASS Be on time. If you must miss class, be sure to talk with your instructor or a classmate about what was covered.

MANAGE YOUR TIME School, work, family, and other commitments place a lot of demand on your time. To be successful, you must be able to devote time to study math every day. Writing out a weekly schedule that lists your class schedule, work schedule, and all other commitments with times that are not flexible will help you to determine when you can study. Use the companion resources that accompany the book, such as MyMathLab, so you can study online with tutoring help whenever you have the time.

Instructor Contact Information

Name:

Office Hours:

Office Location:

Phone Number:

E-mail Address:

Campus Tutoring Center

Location:

Hours:

DO NOT WAIT TO GET HELP If you are having difficulty, get help immediately. Since the material presented in class usually builds on previous material, it is very easy to fall behind. Ask your instructor if he or she is available during office hours or get help at the tutoring center on campus.

FORM A STUDY GROUP A study group provides an opportunity to discuss class material and homework problems. Find at least two other people in your class who are committed to being successful. Exchange contact information and plan to meet or work together regularly throughout the semester either in person or via e-mail, MyMathLab, or phone.

USE YOUR BOOK'S STUDY RESOURCES There are additional resources and support materials available with this book to help you succeed. See the list below and in the preface.

Notebook and Note Taking

Taking good notes and keeping a neat, well-organized notebook are important factors in being successful. If you do your homework online through MyMathLab, you should still keep a notebook to stay organized.

YOUR NOTEBOOK Use a loose-leaf binder divided into four sections: (1) notes, (2) homework, (3) graded tests (and quizzes), and (4) handouts. Or combine the resources in MyMathLab with the MyWorkBook with Chapter Summaries.

TAKING NOTES

❐ Copy all important information. Also, write all points that are not clear to you so that you can discuss them with your instructor, a tutor, or your study group.

❐ Write explanations of what you are doing in your own words next to each step of a practice problem.

❐ Listen carefully to what your instructor emphasizes and make note of it.

The following resources are available in MyMathLab, through your college bookstore, and at **www.pearsonhighered.com:**

- Student's Solutions Manual
- Video Resources with Chapter Test Prep Videos
- MyMathLab
- MyWorkBook with Chapter Summaries

Full descriptions are available in the preface.

Intermediate Algebra
through Applications

Third Edition

GEOFFREY AKST · SADIE BRAGG
Borough of Manhattan Community College, The City University of New York

PEARSON

Boston   Columbus   Indianapolis   New York   San Francisco   Upper Saddle River
Amsterdam   Cape Town   Dubai   London   Madrid   Milan   Munich   Paris   Montréal   Toronto
Delhi   Mexico City   São Paulo   Sydney   Hong Kong   Seoul   Singapore   Taipei   Tokyo

Editorial Director: Christine Hoag
Editor in Chief: Maureen O'Connor
Executive Content Editor: Kari Heen
Content Editor: Katie DePasquale
Editorial Assistant: Rachel Haskell
Senior Managing Editor: Karen Wernholm
Senior Production Supervisor: Ron Hampton
Senior Cover Designer: Barbara T. Atkinson
Senior Technology/Author Support Specialist: Joe Vetere
Text Design: Leslie Haimes
Composition: PreMediaGlobal
Image Research Manager: Rachel Youdelman
Media Producer: Aimee Thorne
Software Development: TestGen: Mary Durnwald; MathXL: Jozef Kubit
Executive Marketing Manager: Michelle Renda
Marketing Manager: Rachel Ross
Marketing Coordinator: Ashley Bryan
Procurement Manager/Boston: Evelyn Beaton
Procurement Specialist: Debbie Rossi
Media Procurement Specialist: Ginny Michaud
Cover Photo: Bamboo on white © Subotina Anna/Shutterstock

Library of Congress Cataloging-in-Publication Data
Akst, Geoffrey.
   Intermediate algebra through applications / Geoffrey Akst, Sadie Bragg.—3rd ed.
     p. cm
   Includes index.
   ISBN-13: 978-0-321-74671-9   ISBN-10: 0-321-74671-6 (student ed.: alk. paper)
    1. Mathematics—Textbooks.   I. Bragg, Sadie   II. Title.
   QA154.3.A52 2013
   512—dc22   2011005566

1 2 3 4 5 6 7 8 9 10—CRK—16 15 14 13 12

pearsonhighered.com

For
William W. Wilcox

Contents

Preface

FROM THE AUTHORS

Our goal in writing *Intermediate Algebra through Applications* was to create a text that would help students progress and succeed in their college developmental math course. Throughout, we emphasize an applied approach, which has two advantages. First of all, it can help students prepare to meet their future mathematical demands—across disciplines, in subsequent coursework, in everyday life, and on the job. Secondly, this approach can be motivating, convincing students that mathematics is worth learning and more than just a school subject.

We have attempted to make the text readable, with understandable explanations and exercises for honing skills. We have also put together a set of easy-to-grasp features, consistent across sections and chapters.

To address many of the issues raised by national professional organizations, including AMATYC, NCTM, and NADE, we have been careful to stress connections to other disciplines; to incorporate the appropriate use of technology; to integrate quantitative reasoning skills; to include problem sets that facilitate student writing, critical thinking, and collaborative activities; and to emphasize real-world data in examples and exercises.

Above all, we have tried to develop a flexible text that can meet the needs of students in both traditional and redesigned developmental courses.

This text is part of the *through Applications* series that includes the following:

WHAT'S NEW IN THE THIRD EDITION?

Say Why Exercises   New fill-in-the-blank problems, located at the beginning of each chapter review, providing practice in reasoning and communicating mathematical ideas (see page 432).

Updated and Expanded Section Exercise Sets   Additional practice in mastering skills.

Lengthening of Cumulative Review Exercise Sets   Twice as many review exercises in response to user demand (see pages 270–271).

Greater Emphasis on Learning Objectives   End-of-section exercises closely aligned with the learning objectives in order to encourage and facilitate review (see pages 173 and 183–191).

More Examples and Exercises Based on Real Data   Additional and more varied applied problems that are useful, realistic, and authentic (see page 364).

Parallel Paired Exercises   Odd/even pairs of problems that more closely reflect the same learning objective (see page 375–377).

Easy-to-Locate Features   Color borders added for back-of-book answer, glossary, and index pages.

Highlighting of Quantitative Literacy Skills   Additional exercises that provide practice in number sense, proportional reasoning, and the interpretation of tables and graphs (see page 27).

Increased Attention to Photos and Graphics   Carefully selected photos to make problems seem more realistic, and relevant graphics to better meet the needs of visual learners (see page 345).

Newly Expanded and Robust MyMathLab Coverage!   One of *every* problem type is now assignable in MyMathLab.

Now Two MyMathLab Course Options

1. **Standard MyMathLab** allows you to build *your* course *your* way, offering maximum flexibility and complete control over all aspects of assignment creation. Starting with a clean slate lets you choose the exact quantity and type of problems you want to include for your students. You can also select from prebuilt assignments to give you a starting point.
2. **Ready-to-Go MyMathLab** comes with assignments prebuilt and preassigned, reducing start-up time. You can always edit individual assignments, as needed, through the semester.

KEY FEATURES

Math Study Skills Foldout   A full-color foldout with tips on organization, test preparation, time management, and more (see inside front cover).

Chapter Openers   Extended real-world applications at the beginning of each chapter to motivate student interest and demonstrate how mathematics is used (see page 85).

Pretests and Posttests   Chapter tests, which are particularly useful in a self-paced, lab, or digital environment (see pages 86 and 347–348).

Section Objectives   Clearly stated learning objectives at the beginning of each section to identify topics to be covered (see page 544).

Side-by-Side Example/Practice Format   Distinctive side-by-side format that pairs each example with a corresponding practice exercise and gets students actively involved from the start (see page 243).

Tips   Helpful suggestions and cautions for avoiding mistakes (see page 624).

Journal Entries   Writing assignments in response to probing questions interspersed throughout the text (see page 107).

Calculator Inserts   Optional calculator and computer software instruction to solve section problems (see pages 281–282).

Cultural Notes   Glimpses of how mathematics has evolved across cultures and throughout history (see page 503).

For Extra Help   Boxes at the beginning of every section's exercise set that direct students to helpful resources (see page 69).

Mathematically Speaking Exercises   Vocabulary exercises in each section to help students understand and use standard mathematical terminology (see page 375).

Mixed Practice Exercises   Problems in synthesizing section material (see page 71).

Application Exercises   End-of-section problems to apply the topic at hand in a wide range of contexts (see pages 491–492).

Mindstretcher Exercises   Nonstandard section problems in critical thinking, mathematical reasoning, pattern recognition, historical connections, writing, and group work to deepen understanding and provide enrichment (see page 379).

Key Concepts and Skills Summary   With a focus on descriptions and examples, the main points of the chapter organized into a practical and comprehensive chart (see pages 428–431).

Chapter Review Exercises   Problems for reviewing chapter content, arranged by section (see pages 257–263).

Chapter Mixed Application Exercises   Practice in applying topics across the chapter (see pages 345–346).

Cumulative Review Exercises   Problems to maintain and build on the mathematical content covered in previous chapters (see pages 664–665).

Appendixes   Four brief appendixes—Table of Symbols, Introduction to Graphing Calculators, Determinants and Cramer's Rule, and Synthetic Division.

U.S. and Metric Unit Tables   Located opposite the inside back cover for quick reference.

Geometric Formulas   A reference on the inside back cover of the text displaying standard formulas for perimeter, circumference, area, and volume.

Coherent Development   Texts with consistent content and style across the developmental math curriculum.

WHAT SUPPLEMENTS ARE AVAILABLE?

For a complete list of the supplements and study aids that accompany *Intermediate Algebra through Applications*, Third Edition, see p. xi.

ACKNOWLEDGMENTS

We are grateful to everyone who has helped to shape this textbook by responding to questionnaires, participating in telephone surveys and focus groups, reviewing the manuscript, and using the text in their classes. We wish to thank James J. Ball, *Indiana State University;* Norma Bisulca, *University of Maine, Augusta;* Scott Brown, *Auburn University, Montgomery;* Joanne Brunner, *Joliet Junior College;* Kristin Chatas, *Washtenaw Community College;* Edith Cook, *Suffolk University;* Addie L. Davis, Ph.D., *Olive-Harvey College;* Mary Deas, *Johnson County Community College;* James Dressler, *Seattle Central Community College;* Karen Ernst, *Hawkeye Community College;* Sam Evers, *University of Alabama, Tuscaloosa;* Amadou Hama, *Kennedy King College;* Mary Beth Headlee, *State College of Florida;* Dr. Albert Hemenway, *Los Angeles Mission College;* Max Hibbs, *Blinn College;* Nancy R. Johnson, Ph.D., *State College of Florida;* Judy Kasabian, *El Camino College;* Mickey Levendusky, *Pima Community College, Downtown Campus;* Tony Masci, *Notre Dame College;* Marcel Maupin, *Oklahoma State University–Oklahoma City;* Margaret Patin, *Vernon College;* Matthew S. Pitassi, *Rio Hondo College;* Merrel Pepper, *Southeast Technical Institute;* Carol Perezluha, *Seminole State College of Florida;* Lynn Rickabaugh, *Aiken Technical College;* Patricia C. Rome, *Delgado Community College;* May Shaw, *Northcentral Technical College;* and Jackie Wing, *Angelina College.* In addition, we would like to extend our gratitude to our accuracy checkers and to those who helped us perfect the content in many ways: Michael Carlisle; Lisa Collete; Paul Lorczak; Denise Heban; Perian Herring; Christine Verity; Janis Cimperman; Ann Ostberg; and Lenore Parens.

Writing a textbook requires the contributions of many individuals. Special thanks go to Greg Tobin, President, Mathematics and Statistics, Pearson Arts and Sciences, for encouraging and supporting us throughout the entire process. We thank Kari Heen and Katie DePasquale for their patience and tact, Michelle Renda, Rachel Ross, and Maureen O'Connor for keeping us abreast of market trends, Rachel Haskell for attending to the endless details connected with the project, Ron Hampton, Elka Block Laura Osterbrock, Laura Hakala, Marta Johnson, and Rachel Youdelman for their support throughout the production process, Barbara Atkinson for the cover design, and the entire Pearson developmental mathematics team for helping to make this text one of which we are very proud.

Geoffrey Akst

Sadie Bragg

Student Supplements

Student's Solutions Manual
By Beverly Fusfield
- Provides detailed solutions to the odd-numbered exercises in each exercise set and solutions to all chapter pretests and post-tests, practice exercises, review exercises, and cumulative review exercises

ISBN-10: 0-321-75715-7   ISBN-13: 978-0-321-75715-9

New Video Resources on DVD with Chapter Test Prep Videos
- Complete set of digitized videos on DVD for students to use at home or on campus
- Includes a full lecture for each section of the text
- Covers examples, practice problems, and exercises from the textbook that are marked with the ⊙ icon
- Optional captioning in English is available
- Step-by-step video solutions for each chapter test
- Chapter Test Prep Videos are also available on YouTube (search by using author name and book title) and in MyMathLab

ISBN-10: 0-321-78631-9   ISBN-13: 978-0-321-78631-9

MyWorkBook with Chapter Summaries
By Carrie Green
- Provides one worksheet for each section of the text, organized by section objective, along with the end-of-chapter summaries from the textbook
- Each worksheet lists the associated objectives from the text, provides fill-in-the-blank vocabulary practice, and exercises for each objective

ISBN-10: 0-321-75975-3   ISBN-13: 978-0-321-75975-7

MathXL Online Course (access code required)

InterAct Math Tutorial Website
www.interactmath.com
- Get practice and tutorial help online
- Provides algorithmically generated practice exercises that correlate directly to the textbook exercises
- Retry an exercise as many times as desired with new values each time for unlimited practice and mastery
- Every exercise is accompanied by an interactive guided solution that gives the student helpful feedback when an incorrect answer is entered
- View the steps of a worked-out sample problem similar to the one that has been worked on

Instructor Supplements

Annotated Instructor's Edition
- Provides answers to all text exercises in color next to the corresponding problems
- Includes teaching tips

ISBN-10: 0-321-75726-2   ISBN-13: 978-0-321-75726-5

Instructor's Solutions Manual (download only)
By Beverly Fusfield
- Provides complete solutions to even-numbered section exercises
- Contains answers to all Mindstretcher problems

ISBN-10: 0-321-75717-3   ISBN-13: 978-0-321-75717-3

Instructor's Resource Manual with Tests and Mini-Lectures (download only)
By Deana Richmond
- Contains three free-response and one multiple-choice test form per chapter and two final exams
- Includes resources designed to help both new and adjunct faculty with course preparation and classroom management, including sample syllabi, tips for using supplements and technology, and useful external resources
- Offers helpful teaching tips correlated to the sections of the text

ISBN-10: 0-321-63934-0   ISBN-13: 978-0-321-63934-9

PowerPoint Lecture Slides (available online)
- Present key concepts and definitions from the text

TestGen® (available for download from the Instructor's Resource Center)

AVAILABLE FOR STUDENTS AND INSTRUCTORS

MyMathLab® Ready-to-Go Course (access code required)

These new Ready-to-Go courses provide students with all the same great MyMathLab features that you're used to, but make it easier for instructors to get started. Each course includes preassigned homework and quizzes to make creating your course even simpler. Ask your Pearson representative about the details for this particular course or to see a copy of this course.

MyMathLab with Pearson eText—Instant Access for *Intermediate Algebra through Applications*

MyMathLab delivers proven results in helping individual students succeed. It provides engaging experiences that personalize, stimulate, and measure learning for each student. And, it comes from a trusted partner with educational expertise and an eye on the future. To learn more about how MyMathLab combines proven learning applications with powerful assessment, visit www.mymathlab .com or contact your Pearson representative.

MathXL—Instant Access for *Intermediate Algebra through Applications*

MathXL® is the homework and assessment engine that runs MyMathLab. (MyMathLab is MathXL plus a learning management system.) With MathXL, instructors can

- Create, edit, and assign online homework and tests using algorithmically generated exercises correlated at the objective level to the textbook.
- Create and assign their own online exercises and import TestGen tests for added flexibility.
- Maintain records of all student work tracked in MathXL's online gradebook.

With MathXL, students can

- Take chapter tests in MathXL and receive personalized study plans and/or personalized homework assignments based on their test results.
- Use the study plan and/or the homework to link directly to tutorial exercises for the objectives they need to study.
- Access supplemental animations and video clips directly from selected exercises.

MathXL is available to qualified adopters. For more information, visit www.mathxl.com or contact your Pearson representative.

Index of Applications

Flower garden, 806
Football field grounds crew, 491
Height of water in a container, 171
Length of a kite string, 536
Maximum capacity of an elevator, 232
Maximum load on a truck, 125
Members of a professional organization, 222, 698
Nanosecond, 59
Nozzle pressure of a firefighter's hose, 515
Phone number, 155
Photography, 40, 442, 496
Rooms in an apartment building, 68
Rubber hose, 786
Snail's speed, 98
Square parcel of land, 787
Stackable storage cabinets, 417
Temperature of a lightbulb filament, 353
Thickness of coins, 315
Time for a brother to catch up to his sister, 345
Walkie-talkies, 750
Weight of a box of candy, 144
Weight of a person, 147
Weight of a roll of nickels, 141
Whispering gallery, 787
Width of a machine part, 585
Women's clothing sizes, 221

Geometry

Altitude of a triangle, 774
Angles of a triangle, 104
Area of a circular region, 26, 73, 417, 554
Area of a deck, 379
Area of a dog run, 435
Area of a mat around a picture, 417, 436
Area of a newspaper, 374
Area of a parking lot, 353
Area of a rectangular region, 104, 155, 546
Area of a soccer field, 345
Area of a square, 379, 542
Area of a trapezoid, 108
Area of a triangular region, 40, 423
Area of an equilateral triangle, 554
Area of an ice skating rink, 405
Diagonal of a basketball court, 515
Diagonals of a polygon, 392
Diagonals of a television set, 540
Diagonals of a trapezoid, 488
Dimensions of a computer screen, 426
Dimensions of a football field, 348
Dimensions of a garden, 301, 609, 628–629, 652, 762
Dimensions of a movie or TV screen, 536, 598
Dimensions of a patio, 661
Dimensions of a rectangular region, 417, 652
Dimensions of a soccer field, 425
Dimensions of a square, 796
Dimensions of a swimming pool, 395
Dimensions of a walkway, 425
Exterior angles of a regular polygon, 511
Height of a box, 564

Interior angles of a polygon, 111
Length of the side of a box, 527
Length of the side of a square, 527, 542
Length of the side of a triangle, 546
Maximum area, 588, 641, 660, 762
Perimeter of a rectangular region, 72, 96
Perimeter of a triangle, 96
Perimeter of the base of the Great Pyramid at Giza, 542
Perpendicular bisector, 774
Radius of a balloon, 417
Radius of a circular region, 588, 599
Radius of a cone, 747
Radius of a cylinder, 467
Radius of a sphere, 555, 583, 585
Sides of a polygon, 686
Surface area of a box, 40, 395, 446
Surface area of a cylinder, 112, 455
Surface area of a rectangular prism, 109
Surface area of a sphere, 446, 476, 562
Using a rectangle to make a box, 609, 660
Vertical angles, 96
Volume of a box, 36, 364, 378, 417
Volume of a cube, 58
Volume of a cylinder, 58, 665
Volume of a foam block, 417
Volume of a sphere, 476, 551
Volume of buildings, 497–498

Government

Candidate's votes, 288
Electoral votes, 14
Federal budget, 165–166
Federal government income, 59
Federal income tax, 85
Mayor's approval rating, 156
Mayoral election, 144
Number of senators in a state, 165
Officers in the U.S. Department of Defense, 360
Political campaign, 157, 233
Popular votes for U.S. presidential candidate, 246
Revenue and expenditures of the federal government, 26
U.S. House of Representatives, 14
U.S. Postal Service, 60, 315

Health and Medicine

Adult's body temperature, 148
Basal energy expenditure, 111
Birth lengths and weights of babies, 235
Body mass index, 135, 499, 618
Calories, 23, 148
Child's weight, 169, 180
Consultation fee for patients, 220
Dental service expenditures, 513
Drug concentration in the bloodstream, 698, 734, 745
Drug dosage, 494
Eye's focusing ability, 112, 510

Filling prescriptions, 441
Flu epidemic, 179
Height of a child, 189
HIV/AIDS, 83
Hydrocortisone cream, 99–100
Ideal body weight, 83, 111, 686
Lithotripsy, 748
Losing weight, 40
Medical supply manufacturer, 289
Medicine dosage, 27, 83, 95, 155, 250, 479
Nicotine-replacement therapy, 27
Nutritional information, 326, 348
Patient's medical bill, 204
Patient's temperature, 165
Personal trainer recommendations, 491
Private health care expenditure, 733
Recommended daily intake of calcium, 82, 206
Recommended daily fiber intake, 315
Target heart rate, 156
Time of death of a person, 743
Weight-loss clinic, 27, 190
White blood cells, 127

Labor

Advertising a job opening, 155
Annual raises, 189
Commission, 243
Company's employment test, 331
Employee number, 104
Employee satisfaction with benefits, 491
Gross pay, 585
Hourly earnings, 264
Hours worked, 262, 289, 336
Lay-offs, 345
Number of construction workers, 79
Overtime pay, 136
Part-time jobs, 227, 289, 336
Salary, 95, 103, 221, 346, 387, 686, 697, 742
Severance check, 694
Software company employees, 222
Weekly pay, 74, 262
Working alone to complete a job, 477, 492, 511, 614, 618, 663
Working together to complete a job, 487, 491, 510, 585

Physics

Acceleration of an object, 476
Beer-Lambert Law, 726
Boyle's Law, 502
Centrifugal force, 499
Doppler effect, 477
Dropped object, 26, 364, 395, 413, 417, 502, 510, 513, 515, 527, 536, 542, 564, 599, 659, 792–793, 816
Force, 455, 564
Hooke's Law, 441
Illumination, 502, 582
Intensity of sound, 524, 705–706, 719, 733, 745
Length and temperature of an object, 214

Photo Credits

Algebra Basics

Real Numbers and the Periodic Table

Real numbers appear throughout the *periodic table*, a key reference tool that chemists use to organize what is known about the elements that make up all matter.

The elements are arranged in rows (periods) and columns (groups) based on their chemical and physical properties. They are positioned in the table so that their *atomic number*—a whole number that represents the number of protons in an atom of the element—is increasing from left to right and from top to bottom.

The order of the elements also coincides with increasing *atomic mass*—a decimal number that is an average for the various isotopes, or varieties, of the element.

The elements in a column of the table generally have not only similar chemical properties but also the same valence—a positive or negative integer that represents the combining power of an element with other elements.

The periodic table has been used to predict the existence of elements yet to be discovered, as well as their properties. (*Source:* Karen Timberlake, *Chemistry: An Introduction of General, Organic, and Biological Chemistry*, Prentice Hall, 2011)

21	22	23	24	25	26	27	28	29	30
Sc	**Ti**	**V**	**Cr**	**Mn**	**Fe**	**Co**	**Ni**	**Cu**	**Zn**
44.96	47.90	50.94	52.00	54.94	55.85	58.93	58.71	63.55	65.38

39	40	41	42	43	44	45	46	47	48
Y	**Zr**	**Nb**	**Mo**	**Tc**	**Ru**	**Rh**	**Pd**	**Ag**	**Cd**
88.91	91.22	92.91	95.94	98	101.07	102.91	106.4	107.87	112.40

A portion of the periodic table. The atomic number is above the symbol for the element and the atomic mass is below.

To see if you have already mastered the topics in this chapter, take this test.

1. Graph -0.5 on the number line.

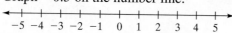

2. Evaluate: $|-8|$

3. Fill in the box with $>$ or $<$ to make a true statement:
   -2.7 ▢ -2.5

4. Identify the property of real numbers illustrated in the statement:

   $$2 \cdot (a + 6) = 2 \cdot (6 + a)$$

Calculate.

5. $-7 - 6$

6. $12 - 18 + (-3) - (-4)$

7. $|5(-2)(4)|$

8. $(-6)^2$

9. 2^0

10. $3(10 - 8) - (7 - 13)^2 \div 12$

11. $6[2(5 - 3) - (4 + 1)]^2$

12. $\dfrac{(-2)^3 - (-8)}{15 - 11}$

13. Simplify: $\dfrac{y^6 \cdot y}{y^5}$

14. Simplify: $-(3a^4b^2)^3$

15. Rewrite using only positive exponents: $-9x^{-7}$

16. Translate to an algebraic expression: Ten more than twice the product of a and b

17. If an item is discounted 10%, then discounted 10% again, and finally discounted 10% a third time, what portion of its original value is it now worth? Express the answer as a decimal in exponential form.

18. Items that cost c dollars apiece to produce are regularly sold at $2c$ dollars each. If 2000 items are produced, write an algebraic expression to represent the amount of profit after 1000 are sold at the regular sale price and 400 at half price.

19. The melting point of hydrogen is $-259.34°C$. The melting point of nitrogen is $-210°C$. Which element melts at a higher temperature?

20. The number (in thousands) of bachelor's degrees earned in the United States in either mathematics or statistics for a given year can be approximated by the algebraic expression $-0.018t^3 + 0.28t^2 - 0.66t + 11.84$, where t represents the number of years since 1999. Using this model, determine, to the nearest thousand, how many bachelor's degrees were earned in mathematics or statistics in the year 2009. (*Source:* nces.ed.gov)

• Check your answers on page A-15.

1.1   Introduction to Real Numbers

OBJECTIVES

Ⓐ To identify different kinds of real numbers: whole numbers, integers, rational numbers, and irrational numbers

Ⓑ To graph a real number on a number line

Ⓒ To find the additive inverse or the absolute value of real numbers

Ⓓ To compare real numbers

Ⓔ To represent a set of real numbers using interval notation

Ⓕ To solve applied problems involving real numbers

What Algebra Is and Why It Is Important

Algebra is a language that allows us to express the patterns and rules of arithmetic. Consider, for instance, the rule for finding the product of two fractions: Multiply the numerators to get the numerator of the product, and multiply the denominators to get the denominator of the product.

$$\frac{2}{3} \cdot \frac{1}{5} = \frac{2 \cdot 1}{3 \cdot 5} = \frac{2}{15}$$

In the language of algebra, we can write the general rule using letters to represent numbers.

$$\frac{a}{b} \cdot \frac{c}{d} = \frac{a \cdot c}{b \cdot d} = \frac{ac}{bd}$$

Algebra allows us not only to communicate rules such as the one just shown but also to solve problems. These problems arise in a variety of disciplines, such as physics, economics, and medicine.

This chapter reviews the basics of algebra. The focus is on two of the key concepts of algebra, namely, *real numbers* and *algebraic expressions*. We begin with a discussion of real numbers, including their operations, properties, and applications.

The Real Numbers

Real numbers are numbers that can be represented as points on a number line. These numbers are an extension of those used in arithmetic and allow us to solve problems that we could not otherwise solve.

There are different kinds of real numbers: whole numbers, integers, rational numbers, and irrational numbers. The relationships among these kinds of numbers are shown in the diagram to the right. Note from the diagram that every real number is either rational or irrational. Furthermore, every integer is a rational number and every whole number is an integer.

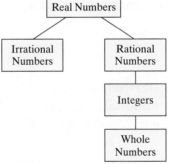

Now, let's consider each of these kinds of real numbers.

Whole Numbers and Integers

In arithmetic, the *whole numbers* consist of 0, 1, 2, 3, 4, 5, 6, 7, 8, 9, . . . ; the three dots mean that these numbers go on forever in the same pattern.

We can represent the whole numbers on a number line, as shown in the following figure. A number to the right of another on a number line is the greater number. A number to the left of another number on the number line is the lesser number.

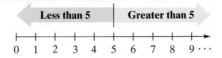

Sometimes, we need to consider numbers that are to the left of 0, that is, numbers that are less than 0. For instance, temperatures below 0 and a company's quarterly loss are values that are less than 0.

More generally, numbers to the *left* of 0 on the number line are said to be *negative*. Note how we write negative numbers: $-1, -2, -3, -4$, and so on. By contrast, the whole numbers to the right of 0 (1, 2, 3, 4, and so on) are said to be *positive*; they can also be written: $+1, +2, +3, +4$, and so on. The positive and negative numbers, together with 0, are called *integers* and are shown on the following number line. The number line extends without end to the right and to the left, as the arrows indicate.

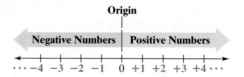

On the preceding number line, note that:

- the point 0 is called the *origin*,
- numbers to the right of 0 are positive, and numbers to the left of 0 are negative,
- the number 0 is neither positive nor negative.

> **DEFINITION**
> The **integers** are the numbers
> $$\ldots, -4, -3, -2, -1, 0, +1, +2, +3, +4, \ldots$$
> continuing infinitely in both directions.

Rational Numbers

A *rational number* is a real number that can be written as the quotient of two integers, where the denominator is not equal to 0. Some examples of rational numbers are:

$$\frac{3}{5}, -\frac{1}{3}, 96, 0, 8.9, -0.03$$

> **DEFINITION**
> **Rational numbers** are real numbers that can be written in the form $\frac{a}{b}$, where a and b are integers and $b \neq 0$.

A rational number can be expressed in a variety of ways. For instance, we can write $-\frac{1}{3}$ as either $\frac{-1}{3}$ or $\frac{1}{-3}$. The numbers 96 and 0 can also be written as $\frac{96}{1}$ and $\frac{0}{1}$, respectively. Likewise, we can write 8.9 as $\frac{89}{10}$ and -0.03 as $-\frac{3}{100}$. All rational numbers have corresponding decimal representations that either *terminate* or *repeat*. For example,

$$\frac{1}{5} = 0.2 \quad \text{and} \quad \frac{7}{8} = 0.875 \qquad \frac{4}{9} = 0.4444\ldots \quad \text{and} \quad \frac{17}{11} = 1.5454\ldots$$

Terminating decimals

Repeating decimals

We can write $0.4444\ldots$ as $0.\overline{4}$ and $1.5454\ldots$ as $1.\overline{54}$, where the bar indicates that the digit or digits repeat infinitely.

Just as with integers, we can visually represent a rational number as a point on a number line. To *graph* a number, we locate the point on the number line the appropriate distance to the left or right of 0 (the origin) and mark it, as shown on the number line below:

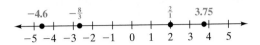

EXAMPLE 1

Graph each number on the same number line.

a. $-\dfrac{3}{2}$ **b.** 3.1

Solution

a. Because $-\dfrac{3}{2}$ can be expressed as $-1\dfrac{1}{2}$, or -1.5, the point $-\dfrac{3}{2}$ can be graphed halfway between -1 and -2.

b. The number 3.1 is between 3 and 4, but closer to 3.

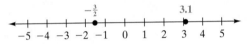

PRACTICE 1

Graph on the same number line.

a. -2.5 **b.** $\dfrac{7}{4}$

Irrational Numbers

We know that any rational number can be written as the quotient (or ratio) of two integers. However, there are other real numbers that cannot be written in this form. These numbers are called *irrational numbers*. For these numbers, the corresponding decimal representations neither terminate nor repeat.

Some examples of irrational numbers are given below:

$$\sqrt{2} = 1.414213\ldots \quad \text{The square root of 2}$$
$$-\sqrt{3} = -1.732050\ldots \quad \text{The negative square root of 3}$$
$$\pi = 3.141592\ldots \quad \text{Pi, the ratio of the circumference of a circle to its diameter}$$

In many computations with irrational numbers, we use decimal approximations that are rounded to a certain number of decimal places. For instance,

$$\sqrt{2} \approx 1.41 \qquad -\sqrt{3} \approx -1.73 \qquad \pi \approx 3.14$$

are decimal approximations of irrational numbers. Recall that the symbol $\approx$ is read, "is approximately equal to."

Irrational numbers, like rational numbers, can be graphed on a number line. The rational numbers and the irrational numbers together make up the real numbers.

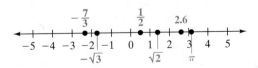

Absolute Value

On a number line, -2 and $+2$ (that is, 2) are opposites of each other. Likewise, $-\dfrac{1}{4}$ and $+\dfrac{1}{4}$ are opposites. Note that to find the opposite of a number, we change the sign of the number.

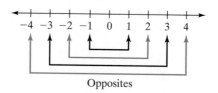

Opposites

From the number line above, we see that each number in a pair of opposites is the same distance from 0 as the other.

> **DEFINITION**
>
> Two real numbers that are the same distance from 0, but on opposite sides of 0, on a number line are called **opposites** or **additive inverses**. For any real number a, the additive inverse of a is written $-a$.

What is the additive inverse of 0?

EXAMPLE 2

Find the additive inverse of each number.

Solution

Number	Additive Inverse
a. 4	-4
b. $-\dfrac{1}{3}$	$\dfrac{1}{3}$
c. $10\dfrac{1}{2}$	$-10\dfrac{1}{2}$
d. -0.2	0.2

PRACTICE 2

Find the additive inverse of each number.

Number	Additive Inverse
a. -41	
b. $-8\dfrac{1}{2}$	
c. 1.7	
d. $-\dfrac{2}{5}$	

Because the number -3 is negative, it lies 3 units *to the left* of 0. The number 3, which is positive, lies 3 units *to the right* of 0.

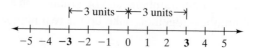

On a number line, the distance a number is from 0 is called its *absolute value*. Thus, the absolute value of $+3$, which we write as $|3|$, is 3. The absolute value of -3, which we write as $|-3|$, is also 3.

> **DEFINITION**
>
> The **absolute value** of a number is its distance from 0 on a number line. The absolute value of the number n is written $|n|$.

Note that this definition is geometric; that is, it is expressed in terms of a number line.

The following properties of absolute value help us to find the absolute value of any number:

- The absolute value of a positive number is the number itself.
- The absolute value of a negative number is its additive inverse.
- The absolute value of 0 is 0.
- The absolute value of a number is always positive or 0.
- Two numbers that are additive inverses have the same absolute value.

EXAMPLE 3

Evaluate.

a. $|-7|$ **b.** $|3.2|$ **c.** $\left|-\dfrac{3}{8}\right|$ **d.** $-|-5|$

Solution

a. $|-7| = 7$ Because the absolute value of a negative number is its additive inverse, the absolute value of -7 is 7.

b. $|3.2| = 3.2$ Because the absolute value of a positive number is the number itself, the absolute value of 3.2 is 3.2.

c. $\left|-\dfrac{3}{8}\right| = \dfrac{3}{8}$ Because $-\dfrac{3}{8}$ is negative, its absolute value is its additive inverse, which is $\dfrac{3}{8}$.

d. $-|-5| = -(5)$ $|-5| = 5$
         $= -5$

PRACTICE 3

Find the value.

a. $\left|\dfrac{2}{5}\right|$

b. $|0|$

c. $|-2.9|$

d. $-|-9|$

Comparing Real Numbers

A number line helps us to compare two real numbers, that is, to decide which number is greater and which is lesser. As previously noted, given two numbers on a number line, the number to the right is greater than the number to the left.

The *equal sign* $(=)$ and the *inequality symbols* $(\neq, <, \leq, >,$ and $\geq)$ are used to compare numbers.

$=$	means *is equal to*	$4.5 = 4\frac{1}{2}$ is read "4.5 is equal to $4\frac{1}{2}$."
$\neq$	means *is not equal to*	$2 \neq -2$ is read "2 is not equal to -2."
$<$	means *is less than*	$-3 < 0$ is read "-3 is less than 0."
$\leq$	means *is less than or equal to*	$5 \leq 7$ is read "5 is less than or equal to 7."
$>$	means *is greater than*	$-1 > -5$ is read "-1 is greater than -5."
$\geq$	means *is greater than or equal to*	$6 \geq 1$ is read "6 is greater than or equal to 1."

The statements $2 \neq -2$, $-3 < 0$, $5 \leq 7$, $-1 > -5$, and $6 \geq 1$ are *inequalities*.

The statement $4.5 = 4\dfrac{1}{2}$ is an *equation*.

When comparing real numbers, remember that:

- 0 is greater than any negative number, because all negative numbers on a number line lie to the left of 0. For example, $0 > -1$.
- 0 is less than any positive number, because all positive numbers lie to the right of 0 on a number line. For example, $0 < 4$.
- any positive number is greater than any negative number, because on a number line all positive numbers lie to the right of all negative numbers. For example, $4 > -1$.

EXAMPLE 4

Indicate whether each inequality is true or false.

a. $0 > -6$ **b.** $0 \geq \dfrac{1}{4}$ **c.** $1.5 < -1.5$ **d.** $-7 \leq -7$

Solution

a. $0 > -6$ True, because 0 is greater than any negative number

b. $0 \geq \dfrac{1}{4}$ False, because 0 is less than any positive number

c. $1.5 < -1.5$ False, because any positive number is greater than any negative number

d. $-7 \leq -7$ True, because -7 is equal to -7

PRACTICE 4

Indicate whether the inequality is true or false.

a. $0 < -6$

b. $-7 \leq 0$

c. $\dfrac{10}{6} \geq \dfrac{5}{3}$

d. $-1 > 0.3$

EXAMPLE 5

Graph the numbers $-2, 0.4, -3,$ and $-\dfrac{1}{3}$ on the same number line.

Then, order them from least to greatest.

Solution

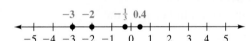

Reading the graph, we see that the numbers in order from least to

greatest are $-3, -2, -\dfrac{1}{3},$ and 0.4.

PRACTICE 5

Graph the numbers $-1.4, 3, -\dfrac{1}{2},$ and

-2.6 on the same number line. Then, rank them in descending order.

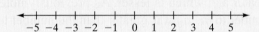

When applied to particular situations, the meaning of *greater* and *lesser* or *positive* and *negative* may vary. For example, in a temperature problem, the greater of two numbers represents a warmer temperature. In a financial problem, a positive number can mean a credit or a profit, whereas a negative number can mean a debt or a loss. In a motion problem, movement to the right or upward is often represented by a positive number, whereas movement to the left or downward is represented by a negative number.

EXAMPLE 6

The following table shows the temperature below which various plants freeze and die:

Plant	Asters	Carnations	Mums
Hardy to	$-20°F$	$-5°F$	$-30°F$

(*Source: The American Horticultural Society, A–Z Encyclopedia of Garden Plants*)

Which of these plants is hardy under the coldest conditions?

Solution   To determine which of these plants is the most hardy, we compare each of the temperatures. Of the three plants, mums are hardy under the coldest conditions because $-30 < -20 < -5$, that is, $-30°F < -20°F < -5°F$.

PRACTICE 6

When fog reached the New York City area, visibility was reduced to $\frac{3}{8}$ mi at LaGuardia Airport, $\frac{3}{4}$ mi at Kennedy Airport, and $\frac{1}{2}$ mi at Newark Airport. Which of these airports had the best visibility?

Sets and Interval Notation

A set is a collection of objects. Each of these objects is said to be an **element** or a **member** of the set.

A set can be represented by listing its elements separated by commas and placing braces around the listing. A set such as $\{2, 4, 6\}$ whose elements we can count is said to be **finite**. By contrast, the set of whole numbers $\{0, 1, 2, 3, \dots\}$, whose elements go on without end, is **infinite**. The set with no elements is called the **empty set**.

We typically use a capital letter to name a set. For instance, we can let $S = \{2, 4, 6\}$.

Sets can be combined in various ways. The **intersection** of sets A and B, written $A \cap B$, is the set of all elements that are in both A and B. So if $A = \{0, 5\}$ and $B = \{1, 5\}$, then $A \cap B = \{5\}$. The **union** of sets A and B, written $A \cup B$, is the set of all elements that are in either A or B or both. So for the given sets A and B, $A \cup B = \{0, 1, 5\}$.

In this text, we mainly consider sets whose elements are real numbers. Such sets can be pictured, or graphed, on a number line. Consider, for instance, the set of all real numbers greater than 3. The graph of this set is:

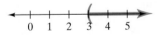

This graph extends without end in the positive direction on the number line. Note also that the left parenthesis on the graph indicates that the left endpoint 3 is not an element of the set and so is not part of the graph. Using **interval notation**, we can also represent this set as $(3, \infty)$, where the symbol ∞, read as "infinity," indicates that the interval extends without bound in the positive direction.

Let's look at another example. Consider the set of all real numbers less than or equal to 4. This set can be graphed as follows:

The graph extends indefinitely in the negative direction, and the right bracket on the graph indicates that the right endpoint 4 is included. In interval notation, the set can be represented as $(-\infty, 4]$, where the symbol $-\infty$, read as "negative infinity," indicates that the interval extends without bound in the negative direction.

Finally, consider the set of all real numbers between −1 and 2, including both −1 and 2. The graph of this set, represented in interval notation by $[-1, 2]$, is:

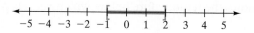

Note that in this graph, the brackets indicate that the two endpoints are included.

EXAMPLE 7

Graph each set on a number line, and represent the set in interval notation.

a. All real numbers less than 2

b. All real numbers greater than or equal to 0

c. All real numbers between 1 and 4 excluding 1 but including 4

Solution

a. We need to graph the set of all real numbers less than 2. The graph of this set is:

The graph extends infinitely in the negative direction, and the right parenthesis on the graph indicates that the right endpoint 2 is excluded. In interval notation, the set can be represented as $(-\infty, 2)$.

b. We are considering the set of all real numbers greater than or equal to 0. The graph of this set is:

Note that the graph extends infinitely in the positive direction, and the left bracket tells us that the left endpoint 0 is included. Using interval notation, we can express the set as $[0, \infty)$.

c. We need to graph the set of all real numbers between 1 and 4 excluding 1 but including 4. The graph of this set is:

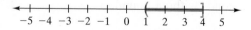

This graph extends from 1 to 4. The left parenthesis tells us that the left endpoint 1 is excluded, and the right bracket tells us that the right endpoint 4 is included. The set can be written as $(1, 4]$.

PRACTICE 7

Graph on the number line, and express in interval notation.

a. The set of all real numbers greater than or equal to 2

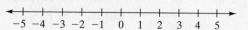

b. The set of all real numbers less than 0

c. The set of all real numbers between −2 and 3 including −2 but excluding 3

Mathematically Speaking

Fill in each blank with the most appropriate term or phrase from the given list.

real numbers	equation	rational numbers
integers	absolute value	infinity
additive inverses	whole numbers	inequality
multiplicative inverses	sign	irrational numbers

1. The _____ consist of 0, 1, 2, 3, 4, 5, 6, 7, 8, 9, ...

2. The _____ are the numbers $-4, -3, -2, -1, 0, +1, +2, +3, +4, \ldots$ continuing infinitely in both directions.

3. Real numbers that can be written as the quotient of two integers, where the denominator is not equal to 0, are called _____.

4. The decimal representations of _____ neither terminate nor repeat.

5. Two real numbers that are the same distance from 0, but on opposite sides of 0, on a number line are called opposites or _____.

6. The _____ of a number is its distance from 0 on the number line.

7. The symbol ∞ is read _____.

8. The statement $8 \geq 6$ is a(n) _____.

A *Classify each number by writing checks in the appropriate boxes.*

	Whole Numbers	Integers	Rational Numbers	Irrational Numbers	Real Numbers
9. -5					
10. $\dfrac{8}{3}$					
11. -3.9					
12. $\sqrt{2}$					
13. 7					
14. $\dfrac{1}{5}$					
15. $\sqrt{3}$					
16. $5.1\overline{6}$					

B *Graph on the number line.*

17. -4

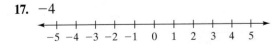

18. -1

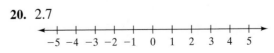

19. 3.5

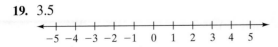

20. 2.7

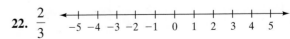

21. $\dfrac{5}{4}$

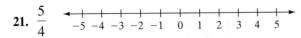

22. $\dfrac{2}{3}$

C *Find the opposite of each number.*

23. -5
24. -32
25. $\dfrac{1}{4}$
26. $\dfrac{3}{4}$

Find the additive inverse of each number.

Number	**27.** -12	**28.** $-\dfrac{5}{6}$	**29.** 2.7	**30.** $-\dfrac{1}{3}$
Additive Inverse				

Find the value.

31. $\left|\dfrac{1}{4}\right|$
32. $\left|\dfrac{1}{3}\right|$
33. $|-9|$
34. $|-5|$

35. $-|-2|$
36. $-|-7|$
37. $-|5.3|$
38. $-|0.5|$

Solve. If impossible, explain why.

39. Name all numbers that have an absolute value of 6.

40. Name all numbers that have an absolute value of 0.2.

41. Name a number whose absolute value is -1.

42. Name three different numbers that have the same absolute value.

D *Indicate whether each inequality is true or false.*

43. $-1 < -5$
44. $-3 < -7$
45. $\dfrac{5}{4} \geq \dfrac{5}{3}$
46. $\dfrac{7}{6} \geq \dfrac{7}{5}$

47. $-1\dfrac{1}{4} \leq -1\dfrac{1}{4}$
48. $-2 \leq -2$
49. $-|3| \geq -3$
50. $-1 \geq |-1|$

51. $|-1.2| \leq |-0.5|$
52. $|-2.6| > |-2.4|$

Replace each ▓ *with* $<$, $>$, *or* $=$ *to make a true statement.*

53. 0 ▓ -7
54. 3 ▓ -5
55. -4 ▓ -6
56. -5 ▓ -1

57. -7.4 ▓ -7
58. -9 ▓ -9.6
59. $|8|$ ▓ $|-8|$
60. $|-6|$ ▓ $|-7|$

61. $-|4.1|$ ▓ $|-4.1|$
62. $-|-1.6|$ ▓ $-|1.6|$

Graph the numbers. Then, name them in order from least to greatest.

63. $-4, \dfrac{3}{4}, 5,$ and $-\dfrac{2}{5}$

64. $-2, \dfrac{5}{2}, 0,$ and $-\dfrac{1}{2}$

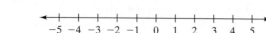

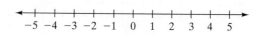

65. $1, 4.3, -1.8,$ and -3.5

66. $3, 3.2, -\dfrac{4}{3},$ and -0.25

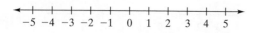

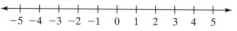

E *Graph on the number line, and express in interval notation.*

67. The set of all real numbers greater than or equal to −4

68. The set of all real numbers greater than −2

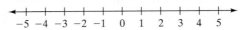

69. The set of all real numbers less than 3

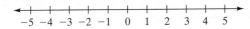

70. The set of all real numbers less than or equal to 4

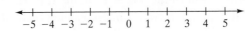

71. The set of all real numbers between −4 and 3 including both −4 and 3

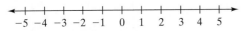

72. The set of all real numbers between 0 and 2 excluding 0 and including 2

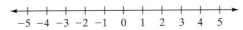

Mixed Practice

Solve.

73. If possible, name a number whose absolute value is −3. If it is impossible, explain why.

74. Graph the numbers 3, −3, 2.1, and 0. Then, name them in order from least to greatest.

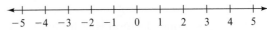

75. Classify the number $-\sqrt{5}$ by writing checks in the appropriate boxes.

Whole Number	Integer	Rational Number	Irrational Number	Real Number

76. Find the additive inverse of 3.5.

77. Graph on the number line the set of all real numbers between −2 and 1 including −2 and excluding 1, and express in interval notation.

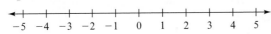

78. Graph the number 4.6 on the number line.

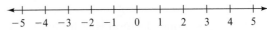

Applications

F *Solve.*

79. The average temperature at the South Pole in the month of January is −16°F. In the month of July, it is −74°F. In which month is it warmer at the South Pole? How do you know? (*Source: Encyclopedia Britannica Almanac 2003*)

80. The mean surface temperature of the planet Saturn is −292°F. The mean surface temperature of the planet Neptune is −238°F. Which planet is colder? How do you know?

81. The Caspian Sea (the lowest point in Europe) lies 27 m below sea level. The Dead Sea (the lowest point in Asia) lies 400 m below sea level. Write an inequality that compares the elevation of the Caspian Sea to that of the Dead Sea. (*Source: National Geographic Family Reference Atlas of the World*)

82. The price of a stock on Monday was $53.42 per share. On Tuesday, the price of the stock was $53.45 per share. Write an inequality that compares the price of the stock on Monday to the price of the stock on Tuesday.

83. The apportionment of the 435 seats in the U.S. House of Representatives is determined by the total population counts in each state obtained through a census that is taken every ten years. The table shows the number of representatives for selected states based on the 2000 census and the 2010 census. (*Source: U.S. Bureau of the Census*)

   a. Which states gained representatives as a result of the 2010 census? Which states lost representatives as a result of the 2010 census?

   b. Was the total population of Florida greater than or less than the total population of Ohio in 2010? Explain how you know.

   c. Which states had the same number of representatives in 2010 as a result of the census? What does this tell you about the total populations of those states?

State	2000	2010
Arizona	8	9
California	53	53
Florida	25	27
Illinois	19	18
New York	29	27
Ohio	18	16
Pennsylvania	19	18
Texas	32	36
Wisconsin	8	8

84. Each state receives one electoral vote for each of its U.S. representatives plus one electoral vote for each of its two senators. The table shows the number of electoral votes for selected states resulting from the 2000 census and the 2010* census. (*Source: U.S. Bureau of the Census*)

   a. Which states gained electoral votes as a result of the 2010 census? Which states lost electoral votes as a result of the 2010 census?

   b. Write an inequality that compares the number of electoral votes held by Connecticut to the number of electoral votes held by Colorado resulting from the 2010 census.

   c. For many states, the candidate that receives the most votes in a state wins the electoral votes for that state. In the 2008 presidential election, the Republican candidate John McCain received 2,048,759 votes in Georgia. The Democratic candidate Barack Obama received 1,844,123 votes in Georgia. Which candidate received that state's electoral votes? (*Source:* National Archives and Records Administration, *Federal Register*)

State	2000	2010
Colorado	9	9
Connecticut	7	7
Georgia	15	16
Illinois	21	20
Michigan	17	16
Nevada	5	6

*Allocations from the 2010 census apply to the 2012, 2016, and 2020 presidential elections.

85. The average price x of a gallon of gasoline this month was down $0.04 from the average price y of a gallon of gasoline last month. Write an inequality that compares the two averages.

86. The enrollment x at a local college this year was up 8% from the enrollment y last year. Write an inequality that compares the college's two enrollments.

87. The base sticker price x of a new SUV was $3565 higher than the base sticker price y of a new minivan. Write an inequality that compares the two base sticker prices.

88. The heating bill x last month was $42.27 lower than the heating bill y this month. Write an inequality that compares the two heating bills.

89. The table shows the final scores with respect to par for selected PGA players in a recent golf tournament. (*Source:* PGA)

    a. Plot the final scores on a number line.

Player	Final Score (with respect to par)
Ian Hoffman	−4
Nick Jones	−13
Tim Lamb	+2
Greg Mahoney	−12
Brian Richey	+8

    b. What does 0 on the number line represent in this context?

    c. Which player's score was the farthest from par?

90. The number of inches the monthly precipitation was above or below the normal monthly precipitation for a recent year in Seattle, Washington, is shown in the table. (*Source:* National Weather Service, Sea-Tac International Airport Station)

Month	Jan.	Feb.	Mar.	Apr.	May	Jun.
Difference in Precipitation from Normal (in inches)	−1.4	0	−0.9	+1.7	−0.7	+0.2

Month	Jul.	Aug.	Sep.	Oct.	Nov.	Dec.
Difference in Precipitation from Normal (in inches)	−0.2	−1	−1.2	−2.5	−2.2	+0.4

    a. Plot the difference in the monthly precipitation from the normal on a number line.

    b. What does 0 on the number line represent in the context of the problem?

    c. In which month was the precipitation farthest from the normal monthly precipitation?

• Check your answers on page A-15.

MINDStretchers

Critical Thinking

1. Suppose a and b represent two real numbers such that $a < b$. Under what circumstances will $|a| > |b|$? Give some examples to support your answer.

Research

2. Either in your college library or on the web, investigate the earliest use of negative numbers. Write a few sentences to summarize your findings.

Writing

3. Explain why every integer is a rational number, but not every rational number is an integer.

1.2 Operations with Real Numbers

In this section, we review the four basic operations on real numbers. First, let's look at *addition*.

A number line gives us a picture of adding real numbers. For instance, to add -1 and -4, we start at the point corresponding to the first number -1. The second number -4 is negative so we move 4 units to the *left*. We end at -5, which is the *sum*. So, $-1 + (-4) = -5$.

Move 4 units to the *left*.

$$
\begin{array}{ccccccccccccc}
-6 & -5 & -4 & -3 & -2 & -1 & 0 & 1 & 2 & 3 & 4 & 5 & 6
\end{array}
$$

End   Start

Note that both -1 and -4 are negative, as is their sum.

A number line gives us a visual way to add two real numbers, but this method is not very efficient, especially for adding large numbers. The following rule is a shortcut for adding real numbers:

> ### To Add Real Numbers
>
> - If the numbers have the same sign, add their absolute values and keep their sign.
>
> - If the numbers have different signs, find the difference between the larger absolute value and the smaller absolute value and take the sign of the number with the larger absolute value.

OBJECTIVES

A To add, subtract, multiply, or divide real numbers

B To solve applied problems involving operations with real numbers

EXAMPLE 1

Add: -3 and -19

Solution   Because both numbers are negative, we find their absolute values, and then add.

$$|-3| + |-19| = 3 + 19 = 22$$

The sum of two negative numbers is negative.

$$-3 + (-19) = -22$$

Negative numbers   Negative sum

PRACTICE 1

Find the sum: $-11 + (-18)$

EXAMPLE 2

Find the sum of -8.1 and 4.2.

Solution   Here we are adding a negative number and a positive number, that is, numbers with *different* signs. First, we find the absolute values.

$$|-8.1| = 8.1 \quad \text{and} \quad |4.2| = 4.2$$

PRACTICE 2

Add: $7.1 + (-5.3)$

Then, we subtract the smaller absolute value from the larger.

$$8.1 - 4.2 = 3.9$$

Because -8.1 has the larger absolute value and its sign is negative, the sum is also negative.

$$-8.1 + (+4.2) = -8.1 + 4.2 = -3.9$$

Negative Positive   The sum takes the sign of the number with
number   number   the larger absolute value.

Subtraction of real numbers is based on addition, and involves finding the additive inverse of a number.

Consider the difference $8 - 3$ and the sum $8 + (-3)$. Computing the value of each, we get $8 - 3 = 5$ and $8 + (-3) = 5$. Note that the result of each computation is the same. So

$$8 - 3 = 8 + (-3).$$

In general, for any real numbers a and b,

$$a - b = a + (-b).$$

In other words, we can change a problem involving the subtraction of real numbers to an equivalent one involving the addition of real numbers by adding the *additive inverse* of the number being subtracted.

To Subtract Real Numbers

- Change the operation of subtraction to addition and change the number being subtracted to its additive inverse.

- Then, follow the rule for adding real numbers.

EXAMPLE 3

Find the difference: $3 - (-5)$

Solution   Change the operation from subtraction to addition.

$$3 - (-5) \quad = \quad 3 + (+5) = 8$$

Change the second number from -5 to $+5$.

PRACTICE 3

Subtract: $8 - (-1)$

EXAMPLE 4

Subtract: $-6.3 - 2$

Solution

$$-6.3 - 2 = (-6.3) - (+2) = (-6.3) + (-2) = -8.3$$

PRACTICE 4

Subtract: $-4 - 10.9$

From arithmetic, we know that the product of two positive numbers is a positive number. In computations with real numbers, we must also consider the product of a negative number and a positive number as well as the product of two negative numbers. For instance, to find the product of 2 and -4, we use the fact that multiplication is repeated addition to get:

$$2(-4) = (-4) + (-4)$$
$$= -8$$

This result suggests that the product of a positive number and a negative number is negative. To multiply two negative numbers, we consider the following pattern:

This number is   The product is increasing
decreasing by 1 each time.   by 4 each time.

$$2(-4) = -8$$
$$1(-4) = -4$$
$$0(-4) = \;\;\,0$$
$$-1(-4) = \;\;\,4$$
$$-2(-4) = \;\;\,8$$

This pattern suggests that the product of two negative numbers is positive.

The following rule, based on the concept of absolute value, is for *multiplying* real numbers:

To Multiply Real Numbers

- Multiply the absolute values of the numbers.

- If the numbers have the same sign, their product is positive; if they have different signs, their product is negative.

EXAMPLE 5

Multiply: -3 by -9

Solution

$|-3| = 3$ and $|-9| = 9$ Find the absolute values.

$3 \cdot 9 = 27$ Multiply the absolute values.

$(-3)(-9) = 27$ The product of two negatives is positive.

PRACTICE 5

Find the product of -2 and -50.

EXAMPLE 6

Multiply -2.5 by 2.

Solution

$|-2.5| = 2.5$ and $|2| = 2$ Find the absolute values.

$(2.5)(2) = 5$ Multiply the absolute values.

$(-2.5)(2) = -5$ The product of a negative and a positive is negative.

PRACTICE 6

Find the product of -0.5 and 6.2.

TIP

Same Sign	
Positive $\cdot$ Positive = Positive	Negative $\cdot$ Negative = Positive
Different Signs	
Positive $\cdot$ Negative = Negative	Negative $\cdot$ Positive = Negative

Some expressions involve repeated multiplication of the same factor. In these expressions, *exponential notation* is often used as a shorthand method for representing this multiplication. For instance, the product $(-5)(-5)$ can be written as $(-5)^2$. This expression is read "-5 *to the second power*" or "-5 *squared.*"

EXAMPLE 7

Evaluate.

a. $(-5)^2$ **b.** -5^2

Solution

a. $(-5)^2 = (-5)(-5) = 25$

b. $-5^2 = -(5 \cdot 5) = -25$

PRACTICE 7

Calculate.

a. $(-7)^2$ **b.** -7^2

We now consider *division*—the last of the four basic operations on real numbers. As with multiplication, division of two real numbers can be expressed in terms of their absolute values.

To Divide Real Numbers

- Divide the absolute values of the numbers.

- If the numbers have the same sign, their quotient is positive; if the numbers have different signs, their quotient is negative.

EXAMPLE 8

Find the quotient.

a. $(-10) \div (-5)$ **b.** $\dfrac{-1.8}{6}$

Solution In each problem, we first find the absolute values. Then, we divide them. Finally, we attach the appropriate sign to this quotient.

a. $(-10) \div (-5)$

     $|-10| = 10$ and $|-5| = 5$

     $10 \div 5 = 2$

Since the numbers have the *same sign*, their quotient is positive. So $(-10) \div (-5) = 2$.

b. $\dfrac{-1.8}{6}$

     $|-1.8| = 1.8$ and $|6| = 6$

     $\dfrac{1.8}{6} = 0.3$

The numbers have *different* signs, so the quotient is negative.

$$\frac{-1.8}{6} = -0.3$$

PRACTICE 8

Divide.

a. $35 \div (-5)$

b. $\dfrac{-4.2}{-6}$

TIP

Same Sign

$$\frac{\text{Positive}}{\text{Positive}} = \text{Positive} \qquad \frac{\text{Negative}}{\text{Negative}} = \text{Positive}$$

Different Signs

$$\frac{\text{Positive}}{\text{Negative}} = \text{Negative} \qquad \frac{\text{Negative}}{\text{Positive}} = \text{Negative}$$

Now, let's consider division involving rational numbers. Just as subtraction is defined in terms of the additive inverse, so division is defined in terms of the *multiplicative inverse* (or *reciprocal*).

DEFINITION

Two real numbers are said to be **multiplicative inverses** (or **reciprocals**) if their product is 1. For any nonzero real number a, the multiplicative inverse of a is written $\frac{1}{a}$.

To divide two real numbers, we multiply the first number by the multiplicative inverse of the second number.

$$6 \div \frac{2}{3} = 6 \cdot \frac{3}{2} = \overset{3}{6} \cdot \frac{3}{\underset{1}{2}} = 9$$

This approach is especially useful if the numbers in the division problem are rational. Note that in this problem the multiplicative inverse of $\frac{2}{3}$ is $\frac{3}{2}$, since their product is 1. We can find the multiplicative inverse of $\frac{2}{3}$ by interchanging its numerator and denominator.

EXAMPLE 9

Divide.

a. $-\frac{1}{2} \div \left(-\frac{1}{8}\right)$ **b.** $-\frac{1}{6} \div 2$ **c.** $-12 \div \left(\frac{2}{3}\right)$

Solution

a. $-\frac{1}{2} \div \left(-\frac{1}{8}\right) = -\frac{1}{2} \cdot \left(-\frac{8}{1}\right) = 4$

b. $-\frac{1}{6} \div 2 = -\frac{1}{6} \div \frac{2}{1} = -\frac{1}{6} \cdot \frac{1}{2} = -\frac{1}{12}$

c. $-12 \div \frac{2}{3} = -12 \cdot \frac{3}{2} = -18$

PRACTICE 9

Divide.

a. $-\frac{3}{4} \div \frac{3}{8}$

b. $-\frac{1}{8} \div (-4)$

c. $10 \div \left(-\frac{1}{5}\right)$

The Order of Operations

Some expressions involve more than one operation, say addition and multiplication. The order in which the operations are performed determines the value of the expression. The *order of operations rule* tells us in which order to carry out the operations in the expression so that its value is unambiguous.

> ## Order of Operations Rule
>
> To evaluate mathematical expressions, carry out the operations in the following order:
>
> **1.** First, perform the operations within any grouping symbols, such as parentheses ( ), brackets [ ], or braces { }. Work from the innermost pair of grouping symbols to the outermost pair.
>
> **2.** Then, evaluate any numbers raised to a power.
>
> **3.** Next, perform all multiplications and divisions as they appear from left to right.
>
> **4.** Finally, do all additions and subtractions as they appear from left to right.

Note that parentheses are understood around the numerator and denominator of any fraction, as well as under a radical sign and within an absolute value sign. So, for instance, $\frac{2+1}{5}$ means $\frac{(2+1)}{5}$, or $\frac{3}{5}$; $\sqrt{2+1}$ means $\sqrt{(2+1)}$, or $\sqrt{3}$; and $|3-1|$ means $|(3-1)|$, that is, $|2|$ or 2.

Following the order of operations rule, we simplify mathematical expressions by first performing the operations within any grouping symbols, carrying out the innermost operations first.

EXAMPLE 10

Simplify: $[8-(2+6)]-7$

Solution

$$[8-(2+6)]-7 = [8-(8)]-7 \quad \text{Add within parentheses, the innermost grouping symbols.}$$
$$= 0-7 \quad \text{Subtract within the brackets.}$$
$$= -7$$

So $[8-(2+6)]-7 = -7$.

PRACTICE 10

Calculate: $9+[(1-7)-5]$

In applying the order of operations rule, we see that multiplication is performed before either addition or subtraction, working from left to right.

EXAMPLE 11

Simplify: $-2(24)-5(-1)$

Solution We use the order of operations rule.

$$-2(24)-5(-1) = -48-(-5) \quad \text{Multiply.}$$
$$= -48+5 \quad \text{Subtract } -5.$$
$$= -43 \quad \text{Add 5.}$$

So $-2(24)-5(-1) = -43$.

PRACTICE 11

Calculate: $3(-10)-(-5)(6)$

Here are some other examples of simplifying expressions by applying the order of operations rule.

EXAMPLE 12

Evaluate.

a. $-8 \div (-2)(-2)$

b. $\dfrac{-5 + (-7)}{2}$

c. $2 - 3^2$

Solution

a. $-8 \div (-2)(-2) = 4(-2)$ Divide -8 by -2.

                   $= -8$ Multiply.

b. $\dfrac{-5 + (-7)}{2} = \dfrac{-12}{2}$ Parentheses are understood to be around the numerator, so add -5 and -7.

           $= -6$ Divide.

c. $2 - 3^2 = 2 - 9$ Square 3.

        $= -7$ Subtract.

PRACTICE 12

Simplify.

a. $(-3)(-4) \div (-2)$

b. $\dfrac{-9 - (-3)}{2}$

c. $(-5)^2 + (-1)^2$

EXAMPLE 13

Find the value of each expression.

a. $|3 - 5|$ **b.** $8 - |-10|$ **c.** $\sqrt{1 + 1}$

Solution

a. $|3 - 5| = |-2|$ Parentheses are understood within an absolute value sign, so subtract first.

           $= 2$ Take the absolute value.

b. $8 - |-10| = 8 - 10$ Take the absolute value.

            $= -2$ Subtract.

c. $\sqrt{1 + 1} = \sqrt{2}$ Parentheses are understood under a radical sign, so add.

PRACTICE 13

Evaluate.

a. $|0 - 3|$

b. $-2 + |-1|$

c. $\sqrt{4 - 1}$

EXAMPLE 14

The following tables show the number of calories in servings of various foods and the number of calories burned by various activities:

Food (per serving)	Number of Calories
Apple	80
Banana	105
Pretzel, stick	30
Ginger ale, can	125
Donut	210

(*Source: The World Almanac and Book of Facts*, 2000)

Activity (1 hr)	Number of Calories Burned*
Swimming	288
Bicycling	612
Football	498
Basketball	450
Scrubbing floors	440

*For a 150 lb person

(*Source: Exercise and Weight Control*, President's Council on Physical Fitness and Sports)

Find the net calories in each situation.

a. Suppose that you eat 3 servings of stick pretzels and then play basketball for $\frac{1}{2}$ hr.

b. Suppose that you eat 3 servings of apples and 2 of donuts. You then go bicycling for 2 hr and then swim for 1 hr.

Solution

a. The number of calories in 3 servings of stick pretzels: $3 \cdot 30 = 90$ calories.

The number of calories burned by playing basketball for $\frac{1}{2}$ hr: $\frac{450}{2} = 225$ calories.

The net calories: $90 - 225 = -135$ calories.

b. The number of calories in 3 servings of apples and 2 of donuts: $(3)(80) + (2)(210) = 660$.

The number of calories burned by bicycling for 2 hr and then swimming for 1 hr: $(2)(612) + (1)(288) = 1512$.

The net calories: $660 - 1512 = -852$ calories.

PRACTICE 14

To discourage guessing on a test, an instructor deducts points for incorrect answers, grading according to the following scheme:

Performance on a Test Item	Score
Correct	5
Incorrect	−5
Blank	0

How would the instructor grade each of the following tests?

Test	Number of Items Correct	Number of Items Incorrect	Number of Items Blank	Test Grade
A	17	1	2	
B	19	1	0	
C	12	7	1	

Mathematically Speaking

Fill in each blank with the most appropriate term or phrase from the given list.

reciprocal	opposite	multiplicative inverse
order of operations	product	negative
rule	additive inverse	difference
positive	distributive property	

1. The sum of two negative numbers is _____.

2. To subtract real numbers, add the _____ of the number being subtracted.

3. If two real numbers have the same sign, their _____ is positive.

4. To divide two real numbers, multiply the first number by the _____ of the second number.

5. The quotient of real numbers with the same signs is _____.

6. The _____ tells us in which order to carry out the operations in an expression.

A **Add or subtract.**

7. $-7 + 9$

8. $-8 + 4$

9. $2 - (-2)$

10. $3 - (-1)$

11. $-1 + (-10)$

12. $-5 + (-6)$

13. $3 - 4$

14. $7 - 11$

15. $-1 - 7$

16. $-4 - 2$

17. $2.9 + (-2.9)$

18. $-1.8 - (-1.8)$

19. $5.6 + (-9.2)$

20. $8 + (-3.1)$

21. $-3.7 - (-4)$

22. $-6.2 - (-7)$

23. $\left(-\dfrac{1}{4}\right) + \left(-\dfrac{3}{4}\right)$

24. $-\dfrac{6}{7} + \left(-1\dfrac{1}{7}\right)$

25. $\dfrac{7}{10} + \left(-\dfrac{9}{10}\right)$

26. $-\dfrac{1}{6} + \dfrac{5}{6}$

27. $-\dfrac{2}{5} - \dfrac{3}{5}$

28. $-1\dfrac{1}{8} - \dfrac{7}{8}$

29. $-4.9 - (-9.3)$

30. $1.7 - (-5.7)$

31. $7.23 - (-7.23)$

32. $-10.34 - 10.34$

Multiply or divide.

33. $(-8)(-5)$

34. $(-7)(-6)$

35. $4(-10)$

36. $(-3)(12)$

37. $(-28) \div (-4)$

38. $(-24) \div (-3)$

39. $45 \div (-9)$

40. $-48 \div 16$

41. $\dfrac{-10}{2}$

42. $\dfrac{30}{-6}$

43. $(-2.3)(5)$

44. $-3.1(8)$

45. $(-2.5)(-3.5)$

46. $(-0.4)(-4.2)$

47. $\dfrac{-8.8}{4}$

48. $\dfrac{3.4}{-2}$

49. $\dfrac{-12}{-0.4}$

50. $\dfrac{-0.5}{-5}$

51. $\left(-\dfrac{3}{4}\right)(-8)$

52. $(15)\left(-\dfrac{4}{5}\right)$

53. $6 \div \left(-\dfrac{1}{4}\right)$

54. $-4 \div \dfrac{1}{3}$

55. $\left(-\dfrac{1}{2}\right)\left(\dfrac{2}{3}\right)$

56. $\left(-\dfrac{3}{8}\right)\left(-\dfrac{16}{21}\right)$

57. $\left(-\dfrac{2}{5}\right) \div \left(-\dfrac{3}{5}\right)$

58. $\left(-\dfrac{4}{7}\right) \div \dfrac{1}{14}$

Evaluate.

59. 4^3 **60.** 8^2 **61.** $(-3)^2$ **62.** $(-10)^4$

63. -2^2 **64.** -3^3

Simplify.

65. $-3(5 - 11) - 6$ **66.** $2(7 - 12) + 11$ **67.** $-3[-5 - (1 + 7)]$ **68.** $4[-8 - (8 - 10)]$

69. $-2(10) - 4(-7)$ **70.** $10(-5) + 9(-2)$ **71.** $-36 \div (-4)(-3)$ **72.** $24 \div 4(-2)$

73. $6(-8) \div (-16)(5)$ **74.** $-2(18) \div (9)(-6)$ **75.** $2[3(5 - 8) + 2(6 - 7)]$

76. $-[6(2 + 1) - 3(2 - 10)]$ **77.** $\dfrac{(-4)(7) + 8(-1)}{5 - 11}$ **78.** $\dfrac{(-3)(8) + (-5)(-6)}{-2 - 7}$

79. $(2)(10) - 3^2$ **80.** $(-1)(3) - 5^2$ **81.** $-3 + 4^2 \div 8(5 - 6)^2$

82. $9(7 - 5)^2 \div (-2)^2 - 7$ **83.** $-5[3^2 - 4(32 \div 16)^3 \div 8] + 25$

84. $-3[2(4^2 - 9) \div 7 + (-2)^3] + 12$ **85.** $\dfrac{12 - 15}{-3} + \dfrac{-5 - 11}{-4}$

86. $\dfrac{17 - 1}{2} - \dfrac{42}{8 - 2}$

Find the value of each expression.

87. $|2 - 9|$ **88.** $|7 - 3|$ **89.** $-|6| + |-15|$

90. $-|-3| - |8|$ **91.** $-|6 - 4^2| - |(2)(-7)|$ **92.** $|1 - 9| + |10 - (-5)^2|$

93. $\sqrt{8 + 1} - 4$ **94.** $\sqrt{7 - 5 + 13}$ **95.** $\sqrt{2(12 - 8 - 1)}$

96. $\sqrt{3(1 - 6 + 12)}$

Mixed Practice

Solve.

97. Subtract -2.1 from -4.

98. Find the product of 3 and -11.

99. Find the sum of -6 and 9.

100. Find the value of $10 - |2 - 8|$.

101. Find the quotient of -3.5 and 0.5.

102. Simplify: $(-4)(-2)^3 \div (3^2 - 5)$

Applications

B *Solve.*

103. To avoid turbulent air, an airplane flying at an altitude of 26,000 ft climbs to an altitude of 31,000 ft. What was the change in altitude?

104. At noon, the temperature was 10°F. By 6:00 P.M. the temperature had dropped 12°. What was the temperature at 6:00 P.M.?

105. In a football game, an offensive player is penalized, resulting in a loss of 10 yd for the offense. On the next play, a defensive player is penalized for holding, which results in a gain of 5 yd for the offense. What was the total gain or loss of yd for the offense?

106. The summit of Mauna Loa, a volcano on the island of Hawaii, is 4,170 m above sea level. The distance from the base of Mauna Loa to the summit is 17,170 m. How far below sea level is the base of Mauna Loa? (*Source:* USGS Hawaiian Volcano Observatory)

107. The height of an object in feet above the ground 3 sec after it is dropped from a height of 160 ft is given by the expression $-16(3)^2 + 160$. What is the height of the object?

108. The approximate area of the passivity zone of the wrestling mat shown in the figure is $3.14[(4.5)^2 - (3.5)^2]$ m². Calculate this area.

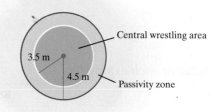

109. The table shows the total revenue and total expenditures of the federal government for the years 2005–2010. (*Source:* Office of Management and Budget)

Year	2005	2006	2007	2008	2009	2010
Receipts (in trillions)	$2.15	$2.41	$2.57	$2.52	$2.10	$2.17
Outlays (in trillions)	$2.47	$2.66	$2.73	$2.98	$3.52	$3.72

a. Complete the following table to show the annual budget surplus or deficit of the federal government.

Year	2005	2006	2007	2008	2009	2010
Receipts-Outlays (in trillions)						

b. What was the average annual surplus or deficit for the six-year period?

$(- |8 - 4^2| - |(2)(-4)|$

$(- |8 - 16| - |(2)(-4)|$

$(- |-8| - |(2)(-4)|$

$(-8 - |(2)(-4)|$

$-8 - |-8|$

$-8 - 8 = -16$

$\begin{array}{r} 2\,768 \\ - 783 \\ \hline 1,985 \end{array}$

$|7 + 31| + (-1) + (-10) + (-9) + (-15) + (-7)$

$48 + 11 + 24 - 7$

$59 + 17$

$76 \div 7$

$17 + 31 + (-1) + (-10) + (-9) + (-15) + (-7) = \dfrac{6}{7}$

WRt 102

$$-1\frac{1}{7} + \frac{2}{7}$$

$$7\overline{\smash{\big)}8} \qquad -1\frac{3}{7}$$
$$\underline{7}$$
$$1 \qquad \frac{-8}{19} \cdot \frac{-19}{1} = 8 \qquad -\frac{5}{13}, -\frac{32}{23} = \frac{10}{9}$$

$$\frac{-15}{6} \quad \frac{-10}{9}$$

$$8 \cdot 8 \cdot 8$$

$$-7(1-8)-7$$
$$-14(4)+16 \qquad -7(-7)-7$$
$$\overline{-4+2} \qquad 49-7 = 42 \qquad 4(-2) = -8$$

$$\frac{-48+16}{-2} \quad \frac{-32}{-2} \; \boxed{16} \qquad \frac{-14(4)+16}{-4+2} \qquad \frac{-56+16}{-2}$$

$$\frac{-40}{-2} = \boxed{-20}$$

$$-|8-4^2| - |(2)(-4)| = \qquad -|8-4^2| -2+8 =$$
$$-|8-16| - |-8| = \qquad -|8-16| + 6$$
$$-|-8| - |8| = \qquad -|-8| + 6$$
$$-8 - 8 = 0 \qquad -8 + 6$$
$$-2$$

110. The table below shows the change in the closing price per share of stock from the previous day for a company over a six-day period in February 2011. The price per share of stock at the close of business after Day 1 was $30.54.

a. Complete the table to show the price per share of stock for the remaining days.

Day	1	2	3	4	5	6
Change from Previous Day (in dollars)	+0.60	−0.04	+0.27	+0.08	−0.45	−0.21
Price per Share (in dollars)	30.54					

b. What do negative changes from the previous day mean?

111. The daily low temperatures in Fargo, North Dakota, for seven consecutive days are given in the table below. Calculate the average daily low temperature for the seven-day period, rounded to the nearest integer. (*Source:* National Oceanic and Atmospheric Administration)

Day	1	2	3	4	5	6	7
Low Temperature	22°F	27°F	−1°F	−9°F	−2°F	−13°F	−9°F

112. A man weighing 218 lb joined a weight-loss program. If he lost an average of 2.6 lb per week, approximately how many weeks did it take him to reach his goal of weighing 175 lb?

113. The dosage instructions on a prescription call for a patient to take $1\frac{1}{2}$ tablets of a 10-mg medication three times a day for 10 days. How many milligrams of the medication will the patient take in one day?

114. A certain type of nicotine-replacement therapy uses patches with varying concentrations of nicotine to gradually decrease a smoker's nicotine dependency. The patches used in the first phase of the therapy each contain 21 mg of nicotine. For each remaining phase of therapy, the concentration of nicotine in the patches is decreased by 7 mg. What is the concentration of nicotine in the patches used in the third phase of therapy?

115. The graph shows the annual revenues in millions of dollars of a small publishing company.

a. Use the graph to complete the table.

Year	2008	2009	2010	2011
Change in Revenue from Previous Year (in millions of dollars)				

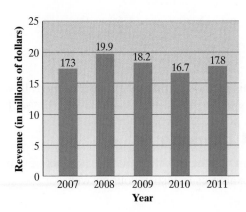

b. What was the average annual change in revenue?

c. What was the average annual revenue over the five-year period?

116. The graph shows the closing price per share of stock for a firm during a week in February 2011.

a. Use the graph to the right to complete the following table.

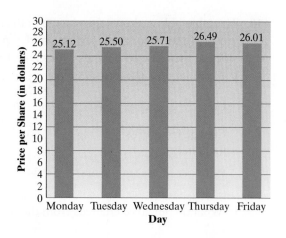

Day	Tuesday	Wednesday	Thursday	Friday
Change in Price from Previous Day (in dollars)				

b. What was the average change in the price of the stock over the week?

c. What was the average price per share of stock for the week?

• Check your answers on page A-15.

MINDStretchers

Mathematical Reasoning

1. For what real numbers a and b is $|a| + |b| = |a + b|$? Justify your answer.

Historical

2. In using the *Russian peasant* method of multiplication, we begin by writing the two numbers to be multiplied under one another, forming two rows. We start the process by halving the number in the top row (ignoring any remainders) while at the same time doubling the number in the bottom row. The process stops when 1 is reached in the top row. The product is the sum of the numbers in the bottom row that are under the odd numbers in the top row. For example, the product of 70 and 95 is found as follows:

70	35	17	8	4	2	1
95	190	380	760	1520	3040	6080

$$190 + 380 \quad + \quad 6080 = 6650 \leftarrow \text{The product is the sum of the three numbers.}$$

Use this method to find the product of the following:
a. 84 and 43   **b.** 115 and 56

Critical Thinking

3. Explain whether it is true that the product of two real numbers is always greater than the quotient of the two numbers. Give examples to justify your answer.

1.3 Properties of Real Numbers

OBJECTIVES

A To use the properties of real numbers

B To solve applied problems involving the properties of real numbers

In this section, we focus on some of the key properties of real numbers—the commutative properties, the associative properties, the identity properties, the inverse properties, the multiplicative property of 0, and the distributive property. These properties are very important to the study of algebra, for they underlie algebraic procedures and, in particular, they help us to simplify complicated expressions.

The Commutative Properties

Let's begin by considering the *commutative property of addition*.

Commutative Property of Addition

For any real numbers a and b,

$$a + b = b + a.$$

In words, this property states that the sum of two numbers is the same, regardless of order. For instance, the sums $3 + 5$ and $5 + 3$ give the same result. Note that in this statement of the commutative property, as well as in the other properties of real numbers, symbols are used to represent real numbers. The letter a represents one number and the letter b represents another number. In algebra, a **variable** is a quantity that is unknown, that is, one that can change or vary in value.

EXAMPLE 1

Rewrite each expression using the commutative property of addition.

a. $7 + (-2)$ **b.** $2x + y$

Solution

a. $7 + (-2) = (-2) + 7$, or $-2 + 7$

b. $2x + y = y + 2x$

PRACTICE 1

Use the commutative property of addition to rewrite each expression.

a. $-3 + 5$

b. $b + 3a$

Now, we look at the *commutative property of multiplication*.

Commutative Property of Multiplication

For any real numbers a and b,

$$a \cdot b = b \cdot a.$$

In words, this property states that the product of two numbers is the same, regardless of order. For instance, the products $3 \cdot 5$ and $5 \cdot 3$ are equal. Note that when writing an expression involving a product, we can either omit the multiplication symbol or use parentheses. For instance, we can write $a \cdot b$ as either ab or $(a)(b)$.

EXAMPLE 2

Rewrite each expression using the commutative property of multiplication.

a. $(-5)(-3)$ **b.** $x(-7)$

Solution

a. $(-5)(-3) = (-3)(-5)$ **b.** $x(-7) = (-7)x,$ or $-7x$

PRACTICE 2

Use the commutative property of multiplication to rewrite each expression.

a. $(-8)(2)$

b. $-4n$

The Associative Properties

Next, let's consider the two associative properties: first, the *associative property of addition* and then, the *associative property of multiplication*.

> **Associative Property of Addition**
>
> For any real numbers a, b, and c,
> $$(a + b) + c = a + (b + c).$$

In words, this property states that when adding three numbers, their sum is the same, regardless of how they are grouped. For instance, the result of adding 2, 3, and 6 is the same whether we add 2 and 3 first and then add 6 to this sum, or we add 3 and 6 first and then add 2 to this sum, that is, $(2 + 3) + 6 = 2 + (3 + 6)$.

EXAMPLE 3

Rewrite each expression using the associative property of addition.

a. $[(-4) + (-3)] + 6$ **b.** $(2p + q) + r$

Solution

a. $[(-4) + (-3)] + 6 = (-4) + [(-3) + 6]$

b. $(2p + q) + r = 2p + (q + r)$

PRACTICE 3

Use the associative property of addition to rewrite each expression.

a. $[8 + (-1)] + 2$

b. $(x + 3y) + z$

Similarly, the associative property of multiplication involves the regrouping of three numbers.

> **Associative Property of Multiplication**
>
> For any real numbers a, b, and c,
> $$(a \cdot b)c = a(b \cdot c).$$

In words, this property states that when multiplying three numbers, their product is the same, regardless of how they are grouped. For instance, $(4 \cdot 2) \cdot 5 = 4 \cdot (2 \cdot 5)$.

EXAMPLE 4

Rewrite each expression using the associative property of multiplication.

a. $[(-5)(-1)](2)$ **b.** $(2)(4x)$

Solution

a. $[(-5)(-1)](2) = (-5)[(-1)(2)]$ **b.** $(2)(4x) = (2 \cdot 4)x$

PRACTICE 4

Use the associative property of multiplication to rewrite each expression.

a. $[(-3)(5)](-2)$

b. $(3)(-6n)$

When adding three or more numbers, their sum is the same regardless of the order or grouping because of the commutative and associative properties. Note that when adding or multiplying three or more positive numbers and negative numbers, it is usually easier to work with the positives separately from the negatives.

EXAMPLE 5

Find the sum: $5 + (-6) + 3 + (-5)$

Solution   Let's group the numbers by sign:

$$\underbrace{5 + 3}_{\text{Positives}} \quad + \quad \underbrace{(-6) + (-5)}_{\text{Negatives}}$$

$$5 + 3 = 8 \qquad \text{Add the positives.}$$
$$-6 + (-5) = -11 \qquad \text{Add the negatives.}$$
$$8 + (-11) = -3 \qquad \text{Find the sum of the positives and the negatives.}$$

So $5 + (-6) + 3 + (-5) = -3$.

PRACTICE 5

Find the sum: $-8 + (-4) + 3 + (-8)$

EXAMPLE 6

Multiply: $5(-2)(-1)(3)$

Solution   We group the numbers by sign.

$$\underbrace{5(3)}_{\text{Positives}} \quad \underbrace{(-2)(-1)}_{\text{Negatives}}$$

$$5(3) = 15 \qquad \text{Multiply the positives.}$$
$$(-2)(-1) = 2 \qquad \text{Multiply the negatives.}$$
$$15 \cdot 2 = 30 \qquad \text{Find the product of the positives and the negatives.}$$

So $5(-2)(-1)(3) = 30$.

PRACTICE 6

Find the product: $(-6)(-1)(4)(-5)$

The Identity Properties

Two other important properties of real numbers are the *additive identity property* (also called the *identity property of 0*) and the *multiplicative identity property* (the *identity property of 1*).

Additive Identity Property

For any real number a,

$$a + 0 = a \quad \text{and} \quad 0 + a = a.$$

In words, this property states that the sum of any number and zero is the original number. For instance, $5 + 0 = 5$ and $0 + 5 = 5$.

Multiplicative Identity Property

For any real number a,

$$a \cdot 1 = a \quad \text{and} \quad 1 \cdot a = a.$$

In words, this property states that the product of any number and one is the original number. For instance, $3 \cdot 1 = 3$ and $1 \cdot 3 = 3$.

EXAMPLE 7

Perform the indicated operation using an identity property.

a. $-6 + 0$ **b.** $0 + 3x$ **c.** $(-4)(1)$ **d.** $(-7n)(1)$

Solution

a. $-6 + 0 = -6$ **b.** $0 + 3x = 3x$

c. $(-4)(1) = -4$ **d.** $(-7n)(1) = -7n$

PRACTICE 7

Use an identity property to simplify.

a. $-5 + 0$

b. $0 + 6y$

c. $(1)(-2)$

d. $(-5x)(1)$

The Inverse Properties

Next, we consider the two inverse properties: first, the *additive inverse property* and then, the *multiplicative inverse property*.

Additive Inverse Property

For any real number a, there is exactly one number, $-a$, such that

$$a + (-a) = 0 \quad \text{and} \quad -a + a = 0,$$

where a and $-a$, are said to be *additive inverses* of each other.

In words, this property states that the sum of any number and its opposite is zero. For instance, $4 + (-4) = 0$ and $-4 + 4 = 0$.

Recall that when finding the additive inverse of a number, we change the sign of the number. Note that this property relates to the additive inverse or opposite of a number, which was discussed in Section 1.1.

EXAMPLE 8

Find the additive inverse of each.

a. 4.5 **b.** -3 **c.** x

Solution

a. The additive inverse of 4.5 is -4.5.

b. The additive inverse of -3 is 3.

c. The additive inverse of x is $-x$.

PRACTICE 8

Find the additive inverse of each.

a. -2 **b.** $-\dfrac{2}{3}$

c. y

Now, let's look at the multiplicative inverse property. Note that this property relates to the multiplicative inverse or reciprocal of a number, discussed in Section 1.2.

Multiplicative Inverse Property

For any nonzero real number a, there is exactly one number $\dfrac{1}{a}$ such that

$$a \cdot \frac{1}{a} = 1 \quad \text{and} \quad \frac{1}{a} \cdot a = 1,$$

where a and $\dfrac{1}{a}$ are said to be *multiplicative inverses* (or *reciprocals*) of each other.

In words, this property states that the product of a number and its multiplicative inverse is one. For instance, $2 \cdot \dfrac{1}{2} = 1$ and $\dfrac{1}{2} \cdot 2 = 1$.

> **TIP** A shortcut for finding the multiplicative inverse of a rational number is to interchange its numerator and denominator. For instance, the multiplicative inverse of $\dfrac{2}{3}$ is $\dfrac{3}{2}$.

EXAMPLE 9

Find the multiplicative inverse of each.

a. $\dfrac{1}{3}$ **b.** 3 **c.** -2 **d.** $-\dfrac{3}{4}$

Solution

a. To find the multiplicative inverse of $\dfrac{1}{3}$, we interchange the numerator and denominator, getting $\dfrac{3}{1}$, or 3. We can check that $\dfrac{1}{3} \cdot 3 = 1$.

b. The multiplicative inverse of $3 \left(\text{or } \dfrac{3}{1} \right)$ is $\dfrac{1}{3}$.

c. The multiplicative inverse of $-2 \left(\text{or } \dfrac{-2}{1} \right)$ is $\dfrac{1}{-2}$, or $-\dfrac{1}{2}$.

d. The multiplicative inverse of $-\dfrac{3}{4} \left(\text{or } \dfrac{-3}{4} \right)$ is $\dfrac{4}{-3}$, or $-\dfrac{4}{3}$.

PRACTICE 9

Find the multiplicative inverse of each.

a. $\dfrac{1}{5}$ **b.** 2

c. -5 **d.** $-\dfrac{2}{7}$

Note that for any nonzero number, the number and its reciprocal have the same sign.

> **TIP** We can rewrite a rational number such as
> $$\frac{1}{-10} \text{ or } \frac{-1}{10} \text{ as } -\frac{1}{10}; \text{ that is, } \frac{1}{-10} = \frac{-1}{10} = -\frac{1}{10}.$$

The Multiplication Property of Zero

The next property of real numbers that we consider is the *multiplication property of zero*.

Multiplication Property of Zero

For any real number a,

$$a \cdot 0 = 0 \quad \text{and} \quad 0 \cdot a = 0.$$

In words, this property states that the product of any number and zero is zero. For instance, $6 \cdot 0 = 0$ and $0 \cdot 6 = 0$.

EXAMPLE 10

Calculate.

a. $(13.4)(0)$ **b.** $(0)(-4.92)$ **c.** $(7x)(0)$

Solution

a. $(13.4)(0) = 0$ **b.** $(0)(-4.92) = 0$ **c.** $(7x)(0) = 0$

PRACTICE 10

Find the value of each.

a. $(-10.36)(0)$

b. $(0)\left(-\dfrac{2}{3}\right)$

c. $(0)(-4n)$

The Distributive Property

The *distributive property of multiplication with respect to addition*, or simply the *distributive property*, involves two operations—multiplication and addition.

> ## The Distributive Property
>
> For any real numbers a, b, and c,
> $$a \cdot (b + c) = a \cdot b + a \cdot c.$$

In words, this property states that a number times the sum of two quantities is equal to the number times one quantity plus the number times the other quantity. For instance, $5 \cdot (2 + 6) = 5 \cdot 2 + 5 \cdot 6$.

Note:

- When we rewrite $a(b + c)$ as $ab + ac$, we think of this as "removing parentheses" or "distributing out."
- Rewriting $a \cdot b + a \cdot c$ as $a \cdot (b + c)$ is sometimes referred to as applying the distributive property "in reverse."
- Another way to express the distributive property is $(b + c)a = ba + ca$.
- Substituting $-c$ for c gives still another form of the distributive property: $a(b - c) = ab - ac$.

EXAMPLE 11

Rewrite each expression using the distributive property.

a. $(-5)(2 + 6)$ **b.** $0.4(x + y)$

c. $(3 - x) \cdot y$

Solution

a. $(-5)(2 + 6) = (-5)(2) + (-5)(6)$

b. $0.4(x + y) = (0.4)x + (0.4)y = 0.4x + 0.4y$

c. $(3 - x) \cdot y = 3 \cdot y - x \cdot y = 3y - xy$

PRACTICE 11

Rewrite each expression using the distributive property.

a. $(-2)(9 + 4.3)$

b. $0.2(a + b)$

c. $(2 - p) \cdot q$

Division Involving Zero

In some division problems, zero is either the dividend or the divisor.

> **Division Involving zero**
>
> For any real number a,
> - $0 \div a = 0$, where a is nonzero;
> - $a \div 0$ is undefined.

In words, these properties state that zero divided by any nonzero number is zero, whereas any number divided by zero is undefined. For instance, $0 \div 2 = 0$ and $2 \div 0$ is undefined. Note that $0 \div 2$ is 0 because 0×2 is 0 and $2 \div 0$ is undefined because there is no number that when multiplied by 0 gives 2.

EXAMPLE 12

Find the value of each of the following expressions, if possible. (Assume that each variable is nonzero.)

a. $\dfrac{0}{5}$

b. $0 \div x$

c. $-3y \div 0$

Solution

a. $\dfrac{0}{5} = 0$

b. $0 \div x = 0$

c. Not possible, because $-3y \div 0$ is undefined

PRACTICE 12

Evaluate each expression.

a. $\dfrac{0}{-2}$

b. $-2n \div 0$

c. $0 \div t$

The properties of real numbers are typically applied not in isolation but in combination. Consider the following example.

EXAMPLE 13

Show that $(x + 0) + (-x) = 0$, justifying each step.

Solution

$(x + 0) + (-x) = (0 + x) + (-x)$ The commutative property of addition

$= 0 + [x + (-x)]$ The associative property of addition

$= 0 + 0$ The additive inverse property

$= 0$ The additive identity property

PRACTICE 13

Show that $4 \cdot \left(\dfrac{1}{4}x\right) = x$, justifying each step.

$4 \cdot \left(\dfrac{1}{4}x\right) = \left(4 \cdot \dfrac{1}{4}\right)x$ a. _____

$= 1x$ b. _____

$= x$ c. _____

Finally, let's examine the basic properties of real numbers in the context of real-world applications.

EXAMPLE 14

The volume of the packing crate shown is the product of its length, width, and height: *lwh*.

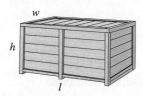

Here are two ways to compute this volume:

- First, find the product of the length and the width and then, multiply this product by the height: $(lw)h$.
- First, find the product of the width and the height and then, multiply the length by this product: $l(wh)$.

Must the two answers be equal? Explain.

Solution   The two answers must be equal. The associative property of multiplication states that $(lw)h = l(wh)$.

PRACTICE 14

A financier receives a 100% return on an investment of *n* dollars. Express this return in dollars, justifying your answer.

Mathematically Speaking

Fill in each blank with the most appropriate term or phrase from the given list.

> reciprocals
>
> associative property of addition
>
> distributive property
>
> commutative property of multiplication
>
> additive inverse property
>
> multiplicative identity property
>
> additive identity property
>
> multiplication property of zero
>
> opposites

1. The _____ states that we get the same product when we multiply two numbers in any order.

2. The _____ states that we get the same result if we regroup the sum of three numbers.

3. The _____ states that the sum of a number and 0 is the number itself.

4. The _____ states that the product of a number and 1 is the number itself.

5. The _____ states that the sum of a number and its additive inverse is the additive identity 0.

6. If $a \cdot \frac{1}{a} = 1$ and $\frac{1}{a} \cdot a = 1$, then a and $\frac{1}{a}$ are multiplicative inverses, or _____, of each other.

7. The _____ states that for any real number a, $a \cdot 0 = 0$ and $0 \cdot a = 0$.

8. The _____ involves two operations, namely, addition and multiplication.

A *Rewrite each expression using the indicated property of real numbers.*

9. The commutative property of addition: $3.7 + 2$

10. The commutative property of addition: $3 \cdot 2 + 1 \cdot (-4)$

11. The associative property of addition: $[(-1) + (-6)] + 7$

12. The associative property of addition:
    $6(8) + [7(-3) + (-1)(2)]$

13. The additive identity property: $-3 + 0$

14. The multiplicative identity property: $1 \cdot (-3)$

15. The distributive property: $3(1 + 9)$

16. The associative property of multiplication:
    $[(-2)(-3)](6)$

17. The commutative property of multiplication:
    $(2 + 7) \cdot 5$

18. The commutative property of multiplication: $5(-8)$

19. The distributive property in reverse:
    $2a + 2b$

20. The distributive property in reverse: $8x + 2x$

Indicate the property that justifies each statement.

21. $n(-3) = -3n$

22. $(n + 2) \cdot 5 = 5 \cdot (n + 2)$

23. $7a + b = b + 7a$

24. $9y + 7x = 7x + 9y$

25. $2s + 2t = 2(s + t)$

26. $8(a + b) = 8a + 8b$

27. $(n + 2) \cdot 0 = 0$

28. $0 \cdot (n + 5) = 0$

29. $-(x + 1) + (x + 1) = 0$

30. $0 = 6n + (-6n)$

31. $5 + (y - x) = (5 + y) - x$

32. $(2 + y) + x = 2 + (y + x)$

33. $\left(8 \cdot \dfrac{1}{8}\right)x = 1x$

34. $1x = \left(10 \cdot \dfrac{1}{10}\right)x$

Calculate, if possible.

35. $(6.1)(0)$

36. $(0)(-2.35)$

37. $(-1\tfrac{1}{5})(0)$

38. $(0)(2\tfrac{3}{4})$

39. $8 + (-3) + (-5)$

40. $12 + 6 + (-10)$

41. $2 + (-3.8) + 9.13 + (-1)$

42. $(-2) + (-3.5) + 7.4 + (-4)$

43. $(-7)(-2)(3)$

44. $(10)(-1)(-6)$

45. $(-2)(-2)(-2)$

46. $(-1)(-1)(-1)$

47. $(-5)(-7)(-2)(10)$

48. $(7)(-1)(-5)(2)$

49. $\dfrac{0}{3}$

50. $0 \div 10$

51. $-2 \div 0$

52. $\dfrac{-7}{0}$

Find each of the following.

53. The additive inverse of 2

54. The additive inverse of 9.1

55. The additive inverse of -7

56. The additive inverse of 0

57. The multiplicative inverse of 7

58. The multiplicative inverse of $\dfrac{1}{2}$

59. The multiplicative inverse of -1

60. The multiplicative inverse of $-\dfrac{2}{3}$

Rewrite each expression without the grouping symbols.

61. $(-4)(2 + 5)$

62. $6(1 - 6)$

63. $(x + 10) \cdot 3$

64. $(y + 4) \cdot 5$

65. $-(a + 6b)$

66. $-(x - 7y)$

67. $n(n - 2)$

68. $a(a + 1)$

To prove the given statement, justify each step.

69. $(2n) \cdot 5 = 10n$

$(2n) \cdot 5 = 5 \cdot (2n)$ **a.** _____

$ = (5 \cdot 2)n$ **b.** _____

$ = 10n$ **c.** _____

70. $t + [5 + (-t)] = 5$

$\begin{aligned} t + [5 + (-t)] &= t + [(-t) + 5] & \text{a.} \underline{\hspace{3cm}} \\ &= [t + (-t)] + 5 & \text{b.} \underline{\hspace{3cm}} \\ &= 0 + 5 & \text{c.} \underline{\hspace{3cm}} \\ &= 5 & \text{d.} \underline{\hspace{3cm}} \end{aligned}$

71. $\left(\dfrac{1}{2}\right)(2p) = p$

$\begin{aligned} \left(\dfrac{1}{2}\right)(2p) &= \left(\dfrac{1}{2} \cdot 2\right)p & \text{a.} \underline{\hspace{3cm}} \\ &= (1)p & \text{b.} \underline{\hspace{3cm}} \\ &= p & \text{c.} \underline{\hspace{3cm}} \end{aligned}$

72. $3x + 2x = 5x$

$\begin{aligned} 3x + 2x &= x \cdot 3 + x \cdot 2 & \text{a.} \underline{\hspace{3cm}} \\ &= x(3 + 2) & \text{b.} \underline{\hspace{3cm}} \\ &= x \cdot 5 & \text{c.} \underline{\hspace{3cm}} \\ &= 5x & \text{d.} \underline{\hspace{3cm}} \end{aligned}$

Mixed Practice

Solve.

73. Find the additive inverse of -0.2.

74. Rewrite the expression $9y + 2y$ using the distributive property.

75. Indicate the definition, property, or number fact that justifies the statement $5(ab) = (5a)b$.

76. Calculate: $(5n)(0)$

77. Justify the statement $\left(\dfrac{1}{4} \cdot 4\right)n = 1n$.

78. Rewrite without the grouping symbols: $t(t - 5)$

Applications

B *Explain each answer in terms of the appropriate property of real numbers or definition.*

79. Each morning, you take a bus from home to a stop, and then you walk to your office. At the end of the day, you make the trip in reverse. Is the distance you travel going to work the same as the distance you travel returning home?

80. A hiker walks m mi west, and then m mi east. How far does she wind up from her original starting point?

81. The rate of sales tax, expressed as a decimal, is r. A shopper buys two items, one costing p dollars and the other q dollars, before tax. One expression for the total tax paid on the two items is $r(p + q)$ and another is $rp + rq$. Will the shopper pay the same amount of taxes regardless of which expression is used?

82. A sales representative makes the trip from Company A to Company B to Company C, and then back to Company A, as shown. The length of this trip can be represented by either $(AB + BC) + CA$ or by $AB + (BC + CA)$. (The notation AB represents the "length of AB.") Must the two representations be equal?

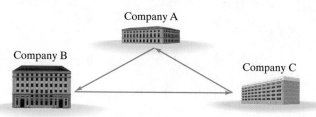

Company A

Company B

Company C

83. An athlete is trying to lose weight. At the beginning of one week, he weighs p lb. At the end of the week, the athlete has neither gained nor lost weight. What is his weight at the end of the week?

84. In a photograph, the pictured height of a child is h in. A photographer enlarges the photograph, multiplying the pictured height by a factor of f. By what factor must the photographer multiply the enlarged picture to shrink the pictured height of the child back to h in.?

85. You calculate the area of the triangle shown using the product $\dfrac{1}{2} \cdot 9 \cdot 12$. Your friend calculates the area of the same triangle using the product $\dfrac{1}{2} \cdot 12 \cdot 9$. Are both calculations correct?

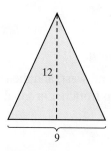

12

9

86. An expression for the surface area of the cereal box shown is given by the expression $2lh + 2wh + 2lw$. Can the surface area also be calculated using the expression $2(lh + wh + lw)$? Explain.

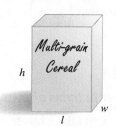

Multi-grain Cereal

h

l

w

• Check your answers on page A-16.

MINDStretchers

Groupwork

1. With a partner, determine whether the commutative and associative properties hold for subtraction and division. Give examples to support your answer.

Critical Thinking

2. Consider the following diagram:

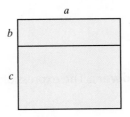

a. Determine the area of the large rectangle formed by the two smaller rectangles in two ways.

b. What property of real numbers does your response in part (a) illustrate?

Writing

3. Explain why 0 has no multiplicative inverse.

1.4 Laws of Exponents and Scientific Notation

OBJECTIVES

A To write or evaluate an expression containing exponents

B To rewrite an expression by using the product or quotient rules of exponents

C To rewrite an expression containing negative exponents

D To rewrite an expression by using the power rule or by raising a product or a quotient to a power

E To change a number written in scientific notation to standard notation, or vice versa

F To solve applied problems involving the laws of exponents and scientific notation

In this section, we focus on exponents with particular attention to the rules known as the *laws of exponents*. We also examine an important and practical application of exponents called *scientific notation*.

Recall from Section 1.2 that *exponential notation* is a shorthand method for representing repeated multiplication of the same factor.

DEFINITION

For any real number x and any positive integer a,

$$x^a = \underbrace{x \cdot x \ldots x \cdot x}_{a \text{ times}}$$

where x is called the **base** and a is called the **exponent** (or **power**). The expression x^a is read "x to the power a."

Note in this definition that the exponent is a positive integer.

Just as we are said to *square* a number when raising it to the power 2, so we speak of *cubing* a number when raising it to the power 3. For instance, the product $(-5) \cdot (-5) \cdot (-5)$, which can be written $(-5)^3$, is read "-5 *to the third power*" or more simply, "-5 *cubed*." Two other exponents are worth singling out:

Exponent 1

For any real number x,

$$x^1 = x.$$

In words, any number raised to the power 1 is the number itself. For instance, $(-5)^1 = -5$.

Exponent 0

For any nonzero real number x,

$$x^0 = 1.$$

In words, any nonzero number raised to the power 0 is 1. For instance, $(-3)^0 = 1$.

Consider the following pattern:

$$4^4 = 256$$
$$4^3 = 64$$
$$4^2 = 16$$
$$4^1 = 4$$
$$4^0 = 1$$

Can you explain why it makes sense for 4^0 to be equal to 1?

Note that throughout this text, we assume that any variable raised to the 0 power represents a nonzero real number.

EXAMPLE 1

Evaluate.

a. $(-2x)^1$ **b.** $(4y)^0$ **c.** $-3n^0$ **d.** $(4x + y)^1$

Solution

a. $(-2x)^1 = -2x$

b. $(4y)^0 = 1$

c. $-3n^0 = -3 \cdot n^0 = -3 \cdot 1 = -3$ **d.** $(4x + y)^1 = 4x + y$

PRACTICE 1

Find the value of each expression.

a. $-3y^1$ **b.** $-(2x)^0$

c. $5x^0$ **d.** $(-7a + b)^1$

Note that in Example 1(c), the exponent only applies to n in the expression $-3n^0$.

Laws of Exponents

The laws of exponents are rules that apply to exponents. These rules are not arbitrary, but follow logically from the definition of an exponent.

First, let's consider the *product rule*. This rule covers the situation in which we are multiplying two exponential expressions with the same base. For instance, consider the product $x^2 \cdot x^3$. From the definition of an exponent, we get:

$$x^2 \cdot x^3 = \overbrace{(x \cdot x)}^{\text{2 factors}} \overbrace{(x \cdot x \cdot x)}^{\text{3 factors}} = x^5$$

$$\underbrace{}_{\text{5 factors}}$$

Since 2 factors of x and 3 additional factors of x make 5 factors of x, it follows that:

$$x^2 \cdot x^3 = x^{2+3} = x^5$$

We can generalize this result as follows:

> ### Product Rule of Exponents
>
> For any nonzero real number x and for any nonnegative integers a and b,
> $$x^a \cdot x^b = x^{a+b}.$$

In words, to multiply powers of the same base, add the exponents and keep the base the same. For instance, $2^3 \cdot 2^5 = 2^{3+5} = 2^8$.

EXAMPLE 2

Rewrite in terms of a base raised to a power, if possible.

a. $x^8 \cdot x^{10}$ **b.** $(-5t)^2 \cdot (-5t)^8$

c. $(m + 1)^4 \cdot (m + 1)$ **d.** $a^5 \cdot b^2$

Solution   Since we are multiplying the same base raised to two exponents, the product rule tells us to add the exponents but *not to change the base.*

a. $x^8 \cdot x^{10} = x^{8+10} = x^{18}$

b. $(-5t)^2 \cdot (-5t)^8 = (-5t)^{2+8} = (-5t)^{10}$

c. $(m + 1)^4 \cdot (m + 1) = (m + 1)^{4+1} = (m + 1)^5$

d. In the expression $a^5 \cdot b^2$, we cannot apply the product rule because the bases are not the same.

PRACTICE 2

Express in terms of a base raised to a power, if possible.

a. $n^3 \cdot n^7$

b. $(-4x)^3 \cdot (-4x)$

c. $(y + 3)^5 \cdot (y + 3)^0$

d. $p^6 \cdot q^3$

Next, let's consider the *quotient rule*. This rule addresses the case when two powers of the same base are divided. For instance, consider the quotient $\dfrac{x^5}{x^2}$. Using the definition of an exponent, we have:

$$\frac{x^5}{x^2} = \frac{\overbrace{x \cdot x \cdot x \cdot \overset{1}{\cancel{x}} \cdot \overset{1}{\cancel{x}}}^{\text{5 factors}}}{\underbrace{\underset{1}{\cancel{x}} \cdot \underset{1}{\cancel{x}}}_{\text{2 factors}}} = \underbrace{x \cdot x \cdot x}_{\text{3 factors}} = x^3$$

Since 5 factors of x divided by 2 factors of x yields 3 factors of x, it follows that:

$$\frac{x^5}{x^2} = x^{5-2} = x^3$$

This result can be generalized as follows:

Quotient Rule of Exponents

For any nonzero real number x and for any nonnegative integers a and b,

$$\frac{x^a}{x^b} = x^{a-b}.$$

In words, to divide powers of the same base, subtract the exponent in the denominator from the exponent in the numerator and keep the base the same. For instance,

$$\frac{10^7}{10^3} = 10^{7-3} = 10^4.$$

EXAMPLE 3

Rewrite each expression as a base to a power, if possible.

a. $\dfrac{p^{12}}{p^7}$ **b.** $\dfrac{(5y)^3}{(5y)^3}$ **c.** $\dfrac{(x+3)^4}{x+3}$ **d.** $\dfrac{x^6}{y^2}$

Solution Since we are dividing powers with the same base, the quotient rule tells us to subtract the exponents but *not to change the base*.

a. $\dfrac{p^{12}}{p^7} = p^{12-7} = p^5$

b. $\dfrac{(5y)^3}{(5y)^3} = (5y)^{3-3} = (5y)^0 = 1$

c. $\dfrac{(x+3)^4}{x+3} = (x+3)^{4-1} = (x+3)^3$

d. In the expression $\dfrac{x^6}{y^2}$, we cannot apply the quotient rule because the bases are different.

PRACTICE 3

Rewrite each expression as a base to a power, if possible.

a. $\dfrac{s^8}{s^2}$

b. $\dfrac{(-3r)^{10}}{(-3r)^9}$

c. $\dfrac{(t-2)^6}{(t-2)^4}$

d. $\dfrac{a^4}{b}$

EXAMPLE 4

Simplify.

a. $n^4 \cdot n^5 \cdot n$ **b.** $(x^3y^2)(x^2y^6)$ **c.** $\dfrac{r^2 \cdot r^2 \cdot r^4}{r^7}$

Solution

a. $n^4 \cdot n^5 \cdot n = n^{10}$ Use the product rule.

b. $(x^3y^2)(x^2y^6) = x^3 \cdot x^2 \cdot y^2 \cdot y^6$ Rearrange the factors.

                   $= x^5y^8$ Use the product rule.

c. $\dfrac{r^2 \cdot r^2 \cdot r^4}{r^7} = \dfrac{r^8}{r^7}$ Use the product rule in the numerator.

           $= r^1$ Then, use the quotient rule.

           $= r$

PRACTICE 4

Simplify.

a. $x^2 \cdot x^8 \cdot x^3$

b. $(a^5b)(a^2b)$

c. $\dfrac{y^{11}}{y^4 \cdot y^3 \cdot y}$

Up to this point, we have only considered exponents that were either positive or 0. However, an exponent may be negative. This case can be demonstrated using the quotient rule. Consider, for instance, the quotient $\dfrac{x^2}{x^5}$. One way to simplify this fraction is by using the definition of an exponent and dividing out common factors, as shown below:

$$\frac{x^2}{x^5} = \frac{\overset{1}{\cancel{x}} \cdot \overset{1}{\cancel{x}}}{\underset{1}{\cancel{x}} \cdot \underset{1}{\cancel{x}} \cdot x \cdot x \cdot x} = \frac{1}{x^3}$$

Another way to simplify this fraction is by using the quotient rule, getting:

$$\frac{x^2}{x^5} = x^{2-5} = x^{-3}$$

Since $\dfrac{x^2}{x^5} = x^{-3}$ and $\dfrac{x^2}{x^5} = \dfrac{1}{x^3}$, we conclude that $x^{-3} = \dfrac{1}{x^3}$. This result is generalized as follows:

Negative Exponent

For any nonzero real number x and for any positive integer a,

$$x^{-a} = \frac{1}{x^a}.$$

In words, to evaluate x^{-a}, take the reciprocal of x^{-a} and change the sign of the exponent. For instance, $3^{-2} = \dfrac{1}{3^2} = \dfrac{1}{9}$.

Note that throughout this text, we assume that any variable raised to a negative exponent is nonzero and any expression with exponents is not considered *simplified* unless it is written with only positive exponents.

EXAMPLE 5

Rewrite with a positive exponent.

a. y^{-7} **b.** $(-5x)^{-1}$ **c.** $(-4+t)^{-2}$

d. $\dfrac{1}{x^{-3}}$ **e.** $p^2 \cdot p^{-3}$ **f.** $\dfrac{n^3}{n^5}$

Solution

a. $y^{-7} = \dfrac{1}{y^7}$

b. $(-5x)^{-1} = \dfrac{1}{-5x} = -\dfrac{1}{5x}$

c. $(-4+t)^{-2} = \dfrac{1}{(-4+t)^2}$

d. $\dfrac{1}{x^{-3}} = 1 \div x^{-3} = 1 \div \dfrac{1}{x^3} = 1 \cdot x^3 = x^3$

e. $p^2 \cdot p^{-3} = p^{2+(-3)} = p^{-1} = \dfrac{1}{p}$

f. $\dfrac{n^3}{n^5} = n^{3-5} = n^{-2} = \dfrac{1}{n^2}$

PRACTICE 5

Rewrite with a positive exponent.

a. a^{-9}

b. $(-4y)^{-1}$

c. $(-5+x)^{-3}$

d. $\dfrac{1}{p^{-4}}$

e. $x^{-5} \cdot x^{-2}$

f. $\dfrac{m}{m^4}$

Note in Example 5(d) that $\dfrac{1}{x^{-3}} = x^3$, which suggests the following:

Reciprocal of x^{-a}

For any nonzero real number x and any integer a,

$$\frac{1}{x^{-a}} = x^a.$$

In words, the reciprocal of x^{-a} is x^a. For instance, $\dfrac{1}{4^{-3}} = 4^3$.

EXAMPLE 6

Simplify by writing each expression using only positive exponents.

a. $\dfrac{1}{x^{-4}}$ **b.** $\dfrac{2}{a^{-6}}$ **c.** $\dfrac{p}{q^{-1}}$ **d.** $\dfrac{3a^2}{b^{-5}}$

Solution

a. $\dfrac{1}{x^{-4}} = x^4$ **b.** $\dfrac{2}{a^{-6}} = 2 \cdot a^6 = 2a^6$

c. $\dfrac{p}{q^{-1}} = p \cdot q^1 = pq$ **d.** $\dfrac{3a^2}{b^{-5}} = 3a^2 \cdot \dfrac{1}{b^{-5}} = 3a^2 \cdot b^5 = 3a^2b^5$

PRACTICE 6

Rewrite as expressions using only positive exponents.

a. $\dfrac{1}{y^{-3}}$ **b.** $\dfrac{5}{x^{-2}}$

c. $\dfrac{m}{n^{-2}}$ **d.** $\dfrac{4r^3}{s^{-3}}$

Now, we consider the *power rule*, which deals with expressions in which a power is raised to a power. For instance, consider the expression $(x^2)^3$. By the definition of an exponent and the product rule, we get:

$$(x^2)^3 = \underbrace{x^2 \cdot x^2 \cdot x^2}_{\text{3 factors of } x^2} = x^{2+2+2} = x^6$$

So we see that $(x^2)^3 = x^6$, which can be generalized as follows:

Power Rule of Exponents

For any nonzero real number x and any integers a and b,

$$(x^a)^b = x^{ab}.$$

In words, to raise a power to a power, multiply the exponents and keep the base. For instance, $(2^3)^4 = 2^{12}$.

EXAMPLE 7

Simplify.

a. $-(x^6)^3$ **b.** $(x^{-5})^2$ **c.** $(n^{-2})^{-1}$

Solution   We apply the rule for raising a power to a power, keeping the base the same and multiplying the exponents.

a. $-(x^6)^3 = -(x^{6 \cdot 3}) = -x^{18}$

b. $(x^{-5})^2 = x^{-5 \cdot 2} = x^{-10} = \dfrac{1}{x^{10}}$

c. $(n^{-2})^{-1} = n^{-2 \cdot (-1)} = n^2$

PRACTICE 7

Simplify.

a. $(y^7)^2$

b. $-(n^{-1})^5$

c. $(p^{-3})^{-3}$

TIP   Don't confuse the *product* rule and the *power* rule.

Product rule: $x^a \cdot x^b = x^{a+b}$

Add the exponents.

Power rule: $(x^a)^b = x^{ab}$

Multiply the exponents.

Another law of exponents has to do with *raising a product to a power*. Consider, for instance, the expression $(2x)^3$.

$$(2x)^3 = \underbrace{(2x)(2x)(2x)}_{\text{3 factors of } 2x} = \underbrace{(2 \cdot 2 \cdot 2)(x \cdot x \cdot x)}_{\text{Rearrange the factors.}} = 2^3 \cdot x^3 = 8x^3$$

We see that $(2x)^3 = 2^3 \cdot x^3$, which we can generalize as follows:

Raising a Product to a Power

For any nonzero real numbers x and y and any integer a,

$$(xy)^a = x^a \cdot y^a.$$

In words, to raise a product to a power, raise each factor to that power, and then simplify. For instance, $(5y)^2 = 5^2 y^2 = 25y^2$.

EXAMPLE 8

Simplify.

a. $(-4p)^3$ **b.** $(2x^4)^4$ **c.** $(-p^2q^3)^6$

d. $-5(m^6n^8)^2$ **e.** $(6x^{-5}y^7)^{-2}$

Solution

a. $(-4p)^3 = (-4)^3(p)^3$ Use the rule for raising a product to a power.

$ = -64p^3$

b. $(2x^4)^4 = 2^4(x^4)^4$ Use the rule for raising a product to a power.

$ = 2^4x^{16}$ Use the power rule.

$ = 16x^{16}$ Simplify.

c. $(-p^2q^3)^6 = (-1)^6(p^2)^6(q^3)^6 = p^{12}q^{18}$

d. $-5(m^6n^8)^2 = -5(m^6)^2(n^8)^2 = -5m^{12}n^{16}$

e. $(6x^{-5}y^7)^{-2} = 6^{-2}(x^{-5})^{-2}(y^7)^{-2}$ Use the rule for raising a product to a power.

$\phantom{(6x^{-5}y^7)^{-2}} = 6^{-2}(x^{10})(y^{-14})$ Use the power rule.

$\phantom{(6x^{-5}y^7)^{-2}} = \dfrac{1}{36} \cdot x^{10} \cdot \dfrac{1}{y^{14}}$ The definition of a negative exponent.

$\phantom{(6x^{-5}y^7)^{-2}} = \dfrac{x^{10}}{36y^{14}}$ Simplify.

PRACTICE 8

Simplify.

a. $(-4x)^2$

b. $(5a^9)^2$

c. $(-q^7r^8)^2$

d. $-6(a^5b^3)^3$

e. $(7x^{-4}y^{-1})^{-2}$

The final law of exponents that we consider here is *raising a quotient to a power*. Consider, for instance, the expression $\left(\dfrac{x}{y}\right)^3$. Using the definition of an exponent, we get:

$$\left(\frac{x}{y}\right)^3 = \underbrace{\frac{x}{y} \cdot \frac{x}{y} \cdot \frac{x}{y}}_{3 \text{ factors of } \frac{x}{y}} = \frac{x \cdot x \cdot x}{y \cdot y \cdot y} = \frac{x^3}{y^3}$$

So we see that $\left(\dfrac{x}{y}\right)^3 = \dfrac{x^3}{y^3}$. We can generalize this result as follows:

Raising a Quotient to a Power

For any nonzero real numbers x and y and any integer a,

$$\left(\frac{x}{y}\right)^a = \frac{x^a}{y^a}.$$

In words, to raise a quotient to a power, raise both the numerator and the denominator to that power, and then simplify. For instance, $\left(\dfrac{2}{3}\right)^4 = \dfrac{2^4}{3^4} = \dfrac{16}{81}$.

EXAMPLE 9

Simplify.

a. $-\left(\dfrac{x}{4}\right)^3$ **b.** $\left(\dfrac{p^2}{q^4}\right)^5$ **c.** $\left(\dfrac{-3x}{4y^7}\right)^2$ **d.** $\left(\dfrac{q^4}{p^2}\right)^{-5}$

Solution

a. $-\left(\dfrac{x}{4}\right)^3 = -\dfrac{x^3}{4^3} = -\dfrac{x^3}{64}$

b. $\left(\dfrac{p^2}{q^4}\right)^5 = \dfrac{(p^2)^5}{(q^4)^5}$ Use the rule for raising a quotient to a power.

$\qquad\quad = \dfrac{p^{10}}{q^{20}}$ Use the power rule.

c. $\left(\dfrac{-3x}{4y^7}\right)^2 = \dfrac{(-3x)^2}{(4y^7)^2}$

$\qquad\quad\ = \dfrac{(-3)^2 x^2}{4^2 (y^7)^2}$

$\qquad\quad\ = \dfrac{9x^2}{16 y^{14}}$

d. $\left(\dfrac{q^4}{p^2}\right)^{-5} = \dfrac{(q^4)^{-5}}{(p^2)^{-5}}$

$\qquad\quad\ = \dfrac{q^{-20}}{p^{-10}}$

$\qquad\quad\ = q^{-20} \cdot \dfrac{1}{p^{-10}}$

$\qquad\quad\ = \dfrac{1}{q^{20}} \cdot p^{10}$

$\qquad\quad\ = \dfrac{p^{10}}{q^{20}}$

PRACTICE 9

Simplify.

a. $\left(\dfrac{y}{2}\right)^5$

b. $\left(\dfrac{u^4}{v^6}\right)^2$

c. $\left(\dfrac{-5a^5}{2b^2}\right)^3$

d. $\left(\dfrac{v^6}{u^4}\right)^{-2}$

Note in Example 9(b) and 9(d) that $\left(\dfrac{p^2}{q^4}\right)^5 = \dfrac{p^{10}}{q^{20}}$ and $\left(\dfrac{q^4}{p^2}\right)^{-5} = \dfrac{p^{10}}{q^{20}}$. So we conclude

that $\left(\dfrac{q^4}{p^2}\right)^{-5} = \left(\dfrac{p^2}{q^4}\right)^5$. This result can be generalized as follows:

Raising a Quotient to a Negative Power

For any nonzero real numbers x and y and any positive integer a,

$$\left(\frac{x}{y}\right)^{-a} = \left(\frac{y}{x}\right)^a.$$

In words, to raise a quotient to a negative power, take the reciprocal of the quotient and change
the sign of the exponent. For instance, $\left(\dfrac{5}{3}\right)^{-2} = \left(\dfrac{3}{5}\right)^2$.

EXAMPLE 10

Simplify: $\left(\dfrac{2a^5}{4b}\right)^{-3}$

Solution

$$\left(\frac{2a^5}{4b}\right)^{-3} = \left(\frac{4b}{2a^5}\right)^{3} = \frac{(4b)^3}{(2a^5)^3} = \frac{64b^3}{8a^{15}}$$

PRACTICE 10

Simplify: $\left(\dfrac{3x^3}{2y^2}\right)^{-4}$

EXAMPLE 11

A computer's memory capacity is often measured in *bits*, *bytes*, and *megabytes*. A bit (short for *binary digit*) is the smallest unit of data in the memory of the computer. A byte is equal to 2^3 bits, whereas a megabyte (MB) is equal to 2^{20} bytes.

a. How many bits are in a megabyte? Write the answer as a power of 2.

b. In some computers, the capacity is expressed in *gigabytes*. A gigabyte (GB) is 2^{10} megabytes. How many bytes are there in a gigabyte? Express the result as a power of 2.

Solution

a. A megabyte is 2^{20} bytes, and a byte is 2^3 bits. So a megabyte is $2^{20} \cdot 2^3 = 2^{23}$ bits.

b. A gigabyte is 2^{10} megabytes, and a megabyte is 2^{20} bytes. So a gigabyte is $2^{10} \cdot 2^{20} = 2^{30}$ bytes.

PRACTICE 11

The concentration of a pollutant in a pond is 30 parts per million (ppm). The pollution level drops by 5% each month, so the amount of pollutant each month is 95% of the amount in the previous month, as shown in the following table:

Month	Pollution Level
1	30 ppm
2	30×0.95 ppm
3	$30 \times (0.95)^2$ ppm
4	$30 \times (0.95)^3$ ppm
5	$30 \times (0.95)^4$ ppm
6	$30 \times (0.95)^5$ ppm
7	$30 \times (0.95)^6$ ppm

a. What will the pollution level be in the twelfth month?

b. The pollution level in the seventh month is how many times that in the second month?

Scientific Notation

In the sciences, an important application of exponents is *scientific notation*. For convenience, scientists commonly use this notation to write very large or very small numbers. Note that scientific notation is based on powers of 10.

Example	Standard Notation	Scientific Notation
The number of hairs on the average human head	150,000	1.5×10^5
The distance between points that an electronic microscope can distinguish	0.0000000002 meter	2×10^{-10} meter

(*Source*: S. Mader, *Inquiry into Life*)

Writing numbers in scientific notation has several advantages over standard notation. For example, when a number contains a long string of 0's, writing it in scientific notation can take fewer digits. Also, numbers written in scientific notation can be relatively easy to multiply or divide.

DEFINITION

A number is expressed in **scientific notation** if it is written in the form

$$a \times 10^n,$$

where n is an integer and a is greater than or equal to 1 but less than 10 ($1 \le a < 10$).

For instance, 7.3×10^5 is written in scientific notation, since 7.3 is between 1 and 10, and 10^5 is a power of 10. Note that for $1 \le a < 10$, the number a must have *one nonzero digit* to the left of the decimal point.

TIP When written in scientific notation, large numbers have positive powers of 10, whereas small numbers have negative powers of 10. For instance, 5×10^{11} is large, and 3×10^{-23} is small.

Now, let's consider how to change a number from scientific notation to standard notation, and vice versa.

EXAMPLE 12

Change each number from scientific notation to standard notation.

a. 3.04×10^5 **b.** 7×10^{-5}

Solution

a. To express this number in standard notation, we need to multiply 3.04 by 10^5. Since $10^5 = 100,000$, multiplying 3.04 by 100,000 gives us:

$$3.04 \times 100,000 = 304,000$$

The number 304,000 is written in standard notation.

 Note that the power of 10 is positive and that the decimal point is moved five places *to the right*. So a shortcut for expressing 3.04×10^5 in standard notation is to move the decimal point in 3.04 five places to the right:

$$3.04 \times 10^5 = 3\underset{\smile}{04000}. = 304,000$$

$$\uparrow\!\!\text{— Move five places to the right.}$$

b. Recall that a decimal point is understood to be at the right end of a whole number. So $7 \times 10^{-5} = 7. \times 10^{-5}$. Using the definition of a negative exponent, we get:

$$7. \times 10^{-5} = 7. \times \frac{1}{10^5}, \text{ or } \frac{7}{10^5}$$

Since $10^5 = 100,000$, we divide 7 by 100,000, giving us:

$$\frac{7}{10^5} = \frac{7}{100,000} = 0.00007$$

The number 0.00007 is written in standard notation.

PRACTICE 12

Express each number in standard notation.

a. 5.193×10^3

b. 3.7×10^{-6}

EXAMPLE 12 (continued)

Note that the power of 10 is negative and that the decimal point is moved five places *to the left*. So a shortcut for expressing 7×10^{-5} in standard notation is to move the decimal point five places to the left:

$$7 \times 10^{-5} = 7. \times 10^{-5} = .0\,0\,0\,0\,7, \text{ or } 0.00007$$

Move five places to the left.

> **TIP** When converting a number from scientific notation to standard notation, move the decimal point to the *left* if the power of 10 is *negative* and to the *right* if the power of 10 is *positive*.

Next, we consider the reverse situation, namely changing a number in standard notation to scientific notation.

EXAMPLE 13

Express each number in scientific notation.

a. 41,000,000,000

b. 0.00000000000000003

Solution

a. For a number to be written in scientific notation, it must be of the form

$$a \times 10^n,$$

where n is an integer and $1 \le a < 10$. We know that 41,000,000,000 and 41,000,000,000. are the same. We move the decimal point *to the left* so that there is one nonzero digit to the left of the decimal point. The number of places moved is the power of 10 that we need to multiply by in order to keep the value of the number fixed.

$$41,000,000,000 = 4.1\,0\,0\,0\,0\,0\,0\,0\,0\,0 \times 10^{10}$$

Move 10 places to the left.

$$= 4.1 \times 10^{10}$$

Note that since 4.1 and 4.1000000000 are equivalent, we could drop the trailing zeros. So 41,000,000,000 expressed in scientific notation is 4.1×10^{10}.

b. We must write the number 0.00000000000000003 in the form

$$a \times 10^n,$$

where n is an integer and $1 \le a < 10$. We move the decimal point *to the right* so that there is one nonzero digit to the left of the decimal point. The number of places moved, preceded by a *negative* sign, is the power of 10 needed.

$$0.00000000000000003 = 0\,0\,0\,0\,0\,0\,0\,0\,0\,0\,0\,0\,0\,0\,0\,3. \times 10^{-17}$$

Move 17 places to the right.

$$= 3 \times 10^{-17}$$

PRACTICE 13

Write each number in scientific notation.

a. 4,000,000,000,000

b. 0.000000000067

When performing calculations involving numbers that are written in scientific notation, we rely on the laws of exponents.

EXAMPLE 14

Calculate, writing the result in scientific notation.

a. $(3 \times 10^{-2})(1.2 \times 10^{7})$

b. $(2.8 \times 10^{3}) \div (4 \times 10^{-5})$

Solution

a. $(3 \times 10^{-2})(1.2 \times 10^{7})$

$$\begin{aligned} &= (3 \times 1.2)(10^{-2} \times 10^{7}) \quad \text{Regroup the factors.} \\ &= (3.6 \times 10^{-2+7}) \quad \text{Use the product rule.} \\ &= 3.6 \times 10^{5} \end{aligned}$$

b. $(2.8 \times 10^{3}) \div (4 \times 10^{-5}) = \dfrac{2.8 \times 10^{3}}{4 \times 10^{-5}}$

$$\begin{aligned} &= \frac{2.8}{4} \times \frac{10^{3}}{10^{-5}} \quad \begin{array}{l}\text{Rewrite the quotient} \\ \text{as a product of quotients.}\end{array} \\ &= 0.7 \times 10^{3-(-5)} \quad \begin{array}{l}\text{Use the quotient} \\ \text{rule.}\end{array} \\ &= 0.7 \times 10^{8} \end{aligned}$$

Note that 0.7×10^{8} is not written in scientific notation, because 0.7 is not a number between 1 and 10, that is, it does not have one nonzero digit to the left of the decimal point. To write 0.7×10^{8} in scientific notation, we convert 0.7 to scientific notation, and then simplify the product.

$$\begin{aligned} 0.7 \times 10^{8} &= 7 \times 10^{-1} \times 10^{8} \\ &= 7 \times 10^{7} \quad \text{Use the product rule.} \end{aligned}$$

So the answer is 7×10^{7}.

PRACTICE 14

Carry out the computation. Express the result in scientific notation.

a. $(2 \times 10^{-5})(5.12 \times 10^{7})$

b. $(1.32 \times 10^{4}) \div (6 \times 10^{-4})$

EXAMPLE 15

The following table gives some characteristics of three planets in the solar system:

Planet	Mean Orbital Radius (in meters)	Mass (in kilograms)	Mean Radius of Planet (in meters)
Mercury	5.80×10^{10}	3.24×10^{23}	2.340×10^{6}
Venus	1.08×10^{11}	4.86×10^{24}	6.10×10^{6}
Jupiter	7.80×10^{11}	1.89×10^{27}	69.8×10^{6}

(*Source:* Peter J. Nolan, *Fundamentals of College Physics*)

a. Rewrite the mass of Venus in standard form.

b. Which has a larger mean orbital radius—Mercury or Venus?

c. Is the mean radius of Jupiter written in scientific notation? Explain.

EXAMPLE 15 (continued)

d. Approximate the ratio of the mean orbital radius of Jupiter to the mean orbital radius of Mercury.

e. What is the difference between the mean orbital radius of Jupiter and that of Venus?

Solution

a. The mass of Venus is 4.86×10^{24} kg. In standard form, the mass is 4,860,000,000,000,000,000,000,000 kg.

b. The mean orbital radius of Mercury is 5.80×10^{10} m, as compared to 1.08×10^{11} m for Venus. The first number has power 10 and the second number has power 11. Since both numbers are written in scientific notation and 11 is the larger exponent, Venus has the larger mean orbital radius.

c. The mean radius of Jupiter is given as 69.8×10^{6} m. This number is not written in scientific notation, since 69.8 is larger than 10.

d. The ratio of the mean orbital radius of Jupiter to that of Mercury is

$$\frac{7.80 \times 10^{11}}{5.80 \times 10^{10}} = \frac{7.8}{5.8} \times 10^{11-10} \approx 1.3 \times 10 = 13.$$

e. The mean orbital radius of Jupiter is 7.80×10^{11} m and that of Venus is 1.08×10^{11} m. Since the powers of 10 are the same, we can apply the distributive property and subtract 1.08 from 7.80.

$$7.80 \times 10^{11} - 1.08 \times 10^{11} = (7.80 - 1.08) \times 10^{11} = 6.72 \times 10^{11}$$

So the difference between the mean orbital radius of each planet is 6.72×10^{11} m.

PRACTICE 15

Physicists use the coefficient of thermal conductivity as a measure of the amount of heat that a material conducts. The following table shows the coefficient for selected materials:

Material	Aluminum	Gold	Brick	Cork Board
Coefficient of Thermal Conductivity	2.34×10^{2}	3.13×10^{2}	6.49×10^{-1}	3.60×10^{-2}

(*Source:* Peter J. Nolan, *Fundamentals of College Physics*)

a. Express in standard form the coefficient for aluminum.

b. Express in standard form the coefficient for cork board.

c. Which material has a higher coefficient—gold or brick?

d. Find the ratio of the coefficient of brick to the coefficient of cork board rounded to the nearest tenth.

Calculators and Scientific Notation

Calculators vary as to how numbers are displayed or entered in scientific notation.

Display

In order to avoid an overflow error, many calculator models change an answer to scientific notation if the answer is either too small or too large to fit into the calculator's display. Many calculators do not display the 10 for a number in scientific notation. Some display an E or e; others show a space, for example, either 3.1E − 4 or 3.1■ −4 instead of 3.1×10^{-4}. *What other differences do you see between written scientific notation and displayed scientific notation?*

EXAMPLE 16

Multiply 4,000,000,000 by 3,000,000,000.

Solution

Press   Display

4000000000 $\boxed{\times}$ 3000000000 $\boxed{\text{ENTER}}$

Check to see if your calculator displays the product in scientific notation, that is, as 1.2E19, 1.2e19, or 1.2■19.

PRACTICE 16

Square 0.000000008. How is the answer displayed?

Enter

Some calculators give the wrong answer to a computation if very large or very small numbers are entered in standard form rather than in scientific notation. To enter a number in scientific notation, many calculators have an exponent key labeled $\boxed{\text{EE}}$, $\boxed{\text{EXP}}$, *or* $\boxed{\text{EEX}}$. *For a negative exponent, a key labeled* $\boxed{+/-}$ *or* $\boxed{(-)}$ *must be pressed either before or after the exponent key, depending on the machine.*

EXAMPLE 17

Multiply 7.2×10^5 by 1.8×10^6 on a calculator.

Solution

Press   Display

7.2 $\boxed{\text{EE}}$ 5 $\boxed{\times}$ 1.8 $\boxed{\text{EE}}$ 6 $\boxed{\text{ENTER}}$

The answer is 1.296×10^{12}. If your calculator has enough places in the display, it may give the answer to this problem in standard form: 1,296,000,000,000.

PRACTICE 17

Use a calculator to divide 8.4×10^{14} by 2.5×10^3.

Mathematically Speaking

Fill in each blank with the most appropriate term or phrase from the given list.

base	added	zero
standard notation	subtracted	exponent
multiplied	divided	scientific notation

1. In the expression x^a, x is called the _____.

2. In the expression x^a, a is called the _____.

3. In the product rule of exponents, the exponents are _____.

4. In the quotient rule of exponents, the exponents are _____.

5. When raising a product to a power, each factor is raised to that power, and then the results are _____.

6. The number 7.1×10^{11} is said to be written in _____.

A *Express in exponential form.*

7. $7 \cdot 7 \cdot 7 \cdot 7$

8. $(-1)(-1)(-1)(-1)(-1)$

9. $a \cdot a \cdot a \cdot a \cdot a$

10. $p \cdot p \cdot p \cdot p \cdot p \cdot p \cdot p$

11. $(-5x)(-5x)(-5x)(-5x)$

12. $(-3n)(-3n)(-3n)$

13. $(a + b)(a + b)(a + b)$

14. $(p - q)(p - q)(p - q)(p - q)$

Evaluate.

15. 4^2

16. 2^5

17. $(-3)^3$

18. $(-5)^3$

19. -3^3

20. -5^3

21. $(-10)^2$

22. $(-9)^2$

23. 7^0

24. 12^0

25. $-4n^0$

26. $6x^0$

27. $(-2y)^0$

28. $-(7n)^0$

29. $(3n)^1$

30. $(-2p)^1$

31. $(3a + b)^0$

32. $(4x - y)^1$

B *Rewrite as a base to a power, if possible.*

33. $3^3 \cdot 3$

34. $2^2 \cdot 2^4$

⊙ 35. $n^8 \cdot n^5$

36. $x^3 \cdot x^6$

⊙ 37. $a^9 \cdot b^4$

38. $p^2 \cdot q^7$

39. $x^0 \cdot x^5$

40. $y^6 \cdot y^0$

41. $(x + 2)^3 \cdot (x + 2)$

42. $(y - 5)^4 \cdot (y - 5)^2$

43. $\dfrac{6^7}{6^3}$

44. $\dfrac{5^{10}}{5^2}$

⊙ 45. $\dfrac{x^{12}}{x^7}$

46. $\dfrac{n^{18}}{n^9}$

⊙ 47. $\dfrac{q^5}{p}$

48. $\dfrac{b^4}{a^2}$

49. $\dfrac{t^8}{t}$

50. $\dfrac{n^{11}}{n^0}$

51. $(2n + 1)^4 \div (2n + 1)$

52. $(3x - 2)^5 \div (3x - 2)^2$

Simplify.

53. $y^4 \cdot y^3 \cdot y^7$

54. $t \cdot t^5 \cdot t^2$

55. $(p^5 q^9)(pq)$

56. $(a^6 b^2)(a^3 b^3)$

57. $(y^2x^2)(-xz^2)(7yz^4)$ **58.** $(-3a^2b^5)(4a^2c)(b^3c)$ **59.** $\dfrac{n^6 \cdot n^2 \cdot n^4}{n^8}$ **60.** $\dfrac{a^3 \cdot a \cdot a^7}{a^{10}}$

61. $\dfrac{r^9 \cdot r^7}{r^{13} \cdot r}$ **62.** $\dfrac{t^2 \cdot t^{11}}{t \cdot t^3}$

C *Express in terms of a base raised to a positive exponent.*

63. 5^{-2} **64.** 10^{-3} **65.** t^{-8} **66.** a^{-7}

67. n^{-10} **68.** x^{-3} **69.** $(-3y)^{-1}$ **70.** $(2x)^{-1}$

71. $(a+6)^{-5}$ **72.** $(x-2)^{-4}$ **73.** $\dfrac{1}{x^{-6}}$ **74.** $\dfrac{1}{y^{-1}}$

Simplify by writing each expression using only positive exponents.

75. $-n^{-8}$ **76.** $-r^{-7}$ **77.** $x^{-4}y^2$ **78.** a^6b^{-3} **79.** $2^{-1}y^2$

80. $-6^{-1}x$ **81.** $-3t^{-2}$ **82.** $4x^{-3}$ **83.** $\dfrac{2}{x^{-4}}$ **84.** $-\dfrac{7}{y^{-2}}$

85. $\dfrac{p^3}{-5q^{-1}}$ **86.** $\dfrac{6n^5}{m^{-2}}$ **87.** $\dfrac{3a^{-2}}{b}$ **88.** $\dfrac{x^4}{3y^{-1}}$ **89.** $\dfrac{n^{-2}}{n^{-5}}$

90. $\dfrac{x^{-1}}{x^{-6}}$ **91.** $\dfrac{(x^{-7})(x^3)}{(x^4)(x^{-1})}$ **92.** $\dfrac{(x^2)(x^{-2})}{(x)(x^{-3})}$

D *Simplify.*

93. $(5^2)^4$ **94.** $(8^5)^2$ **95.** $-(x^3)^5$ **96.** $(p^7)^4$

97. $(n^4)^{-3}$ **98.** $(y^2)^{-6}$ **99.** $-(a^{-5})^{-1}$ **100.** $-(t^{-4})^{-2}$

101. $-(2p)^4$ **102.** $-(2n)^5$ **103.** $(-2n)^5$ **104.** $(-2p)^4$

105. $(-a^5b^2)^2$ **106.** $(pq^3)^2$ **107.** $(-3n^5)^3$ **108.** $(-5x^6)^2$

109. $-2(a^2b^3)^4$ **110.** $5(p^3q)^2$ **111.** $(x^{-9}y^4)^3$ **112.** $(p^{-5}q^3)^4$

113. $(8^{-1}a^8b^{-8})^{-2}$ **114.** $(2^{-1}p^{-7}q)^{-4}$ **115.** $\left(\dfrac{x}{2}\right)^3$ **116.** $-\left(\dfrac{3}{y}\right)^4$

117. $\left(\dfrac{-p^5}{q^3}\right)^4$ **118.** $\left(\dfrac{-a^7}{b^4}\right)^3$ **119.** $-\left(\dfrac{2n^3}{5m^2}\right)^2$ **120.** $-\left(\dfrac{3x^9}{4y^4}\right)^2$

121. $\left(\dfrac{-4r^2}{st^4}\right)^3$ **122.** $\left(\dfrac{-2w}{u^3v^2}\right)^5$ **123.** $\left(\dfrac{3b^{-1}}{a^{-3}}\right)^2$ **124.** $\left(\dfrac{y^{-3}}{2x^{-1}}\right)^3$

125. $\left(\dfrac{3u^4}{4v^3}\right)^{-2}$ **126.** $\left(\dfrac{5a^{10}}{3b^6}\right)^{-3}$ **127.** $-\left(\dfrac{p^{-2}q^2}{r^2t}\right)^{-2}$ **128.** $-\left(\dfrac{a^2b^{-4}}{3c^3}\right)^{-3}$

E *Express in scientific notation.*

129. 400,000,000 **130.** 3,000,000 **131.** 926,000,000,000 **132.** 51,000,000,000

133. 0.0000042 **134.** 0.00079 **135.** 0.00000000074 **136.** 0.000000000313

Express in standard notation.

137. 2.43×10^6

138. 2.64×10^7

139. 3.027×10^{-3}

140. 5.03×10^{-8}

141. 1×10^{-6}

142. 9×10^7

Calculate, writing the result in scientific notation.

143. $(2 \times 10^2)(5 \times 10^4)$

144. $(4 \times 10^4)(7 \times 10^3)$

145. $(2.7 \times 10^{-2})(3 \times 10^{-3})$

146. $(1.4 \times 10^4)(2 \times 10^{-4})$

147. $(6.9 \times 10^5) \div (3 \times 10^2)$

148. $(3.5 \times 10^8) \div (5 \times 10^5)$

149. $(5.4 \times 10^{10}) \div (9 \times 10^4)$

150. $(7.2 \times 10^{-6}) \div (8 \times 10^4)$

151. $(2.613 \times 10^9)(5.391 \times 10^{-12})$

152. $(9.03 \times 10^{-13})(6.775 \times 10^{10})$

153. $(9.821 \times 10^{20}) \div (3.732 \times 10^{12})$

154. $(4.3628 \times 10^6) \div (3.295 \times 10^2)$

Mixed Practice

Solve.

155. Simplify: $\dfrac{x^2 \cdot x^6 \cdot x}{x^3}$

156. Rewrite as a base to a power: $\dfrac{10^7}{10^4}$

157. Evaluate: $-5x^0$

158. Express 0.0000035 in scientific notation.

159. Express in terms of a base raised to a positive exponent: $(n + 1)^{-3}$

160. Rewrite as a base to a power: $n^2 \cdot n^5$

161. Express $(2x)(2x)(2x)$ in exponential form.

162. Express 8.02×10^4 in standard notation.

163. Calculate and write the result in scientific notation: $(2.8 \times 10^4)(3 \times 10^2)$

164. Simplify: $\left(\dfrac{x^2 y^{-5}}{2z}\right)^{-2}$

Applications

F *Solve.*

165. A self-storage facility has large, cube-shaped storage lockers that measure x feet on a side. The owner of the storage facility decides to offer smaller cube-shaped storage lockers. If the dimensions of a small storage locker are half the dimensions of a large storage locker, is the volume of a small storage locker half the volume of a large storage locker? Explain.

166. The volume of a cylindrical can is the product of the area of the circular base and the height of the can. A food company packages its product in two different-sized cylindrical cans. The radius of the base of a large can is double the radius of a small can and the height of each can is the same, as shown. How do the volumes of the cans compare?

167. The number of *E. coli* bacteria in a colony doubles every 20 min. After 2 hours, a colony that starts with n bacteria grows to $(2^3)^2 \cdot n$ bacteria. Simplify this expression.

168. After three years, the value of a car purchased new for x dollars is given by the expression $\left(\dfrac{5}{4}\right)^{-3} \cdot x$. Simplify this expression.

169. The radius of Saturn is approximately 60,000,000 m. Rewrite this radius in scientific notation.

170. The mass of a hydrogen atom is 0.00000000000000000000000017 g. Express this quantity in scientific notation.

171. A nanosecond is a very small unit of time, equal to 1×10^{-9} sec. Express this quantity in standard form.

172. Astronomers use a light-year as a unit of length. If a light-year is approximately 9.5×10^{12} km, write this length in standard form.

173. Every month a bank customer deposits $100 into an account earning 4.5% annual interest compounded monthly.
   a. Calculate the total amount of money she will have deposited into the account after five years.
   b. If the deposits are made at the end of each compounding period, then the total amount in the account after five years is given by the expression $100\left[(1.00375^{12})^5 - 1\right] \div 0.00375$. Calculate this amount.
   c. What was the total interest earned in the five-year period?

174. A couple takes out a 15-yr mortgage in the amount of $180,000. The amount of their monthly payment on the mortgage at 7.5% is given by $\dfrac{1125(1.00625^{12})^{15}}{(1.00625^{12})^{15} - 1}$.
   a. Compute the amount of this monthly payment.
   b. Calculate the total amount of money to the nearest dollar that the couple will pay over the 15-yr period.
   c. What is the total amount of interest paid on the loan over the 15-yr period?

175. The total land area of Japan is approximately 3.78×10^5 km^2. In 2010, the population of Japan was 1.268×10^8 people. Calculate the number of people per square kilometer in Japan that there were in 2010. (*Sources:* U.S. Bureau of the Census, International Database; *Encyclopedia Britannica*)

176. It is estimated that in the year 2050, China will have a population of 1.304×10^9 people. In 1950, China's population was 5.57×10^8 people. Approximately how many times as large as the population in 1950 will the population be in 2050? (*Source:* U.S. Bureau of the Census, International Database, *Encyclopedia Britannica*)

177. The pie graph shows the percent of the total income generated by each source of income for the U.S. federal government in 2010. The total income that year was an estimated 2.162×10^{12}. Calculate the amount of income generated from each source. (*Source:* U.S. Office of Management and Budget)

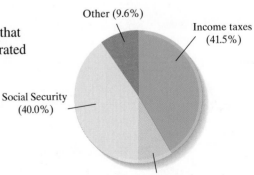

Other (9.6%)

Income taxes (41.5%)

Social Security (40.0%)

Corporate taxes (8.9%)

▦ **178.** The chart shows the amount of revenue brought in to the U.S. Postal Service during 2010 by various types of mail. Calculate the total amount of revenue brought in to the USPS that year. (*Source:* United States Postal Service)

Type of Mail	First class	Advertising	Shipping services	Periodicals	Package services	Other
Revenue (in dollars)	3.4×10^{10}	1.7×10^{10}	8.7×10^{9}	1.9×10^{9}	1.5×10^{9}	3.6×10^{9}

▦ **179.** The Sun converts $\dfrac{2.16 \times 10^{28}}{9 \times 10^{16}}$ kg of matter to energy each minute. Using scientific notation, write the number of kilograms of matter that are converted in an hour.

▦ **180.** The Calvert Cliffs nuclear power plant in Maryland converts $\dfrac{1.45584 \times 10^{14}}{9 \times 10^{16}}$ kg of mass to energy each day. Express this quantity in scientific notation.

• Check your answers on page A-16.

MINDStretchers

Mathematical Reasoning

1. Do you think that 0^0 should be defined to be 0? Should it be defined to be 1? Justify your position.

Research

2. Either in your college library or on the web, investigate the role the mathematician John Wallis played in the development and use of negative exponents. Write a few sentences to summarize your findings.

Writing

3. Recall that a number is said to be written in scientific notation if it is in the form $a \times 10^n$, where n is an integer and $1 \le a < 10$. Explain why we would not want this definition to allow a to be equal to 10.

1.5 Algebraic Expressions: Translating, Evaluating, and Simplifying

OBJECTIVES

A To translate an algebraic expression to a word phrase or a word phrase to an algebraic expression

B To evaluate an algebraic expression

C To simplify an algebraic expression

D To solve applied problems involving algebraic expressions

Recall that a *variable* is a quantity that is unknown, that is, one that can change or *vary* in value. We can use any letter or symbol to represent a variable. By contrast, a *constant* is a known quantity, one whose value does not change. So x is a variable, whereas $\frac{3}{4}$ is a constant.

An *algebraic expression* is an expression in which constants and variables are combined using standard arithmetic operations.

Algebraic expressions consist of one or more *terms*, separated by addition signs. For instance, the algebraic expression $5x + \frac{y}{2} - 1$, which we can think of as $5x + \frac{y}{2} + (-1)$, is made up of three terms:

$$\overset{\text{Terms}}{5x + \frac{y}{2} + (-1)}$$

DEFINITION

A **term** is a number, a variable, or the product or quotient of numbers and variables.

Addition and subtraction signs separate terms. The expression $2y + 5$ consists of two terms, whereas the expression $\frac{3n}{4}$ consists of only one term.

Translating Algebraic Expressions to Word Phrases, and Vice Versa

In applying algebra to practical problems, we frequently need to translate algebraic expressions to words, and vice versa. First, let's consider some of the many ways that we can translate expressions to words.

$x + 3$ translates to	$y - 1$ translates to
• x plus 3	• y minus 1
• x increased by 3	• y decreased by 1
• the sum of x and 3	• the difference between y and 1
• 3 more than x	• 1 less than y

$\frac{2}{3}n$, $\frac{2}{3} \cdot n$, or $\left(\frac{2}{3}\right)(n)$ translates to	$z \div 2$ or $\frac{z}{2}$ translates to
• $\frac{2}{3}$ times n	• z divided by 2
• the product of $\frac{2}{3}$ and n	• the quotient of z and 2
• $\frac{2}{3}$ of n	• the ratio of z and 2
	• $\frac{1}{2}$ of z

EXAMPLE 1

Translate each algebraic expression in the table to words.

Solution

Algebraic Expression	Translation
a. $4x$	4 times x
b. $y - 1$	1 less than y
c. $-2 + z$	The sum of -2 and z
d. $\dfrac{m}{-3}$	m divided by -3
e. $\dfrac{1}{5}r$	$\dfrac{1}{5}$ of r

Note that other translations are also correct.

PRACTICE 1

Translate each algebraic expression to words.

Algebraic Expression	Translation
a. $\dfrac{1}{3}p$	
b. $6 - x$	
c. $s \div (-4)$	
d. $n + (-10)$	
e. $\dfrac{5}{8}m$	

EXAMPLE 2

Translate each algebraic expression to words.

Solution

Algebraic Expression	Translation
a. $5m + 3$	3 more than 5 times m
b. $1 - 2y$	The difference between 1 and $2y$
c. $6(x + y)$	6 times the quantity x plus y
d. $\dfrac{x + y}{4}$	The quantity x plus y divided by 4

PRACTICE 2

Translate each algebraic expression to words.

Algebraic Expression	Translation
a. $6x - 7$	
b. $3 + 2m$	
c. $10(a + b)$	
d. $\dfrac{3}{p - q}$	

Note that in Example 2(c), the sum $(x + y)$ is considered to be a single quantity, because it is enclosed in parentheses.

Now, let's look at some examples of the reverse procedure, that is, translating word phrases to algebraic expressions.

EXAMPLE 3

Translate each phrase to an algebraic expression.

Solution

Phrase	Translation
a. Twice n	$2n$
b. A number decreased by 5	$n - 5$, where n represents the number
c. The quotient of negative 2 and t	$(-2) \div t$
d. $\frac{1}{3}$ of a number	$\frac{1}{3}n$, where n represents the number
e. 4 more than b	$b + 4$

PRACTICE 3

Express each phrase as an algebraic expression.

Phrase	Translation
a. $\frac{1}{2}$ of a number	
b. The sum of a number and negative 1	
c. The difference between x and negative 2	
d. The ratio of 4 and n	
e. The product of negative 3 and d	

EXAMPLE 4

Translate each phrase to an algebraic expression.

Solution

Phrase	Translation
a. The difference between a and the product of 4 and b	$a - 4b$
b. 3 more than 5 times y	$5y + 3$
c. 8 times the quantity r minus s	$8(r - s)$
d. Twice x divided by the sum of x and y	$\dfrac{2x}{x + y}$

PRACTICE 4

Translate each phrase to an algebraic expression.

Phrase	Translation
a. The sum of u and $-v$	
b. 12 less than the product of 3 and y	
c. The quantity a plus b divided by the quantity a minus b	
d. Negative 2 times the difference between x and y	

Evaluating Algebraic Expressions

For an expression to be useful, we need to be able to evaluate it, that is, to find the value of the expression for given values of each variable in the expression.

To Evaluate an Algebraic Expression

- Substitute for each variable the given value.
- Carry out the computation using the order of operations rule.

EXAMPLE 5

Find the value of each expression for $x = -1$, $y = 2$, and $z = 4$.

a. $2x + 1$ **b.** $-z^2$ **c.** $(-z)^2$ **d.** $3x + 4y$

Solution

a. $2x + 1 = 2(-1) + 1$ Replace x with -1.

$\qquad\quad = -2 + 1$ Multiply.

$\qquad\quad = -1$ Add.

b. $-z^2 = -4^2$ Replace z with 4.

$\qquad\;\; = -16$ Square 4.

c. $(-z)^2 = (-4)^2$ Replace z with 4.

$\qquad\quad = 16$ Square -4.

d. $3x + 4y = 3(-1) + 4(2)$ Replace x with -1 and y with 2.

$\qquad\qquad = -3 + 8$ Multiply.

$\qquad\qquad = 5$ Add.

PRACTICE 5

Evaluate each expression for $a = 4$, $b = -1$, $c = -2$, and $d = 3$.

a. $5a - 1$

b. $-c^4$

c. $(-c)^4$

d. $3b - 2d$

Note the difference between the answers in Example 5(b) and (c). In the expression -4^2, the exponent applies only to 4 because the base is 4. By contrast, in the expression $(-4)^2$, the exponent applies to -4, that is, the base is -4.

Some algebraic expressions involve fractions. In evaluating them, recall that parentheses are understood to be around any numerator and any denominator.

EXAMPLE 6

Evaluate each expression if $a = -3$, $b = 4$, and $c = -2$.

a. $\dfrac{2c}{5 + a}$ **b.** $\dfrac{b - c}{b + c}$ **c.** $3a^2 - b^3$ **d.** $5(a + c)$

Solution

a. $\dfrac{2c}{5 + a} = \dfrac{2(-2)}{5 + (-3)} = \dfrac{-4}{2} = -2$

b. $\dfrac{b - c}{b + c} = \dfrac{4 - (-2)}{4 + (-2)} = \dfrac{6}{2} = 3$

c. $3a^2 - b^3 = 3(-3)^2 - (4)^3$

$\qquad\qquad\;\; = 3(9) - (64)$

$\qquad\qquad\;\; = -37$

d. $5(a + c) = 5[-3 + (-2)]$

$\qquad\qquad = 5(-5)$

$\qquad\qquad = -25$

PRACTICE 6

Find the value of each expression when $x = -2$, $y = -1$, and $z = 3$.

a. $\dfrac{x + z}{x - y}$

b. $\dfrac{x - 3z}{y}$

c. $-4z^2 - 4(y - z)$

d. $5y^2 + z^3$

EXAMPLE 7

The weight of an object depends on the gravitational pull of the planet or moon it is on.

a. Write an algebraic expression to represent the weight of an astronaut on Earth that can be approximated by 6 times the astronaut's weight w on the Moon.

b. Find the weight on Earth of an astronaut who weighs 30 lb on the Moon.

Solution

a. We can represent the weight of the astronaut on Earth by the expression $6w$.

b. Since $w = 30$ lb, the weight on Earth is $6 \cdot 35$, or 180 lb.

PRACTICE 7

For each lap of a 1500-meter race, the runner in second place falls 5 m farther behind the runner in first place.

a. How far apart are the runners after L laps?

b. Find the distance between the runners after 3 laps.

Simplifying Algebraic Expressions

Each variable term of an algebraic expression is said to have a *numerical coefficient,* or simply a *coefficient.* For instance, the term $3y$ has coefficient 3 and the term $-4y$ has coefficient -4. Note that the coefficient of y is 1 because $y = 1y$. The terms $3y$, $-4y$, and y are said to be *like* terms.

> **DEFINITION**
> **Like terms** are terms that have the same variables and the same exponents. Terms that are not like are called **unlike terms**.

EXAMPLE 8

For each algebraic expression, identify the terms and indicate whether they are like or unlike.

a. $7a + 6b$ **b.** $n - 9n$ **c.** $3x^2 + x$ **d.** $2a^2b - 2a^2b$

Solution

	Terms	Like or Unlike
a. $7a + 6b$	$7a$ and $6b$ Different variables	Unlike
b. $n - 9n$	n and $-9n$ Same variable	Like
c. $3x^2 + x$	$3x^2$ and x Same variable but different exponents	Unlike
d. $2a^2b - 2a^2b$	$2a^2b$ and $-2a^2b$ Same variables with same exponents	Like

PRACTICE 8

Identify the terms of each algebraic expression and state whether they are like or unlike.

a. $r - 9r$

b. $-x + 5$

c. $2x^2y - xy^2$

d. $m + 5m - m$

We can use the distributive property to simplify algebraic expressions involving like terms. Adding or subtracting like terms using the distributive property is called *combining like terms.*

EXAMPLE 9

Combine like terms.

a. $a - 5a$ **b.** $2y - y + 3$

Solution

a. $a - 5a = 1a - 5a$ Recall that the coefficient of the term a is 1.

$\quad\quad\quad = (1 - 5)a$ Use the distributive property.

$\quad\quad\quad = -4a$ Simplify.

b. $2y - y + 3 = 2y - 1y + 3$ Recall that the coefficient of the term $-y$ is -1.

$\quad\quad\quad\quad\quad = (2 - 1)y + 3$ Use the distributive property.

$\quad\quad\quad\quad\quad = y + 3$ Simplify.

PRACTICE 9

Combine like terms.

a. $-7y - y$

b. $a - 4a + 8b$

EXAMPLE 10

Simplify, if possible.

a. $8x^2 + x$ **b.** $4x^2 - x^3 - 5x^2$ **c.** $-3a^2b + 5a^2b$

Solution

a. $8x^2 + x$ This algebraic expression cannot be simplified because the terms are unlike.

Unlike terms

b. $4x^2 - x^3 - 5x^2 = (4 - 5)x^2 - x^3$

Like terms

$\quad\quad\quad\quad\quad\quad = -1x^2 - x^3$

$\quad\quad\quad\quad\quad\quad = -x^2 - x^3$

c. $-3a^2b + 5a^2b = (-3 + 5)a^2b$

Like terms

$\quad\quad\quad\quad\quad\quad = 2a^2b$

PRACTICE 10

Simplify, if possible.

a. $y^2 - 6y$

b. $2n^2 - 5n^3 + 7n^3$

c. $9xy^2 - xy^2$

Some algebraic expressions involve parentheses. We can use the distributive property to simplify these expressions.

EXAMPLE 11

Simplify: $\dfrac{1}{2}(x + 8) - 4$

Solution

$\dfrac{1}{2}(x + 8) - 4 = \dfrac{1}{2}x + 4 - 4$ Use the distributive property.

$\quad\quad\quad\quad\quad = \dfrac{1}{2}x$ Combine like terms.

PRACTICE 11

Simplify: $5\left(y - \dfrac{2}{5}\right) + 8$

Now, let's consider simplifying algebraic expressions such as $-(2y - 5)$ in which a negative sign precedes an expression in parentheses.

EXAMPLE 12

Simplify: $-(2y - 5)$

Solution

$$-(2y - 5) = -1(2y - 5)$$
$$= -1 \cdot 2y + (-1)(-5) \quad \text{Use the distributive property.}$$
$$= -2y + 5 \quad \text{Simplify.}$$

Because the terms in the expression $-2y + 5$ are unlike, it is not possible to simplify the expression further.

PRACTICE 12

Simplify: $-(4a - 9b)$

EXAMPLE 13

Simplify: $6x - 1 - (3x + 4)$

Solution

$$6x - 1 - (3x + 4) = 6x - 1 - 1(3x + 4)$$
$$= 6x - 1 - 3x - 4$$
$$= 3x - 5$$

PRACTICE 13

Simplify: $3y + 7 - (2y - 5)$

TIP

- When removing parentheses preceded by a *minus sign*, all the terms in parentheses change to the opposite sign.
- When removing parentheses preceded by a *plus sign*, all the terms in parentheses keep the same sign.

EXAMPLE 14

Simplify: $4(8a - 9) - 6(5a - 7)$

Solution

$$4(8a - 9) - 6(5a - 7) = 32a - 36 - 30a + 42$$
$$= 2a + 6$$

PRACTICE 14

Simplify: $5(5y + 8) - 3(7y + 10)$

Some algebraic expressions contain not only parentheses but also brackets. When simplifying expressions containing grouping symbols, we use the distributive property, first removing the innermost grouping symbols and then working outward.

EXAMPLE 15

Simplify: $11 + 3[-2(9x - 5) + 12x]$

Solution

$$11 + 3[-2(9x - 5) + 12x] = 11 + 3[-18x + 10 + 12x]$$
$$= 11 + 3[-6x + 10]$$
$$= 11 - 18x + 30$$
$$= -18x + 41$$

PRACTICE 15

Simplify: $13 - 2[-15y + 4(3y - 1)]$

EXAMPLE 16

Find the total annual interest on $1000 where x dollars is invested at an interest rate of 5% per year and the remainder is invested at 4% per year.

Solution

To compute the annual interest, we multiply the interest rate (changed from a percent to a decimal) by the principal. The 5% interest on x dollars is $0.05x$. On the remainder of $(1000 - x)$ dollars, the 4% interest is $0.04(1000 - x)$. So the total interest generated in one year is

$$0.05x + 0.04(1000 - x) = 0.05x + 40 - 0.04x = 40 + 0.01x,$$

that is, $(40 + 0.01x)$ dollars.

PRACTICE 16

An apartment building has 100 apartments. Of these, y are 3-room apartments and the rest are 4-room apartments. How many rooms are there in the apartment building?

Mathematically Speaking

Fill in each blank with the most appropriate term or phrase from the given list.

expression	coefficient	combining
constant	equal	term
canceling	exponent	like
solve	variable	evaluate

1. A quantity that is unknown, that is, that can change or vary in value, is called a(n) _____.

2. A known quantity, that is, a quantity whose value does not change, is called a(n) _____.

3. In the algebraic expression $7x + \dfrac{y}{3} + 4$, $7x$ is a(n) _____.

4. The term $6t$ has _____ 6.

5. In an algebraic _____, constants and variables are combined using standard arithmetic operations.

6. To _____ an algebraic expression, replace each variable with the given number, and then carry out the computation using the order of operations rule.

7. Terms with the same variables and the same exponents are said to be _____.

8. Adding or subtracting like terms using the distributive property is called _____ like terms.

A *Translate each algebraic expression to words.*

9. $\dfrac{1}{2}n$

10. $\dfrac{2}{3}y$

11. $x - 8$

12. $n - 7$

13. $\dfrac{y}{5}$

14. $a \div (-1)$

15. $a + (-2)$

16. $b + 3$

17. $3x$

18. $8n$

19. $2a - b$

20. $3m - n$

21. $9 + 6n$

22. $4 + 3a$

23. $-3(x + y)$

24. $-2(q - p)$

25. $4(u + v)$

26. $7(a - b)$

27. $\dfrac{r + s}{9}$

28. $\dfrac{a + b}{2}$

29. $\dfrac{-3}{x + y}$

30. $\dfrac{2}{6 - x}$

31. $4p - 5q$

32. $3u - 8v$

33. $\dfrac{2xy}{x - y}$

34. $\dfrac{a + b}{3ab}$

Express each phrase as an algebraic expression.

35. $\dfrac{1}{4}$ of a number

36. $\dfrac{1}{10}$ of a number

37. The sum of a number and -3

38. A number increased by 5

39. The difference between x and -2

40. The difference between -2 and n

41. The ratio of b and a

42. The ratio of x and y

43. The product of -4 and y

44. The product of -2 and t

45. The sum of x and $-y$

46. The sum of p and -3

47. 7 less than the product of h and 3

48. 5 less than twice x

49. The sum of twice c and 4 times d

50. The difference between 6 times s and 5 times t

51. The quantity p minus q divided by the quantity p plus q

52. The quantity a minus $2b$ divided by the quantity a plus $2b$

53. Negative 2 times the quantity a plus 5

54. Negative 3 times the quantity x minus 3

B *Evaluate the given expression for each value of the variable.*

55.

x	-2	-1	0	1	2
$3x + 2$					

56.

x	-2	-1	0	1	2
$2x + 5$					

57.

y	0	1	2	3	4
$2y - 1.5$					

58.

y	0	1	2	3	4
$2y - 3.2$					

59.

x	-4	-2	0	2	4
$-0.5x$					

60.

x	-10	-5	0	5	10
$-0.6x$					

61.

n	-6	-3	0	3	6
$\dfrac{n}{3}$					

62.

n	-12	-4	0	8	12
$\dfrac{n}{4}$					

63.

g	-2	-1	0	1	2
$-2g^2$					

64.

g	-2	-1	0	1	2
$-3g^2$					

65.

a	-2	-1	0	1	2
$a^2 - 3a + 2$					

66.

a	-2	-1	0	1	2
$-a^2 + 3a - 1$					

Evaluate the expression.

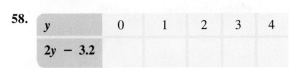

67. For $x = -2$, $y = 4$, $w = -1$, and $z = 5$
  a. $-2x + 3y$
  b. $6x^2$
  c. $3(5w - z)$
  d. $y^2 - 3w^3$

68. For $x = 1$, $y = 2$, $w = -2$, and $z = -1$
  a. $4x + 6y$
  b. $-2w^2$
  c. $-4(w + 6z)$
  d. $2y^2 - w^3$

69. For $a = -1$, $b = 2$, and $c = -4$

   a. $\dfrac{3c}{5 + a}$ **b.** $\dfrac{a + b}{a - b}$

   c. $-c^2$ **d.** $(-a + b)^2$

70. For $a = 4$, $b = -6$, and $c = 3$

   a. $-\dfrac{3c}{a - 2}$ **b.** $\dfrac{b}{2a + b}$

   c. $4c^2$ **d.** $(-a - b)^2$

C *For each algebraic expression, identify the terms and indicate whether they are like or unlike.*

71. $3a + 5b$

72. $2a - 3b$

73. $x - 7x$

74. $2x - 8x$

75. $2x^2 - 6x$

76. $x^2 + 5x^2$

77. $8a^2b - 5a^2b$

78. $3a^2b + 2ab^2$

Combine like terms, if possible.

79. $x + 5x$

80. $n + 4n$

81. $12a^2 - 2a^2$

82. $7x^3 - 13x^3$

83. $9t - 2t + 7$

84. $8p - 8p + 1$

85. $x - 9 + 5x$

86. $2n - 1 + 3n$

87. $9x - 10x^2 + 7x^2$

88. $3x^3 - 2x^2 + x^3$

89. $6x^2 + 6x$

90. $11x^2 + 8x$

91. $-5a^2b + 2a^2b$

92. $-4p^2q + 3p^2q$

Simplify.

93. $\dfrac{1}{2}(x - 2) - 6$

94. $\dfrac{2}{5}(x + 5) - 3$

95. $-(3y - 1)$

96. $-(2y - 4)$

97. $4x - 6 - (x + 7)$

98. $x - 5 - (3x + 2)$

99. $(15a + 2) - 3(5a + 8)$

100. $(5a + 4) - 2(10a - 4)$

101. $9 + 4[-3(y - 1) + 5x]$

102. $1 + 8[-4(y - 1) - 3x]$

103. $7 + 3[-2(x - 7) + 2x]$

104. $7 + 2[-4(x + 6) + 4x]$

105. $3n - \{5[2(9n - 1) - 3(n + 4)] + 2(8 - 5n)\}$

106. $2x - \{4[6(4x - 5) - 2(10x + 1)] + 5(7 - x)\}$

Mixed Practice

Solve.

107. Write as an algebraic expression: -4 is subtracted from a number

108. In the algebraic expression $x^2 - 4x$, identify the terms and indicate whether they are like.

109. Translate the algebraic expression $\dfrac{a - b}{2}$ to words.

110. Combine like terms: $-7x^2 + 2x^2 + 3$

111. Simplify: $9 - 2[(4y - 1) + 8y]$

112. Evaluate the expression $-4(p + 2q)^2$ for $p = 0$ and $q = -1$.

113. Find the value of $|-2.1|$.

114. Replace the   with $<$, $>$, or $=$ to make a true statement: -4.2 -4

Applications

D *Solve.*

115. An anthropologist determines that the bones she examined are those of a female. To determine the approximate height of the female (in inches), she uses the algebraic expression $2.47b + 21.6$, where b is the length of the femur (or thigh bone) in inches. What was the height of the female if the length of the femur is 16.8 in.? (Round to the nearest whole number.)

116. The value of a laptop computer bought new for $1350 decreases by $115 per year. What is the value of the laptop after t years?

117. The pressure (in pounds per square foot) at the bottom of a swimming pool d feet deep is given by the expression $2116.8 + 62d$. What is the pressure at the bottom of a swimming pool that is 8 ft deep?

118. The total revenue (in billions of dollars) for Microsoft Corporation for a given year can be approximated by the algebraic expression $4.6x + 32$, where x represents the number of years since 2003. According to this model, determine the approximate annual revenue for Microsoft in 2007 to the nearest billion dollars. (*Source:* Microsoft Corporation)

119. Write and simplify an algebraic expression for the perimeter of the total fenced-in area shown in the figure.

120. A blank DVD-R holds 4.35 gigabytes (GB) of data, in contrast to a blank CD-R, which holds only 0.7 GB. How much more storage do you get from a spindle of n DVD-Rs than from a spindle of twice as many CD-Rs?

121. In baseball, a player's batting average is the ratio of the number of hits h to the number of times at bat b.

a. Write an algebraic expression that represents a baseball player's batting average.

b. The total number of hits and the total number of times at bat for several major league baseball players during the 2010 season are given in the following table. Complete the table to show each player's batting average, rounded to the nearest thousandth, for the 2010 season. (*Source:* mlb.com)

Player	Number of Hits h	Number of Times at Bat b	Batting Average
Ichiro Suzuki	214	680	
Ryan Braun	188	619	
James Loney	157	588	
Prince Fielder	151	578	

122. A car on the freeway is traveling at a constant rate of 60 mph.

    a. Write an algebraic expression that shows the distance the car travels in t hours.

    b. Use your expression from part (a) to complete the table.

Time t (in hours)	$\frac{1}{2}$	1	$1\frac{3}{4}$	2	$2\frac{1}{2}$
Distance Traveled (in miles)					

123. The total revenue (in millions of dollars) of Netflix, the DVD rental and streaming video service, can be modeled by the algebraic expression $-34x + 625.7$, where x represents the number of quarters since the last quarter of 2009. (*Sources:* netflix.com, finance.yahoo.com)

    a. Complete the following table, rounding the revenue amounts to the nearest whole number:

2010 Quarter	Q1	Q2	Q3	Q4
Revenue (in millions of dollars)				

    b. What trend do you see in the amount of quarterly revenue?

    c. If this trend were to continue, what would the quarterly revenue be in the third quarter of 2011?

124. The number of construction workers (in millions) in the United States for a given year can be approximated by the algebraic expression $-0.15x^2 + 0.63x + 7$, where x represents the number of years since 2004. (*Source:* bls.gov)

    a. Complete the table, rounding to two decimal places.

Year	2007	2008	2009	2010
Number of Construction Workers (in millions)				

    b. Between 2007 and 2010, what was the average annual change in the number of construction workers from each year to the next?

125. An architect is to build a circular fountain with a ledge for sitting, as shown in the figure. The area of the ledge is given by the algebraic expression $\pi(2rx + x^2)$, where r is the radius of the circular fountain and x is the width of the ledge. If $r = 10$ ft and $x = 2$ ft, compute the area of the ledge to the nearest square foot. (Use $\pi \approx 3.14$.)

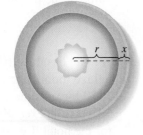

126. A sales representative is reimbursed $50 per day for meals plus $0.24 for each mile he drives. How much is he reimbursed if he drives a total of m miles in d days?

127. A homeowner wants to use part of her yard for a garden. She decides that the garden is to be 5 ft longer than its width w. If she wants to put a fence around the garden, how much fencing does she need to buy?

128. A couple purchases a pair of tickets for the symphony in the orchestra section for d dollars per ticket. Two of their friends decide to attend the concert at the last minute and can only get tickets in the balcony section. Each ticket in the balcony section costs $40 less than a ticket in the orchestra section. Write an expression for the total cost of the four tickets.

129. A pharmacy technician is paid $10.50 per hour.

a. Write an algebraic expression that represents her weekly pay if she works h hours. Then, use your expression to complete the table.

Number of Hours h	30	32.5	35	38	40
Weekly Pay (in dollars)					

b. The pharmacy technician receives overtime pay if she works more than 40 hr/wk. Her weekly pay with overtime can be calculated using the algebraic expression $10.5h + 5.25(h - 40)$. Simplify this expression.

c. Use the expression in part (b) to calculate the pharmacy technician's weekly pay if she works a total of 54 hr.

130. In all, a commuter drives 100 mi a day commuting to and from work. During her daily commute, h mi are on the highway and the rest are on local roads.

a. How many miles per week, that is, on 5 days, does she drive to and from work on local roads?

b. Her highway speed is 60 mph, and her local speed is 30 mph. How long (in hours) does the week's commuting take?

c. Use the expression in part (b) to complete the table, rounding to the nearest tenth.

Daily Travel on Highway h (in miles)	50	40	60	10	90
Weekly Length of Commuting (in hours)					

131. A local cineplex has 15 theaters. Of these, x are 400-seat theaters and the rest are 200-seat theaters. Write and simplify an expression that represents the total number of seats in the cineplex.

132. In a game, a player has a total of 70 chips. Of these, n are 5-dollar chips and the rest are 10-dollar chips. Write and simplify an expression that represents the total value of the player's chips.

133. On a 100-question test that imposes penalties for incorrect answers, scoring is as follows: a correct answer earns 1 point, a blank answer earns 0 points, and an incorrect answer loses $\frac{1}{2}$ point. Write a simplified expression for the total score on the test if c questions were answered correctly and b questions were left blank.

134. A hotel has three types of guest rooms: standard rooms, deluxe rooms, and executive rooms. The daily rate is $120 for a standard room, $165 for a deluxe room, and $230 for an executive room. What is the daily revenue for the hotel if y executive rooms are occupied, x deluxe rooms are occupied, and 75 more standard rooms than deluxe rooms are occupied?

• Check your answers on page A-16.

MINDStretchers

Groupwork

1. The expressions $\dfrac{x^2 - 2x - 8}{x + 2}$ and $x - 4$ seem to be equivalent. Together with a partner, show that for various values of x the expressions are equal in value. Are there any values of x for which the expressions are not equal?

Mathematical Reasoning

2. Suppose that you define a mathematical operation using the symbol ■ as follows:

$$x \ ■ \ y = 3x + 2y$$

Find values of x and y to demonstrate that neither the commutative property nor the associative property is true for the operation of ■.

Writing

3. Describe a situation that could be represented by each algebraic expression.
   a. $1.5x + 10$
   b. $4 + 2(t - 1)$

Cultural Note

Muhammad ibn Musa al-Khwarizmi, a ninth-century Arabic mathematician, wrote *al-Kitab al-mukhtasar fi hisab al-jabr wa'l-muqabala* (*The Compendious Book on Calculation by Completion and Balancing*), one of the earliest treatises on algebra and, in fact, the source of the word *algebra*. This work dealt with solving equations as well as with practical applications of algebra to measurement and legacies. al-Khwarizmi, from whose name the word *algorithm* derives, also wrote influential works on astronomy and on the Hindu numeration system.

(*Sources:* Jan Gullberg, *Mathematics from the Birth of Numbers*, W. W. Norton, 1997; Morris Kline, *Mathematics, A Cultural Approach*, Addison-Wesley, 1962)

Key Concepts and Skills

Concept/Skill	Description	Example
[1.1] **Integers**	The numbers $\ldots, -4, -3, -2, -1, 0, +1, +2, +3, +4, \ldots$ continuing infinitely in both directions.	$-83, 0,$ and 42
[1.1] **Rational numbers**	Real numbers that can be written in the form $\frac{a}{b}$, where a and b are integers and $b \neq 0$.	$\frac{3}{5}, -\frac{1}{3}, 96, 0, 8.9, -0.03,$ and $0.\overline{45}$
[1.1] **Irrational numbers**	Real numbers that cannot be written as the quotient of two integers.	$\sqrt{2}$ and π
[1.1] **Additive inverses (opposites)**	Two real numbers that are the same distance from 0, but on opposite sides of 0, on a number line. For any real number a, the additive inverse of a is written $-a$.	5 and -5 are additive inverses.
[1.1] **The absolute value of a number**	The distance the number is from 0 on a number line. The absolute value of the number n is written $\lvert n \rvert$.	$\lvert -5 \rvert = 5$ $\lvert 8 \rvert = 8$
[1.2] **To add real numbers**	• If the numbers have the same sign, add their absolute values and keep their sign. • If the numbers have different signs, find the difference between the larger and the smaller absolute value, and take the sign of the number with the larger absolute value.	Add: -3 and -19 $\lvert -3 \rvert + \lvert -19 \rvert = 3 + 19 = 22$ The sum of two negative numbers is negative. $-3 + (-19) = -22$
[1.2] **To subtract real numbers**	• Change the operation of subtraction to addition and change the number being subtracted to its additive inverse. • Then, follow the rule for adding real numbers.	$3 - (-5) = 3 + (+5) = 8$
[1.2] **To multiply real numbers**	• Multiply their absolute values. • If the numbers have the same sign, their product is positive; if they have different signs, their product is negative.	$(-3)(-9)$ $\lvert -3 \rvert = 3$ and $\lvert -9 \rvert = 9$ $3 \cdot 9 = 27$ The numbers have the *same sign*, so the product is positive. $(-3)(-9) = 27$
[1.2] **Multiplicative inverses (reciprocals)**	Two real numbers whose product is 1. For any nonzero real number a, the multiplicative inverse of a is written $\frac{1}{a}$.	The multiplicative inverse of 5 is $\frac{1}{5}$.
[1.2] **To divide real numbers**	• Divide their absolute values. • If the numbers have the same sign, their quotient is positive; if the numbers have different signs, their quotient is negative.	Divide: $(-10) \div \left(\frac{1}{5} \right)$ $\lvert -10 \rvert = 10$ and $\left\lvert \frac{1}{5} \right\rvert = \frac{1}{5}$ $10 \div \frac{1}{5} = 10 \cdot \frac{5}{1} = 50$ The numbers have *different signs*, so the quotient is negative. $(-10) \div \left(\frac{1}{5} \right) = -50$

CONCEPT	SKILL

Concept/Skill	Description	Example
[1.2] Order of operations rule	A rule used to evaluate mathematical expressions. Carry out the operations in the following order: 1. First, perform the operations within any grouping symbols, such as parentheses ( ), brackets [ ], or braces { }. Work from the innermost pair of grouping symbols to the outermost pair. 2. Then, evaluate any numbers raised to a power. 3. Next, perform all multiplications and divisions as they appear from left to right. 4. Finally, do all additions and subtractions as they appear from left to right. Parentheses are understood around the numerator and denominator of any fraction, as well as under a radical sign and within an absolute value sign.	$-2(24) - 5(-1)$ $\quad = -48 + 5 = -43$ $2 - 2 \cdot 3^2 = 2 - 2 \cdot 9$ $\qquad\qquad = 2 - 18$ $\qquad\qquad = -16$ $4 + 3(8 - 4) \div 2 \cdot 9$ $\quad = 4 + 3(4) \div 2 \cdot 9$ $\quad = 4 + 12 \div 2 \cdot 9$ $\quad = 4 + 6 \cdot 9$ $\quad = 4 + 54$ $\quad = 58$
[1.3] Commutative property of addition	For any real numbers a and b, $$a + b = b + a.$$	$5 + 8 = 8 + 5$
[1.3] Commutative property of multiplication	For any real numbers a and b, $$a \cdot b = b \cdot a.$$	$5 \cdot 8 = 8 \cdot 5$
[1.3] Associative property of addition	For any real numbers a, b, and c, $$(a + b) + c = a + (b + c).$$	$(2 + 4) + 5 = 2 + (4 + 5)$
[1.3] Associative property of multiplication	For any real numbers a, b, and c, $$(a \cdot b)c = a(b \cdot c).$$	$(2 \cdot 4) \cdot 5 = 2 \cdot (4 \cdot 5)$
[1.3] Additive identity property	For any real number a, $$a + 0 = a \quad \text{and} \quad 0 + a = a.$$	$5 + 0 = 5 \quad \text{and} \quad 0 + 5 = 5$
[1.3] Multiplicative identity property	For any real number a, $$a \cdot 1 = a \quad \text{and} \quad 1 \cdot a = a.$$	$(-3) \cdot 1 = -3 \quad \text{and} \quad 1 \cdot (-3) = -3$
[1.3] Additive inverse property	For any real number a, there is exactly one number, $-a$, such that $$a + (-a) = 0 \quad \text{and} \quad -a + a = 0,$$ where a and $-a$ are said to be additive inverses of each other.	$6 + (-6) = 0$ and $-6 + 6 = 0$
[1.3] Multiplicative inverse property	For any nonzero real number a, there is exactly one number $\dfrac{1}{a}$ such that $$a \cdot \dfrac{1}{a} = 1 \text{ and } \dfrac{1}{a} \cdot a = 1,$$ where a and $\dfrac{1}{a}$ are said to be multiplicative inverses (or reciprocals) of each other.	$5 \cdot \dfrac{1}{5} = 1 \quad \text{and} \quad \dfrac{1}{5} \cdot 5 = 1$

continued

Concept/Skill	Description	Example
[1.3] Multiplication property of zero	For any real number a, $$a \cdot 0 = 0 \quad \text{and} \quad 0 \cdot a = 0$$	$-2 \cdot 0 = 0$ and $0 \cdot (-2) = 0$
[1.3] The distributive property	For any real numbers a, b, and c, $$a \cdot (b + c) = a \cdot b + a \cdot c$$	$2 \cdot (3 + 5) = 2 \cdot 3 + 2 \cdot 5$
[1.3] Division involving zero	For any real number a, • $0 \div a = 0$, where a is nonzero. • $a \div 0$ is undefined.	$0 \div 4 = 0$ $4 \div 0$ is undefined
[1.4] Base and exponent (or power)	For any real number x and any positive integer a, $$x^a = \underbrace{x \cdot x \cdots x \cdot x,}_{a \text{ times}}$$ where x is called the base and a is called the exponent (or power). The expression x^a is read "x to the power a."	$(-5)^2 = (-5)(-5)$ $x^3 = x \cdot x \cdot x$
[1.4] Exponent 1	For any real number x, $$x^1 = x.$$	$(-7)^1 = -7$ $(4x + y)^1 = 4x + y$
[1.4] Exponent 0	For any nonzero real number x, $$x^0 = 1.$$	$(-3)^0 = 1$ $(2a)^0 = 1$
[1.4] Product rule of exponents	For any nonzero real number x and for any integers a and b, $$x^a \cdot x^b = x^{a+b}.$$	$8^4 \cdot 8^2 = 8^6$ $x^8 \cdot x^{10} = x^{18}$
[1.4] Quotient rule of exponents	For any nonzero real number x and for any integers a and b, $$\frac{x^a}{x^b} = x^{a-b}.$$	$\frac{10^5}{10^3} = 10^2 \qquad \frac{p^{12}}{p^7} = p^5$
[1.4] Negative exponent	For any nonzero real number x and for any positive integer a, $$x^{-a} = \frac{1}{x^a}.$$	$5^{-3} = \frac{1}{5^3} \qquad p^{-8} = \frac{1}{p^8}$
[1.4] Reciprocal of x^{-a}	For any nonzero real number x and any integer a, $$\frac{1}{x^{-a}} = x^a.$$	$\frac{1}{x^{-4}} = x^4 \qquad \frac{1}{p^{-7}} = p^7$
[1.4] Power rule of exponents	For any nonzero real number x and any integers a and b, $$(x^a)^b = x^{ab}.$$	$(7^2)^3 = 7^6$ $(p^4)^5 = p^{20}$
[1.4] Raising a product to a power	For any nonzero real numbers x and y and any integer a, $$(xy)^a = x^a \cdot y^a.$$	$(7 \cdot 5)^2 = 7^2 \cdot 5^2$ $(-3a)^2 = (-3)^2(a)^2$

Concept/Skill	Description	Example
[1.4] **Raising a quotient to a power**	For any nonzero real numbers x and y and any integer a, $$\left(\frac{x}{y}\right)^a = \frac{x^a}{y^a}.$$	$$\left(\frac{2}{3}\right)^2 = \frac{2^2}{3^2}$$ $$\left(\frac{a}{b}\right)^4 = \frac{a^4}{b^4}$$
[1.4] **Raising a quotient to a negative power**	For any nonzero real numbers x and y and any integer a, $$\left(\frac{x}{y}\right)^{-a} = \left(\frac{y}{x}\right)^a.$$	$$\left(\frac{2}{3}\right)^{-2} = \left(\frac{3}{2}\right)^2$$ $$\left(\frac{a}{b}\right)^{-4} = \left(\frac{b}{a}\right)^4$$
[1.4] **Scientific notation**	A number is expressed in scientific notation if it is written in the form $$a \times 10^n,$$ where n is an integer and a is greater than or equal to 1 but less than 10 ($1 \le a < 10$).	8.3×10^4 1.24×10^{-7}
[1.5] **Algebraic expression**	An expression in which constants and variables are combined using the standard arithmetic operations.	$5x + \dfrac{y}{2} - 1$
[1.5] **Term**	A number, a variable, or the product or quotient of numbers and variables.	Terms $$5x + \frac{y}{2} + (-1)$$
[1.5] **To evaluate an algebraic expression**	• Substitute for each variable the given value. • Carry out the computation using the order of operations rule.	Find the value of $3x + 4y$ for $x = -1$ and $y = 2$. $3x + 4y = 3(-1) + 4(2)$ $= -3 + 8 = 5$
[1.5] **Like terms**	Terms that have the same variables and the same exponents. Terms that are not like are called *unlike terms*.	$4a$ and $-6a$ Like $2x^2y$ and $3x^2y$ Like $-9s$ and $-9t$ Unlike $5x$ and $3x^2$ Unlike

Say Why
Fill in each blank.

1. The number -0.17 _____ an irrational number

   is/is not

   because _____

   _____ .

2. The additive inverse of $\left|-4\right|$ _____ -4 because

   is/is not

   _____

   _____ .

3. The expression $(-7 + x)3$ _____ be rewritten as

   can/cannot

   $(-7)3 + x \cdot 3$ because _____

   _____ .

4. The number 11.852×10^{-3} _____ written in

   is/is not

   scientific notation because _____

   _____ .

5. The expressions 6^0 and 3^0 _____ the same

   have/do not have

   value because _____

   _____ .

6. The expression $1 + 5x^2$ _____ consist of two

   does/does not

   terms because _____

   _____ .

[1.1]

7. True or false: $\sqrt{2}$ is a rational number.

8. True or false: -5 is an integer.

9. Graph -1.5 on the number line.

10. Find the additive inverse of -8.

11. Find the value of $\left|-6\right|$.

12. True or false: $-2 < -7$.

13. True or false: $0 \geq -6.1$.

14. **a.** Graph the set of all real numbers less than or equal to 0 on the number line.

    b. Express this set in interval notation.

[1.2] *Calculate.*

15. $3.4 + (-5.9)$

16. $5 - (-2)$

17. $(-4.5)(-3)$

18. $\left(\dfrac{1}{4}\right)^2$

19. $(-8) \div (-4)$

20. $-3(4 - 10)$

21. $\left[3(5 + 6) - 3(4 - 5)\right]$

22. $\dfrac{-1 + (-9)}{5}$

23. $(-5)^2 - 3^2$

24. $\left|3 - 9\right|$

25. $-\left|-1\right| - \left|6\right|$

26. $\sqrt{29 - 16}$

[1.3] *Rewrite each expression using the indicated property of real numbers.*

27. Commutative property of addition: $3 + 9$

28. Distributive property: $(-3)(1 + 9)$

Identify the definition or property of real numbers that justifies each statement.

29. $(3x + y) + z = 3x + (y + z)$

30. $-(x + 1) + (x + 1) = 0$

Calculate.

31. $10 + (-2) + (-1)$

32. $(-4)(-5)(-2)(4)$

Find.

33. The additive inverse of 4

34. The multiplicative inverse of $\dfrac{2}{3}$

Rewrite each expression using the distributive property.

35. $3(a - 4b)$

36. $-(x - 5)$

[1.4]

37. Express $(-5x)(-5x)(-5x)$ in exponential form.

Evaluate.

38. $(-4)^3$

39. 13^0

40. $(-3xy)^1$

Express in terms of a base raised to a positive exponent.

41. 7^{-2}

42. $(2y)^{-1}$

Simplify, if possible.

43. $n^8 \cdot n^4$

44. $(ab^2)(a^2b)$

45. $\dfrac{a^5}{b^2}$

46. $(2n + 1)^4 \div (2n + 1)$

47. $(-y^3)^5$

48. $(pq^3)^4$

49. $\left(\dfrac{-u^3}{uv^2}\right)^4$

50. $-4t^{-2}$

51. $\dfrac{2p^3}{q^{-1}}$

52. $x^5 \cdot x^{-4}$

53. $(-2p^2q^{-5})^{-2}$

54. $\left(\dfrac{x^4}{3y}\right)^{-2}$

Express in scientific notation.

55. 200,000,000

56. 0.00031

Express in standard notation.

57. 2.86×10^5

58. 5.02×10^{-3}

Calculate. Write the answer in scientific notation.

59. $(3 \times 10^2)(5 \times 10^4)$

60. $(8 \times 10^5) \div (2 \times 10^3)$

[1.5] *Translate each algebraic expression to words.*

61. $x - 4$

62. $\dfrac{3}{p + q}$

Express each word phrase as an algebraic expression.

63. The difference between x and -5

64. The quantity p plus q divided by the quantity q minus p

65. Evaluate the expression for each value of x.

x	−2	−1	0	1	2
$2x - 5$					

66. Evaluate the expression $3(w - 5z)$ for $w = -3$ and $z = 2$.

Identify the terms in each expression and indicate whether they are like or unlike.

67. $2a + 5b$

68. $3a^2b - a^2b$

Combine like terms.

69. $x - 5x$

70. $-4ab^2 + 7ab^2$

71. $6t - 3t + 2$

72. $5x^2 - 7x + 9x^2 + x$

Simplify.

73. $-(5y - 1)$

74. $6 + 3[-2(x - 6) + 4x]$

Mixed Applications

Solve.

75. The graph shows the daily low temperature (in °F) in Fairbanks, Alaska, for the second week of March in a recent year. (*Sources:* Alaska Climate Research Center; NOAA Weather Service at Fairbanks, Alaska)

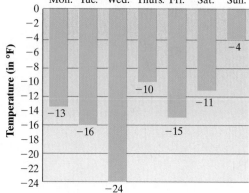

Daily Low Temperature

a. On which day did the lowest daily low temperature occur?

b. Write an inequality that compares the daily low temperature on Tuesday to the daily low temperature on Friday.

c. What was the average daily low temperature for the week?

76. The table shows the quarterly profits in millions of dollars for a small company.

Quarter	1	2	3	4
Profit (in millions of dollars)	1.7	−0.5	3.8	−1.75

a. Plot the profits on a number line.

b. In which quarters did the company show a loss?

c. What was the average quarterly profit?

77. The balance in a checking account with overdraft privileges was $578.37. What is the new balance after a check for $585.00 is cashed?

78. The recommended daily amount of calcium for people age 9 through 18 is 1300 mg. An 8-oz glass of low-fat milk contains about 23% of the recommended daily amount of calcium. How many milligrams of calcium does an 8-oz glass of milk contain to the nearest 100 mg? (*Source:* Institute of Medicine)

79. Death Valley (the lowest point in North America) is 86 m below sea level. The top of Mt. McKinley (the highest point in North America) is 6194 m above sea level. Calculate the difference between these elevations. (*Source: National Geographic Family Reference Atlas of the World*)

80. The table below shows the first-round scores of selected LPGA players in a recent golf tournament. A player's score with respect to par is calculated by subtracting 72 (par score for 18 holes) from the score for the round. (*Source:* LPGA)

a. Complete the table to show each player's score with respect to par.

Player	Christie Kerr	Se Ri Pak	Angela Stanford	Karrie Webb	Michelle Wie
Score	70	69	73	74	72
Score with Respect to Par					

b. What do negative results mean?

81. The projected world population in 2050 is 8,900,000,000. Express this population in scientific notation. (*Source:* un.org)

82. An estimated 3.3×10^7 people worldwide were living with HIV/AIDS in 2009. Of these people, 2,500,000 were children and the rest were adults. How many adults were living with HIV/AIDS in 2009? Express the answer in standard notation. (*Source:* National Institutes of Health)

83. The height in meters (above ground) of an object 2 sec after it is thrown downward from a height of 100 m with an initial velocity of 3 m/sec is given by the expression $-4.9(2)^2 - 3(2) + 100$. What is the height of the object?

84. A clinical dietician calculates the ideal body weight (in pounds) of a man who is 5 ft 11 in. (71 in.) tall using the expression $6(71 - 60) + 106$. Calculate the ideal body weight of this man.

85. A medication is to be administered to a patient intravenously. To calculate the number of drops per minute required to fill an order for a patient, a nurse uses the expression $\frac{A \cdot n}{60}$, where A is the amount of medication to be administered (in mL per hr) and n is the drip factor of the intravenous (IV) tubing (in drips per mL).

a. Calculate the number of drops per minute needed for an order of 100 mL per hr using IV tubing with a drip factor of 15 drips per mL.

b. An order of 1500 mL of an IV medication is to be administered over 12 hr. How many milliliters are to be administered per hour?

c. Calculate the number of drops per minute required to fill the order in part (b) using IV tubing with a drip factor of 15 drips per mL.

86. The height (in feet) at time t (in seconds) of an object that is thrown upward from a height of 300 ft with an initial velocity of 20 ft/sec is given by the algebraic expression $-16t^2 + 20t + 300$.

a. Complete the table.

Time t (in seconds)	0	1	2	3	4	5
Height (in feet)						

b. What does a height of 0 ft represent in the context of this problem?

87. A lottery winner invests $45,000 in two funds earning simple interest. She invests x dollars in an account earning 5% interest and the rest in an account earning 6%. Write an expression for the total amount of annual interest earned on the investment.

88. A parking garage charges $12 for the first hour of parking and $1.50 for each hour (or part of an hour) thereafter. Write an expression for the cost of parking for h hr.

• Check your answers on page A-17.

CHAPTER 1   Posttest

FOR EXTRA HELP

The Chapter Test Prep Videos with test solutions are available on DVD, in MyMathLab, and on YouTube (search "AkstIntermediate Alg" and click on "Channels").

To see if you have mastered the topics in this chapter, take this test.

1. Graph -3.5 on the number line.

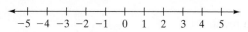

2. Find the value of $|-9|$.

3. True or false: $-2 < -5$

4. Rewrite using the associative property of addition:
$-4 + (x + y)$

Calculate.

5. $3 - (-6)$

6. $10 + |(-2) + (-1)|$

7. $(-3)(5)(-1)(-2)$

8. -13^0

9. $-\left(-\dfrac{1}{2}\right)^2$

10. $(12)(-7) \div (-6)(3) + (-2)^2$

11. $-\left[6(7 + 4) - 3^2(8 - 1)\right]^3$

12. $\dfrac{-4 - (-2)^4}{4(5^2 - 13 \cdot 2)}$

Simplify.

13. $\dfrac{2p^3}{q^{-2}}$

14. $(ab^4)^3(a^6b)$

15. $\left(\dfrac{x^4y^2}{z^{-1}}\right)^{-2}$

16. Translate the expression $4(p - q)$ to words.

17. Evaluate the expression $2(x - 3y)$ for $x = -1$ and $y = 5$.

18. Simplify: $4 + 3\left[-8(x - 1) + 4x\right]$

19. The cost of a wedding reception is $5000 plus $80 per guest. Write an expression to show the total cost if x guests attend a wedding reception.

20. The distance from the Sun to planet Earth is approximately 9.3×10^7 mi. This distance represents one astronomical unit. The distance from the Sun to the planet Venus is approximately 0.7 astronomical units. What is this distance in miles? Express the answer in scientific notation.

• Check your answers on page A-17.

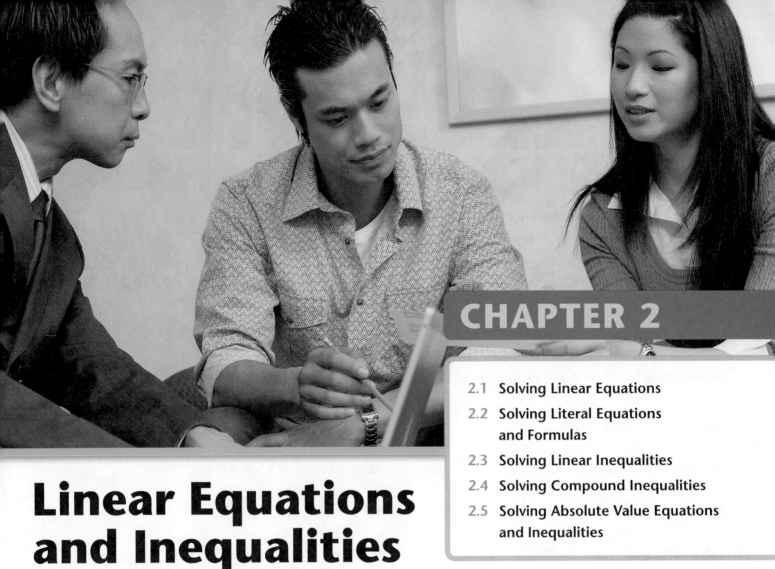

Linear Equations and Inequalities

Taxes and Linear Equations

Linear equations play a key role in determining the amount of federal income tax that Americans pay. According to the tax schedule below, a married couple filing jointly with a taxable income of $110,000 would be in the 25% *tax bracket*. This couple could calculate their taxes by adding $9362.50 to 25% of $42,000—the amount by which their taxable income exceeds $68,000. In this tax bracket, the following linear equation relates taxes T to taxable income I:

$$T = 0.25I - 7637.50$$

The couple would pay $19,862.50 in federal income taxes. (*Source:* irs.gov)

Schedule Y-1—If your filing status is **Married filing jointly** or **Qualifying widow(er)**

If your taxable income is: Over—	But not over—	The tax is:	of the amount over—
$0	$16,750	. . . . **10%**	$0
16,750	68,000	**$1,675.00 + 15%**	16,750
68,000	137,300	**9,362.50 + 25%**	68,000
137,300	209,250	**26,687.50 + 28%**	137,300
209,250	373,650	**46,833.50 + 33%**	209,250
373,650	. . . .	**101,085.50 + 35%**	373,650

1. Is -2 a solution of the equation $\frac{1}{2}x - 6 = 7$?

2. Solve: $2x - 10 = -5$

3. Solve: $5x + 9 = 3x + 13$

4. Solve: $4 - 6n = 2(1 - 2n)$

5. Solve $4x - 3y = 6$ for y in terms of x.

6. Is 5 a solution of the inequality $8 - 2x \leq 7$?

7. Solve and graph: $x - 5 \leq 2x + 1$

8. Solve and graph: $4x + 3 < 7$ and $5x - 1 \geq -16$

9. Solve and graph: $\frac{2}{3}x + 4 \leq 0$ or $10 - 9x \leq 19$

10. Solve: $-17 < 6x + 13 < 25$

11. Find the distance between -3.7 and 2.2.

12. Solve: $|x + 9| = 5$

13. Solve: $|y| - 2 = -1$

14. Solve: $|4x + 3| = |x - 9|$

15. Solve and graph: $|2x - 1| \geq 1$

16. A new copier purchased for $5000 depreciates in value by $840 per year. In how many years will the copier reach its salvage value of $800?

17. A chemist adds 4 L of water to 6 L of a 65% saline solution. What is the saline concentration of the new solution?

18. A sales representative is reimbursed for his expenses each week. His company pays $150 per day for meals and lodging and $0.30 per mile he drives. If his weekly expenses cannot exceed $900, how many miles can he drive in a five-day workweek without going over budget?

19. The owner of a small juice company estimates that her monthly cost is $4500 plus $0.60 per bottle of juice produced. How many bottles of juice should be produced per month if she wants to keep her monthly cost between $7500 and $9000?

20. A contractor estimates that to remodel a kitchen it will cost about $12,000 with a margin of error of $2500. What are the minimum and maximum amounts the homeowner could pay for the kitchen remodeling?

• Check your answers on page A-17.

2.1 Solving Linear Equations

What Equations Are and Why They Are Important

In this chapter, we consider one of the most important concepts in algebra, the *equation*. Equations are important because they are used to solve a wide variety of problems in many fields, such as chemistry, meteorology, and business. For instance, suppose that in a chemistry experiment you use the balance scale shown below to find the mass of the solution in a beaker.

A To determine whether a given number is a solution of a given equation

B To solve a linear equation using the addition or multiplication property

C To solve applied problems involving the addition or multiplication property

We know that if the scale is balanced, we can write the following equation to find the mass of the solution:

$$26 = 9 + x$$

Equations and Solutions

An equation such as $26 = 9 + x$ has two expressions—one on the left side of the equal sign and one on the right:

Equal sign

$$26 = 9 + x$$

Left side   Right side

DEFINITION

An **equation** is a mathematical statement that two expressions are equal.

Some examples of equations are:

$$4 + 5 + 6 = 15 \qquad x - 9 = 12 \qquad -8y = 24 \qquad 2(w - 5) = 10$$

Equations may be true or false. The equation $26 = 9 + x$ is true if 17 is substituted for the variable x.

$$26 = 9 + x$$
$$26 = 9 + 17 \qquad \text{True}$$

However, this equation is false if 7 is substituted for the variable x.

$$26 = 9 + x$$
$$26 = 9 + 7 \qquad \text{False}$$

The number 17 is called a *solution* of the equation $26 = 9 + x$ because when substituting 17 for x, we get a true statement.

DEFINITION

A **solution of an equation** is a value of the variable that makes the equation a true statement.

EXAMPLE 1

Determine whether $-\dfrac{3}{2}$ is a solution of the equation

$4x + 3 = 2(x - 1)$.

Solution

$$4x + 3 = 2(x - 1)$$

$$4\left(-\frac{3}{2}\right) + 3 \stackrel{?}{=} 2\left(-\frac{3}{2} - 1\right) \qquad \text{Substitute } -\frac{3}{2} \text{ for } x.$$

$$-6 + 3 \stackrel{?}{=} 2\left(-\frac{5}{2}\right) \qquad \text{Evaluate each side.}$$

$$-3 = -5 \qquad \text{False}$$

Since $-3 = -5$ is a false statement, $-\dfrac{3}{2}$ is *not* a solution of the equation.

PRACTICE 1

Is 6 a solution of the equation
$y + 2(y - 4) = 10$?

DEFINITION

A **linear equation in one variable** is an equation that can be written in the form

$$ax + b = c,$$

where a, b, and c are real numbers and $a \neq 0$.

Consider the equations $x = 5$ and $2x = 10$.

For $x = 5$, the solution is 5.
For $2x = 10$, we see by inspection that the solution is also 5.

The equations $x = 5$ and $2x = 10$ have the same solution, and so are *equivalent*.

DEFINITION

Equivalent equations are equations that have the same solution.

To *solve* an equation means to find the number or constant that, when substituted for the variable, makes the equation a true statement. This solution is found by changing the equation to an equivalent equation of the following form:

$$x = \boxed{}$$

The variable is isolated (alone   The number, or constant, is
on one side with coefficient 1).   isolated on the other side.

Addition Property

One of the properties that we use in solving equations involves adding. Suppose we have the equation $a = b$. If we add c to each side of the equation, we get $a + c = b + c$.

Addition Property of Equality

For any real numbers a, b, and c, if $a = b$, then $a + c = b + c$.

In words, this property states that when adding any real number to each side of an equation, the result is an equivalent equation. For instance, given that $5 = 5$, it follows that $5 + 3 = 5 + 3$.

Since subtracting a number is the same as adding its opposite, the addition property also allows us to subtract the same number from each side of an equation. For instance, if $a = b$, then subtracting c from each side gives us $a - c = b - c$.

EXAMPLE 2

Solve and check.

a. $x - 6 = -2.1$ **b.** $y + \dfrac{3}{5} = \dfrac{9}{10}$

Solution

a.
$$x - 6 = -2.1$$
$$x - 6 + 6 = -2.1 + 6 \qquad \text{Add 6 to each side of the equation.}$$
$$x + 0 = 3.9 \qquad \text{Simplify.}$$
$$x = 3.9$$

Check $x - 6 = -2.1$
$$3.9 - 6 \stackrel{?}{=} -2.1 \qquad \text{Substitute 3.9 for } x \text{ in the original equation.}$$
$$-2.1 = -2.1 \qquad \text{True}$$

The solution is 3.9.

b.
$$y + \frac{3}{5} = \frac{9}{10}$$
$$y + \frac{3}{5} - \frac{3}{5} = \frac{9}{10} - \frac{3}{5} \qquad \text{Subtract } \frac{3}{5} \text{ from each side of the equation.}$$
$$y + 0 = \frac{9}{10} - \frac{6}{10} \qquad \text{Simplify.}$$
$$y = \frac{3}{10}$$

Check $y + \dfrac{3}{5} = \dfrac{9}{10}$
$$\frac{3}{10} + \frac{3}{5} \stackrel{?}{=} \frac{9}{10} \qquad \text{Substitute } \frac{3}{10} \text{ for } x \text{ in the original equation.}$$
$$\frac{3}{10} + \frac{6}{10} \stackrel{?}{=} \frac{9}{10}$$
$$\frac{9}{10} = \frac{9}{10} \qquad \text{True}$$

The solution is $\dfrac{3}{10}$.

PRACTICE 2

Solve and check.

a. $x - \dfrac{1}{2} = 1\dfrac{1}{2}$

b. $y + 12.3 = -13$

Multiplication Property

Another property that is useful in solving equations involves multiplying. Consider the equation $a = b$. If we multiply each side of the equation by a number c, we get $ac = bc$.

The Multiplication Property of Equality

For any real numbers a, b, and c, if $a = b$, then $a \cdot c = b \cdot c$.

In words, this property states that when multiplying each side of an equation by any real number, the result is an equivalent equation. For instance, given that $5 = 5$, we can conclude that $5 \cdot 3 = 5 \cdot 3$.

Since dividing by a number is the same as multiplying by its reciprocal, the multiplication property also allows us to divide each side of an equation by a nonzero number. For instance, if $a = b$, then multiplying each side by $\frac{1}{c}$ gives us $a \cdot \frac{1}{c} = b \cdot \frac{1}{c}$ or, equivalently, $\frac{a}{c} = \frac{b}{c}$, for $c \neq 0$.

EXAMPLE 3

Solve.

a. $-y = 25$ **b.** $-27 = \dfrac{x}{3}$ **c.** $2.5y = 5$

Solution

a. $-y = 25$

$$\frac{-1y}{-1} = \frac{25}{-1} \qquad \text{Divide each side of the equation by } -1.$$

$$y = -25$$

The solution is -25.

b. $-27 = \dfrac{x}{3}$

$$-27 = \frac{1}{3}x$$

$$3(-27) = 3\left(\frac{1}{3}\right)x \qquad \text{Multiply each side of the equation by 3.}$$

$$-81 = 1x \qquad \text{Simplify.}$$

$$-81 = x$$

$$x = -81$$

The solution is -81.

c. $2.5y = 5$

$$\frac{2.5y}{2.5} = \frac{5}{2.5} \qquad \text{Divide each side of the equation by 2.5.}$$

$$1y = 2 \qquad \text{Simplify.}$$

$$y = 2$$

The solution is 2.

PRACTICE 3

Solve.

a. $8x = -24$

b. $-\dfrac{2}{5}y = 10$

c. $15.9 = 1.5x$

Equations Involving Both the Addition and Multiplication Properties

We now turn our attention to solving equations that involve both the addition and multiplication properties. Let's consider

$$-3w + 25 = 46 \quad \text{and} \quad \frac{s}{3} - 2 = -4.$$

To solve these equations, we first use the addition property to get the variable term alone on one side. Then, we use the multiplication property to isolate the variable.

EXAMPLE 4

Solve and check.

a. $-3w + 25 = 46$ **b.** $\dfrac{s}{3} - 2 = -4$

Solution

a.
$$-3w + 25 = 46$$
$$-3w + 25 - 25 = 46 - 25 \qquad \text{Subtract 25 from each side of the equation.}$$
$$-3w = 21$$
$$\frac{-3w}{-3} = \frac{21}{-3} \qquad \text{Divide each side of the equation by } -3.$$
$$w = -7$$

Check
$$-3w + 25 = 46$$
$$-3(-7) + 25 \stackrel{?}{=} 46 \qquad \text{Substitute } -7 \text{ for } w \text{ in the original equation.}$$
$$21 + 25 \stackrel{?}{=} 46$$
$$46 = 46 \qquad \text{True}$$

The solution is -7.

b.
$$\frac{s}{3} - 2 = -4$$
$$\frac{s}{3} - 2 + 2 = -4 + 2 \qquad \text{Add 2 to each side of the equation.}$$
$$\frac{s}{3} = -2$$
$$3\left(\frac{s}{3}\right) = 3(-2) \qquad \text{Multiply each side of the equation by 3.}$$
$$s = -6$$

Check
$$\frac{s}{3} - 2 = -4$$
$$\frac{-6}{3} - 2 \stackrel{?}{=} -4 \qquad \text{Substitute } -6 \text{ for } s \text{ in the original equation.}$$
$$-2 - 2 \stackrel{?}{=} -4$$
$$-4 = -4 \qquad \text{True}$$

The solution is -6.

PRACTICE 4

Solve and check.

a. $-5m + 20 = 45$

b. $\dfrac{2}{3}x - 9 = 17$

Equations Involving Combining Like Terms

Now, let's consider equations that have like terms on each side of the equation. Here, we combine the like terms on each side.

EXAMPLE 5

Solve and check.

a. $3y - 15 - 5y = 18 + 2y - 9$

b. $30x + 42 - 9x = 11 - x - 2$

Solution

a.
$$3y - 15 - 5y = 18 + 2y - 9$$

$-2y - 15 = 9 + 2y$	Combine like terms.
$-2y - 2y - 15 = 9 + 2y - 2y$	Subtract $2y$ from each side of the equation.
$-4y - 15 = 9$	Combine like terms.
$-4y - 15 + 15 = 9 + 15$	Add 15 to each side of the equation.
$-4y = 24$	
$\dfrac{-4y}{-4} = \dfrac{24}{-4}$	Divide each side of the equation by -4.
$y = -6$	

Check $3y - 15 - 5y = 18 + 2y - 9$

$3(-6) - 15 - 5(-6) \stackrel{?}{=} 18 + 2(-6) - 9$	Substitute -6 for y in the original equation.
$-18 - 15 + 30 \stackrel{?}{=} 18 - 12 - 9$	
$-3 = -3$	True

The solution is -6.

b.
$$30x + 42 - 9x = 11 - x - 2$$

$21x + 42 = 9 - x$	Combine like terms.
$21x + x + 42 = 9 - x + x$	Add x to each side of the equation.
$22x + 42 = 9$	Combine like terms.
$22x + 42 - 42 = 9 - 42$	Subtract 42 from each side of the equation.
$22x = -33$	
$\dfrac{22x}{22} = \dfrac{-33}{22}$	Divide each side of the equation by 22.
$x = -\dfrac{3}{2}$	Simplify.

Check $30x + 42 - 9x = 11 - x - 2$

$30\left(-\dfrac{3}{2}\right) + 42 - 9\left(-\dfrac{3}{2}\right) \stackrel{?}{=} 11 - \left(-\dfrac{3}{2}\right) - 2$	Substitute $-\dfrac{3}{2}$ for x in the original equation.
$-45 + 42 + \dfrac{27}{2} \stackrel{?}{=} 11 + \dfrac{3}{2} - 2$	
$-3 + \dfrac{27}{2} \stackrel{?}{=} 9 + \dfrac{3}{2}$	
$\dfrac{-6}{2} + \dfrac{27}{2} \stackrel{?}{=} \dfrac{18}{2} + \dfrac{3}{2}$	
$\dfrac{21}{2} = \dfrac{21}{2}$	True

The solution is $-\dfrac{3}{2}$.

Solve and check.

a. $10n + 15 + 4n = -13 - n - 2$

b. $6 + 15y - 18 = 8 - y - 9y$

Equations Involving Parentheses

Some equations contain parentheses. To solve these equations, we first use the distributive property to remove parentheses. Then, we proceed as in the previous examples.

EXAMPLE 6

Solve.

a. $2y + 3(y - 9) = -3(y + 6) - y$

b. $5[2 - (2n - 4)] = 2(5 - 3n)$

Solution

a. $2y + 3(y - 9) = -3(y + 6) - y$

$$
\begin{array}{ll}
2y + 3y - 27 = -3y - 18 - y & \text{Use the distributive property.} \\
5y - 27 = -4y - 18 & \text{Combine like terms.} \\
5y + 4y - 27 = -4y + 4y - 18 & \text{Add } 4y \text{ to each side of the equation.} \\
9y - 27 = -18 & \text{Combine like terms.} \\
9y - 27 + 27 = -18 + 27 & \text{Add 27 to each side of the equation.} \\
9y = 9 & \text{Combine like terms.} \\
\dfrac{9y}{9} = \dfrac{9}{9} & \text{Divide each side of the equation by 9.} \\
y = 1 & \text{Simplify.}
\end{array}
$$

The solution is 1.

b. $5[2 - (2n - 4)] = 2(5 - 3n)$

$$
\begin{array}{ll}
5[2 - 2n + 4] = 10 - 6n & \text{Use the distributive property.} \\
5[6 - 2n] = 10 - 6n & \text{Combine like terms.} \\
30 - 10n = 10 - 6n & \text{Use the distributive property.} \\
30 - 10n + 6n = 10 - 6n + 6n & \text{Add } 6n \text{ to each side of the equation.} \\
30 - 4n = 10 & \text{Combine like terms.} \\
30 - 30 - 4n = 10 - 30 & \text{Subtract 30 from each side of the equation.} \\
-4n = -20 & \text{Combine like terms.} \\
\dfrac{-4n}{-4} = \dfrac{-20}{-4} & \text{Divide each side of the equation by } -4. \\
n = 5 & \text{Simplify.}
\end{array}
$$

The solution is 5.

PRACTICE 6

Solve.

a. $9x + 5(x + 3) = -(x + 13) + x$

b. $-8t - [3(11 - 2t) + 11] = 9t$

In general, to solve linear equations, we follow this procedure:

To Solve a Linear Equation

- Use the distributive property to clear the equation of parentheses, if necessary.
- Combine like terms.
- Use the addition property to isolate the term with the variable.
- Use the multiplication property to isolate the variable.
- Check by substituting the solution in the original equation.

Equations are often useful *mathematical models* that represent real-world situations. Although there is no formula for solving applied problems in algebra, it is a good idea to keep the following problem-solving steps in mind:

> ## To Solve a Word Problem Using an Equation
> - Read the problem carefully.
> - Translate the word problem into an equation.
> - Solve the equation.
> - Check the solution in the original problem.
> - State the conclusion.

In Section 1.5, we translated word phrases to algebraic expressions. Similarly, we can translate word sentences to equations. In solving applied problems, first we translate the given word sentence to an equation, and then we solve the equation.

Now, let's take a look at several applied problems involving linear equations.

EXAMPLE 7

In an electrical engineering lab, a group of ten 6-volt and 12-volt batteries are wired in a series as shown.

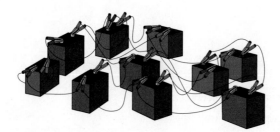

The sum of their voltages produces a power supply of 96 V. Find how many of each type of battery are used.

Solution   There are ten 6- and 12-V batteries in series. Let n represent the number of 6-V batteries. Then, $10 - n$ represents the number of 12-V batteries. Since the sum of their voltages produces a power supply of 96 V, we can write the following:

Word sentence   Together the 6-V batteries and the 12-V batteries produce 96 V.

Equation $6n$ $+$ $12(10 - n)$ $=$ 96

Now, we solve for n.

$$6n + 12(10 - n) = 96$$
$$6n + 120 - 12n = 96$$
$$-6n = -24$$
$$n = 4$$

Then, $10 - n = 10 - 4 = 6.$

Check

$$6n + 12(10 - n) = 96$$
$$6(4) + 12(10 - 4) \stackrel{?}{=} 96$$
$$24 + 12(6) \stackrel{?}{=} 96$$
$$24 + 72 = 96 \quad \text{True}$$

We conclude that there are four 6-V batteries and six 12-V batteries wired in series.

PRACTICE 7

A company planning an advertising campaign must pay $4000 for each newspaper ad and $2000 for each radio commercial. The company plans to use 30 fewer radio commercials than newspaper ads and allocates $600,000 for the ads and the commercials. How many radio commercials will be in the advertising campaign?

Some applications involving linear equations are modeled using *consecutive integers*. Examples of consecutive integers are 13, 14, and 15, as well as -20, -19, and -18. If x is an integer, a general representation of three consecutive integers is x, $x + 1$, and $x + 2$.

Examples of *consecutive even integers* are 6, 8, and 10. If x is an even integer, a general representation of three consecutive even integers is x, $x + 2$, and $x + 4$.

Beginning with an odd integer and counting by 2's gives *consecutive odd integers*. For example, three consecutive odd integers are 19, 21, and 23. If x is an odd integer, a general representation of three consecutive odd integers is x, $x + 2$, and $x + 4$.

EXAMPLE 8

A patient's dosage of a medication had been 50 mg. It is to be reduced to 20 mg in three administrations. If the dosage is reduced by consecutive even integer amounts, how many milligrams of medication are administered to the patient each time?

Solution   The dosage is to be reduced from 50 mg to 20 mg, that is, by a total of 30 mg. The dosages are reduced by consecutive even integer amounts, so that the three reductions can be represented by x, $x + 2$, and $x + 4$. We can set the total reduction equal to the sum of these reductions.

$$x + (x + 2) + (x + 4) = 30$$
$$x + x + 2 + x + 4 = 30$$
$$3x + 6 = 30$$
$$3x = 24$$
$$x = 8$$

The reductions from the original 50-mg dosage are, therefore, 8, 10, and 12 mg, resulting in dosages of 42, 32, and 20 mg. Note that the third dosage is 20 mg, as required.

PRACTICE 8

A prospective employee is considering a job offer in which his salary will increase by $1000 each year. If he wants the total of his earnings over four years to equal $200,000, what starting salary must he negotiate?

Some problems involving linear equations deal with *geometric* concepts.

EXAMPLE 9

The measures of vertical angles are equal. If $\angle AOB$ and $\angle COD$ are vertical angles, find the measure of each angle.

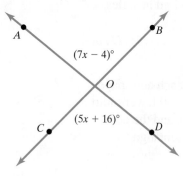

Solution Since the measures of vertical angles are equal, we can solve the following equation for x:

$$7x - 4 = 5x + 16$$
$$2x - 4 = 16$$
$$2x = 20$$
$$x = 10$$

Since $x = 10$, the measure of $\angle AOB$ is
$(7x - 4)° = (7 \cdot 10 - 4)° = (70 - 4)° = 66°$.
Similarly, the measure of $\angle COD$ is
$(5x + 16)° = (5 \cdot 10 + 16)° = (50 + 16)° = 66°$.

So the measure of each angle is $66°$.

PRACTICE 9

The perimeters of the rectangle and triangle shown are equal. Find the perimeter of each.

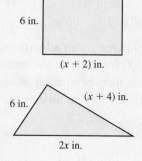

Let's now turn our attention to another kind of word problem, namely one involving *uniform* or *average motion*. To solve such problems, we use the equation $d = rt$, where d is the distance traveled, r is the rate or speed, and t is time.

EXAMPLE 10

A group of students took a trip from Washington, D.C., to New York City as shown on the map. If the driver averaged 60 mph, how long to the nearest hour did it take them to get to New York City?

PRACTICE 10

The bus trip from Miami to San Francisco—a distance of 3348 mi—takes 70 hr of driving time. What is the average speed of the bus, rounded to the nearest mile per hour? (*Source:* Greyhound)

Solution   The distance between Washington, D.C., and New York City is 230 mi and the rate of the car is 60 mph. Using the equation $d = rt$, we can find the time.

$$d = rt$$
$$230 = 60t \qquad \text{Substitute 230 for } d \text{ and 60 for } r.$$
$$\frac{230}{60} = \frac{60t}{60}$$
$$t = 3\tfrac{5}{6}$$

Check

$$230 = 60t$$
$$230 \stackrel{?}{=} 60(3\tfrac{5}{6}) \qquad \text{Substitute } 3\tfrac{5}{6} \text{ for } t \text{ in the original equation.}$$
$$230 \stackrel{?}{=} 60\left(\frac{23}{6}\right)$$
$$230 = 230 \qquad \text{True}$$

So it took about 4 hr to get from Washington, D.C., to New York City.

There are other types of uniform motion problems that lead to more complicated equations. Some of these problems involve one or more objects traveling the same distance, at different rates, and for different lengths of time.

EXAMPLE 11

Twenty minutes after a wife left for work on the bus, her husband noticed that she had left her briefcase at home. The husband left home, driving at 36 mph, to catch up with the bus that was traveling at 24 mph. How long did it take him to catch up with the bus?

Solution   Let t represent the number of hours traveled by the husband. Since his wife left 20 min earlier, $t + \frac{1}{3}$ represents the number of hours she traveled. Notice that because the time is in hours, we express 20 min as $\frac{20}{60}$ hr, or $\frac{1}{3}$ hr. The bus and the car are each traveling at a constant speed. So the distance that it travels is a product of its rate of travel (speed) and the time it has traveled. Putting these quantities in a table clarifies their relationship.

	Rate	·	Time	=	Distance
Husband	36		t		$36t$
Wife	24		$t + \frac{1}{3}$		$24\left(t + \frac{1}{3}\right)$

PRACTICE 11

A car leaves Atlanta traveling at 45 mph. Forty minutes later, a second car leaves the same city traveling at 55 mph in the same direction. In how many hours will the second car overtake the first car?

r	t	= D
45	t	$45t$
55	$t + \frac{4}{6}$	$55\left(t + \frac{2}{3}\right)$

$$\frac{40}{60} = \frac{4}{6} \quad \frac{2}{3}$$

$$45t = 55\left(t + \frac{2}{3}\right)$$
$$45t = 55t + \frac{110}{3}$$
$$\frac{-10t}{-10} = \frac{110}{3} \cdot \frac{1}{10}$$

$$\frac{11}{3}$$

EXAMPLE 11 (continued)

Since the husband and wife travel the same distance to the point where they meet, we can write the following equation, and then solve:

$$36t = 24\left(t + \frac{1}{3}\right)$$
$$36t = 24t + 8$$
$$12t = 8$$
$$t = \frac{8}{12}$$
$$t = \frac{2}{3}$$

So it will take the husband $\frac{2}{3}$ hr, or 40 min, to catch up with the bus.

EXAMPLE 12

Two snails, 432 cm apart, crawl toward each other at rates that differ by 2 cm/min. If it takes the snails 27 min to meet, how fast is each snail crawling?

Solution   Let x represent the rate of the snail crawling in one direction and $x + 2$ represent the rate of the snail crawling in the opposite direction.

	Rate ·	Time =	Distance
Snail crawling in one direction	x	27	$27x$
Snail crawling in opposite direction	$x + 2$	27	$27(x + 2)$

The total distance is 432 cm. So we can solve the following equation to find the rate of each snail:

$$27x + 27(x + 2) = 432$$
$$27x + 27x + 54 = 432$$
$$54x + 54 = 432$$
$$54x = 378$$
$$x = 7$$

Then, $x + 2 = 7 + 2 = 9$. So the rates of the snails are 7 centimeters per minute and 9 cm/min.

PRACTICE 12

Two cyclists simultaneously start traveling from the same point, but in opposite directions, down a straight road, one at 28 km/hr and the other at 36 km/hr. How long will they have traveled by the time they are 96 km apart?

Now, we will focus on two types of problems involving *percent*. These are *investment* or *simple interest* problems, and *mixture* or *solution* problems.

First, we consider investment or simple interest problems. Interest depends on the amount of money borrowed (the principal, p), the annual rate of interest (r, usually expressed as a percent), and the length of time the money is borrowed (t, generally expressed in years). We can compute the amount of interest by multiplying the principal by the rate of interest and the number of years. This is given by the equation $I = prt$. In the same way, investment return depends on the amount of money invested p, the annual rate of return r, and the length of time the money is invested t. Note that whether we are considering investment or interest, the formula $I = prt$ simplifies to $I = pr$ if t is given to be 1 yr.

EXAMPLE 13

The amount of money that a broker invested in bonds was double what she invested in a mutual fund. The profit on the bonds was 4% and the mutual fund lost 4% in value. If the net profit for 2 yr was $800, how much was invested in each?

Solution   We set up a table to organize the given information, letting x represent the amount of money invested in the mutual fund. Note that the profit for each investment is the product of each principal, the rate of profit, and the time.

	Principal $\cdot$	Rate of Profit $\cdot$	Time $=$	Profit
Mutual Fund Amount	x	-0.04	2	$(-0.04)(2)x$
Amount of Bonds	$2x$	0.04	2	$0.04(2)(2x)$
Total Investment				800

The profit on the total investment is the sum of the profits on the mutual fund and the bonds:

$$(-0.04)(2)x + (0.04)(2)(2x) = 800$$
$$-0.08x + 0.16x = 800$$
$$0.08x = 800$$
$$\frac{0.08x}{0.08} = \frac{800}{0.08}$$
$$x = 10,000$$

Then, $2x = 2(10,000) = 20,000$. So $10,000 was invested in mutual funds and $20,000 in bonds.

PRACTICE 13

A beneficiary split her $200,000 inheritance into two investments. One of the investments gained 12% and the other lost 8%. If she broke even on the two investments, how much money did she invest in each?

In mixture problems, the amount of a particular ingredient in the solution or mixture is often expressed as a percent of the total. Chemists and pharmacists often deal with mixture and solution problems.

EXAMPLE 14

A prescription calls for 150 g of 2.5% hydrocortisone cream, meaning 150 g of cream, 2.5% of which is hydrocortisone. The local pharmacist has in stock 0.5% and 3% hydrocortisone creams. How many grams of each must she mix to fill the prescription?

Solution   The pharmacist has available two different strengths of hydrocortisone creams to fill the prescription for 150 g of 2.5% hydrocortisone cream. Let n represent the number of grams of the 0.5% hydrocortisone cream to be used in the mixture. Then, $150 - n$ represents the number of grams of the 3% hydrocortisone cream.

PRACTICE 14

A mechanic added 3 qt of pure antifreeze to 9 qt of a 40% antifreeze solution. What is the percent of antifreeze in the new solution?

EXAMPLE 14 (continued)

We set up the table below to organize the given information.

Action	Percent of Hydrocortisone	Amount of Cream (g)	Amount of Hydrocortisone (g)
Start with	0.5%	n	$0.005n$
Add	3%	$150 - n$	$0.03(150 - n)$
Finish with	2.5%	150	$0.025(150)$

The amount of hydrocortisone that she started with plus the amount of hydrocortisone that she added equals the amount of hydrocortisone that she finished with, which has to be 2.5% of 150 g. To find how many grams of each strength of hydrocortisone cream the pharmacist must mix to fill the prescription, we solve the following equation:

$$0.005n + 0.03(150 - n) = 0.025(150)$$
$$0.005n + 4.5 - 0.03n = 3.75$$
$$-0.025n + 4.5 = 3.75$$
$$-0.025n = -0.75$$
$$n = 30$$

Then, $150 - n = 150 - 30 = 120$. So the pharmacist must mix 30 g of the 0.5% hydrocortisone cream with 120 g of the 3% hydrocortisone cream.

Do you see that we could have solved the equation in Example 14 by first multiplying each side by 1000? What is the advantage in this approach?

Mathematically Speaking

Fill in each blank with the most appropriate term or phrase from the given list.

linear equation	multiplicative inverse	multiplication property
evaluation	property	of equality
inequality	equivalent equation	distributive property
equation	solution	

1. A mathematical statement that two expressions are equal is called a(n) _____.

2. A(n) _____ of an equation is a value of the variable that makes the equation a true statement.

3. An equation in one variable that can be written in the form $ax + b = c$, where a, b, and c are real numbers and $a \neq 0$, is called a(n) _____.

4. According to the _____, if each side of an equation is multiplied by any real number, the result is an equivalent equation.

A *Determine whether the given value is a solution of the equation.*

5. $2x - 1 = 5; 2$

6. $6 - 3x = 9; 1$

7. $1 - \dfrac{t}{4} = -2; -4$

8. $-11 = \dfrac{2n}{3} - 7; -6$

9. $5y - 6 = 7y - 5; -\dfrac{1}{2}$

10. $8 - 3x = 2x + 7; \dfrac{1}{5}$

11. $2n - 10 = 4(3n - 15); 7$

12. $-(4x - 7) = 17 - 6x; 5$

13. $\dfrac{1}{2}\left(9 - \dfrac{x}{4}\right) + x = \dfrac{x}{3} - 2; -12$

14. $\dfrac{2}{3}y - 7 = \dfrac{1}{4}(3y + 5) - 8; -3$

B *Solve and check.*

15. $x + 4 = 2$

16. $x + 5 = 1$

17. $x - 3.7 = -2$

18. $n - 8 = -6.4$

19. $\dfrac{3}{4} + y = -\dfrac{5}{8}$

20. $-\dfrac{1}{6} + t = 2\dfrac{2}{3}$

21. $\dfrac{n}{5} = -1$

22. $\dfrac{x}{2} = 7$

23. $16 - 4y = 0$

24. $42 - 6x = 0$

25. $\dfrac{2}{3}n = \dfrac{4}{9}$

26. $\dfrac{5}{7}y = -\dfrac{15}{28}$

27. $-8.1 = 0.9a$

28. $-8 = -1.6t$

29. $5x + 1 = -4$

30. $3x + 2 = -7$

31. $12 - x = -10$

32. $7 - y = 5$

33. $-4n - 6 = 10$

34. $-2y + 3 = -5$

35. $9 + 8n = 9$

36. $7y - 15 = -15$

37. $\dfrac{t}{3} - 5 = 2$

38. $-\dfrac{n}{2} + 8 = -3$

39. $4 - \dfrac{3}{5}x = 13$

40. $\dfrac{3}{4}x - 10 = -16$

41. $0.2 = -1.2n - 7$

42. $-6.3 = 5.7 - 2x$

43. $8y = y$

44. $-9x = 2x$

45. $10 - 5x = x + 18$

46. $9n - 13 = 6n - 11$

47. $7y - 8 = 12y - 8$

48. $13 + 4x = 13 - 3x$

49. $8a - 3 - 5a = 15$

50. $2t + 4 + t = -5$

51. $16n = 7n - 15 - 6n$

52. $-13x = 19 - 8x - 14$

53. $2.4 - 0.6x + 3.3 = 1.3x$

54. $1.7n + 1 - 2.3n = -1.4$

55. $18 - 12n = 16 + 3n - 11$

56. $9y + 4 - 14y = 2 - y$

57. $23t + 11 - 15t = 6t - 18 + 7$

58. $14n - 6 - 17 = 9n + 12 - 2n$

59. $11y + 24 - 18y = 13y - 21 - 10y$

60. $9y - 13y + 5 = 17 + 4y - 14$

61. $5.7 + 3.6x - 7.2 = 0.6x + 2.5 - 2x$

62. $4n + 1.3 - 3.2n = 1 - 5.2n - 3.9$

63. $-2(x - 6) = 4$

64. $6(x - 5) = -12$

65. $7 - (3n - 8) = -6$

66. $-10 - (5x + 2) = 8$

67. $-4(7 + 3x) = -5(2x + 8)$

68. $-2(9 - 7x) = 3(8x - 6)$

69. $\dfrac{1}{2}(16n - 12) = 9n + 11$

70. $\dfrac{2}{3}(9n + 3) = 2n - 14$

71. $5x - 2(x + 6) = 6(x - 1) - 8$

72. $4(4 - x) - 1 = 7(x + 3) - 14$

73. $13 - 9(2n + 3) = 4(6n + 1) - 15n$

74. $5(5x - 7) + 40 = 2x + 3(8x + 5)$

75. $3[1 - (4x - 5)] = 5(6 - 2x)$

76. $-4(3n + 7) = 2[9 - (7n + 10)]$

77. $\dfrac{1}{2}(18 - 6n) + 5n = 10 - \dfrac{1}{4}(16n + 20)$

78. $\dfrac{2}{3}(9y - 6) - 5 = \dfrac{2}{5}(30y - 25) - 7y$

79. $12y - [9(2 - y) - 8] = 5y + 3(6 - 4y)$

80. $15x + [4(7 - 2x) - 12] = 10x - (5x - 8)$

81. $6(3x - 8) - 4(4x - 9) = 3(5x + 3) - (7x + 1)$

82. $2(5x + 4) - 3(3x - 7) = 6(2x + 5) - (6x - 9)$

83. $\dfrac{1}{2}x - \dfrac{3}{4}x + 5 = \dfrac{1}{6}(10 - 3x) - \dfrac{2}{3}$

84. $2\left(\dfrac{3}{5}n - \dfrac{1}{8}\right) - \dfrac{7}{10}n = 2 + \dfrac{4}{5}n$

Mixed Practice

Solve.

85. Is -26 a solution of the equation $\dfrac{-t}{2} + 8 = -5$?

Solve and check.

86. $0.5 = 0.5x - 3$

87. $n - 6 = 3.2$

88. $6t - 2t + 3 = 10 + 3t - t$

89. $3(4n - 1) - 2(n + 4) = 2(3n + 5)$

90. $\dfrac{3}{4}y = \dfrac{5}{8}$

91. $14x - [7 - 3(9x - 4)] = 22$

92. $2[6 - 5(1 - y)] = 3(y - 2) - 5y$

Applications

C *Solve.*

93. The cost of a book at an online discount book retailer is 20% less than the retail price of the book. The total cost to purchase the book online is $33.59, which includes a shipping fee of $3.99.
  a. What is the original retail price of the book?
  b. Excluding the shipping fee, how much money was saved purchasing the book online?

94. The metered rate for a taxi in Boston is $2.60 plus $0.40 for each $\frac{1}{7}$ mi traveled. (*Source:* taxifarefinder.com)
  a. How many miles, to the nearest mile, did a passenger travel if the total fare was $19?
  b. A passenger taking a taxi from Logan Airport was charged $7.50 in addition to the regular rate. How many miles, to the nearest mile, did the passenger travel if the total fare was $40.50?

95. There are 10 equally spaced hurdles in the women's 400-meter hurdle track event. The distance from the starting line to the first hurdle is 45 m, and the distance from the last hurdle to the finish line is 40 m. What is the distance between the hurdles?

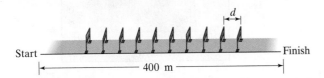

96. The bill for a two-night hotel stay was $280.69, which included a room service charge of $23.15, a hotel restaurant charge of $47.50, and a room tax charge of $32.04. What was the room rate charged per night?

97. A telephone company offers two plans for local and long-distance calling. Plan A costs a flat fee of $39.95 per month for unlimited local and long-distance calling. Plan B costs $14.95 for unlimited local calling and $4 plus an additional $0.07 per minute for long-distance calling per month. For what number of minutes will the monthly cost of Plan B be the same as the monthly cost of Plan A?

98. A technology company offers an employee two salary options. The first option is an annual salary of $60,000 plus 0.5% of the company's annual revenue. The second option is an annual salary of $36,000 plus 1.5% of the company's annual revenue. For what revenue amount will the salary for both options be the same?

99. An interior designer has small and large throw pillows made for a family room. The small pillows require $\frac{1}{2}$ yd^2 of fabric and the large pillows require 2 yd^2 of fabric. If 18 yd^2 of fabric are used to make 18 pillows, how many of each type of pillow are made for the family room?

100. During a clothing drive, a local community center used two different-size storage bins to sort and carry donated items. One type of bin has a storage capacity of 15 ft^3 and the other has a storage capacity of 27 ft^3. The center used 26 bins with a total storage capacity of 486 ft^3. How many of each type of bin were used for the clothing drive?

101. An academic organization holds its national conference annually. This year, the conference registration fee was $120 for members and $200 for nonmembers. A total of $268,120 was collected in registration fees from 2017 registrants. How many nonmembers registered for the conference?

102. A florist making bouquets for the bridesmaids in a wedding party used a combination of oceana roses and white roses. He charged $6 per stem for the oceana roses and $9 per stem for the white roses. The bill for the bouquets was $840. If the florist used a total of 120 roses, how many of each type were used to make the bouquets?

103. Each employee of a company is assigned an employee number. The three newest employees were assigned employee numbers in consecutive order. If the sum of the first two employee numbers assigned is 2407, what was the third number assigned?

104. The inventory manager for a women's catalog clothing company is responsible for assigning item numbers to the clothes for inventory purposes. Each item number consists of a four-digit style number followed by a two-digit size number. The item number for a new skirt in inventory is assigned the same style number, but three different size numbers—to identify petite, regular, and tall sizes. If the style number is 4731 and the size numbers are consecutive odd integers whose sum is 51, what are the three item numbers assigned to the skirt?

105. The measure of an exterior angle of a triangle is equal to the sum of the measures of the two nonadjacent interior angles. What are the measures of the nonadjacent interior angles of the triangle shown?

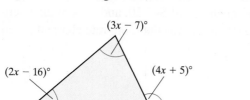

106. The perimeter of the rug shown is 24 ft less than the perimeter of the rectangular floor. Find the area of the floor not covered by the rug.

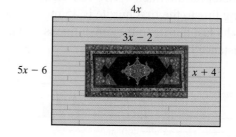

107. The AIDS Walk is an annual 10-km walkathon held in various cities throughout the United States. If a participant walks at an average rate of 4 km/hr, how long will it take her to finish the walkathon?

108. The driving time for a road trip from San Francisco to Phoenix, a distance of 751 mi, is about $11\frac{3}{4}$ hr. What is the average speed of the car, rounded to the nearest mile per hour? (*Source: Rand McNally Atlas*)

109. Fifteen minutes after a boy left for school on his bike, his mother noticed that he had left his term paper on the kitchen table. His mother left home, driving at a rate of 32 mph, to catch up with him. If he had been bicycling at a rate of 8 mph, how long did it take his mother to catch up with him?

110. A car leaves Las Vegas traveling at a rate of 50 mph. Half an hour later, a second car leaves the same city traveling at a rate of 70 mph in the same direction. In how many hours will the second car catch up to the first car?

111. Two groups of college students decide to take a weekend road trip. The first group leaves at 8:00 A.M. traveling at a rate of 60 mph. The second group leaves 20 min later traveling at a rate of 75 mph. Within the first half hour of the trip, the first group stops for 10 min to get gas for the car. At what time will the second group catch up to the first group?

112. Two cars leave the Clara Barton rest stop on the New Jersey Turnpike at the same time heading south. One car travels at an average speed of 54 mph and the other car travels at an average speed of 66 mph. How long after they leave the rest stop will the two cars be 5 mi apart?

113. Two friends live 2.2 mi apart along the same road. If they walk toward each other at rates that differ by $\frac{1}{2}$ mph and meet 24 min later, how fast is each person walking?

114. A husband and wife leave the same train station at the same time on different trains traveling in opposite directions. One train travels 8 mph faster than the other. If the trains are 182 mi apart after 1 hr 45 min, what is the speed of each train?

115. A sales representative invests his $12,000 bonus in two mutual funds. After one year, one fund made a 7% profit and the other lost 9%. If the total profit on the investments was $200, how much did he invest in each mutual fund?

116. An engineer invests her $25,000 bonus into two funds. After one year, one fund gains 6% and the other fund loses 4%. If she broke even on the investments, how much did she invest in the fund that lost money?

117. A broker invests three times the amount of money in a low-risk mutual fund as in a high-risk mutual fund. The low-risk fund showed a return of 3% and the high-risk fund showed a loss of 12%. If the net loss on the investments was $525, how much money did the broker invest?

118. An investor's stock portfolio includes three investments: a growth-and-income fund, bonds, and a balanced fund. Twice as much is invested in the growth-and-income fund as in the balanced fund, and one-third as much is invested in bonds as in the balanced fund. The growth-and-income fund had a return of 8%, the bonds had a return of 3%, and the balanced fund had a loss of 5%. If the net return on the investments was $1080, how much was invested in each fund?

119. As part of a lab experiment for a chemistry class, a student adds pure alcohol to 24 mL of a 9% alcohol solution. How much pure alcohol must he add to the 9% solution to make a 16% solution?

120. A chemist has a 50-mL solution that is 25% acid. How much water should she add to make a 10% solution?

121. A motorist wants 10 gal of 90-octane gasoline to fill the gas tank in her car. The gas station only has 87 octane and 92 octane available.

a. How much of each grade of gasoline does she need to mix together to make the required amount?

b. If 87 octane costs $3.80 per gallon and 92 octane costs $4.10 per gallon, how much did she pay to fill her gas tank?

122. The manager of the meat department at a grocery store wants to make packages of 90% lean ground beef by mixing 85% lean ground beef with 97% lean ground beef.

a. How many pounds of 85% lean ground beef should be mixed with 97% lean ground beef to make a 2-lb package of 90% lean ground beef?

b. If the 85% lean ground beef sells for $3.59 per pound and the 97% lean ground beef sells for $4.29 per pound, for how much per pound should the manager sell the 90% lean ground beef?

• Check your answers on page A-18.

MIND*Stretchers*

Writing

1. Describe a situation that could be solved using the equation $5(x + 20) = 100$.

Mathematical Reasoning

2. Not all equations in one variable have a single real-number solution. Describe the solutions of the following equations:

a. $2x - 5 = 2x - 5$ **b.** $3x + 2 = 3x - 1$

Is there a way to identify equations of the form $ax + b = dx + c$, where a, b, c, and d are real numbers and $a \neq 0$ and $d \neq 0$, that have the solutions described above? Explain.

(continued)

Patterns

3. A table can be used to solve equations.

   a. Complete the following table. Identify the solution to the equation $-4(2x - 1) = 7 - 5x$.

x	-4	-3	-2	-1	0	1	2	3	4
$-4(2x - 1)$									
$7 - 5x$									

   b. Complete the following table. What can you conclude about the solution of the equation $6x - (2x + 5) = 7 - 4x$?

x	-4	-3	-2	-1	0	1	2	3	4
$6x - (2x + 5)$									
$7 - 4x$									

   c. Fill in the table to solve the equation from part (b). (*Hint:* Use increments of 0.1 for x.)

x	1.1	1.2	1.3	1.4	1.5	1.6	1.7	1.8	1.9
$6x - (2x + 5)$									
$7 - 4x$									

Cultural Note

The Rhind papyrus, held in the British Museum, was written about 1650 B.C. and is the best surviving example of Egyptian mathematics, including algebra. It is a written record, much like a mathematics textbook, that includes problems that reveal the Egyptians' algebraic knowledge. The Egyptians of that time solved exercises equivalent to linear equations in one unknown. They did not use symbols (for instance, $2x + 6 = 8$) like those in common use today. All of their problems were stated and solved verbally.

(*Source:* Jan Gullberg, *Mathematics from the Birth of Numbers*, W.W. Norton & Company, 1997)

2.2 Solving Literal Equations and Formulas

In many situations, equations describe the relationship between two or more variables. Such equations are called *literal* equations.

Consider, for instance, the literal equation

$$Ax + By = C.$$

Since there is more than one variable, we can solve for one of the variables in terms of the others, for instance, x in terms of A, B, C, and y. Solving for x is especially useful if we want to find the value of x for various values of A, B, C, and y.

EXAMPLE 1

Solve $Ax + By = C$ for x in terms of A, B, C, and y.

Solution

$$Ax + By = C$$
$$Ax + By - By = C - By \qquad \text{Subtract } By \text{ from each side of the equation.}$$
$$Ax = C - By \qquad \text{Simplify.}$$
$$\frac{Ax}{A} = \frac{C - By}{A} \qquad \text{Divide each side of the equation by } A.$$
$$x = \frac{C - By}{A} \qquad \text{Simplify.}$$

The solution is $x = \dfrac{C - By}{A}$.

PRACTICE 1

Solve $y = mx + b$ for x in terms of y, m, and b.

Note that we can solve literal equations by using the addition and multiplication properties that we used in Section 2.1. Explain how you would check the solution for the equation $Ax + By = C$ in Example 1.

Sometimes, we use the distributive property to solve certain literal equations.

EXAMPLE 2

Solve $a + b(x - y) = c$ for x in terms of a, b, c, and y.

Solution

$$a + b(x - y) = c$$
$$a + bx - by = c \qquad \text{Use the distributive property.}$$
$$a - a + bx - by = c - a \qquad \text{Subtract } a \text{ from each side of the equation.}$$
$$bx - by = c - a$$
$$bx - by + by = c - a + by \qquad \text{Add } by \text{ to each side of the equation.}$$
$$bx = c - a + by$$
$$\frac{bx}{b} = \frac{c - a + by}{b} \qquad \text{Divide each side of the equation by } b.$$
$$x = \frac{c - a + by}{b}$$

PRACTICE 2

Solve $y - b = m(x - a)$ for x in terms of a, b, m, and y.

A *formula* is a special type of literal equation that provides a symbolic description of the some real-world object or action. We can use these types of equations to describe situations such as how the distance you walk depends on how fast and how long you walk, how the size of a circular pan relates to its diameter, and how the amount of money you pay is determined by the cost of the items you buy.

EXAMPLE 3

In business, $A = P(1 + rt)$ is a formula used for computing the amount in an account that generates simple interest. In the formula, A represents the amount, P the original principal, r the annual rate of interest, and t the time in years.

a. Solve for r in terms of A, P, and t.

b. If $A = \$5600$, $P = \$5000$, and $t = 2$ yr, find the value of r.

Solution

a. Solving $A = P(1 + rt)$ for r gives us

$$A = P(1 + rt)$$
$$A = P + Prt \qquad \text{Use the distributive property.}$$
$$A - P = P - P + Prt \qquad \text{Subtract } P \text{ from each side of the equation.}$$
$$A - P = Prt$$
$$\frac{A - P}{Pt} = \frac{Prt}{Pt} \qquad \text{Divide each side of the equation by } Pt.$$
$$\frac{A - P}{Pt} = r$$

So $r = \dfrac{A - P}{Pt}$.

b. We are given that $A = \$5600$, $P = \$5000$, and $t = 2$ yr. So we substitute in the formula, $r = \dfrac{A - P}{Pt}$, to find r.

$$r = \frac{A - P}{Pt}$$
$$= \frac{5600 - 5000}{5000(2)}$$
$$= \frac{600}{10,000}$$
$$= 0.06$$

So the rate is 6%.

PRACTICE 3

In architecture, many geometry formulas are used. The formula

$$A = \frac{1}{2}h(b_1 + b_2)$$ gives the area A

of a trapezoid, where h stands for the altitude (or height), b_1 for the upper base, and b_2 for the lower base.

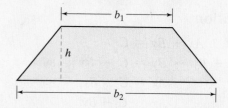

a. Find a formula for b_2 in terms of A, h, and b_1.

b. If $A = 350$ ft^2, $h = 10$ ft, and $b_1 = 20$ ft, find b_2.

EXAMPLE 4

The Fahrenheit temperature F is $32°$ more than nine-fifths of the equivalent Celsius temperature C.

a. Write a formula for this relationship.

b. Solve the formula for C.

c. What Celsius temperature corresponds to the Fahrenheit temperature $86°$?

Solution

a. Fahrenheit temperature F is $32°$ more than nine-fifths of the equivalent Celsius temperature C. To get the corresponding formula, we translate this statement to an equation with mathematical symbols.

$$F = \frac{9}{5}C + 32$$

b. We solve $F = \frac{9}{5}C + 32$ for C.

$$F = \frac{9}{5}C + 32$$

$$F - 32 = \frac{9}{5}C + 32 - 32$$

$$F - 32 = \frac{9}{5}C$$

$$\frac{5}{9}(F - 32) = \frac{5}{9}\left(\frac{9}{5}\right)C$$

$$\frac{5}{9}(F - 32) = C$$

So $C = \frac{5}{9}(F - 32)$.

c. Substitute 86 for F in the formula $C = \frac{5}{9}(F - 32)$.

$$C = \frac{5}{9}(F - 32)$$

$$= \frac{5}{9}(86 - 32)$$

$$= \frac{5}{9}(54)$$

$$= 30, \text{ or } 30°C$$

So $30°C$ corresponds to $86°F$.

PRACTICE 4

The formula for the surface area S of a rectangular prism with a square base is the sum of twice a^2 and four times the product of a and h, where a is the length of a side of the square base and h is the height.

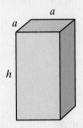

a. Write a formula for this relationship.

b. Solve the formula for h.

c. What is the height of a rectangular prism with surface area 504 in^2. and square base with sides of length 6 in.?

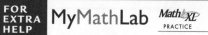
A *Solve for the indicated variable.*

1. $7m + n = 2$ for n

2. $b - 6a = -3$ for b

3. $9a - 3b = 0$ for a

4. $4x + 8y = 0$ for y

5. $\frac{2}{3}xy = z$ for x

6. $-\frac{1}{2}mp = n$ for m

7. $5a + b = c$ for a

8. $4m - y = n$ for m

9. $ax - by = c$ for y

10. $ay - bx = -c$ for y

11. $qrs + t = 0$ for r

12. $uvw - x = 0$ for w

13. $\frac{x}{y} - z = v$ for x

14. $n + \frac{p}{q} = -r$ for p

15. $\frac{a - b}{c} = -d$ for b

16. $\frac{a + b}{-c} = d$ for a

17. $\frac{rs}{t} = v + w$ for s

18. $\frac{ut}{-w} = x - y$ for t

19. $wx - y = u + s$ for w

20. $p - q = rs + t$ for r

21. $a(b + c) = d - g$ for c

22. $x(u - v) + r = y$ for v

Solve the formula for the indicated variable.

23. Ohm's Law: $V = IR$ for R

24. Area of a regular polygon: $A = \frac{1}{2}ap$ for a

25. Centripetal force: $F = \frac{mv^2}{r}$ for m

26. Volume of a cone: $V = \frac{1}{3}\pi r^2 h$ for h

27. Perimeter of a rectangle: $P = 2(l + w)$ for w

28. Investment at simple interest: $A = P(1 + rt)$ for t

29. Lateral area of a frustum: $A = \pi s(r_1 + r_2)$ for r_1

30. Area of a trapezoid: $A = \frac{1}{2}(b_1 + b_2)h$ for b_1

31. The nth partial sum of an arithmetic sequence: $S = \frac{n(a_1 + a_n)}{2}$ for a_n

32. The nth term of an arithmetic sequence: $a_n = a_1 + (n - 1)d$ for n

Mixed Practice

33. Solve for F: $d(F + G) = b - c$

34. Solve for x: $2x + y = z$

35. In the formula $A = 2\pi r(r + h)$ for the surface area of a cylinder, solve for h.

36. Solve for b: $a + \frac{b}{c} = d$

Applications

B *Solve.*

37. The final velocity V of an object under constant acceleration can be found using the formula $V^2 = v^2 + 2as$, where v is initial velocity (in meters per second), a is acceleration (in meters per second^2), and s is distance (in meters).

a. Solve the formula for a.

b. Find the acceleration of a car if it travels 125 m before coming to a complete stop ($V = 0$ m/sec) from an initial velocity of 20 m/sec.

38. The selling price s of an item is $s = c + mc$, where c is the cost of the item and m is the percent markup based on cost (in decimal form).

a. Solve the formula for m.

b. What is the percent markup on an item that costs \$42 and sells for \$73.50?

39. A clinical dietitian can estimate the ideal body weight I for men using the formula $I = 6(h - 60) + 106$, where h is height (in inches).

a. Solve the formula for h.

b. How tall is a man whose ideal body weight is 190 lb?

40. In geometry, the formula for calculating the sum of the measures of the interior angles S (in degrees) of a polygon is $S = 180(n - 2)$, where n is the number of sides of the polygon.

a. Solve the formula for n.

b. If the sum of the measures of the interior angles of a polygon is 1620°, how many sides does the polygon have?

41. The minimum number of calories required by the body for basic life processes is called the *basal energy expenditure*. The basal energy expenditure for women, represented by E, can be determined using the formula

$$E = 9.6w + 1.7h - 4.7y + 655,$$

where w is weight (in kilograms), h is height (in centimeters), and y is age (in years). (*Source: She Does Math!*, Mathematical Association of America, 1995)

a. Solve the formula for y.

b. Approximate the age of a woman who weighs 55 kg, is 160 cm tall, and has a basal energy expenditure of 1305.

42. In order to account for daily physical activity, the basal energy expenditure formula (see Exercise 41) can be adjusted by multiplying the right side by an activity factor. For a moderately active woman, this factor is 1.4.

a. Use the formula in Exercise 41 to write an adjusted formula for y of a moderately active woman.

b. Approximate the age of a moderately active woman who weighs 66 kg, is 170 cm tall, and has an adjusted basal energy expenditure of 1906.

43. In baseball, a pitcher's *earned run average E* is determined by calculating nine times the ratio of the number of earned runs R to the number of innings pitched I.

a. Write a formula for E.

b. Solve this formula for R.

c. The table below shows the earned run average and the total number of innings pitched for selected major league pitchers for the 2010 season. Use the formula to complete the table. (Round your answers to the nearest whole number.)
(*Source:* baseball-reference.com)

Pitcher	E	I	R
Félix Hernández	2.27	249.2	
Josh Johnson	2.30	183.2	
Clay Buchholz	2.33	173.2	
Adam Wainwright	2.42	230.1	

44. The total surface area S of a cylinder, including the top and bottom, is $2\pi r$ times the sum of r and h, where r represents the radius of the cylinder and h its height.

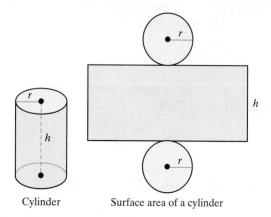

Cylinder   Surface area of a cylinder

a. Write a formula for this relationship.

b. Solve the formula for h.

c. To the nearest inch, calculate the height of a cylindrical bin whose surface area is 924 in.^{2} and radius is 7 in. Use $\pi \approx 3.14$.

45. The eye's focusing ability or *accommodation* is measured in diopters. An ophthalmologist can calculate a person's accommodation A by dividing 100 by the distance d in centimeters that a person can bring an object to the eyes and still see it.

a. Write a formula for accommodation.

b. What is the accommodation of a person who can see an object clearly within 12.5 cm?

46. The equivalent resistance R (in ohms) of two resistors connected in parallel is the product of their resistances, R_1 and R_2, divided by the sum of their resistances.

a. Write a formula for this relationship.

b. What is the equivalent resistance of two resistors connected in parallel if their resistances are 3 ohms and 5 ohms, respectively? (Express your answer as a decimal.)

• Check your answers on page A-18.

MINDStretchers

Mathematical Reasoning

1. Assume that a, b, and c are integers and $a \neq 0$. Prove that the solution to the linear equation $ax + b = c$ is a rational number.

Groupwork

2. Working with a partner, give an example of a situation that the following formula might describe:

$$d = bc + a$$

Explain what each variable represents in your example.

Patterns

3. Identify a pattern in the table and use it to write an equation that relates d and t.

d	1	3	5
t	5	9	13

2.3   Solving Linear Inequalities

OBJECTIVES

Ⓐ To determine whether a given number is a solution of a given inequality

Ⓑ To graph the solutions of linear inequalities on a number line

Ⓒ To solve a linear inequality using the addition or multiplication properties of inequalities

Ⓓ To solve applied problems involving linear inequalities

In Section 2.1, we discussed situations that could be described and solved by a linear equation. In this section, we consider situations that can be described and solved by a linear *inequality*.

For instance, suppose you want to spend no more than $50.00 per month on your smartphone bill for voice calls. How many minutes of calling time per month will you get from a smartphone plan that costs $39.99 per month plus $0.45 per minute for calls after the first 500 free minutes? To determine the number of minutes, you can solve the following inequality:

$$39.99 + 0.45(x - 500) \le 50$$

In this section, we will discuss how to solve such inequalities.

Inequalities and Solutions

Recall from Section 1.1 that we used inequality symbols to compare real numbers. Now, we consider inequalities that compare expressions involving variables, for example,

$$x < 1, \qquad x - 5 \ge 4, \qquad \text{and} \qquad 6y \ne 18.$$

DEFINITION

An **inequality** is any mathematical statement containing $<, \le, >, \ge,$ or $\ne$.

Just as in the case of equations, inequalities may be either true or false. For instance, the inequality $x < 1$ is true if 0 is substituted for the variable x.

$$x < 1$$
$$0 < 1 \quad \text{True}$$

However, this inequality is false if 5 is substituted for the variable x.

$$x < 1$$
$$5 < 1 \quad \text{False}$$

The number 0 is called a *solution* of the inequality $x < 1$ because when substituting 0 for x, we get a true statement. Note that there are many values for x that make $x < 1$ true. Can you name them all? Explain.

DEFINITION

A **solution of an inequality** is any value of the variable that makes the inequality true. To **solve an inequality** is to find all of its solutions.

EXAMPLE 1

Determine whether 2 is a solution of $2x - 11 > -7$.

Solution   We substitute 2 for x in the inequality $2x - 11 > -7$ to determine whether it is a solution.

$$2x - 11 > -7$$
$$2(2) - 11 \overset{?}{>} -7$$
$$4 - 11 \overset{?}{>} -7$$
$$-7 > -7 \quad \text{False}$$

Because $-7 > -7$ is a false statement, 2 *is not* a solution of $2x - 11 > -7$.

PRACTICE 1

Determine whether -4 is a solution of $3y + 8 < -1$.

EXAMPLE 2

Determine whether 7 is a solution of $7y + 4 \leq 39 + 2y$.

Solution   We substitute 7 for y in the inequality $7y + 4 \leq 39 + 2y$.

$$7y + 4 \leq 39 + 2y$$

$$7(7) + 4 \overset{?}{\leq} 39 + 2(7)$$

$$49 + 4 \overset{?}{\leq} 39 + 14$$

$$53 \leq 53 \qquad \text{· True}$$

Because $53 \leq 53$ is a true statement, 7 is a solution of
$7y + 4 \leq 39 + 2y$.

PRACTICE 2

Determine whether 10 is a solution
of $5n - 11 \geq 2n + 28$.

For any inequality, we can draw on a number line a picture of its solutions—the *graph*
of the inequality.

EXAMPLE 3

Graph the following on a number line.

a. $x < 1$ **b.** $x \geq -3$

Solution

a. The graph of $x < 1$ includes all points on the number line to the left of 1.
The right parenthesis on the graph shows that 1 *is not* a solution.

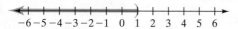

This graph shows that the solutions of $x < 1$ are all real numbers
less than 1, where the endpoint 1 is not included.

Note that for the inequality $x < 1$, its graph extends without
end in the negative direction on the number line. We can also represent the solution of $x < 1$ using interval notation. So for $x < 1$, we
can write $(-\infty, 1)$. Similarly for $x < a$, we write $(-\infty, a)$, where a
is the endpoint.

b. The graph of $x \geq -3$ includes -3 and all points on the number line
to the right of -3. The left bracket on the graph indicates that -3 is
a solution.

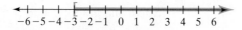

This graph shows that the solutions of $x \geq -3$ are all real numbers
greater than or equal to -3, where the endpoint -3 is included.

For the inequality $x \geq -3$, its graph extends without end in
the positive direction on the number line. We can also represent
the solution of this inequality using interval notation: $[-3, \infty)$.
Similarly for $x \geq a$, we write $[a, \infty)$, where a is the endpoint.

PRACTICE 3 ⊙

Graph the following on a number line.

a. $x > 0$

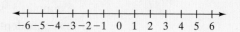

b. $x \leq \dfrac{1}{2}$

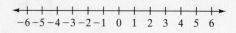

So we can use interval notation, inequality notation, and graphs to represent the solutions of an inequality in one variable. Consider the following table.

INTERVALS, INEQUALITIES, AND GRAPHS

Interval Notation	Inequality Notation	Graph*	Meaning
(a, ∞)	$x > a$		All real numbers greater than a, excluding a
$[a, \infty)$	$x \geq a$		All real numbers greater than or equal to a
$(-\infty, b)$	$x < b$		All real numbers less than b, excluding b
$(-\infty, b]$	$x \leq b$		All real numbers less than or equal to b
$(-\infty, \infty)$			All real numbers

*Some textbooks use the graphical representations ○——→ and ●——→ instead of ←——→ and ⊢——→, respectively.

Solving Inequalities Using the Addition Property

Just as with equations, two inequalities are *equivalent* if they have the same solutions. For example, $x < 3$ is equivalent to $3 > x$. Applying the addition property of inequalities gives us equivalent inequalities.

Addition Property of Inequalities

For any real numbers a, b, and c,

• if $a < b$, then $a + c < b + c$.

• if $a > b$, then $a + c > b + c$.

Similar statements hold for $\leq$ and $\geq$.

Since subtracting c is the same as adding its opposite, $-c$, the addition property of inequalities also allows us to subtract the same number from each side of an inequality.

As with equations, we solve an inequality by expressing it as an equivalent inequality in which the variable term is isolated on one side.

EXAMPLE 4

Solve and graph: $y + 6 > 4$

Solution

$$y + 6 > 4$$
$$y + 6 - 6 > 4 - 6 \quad \text{Subtract 6 from each side of the inequality.}$$
$$y > -2$$

Note that $y + 6 > 4$ and $y > -2$ are equivalent inequalities.
The graph of $y > -2$ is:

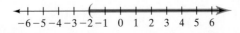

PRACTICE 4

Solve and graph: $x + 3 \leq 1\frac{1}{2}$

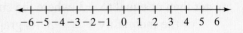

Check Because an inequality has infinitely many solutions, we cannot check them all as we did with an equation. However, we can do a partial check by substituting in the original inequality a couple of points on the graph and a couple not on the graph. So to confirm that all real numbers greater than -2 are the solutions to $y + 6 > 4$, we choose two points on the graph, say, 0 and 3, and two points not on the graph, say, -3 and -5, and substitute them into the original inequality.

Values for y	$y + 6 > 4$	True or False?
0	$0 + 6 \overset{?}{>} 4$	True, since $6 > 4$
3	$3 + 6 \overset{?}{>} 4$	True, since $9 > 4$
-3	$-3 + 6 \overset{?}{>} 4$	False, since $3 < 4$
-5	$-5 + 6 \overset{?}{>} 4$	False, since $1 < 4$

Since the check holds, the solutions are all real numbers greater than -2, but not including -2, or in interval notation, $(-2, \infty)$.

EXAMPLE 5

Solve and graph: $4x + 5 \leq 3x + 8$

Solution

$$4x + 5 \leq 3x + 8$$
$$4x - 3x + 5 \leq 3x - 3x + 8 \qquad \text{Subtract } 3x \text{ from each side of the inequality.}$$
$$x + 5 \leq 8$$
$$x \leq 3 \qquad \text{Subtract 5 from each side of the inequality.}$$

The graph of $x \leq 3$ is:

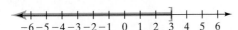

The solutions of $4x + 5 \leq 3x + 8$ are all real numbers that are less than or equal to 3, written in interval notation as $(-\infty, 3]$.

PRACTICE 5

Solve and graph: $5y + 7 \geq 4y + 3$

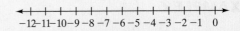

How would you check the solutions of the inequality $4x + 5 \leq 3x + 8$ in Example 5? Explain.

Solving Inequalities Using the Multiplication Property

Consider the true inequality $-2 < 5$. Let's look at what happens when we multiply this inequality by a *positive* number:

$$-2 < 5$$
$$-2 \cdot 3 \overset{?}{<} 5 \cdot 3 \qquad \text{Multiply each side of the inequality by 3.}$$
$$-6 < 15 \qquad \text{True}$$

Now, let's multiply each side of the original inequality by a *negative* number:

$$-2 < 5$$

$$-2(-3) \overset{?}{<} 5(-3) \qquad \text{Multiply each side of the inequality by } -3.$$

$$6 \overset{?}{<} -15 \qquad \text{False, unless the direction of the inequality is reversed}$$

$$6 > -15 \qquad \text{True}$$

Note that multiplying each side of an inequality by a positive number preserves the direction of the inequality, whereas multiplying each side of an inequality by a negative number reverses the direction of the inequality. These examples suggest the *multiplication property of inequalities.*

Multiplication Property of Inequalities

For any real numbers a, b, and c,

- if $a < b$ and c is positive, then $ac < bc$.
- if $a < b$ and c is negative, then $ac > bc$.

Similar statements hold for $>$, $\leq$, and $\geq$.

Does a similar property hold for division? Explain.

EXAMPLE 6

Solve and graph: $\dfrac{2x}{3} < 4$

Solution

$$\frac{2x}{3} < 4$$

$$3 \cdot \frac{2x}{3} < 3 \cdot 4 \qquad \text{Multiply each side of the inequality by 3.}$$

$$2x < 12$$

$$\frac{2x}{2} < \frac{12}{2} \qquad \text{Divide each side of the inequality by 2.}$$

$$x < 6$$

The graph of $x < 6$ is:

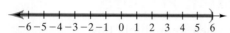

The solutions of $\dfrac{2x}{3} < 4$ are all real numbers less than 6, or, in interval notation, $(-\infty, 6)$.

PRACTICE 6

Solve and graph: $\dfrac{2n}{5} > 1$

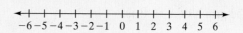

EXAMPLE 7

Solve and graph: $-8z \leq 56$

Solution

$$-8z \leq 56$$

$$\frac{-8z}{-8} \geq \frac{56}{-8} \qquad \text{Divide each side of the inequality by } -8 \text{ and reverse the direction of the inequality.}$$

$$z \geq -7$$

The graph of $z \geq -7$ is:

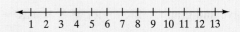

The solutions of $-8z \leq 56$ are all real numbers greater than or equal to -7. We can also write the solutions in interval notation as $[-7, \infty)$.

PRACTICE 7

Solve and graph: $-12x < -108$

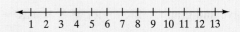

In solving inequalities as well as equations, we sometimes need to use more than one property.

EXAMPLE 8

Solve and graph: $5y - 7 + 2y > 8y - 3$

Solution

$$5y - 7 + 2y > 8y - 3$$

$$7y - 7 > 8y - 3 \qquad \text{Combine like terms.}$$

$$7y - 8y - 7 > 8y - 8y - 3 \qquad \text{Subtract } 8y \text{ from each side of the inequality.}$$

$$-y - 7 > -3 \qquad \text{Combine like terms.}$$

$$-y - 7 + 7 > -3 + 7 \qquad \text{Add 7 to each side of the inequality.}$$

$$-y > 4 \qquad \text{Combine like terms.}$$

$$\frac{-y}{-1} < \frac{4}{-1} \qquad \text{Divide each side of the inequality by } -1 \text{ and reverse the direction of the inequality.}$$

$$y < -4$$

The graph of $y < -4$ is:

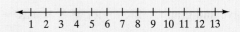

The solutions of $5y - 7 + 2y > 8y - 3$ are all real numbers less than -4, or, in interval notation, $(-\infty, -4)$.

PRACTICE 8

Solve and graph:
$4x - 5x - 10 \geq 10 + x$

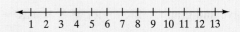

Another way to solve the inequality in Example 8 is as follows:

$$5y - 7 + 2y > 8y - 3$$

$$7y - 7 > 8y - 3 \qquad \text{Combine like terms.}$$

$$7y - 7 + 3 > 8y - 3 + 3 \qquad \text{Add 3 to each side of the inequality.}$$

$$7y - 4 > 8y \qquad \text{Simplify.}$$

$$7y - 7y - 4 > 8y - 7y \qquad \text{Subtract } 7y \text{ from each side of the inequality.}$$

$$-4 > y$$

The solutions are all real numbers less than -4. Recall that $-4 > y$ is equivalent to $y < -4$, which is the inequality we found in Example 8.

EXAMPLE 9

Solve and graph: $2(3 - x) - 8 \geq 12x + 5$

Solution

$2(3 - x) - 8 \geq 12x + 5$

$6 - 2x - 8 \geq 12x + 5$ Use the distributive property.

$-2x - 2 \geq 12x + 5$ Combine like terms.

$-2x - 12x - 2 \geq 12x - 12x + 5$ Subtract $12x$ from each side of the inequality.

$-14x - 2 \geq 5$

$-14x - 2 + 2 \geq 5 + 2$ Add 2 to each side of the inequality.

$-14x \geq 7$ Combine like terms.

$\dfrac{-14x}{-14} \leq \dfrac{7}{-14}$ Divide each side by -14 and reverse the direction of the inequality.

$x \leq -\dfrac{1}{2}$

The graph of $x \leq -\dfrac{1}{2}$ is

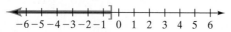

The solutions of $2(3 - x) - 8 \geq 12x + 5$ are all real numbers less than or equal to $-\dfrac{1}{2}$ or, in interval notation, $\left(-\infty, -\dfrac{1}{2}\right]$.

PRACTICE 9

Solve and graph:
$2 - (4y + 5) \leq 2y + 1$

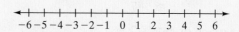

To Solve a Linear Inequality

- Use the distributive property to clear the inequalities of parentheses, if necessary.

- Combine like terms.

- Use the addition property to isolate the variable.

- Use the multiplication property to isolate the variable, being careful to reverse the direction of the inequality symbol when multiplying or dividing by a negative number.

Linear inequalities are used to solve problems in areas such as business, construction and design, science, and linear programming. In applied problems, some of the common word phrases used in translating inequalities are *is less than*, *is less than or equal to*, *is greater than*, and *is greater than or equal to*. The table shows some other word phrases and their translations.

Word Phrase	Translation
x is at most a.	$x \leq a$
x is no more than a.	$x \leq a$
x is at least a.	$x \geq a$
x is no less than a.	$x \geq a$

EXAMPLE 10

The warehouse freight elevator can safely carry a load of at most 4000 lb. A worker needs to move 60-lb crates from the loading dock to the third floor of the building. The worker weighs 180 lb and the cart that he uses weighs 90 lb. What is the greatest number of crates he can move in one trip?

Solution   Let x represent the number of crates that can be moved in one trip. Then, $60x$ represents the total weight of the crates that can be moved in one trip. Since the elevator can safely carry a load of at most 4000 lb, we can write the following inequality:

Weight of Crates		Weight of Worker		Weight of Cart	At Most	Maximum Load Elevator Holds
↓		↓		↓	↓	↓
$60x$	$+$	180	$+$	90	$\leq$	4000

Now, let's solve for x.

$$60x + 180 + 90 \leq 4000$$
$$60x + 270 \leq 4000$$
$$60x + 270 - 270 \leq 4000 - 270$$
$$60x \leq 3730$$
$$\frac{60x}{60} \leq \frac{3730}{60}$$
$$x \leq 62\frac{1}{6}$$

So the greatest number of crates he can move in one trip is 62.

EXAMPLE 11

An account executive invested $10,000 in stocks and bonds for her client. She anticipates that the client will receive a $4\frac{1}{2}\%$ annual return from high-risk stocks and an annual return of $2\frac{1}{2}\%$ from the bonds. The client wants his annual income from stocks and bonds to be at least $300. How much money should the account executive invest in each?

Solution   Let x represent the amount of money invested in stocks at $4\frac{1}{2}\%$. Then, $10,000 - x$ represents the amount of money invested in bonds at $2\frac{1}{2}\%$. To compute the return, we can use the formula $I = prt$. Since $t = 1$, the formula simplifies to $I = pr$. The total amount of the return is the sum of the return from each investment. Since the annual income from both investments is at least $300, we write:

$$0.045x + 0.025(10,000 - x) \geq 300$$
$$0.045x + 250 - 0.025x \geq 300$$
$$0.02x + 250 \geq 300$$
$$0.02x + 250 - 250 \geq 300 - 250$$
$$0.02x \geq 50$$
$$x \geq 2500$$

Then, $10,000 - x = 10,000 - 2500 = 7500$.
So at least $2500 should be invested in stocks and at most $7500 should be invested in bonds.

PRACTICE 10

A student is enrolled in a psychology course. There are four tests in the first marking period of the course. The student's scores on three tests are 85, 91, and 83. What is the lowest score she can get on the fourth test to have an average greater than 89 for the marking period?

PRACTICE 11

An entrepreneur had $54,000 to invest. She invested part of this money in noninsured bonds paying 6% interest annually. She invested the rest in a government-insured certificate of deposit paying only 5% annual interest. If she needs at least $3000 in extra income from these investments, how much money should she place in the bonds?

In Example 11, explain why *at most* $7500 should be invested in bonds.

Mathematically Speaking

Fill in each blank with the most appropriate term or phrase from the given list.

addition property	negative	equation
rational	evaluate	graph
less than	distributive property	less than or equal to
solution	inequality	solve
excludes	includes	positive

1. A(n) _____ is any mathematical statement containing $<, \leq, >, \geq,$ or $\neq$.

2. A(n) _____ of an inequality is any value of the variable that makes the inequality true.

3. To _____ an inequality is to find all of its solutions.

4. On a number line, a picture of the solutions of an inequality is said to be its _____.

5. In interval notation, the left bracket [indicates that the interval _____ the left endpoint.

6. The interval notation $(-\infty, 3]$ means all real numbers _____ 3.

7. The _____ of inequalities states that for any real numbers a, b, and c, if $a \leq b$, then $a + c \leq b + c$.

8. The multiplication property of inequalities states that for any real numbers a, b, and c, if $a \leq b$ and c is _____, then $ac \geq bc$.

A *Determine whether the given value is a solution of the inequality.*

9. $3x - 7 > -2; 2$

10. $5x + 1 < -4; -1$

11. $\dfrac{n}{4} + 10 \leq 8; -8$

12. $\dfrac{y}{2} - 6 \geq -3; 6$

13. $23 - 5x \geq 2x + 9; 10$

14. $4n + 15 > 6n + 20; -7$

15. $\dfrac{2}{3}y + 8 < \dfrac{3}{4}y + 6; -12$

16. $\dfrac{2}{9}x - 11 \leq \dfrac{1}{6}x - 12; -18$

⊙ 17. $4(24n + 1) + 16 > 40n - 2; -\dfrac{3}{8}$

18. $18 - 20y \leq 4y - (28y - 11); \dfrac{3}{4}$

B *Graph the inequality on a number line. Then, write the solution using interval notation.*

19. $x < -1$

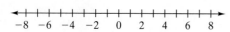

20. $x \leq -4$

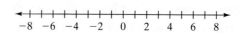

21. $x \geq 2$

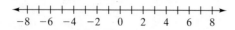

22. $x > 5$

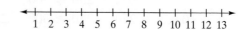

23. $x > -3\frac{1}{2}$

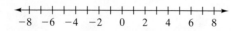

24. $x < 6\frac{1}{2}$

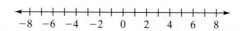

25. $7.5 \geq x$

26. $0.5 \leq x$

C *Solve and graph.*

27. $n + 5 \geq 3$

$$\xleftarrow{\hspace{0.3em}\rule{0pt}{0.6em}\hspace{0.3em}}_{-6\ -5\ -4\ -3\ -2\ -1\ \ 0\ \ 1\ \ 2\ \ 3\ \ 4\ \ 5\ \ 6}\xrightarrow{}$$

28. $x + 6 \leq 10$

$$\xleftarrow{\hspace{0.3em}}_{-6\ -5\ -4\ -3\ -2\ -1\ \ 0\ \ 1\ \ 2\ \ 3\ \ 4\ \ 5\ \ 6}\xrightarrow{}$$

29. $y - 3 < -7$

$$\xleftarrow{\hspace{0.3em}}_{-10\ -9\ -8\ -7\ -6\ -5\ -4\ -3\ -2\ -1\ \ 0\ \ 1\ \ 2}\xrightarrow{}$$

30. $n - 2 > 4$

$$\xleftarrow{\hspace{0.3em}}_{0\ \ 1\ \ 2\ \ 3\ \ 4\ \ 5\ \ 6\ \ 7\ \ 8\ \ 9\ \ 10\ \ 11\ \ 12}\xrightarrow{}$$

31. $x + \dfrac{1}{2} > -1\dfrac{1}{2}$

$$\xleftarrow{\hspace{0.3em}}_{-6\ -5\ -4\ -3\ -2\ -1\ \ 0\ \ 1\ \ 2\ \ 3\ \ 4\ \ 5\ \ 6}\xrightarrow{}$$

32. $n - 2 \geq -3\dfrac{1}{2}$

$$\xleftarrow{\hspace{0.3em}}_{-6\ -5\ -4\ -3\ -2\ -1\ \ 0\ \ 1\ \ 2\ \ 3\ \ 4\ \ 5\ \ 6}\xrightarrow{}$$

33. $-6.5 \geq y - 4.5$

$$\xleftarrow{\hspace{0.3em}}_{-8\ -7\ -6\ -5\ -4\ -3\ -2\ -1\ \ 0\ \ 1\ \ 2\ \ 3\ \ 4}\xrightarrow{}$$

34. $2.5 > n + 5.5$

$$\xleftarrow{\hspace{0.3em}}_{-10\ -9\ -8\ -7\ -6\ -5\ -4\ -3\ -2\ -1\ \ 0\ \ 1\ \ 2}\xrightarrow{}$$

35. $2x - 9 < x - 6$

$$\xleftarrow{\hspace{0.3em}}_{-3\ -2\ -1\ \ 0\ \ 1\ \ 2\ \ 3\ \ 4\ \ 5\ \ 6\ \ 7\ \ 8\ \ 9}\xrightarrow{}$$

36. $7n + 3 > 6n - 1$

$$\xleftarrow{\hspace{0.3em}}_{-10\ -9\ -8\ -7\ -6\ -5\ -4\ -3\ -2\ -1\ \ 0\ \ 1\ \ 2}\xrightarrow{}$$

37. $9x + 5 \geq 8x + 5$

$$\xleftarrow{\hspace{0.3em}}_{-6\ -5\ -4\ -3\ -2\ -1\ \ 0\ \ 1\ \ 2\ \ 3\ \ 4\ \ 5\ \ 6}\xrightarrow{}$$

38. $6y - 4 < 5y - 4$

$$\xleftarrow{\hspace{0.3em}}_{-6\ -5\ -4\ -3\ -2\ -1\ \ 0\ \ 1\ \ 2\ \ 3\ \ 4\ \ 5\ \ 6}\xrightarrow{}$$

39. $1.7 - 2.8x \leq 3.2 - 3.8x$

$$\xleftarrow{\hspace{0.3em}}_{-6\ -5\ -4\ -3\ -2\ -1\ \ 0\ \ 1\ \ 2\ \ 3\ \ 4\ \ 5\ \ 6}\xrightarrow{}$$

40. $4.6 - 6.1y > -7.1y - 0.4$

$$\xleftarrow{\hspace{0.3em}}_{-10\ -9\ -8\ -7\ -6\ -5\ -4\ -3\ -2\ -1\ \ 0\ \ 1\ \ 2}\xrightarrow{}$$

41. $\dfrac{4}{5}x \geq -8$

$$\xleftarrow{\hspace{0.3em}}_{-15\ -14\ -13\ -12\ -11\ -10\ -9\ -8\ -7\ -6\ -5\ -4\ -3}\xrightarrow{}$$

42. $\dfrac{5}{3}x > 10$

$$\xleftarrow{\hspace{0.3em}}_{0\ \ 1\ \ 2\ \ 3\ \ 4\ \ 5\ \ 6\ \ 7\ \ 8\ \ 9\ \ 10\ \ 11\ \ 12}\xrightarrow{}$$

43. $-3n < 15$

$$\xleftarrow{\hspace{0.3em}}_{-10\ -9\ -8\ -7\ -6\ -5\ -4\ -3\ -2\ -1\ \ 0\ \ 1\ \ 2}\xrightarrow{}$$

44. $-6x \geq -24$

$$\xleftarrow{\hspace{0.3em}}_{-2\ -1\ \ 0\ \ 1\ \ 2\ \ 3\ \ 4\ \ 5\ \ 6\ \ 7\ \ 8\ \ 9\ \ 10}\xrightarrow{}$$

45. $-90 \leq -15y$

$$\xleftarrow{\hspace{0.3em}}_{0\ \ 1\ \ 2\ \ 3\ \ 4\ \ 5\ \ 6\ \ 7\ \ 8\ \ 9\ \ 10\ \ 11\ \ 12}\xrightarrow{}$$

46. $77 > -11x$

$$\xleftarrow{\hspace{0.3em}}_{-12\ -11\ -10\ -9\ -8\ -7\ -6\ -5\ -4\ -3\ -2\ -1\ \ 0}\xrightarrow{}$$

47. $-\dfrac{3}{8}x > \dfrac{9}{16}$

$$\xleftarrow{\hspace{0.3em}}_{-6\ -5\ -4\ -3\ -2\ -1\ \ 0\ \ 1\ \ 2\ \ 3\ \ 4\ \ 5\ \ 6}\xrightarrow{}$$

48. $-\dfrac{5}{6}n \leq -\dfrac{5}{12}$

$$\xleftarrow{\hspace{0.3em}}_{-6\ -5\ -4\ -3\ -2\ -1\ \ 0\ \ 1\ \ 2\ \ 3\ \ 4\ \ 5\ \ 6}\xrightarrow{}$$

Match the inequality to its graph.

49. $2 - x \leq 1$　　　　**50.** $7 - 4x > 11$　　　　**51.** $-11 < 2x - 9$　　　　**52.** $-2 \geq 3x - 5$

a.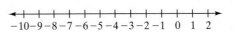
$$\xleftarrow{\hspace{0.3em}}_{-8\ \ -6\ \ -4\ \ -2\ \ \ 0\ \ \ 2\ \ \ 4\ \ \ 6\ \ \ 8}\xrightarrow{}$$

b.
$$\xleftarrow{\hspace{0.3em}}_{-8\ \ -6\ \ -4\ \ -2\ \ \ 0\ \ \ 2\ \ \ 4\ \ \ 6\ \ \ 8}\xrightarrow{}$$

c.
$$\xleftarrow{\hspace{0.3em}}_{-8\ \ -6\ \ -4\ \ -2\ \ \ 0\ \ \ 2\ \ \ 4\ \ \ 6\ \ \ 8}\xrightarrow{}$$

d.
$$\xleftarrow{\hspace{0.3em}}_{-8\ \ -6\ \ -4\ \ -2\ \ \ 0\ \ \ 2\ \ \ 4\ \ \ 6\ \ \ 8}\xrightarrow{}$$

Solve.

53. $2x + 5 \le 13$

54. $4x - 7 > -3$

55. $9 - \dfrac{x}{3} < -5$

56. $12 - \dfrac{2n}{7} \ge 14$

57. $11y + 9 \ge 8y$

58. $-y < -6y + 25$

59. $6y < -18y$

60. $10n > 5n$

61. $3y - 13 > 15 - y$

62. $6n + 10 \le 4n - 10$

63. $\dfrac{2}{3}x + 6 \ge \dfrac{4}{5}x + 9$

64. $\dfrac{x}{4} - 12 \le \dfrac{7}{10}x - 18$

65. $2.8z - 1.3 > -1.6z - 0.2$

66. $-1.5n + 0.7 < 4.9 - 0.8n$

67. $3n - 2 - n > 5n + 19$

68. $23 - 7y > 4y + 3y - 5$

69. $17n - 8n + 14 \le 13n - 10 + 6$

70. $11z - 16 + 5z < 14 - 8z - 21$

71. $3(2x - 1) - 4 \ge 7x - 12$

72. $5(6 - x) + 11 \le 22 - 9x$

73. $15n - 4(5 - 6n) < 32 - (17 - 18n)$

74. $6(7 - 3y) - 12 \ge -2(12 - 8y) - 10y$

75. $\dfrac{3}{4}(12x - 16) + 20 > 25\left(\dfrac{2}{5}x - 1\right) + 9x$

76. $\dfrac{2}{3}(27n + 21) - 8n > 18 - 36\left(\dfrac{n}{6} - 2\right)$

77. $\dfrac{1}{4}\left[2 - 9\left(6 + \dfrac{2}{3}x\right)\right] + 19 \le 17 - \dfrac{1}{2}(8 + 7x)$

78. $37 - \dfrac{3}{5}\left(\dfrac{5}{9}y + 20\right) \le \dfrac{3}{4}[4y - 5(2y - 8)]$

Mixed Practice

79. Determine whether 3 is a solution for the inequality $2x + 7 > 4x - 3$.

80. Solve and graph: $2y - 3 > y + 1$

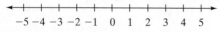

81. Graph the inequality $x \le -1$ on a number line. Then, write the solution using interval notation.

Solve.

82. $5(3y - 1) - 8 > y - 4(2y + 1)$

83. $8n < 10n$

84. $5n - 1 + n \ge 2n + 6$

Applications

D *Solve.*

85. It costs $5 per day for parking at a local parking garage. The owner of the garage offers a monthly parking pass that costs $60. For what number of days is the daily parking rate a better deal?

86. A fare on the Southeastern Pennsylvania Transportation Authority's (SEPTA) transit system is $1.55 per ride if purchased separately. A monthly transit pass costs $83. For what number of rides will the monthly pass be a better deal? (*Source:* septa.org)

87. A cell phone calling plan costs $29.99 per month for unlimited calling within the plan's network and $0.79 per minute (or part thereof) for calls outside the network. How many minutes of calls outside the network can you make if you want your cell phone bill to be less than $50 per month?

88. A film editor has many large audio and video files to transfer from one computer to another. Her USB drive, which holds 7.8 gigabytes (GB), already contains 1.2 GB of data. Each file is 0.35 GB in size. Up to how many files can she move in the first transfer?

89. A nurse planning a seven-day vacation estimates that it will cost $486 for the airfare, $170 for the rental car, and $654 for the hotel. If she wants to spend at most $2500 for the entire vacation, how much can she spend per day on remaining expenses?

90. A hand pallet truck can carry a maximum load of 2200 lb. A worker needs to move cases of paper from the warehouse to a loading dock. He has already stacked 24 cases on a pallet. If the pallet weighs 40 lb and each case of paper weighs 60 lb, how many additional cases could he stack on the pallet before moving it to the loading dock?

91. The profits for a small business in the first three quarters of the year were $120,356, $96,147, and $85,502. What must the company's profit be for the fourth quarter in order to show an average quarterly profit of at least $100,000?

92. A student in an intermediate algebra course must take five exams. His scores on the first four exams this semester were 93, 95, 84, and 80. What is the minimum he must score on the last exam to have an average score of at least 87?

93. Fifteen minutes after a man leaves for his morning run, his wife leaves for a bike ride along the same route. The husband runs at an average rate of 6 mph. At what minimum rate should the wife bike if she wants to catch up with her husband within 20 min?

94. A business executive wants to invest twice the amount of money in a high-risk fund that has a likely return of $5\frac{1}{4}\%$ as in a fund that has a guaranteed return of $4\frac{1}{2}\%$. What is the minimum amount that should be invested in each fund if she hopes to have a return of at least $1200?

95. A retiree invested $30,000 by putting part of it in insured certificates of deposit that paid 6% annually and the rest in a risky stock fund that paid 12% for the year. How much money did the retiree invest in CDs if the total investment return was more than $2460?

96. A young couple took out a mortgage to buy an apartment. Now, they want to sell this apartment for enough money to at least pay off the mortgage, on which they currently owe $18,500. The real estate agent gets 5% of the selling price, and the city has a $500 real estate transfer tax paid by the seller. For how much money should they sell the apartment?

97. A local courier service estimates that its monthly operating cost is $1500 plus $0.85 per delivery. If the service generates revenue of $6 for each delivery, how many deliveries must be made per month in order for the monthly revenue to exceed the monthly cost?

98. During the summer, a college student sells scented glycerin soaps at a kiosk in a local mall. It costs her $2600 per month to rent the kiosk and approximately $1.10 to make each bar of soap. If she charges $4.50 per bar of soap, how many bars must she sell in a month in order to make a profit?

99. Two car dealerships offer lease options on comparable compact cars. One dealership requires $2500 upon signing the lease and $109 per month. The other dealership requires $2000 upon signing the lease and $129 per month. Under what circumstances would you choose the first lease option over the second lease option? Explain.

100. A local bank offers two types of checking accounts. The Basic checking account is $3.85 per month and includes a $0.20-per-check fee. The Basic Plus checking account is $4.95 per month and includes seven free checks and a $0.30-per-check fee for every check thereafter. Under what circumstances should a person select the second checking account option? Explain.

• Check your answers on page A-18.

MINDStretchers

Writing

1. Describe a situation modeled by an inequality whose solutions can be represented by the following graph.

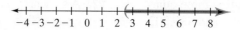

Critical Thinking

2. Suppose that a and b represent two real numbers, where $a > b$.
Consider the following argument: We know that $2 > 1$.

$$2 > 1$$
$$2(b - a) > 1(b - a)$$
$$2b - 2a > b - a$$
$$b - 2a > -a$$
$$b > a$$

So we conclude that b is greater than a. But this is impossible, since we were given that a is greater than b. Find the flaw in the argument.

Mathematical Reasoning

3. An investor has $10,000 to invest in two different funds. Fund A earns 7.5% simple interest, and Fund B earns 5% simple interest annually. He would like to earn at least $800 after the first year. Is this possible? Explain.

2.4 Solving Compound Inequalities

In this section, we look at joining inequalities. Consider this example.

White blood cells, also called leukocytes, help protect the body against disease and infection. A cubic millimeter of blood contains from 4000 to 10,000 leukocytes. The following table shows the condition of white blood cells depending on the number N of leukocytes in 1 mm^3 of blood.

Leukopenia (more at risk of catching an infection)	Normal Range	Leukocytosis (indication of a possible infection in the body)
$N \leq 4000$	$4000 < N < 10,000$	$N \geq 10,000$

(*Source:* Priscilla Lemone and Karen M. Burke, *Medical Surgical Nursing: Critical Thinking and Client Care*)

A normal range of white blood cells in 1 mm^3 of blood can be expressed by two inequalities,

$$4000 < N \quad \text{and} \quad N < 10,000,$$

or by joining these inequalities to get

$$4000 < N < 10,000.$$

DEFINITION

Two inequalities that are joined by the word *and* or the word *or* form a **compound inequality**.

Let's look at the graphs of $4000 < N$, $N < 10,000$, and $4000 < N < 10,000$. For the first graph, recall that $4000 < N$ can be written as $N > 4000$.

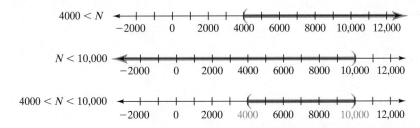

The last graph shows that a solution of $4000 < N < 10,000$ is the intersection of the solutions of the inequalities $4000 < N$ and $N < 10,000$.

We read $4000 < N < 10,000$ as "N is greater than 4000 and less than 10,000." Another way to read this inequality is "N is between 4000 and 10,000, excluding 4000 and 10,000."

We can use interval notation, inequality notation, and graphs to represent the solutions of a compound inequality in one variable, as shown in the table.

INTERVALS, INEQUALITIES, AND GRAPHS

Interval Notation	Inequality Notation	Graph*	Meaning
(a, b)	$a < x < b$		All real numbers between a and b, excluding a and b
$[a, b]$	$a \leq x \leq b$		All real numbers between a and b, including a and b
$[a, b)$	$a \leq x < b$		All real numbers between a and b, including a and excluding b
$(a, b]$	$a < x \leq b$		All real numbers between a and b, excluding a and including b

*Some textbooks use the graphical representations 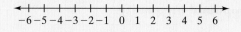 instead of, respectively, and .

Solving Inequalities with the Word *And*

A *solution of a compound inequality* joined by *and* is the intersection of the solutions of the individual inequalities. In other words, a solution is any value of the variable that makes *both* inequalities true.

EXAMPLE 1

Solve and graph: $3x < 12$ and $2x + 1 > -5$

Solution

$3x < 12$	and	$2x + 1 > -5$	
$\dfrac{3x}{3} < \dfrac{12}{3}$	Divide each side by 3.	$2x + 1 - 1 > -5 - 1$	Subtract 1 from each side.
$x < 4$	Simplify.	$2x > -6$	Simplify.
		$\dfrac{2x}{2} > \dfrac{-6}{2}$	Divide each side by 2.
		$x > -3$	Simplify.

$x < 4$ and $x > -3$

Since $x > -3$ is the same as $-3 < x$, we can write $x < 4$ and $x > -3$ as one inequality, getting

$$-3 < x < 4.$$

The solutions of this inequality are all values of x that satisfy both $x < 4$ and $x > -3$. Its graph is the intersection, or overlap, of the graphs of $x < 4$ and $x > -3$, as shown on the following number lines:

Note that the graph includes all points *between* -3 and 4, excluding -3 and 4.

The solutions of $3x < 12$ and $2x + 1 > -5$ are all real numbers between -3 and 4, or, in interval notation, $(-3, 4)$.

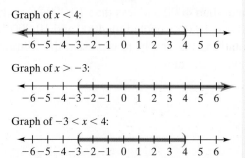

Graph of $x < 4$:

Graph of $x > -3$:

Graph of $-3 < x < 4$:

PRACTICE 1

Solve and graph: $8x \leq 32$ and $5x + 3 \geq 3$

EXAMPLE 2

Solve and graph: $-3 \le 8x + 5 \le 21$

Solution   Write the compound inequality as two inequalities joined by *and*. Then, solve each inequality.

$$-3 \le 8x + 5 \le 21$$

$-3 \le 8x + 5$	and	$8x + 5 \le 21$	
$-3 - 5 \le 8x + 5 - 5$		$8x + 5 - 5 \le 21 - 5$	Subtract 5 from each side.
$-8 \le 8x$		$8x \le 16$	Simplify.
$\dfrac{-8}{8} \le \dfrac{8x}{8}$		$\dfrac{8x}{8} \le \dfrac{16}{8}$	Divide each side by 8.
$-1 \le x$	and	$x \le 2$	

We can combine both inequalities to get:

$$-1 \le x \le 2.$$

The graph of $-1 \le x \le 2$ is

The solutions of $-3 \le 8x + 5 \le 21$ are all real numbers between -1 and 2, including -1 and 2, or, in interval notation, $[-1, 2]$.

Let's consider an alternative way of solving the inequality in Example 2.

$$-3 \le 8x + 5 \le 21$$

$$-3 - 5 \le 8x + 5 - 5 \le 21 - 5 \qquad \text{Subtract 5 from each part of the compound inequality.}$$

$$-8 \le 8x \le 16$$

$$\frac{-8}{8} \le \frac{8x}{8} \le \frac{16}{8} \qquad \text{Divide each part of the compound inequality by 8.}$$

$$-1 \le x \le 2$$

Note that this method has fewer steps and will be useful in the next section.

EXAMPLE 3

Solve and graph: $-2 < -3x + 7 \le 10$

Solution

$$-2 < -3x + 7 \le 10$$

$-2 - 7 < -3x + 7 - 7 \le 10 - 7$	Subtract 7 from each part of the compound inequality.
$-9 < -3x \le 3$	Simplify.
$\dfrac{-9}{-3} > \dfrac{-3x}{-3} \ge \dfrac{3}{-3}$	Divide each part of the compound inequality by -3 and reverse the direction of each inequality.
$3 > x \ge -1$	Simplify.

or alternatively $-1 \le x < 3$.

PRACTICE 2

Solve and graph: $-5 < 6x + 1 < 7$

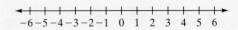

PRACTICE 3

Solve and graph: $-3 < 2 - t < 5$

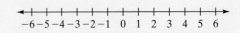

EXAMPLE 3 (continued)

The graph of $-1 \leq x < 3$ is

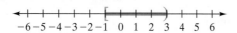

The solutions of $-2 < -3x + 7 \leq 10$ are all the real numbers between -1 and 3, including -1 and excluding 3. The solution written in interval notation is $[-1, 3)$.

EXAMPLE 4

Solve: $1 - 2x \geq 0$ and $2x + 7 \geq 11$

Solution

$$1 - 2x \geq 0 \qquad \text{and} \quad 2x + 7 \geq 11 \qquad \text{Solve each inequality.}$$

$$\begin{array}{ll} -2x \geq -1 & 2x \geq 4 \\ \dfrac{-2x}{-2} \leq \dfrac{-1}{-2} & \dfrac{2x}{2} \geq \dfrac{4}{2} \\ x \leq \dfrac{1}{2} \quad \text{and} & x \geq 2 \end{array}$$

Since there is no real number that is both less than or equal to $\dfrac{1}{2}$ and greater than or equal to 2, there is no solution. By examining the graphs of $x \leq \dfrac{1}{2}$ and $x \geq 2$, we can also see that there are no intersecting points.

PRACTICE 4

Solve: $3x - 2 < 1$ and $4x - 1 > 19$

Solving Inequalities with the Word *Or*

Now, let's focus on compound inequalities that are joined by *or*. A *solution of a compound inequality* joined by *or* is the union of the solutions of the individual inequalities. That is, a solution is any value of the variable that makes *either* inequality true.

EXAMPLE 5

Solve and graph: $7 - 3x \geq 1$ or $5x + 2 \geq 22$

Solution

$$\begin{array}{ll} 7 - 3x \geq 1 & \text{or} \quad 5x + 2 \geq 22 \\ -3x \geq -6 & 5x \geq 20 \\ \dfrac{-3x}{-3} \leq \dfrac{-6}{-3} \quad \begin{array}{l} \text{\footnotesize Divide each side by } -3, \\ \text{\footnotesize and reverse the direc-} \\ \text{\footnotesize tion of the inequality.} \end{array} & \dfrac{5x}{5} \geq \dfrac{20}{5} \\ x \leq 2 & \text{or} \quad x \geq 4 \end{array}$$

The solutions of this inequality are all values of x that satisfy either $x \leq 2$ or $x \geq 4$. Its graph is the union of the graphs of $x \leq 2$ and $x \geq 4$.

PRACTICE 5

Solve and graph: $4x - 9 > 3$ or $8 - x > 10$

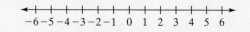

Graph of $x \le 2$:

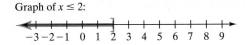

Graph of $x \ge 4$:

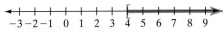

Graph of $x \le 2$ or $x \ge 4$:

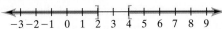

The solutions of $7 - 3x \ge 1$ or $5x + 2 \ge 22$ are all real numbers less than or equal to 2 *or* greater than or equal to 4, or in interval notation, the union of the intervals $(-\infty, 2]$ and $[4, \infty)$. We can also write this as $(-\infty, 2] \cup [4, \infty)$, where the symbol $\cup$ denotes the union of the solution intervals.

Are numbers in the interval $(2, 4)$ solutions of the compound inequality $7 - 3x \ge 1$ or $5x + 2 \ge 22$ in Example 5? Explain.

EXAMPLE 6

Solve and graph: $3x - 4 < 11$ or $-2x + 5 \le 3$

Solution

$$
\begin{array}{cc}
3x - 4 < 11 & \text{or} \quad -2x + 5 \le 3 \\
3x < 15 & -2x \le -2 \\
\dfrac{3x}{3} < \dfrac{15}{3} & \dfrac{-2x}{-2} \ge \dfrac{-2}{-2} \\
x < 5 \quad \text{or} & x \ge 1
\end{array}
$$

Divide each side by -2, and reverse the direction of the inequality.

The solutions of this inequality are all values of x that satisfy either $x < 5$ or $x \ge 1$. By looking at the graphs of $x < 5$ and $x \ge 1$, we see that the union of the graphs is the entire real-number line.

Graph of $x < 5$:

Graph of $x \ge 1$:

Graph of $x < 5$ or $x \ge 1$:

The solutions of $3x - 4 < 11$ or $-2x + 5 \le 3$ are all real numbers less than 5 *or* greater than or equal to 1. Another way to express the solution is the union of the intervals $(-\infty, 5)$ and $[1, \infty)$. Since every real number is either less than 5 or greater than or equal to 1, the solutions are all real numbers, or $(-\infty, \infty)$ in interval notation.

PRACTICE 6

Solve and graph: $-3x - 2 < 7$ or $5x + 1 \le 11$

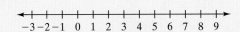

Compound inequalities are used in solving problems in such areas as science, medicine, and sports.

EXAMPLE 7

A runner trains for the New York City Marathon only on weekdays. Each week, she plans to run between 20 mi and 25 mi, inclusive. On Monday she ran 4 mi, and on Tuesday $4\frac{1}{2}$ mi. Find the least and the greatest numbers of miles she needs to run on each of the next three days to maintain her training schedule.

Solution   Let x represent the number of miles run each day. Then $3x$ represents the number of miles run for three days. Since each week the runner plans to run 20 mi to 25 mi, we can write the following compound inequality:

Least Number of Miles Run	Number of Miles Run for 3 Days	Miles Run on Monday	Miles Run on Tuesday	Greatest Number of Miles Run
↓	↓	↓	↓	↓
20 ≤	$3x$ +	4 +	$4\frac{1}{2}$ ≤	25

Now let's solve for x.

$$20 \le 3x + 4 + 4\tfrac{1}{2} \le 25$$
$$20 \le 3x + 8\tfrac{1}{2} \le 25$$

$$20 \le 3x + 8\tfrac{1}{2} \qquad \text{and} \qquad 3x + 8\tfrac{1}{2} \le 25$$
$$20 - 8\tfrac{1}{2} \le 3x + 8\tfrac{1}{2} - 8\tfrac{1}{2} \qquad 3x + 8\tfrac{1}{2} - 8\tfrac{1}{2} \le 25 - 8\tfrac{1}{2}$$
$$11\tfrac{1}{2} \le 3x \qquad\qquad\qquad 3x \le 16\tfrac{1}{2}$$
$$3\tfrac{5}{6} \le x \qquad \text{and} \qquad x \le 5\tfrac{1}{2}$$

Writing this as a combined inequality, we get $3\frac{5}{6} \le x \le 5\frac{1}{2}$. So the runner will need to run between $3\frac{5}{6}$ mi and $5\frac{1}{2}$ mi, inclusive, each day.

PRACTICE 7

The acidity of the water in a college's swimming pool is considered normal if the average of three pH readings is between 7.1 and 7.7, including both 7.1 and 7.7. The first two readings for the swimming pool are 7.3 and 7.8. What possible value for the third reading will make the average pH normal?

Mathematically Speaking

Fill in each blank with the most appropriate term or phrase from the given list.

intersection	union	expressions	inequalities
equations	and	or	

1. A compound inequality is formed by joining two _____ with the word *and* or the word *or*.

2. The graph of a compound inequality joined by the word _____ is the intersection, or overlap, of the graphs of each of the individual inequalities.

3. The graph of a compound inequality joined by the word _____ is the union of the graphs of each of the individual inequalities.

4. In interval notation, the symbol ∪ denotes a(n) _____.

Ⓐ *Rewrite the inequality using interval notation. Then, graph the inequality.*

5. $2 < x < 5$

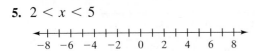

6. $1 \le x \le 6$

7. $-3 \le x < 0$

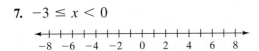

8. $-5 < x \le -0.5$

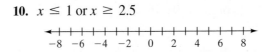

9. $x < -4$ or $x > 3$

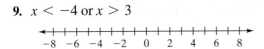

10. $x \le 1$ or $x \ge 2.5$

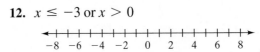

11. $x < -2$ or $x \ge 0.5$

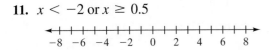

12. $x \le -3$ or $x > 0$

Solve and graph the inequality.

13. $2x > -10$ and $3x - 1 < 8$

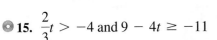

14. $5x - 7 > 8$ and $4x < 16$

Ⓞ 15. $\dfrac{2}{3}t > -4$ and $9 - 4t \ge -11$

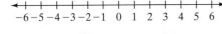

16. $-\dfrac{n}{4} \le 1$ and $2n - 17 < -17$

17. $7 \le 6x + 1$ and $-4x \ge 28$

18. $6 < -5x - 4$ and $2x > 11$

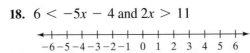

19. $-41 < 9a - 5 \le 13$

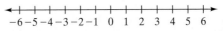

20. $-17 \le 3y + 10 < 1$

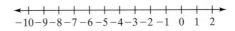

21. $-13 < \dfrac{4x}{5} - 9 < -7$

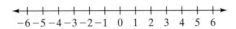

22. $-13 \leq \dfrac{2n}{3} - 7 \leq -10$

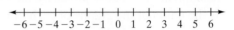

23. $8 \leq 12 - h$ and $-\dfrac{2}{3}h > 1$

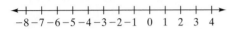

24. $-6 > -a - 1$ and $\dfrac{2}{5}a > 4$

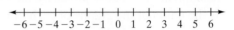

25. $0 \leq -8x - 20 \leq 24$

26. $-9 < 15 - 6x < 30$

27. $2x - 3 < -5$ or $3x + 4 > 10$

28. $5x + 2 \leq -18$ or $4x - 1 \geq 15$

29. $6 - r > 14$ or $-11 \leq 3r - 8$

30. $-16 \geq 19 - 7n$ or $11n + 23 < -10$

31. $18 - 10x \geq 23$ or $26 + 9x \geq -28$

32. $8x + 13 < 25$ or $-9 - 6x < 33$

33. $-17 \leq 16 - 11a$ or $-5 < -7a - 12$

34. $19 > -5t + 24$ or $-27 \leq 13t - 14$

35. $\dfrac{n}{2} + 8 > 11$ or $9 - \dfrac{2n}{3} \geq 6$

36. $\dfrac{x}{7} - 10 \leq -10$ or $\dfrac{3x}{4} + 5 > 8$

37. $4x + 19 > -19$ or $15 - 3x \geq 26$

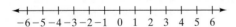

38. $-7 < 2x + 7$ or $6x + 21 \leq 30$

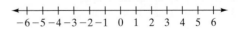

Solve.

39. $7x - 9 > 4x - 18$ and $6x + 5 < x$

40. $2x + 11 \leq 3x + 17$ and $8x - 9 \leq x + 12$

41. $5x + 9 \leq 6x - 3$ or $10x - 7 < 4x + 5$

42. $12x < 14x - 10$ or $3x + 8 \geq 7x - 8$

43. $3x - 9 \leq -2 - \dfrac{x}{2}$ and $15x - 20x < -40$

44. $13x + 2 > 16x$ and $4 - \dfrac{x}{3} \leq x - 2$

45. $\frac{1}{2}x - 21 < \frac{1}{4}x - 10$ or $6x + 19 > 4x + 5$

46. $x + 1 > \frac{7}{8}x + \frac{3}{4}$ or $12x - 23 < 8x - 15$

47. $2(5x + 1) > 9x + 1$ and $11x - 7 \geq 4x$

48. $13 - 3x > -8$ and $12x + 7 < -(1 - 10x)$

49. $-15 \leq 5(x - 7) < 20$

50. $-16 < -4(9 - x) \leq 24$

51. $1 \leq -\frac{1}{3}(4x - 27) < 17$

52. $-7 < \frac{1}{2}(16 - 5x) < 23$

53. $13 \leq 8x - 3(4x + 1) \leq 25$

54. $-15 < 21 - 2(3x + 9) < 0$

55. $-8.3 < 1.7 - 0.5(7x - 1) < 3.95$

56. $-10.3 < 2(1.4x - 5.1) - 0.1 \leq 6.5$

Mixed Practice

Rewrite the inequality using interval notation. Then, graph the inequality.

57. $x > 1$ or $x \leq -2$

$-6\,-5\,-4\,-3\,-2\,-1\ \ 0\ \ 1\ \ 2\ \ 3\ \ 4\ \ 5\ \ 6$

58. $3 \leq x < 6$

$3\ \ 4\ \ 5\ \ 6\ \ 7\ \ 8\ \ 9\ \ 10\ \ 11\ \ 12\ \ 13\ \ 14\ \ 15$

Solve and graph the inequality.

59. $7 < 2x + 1$ and $2x \leq 8$

$1\ \ 2\ \ 3\ \ 4\ \ 5\ \ 6$

60. $-8 < 4 + 3(2y + 1) < 0$

$-4\,-3\,-2\,-1\ \ 0\ \ 1$

61. $6 \geq 2x - 5 > 0$

$1\ \ 2\ \ 3\ \ 4\ \ 5\ \ 6$

62. $10 - s \geq 4$ or $-1 < 2s + 3$

$-6\,-5\,-4\,-3\,-2\,-1\ \ 0\ \ 1\ \ 2\ \ 3\ \ 4\ \ 5\ \ 6$

Applications

B *Solve.*

63. To receive a B in an intermediate algebra course, a student's exam average must be between 83 and 87. A student scored 87, 82, and 80 on the first three exams of the semester. What are the possible scores she can get on the fourth exam to have a B average for the semester?

64. On a cross-country move, a couple plans to drive between 450 and 600 miles per day. If they estimate that their average driving speed will be 60 mph, how many hours per day will they be driving?

65. A person's *body mass index* (BMI) is used to help assess that person's risk for diseases associated with an unhealthy weight. BMI is calculated using the expression $\frac{703w}{h^2}$, where w is a person's weight in pounds and h is his or her height in inches. People who are considered to be at an unhealthy weight for their height have BMIs less than 18.5 or greater than 24.9. Find all of the unhealthy weights for a person who is 5 ft 8 in. (or 68 in.) tall, rounded to the nearest pound.

66. Matter is in a liquid state between its melting point and its boiling point. The melting point of the element mercury is about $-38.8°C$ and its boiling point is about $356.7°C$. The formula $C = \frac{5}{9}(F - 32)$ is used to convert a temperature in degrees Fahrenheit (°F) to the equivalent temperature in degrees Celsius (°C). To the nearest tenth of a degree, determine the temperatures in °F for which mercury is not in the liquid state.

67. Named after the Austrian physicist Ernst Mach, Mach numbers are associated with aircraft traveling at supersonic speeds. The Mach number of an aircraft is the quotient of its speed and the speed of sound, which is approximately 740 mph. What is the range of speeds for aircrafts whose Mach numbers are between 1.0 and 2.0? (*Source:* NASA, Glenn Research Center)

68. According to the Americans with Disabilities Act *Accessibility Guidelines*, the number of handicap seats required for all new construction in places of assembly with over 500 fixed seats is 6 plus 1 additional handicap seat for each 100 fixed seats above 500. How many handicap seats are required for a new auditorium that is to have between 800 and 1200 fixed seats?

69. The cost of a wedding reception is $2500 plus $50 for each guest. If a couple would like to keep the cost of the reception between $7500 and $10,000, how many guests can the couple invite?

70. A student's cell phone plan costs $29 per month, with the first 200 min free. Additional minutes (or part thereof) are charged at a rate of $0.49 per min. How many minutes does she use her cell phone if her average monthly bill is between $78 and $127?

71. The owner of a cookie shop plans to sell gift baskets for special occasions. He estimates that the monthly cost to make the gift baskets is $200 plus $5 per basket. Each basket will sell for $45. How many gift baskets must be sold per month if the owner wants to make a monthly profit between $1000 and $2000? (*Hint:* Profit = Revenue − Cost.)

72. A factory worker earns $12 per hour plus an overtime rate of $16 for every hour over 40 hr she works per week. How many hours of overtime must she work if she wants to make between $600 and $800 per week?

• Check your answers on page A-19.

MINDStretchers

Mathematical Reasoning

1. Consider the equation $x + 1 = 8$.
  a. Express this equation as a compound inequality.
  b. Explain how you know that the compound inequality is equivalent to the equation.

Writing

2. Suppose that two inequalities are joined by *and* to form a compound inequality, and the same two inequalities are joined by *or* to form a second compound inequality. Is it possible for the two compound inequalities to have the same solutions? Explain.

Groupwork

3. With a partner, discuss how you can solve the following inequalities. Then, solve each inequality.
  a. $-4x < 2x - 18 < -x$
  b. $7x - 5 \le 4x - 3 \le 8x - 3$
  c. $3(2x + 1) < 9(2 - x) \le 20 - 5x$

2.5 # Solving Absolute Value Equations and Inequalities

Recall from Section 1.1 that the *absolute value* of a number is its distance from 0 on a number line.

A To find the distance between numbers on a number line

DEFINITION
For any real number x, the **absolute value** of x, written as $|x|$, is
$$|x| = x \text{ when } x \geq 0 \quad \text{and} \quad |x| = -x \text{ when } x < 0.$$

B To solve an equation involving absolute value

C To solve an inequality involving absolute value

Let's look at $|4|$ and $|-4|$ on the number line.

D To solve applied problems involving absolute value equations or inequalities

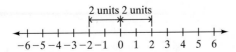

The distance from 0 to 4 or from 0 to -4 is 4 units. So we write:
$$|4| = 4 \quad |-4| = 4$$

Now, let's look at the distance between -2 and 4 on the number line.

The distance from 0 to 4 or from 0 to -4 is 4 units. So we write:

Another way to find the distance between -2 and 4 on the number line is to find the absolute value of the difference of the numbers, as shown:
$$|-2 - 4| = |-6| = 6 \quad \text{or} \quad |4 - (-2)| = |6| = 6$$

Explain why the order in which we subtract does not matter when finding the distance between two points on a number line.

DEFINITION
For any real numbers a and b, the **distance** between them on a number line is $|a - b|$ or $|b - a|$.

Solving Absolute Value Equations

We can use a number line to solve an *absolute value equation*.

EXAMPLE 1
Solve $|x| = 2$ using a number line.

Solution Using a number line, we find all the numbers that are 2 units from 0.

Both -2 and 2 are two units away from 0 on the number line.

PRACTICE 1
Solve $|x| = 5$ using a number line.

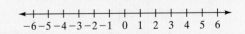

EXAMPLE 1 (continued)

Check

$$|x| = 2$$ $$|x| = 2$$
$$|-2| \stackrel{?}{=} 2$$ Substitute -2 for x. $$|2| \stackrel{?}{=} 2$$ Substitute 2 for x.
$$2 = 2$$ True $$2 = 2$$ True

So the solutions are -2 and 2.

EXAMPLE 2

Solve: **a.** $|x| = 0$ **b.** $|x| = -1$

Solution

a. $|x| = 0$
    $x = 0$

Zero is the only number with absolute value 0. So the solution is 0.

b. $|x| = -1$
No solution

Since the absolute value represents a distance, it cannot be negative. So there is no solution.

PRACTICE 2

Solve.

a. $0 = |y|$

b. $|y| = -\dfrac{1}{2}$

The previous examples suggest the following rule for solving linear absolute value equations:

To Solve an Absolute Value Equation

For any positive number a and any algebraic expression X,

- if $|X| = a$, then $X = a$ or $X = -a$.
- if $|X| = 0$, then $X = 0$.
- if $|X| = -a$, then the equation has no solution.

We can use this rule and the properties of equality to solve absolute value equations.

EXAMPLE 3

Solve: $5|x| - 4 = 11$

Solution

$$5|x| - 4 = 11$$
$$5|x| - 4 + 4 = 11 + 4$$ Add 4 to each side of the equation.
$$5|x| = 15$$ Simplify.
$$\frac{5|x|}{5} = \frac{15}{5}$$ Divide each side of the equation by 5.
$$|x| = 3$$
$$x = 3 \quad \text{or} \quad x = -3$$

The solutions are 3 and -3. Note that since the absolute value is equal to a positive number, 3, there are two solutions.

PRACTICE 3

Solve: $2|y| - 9 = 1$

EXAMPLE 4

Solve: $|x - 8| = 5$

Solution

$$|x - 8| = 5$$

$$|X| = a$$

Since 5 is positive, $x - 8 = 5$ or $x - 8 = -5$. Solving each equation, we get:

$$x - 8 = 5 \quad \text{or} \quad x - 8 = -5$$
$$x = 13 \quad \text{or} \quad x = 3$$

The solutions are 13 and 3.

PRACTICE 4

Solve: $|x + 6| = 4$

EXAMPLE 5

Solve: $|2y + 5| = 9$

Solution

$$|2y + 5| = 9$$

$$|X| = a$$

Since 9 is positive, $2y + 5 = 9$ or $2y + 5 = -9$.

$$2y + 5 = 9 \quad \text{or} \quad 2y + 5 = -9$$
$$2y = 4 \qquad\qquad 2y = -14$$
$$y = 2 \quad \text{or} \qquad y = -7$$

The solutions are 2 and -7.

PRACTICE 5

Solve: $|3x - 1| = 8$

EXAMPLE 6

Solve: $-3|x - 3| = 9$

Solution

$$-3|x - 3| = 9$$
$$|x - 3| = -3 \qquad \text{Divide each side of the equation by } -3.$$

$$|X| = -a$$

Since the absolute value cannot be negative, the equation has no solution.

PRACTICE 6

Solve: $-\dfrac{1}{2}|x - 1| = 4$

In some equations, there are two absolute value expressions. Consider, for instance, the equation $|x| = |y|$. We are told that x and y are equal in absolute value. This means that they are the same distance from 0 on the number line. So they must be either equal to each other or opposites of each other. We can express this as follows:

$$|x| = |y| \quad \text{means} \quad x = y \quad \text{or} \quad x = -y$$

Equal in absolute value   Equal to each other   Opposites of each other

EXAMPLE 7

Solve and check: $|2y + 1| = |y + 2|$

Solution   Since the expressions have the same absolute value, either $2y + 1 = y + 2$ or $2y + 1 = -(y + 2)$. Solving each equation, we get:

$$|2y + 1| = |y + 2|$$

$2y + 1 = y + 2$ or $2y + 1 = -(y + 2)$
$y = 1$ | $2y + 1 = -y - 2$
   | $3y = -3$
or $y = -1$

Check

Substitute 1 for y. | Substitute -1 for y.

$|2y + 1| = |y + 2|$ | $|2y + 1| = |y + 2|$
$|2(1) + 1| \overset{?}{=} |1 + 2|$ | $|2(-1) + 1| \overset{?}{=} |-1 + 2|$
$|2 + 1| \overset{?}{=} |1 + 2|$ | $|-2 + 1| \overset{?}{=} |-1 + 2|$
$|3| = |3|$ | $|-1| = |1|$
$3 = 3$ True | $1 = 1$ True

The solutions are 1 and -1.

PRACTICE 7

Solve and check: $|5x + 2| = |4x + 1|$

EXAMPLE 8

Solve: $|x - 2| = |x - 1|$

Solution   We know that $x - 2 = x - 1$ or $x - 2 = -(x - 1)$. Solving each equation, we get:

$$|x - 2| = |x - 1|$$

$x - 2 = x - 1$ or $x - 2 = -(x - 1)$
$-2 = -1$ False | $x - 2 = -x + 1$
   | $2x = 3$
   | $x = \dfrac{3}{2}$

Since $-2 = -1$ is a false statement, there is no solution to the equation $x - 2 = x - 1$. So the only solution to the original equation is $\dfrac{3}{2}$.

PRACTICE 8

Solve: $|3 - x| = |x + 4|$

Absolute value equations are used in solving problems in areas such as manufacturing, banking, and voting.

The symbol $\pm$ (read "plus or minus") is used to represent both the positive and negative values of a number. For example, ± 3 represents both 3 and -3. We often use this symbol to indicate *tolerance*—the allowable error above and below an ideal measurement. For example, a machine that cuts wood to be made into metersticks might be set to allow for a tolerance of ± 0.001 m. The maximum length of the meterstick is 1 m $+ 0.001$ m $= 1.001$ m. The minimum length is 1 m $- 0.001$ m $= 0.999$ m.

Alternatively, the maximum and minimum lengths of the meterstick can be found by solving the absolute value equation $|x - 1| = 0.001$, where x is the length of the meterstick.

$$|x - 1| = 0.001$$

$$x - 1 = 0.001 \quad \text{or} \quad x - 1 = -0.001$$
$$x = 1.001 \quad \text{or} \quad x = 0.999$$

So again we see that the maximum and minimum lengths are 1.001 m and 0.999 m, respectively.

EXAMPLE 9

A juice manufacturer claims that each of its juice cartons contains 64 oz of juice. The manufacturer allows for a tolerance of ± 2.5 oz per case of 12 cartons. Find the maximum and minimum acceptable amounts of juice per case.

Solution   Let x represent the amount of juice in a case. To find the maximum and minimum acceptable values of x, we solve the following absolute value equation:

$$|x - \underbrace{12(64)}| = 2.5$$

$$\downarrow$$

Amount of juice the
manufacturer claims
per case

$$x - 12(64) = 2.5 \quad \text{or} \quad x - 12(64) = -2.5$$
$$x - 768 = 2.5 \qquad\qquad x - 768 = -2.5$$
$$x = 770.5 \quad \text{or} \qquad x = 765.5$$

So the maximum and minimum acceptable amounts of juice per case are 770.5 oz and 765.5 oz, respectively.

PRACTICE 9

A nickel weighs approximately 0.18 oz. To check that there are 40 nickels in a roll, a teller weighs the roll. The bank allows an error of 0.0015 oz in the total weight. What are the maximum and minimum weights of an acceptable nickel roll given that the wrapper weighs 0.05 oz?

Solving Absolute Value Inequalities

As with absolute value equations, we can use a number line to solve an *absolute value inequality*.

EXAMPLE 10

Solve $|x| < 2$ using a number line.

Solution   The inequality $|x| < 2$ represents all real numbers whose distance from 0 is less than 2 units. The graph of $|x| < 2$ is

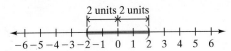

PRACTICE 10

Solve $|x| < 5$ using a number line.

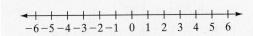

EXAMPLE 10 (continued)

We can check the graph by substituting any number between -2 and 2, say -1, in the original inequality.

$$|x| < 2$$

$$\overset{?}{|-1| < 2}$$

$$1 < 2 \quad \text{True}$$

The solutions of $|x| < 2$ are all real numbers between but not including -2 and 2, that is, all real numbers that satisfy the compound inequality $-2 < x < 2$. We can also express the solutions in interval notation as $(-2, 2)$.

EXAMPLE 11

Solve $|x| > 2$ using a number line.

Solution   The inequality $|x| > 2$ represents all real numbers whose distance from 0 is greater than 2 units. The graph of $|x| > 2$ is

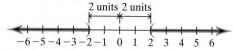

The solutions of $|x| > 2$ are all real numbers that are less than -2 or greater than 2, which can be written as the compound inequality $x < -2$ or $x > 2$. We can express the solutions in interval notation as the union of $(-\infty, -2)$ and $(2, \infty)$; that is, $(-\infty, -2) \cup (2, \infty)$.

PRACTICE 11

Solve $|x| > 5$ using a number line.

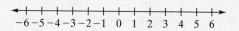

The preceding examples suggest the following rule for solving absolute value inequalities:

> ### To Solve an Absolute Value Inequality
>
> For any positive number a and any algebraic expression X,
>
> - if $|X| < a$, then $-a < X < a$.
> - if $|X| > a$, then $X < -a$ or $X > a$.
>
> Similar rules hold for $|X| \leq a$ and $|X| \geq a$.

We can use this rule to solve absolute value inequalities.

EXAMPLE 12

Solve and graph: $|x - 3| \leq 4$

Solution

$\lvert x - 3 \rvert \leq 4$	
$-4 \leq x - 3 \leq 4$	Write a compound inequality.
$-4 + 3 \leq x - 3 + 3 \leq 4 + 3$	Add 3 to each part of the compound inequality.
$-1 \leq x \leq 7$	Simplify.

PRACTICE 12

Solve and graph: $|x + 2| < 5$

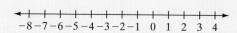

The graph of $|x - 3| \le 4$ is:

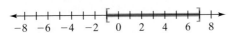

The solutions of $|x - 3| \le 4$ are all real numbers between -1 and 7, including -1 and 7, written in interval notation as $[-1, 7]$.

EXAMPLE 13

Solve and graph: $|x + 5| > 3$

Solution

$$|x + 5| > 3$$

$x + 5 < -3$ or $x + 5 > 3$ Write a compound inequality.

$\quad x < -8$ or $\quad x > -2$ Subtract 5 from each side.

The graph of $|x + 5| > 3$ is:

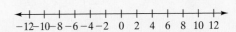

The solutions of $|x + 5| > 3$ are all real numbers less than -8 or greater than -2, written in interval notation as the union of $(-\infty, -8)$ and $(-2, \infty)$; that is, $(-\infty, -8) \cup (-2, \infty)$.

EXAMPLE 14

Solve and graph: $|2x - 5| < 7$

Solution

$$|2x - 5| < 7$$

$-7 < 2x - 5 < 7$ Write a compound inequality.

$-2 < 2x < 12$ Add 5 to each part of the compound inequality.

$-1 < x < 6$ Divide each part of the compound inequality by 2.

The graph of $-1 < x < 6$ is:

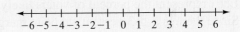

The solutions of $|2x - 5| < 7$ are all real numbers between -1 and 6, written in interval notation as $(-1, 6)$.

EXAMPLE 15

Solve and graph: $|1 - 2x| \ge 2$

Solution

$$|1 - 2x| \ge 2$$

$1 - 2x \le -2$ or $1 - 2x \ge 2$ Write a compound inequality.

$\quad -2x \le -3 \qquad\qquad -2x \ge 1$ Subtract 1 from each side.

$\quad x \ge \dfrac{3}{2} \qquad\qquad\quad x \le -\dfrac{1}{2}$ Divide each side by -2 and reverse the inequality.

PRACTICE 13

Solve and graph: $|x - 2| \ge 4$

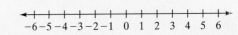

PRACTICE 14

Solve and graph: $|3x - 1| \le 4$

PRACTICE 15

Solve and graph: $|3 - 4x| > 1$

EXAMPLE 15 (continued)

The graph of $x \geq \dfrac{3}{2}$ or $x \leq -\dfrac{1}{2}$ is:

$$\overset{\longleftarrow \; | \; | \; | \; | \; | \; | \; | \;] \; | \; | \; [\; | \; | \; | \; | \; | \; \longrightarrow}{\underset{-6\,-5\,-4\,-3\,-2\,-1\;\;0\;\;1\;\;2\;\;3\;\;4\;\;5\;\;6}{}}$$

The solutions of $|1 - 2x| \geq 2$ are all real numbers less than or equal to $-\dfrac{1}{2}$ or greater than or equal to $\dfrac{3}{2}$. We can express the solutions in interval notation as the union of $\left(-\infty, -\dfrac{1}{2}\right]$ and $\left[\dfrac{3}{2}, \infty\right)$, that is, $\left(-\infty, -\dfrac{1}{2}\right] \cup \left[\dfrac{3}{2}, \infty\right)$.

As with absolute value equations, we can use absolute value inequalities to solve applied problems.

EXAMPLE 16

According to a poll, 45% of voters in an upcoming mayoral election are likely to vote for the incumbent.

a. If the poll has a margin of error of no more than ± 3 percentage points, write an absolute value inequality to represent this situation.

b. Use the inequality found in part (a) to find the least and greatest percent of voters likely to vote for the incumbent according to this poll.

Solution

a. Let x represent the percent of voters likely to vote for the incumbent. Since there is a ± 3 percentage point margin of error, the inequality describing this situation is $|x - 45| \leq 3$.

b. Solving $|x - 45| \leq 3$ for x, we get:

$$|x - 45| \leq 3$$
$$-3 \leq x - 45 \leq 3$$
$$42 \leq x \leq 48$$

So the least and greatest percent of voters likely to vote for the incumbent are 42% and 48%, respectively.

PRACTICE 16

A box of a particular brand of candy is supposed to weigh 454 g. The quality control inspector randomly selects boxes to weigh, sending back any box that is not within 5 g of the ideal weight.

a. Write an absolute value inequality for this situation.

b. What is the range of acceptable weights for a box of candy?

Mathematically Speaking

Fill in each blank with the most appropriate term or phrase from the given list.

plus or minus	the opposite of that number	that number
two solutions	add and then subtract	no solution

1. The absolute value of a negative number is _____.

2. For any positive number a and any algebraic expression X, the equation $|X| = -a$ has _____.

3. For any positive number a and any algebraic expression X, the equation $|X| = a$ has _____.

4. The symbol $\pm$ means _____.

A *Find the distance between the given numbers on a number line.*

5. -3 and 3

6. -5 and 6

7. -11 and -4

8. -13 and -2

9. 1.5 and 9

10. 0.6 and 8.3

B *Solve.*

11. $|x| = 7$

12. $|x| = 1$

13. $|n| = \dfrac{2}{3}$

14. $|y| = \dfrac{5}{2}$

15. $|6x| = 24$

16. $|5n| = 10$

17. $2|y| = -1.6$

18. $-3|x| = 12.3$

19. $|x| + 1 = 1$

20. $|x| + 7 = 7$

21. $|x| - 4 = -2$

22. $|z| - 5 = 6$

23. $3|n| + 10 = 7$

24. $2|x| - 4 = 8$

25. $12 - 4|y| = -16$

26. $-6|x| - 3 = -33$

27. $|4x| + 9 = 15$

28. $|3x| + 11 = 12$

29. $\left|\dfrac{2}{3}n\right| - 5 = -1$

30. $\left|\dfrac{1}{2}x\right| - 5 = 0$

31. $|y + 7| = 4$

32. $|n - 9| = 12$

33. $|x - 3| = -1$

34. $|x + 6| = 10$

35. $|2z + 13| = 21$

36. $|3n - 8| = -5$

37. $|4x - 11| = 17$

38. $|5x + 3| = 2$

39. $2|3n - 1| = 16$

40. $4|4 - 2x| = 24$

41. $-|10 - 6z| = 7$

42. $3|3n + 2| = -9$

43. $-\dfrac{1}{3}|8 - 7n| = -12$

44. $-\dfrac{1}{4}|5x + 6| = -4$

45. $|5x| = |7x - 24|$

46. $|4y + 1| = |6y|$

47. $\left|\dfrac{1}{2}x + 3\right| = \left|\dfrac{1}{4}x\right|$

48. $\left|\dfrac{1}{3}n - 8\right| = \left|\dfrac{1}{5}n\right|$

49. $|2x - 3| = |x + 9|$

50. $|n - 4| = |3n + 2|$

51. $|5y + 1| = |6y - 1|$

52. $|4z - 10| = |3z + 10|$

53. $|11 - 7x| = |5 - 9x|$

54. $|12 - 3z| = |15 - 6z|$

55. $|x + 7| = |x + 1|$

56. $|x - 8| = |x + 6|$

57. $|4 - n| = |n - 2|$

58. $|y - 9| = |5 - y|$

59. $-|3t + 4| = -|2t - 1|$

60. $-|5n - 3| = -|n - 4|$

61. $|4a + 7| = -|8 - a|$

62. $-|2y + 6| = |3y - 2|$

C *Solve, and then graph.*

63. $|x| > 3$

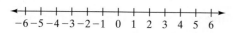

64. $|x| \geq 4$

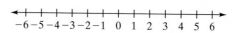

65. $|x| \leq \dfrac{1}{2}$

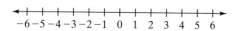

66. $|x| < \dfrac{3}{2}$

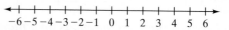

67. $|7x| \geq 21$

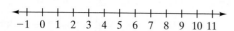

68. $|5x| \geq 10$

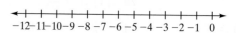

69. $|x + 4| < 4$

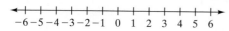

70. $|x - 1| \leq 1$

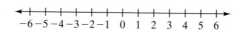

71. $|x - 5| \geq 3$

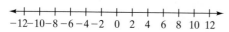

72. $|x + 7| > 1$

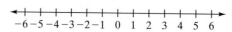

73. $|x| + 5 \geq 6$

74. $|x| + 2 > 4$

75. $|x| - 3 < -2$

76. $|x| - 5 < -1$

77. $|2x + 3| \geq 11$

78. $|4x + 5| > 7$

79. $|6 - 3x| < 9$

80. $|7 - 2x| \leq 1$

81. $\left|\dfrac{2}{3}x + 4\right| \geq 0$

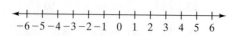

82. $\left|\dfrac{1}{2}x - 1\right| \geq 0$

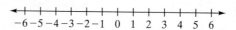

83. $-|2x - 7| < -6$

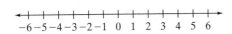

84. $-|5x - 8| \geq -2$

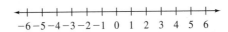

85. $|4x - 3| - 6 \leq 5$

$$\xleftarrow{+\ +\ +\ +\ +\ +\ +\ +\ +\ +\ +\ +\ +\ }\rightarrow$$
$$\quad -6\ -5\ -4\ -3\ -2\ -1\ \ 0\ \ 1\ \ 2\ \ 3\ \ 4\ \ 5\ \ 6$$

86. $|3x + 1| + 2 \geq 9$

$$\xleftarrow{+\ +\ +\ +\ +\ +\ +\ +\ +\ +\ +\ +\ +\ }\rightarrow$$
$$\quad -6\ -5\ -4\ -3\ -2\ -1\ \ 0\ \ 1\ \ 2\ \ 3\ \ 4\ \ 5\ \ 6$$

Mixed Practice

D *Solve.*

87. Find the distance between -1 and 4 on a number line.

88. $|x + 4| = 6$

89. $|10 - x| = |2x + 1|$

90. $7 + 3|n| = -2$

Solve, and then graph.

91. $11 < |4x - 1|$

$$\xleftarrow{+\ +\ +\ +\ +\ +\ +\ +\ +\ +\ +\ +\ +\ }\rightarrow$$
$$\quad -6\ -5\ -4\ -3\ -2\ -1\ \ 0\ \ 1\ \ 2\ \ 3\ \ 4\ \ 5\ \ 6$$

92. $|10y| \leq 30$

$$\xleftarrow{+\ +\ +\ +\ +\ +\ +\ +\ +\ +\ +\ +\ +\ }\rightarrow$$
$$\quad -6\ -5\ -4\ -3\ -2\ -1\ \ 0\ \ 1\ \ 2\ \ 3\ \ 4\ \ 5\ \ 6$$

Applications

Solve.

93. A digital thermometer with a temperature probe is used in a laboratory experiment. The thermometer has an accuracy of $\pm 0.4°C$. If the reading on the thermometer is 36.8°C, what are the maximum and minimum possible temperatures?

94. A bathroom scale has a tolerance of ± 1.5 lb. If a person standing on the scale sees a reading of 147 lb, what are the maximum and minimum possible weights of the person?

95. A police radar gun is calibrated to have an allowable maximum error of ± 1 mph. If a car is clocked at 37 mph, what are the maximum and minimum possible speeds that the car was traveling?

96. The average of the scores on a math exam was 68. If the maximum and minimum scores on the exam differed from the average by 18 points, what were the maximum and minimum exam scores?

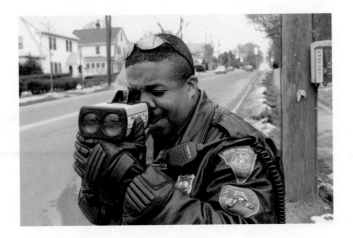

97. A survey conducted in a town showed that 68% of the residents were opposed to the construction of an additional runway at the nearby airport. The survey has a margin of error of no more than $\pm 4\%$.

    a. Write an absolute value inequality to represent this situation.

    b. Use the inequality in part (a) to find the possible values for the actual percentage of residents opposed to the construction of the runway.

98. A patient placed on a restricted diet is allowed 1300 calories per day with a tolerance of no more than 50 calories.

    a. Write an inequality to represent this situation.

    b. Use the inequality in part (a) to find the range of allowable calories per day in the patient's diet.

99. The average annual income of residents in an apartment house is $39,000. The income of a particular resident is *not* within $5000 of the average.

    a. Write an absolute value inequality that describes the income I of the resident.

    b. Solve the inequality in part (a) to find the possible income of the resident.

100. The daily production P of oil at a refinery is within 50,000 of 700,000 barrels.

    a. Express this statement as an absolute value inequality.

    b. Solve the inequality in part (a) to find the possible daily production of the refinery.

101. On a highway, the speed limit is 55 mph, and slower cars are required to travel at least 35 mph. What absolute value inequality must any legal speed s satisfy?

102. An adult's body temperature t is considered to be normal if it is at least 97.6°F and at most 99.6°F. Express this statement as an absolute value inequality.

                       • Check your answers on page A-19.

MINDStretchers

Critical Thinking

  1. What are the solutions of each inequality?

    a. $|x| < a$, where $a < 0$

    b. $|x| > a$, where $a < 0$

Mathematical Reasoning

  2. If you solve an inequality of the form $|x - a| < b$, where a is a real number and b is a positive real number, explain how you can write the solutions of the inequality $|x - a| \geq b$ without solving the inequality. Justify your answer.

Writing

  3. Write a situation that can be represented by each of the following.

    a. $|x - 15| = 2$ **b.** $|x - 30| < 10$

Key Concepts and Skills

Concept/Skill	Description	Example
[2.1] **Equation**	A mathematical statement that two expressions are equal.	$6 + 1 = 7$ $x + 1 = 7$
[2.1] **Solution of an equation**	A value of the variable that makes the equation a true statement.	$5x = 10$ Is $x = 2$ a solution? Yes, because $5 \cdot 2 = 10$ is true.
[2.1] **Linear equation in one variable**	An equation that can be written in the form $ax + b = c$, where a, b, and c are real numbers and $a \neq 0$.	$2x + 5 = 6$ $-13y = 3$
[2.1] **Equivalent equations**	Equations that have the same solution.	$2x + 5 = 8$ and $2x = 3$
[2.1] **Addition property**	For any real numbers a, b, and c, if $a = b$, then $a + c = b + c$.	If $6 = 6$, then $6 + 1 = 6 + 1$. If $x - 2 = 10$, then $x = 12$.
[2.1] **Multiplication property**	For any real numbers a, b, and c, if $a = b$, then $ac = bc$.	If $5 = 5$, then $(5)(2) = 5(2)$. If $\dfrac{x}{3} = 6$, then $x = 18$.
[2.1] **To solve a linear equation**	• Use the distributive property to remove parentheses, if necessary. • Combine like terms. • Use the addition property to isolate the term with the variable. • Use the multiplication property to isolate the variable. • Check by substituting the solution in the original equation.	$4(2x + 3) = -3(x - 1) + 20$ $8x + 12 = -3x + 3 + 20$ $8x + 12 = -3x + 23$ $11x + 12 = 23$ $11x = 11$ $x = 1$ **Check** $4[2(1) + 3] \overset{?}{=} -3(1 - 1) + 20$ $4(5) \overset{?}{=} -3(0) + 20$ $20 = 20$ True
[2.1] **To solve a word problem using an equation**	• Read the problem carefully. • Translate the word problem into an equation. • Solve the equation. • Check the solution in the original problem. • State the conclusion.	The sum of two consecutive even integers is equal to 122. Find the integers. Let $x =$ the first integer. Then $x + 2 =$ the second integer. $x + (x + 2) = 122$ $2x + 2 = 122$ $2x = 120$ $x = 60$ So $x + 2 = 62$. **Check** $60 + 62 \overset{?}{=} 122$ $122 = 122$ True The first integer is 60, and the second is 62.

continued

Concept/Skill	Description	Example
[2.3] **Inequality**	Any mathematical statement containing $<, \leq, >, \geq,$ or $\neq$.	$6 + 5 > 10$ $x + 4 \leq 12$
[2.3] **Solution of an inequality**	Any value of the variable that makes the inequality true. To **solve** an inequality is to find all of its solutions.	$3x + 1 > 10$ Is $x = 4$ a solution? Yes, because $13 > 10$ is true.
[2.3] **Addition property of inequalities**	For all real numbers a, b, and c, • if $a < b$ then $a + c < b + c$. • if $a > b$ then $a + c > b + c$. Similar statements hold for $\leq$ or $\geq$.	If $x + 6 > 4$, then $x > -2$.
[2.3] **Multiplication property of inequalities**	For all real numbers a, b, and c, • if $a < b$ and $c > 0$, then $ac < bc$. • if $a < b$ and $c < 0$, then $ac > bc$. Similar statements hold for $>$, $\leq$, or $\geq$.	If $\dfrac{x}{3} < 2$, then $x < 6$. If $-2x < 8$, then $x > -4$.
[2.3] **To solve a linear inequality**	• Use the distributive property to remove parentheses, if necessary. • Combine like terms. • Use the addition property to isolate the variable. • Use the multiplication property to isolate the variable, being careful to reverse the direction of the inequality symbol when multiplying or dividing by a negative number.	$3(1 - 2x) + 5 < 3x - 1$ $3 - 6x + 5 < 3x - 1$ $-6x + 8 < 3x - 1$ $-9x + 8 < -1$ $-9x < -9$ $x > 1$ interval notation: $(1, \infty)$ $-6\ -5\ -4\ -3\ -2\ -1\ \ 0\ \ 1\ \ 2\ \ 3\ \ 4\ \ 5\ \ 6$
[2.4] **Compound inequality**	An inequality formed by two inequalities joined by the word *and* or the word *or*.	$x > 5$ *and* $x < 6$, or $5 < x < 6$ $x < 200$ *or* $x > 300$
[2.5] **Absolute value of x**	For any real number x, $|x| = x$ when $x \geq 0$ and $|x| = -x$ when $x < 0$.	$|6| = 6$ $|0| = 0$ $|-5| = 5$
[2.5] **Distance**	For any real numbers a and b, the distance between them is $|a - b|$, or $|b - a|$.	The distance between 8 and 5: $|8 - 5| = 3$ or $|5 - 8| = 3$
[2.5] **To solve an absolute value equation**	For any positive number a and any algebraic expression X, • if $|X| = a$, then $X = a$ or $X = -a$. • if $|X| = 0$, then $X = 0$. • if $|X| = -a$, then the equation has no solution.	$|x + 2| = 5$ $x + 2 = 5$ or $x + 2 = -5$ $x = 3$ $x = -7$ $|x + 6| = 0$ $x + 6 = 0$ $x = -6$ $|x + 2| = -5$ No solution

CONCEPT   SKILL

Concept/Skill	Description	Example
[2.5] To solve an absolute value inequality	For any positive number a and any algebraic expression X, • if $|X| < a$, then $-a < X < a$. • if $|X| > a$, then $X < -a$ or $X > a$. Similar rules are true for $|X| \le a$ and $|X| \ge a$.	$|2x - 3| < 7$ $-7 < 2x - 3 < 7$ $-4 < 2x < 10$ $-2 < x < 5$ interval notation: $(-2, 5)$ $|2x - 3| \ge 7$ $2x - 3 \le -7$ or $2x - 3 \ge 7$ $x \le -2$ or $x \ge 5$ interval notation: $(-\infty, 2] \cup [5, \infty)$

INTERVALS, INEQUALITIES, AND GRAPHS

Interval Notation	Inequality Notation	Graph*	Meaning
(a, ∞)	$x > a$		All real numbers greater than a, excluding a
$[a, \infty)$	$x \ge a$		All real numbers greater than or equal to a
$(-\infty, b)$	$x < b$		All real numbers less than b, excluding b
$(-\infty, b]$	$x \le b$		All real numbers less than or equal to b
$(-\infty, \infty)$			All real numbers

*Some textbooks use the graphical representations ○——→ and ●——→ instead of (——→ and [——→, respectively.

Interval Notation	Inequality Notation	Graph*	Meaning
(a, b)	$a < x < b$		All real numbers between a and b, excluding a and b
$[a, b]$	$a \le x \le b$		All real numbers between a and b, including a and b
$[a, b)$	$a \le x < b$		All real numbers between a and b, including a and excluding b
$(a, b]$	$a < x \le b$		All real numbers between a and b, excluding a and including b

*Some textbooks use the graphical representations —○——○— and —●——●— instead of —(——)— and —[——]—, respectively.

CHAPTER 2   Review Exercises

Say Why

Fill in each blank.

1. The number 5 _____ a solution of the equation
   is/is not
   $2x - 4 = 14$ because _____
   _____.

2. The statement $1 - 6c = 2c$ _____ a linear equation
   is/is not
   in one variable because _____
   _____.

3. The equations $3x - 7 = 20$ and $4x = 36$ _____
   are/are not
   equivalent because _____
   _____.

4. $3x + 2y = 6$ _____ a literal equation because
   is/is not
   _____
   _____.

5. The equation $|x| + 1 = 0$ _____ have a solution
   does/does not
   because _____
   _____.

6. The number -3 _____ in the interval $(-\infty, \infty)$
   is/is not
   because _____
   _____.

[2.1]   *Determine whether the given value is a solution of the equation.*

7. $2x + 5 = 6x - 11; 4$

8. $12 - 7x = 4(3x + 1); -1$

9. $\frac{2}{3}(9x - 12) = -(8 - 3x); -2$

10. $3x + \frac{1}{2}(8x + 12) = 6 - (1 - 4x); -\frac{1}{3}$

Solve and check.

11. $-\frac{2}{3}x = 4$

12. $3n + 4 = -8$

13. $7x - 16 = 9x$

14. $17 - 2t = 11 - 10t$

15. $4n - 3 - 8n + 15 = 0$

16. $3.5 - 1.9a - 4 = 1.1a - 2.6$

17. $\frac{1}{2}(3x - 16) = -15$

18. $3(4y + 8) = -4(7 - 9y)$

19. $6y - 2(1 - y) = 12y + 18$

20. $4[1 - (5x - 2)] + 11x = 7 - (7x - 7)$

[2.2]   *Solve for the indicated variable.*

21. $10x + 2y = -2$ for y

22. $y - 4x = 8$ for x

23. $ab + cd = 0$ for c

24. $6xy - v = z$ for x

25. $a(b - c) = d$ for b

26. $q - r(s + t) = p$ for t

Solve the formula for the indicated variable.

27. Surface area of a cylinder: $A = 2\pi r^2 + 2\pi rh$ for h

28. Fried's rule (dosages): $D = \dfrac{Ay}{150}$ for A

29. Ideal gas law: $PV = nRT$ for T

30. Volume of a rectangular prism: $V = lwh$ for w

[2.3] *Determine whether the given value is a solution of the inequality.*

31. $4n - 7 > 3n - 6; -2$

32. $\frac{1}{4}(y + 5) \leq 10 - \frac{1}{3}y; 15$

Match the inequality to its graph.

33. $x < 2$

34. $x \geq 2$

35. $x \leq 2$

36. $x > 2$

a.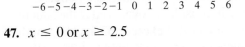

b.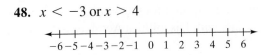

c.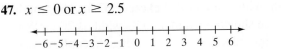

d.

Solve and graph.

37. $-\frac{3}{5} < \frac{1}{10}x$

38. $2n - 7 \leq 9$

39. $11 - 5y \leq 11 - 4y$

40. $-3x > x + 1$

Solve. Then, write the solution in interval notation.

41. $16y - 12y + 6 \geq 13 + 9y - 22$

42. $6(2 - 0.5x) \leq -12$

43. $10 - (4x + 14) < -2(5x + 8)$

44. $\frac{1}{3}[7 - (24n - 11)] > 30 - 13n - 4$

[2.4] *Rewrite the inequality using interval notation. Then, graph the inequality.*

45. $-4 < x \leq 1$

46. $-2 \leq x < 3$

47. $x \leq 0$ or $x \geq 2.5$

48. $x < -3$ or $x > 4$

Solve and graph.

49. $2x - 3 > 1$ and $x + 7 < 12$

50. $\frac{1}{2}x \geq -2$ and $1 - 3x > -8$

51. $-11 < 5x + 4 \leq 14$

52. $0 \leq 2 - 4x \leq 18$

53. $3x + 1 < -2$ or $4x - 7 \geq 7$

54. $\frac{1}{4}x + 3 < 6$ or $2 - 5x < 12$

Solve.

55. $8x + 12 \le 2x$ or $x < 3x$

56. $7x - 6 \le 8x - 4$ and $19 - x \le 10 - 4x$

[2.5] *Find the distance between the two real numbers on the number line.*

57. -2 and 9

58. -6.4 and -5.7

Solve.

59. $|x| = 15$

60. $|4y| = 24$

61. $|n| - 12 = -3$

62. $11 - 2|x| = 9$

63. $|y + 2| = -5$

64. $13 - |5 - 6y| = 12$

65. $|3x - 9| = |4x + 15|$

66. $|x - 1| = |x + 1|$

Solve, then graph.

67. $|x| < 1.5$

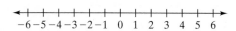

68. $|x - 4| \le 3$

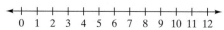

69. $|x| + 2 \ge 2$

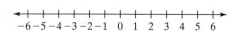

70. $|2x - 5| > 4$

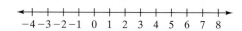

71. $-|4x - 9| \le -1$

72. $|2x - 7| - 1 < 4$

Mixed Applications

73. A minimum floor space of 1440 in² is required for a stationary wheelchair. If the space is required to be 48 in. long, how many inches are required for the width?
(*Source:* Americans with Disabilities Act *Accessibility Guidelines*)

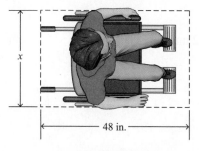

Clear Floor Space

74. According to American Airlines' baggage policies, checked baggage is considered oversized if the sum of its length, width, and height exceeds 62 in. The length of the suitcase shown is 30 in. and its width is 20 in. What is its maximum height for the luggage to conform to the baggage restriction? (*Source:* aa.com)

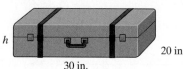

75. The cost of a speeding ticket is $50 up to the first 10 mph over the speed limit plus $10 for every mile per hour over that amount. How fast was a person driving if a $120 speeding ticket was issued on a road with a 55-mph speed limit?

76. A small quilt-making company estimates that its monthly cost is $2450 plus $20 for each quilt produced. Each quilt sells for $90. How many quilts must be sold in order for the cost to be equal to the revenue?

77. A student notices that the last four digits of her phone number are consecutive integers. If the sum of the digits is 18, what are those digits?

78. A poster is twice as long as it is wide. What is the area of the poster if the perimeter is 24 ft?

79. Ten minutes after a bus left a highway rest stop, a car left the same rest stop traveling in the same direction as the bus. If the car traveled at a rate of 60 mph and the bus traveled at a rate of 50 mph, how long did it take for the car to catch up to the bus?

80. Two friends leave work at the same time driving in opposite directions. One friend is driving 6 mph slower than the other. After 5 min, they are 5.5 mi apart. At what rate is each person driving?

81. A chemist has available a 15% alcohol solution and a 75% alcohol solution. How many liters of each solution does she need to make 8 L of a 39% alcohol solution?

82. To help save for retirement, a couple invests three times the amount of money in a low-risk fund as in a high-risk fund. After one year, the low-risk fund showed a 4.5% return and the high-risk fund showed an 8% return. If the total return on the investments was $1075, how much did the couple invest in each fund?

83. The formula $I = Prt$ is used to calculate the simple interest I earned in t years on an investment of P dollars at an interest rate r (expressed as a decimal).

   a. Solve the formula for r in terms of I, P, and t.

   b. Find the interest rate if an investment of $1000 earns $275 in interest in five years.

84. Young's rule is used to calculate the appropriate dose D of a medicine to be administered to a child. To apply this rule, divide the product of the adult dose A and the age of the child y (in years) by the sum of the age of the child and 12.

   a. Write a formula for Young's rule.

   b. Calculate the correct dose of a medication for a 4-year-old child if the adult dose is 120 mg.

85. High-profile award shows hire seat fillers to occupy the seats of celebrities who leave, so that the ceremony always appears full on TV. It is expected that 5%–10% of the seats will be empty at any given time. In addition, there must be 1 to 3 ushers for each 5 rows of seats. If the ceremony's auditorium has 50 rows of seats, with 20 seats in each row, what is the minimum combined number of seat fillers and ushers needed for the event?

86. At the end of a long-range mission, a fighter jet pilot realizes that he does not have enough fuel to reach the base 1000 mi away. A refueling tanker, in the air between the base and the jet, is 900 mi from the jet. The tanker starts heading toward the jet to refuel it, traveling at half its speed. If they meet in 36 min, what was the jet's speed?

87. Moving company A offers truck rentals for $15 per day plus an additional $0.20 for each mile driven. Moving company B offers truck rentals for $35 per day plus an additional $0.10 for each mile driven. A truck has to be driven how many miles if moving company B is a better deal?

88. A local newspaper charges $45 per day for ads that are 25 words or less. Each word over that amount is charged at a rate of $1.20 per word. A company wants to advertise a job opening in the paper for seven days. What is the maximum number of words the company can have in the ad if it does not want to spend more than $420?

89. The minimum speed posted on a highway is 45 mph and the maximum speed is 65 mph. If a driver stays within the posted speed limits, what is the range of distances that can be driven in 3 hr?

90. The *target heart rate* is a desired range of heart rates reached during physical activity. For a moderate-intensity workout, it is recommended that a person's target heart rate is between 50% and 70% of the person's maximum heart rate. Target heart rate can be determined by solving the compound inequality

$$220 - 2R \leq a \quad \text{and} \quad 220 - \frac{10}{7}R \geq a,$$

where R is the heart rate in beats per minute and a is the person's age in years. What is the target heart rate for a 35-year-old person?

91. A clothing company manufactures women's pants. If the inseam for regular-length pants is 30 in. with a tolerance of $\pm \frac{1}{2}$ in., what are the maximum and minimum lengths of a pair of pants the company manufactures?

92. A recent poll showed that the mayor's approval rating was 49%. If the margin of error was no more than ± 2 percentage points, what are all the possible actual percentages for the mayor's approval rating?

• Check your answers on page A-20.

CHAPTER 2   Posttest

FOR EXTRA HELP

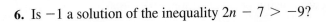

CHAPTER
Test Prep VIDEOS

The Chapter Test Prep Videos with test solutions are available on DVD, in MyMathLab, and on YouTube ™ (search "AkstIntermediate Alg" and click on "Channels").

To see if you have mastered the topics in this chapter, take this test.

1. Is -3 a solution of the equation $\frac{1}{3}x - 8 = -9$?

2. Solve: $4x + 7 = 5$

3. Solve: $2n - 11 = 6n + 17$

4. Solve: $3(1 + 3t) = -(5t - 17)$

5. Solve $a(x + y) - b = c$ for y in terms of a, x, b, and c.

6. Is -1 a solution of the inequality $2n - 7 > -9$?

7. Solve and graph: $5x + 9 > 3(x + 9)$

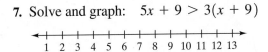

8. Solve and graph: $2x - 7 < 2$ and $x - 1 \geq -x$

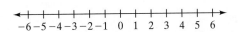

9. Solve and graph: $13 - \frac{3}{4}n > 16$ or $15 > 20 - 5n$

10. Solve: $-10 \leq 14 - 12x < 8$

11. Find the distance between -9.4 and -5.6.

12. Solve: $|3x - 5| = 4$

13. Solve: $-|2y + 1| + 2 = -9$

14. Solve: $|x - 6| = |x - 7|$

15. Solve and graph: $|3x - 5| < 4$

16. A day spa offers two types of massage: relaxation and deep tissue. The spa charges $70 for a relaxation massage and $90 for a deep-tissue massage. In a particular week, the spa provided 53 massages and earned $3930. How many of each type of massage did the spa provide?

17. A beneficiary invests $50,000 of his inheritance in two funds. After one year, one of the funds lost 8% and the other gained 6%. If lost $850 on the investments overall, how much did he invest in each fund?

18. A local health club offers two membership options: Option A costs $50 per month for unlimited use of the gym, and Option B costs $20 per month plus an additional $1.50 per hour for use of the gym. For what number of hours of use per month will Option A be a better deal?

19. The office manager of a local political campaign needs to have postcards and flyers printed. The cost to print one postcard is $1.40 and the cost to print each flyer is $0.10. Each of the 1000 households with registered voters is to receive a postcard by mail. The postage for one postcard is $0.29. If she must keep the total costs between $1700 and $1800, how many flyers can she have printed?

20. A paper mill's cutting machine has a tolerance of ± 0.001 in. in either direction. If the machine is set to cut 48-in. by 48-in. sheets, what is the range of areas of a sheet of paper?

• Check your answers on page A-20.

Cumulative Review Exercises

To help you review, solve the following.

1. Is the statement $|-1.4| > |-1.3|$ true or false?

2. Calculate: $7 - [12 - (1 + 3)^2] \div -4$

3. Rewrite $2 \cdot (3 + 5)$ using the commutative property of addition.

4. Simplify: $(a^3b)^2$

5. Rewrite using positive exponents: $6n^{-1}$

6. Translate the phrase to an algebraic expression: One-third the sum of twice a number and nine.

7. Evaluate $4x^2y - (z + 2y)$ for $x = -2$, $y = 0.5$, and $z = -1$.

8. Combine: $7x + 8 - 9x + 5 + 3(x - 7)$

9. Solve and check: $-5 = 1.3a + 0.2$

10. Solve for c: $a - \dfrac{bc}{d} = e$

11. Is -24 is a solution of $\dfrac{1}{4}x + 3 < \dfrac{1}{6}x + 1$?

12. Solve and graph: $9x + 4 \geq 7x - 2$

$$\overset{\longleftrightarrow}{\underset{-4 \quad -3 \quad -2 \quad -1 \quad 0 \quad 1 \quad 2 \quad 3 \quad 4}{}}$$

13. Solve: $x - 7 \geq -2x + 2$ or $4x - 5 < 2x - 3$

14. Solve: $|x + 2| = -3$

15. The change in the price of a stock for a five-day period is shown in the table. Calculate the average daily change in price for the five-day period.

Day	1	2	3	4	5
Change (in dollars)	+0.12	−0.23	−0.14	+0.13	−0.08

16. One astronomical unit is approximately 9.3×10^7 mi. The distance from the Sun to the planet Neptune is 2.793×10^9 mi. Compute the distance from the Sun to the planet Neptune in astronomical units. (Round your answer to the nearest tenth.)

17. The prime factorization of a whole number is the unique way to write it as a product of prime numbers. For instance, the prime factorization of $24 = 2^3 \times 3^1$, and the prime factorization of $100 = 2^2 \times 5^2$. Verify, through their prime factorizations, that $24 \times 100 = 2400$.

18. During a storm, you can estimate the number of miles m away a bolt of lightning strikes by first counting the number of seconds s between the bolt of lightning and the associated clap of thunder and then dividing by 5.

   a. Translate this relationship to a formula.

   b. If lightning strikes 2.5 mi away, how many seconds will elapse before you hear the thunder?

19. For a given year, the percent of people in the United States who used the Internet can be approximated by the algebraic expression $1.9x + 71.2$, where x is the number of years since 2006. According to this model, what percent of the U.S. population, rounded to the nearest whole percent, used the Internet in the year 2008? (*Sources:* Pew Internet and American Life Project)

20. According to a recent report on U.S. males, aged 20−29, 80% of them have a height within 3.85 in. of 69.65 in. Write an inequality to represent the situation of the 20% of the males in this age group who fall outside this height range. (*Source:* cdc.gov)

• Check your answers on page A-20.

Graphs, Linear Equations and Inequalities, and Functions

Graphs and Inflation: Buying Power

The U.S. Department of Labor's Bureau of Labor Statistics provides national data on inflation. The bureau uses the average consumer price index for a given calendar year to calculate inflation. A good way to understand the impact of inflation on our buying power is to calculate how much $100 worth of goods purchased in 1967 would cost in successive years. The table to the right shows a comparison of costs from 1970 to 2010. Although the table is more detailed, the graph below gives a clearer picture of inflation. (*Source:* bls.gov)

Year	Dollars
1970	$116
1975	$161
1980	$249
1985	$322
1990	$391
1995	$457
2000	$516
2005	$585
2010	$653

1. Plot the points $A(2, 5)$ and $B(-3, -1)$ on the coordinate plane.

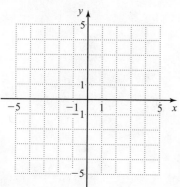

2. Calculate the slope of the line that passes through the points $(-6, 1)$ and $(3, -11)$.

3. Graph the line that passes through the point $(0, -4)$ and has slope $m = \dfrac{3}{4}$ on the coordinate plane.

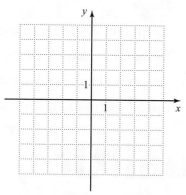

4. Given $P(1, 7)$, $Q(5, 3)$, $R(-8, -10)$, and $S(-15, -3)$, determine whether $\overleftrightarrow{PQ}$ and $\overleftrightarrow{RS}$ are parallel, perpendicular, or neither.

5. Is $(3, -2)$ a solution of the equation $3x + 2y = 5$?

6. Find the *x*- and *y*-intercepts of the graph of $4y - 6x = 12$.

Graph the following equations.

7. $y = 3x + 1$

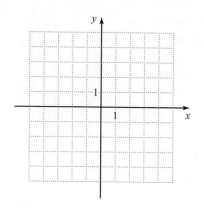

8. $12x - 6 = 0$

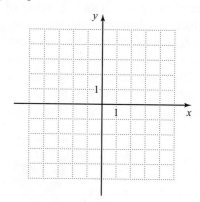

9. $4x + 3y = -9$

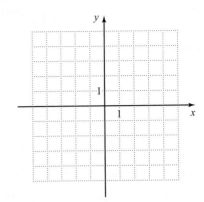

10. Write an equation of the line that has slope 2 and that passes through the point $(-1, 5)$.

11. Write an equation of the line that passes through the points $(0, -1)$ and $(-2, 2)$.

12. Determine if $(-3, 4)$ is a solution of $4x - 2y > -2$.

Graph the following inequalities.

13. $y < \dfrac{1}{2}x + 3$

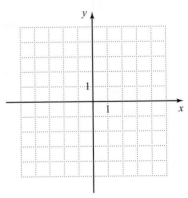

14. $-2x - 6y \le 6$

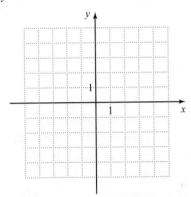

15. If $f(x) = |2x - 8|$, evaluate $f(-6)$.

16. Graph the function $f(x) = 1 - 1.5x$ for $x \ge -2$, and identify its domain and range.

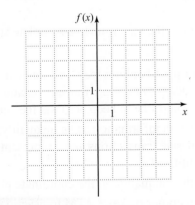

17. The total number of takeoffs and landings at Chicago's O'Hare Airport for the years from 2007 to 2009 are shown in the table to the right:

Point	Year t	Total Takeoffs and Landings M
A	2007	926,973
B	2008	881,566
C	2009	827,899

(*Source:* airports.org)

a. Choose an appropriate scale and plot and label the points on the coordinate plane. Then sketch $\overline{AB}$ and $\overline{BC}$.

b. Compute the slopes of $\overline{AB}$ and $\overline{BC}$.

c. Did the total number of takeoffs and landings change at the same rate over the three years? Explain.

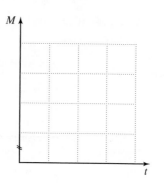

18. A utility company charges its residential customers a flat monthly fee of $20 for electricity service plus $0.15 per kilowatt-hour used during the month.

a. Write a linear equation that shows the monthly electric bill B (in dollars) in terms of the number of kilowatt-hours k of electricity used.

b. Graph this equation on the coordinate plane.

c. Use the graph to estimate the monthly bill if 300 kWh of electricity are used.

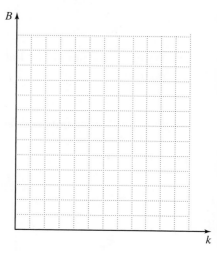

19. When the brakes of a car are applied, the speed of the car decreases by the same amount each minute. Two seconds after the brakes are applied, the speed of the car was 24 mph. After 4 sec, the speed of the car was 8 mph.

a. Write an equation that shows the speed s in miles per hour of the car t sec after the brakes are applied.

b. What is the slope of the line? Explain its meaning in this situation.

c. What is the s-intercept of the graph of the equation? What does it represent in this situation?

d. When does the car come to a complete stop?

20. A customer pays banking fees for using her check/cash card. The bank charges $1.50 for each point-of-sale transaction and $1 for each ATM transaction. The customer would like to keep her total fees under $20 per month.

a. Express this information as an inequality, letting x represent the number of point-of-sale transactions and y represent the number of ATM transactions per month.

b. Graph the inequality on the coordinate plane.

c. If the customer makes 8 ATM transactions, can she make 10 point-of-sale transactions? Explain.

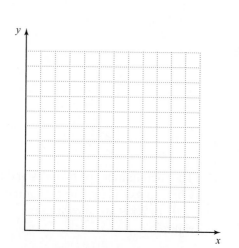

● Check your answers on page A-20.

3.1 The Rectangular Coordinate System

OBJECTIVES

A To plot points on a coordinate plane

B To identify the quadrants of a coordinate plane

C To interpret graphs in applied problems

What Graphing Is and Why It Is Important

A graph is a picture of a mathematical relationship such as an equation or an inequality. Graphs help us to visualize problems in traditional algebra and allow us to focus on the geometry of a problem. Although lacking the precision of an equation, a graph can clarify at a glance patterns and trends in a relationship.

In the past, the graphing approach to problem solving was usually more time-consuming than the traditional algebraic approach. Today, the use of graphing calculators and computer software packages has made graphing easier. But to utilize these electronic tools in graphing, you must first understand the concepts and skills involved in graphing that are discussed in this chapter.

In this chapter, the relationships that we graph are relatively simple. In subsequent chapters, however, we will apply graphing techniques to more complex relationships.

A Coordinate Plane

The flat surface on which we draw graphs is called a **coordinate plane**. To create a coordinate plane, we first sketch two perpendicular number lines—one horizontal, the other vertical—intersecting at their zeros. The point where they intersect is called the **origin**, and each number line is called an *axis*. It is common practice to refer to the horizontal number line as the **x-axis** and the vertical number line as the **y-axis**.

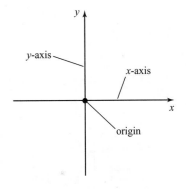

The location of each point on a coordinate plane is represented by a pair of coordinates called an **ordered pair**. The first coordinate in an ordered pair represents a horizontal distance and is called the **x-coordinate**. The second coordinate represents a vertical distance and is called the **y-coordinate**.

To *plot* (or graph) a point on a coordinate plane, we find its location represented by its ordered pair. For example, to plot $(3, 2)$, we start at the origin, go 3 units *to the right,* then go *up* 2 units. For this point, we say that $x = 3$ and $y = 2$.

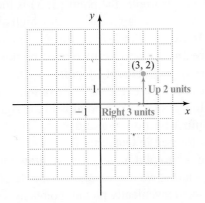

When an ordered pair has a negative x-coordinate, the corresponding point is to the left of the y-axis, as shown on the coordinate plane to the right. When an ordered pair's y-coordinate is negative, the point is below the x-axis.

Any ordered pair whose y-coordinate is 0 will correspond to a point that is on the x-axis. For instance, the points $(-4, 0)$ and $(2, 0)$ are on the x-axis, as shown in the coordinate plane shown on the next page to the left. Similarly, any ordered pair whose x-coordinate

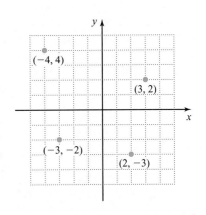

is 0 will correspond to a point that is on the y-axis. For instance, the points $(0, 2)$ and $(0, -3)$ are on the y-axis, as shown in the coordinate plane shown below to the right.

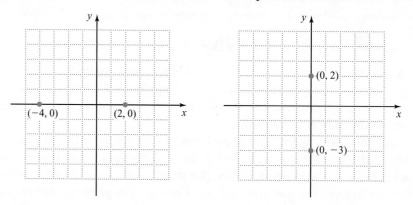

Sometimes, we name points with letters. We might refer to a point as P or A or any other letter that we choose. Generally, a capital letter is used. If we want to emphasize that point P has coordinates $(2, 8)$, we may write it as $P(2, 8)$. If there is a point whose coordinates we do not know, we may refer to it as (x, y) or $P(x, y)$, where x and y are the unknown coordinates.

Sometimes to distinguish one point from another, we use *subscripts* to name them. We may refer to a pair of points as

- P_1 and P_2 (read "P sub one" and "P sub two"),

- (x_1, y_1) and (x_2, y_2), or

- $P_1(x_1, y_1)$ and $P_2(x_2, y_2)$.

Notice that y_1 is the y-coordinate that corresponds to the x-coordinate x_1, and y_2 corresponds to x_2.

The x- and y-axes are boundaries that separate a coordinate plane into four regions called **quadrants**. These quadrants are named in counterclockwise order starting with Quadrant I, as in the figure to the left.

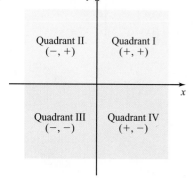

The quadrant in which a point is located tells us something about its coordinates. For instance, in Quadrant I, any point is to the right of the y-axis and above the x-axis, so both its coordinates must be positive. For example, the point $(3, 2)$ is located in Quadrant I. Points in this quadrant are of particular interest in applications in which all quantities must be positive.

A point in Quadrant II lies to the left of the y-axis and above the x-axis, so its x-coordinate must be negative and its y-coordinate must be positive. For example, the point $(-4, 4)$ is located in Quadrant II. In Quadrant III, points are to the left of the y-axis and below the x-axis, so both coordinates must be negative. For example, the point $(-3, -2)$ is located in Quadrant III. And finally, in Quadrant IV, a point is to the right of the y-axis and below the x-axis, so its x-coordinate must be positive and its y-coordinate must be negative. For example, the point $(2, -3)$ is located in Quadrant IV.

Often the points that we are to plot affect how we draw the axes on a coordinate plane. For instance, in the following planes, we choose an appropriate *scale* for each axis—the length between adjacent tick marks—to conveniently plot all points in question. And depending on the location of the points to be plotted, we may choose to show only part of a coordinate plane.

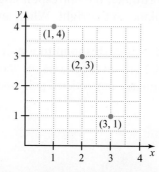

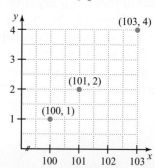

Frequently, we plot data points to make a prediction, that is, to extend the observed pattern in order to estimate the missing data. We do so by connecting the plotted points with line segments or by drawing one line through the plotted points. We can use coordinates from these lines or line segments to help us make predictions.

EXAMPLE 1

The temperature chart of a patient is shown here.

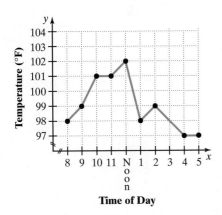

Time of Day

a. What is a reasonable estimate of the temperature of the patient at 3 P.M.?

b. Can you be sure that this estimate is correct? Explain.

Solution

a. At 3 P.M., the temperature of the patient seems to have been approximately 98°F.

b. This estimate is not certain, since the patient's temperature was not observed and recorded at that time. The prediction is based on drawing line segments connecting adjacent points.

PRACTICE 1

On the following coordinate plane, the *x*-coordinate represents a state's population according to a recent U.S. census. The *y*-coordinate stands for the corresponding number of senators from the state. (*Source:* census.gov)

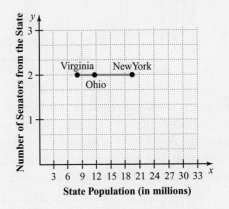

State Population (in millions)

a. How many senators does Ohio have?

b. Describe the pattern that you observe.

EXAMPLE 2

The following table indicates the size of the U.S. population according to the census in the past five decades.

Year	1970	1980	1990	2000	2010
U.S. Population (in millions)	203	227	249	281	309

(*Source:* census.gov)

a. Plot this information on a coordinate plane.

b. Describe any trends that you observe.

PRACTICE 2

The following table shows the amount of total federal budget outlays in three selected years.

Year	1990	2000	2010
Federal Budget Outlays (in billions)	$1253	$1790	$3352

(*Source:* U.S. Office of Management and Budget, 2010)

EXAMPLE 2 (continued)

Solution

a. After labeling the axes, we plot the points. Since all numbers are positive, we only show the first quadrant. We plot the first coordinate from the top row of the given table against the horizontal axis, and plot the second coordinate from the bottom row against the vertical axis. The points are as shown.

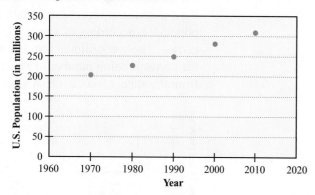

b. The points show the population is increasing over time. They lie approximately in a straight line. We can use this line to estimate the size of the population in years for which we do not have definite information. For instance, we might estimate the population to be about 180 million in 1960 and about 340 million in 2020.

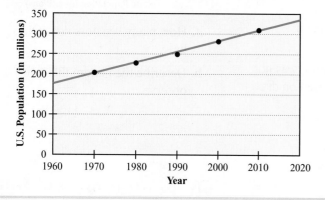

a. Plot this information on the following coordinate plane.

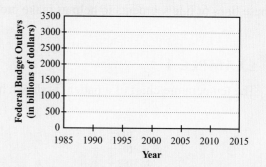

b. Describe any trends that you observe.

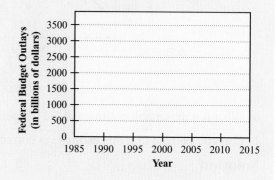

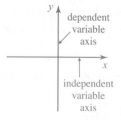

We end this discussion of plotting points with a final comment about variables. On a coordinate plane, the first coordinate of each point is a value of one quantity (or variable), and the second coordinate is the corresponding value of another. For instance, in Example 1, we considered a patient's temperature at various times throughout the day. In this situation, one quantity represents time and the other quantity represents temperature. Notice that the patient's temperature depends on the time of day, rather than the other way around. So we refer to time as the **independent** variable and the temperature as the **dependent** variable. It is customary when plotting points to measure the independent variable along the horizontal axis and the dependent variable along the vertical axis.

Mathematically Speaking

Fill in each blank with the most appropriate term or phrase from the given list.

horizontal	origin	vertical
ordinate	*y*-axis	first
second	Quadrant II	Quadrant IV
coordinate axes	*x*-axis	independent
dependent	ordered pair	

1. The horizontal number line on the coordinate plane is called the _____.

2. The point where the horizontal and vertical axes intersect is called the _____.

3. A(n) _____ is a pair of coordinates representing the location of a point on a coordinate plane.

4. The _____ coordinate in an ordered pair is the *x*-coordinate.

5. The *y*-coordinate in an ordered pair represents a(n) _____ distance.

6. Any point on the _____ has an *x*-coordinate equal to 0.

7. Points in _____ lie to the right of the *y*-axis and below the *x*-axis.

8. The _____ variable is customarily measured along the vertical axis.

A *On the coordinate plane to the right, plot the points with the given coordinates.*

9. $A(2, 1)$
    $B(-3, -4)$
    $C(0, -1)$
    $D(-1, 0)$
    $E(3, -2)$
    $F(-2, 3)$

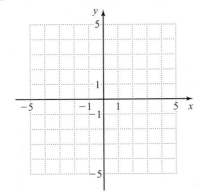

10. $A(2, 0)$
    $B(4, -1)$
    $C(-1, 4)$
    $D(0, -2)$
    $E(1, 2)$
    $F(-2, -1)$

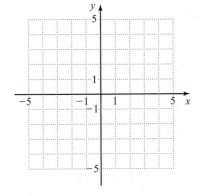

Next to each point, write its coordinates.

11.

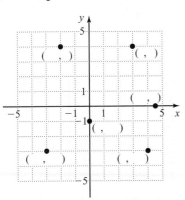

12.

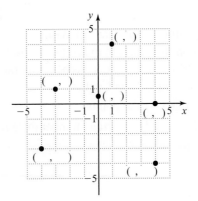

B *Identify the quadrant in which each point is located.*

◉ 13. $(-9, 7)$

14. $(-8, 8)$

◉ 15. $(6, 8)$

16. $(0.1, 4)$

17. $(-2.4, -1.6)$

18. $(-7, -7)$

167

19. $(5.3, -5)$ **20.** $(4.1, -10)$ **21.** (x, y) if $x < 0$ and $y > 0$

22. (x, y) if $x > 0$ and $y < 0$ ⊙ **23.** (x, y) if $x > 0$ and $y = x$ **24.** (x, y) if $x < 0$ and $y = -x$

Mixed Practice

On the given coordinate plane, do the following.

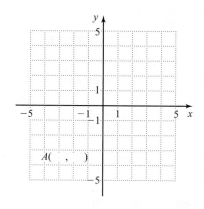

25. Write the coordinates of point A.

26. Plot the point $B(-3, 2)$.

Identify the quadrant in which each point is located.

27. $(1.1, 4)$

28. (x, y) if $x > 0$ and $y = -x$.

Applications

C   *Solve.*

⊙ **29.** On the SAT, there are three parts—critical reading, mathematics, and writing—each with a score ranging from 200 to 800. The graph to the right shows the writing and the math scores for college applicants, coded A, B, C, D, and E.

    a. Estimate the coordinates of each point.

    b. Which students scored higher in math than in writing?

    c. If students A, B, C, D, and E scored 500, 450, 500, 510, and 600, respectively, on critical reading, which student's score is nearest to the maximum total score of 2400?

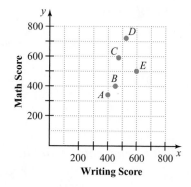

30. The average temperatures in January in Wisconsin for the years from 2005 to 2011 are shown on the graph to the right, where the x-value represents the year ($x = 0$ is 2008) and the y-value is the temperature in degrees Fahrenheit. (*Source:* National Climatic Data Center)

    a. Fill in the coordinates of the points.

    b. Which year had the coldest average temperature in January?

    c. Which year had the warmest average temperature in January?

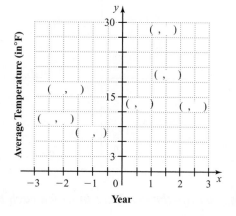

31. The table shows a child's weight (in pounds) recorded at each of his annual physical examinations from birth to age 4. Plot the information on the coordinate plane.

Age a	0	1	2	3	4
Weight w	9	21	28	33	42

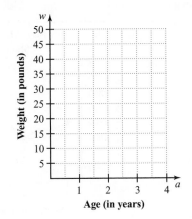

32. The table shows the change from the previous day in the price per share of a retail company's stock for a five-day period. Plot the data on the coordinate plane.

Day x	1	2	3	4	5
Change in Price y (in dollars)	+0.25	−0.70	−0.28	+0.13	+0.60

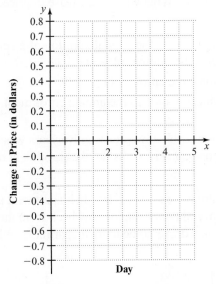

33. The table shows the estimated and projected number of Internet users (in millions) in the United States for various years.

Year	2000	2005	2010	2015
Number of Internet Users (in millions)	135	198	254	288

(*Source:* etforecasts.com)

a. Choose an appropriate scale and plot this information on the coordinate plane.

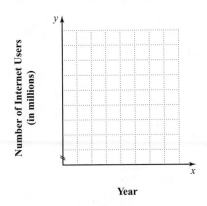

b. Describe any trends you observe.

34. The number of top 40 radio stations in the United States for various years is shown in the following table:

Year	2005	2006	2007	2008	2009	2010
Number of Top 40 Radio Stations	502	485	473	467	483	487

(*Source:* mstreet.net)

a. Choose an appropriate scale and plot the data on the coordinate plane.

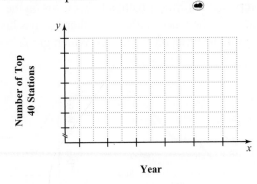

b. Describe the trend that you observe.

35. The graph shows the monthly average for the price of a gallon of regular unleaded gasoline (in cents) in the United States for the 12-month period from
January to December 2008. (*Source:* bls.gov)

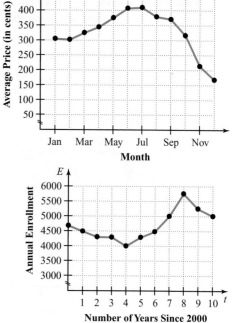

a. Estimate the average price for a gallon of regular unleaded gasoline in
February.

b. In which month was the average price of a gallon of regular unleaded
gasoline below $2?

c. Between which two consecutive months was the greatest decrease in the price
per gallon? What was the decrease in price?

36. The graph shows the annual enrollment *E* for a community college, where *t*
represents the number of years since 2000.

a. Estimate the enrollment in 2004.

b. In which years was the annual enrollment greater than 5000 students?

c. Between which two consecutive years was the greatest decline in enrollment?
By how many students did the enrollment decrease?

37. On Saturday, two friends left their apartment and drove to the beach, where they
spent several hours. On the way back from the beach, they stopped for dinner. After
dinner they drove home to their apartment. Which graph best illustrates the trip? Explain.

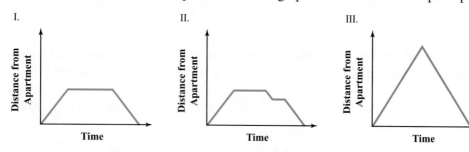

38. In preparing his child's bath, a father turns on the faucet of a bathtub. When the phone rings,
he turns off the faucet. After the father finishes his call, he again turns on the faucet. Once the
tub is full, he turns off the faucet. Which graph best represents the filling of the bathtub? Explain.

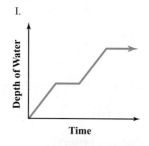
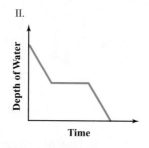
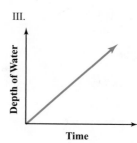

• Check your answers on page A-21.

MINDStretchers

Writing

1. Explain why the order of the coordinates in an ordered pair is important.

Patterns

2. The table lists a given point, the coordinates of its reflection about the y-axis, and the coordinates of its reflection about the x-axis. Study the table.

Point	Reflection about the y-axis	Reflection about the x-axis
$(2, 3)$	$(-2, 3)$	$(2, -3)$
$(4, 8)$	$(-4, 8)$	$(4, -8)$
$(0, 1)$	$(0, 1)$	$(0, -1)$
$(1, 0)$	$(-1, 0)$	$(1, 0)$
$(-3, 5)$	$(3, 5)$	$(-3, -5)$
$(-7, 2)$	$(7, 2)$	$(-7, -2)$
$(-5, -1)$	$(5, -1)$	$(-5, 1)$
$(-2, -2)$	$(2, -2)$	$(-2, 2)$
$(3, -7)$	$(-3, -7)$	$(3, 7)$
$(1, -2)$	$(-1, -2)$	$(1, 2)$

a. What pattern do you observe between the coordinates of the given point and its reflection about the y-axis?

b. What pattern do you observe between the coordinates of the given point and its reflection about the x-axis?

c. Given any point (x, y), what are the coordinates of its reflection about the y-axis? What are the coordinates of its reflection about the x-axis?

Mathematical Reasoning

3. Water is poured at a constant rate into each of the containers shown. For each container, sketch a graph that shows the height of the water in the container over time.

Container A   Container B

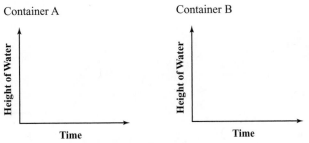

Cultural Note

The seventeenth-century Frenchman René Descartes was a renowned mathematician, philosopher, and scientist. He developed the concepts that underlie graphing on a coordinate plane (or Cartesian system). Among these concepts are the correspondence between ordered pairs of numbers and points on the plane and the correspondence between equations and their graphs. In the poster displayed here, Descartes' spirit is invoked to encourage students at a French university to vote.

3.2 Slope

On the coordinate plane, we are interested not only in points but also in the lines that pass through them. We know that any two points P and Q determine a unique line, which we write as $\overleftrightarrow{PQ}$, (read "line PQ").

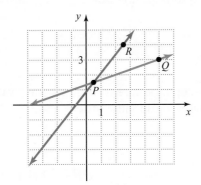

If we take the point P and choose another point R, we get a different line. Note that $\overleftrightarrow{PQ}$ and $\overleftrightarrow{PR}$ have different slants.

In this section, we focus on an important characteristic of a line, namely its *slope*.

Slope

The **slope** of a line, also called its **rate of change**, is a measure of the line's slant. Examining the slope of a line can tell us if the quantity being graphed increases or decreases, as well as how fast that quantity is changing.

Suppose that a straight line on a coordinate plane passes through the points with coordinates (x_1, y_1) and (x_2, y_2), as shown below.

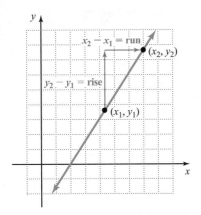

The slope of the line, generally represented by the letter m, is defined as the ratio of the change in y-values to the change in x-values. For the points (x_1, y_1) and (x_2, y_2), we use the following formula:

$$m = \frac{\text{Change in } y\text{-values}}{\text{Change in } x\text{-values}} = \frac{y_2 - y_1}{x_2 - x_1}, \quad \text{where } x_1 \neq x_2$$

In this formula, the numerator of the fraction is the vertical change called the **rise** and the denominator is the horizontal change called the **run**. So another way of writing the formula for slope is $m = \dfrac{\text{rise}}{\text{run}}$.

DEFINITION

The **slope** m of a line passing through the points (x_1, y_1) and (x_2, y_2) is defined to be

$$m = \frac{y_2 - y_1}{x_2 - x_1}, \quad \text{where } x_1 \neq x_2.$$

TIP When computing the slope, be sure to subtract the y-values in the numerator and the x-values in the denominator in the same order.

EXAMPLE 1

Find the slope of the line that passes through the points $(2, 1)$ and $(5, 3)$. Plot the points and then sketch the line.

Solution We are given the coordinates of two points. Let $(2, 1)$ stand for (x_1, y_1) and $(5, 3)$ for (x_2, y_2). Substituting into the slope formula, we get:

$$m = \frac{y_2 - y_1}{x_2 - x_1} = \frac{3 - 1}{5 - 2} = \frac{2}{3}$$

We can also find the slope of a line from its graph. Let's plot the points $(2, 1)$ and $(5, 3)$ and then sketch the line passing through them.

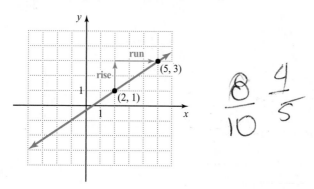

The change in y-values (the rise) is $3 - 1$, or 2. The change in x-values (the run) is $5 - 2$, or 3. So $m = \dfrac{\text{rise}}{\text{run}} = \dfrac{2}{3}$. Therefore, we get the same answer, no matter which formula we use: $m = \dfrac{y_2 - y_1}{x_2 - x_1}$, or $m = \dfrac{\text{rise}}{\text{run}}$.

PRACTICE 1

Find the slope of the line that passes through the points $(4, 7)$ and $(1, 2)$. Sketch the line after plotting these points.

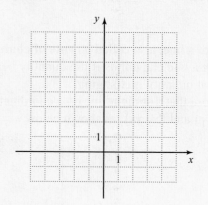

A line rising from left to right like the one shown in Example 1 has a *positive* slope. We say that such a line is *increasing* because as the x-values get larger, the corresponding y-values get larger.

EXAMPLE 2

Sketch the line containing the points $(-2, 1)$ and $(4, -2)$. Find the slope.

Solution First, we plot the points $(-2, 1)$ and $(4, -2)$. Then, we draw a line passing through them.

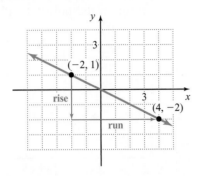

Finally, we find the slope. Starting at $(-2, 1)$, we move 3 units down (for a rise of -3), and then 6 units to the right (for a run of 6) to arrive at the point $(4, -2)$. So the slope $= \dfrac{\text{rise}}{\text{run}} = \dfrac{-3}{6} = -\dfrac{1}{2}$.

PRACTICE 2

Sketch the line passing through the points $(-3, 2)$ and $(1, -1)$. Find the slope.

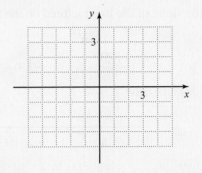

A line falling from left to right as shown in Example 2 has a *negative* slope. We say that such a line is *decreasing* because as the *x*-values get larger, the corresponding *y*-values get smaller.

EXAMPLE 3

Find the slope of the line that passes through the points $(6, 2)$ and $(-2, 2)$. Plot the points, and then sketch the line.

Solution First, we find the slope:

$$\underbrace{(6, 2)}_{x_1 \ y_1} \quad \underbrace{(-2, 2)}_{x_2 \ y_2}$$

$$m = \frac{y_2 - y_1}{x_2 - x_1} = \frac{2 - 2}{-2 - 6} = \frac{0}{-8} = 0$$

So the slope of the line passing through these points is 0.
   Next, we plot the points and then sketch the line.

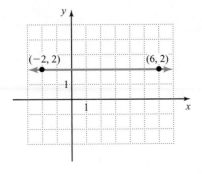

PRACTICE 3

On the following coordinate plane, plot the points $(4, -3)$ and $(-1, -3)$. Sketch the line passing through those points, and then compute its slope.

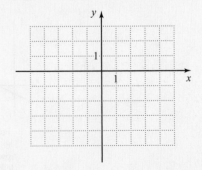

When the slope of a line is 0, its graph is a *horizontal* line, as shown in Example 3. All points on a horizontal line have the same y-coordinate; that is, the y-values of such a line are constant for all x-values.

EXAMPLE 4

What is the slope of the line pictured on the following coordinate plane?

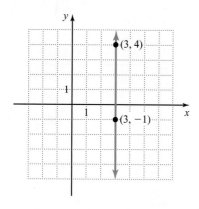

Solution

$$(3, -1) \quad (3, 4)$$
$$\downarrow \quad \downarrow \qquad \downarrow \quad \downarrow$$
$$x_1 \quad y_1 \qquad x_2 \quad y_2$$

$$m = \frac{y_2 - y_1}{x_2 - x_1} = \frac{4 - (-1)}{3 - 3} = \frac{5}{0}$$

Since division by 0 is undefined, the slope of this line is also undefined.

PRACTICE 4

Find the slope of the line that passes through the points $(-1, 5)$ and $(-1, 0)$.

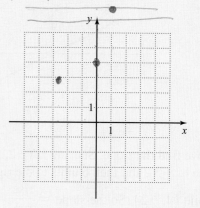

When the slope of a line is undefined, the graph is a *vertical* line, as shown in Example 4. All points on a vertical line have the same x-coordinate; that is, the x-values are constant for all y-values.

As shown in the last few examples, the sign of the slope of a line tells us a good deal about the line. As we continue our discussion of graphing lines, it will be helpful to keep the following graphs in mind.

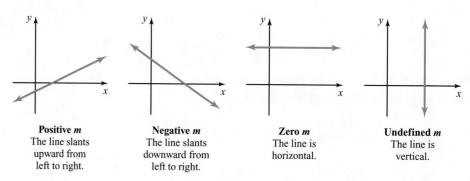

Positive *m*
The line slants
upward from
left to right.

Negative *m*
The line slants
downward from
left to right.

Zero *m*
The line is
horizontal.

Undefined *m*
The line is
vertical.

EXAMPLE 5

Calculate the slopes for the lines shown.

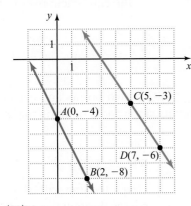

Solution For $\overleftrightarrow{AB}$, passing through $A(0, -4)$ and $B(2, -8)$,

$$m = \frac{y_2 - y_1}{x_2 - x_1} = \frac{-8 - (-4)}{2 - 0} = \frac{-4}{2} = -2.$$

For $\overleftrightarrow{CD}$, passing through $C(5, -3)$ and $D(7, -6)$,

$$m = \frac{y_2 - y_1}{x_2 - x_1} = \frac{-6 - (-3)}{7 - 5} = \frac{-3}{2} = -\frac{3}{2}.$$

Note that both lines have negative slopes and slant downward from left to right.

PRACTICE 5

Compute the slopes for the lines shown.

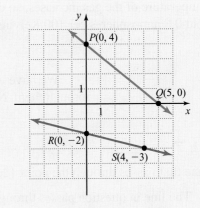

EXAMPLE 6

Students conduct an experiment on the gas contained in a sealed tube in a chemistry lab. The experiment is to heat the gas and then to measure the resulting pressure in the tube. In the lab manual, students plot points to show the gas pressure for various temperatures.

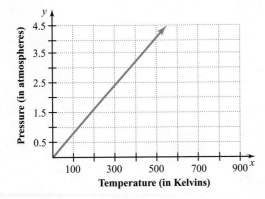

a. Is the slope of this line positive, negative, zero, or undefined?

b. In a sentence, explain the significance of your answer to part (a) in terms of temperature and pressure.

PRACTICE 6

To reduce their taxes, many businesses use *the straight-line method of depreciation* to estimate the change in the value over time of equipment that they own. The graph shows the value of equipment owned from the time of purchase to seven years later.

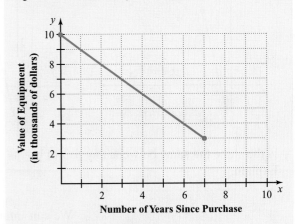

a. Is the slope of this line positive, negative, zero, or undefined?

EXAMPLE 6 (continued)

Solution

a. The slope of the line is positive, since it slants upward from left to right.

b. As the temperature of the gas increases, so does the pressure in the tube (about 0.8 atmospheres per 100 Kelvins).

b. In a sentence or two, explain the significance of your answer to part (a) in terms of the value of the equipment over time.

We have already graphed a line by plotting two points and drawing the line passing through them. Now, let's graph a line given its slope and a point on the line.

EXAMPLE 7

Graph the line that passes through the point $(2, 5)$ and has slope 4.

Solution The line in question passes through the point $(2, 5)$. But there are infinitely many such lines—which is the right one? We use the given slope 4 to find a second point through which the line also passes. Since 4 can be written as $\frac{4}{1}$, we have: Slope $= \dfrac{\text{rise}}{\text{run}} = \dfrac{4}{1}$. Starting at $(2, 5)$, we move 4 units up (for a rise of 4) and then 1 unit to the right (for a run of 1). We arrive at the point $(3, 9)$. Finally, we sketch the line passing through the points $(2, 5)$ and $(3, 9)$, as shown in the first graph that follows.

Since $\dfrac{4}{1} = \dfrac{-4}{-1}$, we could have started at $(2, 5)$ and moved down 4 units (for a rise of -4) and then 1 unit to the left (for a run of -1). In this case, we would arrive at $(1, 1)$, which is another point on the same line, as shown in the graph on the right.

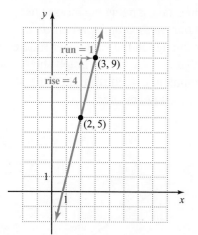

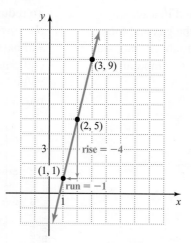

Is the slope the same between any pair of points on the graphed line? Explain.

PRACTICE 7

Graph the line with slope 5 that passes through the point $(1, -2)$.

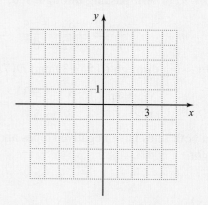

We can interpret slope as the rate of change of one variable with respect to another variable, typically y with respect to x. For example, if a graph shows the number of copies y a photocopier can produce during different times in x minutes, then that graph's slope represents the number of copies per minute. For a graph that shows the cost y of a taxi ride for trips of different lengths in x miles, the slope represents the cost per mile.

EXAMPLE 8

After a stock split, the value of the stock increased, as shown in the graph. Did the rate of increase change over time? Explain.

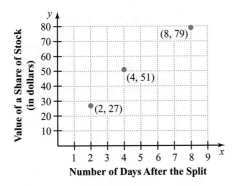

a. Find the rate of increase between the second and fourth days after the split.

b. Find the rate of increase between the fourth and eighth days after the split.

c. Were the two rates of increase the same?

Solution   The rate of increase is represented by the slope of the line passing through each pair of points.

a. The slope of the line that passes through the points $(2, 27)$ and $(4, 51)$ is $\dfrac{51 - 27}{4 - 2} = \dfrac{24}{2} = 12$. So the rate of increase between the second and fourth days after the split is $12 per day.

b. The slope of the line that passes through the points $(4, 51)$ and $(8, 79)$ is $\dfrac{79 - 51}{8 - 4} = \dfrac{28}{4} = 7$. So the rate of increase between the fourth and eighth days after the split is $7 per day.

c. Since $12 > 7$, the two rates of increase were not the same.

PRACTICE 8

A town was struck by a flu epidemic. Use the graph below to determine each of the following:

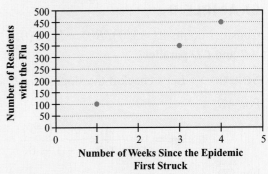

a. The rate of infection between the first and third weeks

b. The rate of infection between the third and fourth weeks

c. Which rate of infection was greater

Parallel and Perpendicular Lines

By examining the slopes of given lines on a coordinate plane, we can determine if the lines are

- parallel or
- perpendicular.

Let's consider *parallel lines* first.

Since the slope of a line is a measure of its slant, lines with equal slopes are parallel. On the coordinate plane shown here, if we know the coordinates of points P, Q, R, and S, we can verify that $\overleftrightarrow{PQ}$ and $\overleftrightarrow{RS}$ are parallel by computing their slopes and then checking whether they are equal.

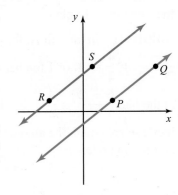

> **DEFINITION**
> Two nonvertical lines are **parallel** if and only if their slopes are equal. In other words, if two lines have slopes m_1 and m_2, they are parallel if and only if $m_1 = m_2$.

EXAMPLE 9

Consider points $A(-3, 0)$, $B(0, -5)$, $C(0, 5)$, and $D(3, 0)$. Are lines $\overleftrightarrow{AB}$ and $\overleftrightarrow{CD}$ parallel?

Solution   Let's check if the slopes of $\overleftrightarrow{AB}$ and $\overleftrightarrow{CD}$ are equal.
For $\overleftrightarrow{AB}$,

$$m = \frac{y_2 - y_1}{x_2 - x_1} = \frac{0 - (-5)}{-3 - 0} = \frac{5}{-3} = -\frac{5}{3}.$$

For $\overleftrightarrow{CD}$,

$$m = \frac{y_2 - y_1}{x_2 - x_1} = \frac{5 - 0}{0 - 3} = \frac{5}{-3} = -\frac{5}{3}.$$

Since the slopes of $\overleftrightarrow{AB}$ and $\overleftrightarrow{CD}$ are equal, the lines are parallel.

PRACTICE 9

Given points $E(0, 5)$, $F(4, -1)$, $G(0, 2)$, and $H(2, -2)$, decide whether lines $\overleftrightarrow{EF}$ and $\overleftrightarrow{GH}$ are parallel.

EXAMPLE 10

The following graph records the amount of garbage deposited in landfills A and B after they are opened.

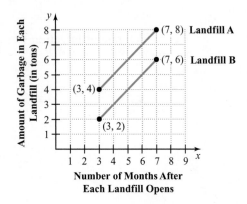

Are the amounts of garbage deposited at the two landfills growing at the same rate? Explain.

Solution   In landfill A, the garbage grew at the rate
$\dfrac{8 - 4}{7 - 3} = \dfrac{4}{4} = 1$, or 1 ton per month. The garbage in landfill B grew
at the rate $\dfrac{6 - 2}{7 - 3} = \dfrac{4}{4} = 1$, or 1 ton per month. Since the rates of
increase are equal, the amount of garbage in each landfill is growing at the same rate.

PRACTICE 10

The weights of a brother and of a sister at age 3 and at age 7, respectively, are recorded as shown. Did their weights increase at the same rate? Explain your response.

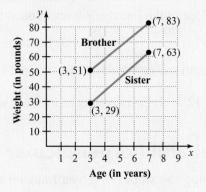

Now, let's turn to the question of *perpendicular lines*. Since the earliest days of building construction, people have been interested in determining whether two lines meet at right angles. When considering this problem on a coordinate plane, the key is to check the slopes of the two lines.

On a coordinate plane, it can be shown that two lines are perpendicular to one another when the product of their slopes is -1. For instance, $\overleftrightarrow{PQ}$ on the coordinate plane to the right has slope $\dfrac{7-2}{3-1} = \dfrac{5}{2}$, and $\overleftrightarrow{QR}$ has slope $\dfrac{2-0}{1-6} = \dfrac{2}{-5} = -\dfrac{2}{5}$. The product of these slopes is $\left(\dfrac{5}{2}\right)\left(-\dfrac{2}{5}\right) = -1$. Note that the slopes are *negative reciprocals* of each other.

Therefore, the two lines must be perpendicular to one another.

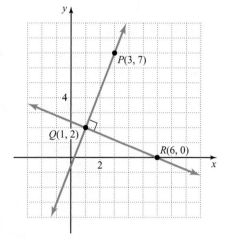

DEFINITION

Two nonvertical lines are **perpendicular** if and only if the product of their slopes is -1. In other words, if the two lines have slopes m_1 and m_2, they are perpendicular if and only if $m_1 \cdot m_2 = -1$.

Note that for perpendicular lines, the slopes m_1 and m_2 are negative reciprocals of each other, so $m_1 = -\dfrac{1}{m_2}$.

Can you explain why we added the condition that the lines be nonvertical in considering the slopes of perpendicular lines?

EXAMPLE 11

Determine if lines $\overleftrightarrow{AB}$ and $\overleftrightarrow{BC}$ are perpendicular to one another.

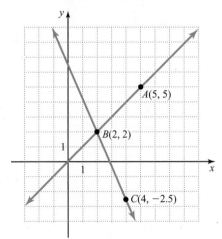

Solution   First, let's compute the slopes of the two lines in question. For $\overleftrightarrow{AB}$,

$$m = \frac{y_2 - y_1}{x_2 - x_1} = \frac{5 - 2}{5 - 2} = \frac{3}{3} = 1.$$

For $\overleftrightarrow{BC}$,

$$m = \frac{y_2 - y_1}{x_2 - x_1} = \frac{2 - (-2.5)}{2 - 4} = \frac{4.5}{-2} = -2.25.$$

To check if $\overleftrightarrow{AB}$ and $\overleftrightarrow{BC}$ are perpendicular, we find the product of their slopes:

$$(1)(-2.25) = -2.25$$

Since this product is not equal to -1, the lines are not perpendicular to one another.

PRACTICE 11

Consider points $A(1, 4)$, $B(2, 5)$, and $C(-1, 5)$. Plot these points on the following coordinate plane. Decide if line $\overleftrightarrow{AB}$ is perpendicular to line $\overleftrightarrow{AC}$.

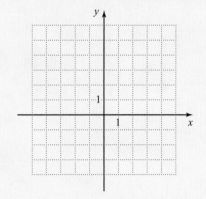

EXAMPLE 12

As shown below, a hiker is at point A and wants to take the shortest route through a field to reach a nearby road, represented by $\overline{BC}$ (read "line segment BC"). The shortest route will be to head perpendicular to the road. Is $\overline{AD}$ the shortest route? Explain.

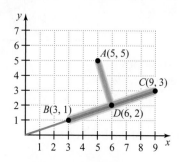

Solution   The slope of $\overline{AD}$ is $\dfrac{5-2}{5-6} = \dfrac{3}{-1} = -3$, and the slope of $\overline{BC}$ is $\dfrac{3-1}{9-3} = \dfrac{2}{6} = \dfrac{1}{3}$. Since these slopes are negative reciprocals, the line segments are perpendicular, so $\overline{AD}$ is the shortest route to the road.

PRACTICE 12

Consider the triangular shelf ABC shown below. For the shelf to fit in a closet, it must be shaped like a right triangle. Will the shelf fit? Explain.

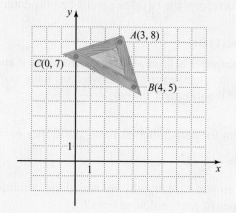

Mathematically Speaking

Fill in each blank with the most appropriate term or phrase from the given list.

vertical	parallel	decreasing	*y*-values to the change in *x*-values
perpendicular	horizontal	negative	*x*-values to the change in *y*-values
rise	slant	increasing	positive

1. The slope of a line is the ratio of the change in _____.

2. Slope can be calculated by dividing the _____ by the run.

3. A line falling from left to right is _____.

4. A line slanting up from left to right has a(n) _____ slope.

5. A line with a slope of 0 is _____.

6. A line with undefined slope is _____.

7. Two nonvertical lines are _____ if and only if their slopes are equal.

8. Two nonvertical lines are _____ if and only if the product of their slopes is −1.

Ⓐ *Compute the slope m of the line that passes through the given points. Then, plot the points and sketch the line.*

9. $(4, 0)$ and $(2, 2)$ $m =$

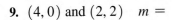
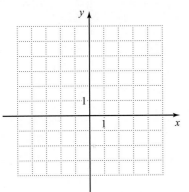

10. $(0, 0)$ and $(-4, 3)$ $m =$

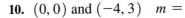
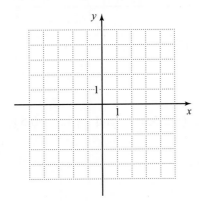

11. $(5, 3)$ and $(4, 1)$ $m =$

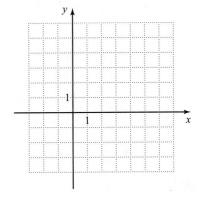

12. $(3, -1)$ and $(-1, -2)$ $m =$

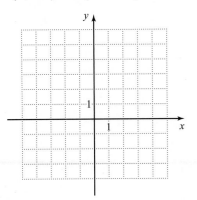

13. $(0, -4)$ and $(5, 0)$ $m =$

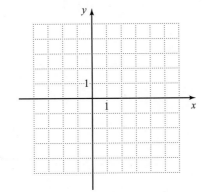

14. $(6, 3)$ and $(3, -2)$ $m =$

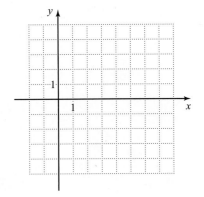

15. $(-1, 0)$ and $(-3, 0)$ $m =$

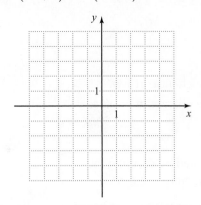

16. $(0, 1)$ and $(2, 1)$ $m =$

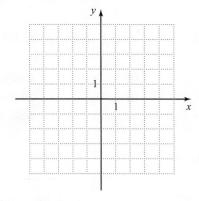

17. $(-4, 2)$ and $(-4, -2)$
$m =$

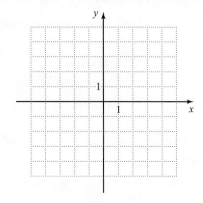

18. $(3, -3)$ and $(3, 1)$
$m =$

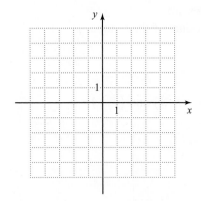

19. $(-1, 5)$ and $(3, -1)$
$m =$

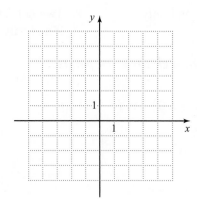

20. $(2, -3)$ and $(-2, 1)$
$m =$

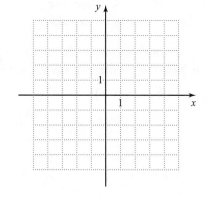

21. $(-0.5, -1.5)$ and $(0.5, -3)$
$m =$

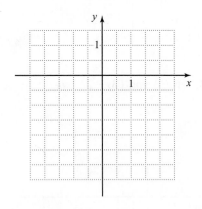

22. $(-2.5, -1)$ and $(-3.5, -0.5)$
$m =$

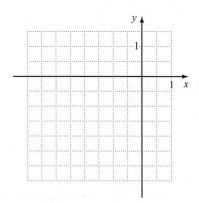

On each graph, calculate the slopes of the lines shown.

23.

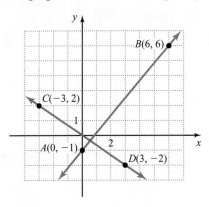

24.

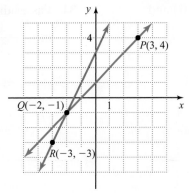

B *For each graph, identify whether the slope of the line is positive, negative, zero, or undefined.*

25.

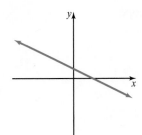

26.

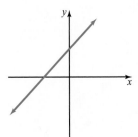

27.

28.

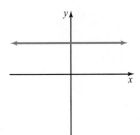

29.

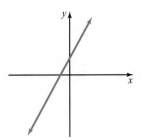

30.

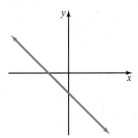

31.

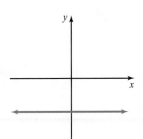

32.

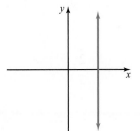

C *Graph the line on the coordinate plane using the given information.*

33. Passes through $(0, 0)$ and $m = -3$

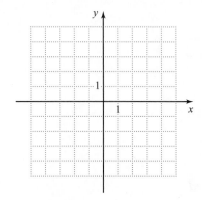

34. Passes through $(-1, -1)$ and $m = -\dfrac{3}{2}$

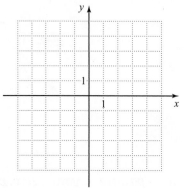

35. Passes through $(0, -3)$ and $m = 1$

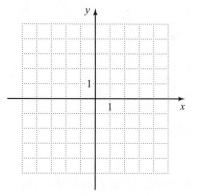

36. Passes through $(-2, 4)$ and $m = \dfrac{1}{4}$

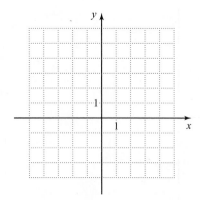

37. Passes through $(-3, -2)$ and $m = 0$

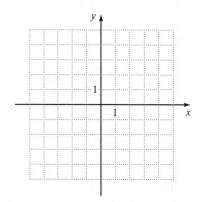

38. Passes through $(2, 2)$ and has undefined slope

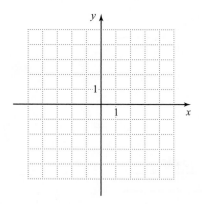

Find two additional points that lie on the line that passes through the given point and has slope m.

39. $(0, 4)$ $m = 2$

40. $(3, 0)$ $m = 1$

41. $(-5, 2)$ $m = -4$

42. $(6, -3)$ $m = -2$

43. $(-4, 1)$ $m = \dfrac{1}{2}$

44. $(-6, -5)$ $m = \dfrac{1}{3}$

45. $(-8, -1)$ $m = -\dfrac{3}{4}$

46. $(-10, 10)$ $m = -\dfrac{2}{3}$

47. $\left(\dfrac{1}{2}, -7\right)$ $m = 0$

48. $\left(-\dfrac{1}{2}, 9\right)$ $m = 0$

49. $(-2.4, 1)$ m is undefined.

50. $(3.6, -1.2)$ m is undefined.

D *Given points P, Q, R, and S, determine whether $\overleftrightarrow{PQ}$ and $\overleftrightarrow{RS}$ are parallel, perpendicular, or neither.*

51. $P(5, 1), Q(3, 2), R(8, 4),$ and $S(9, 6)$

52. $P(2, -2), Q(5, 2), R(7, 0),$ and $S(11, -3)$

53. $P(7, 2), Q(-9, 2), R(1, -4),$ and $S(4, -4)$

54. $P(0, -4), Q(-1, -10), R(-12, 6),$ and $S(-13, 0)$

55. $P(1.6, 3.2), Q(-2.4, -4.8), R(5.9, 0.5),$ and $S(0.9, 10.5)$

56. $P(2.7, -2.7), Q(6.7, 1.3), R(7.5, 5.5),$ and $S(2.5, 0.5)$

Mixed Practice

Solve.

57. Identify whether the slope of the line shown is positive, negative, zero, or undefined.

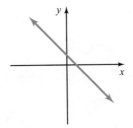

58. Compute the slope m of the line that passes through the points $(2, 5)$ and $(1, -4)$. Then, plot the points and sketch the line that passes through them.

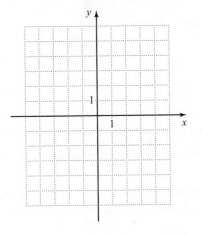

59. Given points $P(2, 1), Q(-2, 5), R(0, 3),$ and $S(1, 4),$ determine whether $\overleftrightarrow{PQ}$ and $\overleftrightarrow{RS}$ are parallel, perpendicular, or neither.

60. Consider the line that passes through $(0, 2)$ and has slope $m = 3$. Find two additional points that lie on the line.

61. On the coordinate plane, graph the line that passes through $(3, 1)$ and has $m = 0$.

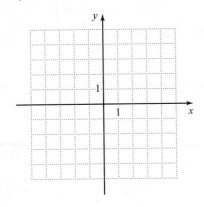

62. Calculate the slope of the line shown.

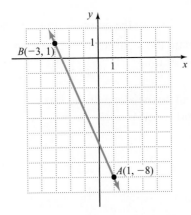

Applications

E *Solve.*

63. According to the *Accessibility Guidelines of the Americans with Disabilities Act*, the slope of a curb ramp adjoining a sidewalk and street cannot exceed $\frac{1}{20}$. Does the curb ramp shown in the figure meet this guideline? Explain.

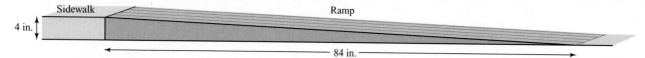

64. Approximate the slope of the roller coaster track shown.

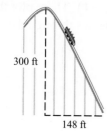

65. Most states have a standard sales tax rate that is applied to purchases. The graph displays the amount of sales tax charged on taxable purchases in Georgia.

   a. Is the slope of the line positive, negative, zero, or undefined?

   b. Explain the significance of your answer to part (a) in terms of the amount of a purchase and the amount of sales tax charged.

   c. Calculate the slope of the line and explain what it represents in the context of the problem.

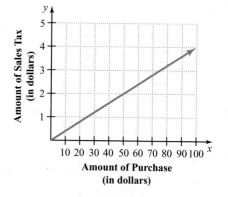

66. The graph shows the speed of a car with a constant acceleration after the brakes are applied.

   a. Is the slope of the graph positive, negative, zero, or undefined?

   b. Explain the significance of your answer to part (a) in terms of the velocity of the car over time.

   c. Calculate the slope of the line and explain its meaning in terms of the context of the problem.

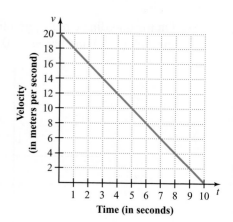

67. Two coffee shops sell a Kona blend coffee. The graph shows the cost of x pounds of Kona coffee at each coffee shop.

    a. What does the slope of each line represent?

    b. Which coffee shop charges more for the Kona coffee? Explain.

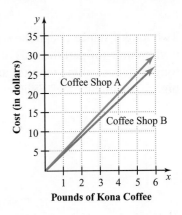

68. Two copy machines started a copy job at the same time. The lines show the number of copies produced over t minutes.

    a. What does the slope of each line represent?

    b. Which copier was slower? Explain.

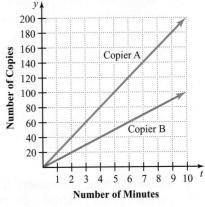

69. The height of a child is recorded at each of her annual checkups. The chart shows her height at various ages.

Point	A	B	C
Age (in years)	1	3	5
Height (in inches)	32	38	42

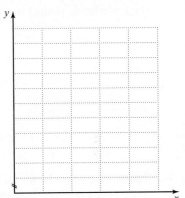

    a. Choose an appropriate scale and plot the points on the coordinate plane. Then, sketch line segments $\overline{AB}$ and $\overline{BC}$.

    b. Compute the slopes of $\overline{AB}$ and $\overline{BC}$.

    c. Was the child's rate of growth constant over the four years? Explain.

70. A company awards its employees annual raises based on seniority. The annual salary of an employee is shown in the following table.

Point	P	Q	R
Years Employed	2	4	6
Annual Salary (in dollars)	$34,200	$37,278	$40,796

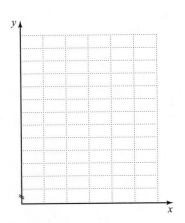

    a. Choose an appropriate scale and plot the points on the coordinate plane. Then, sketch $\overline{PQ}$ and $\overline{QR}$.

    b. Calculate the slopes of $\overline{PQ}$ and $\overline{QR}$.

    c. Was the amount of the annual raise constant over the four-year period? Explain.

71. Two long-distance telephone service providers charge a flat monthly fee plus a fee for each minute (or part thereof) of a long-distance call. The graph shows the number of long-distance calling minutes and the monthly phone bill. Which provider charges the higher per-minute fee? Explain.

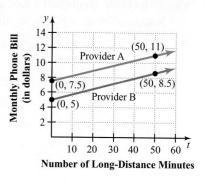

72. A weight-loss clinic provides diet and exercise recommendations for its clients. Two clients are placed on similar weight-loss plans. Their weights over a 12-week period are shown on the graph. Did the two clients lose weight at the same rate? Explain.

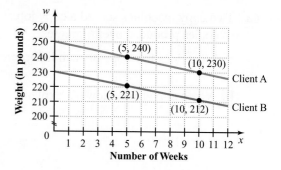

73. The rectangular glass insert of a patio table has to be replaced because of a crack. To ensure a perfect fit, the glass cutter must check that the new piece of glass (shown on the coordinate plane) is rectangular. Will the glass insert fit?

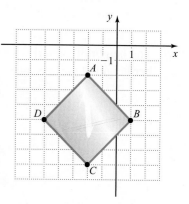

74. Three roads are shown on the map to the right. The road passing through points A and C goes due north–south. Does the road passing through points B and C go due east–west?

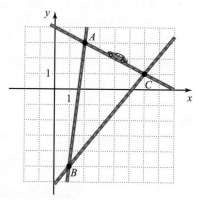

75. The graphs represent two different calling plans offered by a cell phone company. Use slope to describe each plan.

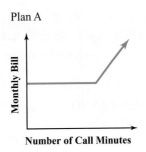

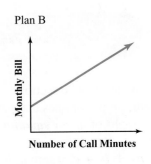

76. The graph shows an ISP (Internet service provider) plan. What does the slope tell you about the plan?

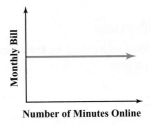

• Check your answers on page A-22.

MINDStretchers

Research

1. No one is certain why the letter *m* is used to represent slope. In your college library or on the web, identify a couple of theories that explain this notation. Summarize your findings.

Groupwork

2. Working with a partner, plot the points $A(2, 3)$, $B(-1, -1)$, and $C(-2, 4)$. Then, sketch $\overleftrightarrow{AB}$. Find a point D such that:

a. $\overleftrightarrow{CD}$ is parallel to $\overleftrightarrow{AB}$.

b. $\overleftrightarrow{CD}$ is perpendicular to $\overleftrightarrow{AB}$.

Compare your answers to those of other groups. Are your answers the same? How many choices are there for the point D in each case?

Mathematical Reasoning

3. Recall that slope can be defined as $m = \dfrac{\text{rise}}{\text{run}}$. Consider a line that has slope $m = 1$.

a. How does the slope of the line change if the rise is increased and the run remains unchanged?

b. How does the slope of the line change if the run is increased and the rise remains unchanged?

c. If the run is decreased and the slope remains unchanged, how does the rise change?

3.3 Graphing Linear Equations

In the first two sections of this chapter, we discussed plotting points on a coordinate plane and the slopes of lines that pass through a given pair of points. In this section, we will focus on the graphs of *linear equations in two variables*.

Recall from Chapter 2 that a linear equation in one variable is an equation of the form $ax + b = c$, where a, b, and c are real numbers and $a \neq 0$. We now consider *linear equations in two variables*.

> **DEFINITION**
>
> A **linear equation in two variables**, x and y, is an equation that can be written in the *general form* $Ax + By = C$, where A, B, and C are real numbers and A and B are not both 0.

Some examples of linear equations in two variables are

$$4x + 3y = 6 \qquad y = 2x + 1 \qquad 5x = y - 1$$

Solutions of Linear Equations in Two Variables

We know that a linear equation in one variable generally has a *single* real-number solution. In the equation $2x + 1 = 7$, for instance, the real number 3 is the only solution since it is the only number that satisfies the equation. Similarly, a *solution* of an equation in two variables is an ordered pair of numbers (x, y) that when substituted into the equation results in a true statement.

> **DEFINITION**
>
> A **solution** of an equation in two variables is an ordered pair of numbers that makes the equation true.

For instance, consider the equation $y = 2x + 1$. Here, the ordered pair $(1, 3)$ is a solution, since substituting 1 for x and 3 for y results in a true statement.

$$y = 2x + 1$$
$$3 = 2(1) + 1$$
$$3 = 3 \qquad \text{True}$$

Equations in two variables generally have an *infinite* number of solutions. For example, the ordered pairs $(2, 5)$ and $(0, 1)$ are also solutions of $y = 2x + 1$.

EXAMPLE 1

Determine whether each ordered pair is a solution of the equation $x + 4y = 6$.

a. $(2, 1)$ **b.** $(-3, 0)$

Solution

a. Substituting 2 for x and 1 for y, we get:

$$x + 4y = 6$$
$$2 + 4(1) \stackrel{?}{=} 6$$
$$6 = 6 \qquad \text{True}$$

So $(2, 1)$ is a solution.

PRACTICE 1

Determine whether each ordered pair is a solution of the equation $2x - 3y = 5$.

a. $(3, 2)$

b. $(1, -1)$

b. Substituting -3 for x and 0 for y gives us:

$$x + 4y = 6$$
$$-3 + 4(0) \stackrel{?}{=} 6$$
$$-3 = 6 \quad \text{False}$$

So $(-3, 0)$ is not a solution.

Now, let's consider how to find solutions of an equation in two variables. If we know the value of one of the variables, we can compute the value of the other. For instance, suppose that we want to find the solution of the equation $3x + y = 8$ for $x = 4$.

$$3x + y = 8$$
$$3 \cdot 4 + y = 8$$
$$12 + y = 8$$
$$y = 8 - 12$$
$$y = -4$$

So $x = 4$ and $y = -4$, or $(4, -4)$, is a solution.

Since an equation in two variables can have an infinite number of solutions, a table is a good way to organize and keep track of ordered pair solutions. For example, the table shows some additional solutions of the equation $3x + y = 8$.

x	y
-2	14
0	8
2	2
4	-4

Graphing Linear Equations

The graph of an equation, or, more precisely, the graph of its solutions, is a kind of picture of the equation.

> **DEFINITION**
>
> The **graph** of a linear equation in two variables consists of all points whose coordinates make the equation true.

To graph a linear equation in two variables, say x and y, we first solve the equation for y. Next, we find several solutions, keeping track in a table of the x- and y-values. Then, we plot the points on a coordinate plane. Finally, we sketch the line passing through the points.

EXAMPLE 2

Sketch the graph of $-3x + y = 1$.

Solution   We begin by solving the equation $-3x + y = 1$ for y.

$$-3x + y = 1$$
$$y = 3x + 1$$

Because y is isolated, we can easily compute y-values by substituting arbitrary values of x. For instance, suppose we let x be 0. To compute y, we substitute 0 for x:

$$y = 3x + 1 = 3 \cdot 0 + 1 = 0 + 1 = 1$$

So $x = 0$ and $y = 1$, or $(0, 1)$, is a solution to this equation.

PRACTICE 2

On the given coordinate plane, graph the equation $3x + y = -1$.

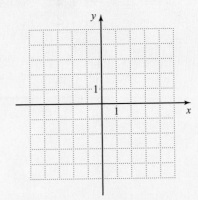

EXAMPLE 2 (continued)

In the same way, we can conclude that $(-1, -2)$ and $(1, 4)$ are solutions to the equation. Next, we enter these results in a table.

x	-1	0	1
y	-2	1	4

Then, we plot on a coordinate plane the three points $(-1, -2)$, $(0, 1)$, and $(1, 4)$, which should all lie on the same line. The graph of the equation $y = 3x + 1$ is the line passing through these points. So any point on this line satisfies the equation $-3x + y = 1$.

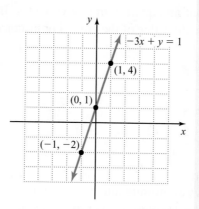

Example 2 suggests the following procedure for graphing a linear equation.

To Graph a Linear Equation in Two Variables

- Isolate one of the variables (usually y) if it is not already done.
- Choose three x-values, entering them into a table.
- Complete the table by finding the corresponding y-values.
- Plot the three points—two to draw the line and the third to serve as a *checkpoint*.
- Draw the line passing through the points.

EXAMPLE 3

Graph the equation $x - 2y = 4$.

Solution Solving the equation for y, we get:

$$x - 2y = 4$$
$$-2y = -x + 4 \qquad \text{Subtract } x \text{ from each side.}$$
$$y = \frac{-x + 4}{-2} \qquad \text{Divide each side by } -2.$$
$$y = \frac{1}{2}x - 2 \qquad \text{Simplify.}$$

PRACTICE 3

Graph the equation $x + 3y = -9$.

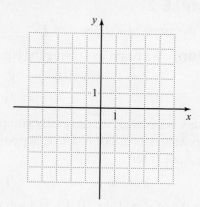

Now, we choose x-values and find the corresponding y-values.

x	$y = \frac{1}{2}x - 2$	(x, y)
0	$y = \frac{1}{2}(0) - 2 = -2$	$(0, -2)$
2	$y = \frac{1}{2}(2) - 2 = -1$	$(2, -1)$
4	$y = \frac{1}{2}(4) - 2 = 0$	$(4, 0)$

Finally, we plot the points and draw the line that passes through them.

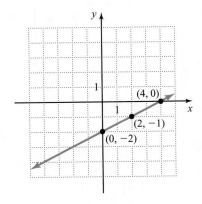

Another way to sketch the graph of an equation written in general form is to plot the points where the graph intersects the x- and y-axes. These points are called the *intercepts* of the graph.

The following graph shows a line passing through two points, $(0, 4)$ and $(3, 0)$, which are both intercepts.

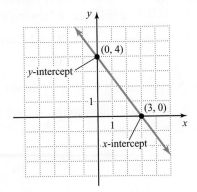

Intercepts are easy to plot and can save work in drawing the graph of a line. Note that an x-intercept lies on the x-axis, so that its y-value must be 0. Similarly, the x-value of a y-intercept must be 0.

EXAMPLE 4

Find the intercepts of the graph of the equation $x + 3y = 6$. Then, graph the equation.

Solution   Since the y-intercept has x-value 0, we let $x = 0$, and then solve the given equation for y.

$$x + 3y = 6$$
$$0 + 3y = 6$$
$$y = 2$$

So the y-intercept is $(0, 2)$.

Similarly, the x-intercept has y-value 0. So we let $y = 0$, and then solve for x.

$$x + 3y = 6$$
$$x + 3(0) = 6$$
$$x = 6$$

So the x-intercept is $(6, 0)$.

We find one additional point, $(-3, 3)$, as a checkpoint. Can you show that $(-3, 3)$ is a point on the graph of $x + 3y = 6$?

We plot the points $(0, 2)$, $(6, 0)$, and $(-3, 3)$ on a coordinate plane and draw a line through the points, getting the desired graph.

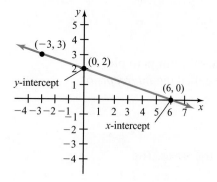

PRACTICE 4

Find the x- and y-intercepts of the graph of the equation $2x - 4y = -8$. Then, graph the equation.

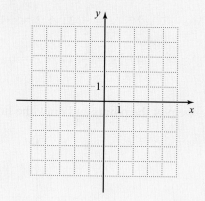

We know that in its general form a linear equation in two variables is written as $Ax + By = C$. In some linear equations, one of the two variables is missing; that is, the coefficient of one of the two variables is 0. Consider the following equations:

$$0x + y = 6, \quad \text{where } A = 0, B = 1, \text{ and } C = 6$$

and

$$x + 0y = 5.8, \quad \text{where } A = 1, B = 0, \text{ and } C = 5.8$$

Note that the equation $0x + y = 6$ is equivalent to the equation $y = 6$. Likewise, the equation $x + 0y = 5.8$ is equivalent to the equation $x = 5.8$. Let's look at the graphs of these equations.

EXAMPLE 5

Graph and find the slope for each.

a. $y = 6$ **b.** $x = 5.8$

Solution

a. For the line $y = 6$, the coefficient of the x-term is 0. All points on this line will have the same y-value, 6, but different x-values. The graph is as shown.

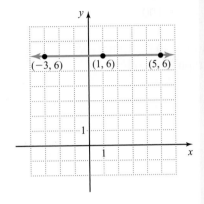

Since the graph of $y = 6$ is a horizontal line, recall that its slope is 0.

b. For the line $x = 5.8$, the coefficient of the y-term is 0. All points on this line will have the same x-value, 5.8, but different y-values. The graph is as shown.

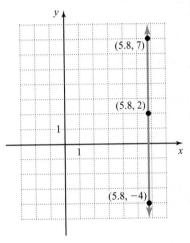

Since the graph of $x = 5.8$ is a vertical line, its slope is undefined.

PRACTICE 5

On the given coordinate plane, graph

a. $y = -1$ **b.** $x = -2.5$

and find the slope for each.

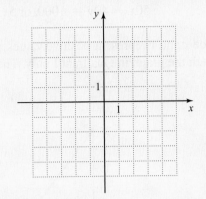

EXAMPLE 6

On the first leg of a trip, a truck driver drove for x hr at a constant speed of 50 mph. On the second leg, he drove for y hr consistently at 40 mph. In all, he drove 1000 mi.

a. Translate this information to an equation.

b. Choose an appropriate scale and graph the equation found in part (a).

c. What are the x- and y-intercepts of this graph? Explain their significance in terms of the trip.

d. Find the slope of the line. Explain whether you would have expected the slope to be positive or negative, and why.

PRACTICE 6

The coins in a cash register, with a total value of $2, consist of d dimes and n nickels.

a. Represent this relationship as an equation.

b. On the next page, choose an appropriate scale and graph the equation found in part (a).

EXAMPLE 6 (continued)

Solution

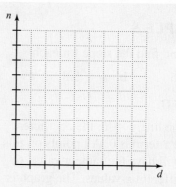

a. Since Rate · Time = Distance, we know that the distance driven in the first leg is $50x$ mi and the distance driven in the second leg is $40y$ mi. The total distance driven is 1000 mi. So we conclude:

$$50x + 40y = 1000, \text{ or } 5x + 4y = 100$$

b. Since x and y represent positive values, we need to plot points and sketch the graph in Quadrant I only. The graph is as follows:

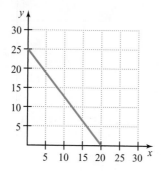

c. Explain the significance of the two intercepts in this context.

c. For the x-intercept, $y = 0$. Substituting in the equation and solving for x gives us $x = 20$. For the y-intercept, $x = 0$. Substituting and solving for y gives us $y = 25$. The x-intercept $(20, 0)$ indicates that the trip would have taken 20 hr if the truck driver had driven the entire trip at 50 mph. The y-intercept $(0, 25)$ indicates that the trip would have taken 25 hr if the truck driver had driven the entire trip at 40 mph.

d. What is the slope of the line? Would you have expected the slope to be positive or negative? Explain.

d. We know that the points $(20, 0)$ and $(0, 25)$ lie on the graph. The slope is the difference in y-values divided by the difference in x-values:

$$m = \frac{25 - 0}{0 - 20} = \frac{25}{-20} = -\frac{5}{4}$$

We could have predicted that the slope would be negative because the given conditions imply that for larger x-values we get smaller y-values; that is, the line has a negative slope.

Mathematically Speaking

Fill in each blank with the most appropriate term or phrase from the given list.

x-value	three x-values	three points
graph	y-intercept	x-intercept
y-value	set	
ordered pair	table	

1. A solution of an equation in two variables is a(n) _____ of numbers that makes the equation true.

2. The _____ of a linear equation in two variables consists of all points whose coordinates make the equation true.

3. One way of graphing a linear equation in two variables is to plot _____.

4. The _____ of a graph is the point where the graph intersects the x-axis.

5. The _____ of a graph is the point where the graph intersects the y-axis.

6. The _____ of an x-intercept must be 0.

A *Determine if the given ordered pair is a solution of the given equation.*

7. $(4, -1); y = \frac{1}{4}x$

8. $(1, 3); y = 3x$

9. $(-2, 5); y = -3x - 1$

10. $(-2, 2); y = \frac{1}{2}x + 3$

11. $(-3, -7); 2x - 3y = 15$

12. $(4, -5); 3x - 2y = 2$

13. $(0.5, -1.4); 10y + 6x = 11$

14. $(2.6, 3.5); 4y - 5x = 1$

15. $(-0.1, 8); y = 8$

16. $(-6, 1.2); y = -6$

Match each equation to its graph.

17. $y = 2x - 1$

18. $y = -2x + 1$

19. $y = -2x - 1$

20. $y = 2x + 1$

a.

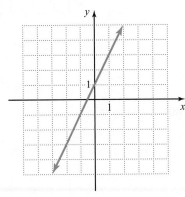

b.

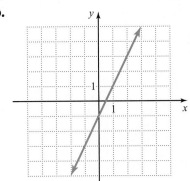

(continued)

199

c.

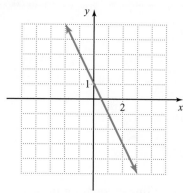

d.

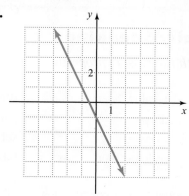

B *Find the x- and y-intercepts of the graph of each equation.*

21. $y = x - 4$

22. $y = x + 2$

23. $y = \frac{1}{3}x + 3$

24. $y = -\frac{1}{2}x - 1$

25. $4x + 2y = -8$

26. $3x + 5y = 30$

27. $5x - 2y = 10$

28. $4x - 3y = -12$

29. $8y - 6x + 2 = 0$

30. $6y - 9x - 12 = 0$

31. $10y = -15$

32. $-7y = 21$

33. $-3x = -18$

34. $16x = 12$

C *Graph each equation.*

35. $y = x + 2$

36. $y = x - 1$

37. $y = -3x - 4$

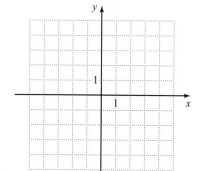

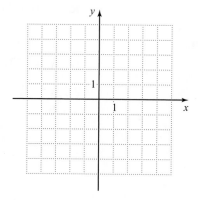

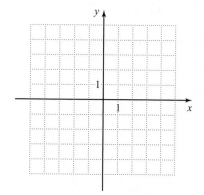

38. $y = -2x + 3$

39. $2x - y = 4$

40. $4x - y = 5$

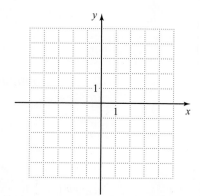

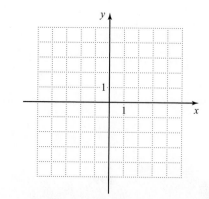

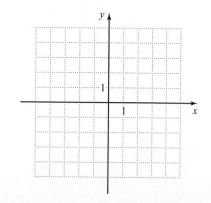

41. $x + 4y = 12$

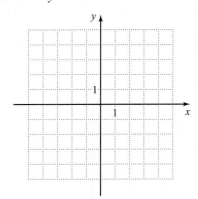

42. $x + 2y = 8$

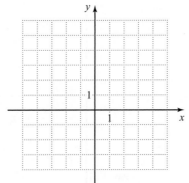

43. $3x + 3y = -6$

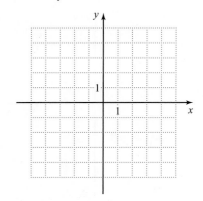

44. $5x + 5y = -20$

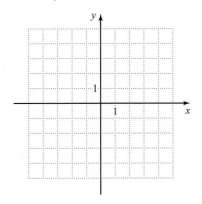

45. $2x - 3y = -3$

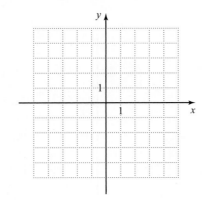

46. $3x - 2y = -4$

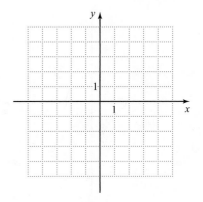

47. $4y - 2x = -8$

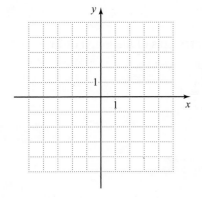

48. $9y - 3x = -9$

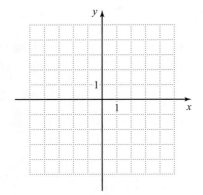

49. $4x + 5y - 10 = 0$

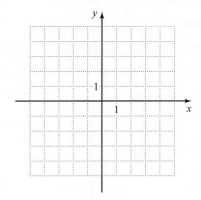

50. $5x + 4y - 16 = 0$

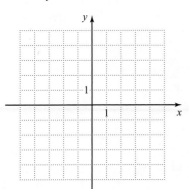

51. $20 - 10y = 0$

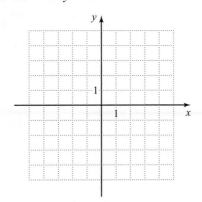

52. $6y - 18 = 0$

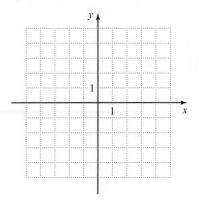

53. $9x = -27$

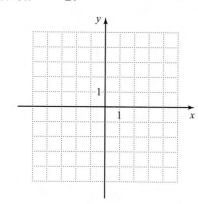

54. $-7x = -7$

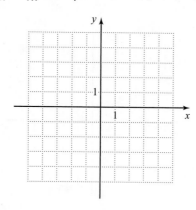

55. $1.4x - 0.35 = 0.7y$

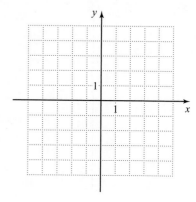

56. $1.6y + 8 = 4.8x$

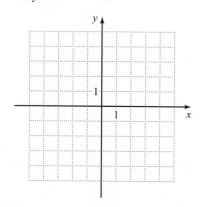

57. $0.5y + 1 = -0.25x$

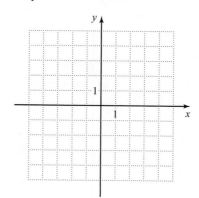

58. $0.5x = 0.2y - 0.6$

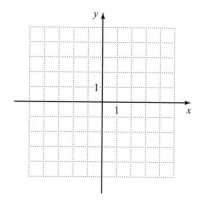

Mixed Practice

Solve.

59. Determine if $(3, -5)$ is a solution of $y = x - 2$.

60. Find the *x*- and *y*-intercepts of the graph of $2x - y = 12$.

61. Graph the equation $y = -2x + 3$.

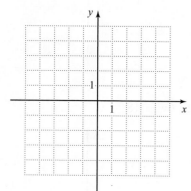

62. Graph $-4x + 4y = 8$.

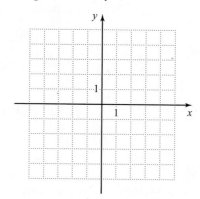

Applications

D *Solve.*

63. One thousand dollars is deposited into a savings account earning 4% simple interest annually. If no additional deposits are made, the amount in the account A (in dollars) after t years is given by

$$A = 40t + 1000.$$

a. Complete the table.

t	0	2	4	6
A				

b. Choose an appropriate scale for the axes and then graph the equation.

c. What does the A-intercept represent in the context of this problem?

d. From the graph, find the number of years it takes for the amount in the account to grow to $1200.

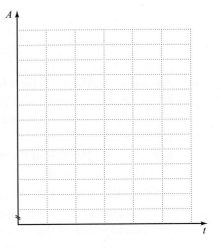

64. A manufacturing company has fixed monthly costs of $500 plus $0.75 for each unit x of a product it produces. If the revenue from each unit sold is $3.25, then the manufacturer's profit P (in dollars) for selling x units is given by

$$P = 2.5x - 500.$$

a. Complete the table.

x	0	100	200	300	400
P					

b. Choose an appropriate scale and then graph the equation.

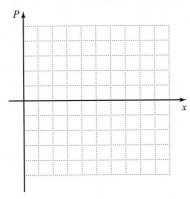

c. What do the x- and P-intercepts represent in the context of this problem?

d. From the graph, determine the number of units that must be sold per month in order to have a profit of $400.

65. A daycare center charges a $50 enrollment fee plus $200 per week for childcare services.

   a. Express the amount A paid to the daycare center in terms of the number of weeks w of childcare services.

   b. Graph the equation on the coordinate plane.

   c. Explain why it makes sense to consider the portion of the graph that lies only in Quadrant I.

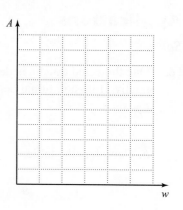

66. A patient's medical bill for a hospital stay totaled $12,000. Of this amount, her insurance company paid $9000. To pay the balance, she works out a payment plan with the hospital's billing office in which she makes monthly payments of $250.

   a. Express the balance of the hospital bill B in terms of the number of months m of payments made.

   b. Graph the equation on the coordinate plane.

   c. What are the m- and B-intercepts of the graph? Explain their significance in terms of the balance of the hospital bill.

67. A subscription for an online news magazine costs $10 per month. Write an equation that represents the cost c of the subscription for m months. Then, graph the equation on the coordinate plane.

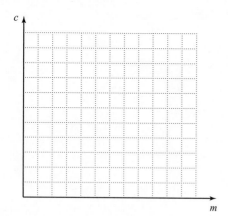

68. A driver sets his cruise control to 65 mph. Write an equation that represents the driver's speed s in terms of the length of time t driven. Then, graph the equation on the coordinate plane.

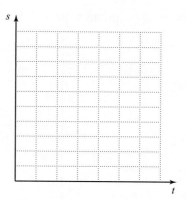

69. A local cable company charges $99 to install a high-speed Internet connection plus a $45 monthly service fee.

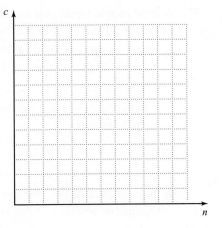

   a. Express the cost c of a high-speed Internet connection (in dollars) in terms of the number of months n of service.

   b. Choose an appropriate scale and graph the equation on the coordinate plane.

   c. From the graph, estimate the total cost for 6 mo of high-speed Internet service.

70. A businessman invested x dollars in a fund earning 6% simple annual interest and y dollars in a fund earning 5% simple annual interest. After one year, the total interest earned on the investments was $1500.

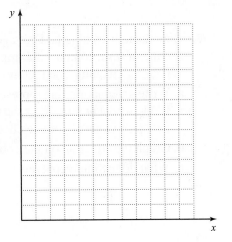

   a. Translate this information into an equation.

   b. Graph the equation on the coordinate plane.

   c. Identify the x- and y-intercepts of the graph. What do the intercepts represent in the context of this problem?

   d. From the graph, estimate the amount invested at 6% if $12,000 is invested at 5%.

71. A college awards associate's degrees and bachelor's degrees. At last year's commencement, the college awarded A associate's degrees and B bachelor's degrees. A total of 1800 degrees were awarded.

   a. Translate this information into an equation.

   b. Choose an appropriate scale and graph the equation.

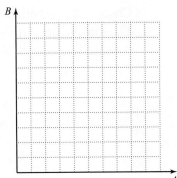

   c. Identify the A- and B-intercepts of the graph. What do they represent in terms of the number of degrees awarded?

   d. In a sentence or two, explain why only some of the points on the graph in Quadrant I are reasonable solutions of this problem.

72. A local movie theater charges $6 per ticket for all shows before 5:00 P.M. and $10 per ticket for all shows after 5:00 P.M. Ticket sales on Saturday amounted to $31,500.

    a. Translate this information into an equation, letting x represent the number of $6 tickets sold and y represent the number of $10 tickets sold.

    b. Graph the equation on the coordinate plane.

    c. What are the x- and y-intercepts of the graph? Explain their significance in terms of the ticket sales.

    d. In a sentence or two, explain why every point on the graph in Quadrant I is not a reasonable solution of this problem.

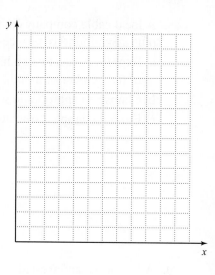

73. A dietician uses milk and cottage cheese as the sources of calcium in her diet. One serving of milk contains 300 mg of calcium, and one serving of cottage cheese contains 100 mg of calcium. The recommended daily amount (RDA) of calcium is 1000 mg.

    a. Write an equation that represents the total calcium from both sources to reach the RDA, where m represents the number of servings of milk and c represents the number of servings of cottage cheese.

    b. Graph the equation on the coordinate plane.

    c. What do the intercepts represent in the context of this problem?

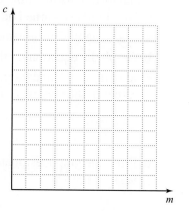

74. A gym offers a special membership price of $35 per month for unlimited use of the gym. The initiation fee for the gym membership is $300.

    a. Express the total cost c of a gym membership for t mo.

    b. Choose an appropriate scale and then graph the equation.

    c. From the graph, estimate the number of months of membership if the total cost is $650.

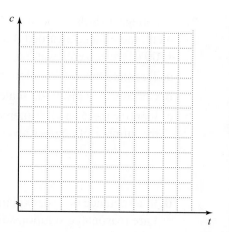

• Check your answers on page A-24.

MINDStretchers

Writing

1. Explain the difference between the solutions of a linear equation of the form $ax + b = c$, where $a \neq 0$, and a linear equation of the form $Ax + By = C$, where A and B are not both 0. Discuss how the graphs of the solutions of each equation compare.

Groupwork

2. Some equations in two variables are not linear. Below are graphs of two nonlinear equations. Working with a partner, identify the x- and y-intercepts of each of the following graphs and show that they satisfy the equation.

   a. $y = -x^2 + 4$ **b.** $x^2 + y^2 = 16$

   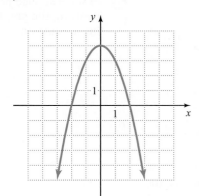 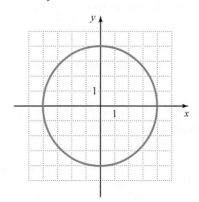

Mathematical Reasoning

3. For each statement, determine whether it is always true, sometimes true, or never true. Justify your answer.

   a. The graph of a linear equation has both an x- and a y-intercept.

   b. The lines $x = a$ and $y = b$, where a and b are real numbers, are perpendicular.

   c. The graph of a linear equation has more than one x-intercept.

   d. It's possible for the graph of a linear equation to have neither an x-intercept nor a y-intercept.

3.4 More on Graphing Linear Equations

Having discussed the general form of a linear equation in the previous section, we now move on to consider two other forms: the *slope-intercept form* and the *point-slope form*.

Slope-Intercept Form

Recall from Section 3.3 that one approach to graphing a linear equation written in general form is to isolate y:

$$-3x + y = 1 \qquad \text{General form}$$
$$y = 3x + 1 \qquad \text{Solve for } y.$$

The linear equation $y = 3x + 1$ is said to be in *slope-intercept form*.

> **DEFINITION**
>
> A linear equation in two variables is in **slope-intercept form** if it is written as
>
> $$y = mx + b,$$
>
> where m and b are real numbers. In this form, m is the slope of the line and $(0, b)$ is the y-intercept of the graph of the equation.

Slope-intercept is the most commonly used form in graphing a linear equation because the slope and y-intercept of the graph can be directly identified from the equation. That is, the slope m is the coefficient of x, and the y-coordinate of the y-intercept is the constant term b in the equation. This form is also convenient in finding the equation of a line when we know both its slope and y-intercept.

To derive the slope-intercept form, consider the line passing through an arbitrary point (x, y) with y-intercept $(0, b)$:

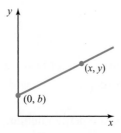

Since slope is the ratio of the change in y-values to the change in x-values, the slope formula gives us:

$$m = \frac{y - b}{x - 0} \quad \begin{matrix} \longleftarrow \text{Change in } y\text{-values} \\ \longleftarrow \text{Change in } x\text{-values} \end{matrix}$$

Solving for y, we get:

$$m = \frac{y - b}{x}$$

$$mx = y - b \qquad \text{Multiply each side by } x.$$

$$mx + b = y \qquad \text{Add } b \text{ to each side.}$$

or

$$y = mx + b,$$

which is the slope-intercept form of the equation corresponding to the desired graph.

The following table gives examples of equations in slope-intercept form:

Equation	Slope m	y-intercept $(0, b)$
$y = \dfrac{1}{2}x + 1$	$\dfrac{1}{2}$	$(0, 1)$
$y = 3x - 1$, or $y = 3x + (-1)$	3	$(0, -1)$
$y = 4x$, or $y = 4x + 0$	4	$(0, 0)$
$y = 2$, or $y = 0x + 2$	0	$(0, 2)$

EXAMPLE 1

Express $5x + 3y = 2$ in slope-intercept form.

Solution Since the slope-intercept form of an equation is $y = mx + b$, we need to solve the given equation for y.

$$5x + 3y = 2$$

$$3y = -5x + 2$$

$$\frac{3y}{3} = \frac{-5}{3}x + \frac{2}{3}$$

$$y = -\frac{5}{3}x + \frac{2}{3}$$

So $y = -\dfrac{5}{3}x + \dfrac{2}{3}$ is the equation in slope-intercept form,

where m is $-\dfrac{5}{3}$ and b is $\dfrac{2}{3}$.

PRACTICE 1

Express $3x - 2y = 1$ in slope-intercept form.

When the equation of a line is written in slope-intercept form, we can sketch the graph of the equation using the slope and y-intercept.

To Graph a Linear Equation in Two Variables Using the Slope and y-Intercept

- Plot the y-intercept.
- Use the slope to find a second point on the line.
- Draw a line through the two points.

EXAMPLE 2

Graph $y = -\frac{3}{2}x + 4$ using the slope and y-intercept.

Solution Since $y = -\frac{3}{2}x + 4$ is in slope-intercept form, the slope

is $-\frac{3}{2}$, and the y-intercept is $(0, 4)$. To graph the equation, we first plot

the y-intercept $(0, 4)$. Since the slope $-\frac{3}{2}$ equals $\frac{-3}{2}$, from the point

$(0, 4)$ we move *down* 3 units and then 2 units to the *right* to find the
second point $(2, 1)$. Then, we draw the line through the points $(0, 4)$
and $(2, 1)$, as shown in the following graph:

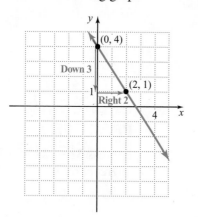

PRACTICE 2

Use the slope and y-intercept to graph
$y = -\frac{1}{2}x - 2$.

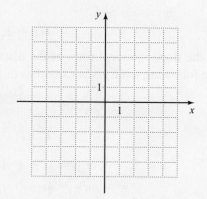

Note that the slope in Example 2 could also be written as $\frac{3}{-2}$, so from the point $(0, 4)$
we could have moved *up* 3 units and then 2 units to the *left*. We would have gotten the second
point $(-2, 7)$, and then plotted and drawn a line through $(0, 4)$ and $(-2, 7)$ to obtain the graph.
Sometimes we need to find the equation of a line when given its slope and y-intercept.

EXAMPLE 3

Find the equation of the line with slope $-\frac{2}{5}$ and with y-intercept
$(0, -1)$.

Solution We are given that the slope is $-\frac{2}{5}$ and the y-intercept

is $(0, -1)$. So $m = -\frac{2}{5}$ and $b = -1$. Substituting in the slope-intercept

form, we get

$$y = \quad mx \ + \quad b$$
$$\qquad \updownarrow \qquad\quad \updownarrow$$
$$y = -\frac{2}{5}x + (-1), \text{ or } y = -\frac{2}{5}x - 1.$$

PRACTICE 3

A line on a coordinate plane has slope
-2 and intersects the y-axis 3 units
above the origin. Write its equation in
slope-intercept form.

EXAMPLE 4

Find the equation of the line that is parallel to the graph of $y = 6x + 1$ and that has a y-intercept of $(0, 3)$.

Solution   The given line, $y = 6x + 1$, is in slope-intercept form. Its slope is, therefore, 6. We know that parallel lines have the same slope, so the line we want also has slope m equal to 6. Since its y-intercept is $(0, 3)$, b equals 3. Therefore, the desired equation is

$$y = 6x + 3.$$

PRACTICE 4

What is the equation of the line parallel to the graph of $y = -6x + 4$ with y-intercept $(0, -2)$?

EXAMPLE 5

What is the equation of the line that is perpendicular to the graph of $y = 8x - 1$ and that has a y-intercept of $(0, 1)$?

Solution   The given line $y = 8x - 1$ is in slope-intercept form, so its slope is 8. We know that if two lines are perpendicular, their slopes are negative reciprocals of each other. Therefore, the slope of any line perpendicular to the graph of $y = 8x - 1$ is $-\dfrac{1}{8}$. So m equals $-\dfrac{1}{8}$.

We are given a y-intercept of $(0, 1)$, so we know $b = 1$. The desired equation is

$$y = -\frac{1}{8}x + 1.$$

PRACTICE 5

Find the equation of the line that is perpendicular to the graph of $y = 3x + 2$ and that has a y-intercept of $(0, 0)$.

EXAMPLE 6

Students studying archaeology know that when the femur bone of an adult female is unearthed, a good estimate of her height is 73 more than double the length of the bone. (Both the length of the femur f and the height h are in centimeters.)

a. Express this relationship as a formula.

b. Graph this relationship.

c. Use the graph to estimate the height of an adult female whose femur was 35 cm in length.

Solution

a. The height h can be approximated by 73 more than double the length of the femur f. So we can write the formula

$$h = 2f + 73.$$

PRACTICE 6

A young couple buys furniture for $2000, agreeing to pay $200 down and $100 at the end of each month until the entire debt is paid off.

a. Write an equation to express the amount that remains to be paid off P in terms of the number of monthly payments m.

EXAMPLE 6 (continued)

b. Using the *y*-intercept $(0, 73)$ and the slope 2, we graph the line.

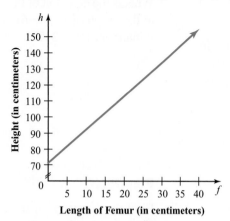

Length of Femur (in centimeters)

c. We see that the point on the graph with first coordinate 35 has second coordinate approximately 145. So our estimate is that a woman with a 35-centimeter femur was about 145 cm tall.

b. Graph the equation found in part (a).

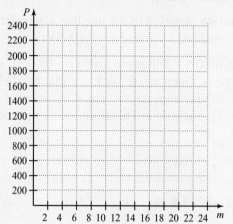

c. Use the graph to estimate how long it will take to pay off the debt.

Point-Slope Form

The third form of a linear equation in two variables that we will discuss is called the *point-slope form*.

> **DEFINITION**
>
> The **point-slope form** of a linear equation in two variables is written as
>
> $$y - y_1 = m(x - x_1),$$
>
> where x_1, y_1, and m are real numbers. In this form, m is the slope, and (x_1, y_1) is a point that lies on the graph of the equation.

This form, although used less frequently than other forms of linear equations, is particularly useful for finding the equation of a line in two situations:

- when the slope and a point on the line are known.
- when two points on the line are known.

To derive the point-slope form, consider the line passing through the points (x, y) and (x_1, y_1).

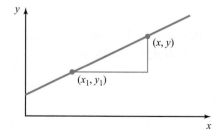

Since slope is the ratio of the change in y-values to the change in x-values, the slope formula gives us

$$m = \frac{y - y_1}{x - x_1}$$

$$m(x - x_1) = y - y_1 \qquad \text{Multiply each side by } (x - x_1).$$

or

$$y - y_1 = m(x - x_1),$$

which is the point-slope form of the equation corresponding to the graph.

EXAMPLE 7

A line with slope -3 passes through the point $(-1, 8)$. Write the equation of this graph in point-slope form.

Solution We are given a point on the line and its slope. An easy way to find the equation of the line is to substitute directly into the point-slope formula, where $x_1 = -1$, $y_1 = 8$, and $m = -3$:

$$y - y_1 = m(x - x_1)$$
$$y - 8 = -3[x - (-1)]$$
$$y - 8 = -3(x + 1)$$

We can leave this equation in point-slope form, or we can simplify the equation and write it in general form:

$$y - 8 = -3x - 3$$
$$3x + y = 5 \qquad \text{General form}$$

Or if we like, we can change the form of this equation to the more common slope-intercept form:

$$y = -3x + 5 \qquad \text{Slope-intercept form}$$

PRACTICE 7

A line passing through the point $(2, 5)$ has slope 4. Write its equation in point-slope form.

EXAMPLE 8

What is the equation of the line passing through the points $(3, -2)$ and $(-3, 0)$?

Solution Since we know the coordinates of two points on the line, we can find its slope:

$$m = \frac{y_2 - y_1}{x_2 - x_1} = \frac{-2 - 0}{3 - (-3)} = \frac{-2}{6} = -\frac{1}{3}$$

The line with slope $-\frac{1}{3}$ passing through point $(-3, 0)$ is:

$$y - y_1 = m(x - x_1)$$

$$y - 0 = -\frac{1}{3}[x - (-3)]$$

$$y - 0 = -\frac{1}{3}(x + 3) \qquad \text{Point-slope form}$$

$$y = -\frac{1}{3}x - 1 \qquad \text{Slope-intercept form}$$

PRACTICE 8

Find the equation in point-slope form of the line passing through $(-6, -2)$ and the origin.

EXAMPLE 9

Pressure under water increases with depth. The pressure P on an object and its depth d are modeled by a linear equation. The pressure at sea level is 1 atmosphere (atm), whereas the pressure 33 ft below sea level is 2 atm. Write the equation relating P in atmospheres and d in feet.

Solution   We know that P and d are related by a linear equation. At sea level, that is, at $d = 0$, $P = 1$. At $d = -33$, $P = 2$. So the points $(0, 1)$ and $(-33, 2)$ lie on the graph of the desired equation. Since we know two points on the graph, let's find the equation in point-slope form. The slope of the line can be approximated as follows:

$$m = \frac{2 - 1}{-33 - 0} = \frac{1}{-33} \approx -0.03$$

Using the point-slope form $P - P_1 = m(d - d_1)$ with the point $(0, 1)$ gives us

$$P - 1 = -0.03(d - 0),$$

which can also be written as $P - 1 = -0.03d$, or, alternatively, as $P = -0.03d + 1$.

PRACTICE 9

The length of a heated object and the temperature of the object are related by a linear equation. A rod at 0°C is 10 m long, and at 25°C is 10.1 m long. Write an equation for length l in terms of temperature T.

Using a Calculator or Computer to Graph Linear Equations

Calculators with graphing capabilities and computers with graphing software allow us to graph equations, even those with complicated coefficients, at the push of a key. Although they vary somewhat in terms of features and commands, these machines all graph the equation of your choice on a coordinate plane.

  To graph an equation, begin by making certain that the equation is in slope-intercept form. On many "graphers," pressing the $\boxed{Y =}$ *key results in a window being displayed on which you enter the equation. For instance, if you wanted to graph* $3x - y = 4$, *you would first solve for y, resulting in* $y = 3x - 4$, *and then enter* $3x - 4$ *to the right of* $\backslash Y1 =$ *on the screen. Pressing the* $\boxed{GRAPH}$ *key displays a coordinate plane in which the graph of* $y = 3x - 4$ *is sketched.*

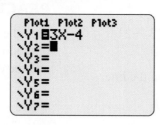

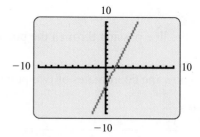

  Many graphers have a TRACE feature that highlights a point on the graph and displays its coordinates. As you hold down an arrow key, the coordinates change as the highlighted point moves along the graph.

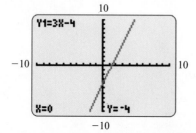

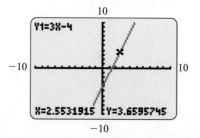

TIP The $\boxed{\textbf{WINDOW}}$ key allows you to set the range and scales for the axes. Before you graph an equation, be sure to set the viewing window in which you would like to display the graph.

EXAMPLE 10

Graph $x + 2y = 6$. Then, use the **TRACE** feature to identify the y-intercept.

Solution First, solve for y: $y = -\dfrac{1}{2}x + 3$. Next, press $\boxed{\textbf{Y} =}$ and enter $-(1/2)x + 3$ to the right of $\backslash\textbf{Y1} =$. Set the viewing window in which you want to display the graph. Then, press $\boxed{\textbf{GRAPH}}$ to display the graph of the equation. If the graph of $y = -\dfrac{1}{2}x + 3$ does not appear, check your grapher's instruction manual.

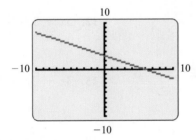

With the **TRACE** feature, move the cursor along the graph until it rests on the y-axis. The displayed coordinates of this y-intercept are $x = 0$ and $y = 3$.

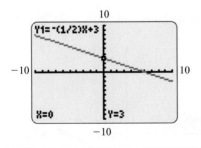

PRACTICE 10

Graph $x - 3y = 3$, and then find the y-intercept with the **TRACE** feature.

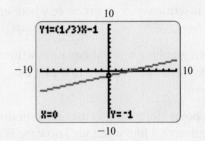

Mathematically Speaking

Fill in each blank with the most appropriate term or phrase from the given list.

point-slope form	y-intercept	x-intercept
slope	two points on the line	
slope-intercept form	one point on the line	

1. A linear equation in two variables is in _____ if it is written as $y = mx + b$, where m and b are real numbers.

2. In the slope-intercept form of an equation, m represents the _____.

3. When graphing a linear equation written in the slope-intercept form, we first plot the _____.

4. The _____ of a linear equation in two variables is written as $y - y_1 = m(x - x_1)$, where x_1, y_1 and m are real numbers.

5. The point-slope form is useful for finding the equation of a line when we know the line's slope and _____.

6. The point-slope form is useful for finding the equation of a line when we do not know its slope, but we do know _____.

A *Write each equation in slope-intercept form. Then, identify the slope and y-intercept.*

7. $y - 2x = 0$

8. $y + 4x = 0$

9. $12x + 3y = -18$

10. $14x - 7y = 28$

11. $3x - 6y = 6$

12. $5x + 15y = -30$

13. $10y - 8x = 100$

14. $12y - 9x = 60$

15. $3x + 2y - 1 = 0$

16. $4x - 3y - 2 = 0$

17. $y - 5 = -3(x - 8)$

18. $y + 3 = 3(x - 6)$

Match each equation to its graph.

19. $3x - 6y = 12$

20. $6y - 3x = 12$

21. $6x + 2y = -4$

22. $6x - 2y = 4$

a.
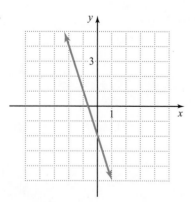

b.
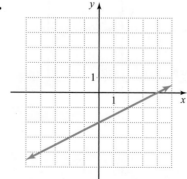

c.

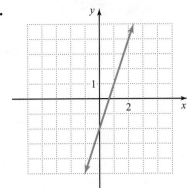

d.

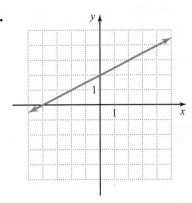

B *Graph each equation using the slope and y-intercept.*

23. $y = 2x + 3$

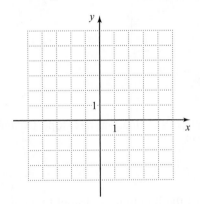

24. $y = 3x - 2$

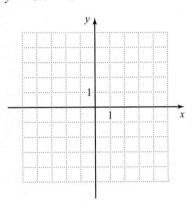

25. $y = -\dfrac{1}{3}x - 1$

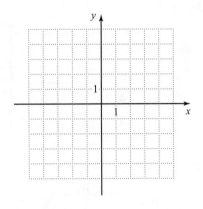

26. $y = -\dfrac{1}{2}x + 4$

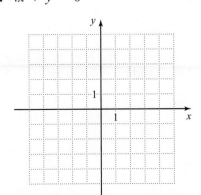

27. $2x - 3y = 9$

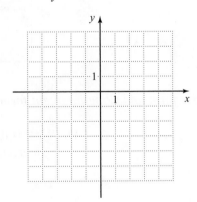

28. $3x - 2y = 4$

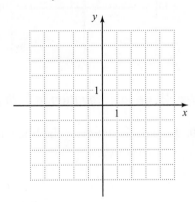

29. $4x + y = 0$

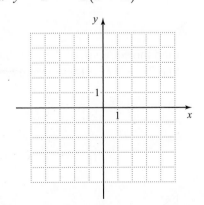

30. $x - 3y = 0$

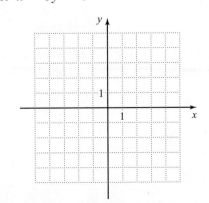

31. $y - 2 = -2(x + 2)$

32. $y + 1 = -(x - 3)$

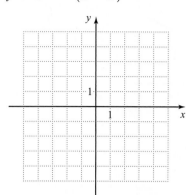

33. $y = 0.5x + 1.5$

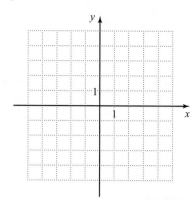

34. $y = -0.5x - 2.5$

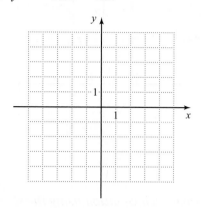

Graph each equation using a graphing calculator.

35. $y = -1.2x - 0.357$

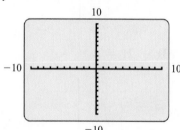

36. $y = -3.4x + 1.12$

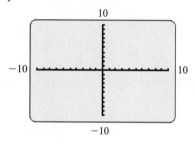

37. $y = 0.875x + 2.013$

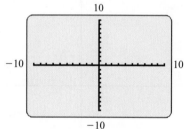

38. $y = 2.044x - 0.0362$

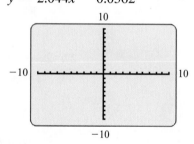

C *Solve. Write each equation in both point-slope form and slope-intercept form.*

39. Find the equation of the line with slope $-\dfrac{3}{5}$ and y-intercept $(0, 8)$.

40. What is the equation of the line that has slope 1.6 and y-intercept $(0, -7)$?

41. What is the equation of the line that is parallel to the line $3x + 9y = 18$ and has y-intercept $(0, -4)$?

42. Find the equation of the line that is parallel to the line $5y = -10x$ and has x-intercept $(1, 0)$.

43. Find the equation of the line that is perpendicular to the line $2y - 4x - 14 = 0$ and has x-intercept $(-5, 0)$.

44. What is the equation of the line that is perpendicular to the line $x - 3y + 3 = 0$ and has y-intercept $(0, -6)$?

45. What is the equation of the vertical line that passes through the point $(-2.3, 0.7)$?

46. What is the equation of the horizontal line that passes through the point $(3.1, -0.7)$?

47. Find the equation of the line with slope -1 that passes through the point $(3, 2)$.

48. Find the equation of the line with slope 3 that passes through the point $(-1, 4)$.

49. What is the equation of the line that has slope $\frac{3}{2}$ and passes through the point $(-6, 5)$?

50. What is the equation of the line that has slope $-\frac{1}{5}$ and passes through the point $(15, -8)$?

51. Find the equation of the line that is parallel to $2x + 4y = 1$ and passes through the point $(2, -1)$.

52. What is the equation of the line that is parallel to $9x - 3y = 2$ and passes through the point $(-1, -4)$?

53. What is the equation of the line that is perpendicular to $5x - 4y = 12$ and passes through the point $(-10, -6)$?

54. Find the equation of the line that is perpendicular to $7 - 8y = 6x$ and passes through the point $(3, -5)$.

55. Find the equation of the line that passes through the points $(3, 7)$ and $(5, 9)$.

56. Find the equation of the line that passes through the points $(1, 4)$ and $(4, 1)$.

Find the equation of each graph.

57.

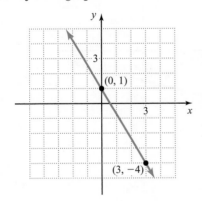

58.

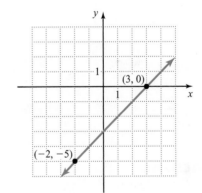

59.

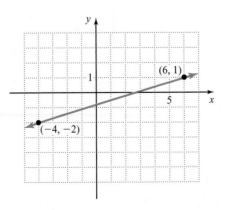

60.

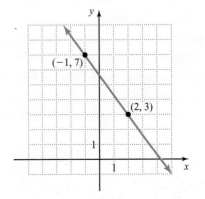

Mixed Practice

Solve.

61. What is the equation of the line that is perpendicular to $2x + y - 5 = 0$ and passes through the point $(2, 4)$?

62. Graph $y - 1 = -3(x + 1)$ using the slope and y-intercept.

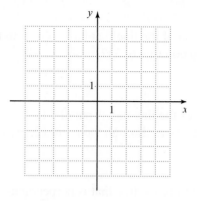

63. Write the equation $x + 4y = 12$ in slope-intercept form. Then, identify the slope and y-intercept.

64. Find the equation of the line that is parallel to $5x - 2y = 12$ and passes through the point $(6, 4)$.

65. Find the equation of the line with slope $-\dfrac{2}{3}$ and y-intercept $(0, -1)$.

66. Find the equation of the following graph.

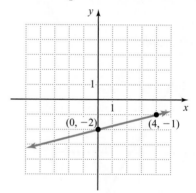

Applications

D *Solve.*

67. A medical group charges patients an initial consultation fee plus \$40 for each additional consultation. A patient pays a total of \$220 for three additional consultations.

a. Write an equation in slope-intercept form to show the total amount A (in dollars) that a patient pays for x additional consultations.

b. Graph the equation on the coordinate plane.

c. What does the A-intercept represent in this situation?

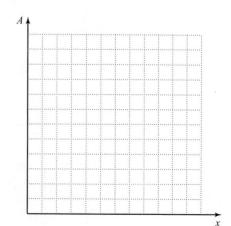

68. A real-estate agent receives a flat monthly salary plus a 0.5% commission on her monthly home sales. In a particular month, her home sales were $500,000 and her total monthly income was $4300.

a. Write an equation in slope-intercept form that shows the real-estate agent's total monthly income I in terms of her monthly home sales s.

b. Graph the equation on the coordinate plane.

c. What does the I-intercept represent in the context of the problem?

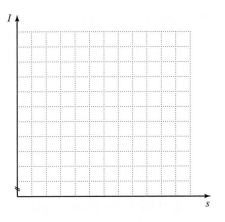

69. The following graph shows the relationship between a Fahrenheit temperature F and the equivalent Celsius temperature C.

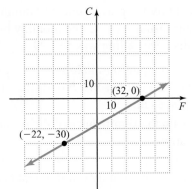

a. Find the equation of the line in point-slope form.

b. If hydrogen chloride boils at $-85°C$, use the equation from part (a) to find the equivalent Fahrenheit temperature.

70. The following graph shows the relationship between women's clothing sizes in the United States x and women's clothing sizes in France y.

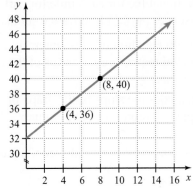

a. Find an equation of the line in slope-intercept form.

b. The size on the label of a dress at a clothing store in France is 44. What is the equivalent dress size in the United States?

71. An object is tossed upward with an initial velocity of 84 ft/sec. Two seconds later, the velocity of the object is 20 ft/sec. After 4 sec, its velocity is -44 ft/sec. The velocity v and the time t are modeled by a linear equation.

a. Write an equation in slope-intercept form that shows the velocity of the object v after t sec.

b. Graph the equation on the coordinate plane.

c. What is the meaning of the slope in this situation?

d. Explain the significance of the t-intercept in the context of the problem.

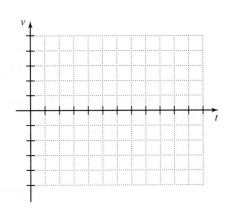

72. One year after purchase, a piece of machinery in a factory has an assessed value of $13,200. After 4 yr, the value of the machinery is assessed at $7800. The value of the machinery V and the time t are modeled by a linear equation.

    a. Write an equation in point-slope form to show the value V of the machinery after t yr. Then, express the equation in slope-intercept form.

    b. Graph the equation on the coordinate plane.

    c. Explain the meaning of the slope in this situation.

    d. What is the V-intercept of the graph? Explain its meaning in the context of the problem.

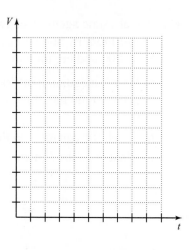

73. A taxi company charges a flat meter fare of $2.50 plus an additional fee for each mile (or part thereof) driven. A passenger pays $11.50 for a 6-mi taxi ride.

    a. Find an equation in slope-intercept form that models the total meter fare f in terms of the number of miles driven m.

    b. What does the slope of the graph of the equation in part (a) represent in this situation?

    c. How many miles were driven if a passenger pays $22?

74. A long-distance telephone service provider charges a flat monthly fee plus 8 cents per minute (or part thereof) for long-distance calls. The monthly bill for 100 min of long-distance calling was $17.99.

    a. Find an equation in slope-intercept form that shows the total monthly bill b in dollars for x min of long-distance calls.

    b. Explain the significance of the b-intercept in this situation.

    c. What is the monthly bill if a person makes 180 min of long-distance calls?

75. Thirty minutes after a truck driver passes the 142-mile marker on a freeway, he passes the 170-mile marker. Find an equation that shows the distance d he drives in t hr.

76. After 2 yr, the amount in a savings account earning simple interest was $1070. After 5 yr, the amount in the account was $1175. Find an equation that represents the amount in the account A after t yr.

77. At its inception, a professional organization had 26 members. Three years later, the organization had grown to 83 members. If membership continues to grow at the same rate, find a linear equation that represents the number of members in the organization n after t years.

78. Twelve years after opening, a software company had grown from 8 employees to 112 employees. If the company continues to grow at the same rate, find a linear equation that shows the number of employees y after t yr.

• Check your answers on page A-25.

MINDStretchers

Groupwork

1. The following graph is composed of line segments $\overline{AB}$, $\overline{BC}$, $\overline{CD}$, and $\overline{DE}$. Working with a partner, find the equation of the line on which each segment lies. Then, state the values of x for which the line segment is graphed.

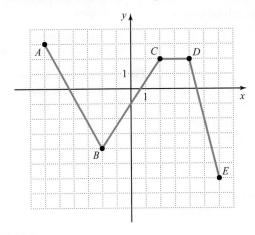

Technology

2. Consider the linear equation $4x - 8 = 0$ in one variable.

   a. Solve the equation for x.

   b. Using a graphing calculator or computer with graphing software, graph $y = 4x - 8$. Then, identify the x-intercept of the graph.

   c. What conclusion can you draw from your answers to parts (a) and (b)?

   d. Solve the equation $1 - \dfrac{1}{3}x = 0$ both algebraically and graphically.

Mathematical Reasoning

3. The general form of a linear equation is $Ax + By = C$.

   a. Write this equation in slope-intercept form.

   b. Identify the slope and y-intercept of the graph of this equation in terms of A, B, and C.

3.5   Graphing Linear Inequalities

Recall that in Section 2.3, we discussed how to graph inequalities in *one variable* on a number line. We saw that each solution to an inequality such as $x \leq 2$ is a real number that when substituted for the variable makes the inequality true. We also saw that the graph of this inequality, the set of all solutions to the inequality, is an interval on the number line.

In this section, we discuss graphing inequalities in *two variables*, such as $3x + y \geq 2$, on a coordinate plane.

DEFINITION

A **linear inequality in two variables** is an inequality that can be written in the form $Ax + By < C$, where A, B, and C are real numbers and A and B are not both 0. The inequality symbol can be $<$, $>$, $\leq$, or $\geq$.

Inequalities in two variables arise from many situations. For example, suppose that the number of full-time students f and part-time students p enrolled in a college is capped at 2500. We can represent this situation by the following inequality:

$$f + p \leq 2500$$

As with solutions of linear equations in two variables, the solutions of linear *inequalities* in two variables are ordered pairs.

DEFINITION

A **solution of a linear inequality in two variables** is an ordered pair of real numbers that when substituted for the variables makes the inequality a true statement.

EXAMPLE 1

Determine whether $(-2, 4)$ is a solution of $2x + y < 7$.

Solution   Substituting -2 for x and 4 for y, we get:

$$2x + y < 7$$
$$2(-2) + 4 \stackrel{?}{<} 7$$
$$0 < 7 \quad \text{True}$$

Since the statement $0 < 7$ is true, $(-2, 4)$ is a solution of the inequality.

PRACTICE 1

Is $(3, -1)$ a solution of $4x - 5y > 2$?

The graph of an inequality is the set of all points in the coordinate plane whose coordinates satisfy the inequality. For instance, consider the inequality $y \geq x + 2$. To graph this inequality, we first graph the corresponding equation $y = x + 2$. Since the inequality involves the $\geq$ symbol, we graph this equation as a solid line. The graph of this equation is called the *boundary line*. This line cuts the plane into two *half-planes*.

Next, we take an arbitrary point on either side of the boundary line—a *test point*. If the coordinates of the test point satisfy the inequality, then the half-plane in which the test point lies is the desired graph. Otherwise, the desired graph is the other half-plane. Can you explain why the test point must *not* be on the boundary line?

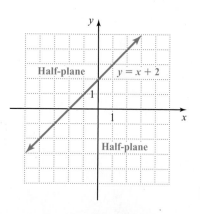

Suppose that we take $(4, 0)$ as our test point:

$$y \geq x + 2$$
$$0 \geq 4 + 2$$
$$0 \geq 6 \qquad \text{False}$$

Since the inequality results in a false statement for the point $(4, 0)$, the half-plane that we want is the region above the graph of $y = x + 2$. So the graph of $y \geq x + 2$ is the boundary line and the shaded region above. See the graph below on the left.

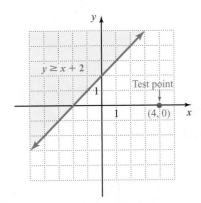 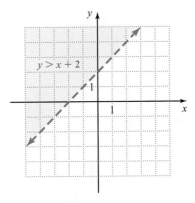

Suppose the inequality was $y > x + 2$. Since the inequality involves the $>$ symbol, which means *strictly* greater than and never equal to, the boundary line is not part of the graph. We indicate that the points on the boundary line are not solutions by drawing a broken line. See the graph above on the right.

To Graph a Linear Inequality in Two Variables

- Graph the corresponding linear equation. For an inequality that involves either the symbol $\leq$ or the symbol $\geq$, draw a solid line; for an inequality with either the symbol $<$ or the symbol $>$, draw a broken line. This line is the boundary between two half-planes.

- Choose a test point in either half-plane, and substitute the coordinates of this point in the inequality. If the resulting inequality is satisfied, then the graph of the inequality is the half-plane containing the test point. If it is not satisfied, then the other half-plane is the graph. In either case, a solid line is part of the graph, a broken line is not. Shade the appropriate region.

EXAMPLE 2

Graph the inequality $y - x < 3$.

Solution First, we graph the corresponding equation $y - x = 3$. Solving for y gives $y = x + 3$. Since the original inequality involves the symbol $<$, we draw a broken line. Then, we choose a test point in either half-plane, say $(0, 0)$. We substitute $x = 0$ and $y = 0$ into the inequality:

$$y - x < 3$$
$$0 - 0 < 3$$
$$0 < 3 \qquad \text{True}$$

PRACTICE 2

Graph the inequality $y - x \geq 6$.

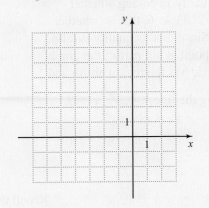

EXAMPLE 2 (continued)

Since the inequality results in a true statement, the half-plane containing the test point is part of the graph. So the half-plane below the line is our graph, and we shade it in. Note that the graph does not include the boundary line since the given inequality involves the symbol $<$.

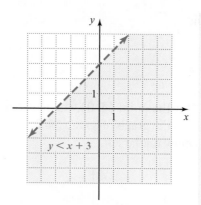

EXAMPLE 3

Graph: $x < 1$

Solution The boundary line, $x = 1$, is the vertical line 1 unit to the right of the y-axis. We need to select a test point on either side of the boundary line. Let's take the point $(0, 0)$ as the test point. Substituting into the inequality $x < 1$, we get $0 < 1$, which is true. So the graph is the half-plane to the left of the line $x = 1$, not including the line.

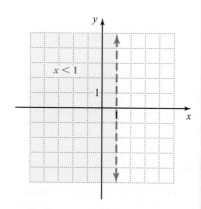

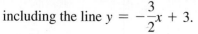

PRACTICE 3

Find the graph of $y \geq -2$.

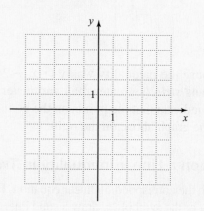

EXAMPLE 4

Graph the inequality $3x + 2y \geq 6$.

Solution The corresponding equation is $3x + 2y = 6$. Solving for y gives us $y = -\dfrac{3}{2}x + 3$. Next, we graph this line. We draw a solid line, since the original inequality symbol is $\geq$.

Taking $(0, 0)$ as the test point, we substitute into the original inequality, checking whether $3(0) + 2(0) \geq 6$, that is, whether $0 \geq 6$. Since this inequality is false, the test point is not part of the graph. So the graph is the region above and including the line $y = -\dfrac{3}{2}x + 3$.

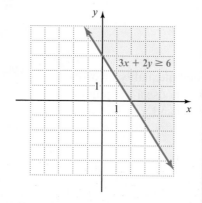

PRACTICE 4

Graph the inequality $2x + 4y < 8$.

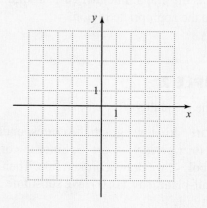

Recall that in Section 2.3 we discussed common word phrases such as "at least" and "at most" that translate to inequalities.

EXAMPLE 5

A plane is carrying bottled water and medicine to victims of an earth-quake. Each bottle of water weighs 5 lb and each container of medicine weighs 20 lb. The plane can carry at most 60,000 lb in cargo.

a. Express this weight limitation as an inequality in terms of the number of bottles of water w and the number of medicine containers m on the plane.

b. Graph this inequality.

c. Use this graph to decide whether the plane can carry 1000 containers of medicine and 2000 bottles of water.

Solution

a. The bottles of water on the plane will weigh $5w$ lb, and the containers of medicine will weigh $20m$ lb. The plane's cargo can weigh at most 60,000 lb; that is, the weight of the cargo is less than or equal to this limitation, which can be represented by the following inequality:

$$5w + 20m \leq 60{,}000$$

b. The corresponding equation is $5w + 20m = 60{,}000$. To graph this equation, we solve for w in terms of m so as to avoid fractions:

$$w = -4m + 12{,}000$$

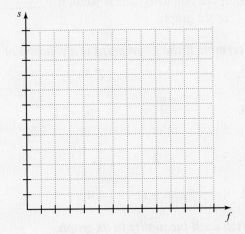

In this situation, all values of w and m must be nonnegative, so we graph the inequality only in the first quadrant. Next, we substitute the test point $(0, 0)$ for (m, w) in the original inequality:

$$5w + 20m \leq 60{,}000$$
$$5 \cdot 0 + 20 \cdot 0 \leq 60{,}000$$

Since it is true that $0 \leq 60{,}000$, the following triangular region including its three sides is the graph of the inequality.

c. To determine whether the plane can carry 1000 containers of medicine and 2000 bottles of water, we plot the point $(1000, 2000)$. Since this point lies within the graph of the inequality, the plane can carry this cargo.

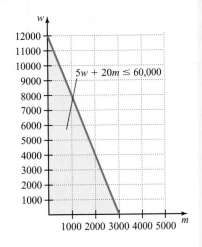

PRACTICE 5

A student has two part-time jobs. At the first job, the student makes $8 per hr and works for f hr in a week. At the second job, he makes $10 per hr and works for s hr in a week. Between the two jobs, he needs to earn at least $200 per wk to pay expenses.

a. Write an inequality that shows the number of hours that the student can work at each job in a week.

b. Graph this inequality.

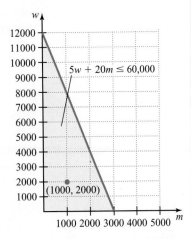

c. Give a few examples of the number of hours that the student can work at each job.

Mathematically Speaking

Fill in each blank with the most appropriate term or phrase from the given list.

solid	is not	test point	is
linear inequality	broken	linear equation	

1. An example of a(n) _____ in two variables is $2x + 3y \leq 6$.

2. To graph a linear inequality in two variables, we first graph the corresponding _____.

3. A(n) _____ boundary line is drawn when graphing a linear inequality that involves the symbol $\leq$ or the symbol $\geq$.

4. A(n) _____ boundary line is drawn when graphing a linear inequality that involves the symbol $<$ or the symbol $>$.

5. If the boundary line is solid, it _____ part of the graph.

6. If the boundary line is broken, it _____ part of the graph.

Ⓐ *Determine if the given point is a solution of the given inequality.*

7. $y > 5 - 2x$ $(2, 3)$

8. $y < 3x - 8$ $(4, -5)$

9. $y \leq \dfrac{1}{2}x + 1$ $(-6, -2)$

10. $y \geq -\dfrac{1}{3}x + 4$ $(-3, 5)$

11. $3x + 2y < 2$ $\left(\dfrac{2}{3}, \dfrac{1}{2}\right)$

12. $4x - 3y > -1$ $\left(\dfrac{3}{4}, 2\right)$

13. $8y - 5x \geq -11$ $(4, 1)$

14. $6y - 9x \leq 13$ $(-4, -6)$

Ⓑ *Match each inequality to its graph.*

15. $x - 3y > 9$

16. $3x - 9y < 27$

17. $6x + 9y \geq -9$

18. $2x + 3y \leq 3$

a.

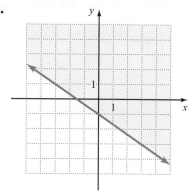

b.

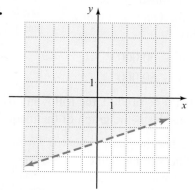

c.

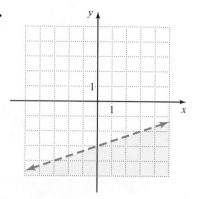

d.

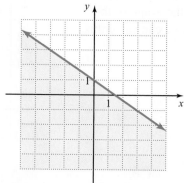

Graph each inequality.

19. $y > 2x$

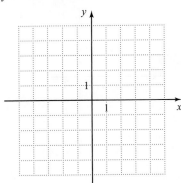

20. $y > -2x$

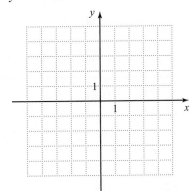

21. $y < x - 4$

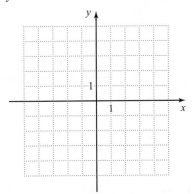

22. $y < 1 - x$

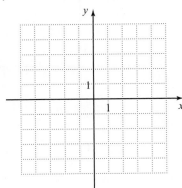

23. $y \geq \dfrac{1}{2}x + 3$

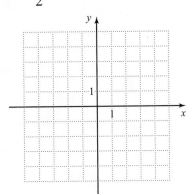

24. $y \geq -\dfrac{1}{4}x - 1$

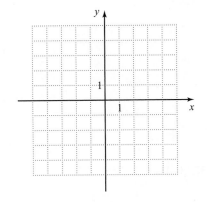

25. $y \leq 2 - 3x$

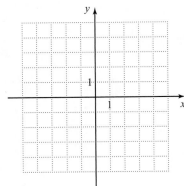

26. $y \leq 3x - 4$

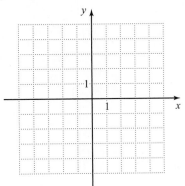

27. $x < -1.5$

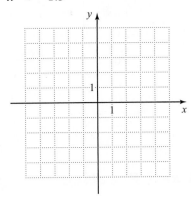

28. $x \geq 2$

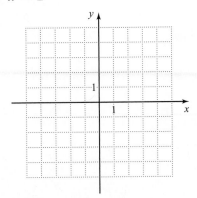

29. $y \leq 4$

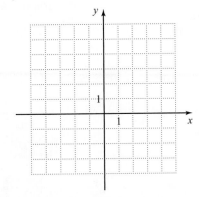

30. $y > -2.5$

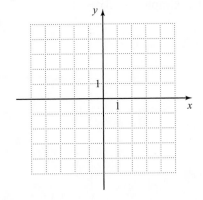

31. $4x + 2y \geq -2$

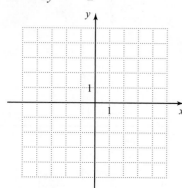

32. $6x + 3y > 9$

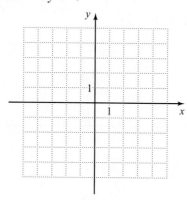

33. $2x - 3y \leq -6$

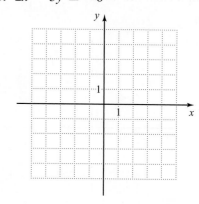

34. $4x - 6y \geq 24$

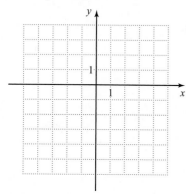

35. $5y - 3x + 20 > 0$

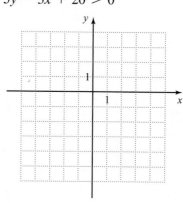

36. $3x + 2y - 8 < 0$

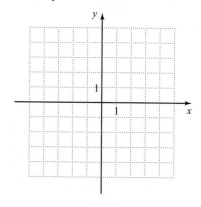

37. $0.4x - 0.3y < 1.2$

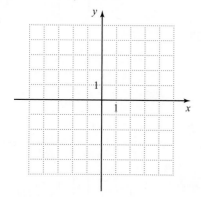

38. $1.2x - 0.9y \geq 3.6$

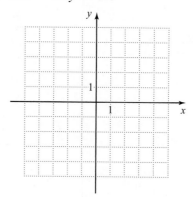

Mixed Practice

Solve.

39. Graph $y < 3x + 1$.

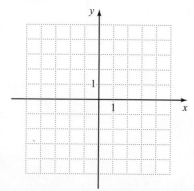

40. Graph $y \geq -3$.

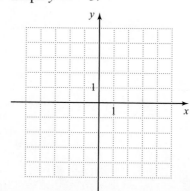

41. Determine if $(3, -1)$ is a solution of $y \le x - 4$.

42. Determine if $\left(\dfrac{1}{2}, -\dfrac{3}{4}\right)$ is a solution of $4x + 8y > -3$.

Applications

C *Solve.*

43. An investor wants to purchase x shares of a technology stock and y shares of a media stock. The technology stock is $30 per share and the media stock is $21 per share. The investor has a maximum of $5000 with which to purchase stock.

  a. Express this information as an inequality.

  b. Graph the inequality.

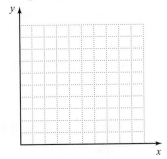

  c. Use the graph to determine if the investor can purchase 110 shares of each stock.

44. A homeowner has 100 ft of fencing with which to enclose a rectangular dog pen in her backyard.

  a. Write an inequality that represents the possible dimensions of the dog pen, where l represents the length and w represents the width of the dog pen.

  b. Graph the inequality on the coordinate plane.

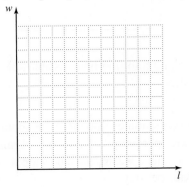

  c. Give one set of dimensions that satisfy the inequality.

45. A credit card company requires a minimum monthly payment of at least 4% of the credit card balance.

  a. Write an inequality that shows the monthly payment p in terms of the balance b.

  b. Graph the inequality for balances up to $1000.

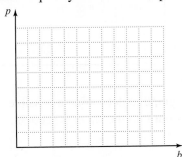

  c. Use the graph to determine a possible monthly payment on a credit card with a balance of $650.

46. A candy shop makes small and large gift baskets for special occasions. The profit from selling a small gift basket is $10 and the profit from selling a large gift basket is $23. The owner of the candy shop wants to make a profit of more than $1000 a month on the sale of gift baskets.

  a. Express this information as an inequality, letting x represent the number of small gift baskets and y represent the number of large gift baskets.

  b. Graph the inequality on the coordinate plane.

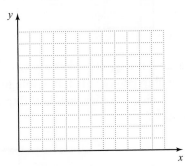

  c. Use the graph to determine if the profit requirement is met if 45 small gift baskets and 25 large gift baskets are sold in a particular month.

47. An elevator has a maximum capacity of 2000 lb. Suppose the average weight of an adult is 160 lb and the average weight of a child is 60 lb.

   a. Write an inequality that relates the number of adults a and children c who can ride in the elevator without overloading it.

   b. Graph the inequality.

   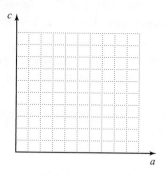

   c. Use the graph to find one possible solution of the inequality.

48. A lottery winner invests x dollars of his winnings into a fund earning 6.25% simple interest and y dollars of his winnings in a fund earning 8.5% simple interest. He would like to earn more than $5000 on the investments.

   a. Express this information as an inequality.

   b. Graph the inequality.

   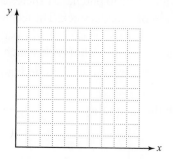

   c. Identify one possible solution of the inequality.

49. A bakeshop estimates that there are about 12 chocolate chip cookies per pound and 8 oatmeal raisin cookies per pound. To reduce waste, the bakeshop makes at most 360 cookies per day.

   a. Express this information as an inequality, letting c represent the number of pounds of chocolate chip cookies and r represent the number of pounds of oatmeal raisin cookies baked per day.

   b. Graph the inequality.

   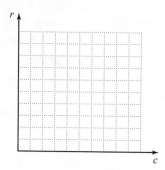

   c. On a particular day, the bakeshop bakes 20 lb of chocolate chip cookies. Is it possible to bake 15 lb of oatmeal raisin cookies that day? Explain.

50. A student wants to put at least 10% of her weekly income into her savings account.

   a. Write an inequality that shows the weekly savings s in terms of her weekly income i.

   b. Graph the inequality for a weekly income up to $500.

   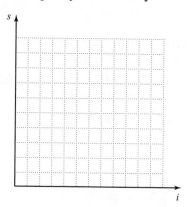

   c. Use the graph to determine a possible weekly income if $50 is put into the savings account.

51. During a political campaign, a candidate makes television and radio commercials. A television commercial requires 30 sec of airtime and a radio commercial requires 60 sec of airtime. Based on the budget, the campaign manager plans for the candidate to get at most 60 min of airtime per week.

a. Write an inequality that relates the number of television commercials T and the number of radio commercials R that can be aired per week.

b. Graph the inequality on the coordinate plane.

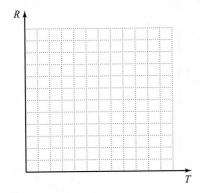

c. Is it possible to air 50 television commercials and 40 radio commercials? Explain.

52. A small perfume company has daily fixed costs of $100 plus $4 for each of the x 1-ounce bottles and $9 for each of the y 2-ounce bottles of perfume produced in a day. The company would like to keep the total daily cost below $600.

a. Write an inequality to represent the situation.

b. Graph the inequality.

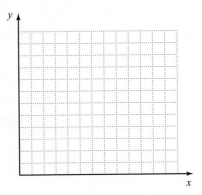

c. Use the graph to find two solutions that satisfy the inequality.

• Check your answers on page A-26.

MINDStretchers

Critical Thinking

1. Not all graphs of inequalities in two variables are linear. Use your knowledge of graphing linear inequalities in two variables to select the correct graph for each nonlinear inequality.

   a. $x^2 + y^2 \geq 1$

   I.
   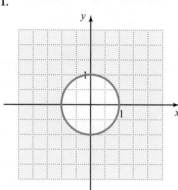

   II.
   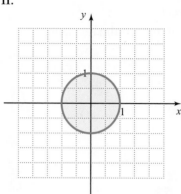

   b. $y \geq x^2 - 1$

   I.
   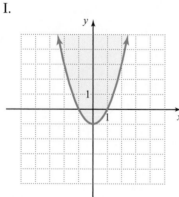

   II.
   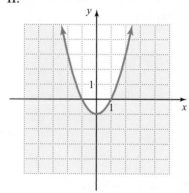

Writing

2. Explain the difference between the solutions of a linear inequality in one variable and the solutions of a linear inequality in two variables. Discuss how the graphs of the solutions compare.

Mathematical Reasoning

3. Consider the graphs: $Ax + By > C$ and $Ax + By < C$. If the two graphs were combined, what would you get?

3.6 Introduction to Functions

OBJECTIVES

Ⓐ To identify a function

Ⓑ To determine the domain or range of a function

Ⓒ To evaluate a function written in function notation

Ⓓ To graph a function

Ⓔ To recognize the graph of a function

Ⓕ To solve applied problems involving functions

In this section, we introduce the concept of a *function*. This concept is one of the most important in mathematics and will be developed more fully in later courses. We begin the section by discussing the idea of a *relation*—a more general concept than function, and then explain why some relations are functions and others are not. Next, we consider several topics related to functions—the domain and the range of a function, the system of notation based on functions, and a few of the many different types of functions. The section concludes with some applications of the concept of function to a variety of situations.

Functions

Let's consider a specific example that will lead us to the concept of function. Suppose that a hospital keeps a record of the birth lengths and birth weights of babies born in the maternity ward. One night, five babies were born with length and weight as shown in the following table:

Birth Length (in centimeters)	46	58	48	30	52
Birth Weight (in kilograms)	3.0	4.0	3.5	2.3	4.5

We can plot these values on a coordinate plane to get a better feel for any trends.

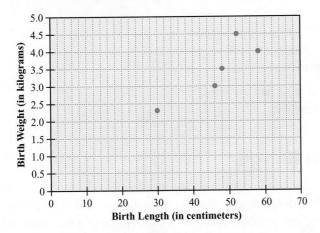

Whether expressed in a table or in a graph, the given information defines a *relation* between birth length and birth weight. We can think of this relation as a set of ordered pairs: $\{(46, 3.0), (58, 4.0), (48, 3.5), (30, 2.3), (52, 4.5)\}$. In general, we can define a relation as follows:

DEFINITION

A **relation** is a set of ordered pairs.

The hospital example defines a special kind of relation called a *function*. We know that the relation is a function because all the first coordinates, here birth lengths, are different from one another. If two babies had been born with the same birth length but with different birth weights, which is certainly possible, the relation would not have been a function.

> **DEFINITION**
>
> A **function** is a relation in which no two ordered pairs have the same first coordinates.

In any function, the first coordinates will be values of one variable (the independent variable) and the second coordinates will be values of another variable (the dependent variable). So in our example, the first coordinates are values of birth length and the second coordinates are values of birth weight. We say that the dependent variable *is a function of* the independent variable: Birth weight depends on birth length. If we write the ordered pairs in the reverse order, how will the dependent and independent variables be affected?

For any function, there is one and only one value of the dependent variable for each value of the independent variable. For instance, your income, if you worked at a fixed hourly wage, is a function of the number of hours you work and is uniquely determined by that number.

EXAMPLE 1

Determine whether each relation represents a function.

a. $\{(-3, 1), (0, 5), (1, 1), (3, 2)\}$

b.

x	4	1	0	4
y	−2	−1	0	2

Solution

a. In this relation, the ordered pairs $(-3, 1)$ and $(1, 1)$ have the same second coordinate, 1. However, all the first coordinates are different from one another. So this relation represents a function.

b. From the table we see that the x-value 4 is paired with two y-values, −2 and 2. So this relation does not represent a function.

PRACTICE 1

Decide whether each relation represents a function.

a. $\{(-5, 4), (-2, 0), (0, 2), (3, 7), (0, 8)\}$

b.

x	4	2	5	−1
y	1	3	0	1

Domain and Range

Each function has a domain and a range. By the *domain* of a function, we mean the set of all first coordinates of the ordered pairs that make up the function. These first coordinates are the possible values of the independent variable. By the *range* of a function, we mean the set of all second coordinates, that is, the values of the dependent variable that the function assigns to the first coordinates.

In the hospital example, recall that the function is the set of ordered pairs $(46, 3.0)$, $(58, 4.0)$, $(48, 3.5)$, $(30, 2.3)$, and $(52, 4.5)$. The following diagram shows the domain and range of this function:

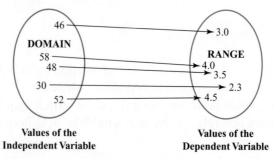

Values of the
Independent Variable

Values of the
Dependent Variable

DEFINITION

The **domain** of a function is the set of all values of the independent variable.
The **range** of a function is the set of all values of the dependent variable.

EXAMPLE 2

Find the domain and range of each function.

a. The function defined by the set of ordered pairs.
$\{(1, 5), (8, -2), (0, 4)\}$

b. The function defined by the following graph:

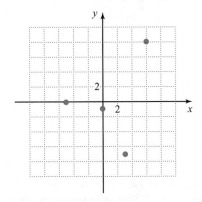

c. The function defined by the following table:

x	0	5	3
y	8	-2	-1

Solution

a. The domain is the set of first coordinates among the ordered pairs that define the function: $\{0, 1, 8\}$. The range is the set of second coordinates: $\{-2, 4, 5\}$.

b. The points plotted on the graph are $(-5, 0), (0, -1), (3, -7)$, and $(6, 8)$. The domain is the set of first coordinates: $\{-5, 0, 3, 6\}$. The range is the set of second coordinates: $\{-7, -1, 0, 8\}$.

c. The domain is the set of x-values in the top row: $\{0, 3, 5\}$. The range is the set of y-values in the bottom row: $\{-2, -1, 8\}$.

PRACTICE 2

Determine the range and domain of each of the following functions.

a. The function defined by the ordered pairs: $(2, 0), (8, -6), (5, 3)$

b. The function defined by the following graph:

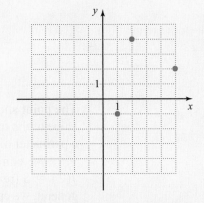

c. The function defined by the following table:

Day of the Month	1	2	3	4	5
Hours of Sleep	6.5	9	7	8	7.5

Function Notation

Another way to think of a function is as a rule (or correspondence) that assigns to each value of the independent variable a single value of the dependent variable. From this point of view, a function is a kind of input-output machine, where the independent variable is the input and the dependent variable is the output.

The rule that defines a function is often written as an equation. Consider, for instance, the equation $y = 4x$, which relates the length x of a side of a square and its perimeter y. In this equation, the value of y depends on x, and we say that y is a function of x. Using *function notation*, we can rewrite this equation as $f(x) = 4x$. The expression $f(x)$ is read "f of x."

The length x of a side of the square is the input, and the perimeter $f(x)$ of the square is the output. The function rule $4x$ assigns to each value x one value $f(x)$. Since both y and $f(x)$ represent $4x$, we can write $y = f(x)$.

> **TIP**   In function notation, the parentheses enclose the quantity that is the independent variable. They do not indicate multiplication.

A function, such as $f(x) = 4x$, can also be represented by a graph. On the following coordinate plane, we can determine the value of $f(x)$ for a particular x-value by finding the y-coordinate of the point on the graph that has that x-value:

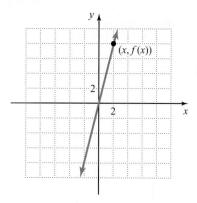

The point shown on the graph has coordinates $(2, 8)$. Therefore we can conclude that the value of the function is 8 when the value of x is 2, that is, a square with side length 2 units has a perimeter of 8 units.

We can express point $(2, 8)$ using function notation. Since $f(x) = 4x$, we write $f(2) = 8$ (read "f of 2 is 8") to indicate that when $x = 2$ the function has value $4 \cdot 2$, or 8. In general, we can evaluate a function by substituting the specific value of the independent variable given in parentheses in the rule that defines the function.

EXAMPLE 3

Let $f(x) = -2x + 4$. Find:

a. $f(5)$ **b.** $f(-3)$ **c.** $f(a)$ **d.** $f(2a)$

Solution

a. By $f(5)$, we mean the value of $f(x)$ when $x = 5$.

$$f(x) = -2x + 4$$
$$f(5) = -2(5) + 4 = -10 + 4 = -6$$

b. $f(-3) = -2(-3) + 4 = 6 + 4 = 10$

c. $f(a) = -2(a) + 4 = -2a + 4$

d. $f(2a) = -2(2a) + 4 = -4a + 4$

PRACTICE 3

Given that $g(x) = 3x - 1$, evaluate each of the following:

a. $g(2)$ **b.** $g(-1)$

c. $g(n)$ **d.** $g(n + 1)$

Note that while the letters f and g are frequently used to represent functions, we can use any letter to represent either a function or its independent variable. So $f(x) = 2x$ and $g(t) = 2t$ both indicate the same function whose value is twice the value of the independent variable.

Types of Functions

Now, we consider the graphs of a few different types of functions, some linear and others nonlinear. Then, we examine their domain and range.

First, we look at the **linear function**, which is a function that can be defined by $f(x) = mx + b$, where m and b are real numbers. An example of a linear function is $f(x) = 3x - 1$, where $m = 3$ and $b = -1$. Let's take a look at the graph of this function.

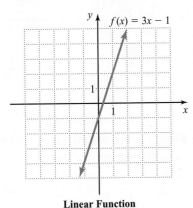

Linear Function

Recall that the graph of a line goes on indefinitely in both directions. We can see from the graph of $f(x) = 3x - 1$ that x can take on any real value, as can y. So the domain of this function is the set of all real numbers, or, in interval notation, $(-\infty, \infty)$. The range of this function is also $(-\infty, \infty)$ since the graph extends as far above or below the x-axis as desired.

Next, we consider the **constant function**, which is a linear function that can be defined by $f(x) = k$, where k is a real number. For instance, consider the constant function $f(x) = 3$, whose graph is shown on the coordinate plane to the right. We see from the graph that the domain of the function is the set of all real numbers, or, in interval notation, $(-\infty, \infty)$. However, the range is $\{3\}$, since the function only takes on the value 3.

Now, let's examine some nonlinear functions. First, we consider the **rational function** $f(x) = \dfrac{1}{x}$. The graph of this function is shown on the following coordinate plane:

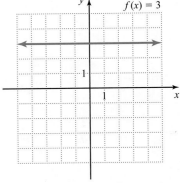

Constant Function

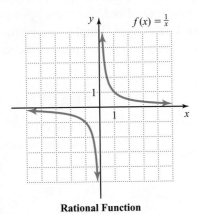

Rational Function

Note that there is no point on the graph with x-coordinate equal to 0. This omission makes sense, since substituting 0 for x in the equation of the graph leads to the expression $\dfrac{1}{0}$, which is undefined. Similarly, there is no point on the graph with y-coordinate equal to 0. We see from the graph that the domain and the range are each the set of all real numbers except for 0.

Next, let's consider the **quadratic function** $f(x) = x^2$. The graph of this function is shown on the coordinate plane on the next page. We see from the graph that the domain of this function is the set of all real numbers, or in interval notation $(-\infty, \infty)$. Since the y-values are all nonnegative numbers, the range is $[0, \infty)$, that is, the set including all positive numbers and 0. We will discuss the quadratic function in more detail in Chapter 8.

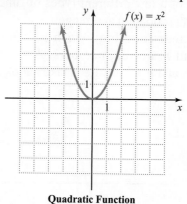

Quadratic Function

Finally, we examine the **absolute value function**, which is the function $f(x) = |x|$, shown in the following graph. Note that except for the origin, the graph of the absolute value function lies entirely above the x-axis.

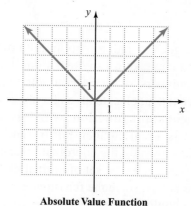

Absolute Value Function

We see from the graph of this function that its domain is $(-\infty, \infty)$. Since the y-values are all nonnegative numbers, the range is $[0, \infty)$, that is, the set including all positive numbers and 0.

To graph a function, we find several ordered pairs by evaluating the function. Then, we plot the ordered pairs in a coordinate plane. Finally, we connect the points. Note that since $y = f(x)$, the ordered pairs (x, y) and $(x, f(x))$ represent the same point.

EXAMPLE 4

Graph $f(x) = -|x|$. Identify the function's domain and range.

Solution First, we find some ordered pairs to plot. Choosing some values of x, we get:

x	$f(x) = -|x|$	(x, y)
-2	$f(-2) = -|-2| = -2$	$(-2, -2)$
-1	$f(-1) = -|-1| = -1$	$(-1, -1)$
0	$f(0) = -|0| = 0$	$(0, 0)$
1	$f(1) = -|1| = -1$	$(1, -1)$
2	$f(2) = -|2| = -2$	$(2, -2)$

PRACTICE 4

Graph $f(x) = |x + 1|$. Identify the function's domain and range.

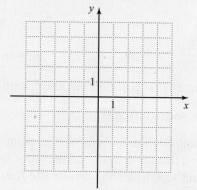

Next, we plot these points, and then sketch the graph. The domain is $(-\infty, \infty)$. We see from the graph that the dependent variable y can take on any nonpositive value, so the range is $(-\infty, 0]$.

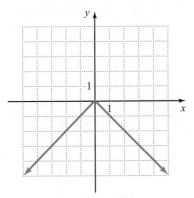

EXAMPLE 5

Graph $f(x) = x - 1$, for $x \le 0$. What are the domain and range of this function?

Solution   Let's begin by identifying some ordered pairs to plot.

x	$f(x) = x - 1$	(x, y)
-4	$f(-4) = -4 - 1 = -5$	$(-4, -5)$
-3	$f(-3) = -3 - 1 = -4$	$(-3, -4)$
-2	$f(-2) = -2 - 1 = -3$	$(-2, -3)$
-1	$f(-1) = -1 - 1 = -2$	$(-1, -2)$
0	$f(0) = 0 - 1 = -1$	$(0, -1)$

Next, we plot the points and draw the line for all x-values less than or equal to 0, as shown in the following graph:

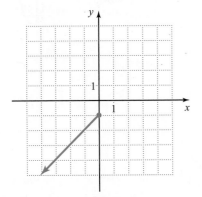

From the graph we see that the domain is $(-\infty, 0]$. The graph indicates that the dependent variable y is any negative number less than or equal to -1. So the range is $(-\infty, -1]$.

PRACTICE 5

Graph $f(x) = 2x$, for $x \ge 0$, and determine the domain and range of this function.

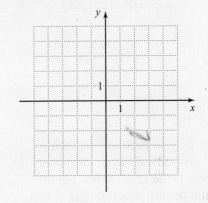

The Vertical-Line Test

One way of representing a function is as a graph. However, some graphs, while representing a relation, do not represent a function. Recall that a relation is only a function if no two distinct ordered pairs have the same first coordinate. It follows that the graph of a function cannot have two points with the same x-coordinate but with different y-coordinates. This suggests a test for determining if a graph represents a function.

> **THE VERTICAL-LINE TEST**
>
> If any vertical line intersects a graph at more than one point, the graph does not represent a function. If no such line exists, then the graph represents a function.

EXAMPLE 6

Consider the following graphs. Determine if each graph represents a function.

a.

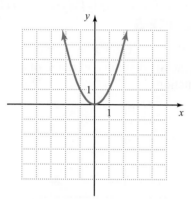

b.

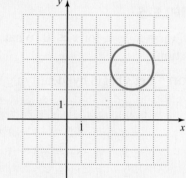

Solution

a. We see by inspecting the graph that any vertical line on the plane intersects the graph at one point. So the vertical-line test tells us that this graph represents a function.

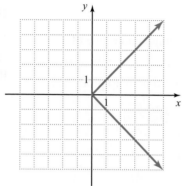

b. Since any vertical line in Quadrants I and IV intersects the graph more than once, the graph fails the vertical-line test. So this graph does not represent a function.

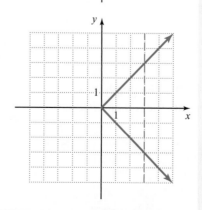

PRACTICE 6

Determine whether each graph represents a function.

a.

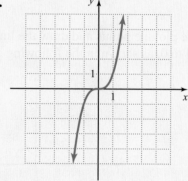

b.

Note that in the following example, only nonnegative values of the two variables make sense, restricting the domain and the range.

EXAMPLE 7

A salesperson earns $20,000 a year plus 10% commission on total sales.

a. Use function notation to express the relationship between the salesperson's yearly earnings $E(s)$ and her total sales s.

b. Express in function notation the annual earnings made on the sale of merchandise worth $30,000. Under these conditions, how much money was earned?

c. Graph the relationship in part (a).

d. Find the domain and the range of the function.

e. Explain how you could conclude that the relationship is a function by examining the graph in part (c).

Solution

a. We know that the salesperson's earnings $E(s)$ amount to $20,000 a year plus an additional 10% commission on her sales s, that is, $E(s) = 20,000 + 0.1s$. Alternatively, we can write this relation as $E(s) = 0.1s + 20,000$.

b. Using function notation, we can represent the total annual earnings on sales of $30,000 by $E(30,000)$. To evaluate, we substitute 30,000 for s in the expression $0.1s + 20,000$. So $E(30,000) = 0.1(30,000) + 20,000 = 23,000$, that is, $23,000.

c. We graph $E(s) = 0.1s + 20,000$.

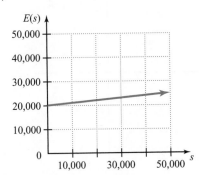

d. Inspecting the graph leads us to conclude that the domain is all real numbers greater than or equal to 0, that is, $[0, \infty)$. The range is all real numbers greater than or equal to 20,000, or in interval notation, $[20,000, \infty)$.

e. The graph in part (c) passes the vertical-line test, showing that the relation is a function.

PRACTICE 7

The cost of renting a car at a local dealer is $40 per day plus $0.10 a mile.

a. Express in function notation the relationship between the daily rental cost in dollars $C(m)$ and the number of miles driven m.

b. Express in function notation the cost of renting a car for a day and driving 200 mi. Then, evaluate the function.

c. Graph the relationship.

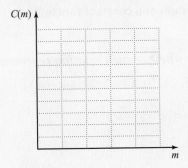

d. Find the domain and the range of the function.

e. Explain how you could decide whether the relationship is a function by examining the graph in part (c).

Mathematically Speaking

Fill in each blank with the most appropriate term or phrase from the given list.

output	range	represents	vertical line
does not represent	function	dependent	independent
horizontal line	domain	relation	input

1. A(n) _____ is a set of ordered pairs.

2. A function's first coordinates are values of the _____ variable.

3. For any function, there is one and only one value of the _____ variable for each value of the other variable.

4. The _____ of a function is the set of all values of the independent variable.

5. The _____ of a function is the set of all values of the dependent variable.

6. If we think of a function as an input-output machine, the independent variable is the _____.

7. The graph of a constant function is a(n) _____.

8. If any vertical line intersects a graph at more than one point, the graph _____ a function.

Ⓐ *Determine whether each relation represents a function.*

9. $\{(-3, 2), (0, 1), (3, 2), (4, -4)\}$

10. $\{(1, 1), (1, -1), (2, -4), (2, 4)\}$

11.

x	−4	−2	−2	1	2
y	7	−3	3	1	6

12.

x	−3	−2	0	2	3
y	5	0	−4	0	5

13.

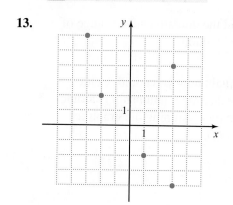

14.

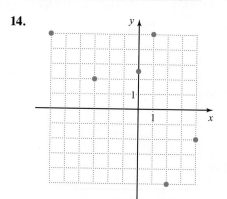

15.

Radius of a Circle r	1	2	3	4	5
Circumference C	2π	4π	6π	8π	10π

16.

Time t (in seconds)	0	5	10	15
Distance d (in feet)	0	20	40	60

B *Find the domain and range of each function.*

17. $\{(-2, 6), (-1, 8), (0, 10), (1, 12), (2, 14)\}$

18. $\{(-6, 3), (-4, 2), (0, 0), (4, -2), (6, -3)\}$

19. $\{(-3, -27), (-1, -1), (0, 0), (1, 1), (3, 27)\}$

20. $\{(-4, 7), (-3, 0), (-2, -5), (-1, -8), (0, -9)\}$

21.

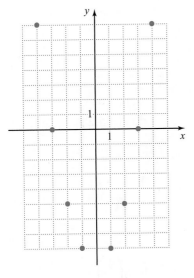

22.

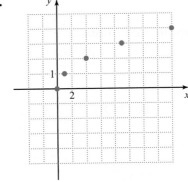

23.

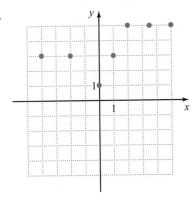

24.

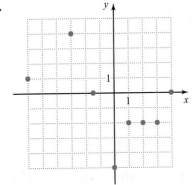

25.

x	−7	−3	−1	2	4
y	7	3	1	−2	−4

26.

x	0	1	2	3
y	9	9	9	9

27.

Year	2007	2008	2009	2010	2011
Score of Winning Superbowl Team	29	17	17	31	31

(*Source:* NFL.com)

28.

Election Year	1996	2000	2004	2008
Popular Votes Cast for Winning U.S. Presidential Candidate	45,590,703	50,456,062	62,040,610	69,498,459

(*Sources: The World Almànac and Book of Facts, 2011*)

C *Evaluate each function for the given values.*

29. $f(x) = 8 - 5x$

    a. $f(2)$ **b.** $f(-1)$

    c. $f\left(\dfrac{3}{5}\right)$ **d.** $f(1.8)$

30. $g(x) = \dfrac{1}{2}x - 1$

    a. $g(0)$ **b.** $g(-6)$

    c. $g(14)$ **d.** $g(-5.2)$

31. $g(x) = 2.4x - 7$

    a. $g(5)$ **b.** $g(-2)$

    c. $g(a)$ **d.** $g(a^2)$

32. $f(x) = -3.5x + 1.5$

    a. $f(3)$ **b.** $f(-1)$

    c. $f(n)$ **d.** $f(n^3)$

33. $f(x) = \left|\dfrac{1}{2}x + 3\right|$

    a. $f(0)$ **b.** $f(-8)$

    c. $f(-4t)$ **d.** $f(t - 6)$

34. $h(x) = 1 - |2x|$

    a. $h(-5)$ **b.** $h\left(\dfrac{1}{2}\right)$

    c. $h(-a)$ **d.** $h(a + 1)$

35. $h(x) = 3x^2 - 6x - 9$

    a. $h(2)$ **b.** $h(-1)$

    c. $h(-n)$ **d.** $h(2n)$

36. $f(x) = 2x^2 + x$

    a. $f(-2)$ **b.** $f\left(-\dfrac{1}{2}\right)$

    c. $f(a)$ **d.** $f(-3a)$

37. $g(x) = 10$

    a. $g(7)$ **b.** $g(-150)$

    c. $g(t)$ **d.** $g(5 - 9t)$

38. $g(x) = -4$

    a. $g(23)$ **b.** $g(-41)$

    c. $g(5n)$ **d.** $g(2n^2)$

D *Graph each function. Then, identify its domain and range.*

39. $f(x) = 5x - 4$

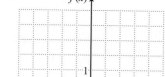
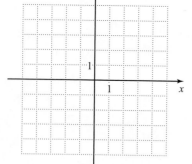

40. $f(x) = -3x + 1$

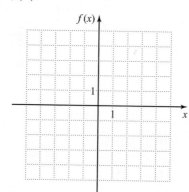

41. $g(x) = |x| + 2$

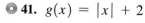
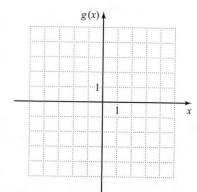

42. $g(x) = -|x| + 3$

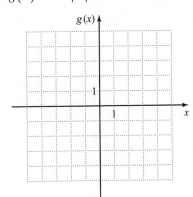

43. $f(x) = -5$

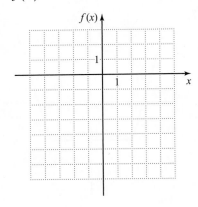

44. $h(x) = 2$

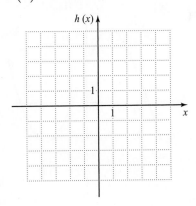

45. $f(x) = -\dfrac{1}{4}x - 1$, for $x \leq 0$

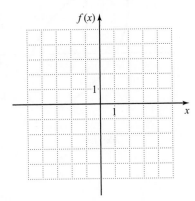

46. $f(x) = \dfrac{3}{2}x + 4$, for $x \geq -4$

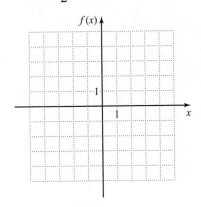

47. $h(x) = |x + 1|$, for $x \geq -2$

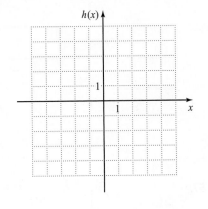

48. $g(x) = |x - 4|$, for $x \leq 4$

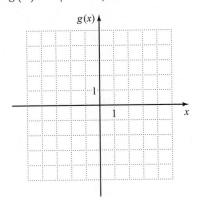

49. $f(x) = x^2$, for $x \leq 0$

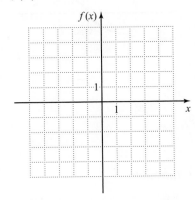

50. $f(x) = 1 - x^2$, for $x \geq 0$

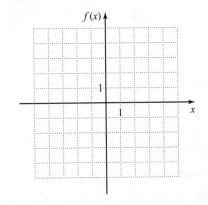

E *Determine whether each graph represents a function.*

51.

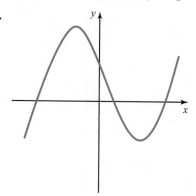

52.

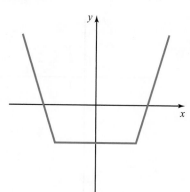

○ 53.

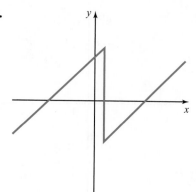

54.

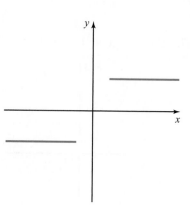

55.

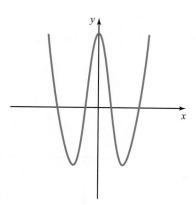

56.

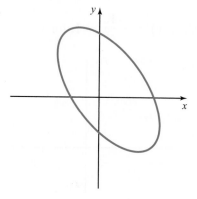

57.

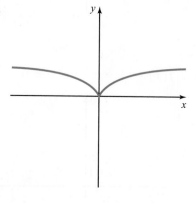

58.

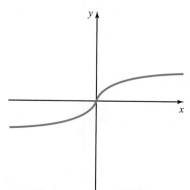

Mixed Practice

Determine whether each relation represents a function.

59. $\{(-2, 6), (-1, 4), (5, 8),$
$(10, -6), (11, 4)\}$

60.

x	−2	0	0	1	2
y	1	3	5	−3	4

61.

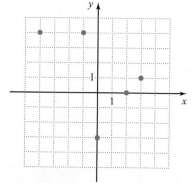

Find the domain and range of each function.

62. $\{(-5, 2), (-2, 7), (0, -6), (4, 1), (7, 33)\}$

63.

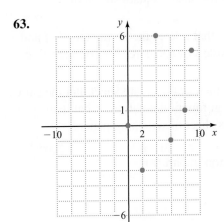

Solve.

64. Evaluate $f(x) = 3x^2 + x$ for the given values.

   a. $f(0)$ **b.** $f(-2)$

   c. $f(n)$ **d.** $f(-4n)$

65. Graph the function $f(x) = 3x^2$ for $x \geq 0$.

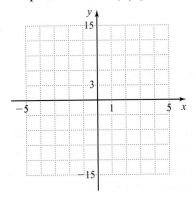

66. Determine whether the graph to the right represents a function.

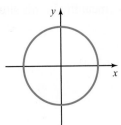

Applications

Solve.

67. A department store is running a sale that gives customers 20% off the total amount of their purchases.

   a. Use function notation to express the relationship between the total discount $d(a)$ and the amount of a purchase a before the discount.

   b. Use function notation to express the total amount of the discount on a purchase of $150. How much money was saved on the purchase?

   c. Graph the function in part (a).

   d. Identify the domain and range of the function.

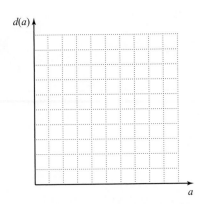

68. A paper manufacturer makes square sheets of wrapping paper.

   a. Write a function that relates the perimeter $p(s)$ and the length of a side s of a square sheet of wrapping paper.

   b. Express the perimeter of a square sheet that measures 12 in. on a side in function notation. What is the perimeter of this sheet of wrapping paper?

   c. Graph the function in part (a).

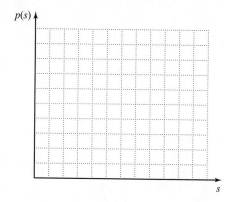

   d. Find the domain and range of the function.

69. A car that is purchased new for $22,500 depreciates in value by $1875 per year.

   a. Use function notation to show the value $V(t)$ of the car (in dollars) t yr after it is purchased.

   b. What is the meaning of $V(6)$? Find this value.

   c. Graph the function in part (a).

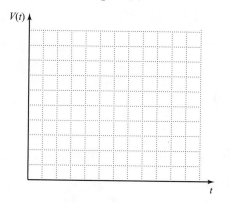

70. A cell phone company charges a $50 activation fee plus $45 per month for unlimited calling within the network.

   a. Use function notation to represent the total amount paid $A(x)$ for x mo of service.

   b. Find $A(4)$. Interpret its meaning in this situation.

   c. Graph the function in part (a).

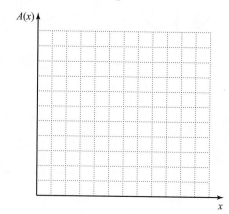

71. A patient's weekly dosage of 500 mg of a medication is reduced by 50 mg/wk.

   a. Express in function notation the relationship between the patient's weekly dosage $d(x)$ and the number of weeks x.

   b. Find $d(2)$ and interpret its meaning in this situation.

   c. Graph the function in part (a).

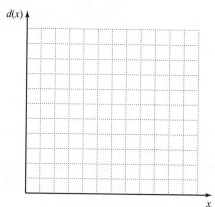

72. A manufacturing plant has monthly fixed costs of $10,000 plus $1.20 for each unit it produces.

    a. Write the relationship between the number of units produced x and the manufacturing plant's monthly cost $C(x)$.

    b. What is $C(500)$? Explain its meaning in the context of the situation.

    c. Graph the function in part (a).

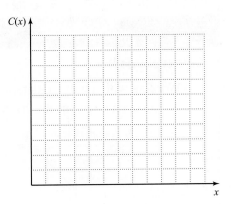

73. The height of an object t sec after it is thrown straight upward from a height of 160 ft with an initial velocity of 48 ft/sec is given by the function

    $$h(t) = -16t^2 + 48t + 160.$$

    a. The graph of this function is given to the right. Use the graph to find $h(0.5)$ and $h(3)$.

    b. Is it possible to consider values of $t > 5$ in this situation? Explain.

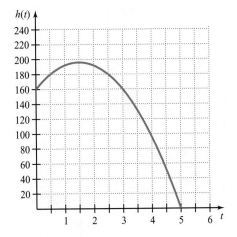

74. A box manufacturer makes open boxes by cutting x-inch by x-inch squares from each corner of a 50-inch by 40-inch piece of cardboard (see figure). The volume of the box is given by the function

    $$V(x) = 4x^3 - 180x^2 + 2000x.$$

    a. The graph of this function is given here. Use the graph to find $V(5)$ and $V(15)$.

    b. Is it reasonable to consider values of $x > 20$ in this situation? Explain.

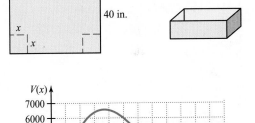

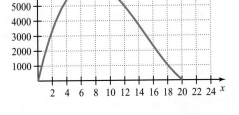

• Check your answers on page A-27.

MINDStretchers

Groupwork

1. Working with a partner, determine the domain and range of each of the following functions. (Recall from Chapter 2 that an open circle means that the point is not part of the graph.)

a.

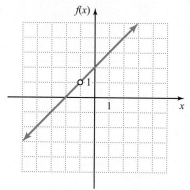

b.

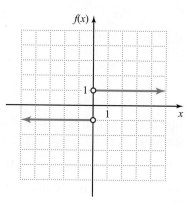

c.

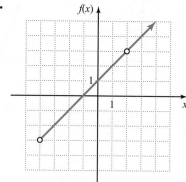

d.

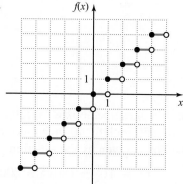

Critical Thinking

2. Some functions are described by two or more rules of correspondence that are evaluated for specific values of the domain. These functions are called *piecewise defined* functions. One such function is:

$$f(x) = \begin{cases} 2x + 5, & x \le -1 \\ -x - 2, & x > -1 \end{cases}$$

   a. Which rule is evaluated if $x = 2$? Explain.
   b. Which rule is evaluated if $x = -3$? Explain.
   c. Find the following function values:

   i. $f(-4)$ ii. $f(-1)$ iii. $f(0)$ iv. $f(3)$

Mathematical Reasoning

3. Can a relation whose domain contains fewer values than its range ever define a function? Explain.

Key Concepts and Skills

Concept/Skill	Description	Example
[3.2] **Slope** *m*	The **slope** of a line passing through the points (x_1, y_1) and (x_2, y_2) is defined to be $$m = \frac{y_2 - y_1}{x_2 - x_1},$$ where $x_1 \neq x_2$.	The slope of the line that passes through the points $(3, 1)$ and $(4, -2)$ is: $$m = \frac{1 - (-2)}{3 - 4} = \frac{3}{-1} = -3$$
[3.2] **Parallel lines**	Two nonvertical lines are **parallel** if and only if their slopes are equal. In other words, if two lines have slopes m_1 and m_2, they are parallel if and only if $m_1 = m_2$.	
[3.2] **Perpendicular lines**	Two nonvertical lines are **perpendicular** if and only if the product of their slopes is -1. In other words, if the two lines have slopes m_1 and m_2, they are perpendicular if and only if $m_1 \cdot m_2 = -1$.	
[3.3] **Linear equation in two variables, *x* and *y***	An equation that can be written in the *general form* as $Ax + By = C$, where A, B, and C are real numbers and A and B are not both 0.	$2x + y = 6$
[3.3] **Solution of an equation in two variables**	An ordered pair of numbers that makes the equation true.	$(1, -2)$ is a solution of $y = 3x - 5$.
[3.3] **Graph of an equation in two variables**	All points whose coordinates make the equation true.	The graph of $y = x$.

continued

Concept/Skill	Description	Example
[3.3] To graph a linear equation in two variables	• Isolate one of the variables (usually y) if it is not already done. • Choose three x-values, entering them into a table. • Complete the table by finding the corresponding y-values. • Plot the three points—two to draw the line and the third to serve as a *checkpoint*. • Draw the line passing through the points.	$x - 2y = 4$ $y = \dfrac{1}{2}x - 2$ 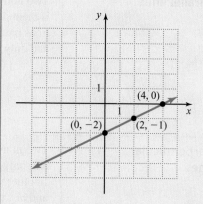 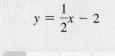 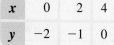
[3.3] Intercepts of a line	The **x-intercept**: the point where the graph intersects the x-axis. The **y-intercept**: the point where the graph intersects the y-axis.	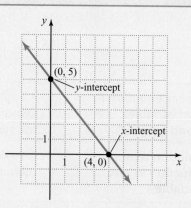
[3.4] Slope-intercept form	A linear equation in two variables written as $y = mx + b$, where m and b are real numbers. In this form, m is the slope of the line, and $(0, b)$ is the y-intercept of the graph of the equation.	In $y = \dfrac{1}{2}x - 1$, the slope is $\dfrac{1}{2}$ and the y-intercept is $(0, -1)$.
[3.4] To graph a linear equation in two variables using the slope and y-intercept	• Plot the y-intercept. • Use the slope to find a second point on the line. • Draw a line through the two points.	Graph: $y = 3x + 1$ 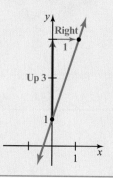 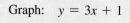

Concept/Skill	Description	Example
[3.4] **Point-slope form**	A linear equation in two variables written as $y - y_1 = m(x - x_1)$, where $x_1, y_1,$ and m are real numbers.	$y - 3 = 5(x + 2)$
[3.5] **Linear inequality in two variables**	An inequality that can be written in the form $Ax + By < C$, where A, B, and C are real numbers and A and B are not both 0. The inequality symbol can be $>, <, \geq,$ or $\leq$.	$2x + 3y < 6$ $x - y \geq 4$
[3.5] **Solution of a linear inequality in two variables**	An ordered pair of real numbers that when substituted for the variables makes the inequality a true statement.	$(3, 1)$ is a solution of $2x + y \geq 7$. $2(3) + 1 \overset{?}{\geq} 7$ $7 \geq 7$ True
[3.5] **To graph a linear inequality in two variables**	• Graph the corresponding linear equation. For an inequality that involves either the symbol $\geq$ or the symbol $\leq$, draw a solid line; for an inequality with either the symbol $>$ or the symbol $<$, draw a broken line. This line is the boundary between two half-planes. • Choose a test point in either half-plane and substitute the coordinates of this point into the inequality. If the resulting inequality is satisfied, then the graph of the inequality is the half-plane containing the test point. If it is not satisfied, then the other half-plane is the graph. In either case, a solid line is part of the graph; a broken line is not. Shade the appropriate region.	$y > x + 3$ $2x + 3y \leq 6$
[3.6] **Relation**	A set of ordered pairs.	$\{(0, 1), (3, 4)\}$
[3.6] **Function**	A relation in which no two ordered pairs have the same first coordinates.	A function: $\{(1, 4), (2, 4)\}$ Not a function: $\{(4, 1), (4, 2)\}$
[3.6] **Domain of a function; range of a function**	Domain: The set of all values of the independent variable. Range: The set of all values of the dependent variable.	Function: $\{(1, 1), (2, 4)(3, 9)\}$ $1 \longrightarrow 1$ $2 \longrightarrow 4$ $3 \longrightarrow 9$ DOMAIN RANGE

continued

Concept/Skill	Description	Example
[3.6] **Vertical-line test**	If any vertical line intersects a graph at more than one point, the graph does not represent a function. If no such line exists, then the graph represents a function.	A function: $y = x^2$ 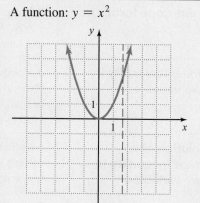 Not a function: $x^2 + y^2 = 4$ 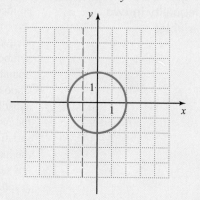

CHAPTER 3   Review Exercises

Say Why
Fill in each blank.

1. Two lines with slopes $\dfrac{-8}{12}$ and $\dfrac{2}{-3}$ $\underline{}$ $\underset{\text{are/are not}}{}$

   parallel because _____

   _____.

2. Two lines with slopes -3 and $-\dfrac{1}{3}$ $\underline{}$ $\underset{\text{are/are not}}{}$

   perpendicular because _____

   _____.

3. The equation $3x^2 + 7y = 1$ $\underset{\text{is/is not}}{\underline{}}$ linear because

   _____.

4. The x-intercept of an equation $\underset{\text{can/cannot}}{\underline{}}$ be $(0, -4)$

   because _____.

5. One solution of $3x - 4y \le -12$ $\underset{\text{is/is not}}{\underline{}}$ $(2, 5)$

   because _____

   _____.

6. The relation defined by $\{(1, 3), (2, 5), (1, 6)\}$

   $\underset{\text{does/does not}}{\underline{}}$ represent a function because

   _____

   _____.

[3.1]

7. Plot each point on the coordinate plane.
   $A(-3, 3), B(0, 0.5), C(1, -4), D(-2, -5)$

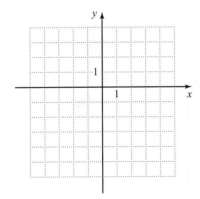

8. Fill in the coordinates of each point.

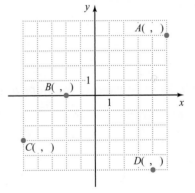

Identify the quadrant(s) in which each point is located.

9. $(1.3, -7)$

10. $(-8, -10)$

11. (x, y) if $xy > 0$

12. (x, y) if $xy < 0$

[3.2] *Compute the slope m of the line that passes through the given points.*
Then, plot the points and sketch the line that passes through them.

13. $(0, 4)$ and $(-3, -2)$ $m =$

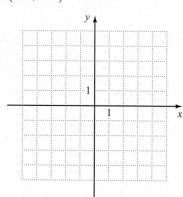

14. $(-3, 3)$ and $(-6, 2)$ $m =$

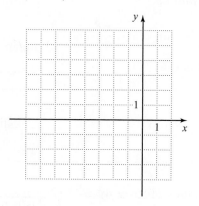

15. $(1.5, -1)$ and $(1.5, 3)$ $m =$

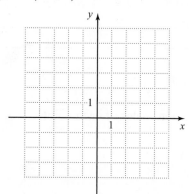

16. $(5, -1)$ and $(2, 4)$ $m =$

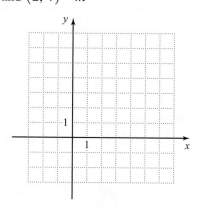

For each graph, identify whether the slope of the line is positive, negative, zero, or undefined.

17.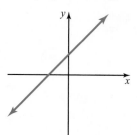

18.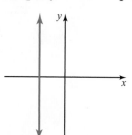

19.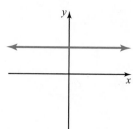

20.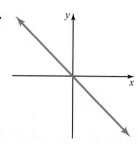

Graph the line on the coordinate plane using the given information.

21. Passes through $(2, 3)$ and $m = 1$

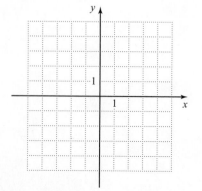

22. Passes through $(0, 1)$ and $m = -\dfrac{2}{3}$

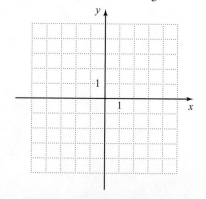

23. Passes through $(-3, -1)$ and $m = -\dfrac{1}{2}$

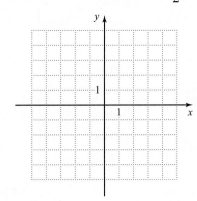

24. Passes through $(-4, 4)$ and $m = 0$

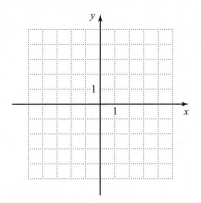

Find two additional points that lie on the line that passes through the given point and has slope m.

25. $(-5, 3)$ $m = -4$

26. $(6, -9)$ $m = 3$

27. $(1.6, -1)$ $m = 0$

28. $(-4, -7)$ $m = -\dfrac{3}{5}$

Given points P, Q, R, and S, determine whether $\overleftrightarrow{PQ}$ and $\overleftrightarrow{RS}$ are parallel, perpendicular, or neither.

29. $P(1, 4), Q(-2, 6), R(9, -5),$ and $S(11, -2)$

30. $P(3.2, -3), Q(-1.8, 7), R(5, 4),$ and $S(1, 12)$

[3.3] *Determine if the given ordered pair is a solution of the given equation.*

31. $y = \dfrac{1}{3}x - 2$ $(6, 0)$

32. $5x - 2y = -1$ $(-1, -3)$

Find the x- and y-intercepts of the graph of each equation.

33. $y = 3x - 9$

34. $2x + 8 = 0$

35. $24 = 12y$

36. $4y - 9x = 6$

Graph each equation.

37. $y = 2x - 1$

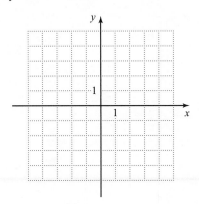

38. $y = -\dfrac{1}{3}x + 2$

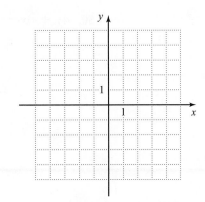

39. $3x + 2y = 6$

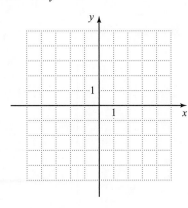

40. $12x - 4y = -8$

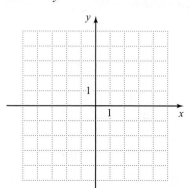

41. $6y = -12$

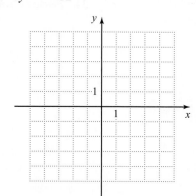

42. $-10x = -25$

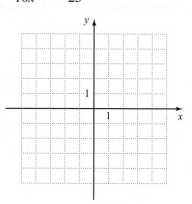

[3.4] *Write each equation in slope-intercept form. Then, identify the slope and y-intercept.*

43. $5x - y = 1$

44. $6y - 4x = -18$

45. $7y + 2x - 14 = 0$

46. $y - 8 = -\dfrac{3}{2}(x - 2)$

Graph each equation using the slope and y-intercept.

47. $y = -\dfrac{1}{4}x + 3$

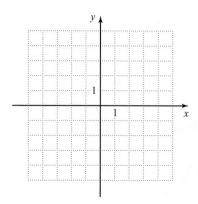

48. $y = 4x - 5$

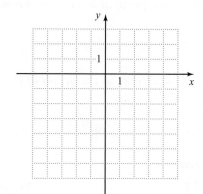

49. $3x - 5y = 20$

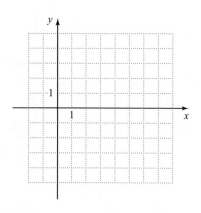

50. $4y + 6x = -24$

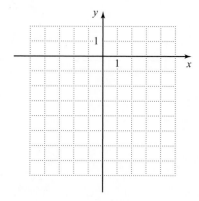

51. Write an equation of the line that has slope -6 and passes through the point $(-1, 3)$.

52. Find the equation of the line that has slope $\dfrac{2}{5}$ and passes through the point $(5, 7)$.

53. What is the equation of the horizontal line that passes through the point $(1.2, -2.4)$?

54. Find the equation of the line that is parallel to $2x - 8y = 40$ and passes through the point $(-8, 1)$.

55. What is the equation of the line that is perpendicular to $9 - 3y = 15x$ and passes through the point $(0, 7)$?

56. The points $(-2, -5)$ and $(-1, -8)$ lie on a line. Find the equation of the line.

[3.5] *Determine if the given point is a solution of the given inequality.*

57. $5x - 6y > 20 \qquad \left(2, -\dfrac{2}{3}\right)$

58. $4y - 2x \le -1 \qquad (-3, -2)$

Graph each inequality.

59. $y \ge -\dfrac{2}{3}x + 3$

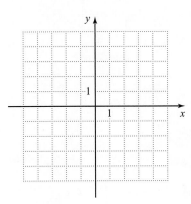

60. $y < 1$

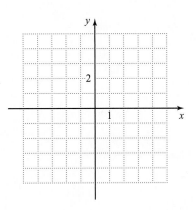

61. $4x - 4y \ge 12$

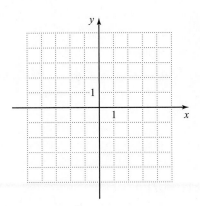

62. $3x + 5y > -10$

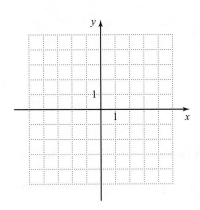

[3.6] *Determine whether each relation represents a function.*

63. $\{(1, 5), (4, 9), (6, 11), (8, 5)\}$

64.

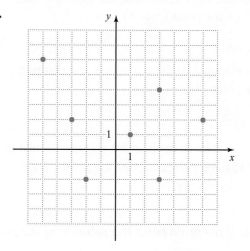

Find the domain and range of each function.

65. $\{(-7, 3), (-5, 3), (-3, 3), (-1, 3)\}$

66.

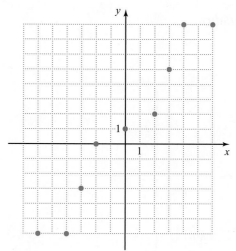

67.

x	-27	-8	-1	0	1	8	27
y	-3	-2	-1	0	1	2	3

68.

Hours Worked	32	26	38	20	24	30
Weekly Pay	$304	$247	$361	$190	$228	$285

Evaluate each function for the given values.

69. $f(x) = \dfrac{1}{3}x + 6$

  a. $f(-9)$ **b.** $f(3.6)$

  c. $f(3a)$ **d.** $f(6a - 12)$

70. $g(x) = |4x - 7|$

  a. $g(3)$ **b.** $g\left(-\dfrac{3}{4}\right)$

  c. $g(2n)$ **d.** $g\left(\dfrac{1}{4}n + 1\right)$

Graph each function. Then, identify its domain and range.

71. $f(x) = 4 - \dfrac{1}{2}x$

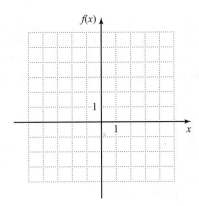

72. $f(x) = -x^2$

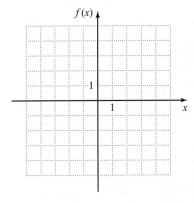

73. $g(x) = -|x| + 5$

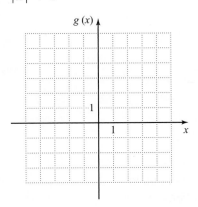

74. $g(x) = 4x - 9$, for $x \geq 1$

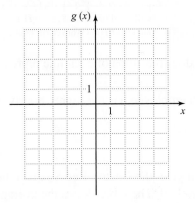

Determine whether each graph represents a function.

75.

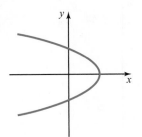

76.

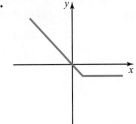

Mixed Applications

Solve.

77. Two employees at a company receive hourly pay. The lines in the graph show the total pay p of each employee for t hr of work.

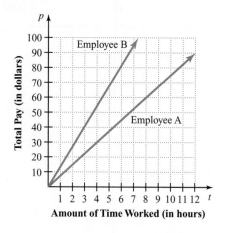

Amount of Time Worked (in hours)

a. What does the slope of each line represent?

b. Which employee has the higher hourly pay? Explain.

78. Many childcare centers charge parents additional fees when they arrive late to pick up their children. The graph shows the late-fee structure for a childcare center. Use slope to describe the center's late fee structure.

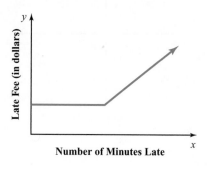

Number of Minutes Late

79. In a recent year, the Boston Red Sox had the highest single-game ticket prices in major league baseball. The table shows the average price of a ticket for a game at Fenway Park for the years from 2008 to 2010, where $t = 0$ represents 2006. (*Source: The Morning Sentinel*)

Point	Year t	Average Ticket Price p (in dollars)
A	2	48.80
B	3	50.24
C	4	52.32

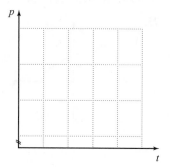

a. Choose an appropriate scale and plot the points on the coordinate plane. Then, sketch $\overline{AB}$ and $\overline{BC}$.

b. Compute the slopes of $\overline{AB}$ and $\overline{BC}$.

c. Was the increase in the average price of a ticket constant over the three years? Explain.

80. For monthly parking, a garage charges an activation fee plus a monthly fee. The total amount A (in dollars) that a new customer pays after parking for m mo is given by:

$$A = 50m + 20$$

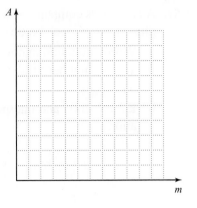

a. Choose an appropriate scale and then graph the equation.

b. What is the slope of the line? What does it represent in the context of this problem?

c. What is the A-intercept? Explain its significance in this situation.

d. From the graph, estimate the total amount a new customer pays for 7 mo of parking.

81. A high school football concession stand sells coffee for $1.50 per cup and hot chocolate for $1.25 per cup. The concession stand collected $1200 from the sale of coffee and hot chocolate during Friday night's game.

a. Write this information as an equation, letting x represent the number of cups of coffee sold and y represent the number of cups of hot chocolate sold.

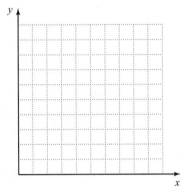

b. Choose an appropriate scale and graph the equation.

c. Identify the x- and y-intercepts of the graph. What do they represent in terms of the sales of the concession stand?

d. From the graph, find the number of cups of coffee sold if 300 cups of hot chocolate were sold.

82. An insurance company charges customers an annual premium for car insurance. For a particular policy, a customer pays $170 per month. After 3 mo, the balance to be paid was $850.

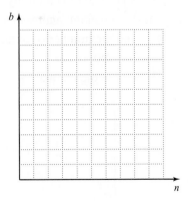

a. Write an equation in slope-intercept form that shows the balance to be paid b in terms of the number of months of payments n.

b. Graph the equation.

c. Identify the intercepts of the graph. Explain their significance in the context of the situation.

83. A machine at a beverage plant fills 165 bottles in 3 min and 330 bottles in 6 min.

a. Find an equation in point-slope form that represents the number of bottles b filled by the machine in m min.

b. What does the slope of the line represent?

c. How many bottles can the machine fill in 1 hr?

84. A cosmetics company makes a monthly profit of $300 plus $1.20 for each bottle of nail polish it sells.

   a. Write a function that relates the monthly profit $P(x)$ and the number of bottles of nail polish x the company sells.

   b. Find $P(200)$ and interpret its meaning in this situation.

   c. Graph the function in part (a).

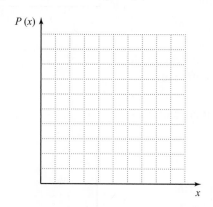

   d. Find the domain and range of the function.

85. A media store sells previously owned CDs and DVDs for $5 and $10 each, respectively. The store would like combined sales of the CDs and DVDs to be at least $500 per month.

   a. Express this information as an inequality, letting c represent the number of CDs sold and d represent the number of DVDs sold per month.

   b. Graph the inequality.

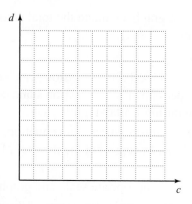

   c. Will the media store meet its sales expectation if 30 CDs and 25 DVDs are sold? Explain.

86. A student has a cell phone plan that charges $0.59 per minute for calls within the service provider's network and $1.49 per minute for calls outside the network. The student does not want his monthly bill to exceed $100 per month.

   a. Write an inequality that shows the number of minutes of calls within the network x and the number of minutes of calls outside the network y that the student can make in 1 mo.

   b. Graph the inequality.

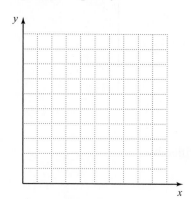

   c. Use the graph to find one possible solution.

• Check your answers on page A-28.

FOR
EXTRA
HELP

CHAPTER
Test Prep
VIDEOS

The Chapter Test Prep Videos with test solutions are available on DVD, in MyMathLab, and on YouTube (search "AkstIntermediate Alg" and click on "Channels").

To see if you have mastered the topics in this chapter, take this test.

1. Plot the points $P(-4, 3)$ and $Q(6, -2)$ on the coordinate plane.

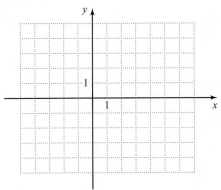

2. Compute the slope of the line that passes through the points $(4.2, -1)$ and $(1.7, 4)$.

3. Graph the line that passes through the point $(2, -3)$ and has slope $m = -4$ on the coordinate plane.

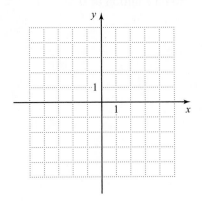

4. Given $P(-6, 9)$, $Q(-8, -5)$, $R(4, 10)$, and $S(-3, 11)$, determine whether $\overleftrightarrow{PQ}$ and $\overleftrightarrow{RS}$ are parallel, perpendicular, or neither.

5. Determine if $\left(\dfrac{1}{2}, -\dfrac{1}{4}\right)$ is a solution of the equation $3x - 2y = 1$.

6. Find the x- and y-intercepts of the graph of $6x - 9y + 3 = 0$.

Graph each equation.

7. $y = -\dfrac{3}{5}x + 3$

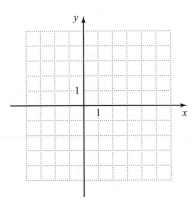

8. $24 = -8y$

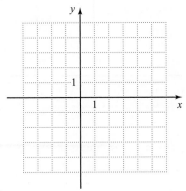

9. $7x - 14y = 56$

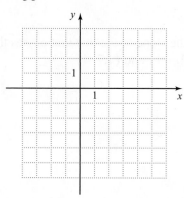

10. What is the equation of the line that is parallel to $y = 4x + 7$ and passes through the point $(0, -1)$?

11. A line passes through the points $(2, 5)$ and $(0, -2)$. What is its equation?

12. If $g(x) = -2x + 11$, evaluate $g(a + 3)$.

13. Graph the function $f(x) = -|x| + 1$ for $x \le 5$, and identify its domain and range.

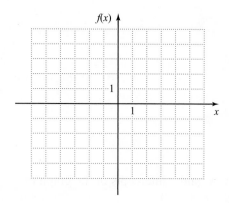

14. Is $(-10, -20)$ a solution of $0.4x - 1.2y \le -15$?

Graph each inequality.

15. $y > -3x + 3$

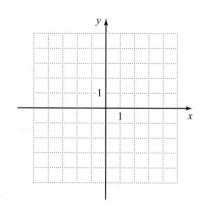

16. $2x - 8y \ge -16$

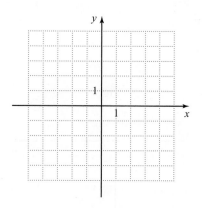

17. The table shows the price per troy ounce of silver for various years. (*Source:* wikipedia.org)

Point	Year t	Price p (in dollars)
A	2000	4.95
B	2005	7.31
C	2010	20.19

a. Choose an appropriate scale and plot the points on the coordinate plane. Then, sketch $\overline{AB}$ and $\overline{BC}$.

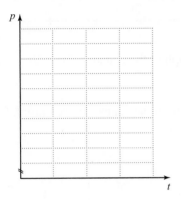

b. Calculate the slopes of $\overline{AB}$ and $\overline{BC}$.

c. Did the price per troy ounce of silver increase at the same rate over the ten years? Explain.

18. A florist charges a bride $4 for each of the r roses and $8 for each of the l lilies used in the floral arrangements for her wedding. The total bill for the floral arrangements was $1200.

a. Express this information as a linear equation.

b. Graph this equation on the coordinate plane.

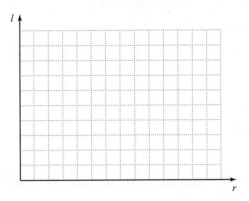

c. Identify the intercepts. Explain the significance of the r- and l-intercepts in this situation.

d. Use the graph to find the number of lilies in a floral arrangement with 150 roses.

19. Five minutes after starting her run, a jogger traveled $\frac{5}{8}$ mi. After 20 min of running at a constant speed, she traveled a distance of $2\frac{1}{2}$ mi.

a. Write an equation that shows the distance d the jogger traveled t min after she starts her run.

b. What is the slope of the line? Explain its meaning in this situation.

c. How long would it take her to run 10 mi?

20. A local farm allows people to pick apples from its orchards. The farm charges $8 for a 10-pound bag and $14 for a 20-pound bag. The farm would like to generate revenue of at least $4000 per week.

a. Express this information as an inequality, letting x represent the number of 10-pound bags and y represent the number of 20-pound bags sold per week.

b. Graph the inequality on the coordinate plane.

c. In a particular week, the farm sells 200 10-pound bags and 250 20-pound bags. Did the farm meet its revenue goal? Explain.

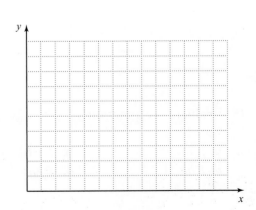

• Check your answers on page A-31.

Cumulative Review Exercises

To help you review, solve these problems.

1. Replace the ▢ with $<$, $>$, or $=$ to make a true statement: $-|3.2|$ ▢ $-|-3.2|$

2. Calculate: $4[25 - 3(8 - 11)^2] \div 2 + (-6)$

3. What property of real numbers is illustrated by the statement $4 \cdot a + 4 \cdot b = 4(a + b)$?

4. Simplify: $\left(\dfrac{2x^2}{y^{-1}}\right)^{-3}$

5. Simplify: $-5n - 2(1 - n) + 7$

6. Solve: $12 - (4x + 9) = 8 - 7x$

7. Solve: $\dfrac{1}{2}x + 2 \geq \dfrac{2}{3}x + 3$

8. Solve and graph: $-6 < 14 - 2x \leq 0$

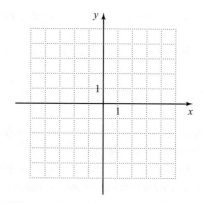

9. Solve: $10 - |3x + 8| = 4$

10. For the graph of the equation $2x - 8y + 4 = 0$, find the x- and y-intercepts.

11. Graph the line that passes through the point $(2, 5)$ with slope 1.

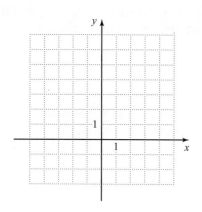

12. Write the coordinates next to each point.

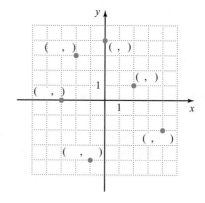

13. Graph the equation $y = -\frac{1}{2}x + 2$ using the slope and y-intercept.

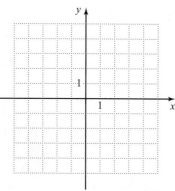

14. Graph the inequality $y > 2x - 1$.

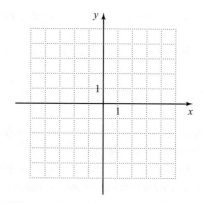

15. Evaluate the function $f(x) = 1.2x - 4$ for the given values.
    a. $f(7)$
    b. $f(-6)$
    c. $f(a)$
    d. $f(3a)$

16. In the human body, there are approximately 10 trillion cells and 1×10^{14} bacteria. What is the ratio of bacteria to cells?

17. A mail-order company charges $5 for shipping on all orders up to $100. For each dollar over that amount, the company charges 5% for shipping. If a customer pays a total of $8.75 for shipping, what was the amount of the order?

18. A chemist needs 20 L of a 45% saline solution. She has an 80% saline solution and a 30% saline solution available in the lab. How much of each solution does she need to mix in order to make the required solution?

19. For next month, you budget at most $50 for downloading movies onto your computer. Your favorite online movie store charges $15 to download a new movie and $5 to download an old movie.

  a. Express this information as an inequality, letting *x* represent the number of new movies and *y* the number of old movies that you download.

  b. Graph the inequality.

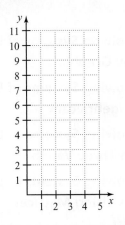

  c. Use the graph to find a solution of the inequality.

20. On a particular day, pure gold was valued at about $48 per gram. Pure gold jewelry is said to be 24 karat. (*Sources:* goldprice.org and wikipedia.org)

  a. Write the relationship between the weight *w* (in grams) of a 24 karat gold ring and the value $V(w)$ of the gold (in dollars).

  b. Find $V(5.1)$ to the nearest whole number, and interpret its meaning.

  c. Graph the function in part (a).

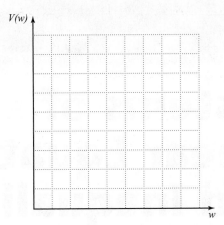

• Check your answers on page A-31.

Systems of Linear Equations and Inequalities

Graphing and College Enrollments

In recent decades, both male and female college enrollments have been growing. However, the number of female college students has been increasing at a significantly higher rate. In the quadrant to the right, the line $y = 0.19x + 4.1$ represents the male enrollment, and the line $y = 0.26x + 3.4$ represents the female enrollment for various years. The graphs of these equations intersect approximately at the point $(10, 6)$, indicating that in 1980 the numbers of men in college and of women in college were very close in value—about 6 million.

Historically, in the 1970s, more men than women had attended college. However, starting in the next decade, the trend reversed, with more females than males in attendance. Furthermore, this "college gender gap" has continued to widen. The number by which the female enrollment exceeds the male enrollment (about 1 million in 1990) is projected to reach some 3 million by 2020. (*Source:* National Center for Educational Statistics)

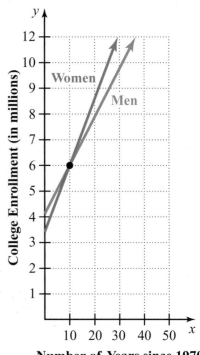

Number of Years since 1970

CHAPTER 4   PRETEST

To see if you have already mastered the topics in this chapter, take this test.

1. Is $(1, 4)$ a solution of the following system?

$$2x - 3y = -10$$
$$5x + 2y = 13$$

Solve by graphing.

2. $y = 3x - 4$
   $y = -(2x - 1)$

3. $9x - 4y = -8$
   $-3x + 2y = 4$

4. $5y - 20 = 0$
   $6x + 8y = 8$

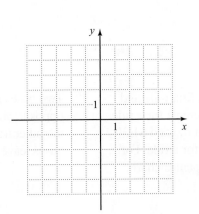

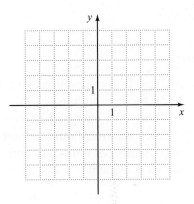

Solve by substitution.

5. $y = 8x + 5$
   $x = \dfrac{1}{2}y - 1$

6. $4x - 2y = 9$
   $y = 3 + 2x$

7. $u + 7v = 8$
   $6u - v = 5$

Solve by elimination.

8. $x + y = 10$
   $3x - y = 6$

9. $5a - 3b = -1$
   $13a - 3b = 1$

10. $0.5x - 1.5y = 1.25$
    $-6x + 18y = -15$

11. Is $(2, -3, 1)$ a solution of the following system?

$$2x - y + z = 8$$
$$x + 3y - 3z = 10$$
$$3x + 2y + z = 1$$

Solve.

12. $x + y + z = 0$
    $2x - y + 3z = 3$
    $x + 2y - z = -3$

13. $3x - 2y = -13$
    $\dfrac{1}{2}y - \dfrac{1}{4}z = 1$
    $x - 2z = 2$

14. Solve the following system using matrices:

$$4x + 5y - z = -4$$
$$x - 2y + z = 5$$
$$2x + y = 5$$

15. Graph the solutions of the system of inequalities:

$$y < \frac{1}{2}x - 1$$
$$y \geq -3x + 1$$
$$y \geq -4$$

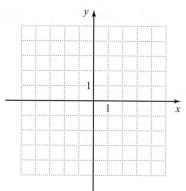

Solve.

16. Red and Black Cab charges $2.50 for the first mile and $1.50 for each additional mile (or part thereof) for a ride. Midtown Taxi charges $2.80 for the first mile and $1.30 for each additional mile (or part thereof) for a ride. For what number of miles will both taxi companies charge the same amount?

17. Two buses leave a bus depot at the same time traveling in opposite directions. One bus travels at a rate of 54 mph and the other travels at a rate of 60 mph. How long after the buses leave will they be 171 mi apart?

18. A lab technician needs 20 L of an antiseptic solution that is 40% alcohol. How can she combine an antiseptic that is 25% alcohol with an antiseptic that is 50% alcohol to obtain the required amount?

19. Popcorn prices at a movie theater's concession stand are shown in the table below.

Size	Price
Small	$6
Medium	$7
Large	$8

On Wednesday, the concession stand sold 807 bags of popcorn. The number of large bags sold was 10 more than twice the number of medium bags sold. If the concession stand collected a total of $5282, how many of each size bag of popcorn were sold?

20. A manager is planning to buy turkey sandwiches and roast beef sandwiches for the annual company party. Due to budget constraints, the most he can spend on this food is $400. A turkey sandwich costs $4.50 and a roast beef sandwich $6.50. He needs to buy at least 50 turkey sandwiches and 20 roast beef sandwiches.

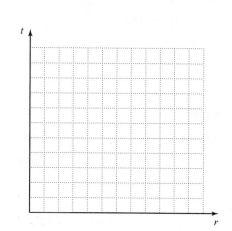

  a. Express this information as a system of inequalities, letting *t* represent the number of turkey sandwiches and *r* the number of roast beef sandwiches.

  b. Graph the solutions of the system.

  c. Determine one example of a possible purchase that he could make.

• Check your answers on page A-32.

4.1 Solving Systems of Linear Equations by Graphing

OBJECTIVES

A To determine whether an ordered pair is a solution of a system of linear equations

B To solve a system of linear equations by graphing

C To solve applied problems involving systems of linear equations

What Systems of Linear Equations Are and Why They Are Important

In Chapter 3, we discussed single linear equations in two variables that describe many situations. Now, we consider more complex situations in which the relationship between two variables is described by a *pair* of linear equations called a *system*.

Consider the following problem: A video shop sold a total of 80 DVDs during a one-day sale. If the shop sold three times as many \$10 DVDs as those priced at \$12, how many of each did the shop sell by the end of the day?

We can solve this problem using two equations. Let x represent the number of \$10 DVDs sold and y represent the number of \$12 DVDs sold.

Word Sentence:	The number of \$10 DVDs	plus	The number of \$12 DVDs	is	The total number of DVDs
Equation:	x	$+$	y	$=$	80

Word Sentence:	The number of \$10 DVDs	is	3 times the number of \$12 DVDs
Equation:	x	$=$	$3y$

The two equations are:

$$x + y = 80$$
$$x = 3y$$

Introduction to Systems of Linear Equations

In the preceding example, we need to consider two equations in order to solve the problem.

> **DEFINITION**
>
> A **system of equations** consists of two or more equations considered simultaneously, that is, together.

If we graph the equations

$$x + y = 80$$
$$x = 3y$$

on the same coordinate plane, we see that the graphs intersect in a single point, namely $(60, 20)$. This suggests that the point lies on the graph of both equations and is, therefore, a *solution* of both equations.

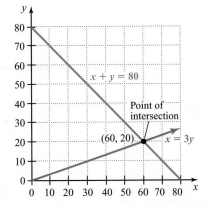

We can confirm that $(60, 20)$ is a solution of both equations in the system

$$x + y = 80$$
$$x = 3y$$

by substituting 60 for x and 20 for y in the original equations, and then checking if *both* equations are true.

$$x + y = 80 \rightarrow 60 + 20 \stackrel{?}{=} 80 \quad \rightarrow 80 = 80 \quad \text{True}$$
$$x = 3y \rightarrow \quad 60 \stackrel{?}{=} 3(20) \rightarrow 60 = 60 \quad \text{True}$$

Since the ordered pair $(60, 20)$ satisfies both equations, we say that it is the solution of the system. So the video shop sold 60 $10 DVDs and 20 $12 DVDs during the one-day sale.

> **DEFINITION**
>
> A **solution** of a system of equations in two variables is an ordered pair of numbers that makes both equations in the system true.

Note that systems of equations are sometimes written with large braces, as in

$$\begin{Bmatrix} x + y = 80 \\ x = 3y \end{Bmatrix} \quad \text{or} \quad \begin{cases} x + y = 80 \\ x = 3y \end{cases}$$

to emphasize that a number must satisfy both of the equations in the system to be a solution of the system.

EXAMPLE 1

Consider the following system:

$$2x - y = 9$$
$$x + 4y = 0$$

a. Is $(5, 1)$ a solution of the system?

b. Is $(4, -1)$ a solution of the system?

Solution

a. To determine if the ordered pair $(5, 1)$ is a solution of this system, we substitute 5 for x and 1 for y in the equations and check to see if both equations are true.

$$2x - y = 9 \rightarrow 2(5) - (1) \stackrel{?}{=} 9 \xrightarrow{\text{Simplifies to}} 9 = 9 \quad \text{True}$$
$$x + 4y = 0 \rightarrow \quad 5 + 4(1) \stackrel{?}{=} 0 \xrightarrow{\text{Simplifies to}} 9 = 0 \quad \text{False}$$

The ordered pair $(5, 1)$ is not a solution of the system because it does not satisfy both equations.

b. To determine if $(4, -1)$ is a solution, we substitute 4 for x and -1 for y in the equations and check to see if both equations are true.

$$2x - y = 9 \rightarrow 2(4) - (-1) \stackrel{?}{=} 9 \xrightarrow{\text{Simplifies to}} 9 = 9 \quad \text{True}$$
$$x + 4y = 0 \rightarrow \quad 4 + 4(-1) \stackrel{?}{=} 0 \xrightarrow{\text{Simplifies to}} 0 = 0 \quad \text{True}$$

The ordered pair $(4, -1)$ is a solution of the system because it satisfies both equations.

PRACTICE 1

Determine if each ordered pair is a solution of the system.

$$3x - 2y = 3$$
$$-x + 3y = 6$$

a. $(3, 3)$

b. $(-1, -2)$

Solving Systems by Graphing

In solving a system of linear equations, a question that immediately comes to mind is how many solutions the system has. Let's consider this question graphically by looking at some examples.

EXAMPLE 2

Solve the following system graphically:

$$x + y = 3$$
$$2x - y = 6$$

Solution   Begin by writing each equation in slope-intercept form to make it easy to graph.

$$x + y = 3 \rightarrow y = -x + 3$$
$$2x - y = 6 \rightarrow y = 2x - 6$$

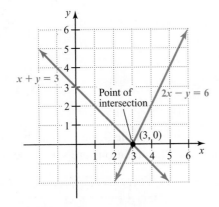

The lines appear to intersect at the point $(3, 0)$.

Check

Since solving systems of equations by graphing may result in approximate solutions, we can confirm that $(3, 0)$ is the exact solution by substituting 3 for x and 0 for y into the original equations.

$$x + y = 3 \rightarrow \quad 3 + 0 \stackrel{?}{=} 3 \quad \xrightarrow{\text{Simplifies to}} \quad 3 = 3 \quad \text{True}$$
$$2x - y = 6 \rightarrow 2(3) - 0 \stackrel{?}{=} 6 \quad \xrightarrow{\text{Simplifies to}} \quad 6 = 6 \quad \text{True}$$

So the solution of the system is $(3, 0)$.

PRACTICE 2

Solve the following system graphically:

$$x - y = 6$$
$$2x + y = 3$$

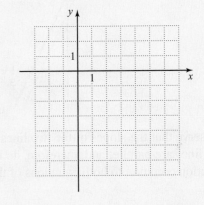

When the graphs of the equations intersect in exactly one point, as in Example 2, the system has just *one solution*. In this case, we say that the system is *consistent* and the equations are *independent*.

EXAMPLE 3

Solve graphically: $2x + y = 4$
$2x + y = 3$

Solution

$$2x + y = 4 \rightarrow y = -2x + 4$$
$$2x + y = 3 \rightarrow y = -2x + 3$$

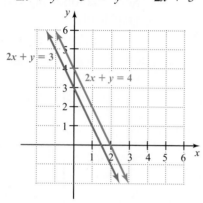

We observe that the slopes of the two lines are the same, namely -2.
So the lines are parallel and, therefore, do not intersect. That is, there is
no solution common to both equations of the system.

PRACTICE 3

Solve graphically: $3x - y = 2$
$3x - y = 4$

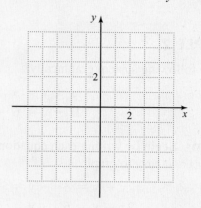

When the graphs are parallel lines, as in Example 3, the system has *no solution*. In this
case, the system is said to be *inconsistent* and the equations are said to be *independent*.

EXAMPLE 4

Solve the following system by graphing: $2x - 3y = -6$
$-6x + 9y = 18$

Solution

$$2x - 3y = -6 \rightarrow y = \frac{2}{3}x + 2$$

$$-6x + 9y = 18 \rightarrow y = \frac{2}{3}x + 2$$

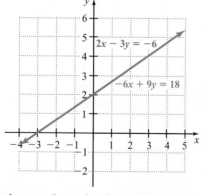

We graph the equations and see that both equations have the same
graph. Any point on this line—for instance, $(0, 2)$—will satisfy both
equations. So there are an infinite number of solutions.

PRACTICE 4

Solve the following system by
graphing:

$$-15x + 5y = 10$$
$$-3x + \ \ y = 2$$

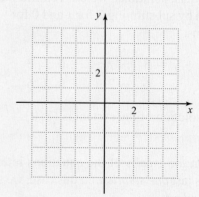

When the graphs are the same line, as in Example 4, the system has *infinitely many solutions*. In this case, the system is said to be *consistent* and the equations are said to be *dependent*, that is, any solution of one equation of the system is also a solution of the other.

The following table summarizes the three types of systems shown in Examples 2 through 4:

Number of Solutions	Description of the System and Its Equations	Description of the System's Graphs	Possible Graphs
One solution	The system is **consistent** and the equations are **independent**.	The graphs intersect at exactly one point.	
No solution	The system is **inconsistent** and the equations are **independent**.	The graphs are parallel lines.	
Infinitely many solutions	The system is **consistent** and the equations are **dependent**.	The graphs are the same line.	

EXAMPLE 5

Two taxi companies compete for business in the same city. Checker Taxi Company charges $3 for the taxi drop plus $1.50 for each mile driven, while Blacktop Taxi Company charges $2.50 for the taxi drop plus $1.75 for each mile driven.

a. Express these relationships as a system of equations.

b. Use graphing to solve the system and interpret the results.

Solution

a. Let C represent the total charge for each company and x the number of miles driven.

Checker Taxi:

Total charge equals $3 for the taxi drop plus $1.50 for each mile driven.

$$C = 3 + 1.5x$$

Blacktop Taxi:

Total charge equals $2.50 for the taxi drop plus $1.75 for each mile driven.

$$C = 2.5 + 1.75x$$

The system of equations is:

$$C = 3 + 1.5x$$
$$C = 2.5 + 1.75x$$

EXAMPLE 5 (continued)

b. Graphing each equation, we get:

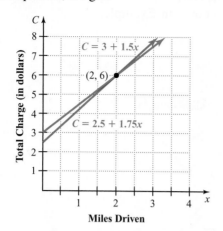

Miles Driven

The lines appear to intersect at $(2, 6)$; that is, $x = 2$ and $C = 6$. This means that the two taxi companies charge the same amount—namely $6—when the taxi trip is 2 mi.

b. Use graphing to solve the system and interpret the results.

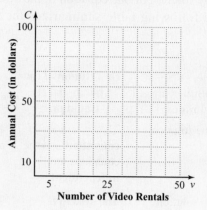

Number of Video Rentals

Businesses often use cost and revenue equations (or functions) to predict sales, adjust prices, and determine profit and loss. The point at which revenue (income) equals cost (expenses) is called the *break-even point*.

EXAMPLE 6

A clothing company makes jackets and sells them for $150 apiece. The company has fixed costs of $10,000 and variable costs of $50 per jacket.

a. Write the revenue function $R(x)$ from the sale of x jackets.

b. Write the cost function $C(x)$ for the production of x jackets.

c. Graph the functions as a system to determine the break-even point for production.

Solution

a. Since the company sells x jackets at $150 per jacket, the income from sales is $150x$. So the revenue function is $R(x) = 150x$.

b. The company's costs are $50 per jacket plus $10,000. So the cost function is $C(x) = 50x + 10{,}000$.

c. Graphing the system, we have:

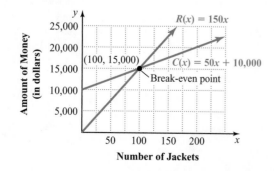

Number of Jackets

PRACTICE 6

A musical production costs $80,000 plus $5900 per performance. A sold-out performance makes $7500.

a. What is the revenue function $R(x)$ from x sold-out performances?

b. What is the cost function $C(x)$ for the production of x sold-out performances?

From the graph, it appears that the lines intersect at the point $(100, 15{,}000)$. So the ordered pair $(100, 15{,}000)$ represents the point at which the revenue is equal to the cost, that is, the break-even point. This means that if the company sells 100 jackets, it will break even with $15,000.

c. Graph the system to determine the number of sold-out performances the production will need in order to break even.

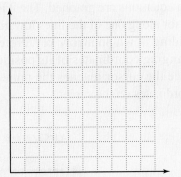

Solving Systems of Linear Equations on a Grapher

Both graphing calculators and computers with graphing software can help facilitate the process of solving systems of linear equations. As with the paper-and-pencil approach, a grapher displays the graphs of the equations that make up a system on the same coordinate plane. We can then use one of the special features of the grapher to read the coordinates of the point at which the graphed lines intersect, that is, the solution of the system.

The most common features of a grapher that help us to read the coordinates of the point of intersection are TRACE, ZOOM, and INTERSECT.

- *With the **TRACE** feature, the cursor runs along either of the graphed lines until it is positioned on or near the point of intersection; the coordinates of that point are then displayed.*
- *The **ZOOM** feature lets us position the cursor as close as we want to the point of intersection.*
- *The **INTERSECT** feature automatically calculates the point of intersection.*

Note that each of the features may give only an approximation for the point of intersection. However, the most accurate approximation of the point of intersection is given by the INTERSECT feature.

EXAMPLE 7

Use either a graphing calculator or graphing software to solve

$$2x - 3y = 6$$
$$3x = 1 - y$$

Solution Begin by solving each equation for y.

$$2x - 3y = 6 \xrightarrow{\text{Isolate } y.} y = \frac{2}{3}x - 2$$

$$3x = 1 - y \xrightarrow{\text{Isolate } y.} y = -3x + 1$$

PRACTICE 7

Use a grapher to solve the following system of equations:

$$5x + 2y = -2$$
$$y = 2x - 3$$

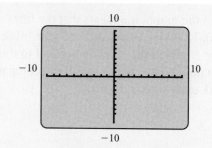

EXAMPLE 7 (continued)

Then, press the $\boxed{Y=}$ key, and enter $(2/3)x - 2$ to the right of **Y1 =** and $-3x + 1$ to the right of **Y2 =** . Set the viewing window. Then, press the $\boxed{GRAPH}$ key to display the coordinate plane on which the two equations are graphed. The **TRACE** feature can be used to move a cursor along one of the lines toward the intersection of the graphs by holding down an arrow key. Note that as the cursor is moved, the changing coordinates of its position will be displayed on the screen. Once the cursor reaches the point of intersection, you can read the coordinates on the screen.

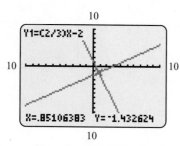

Using the TRACE feature

To get a better approximation of the solution, you can either activate the **ZOOM** feature to zoom in on the intersection point or activate the **INTERSECT** feature, often found in the **CALC** menu.

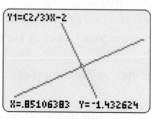

Using the ZOOM feature

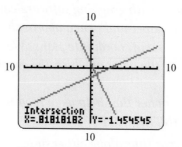

Using the INTERSECT feature

So the approximate solution here is $(0.818, -1.455)$.

Mathematically Speaking

Fill in each blank with the most appropriate term or phrase from the given list.

inconsistent	set	graph
solution	infinitely many	system
no	independent	one

1. A(n) _____ of equations is two or more equations considered simultaneously, that is, together.

2. A(n) _____ of a system of equations in two variables is an ordered pair of numbers that makes both equations in the system true.

3. If the graphs of a system of linear equations intersect at exactly one point, the system has _____ solution(s).

4. If the graphs of a system of linear equations are parallel lines, the system has _____ solution(s).

5. If the graphs of a system of linear equations are the same line, the system has _____ solution(s).

6. Systems that are _____ have no solution.

A *Determine whether the ordered pair is a solution of the system of equations.*

7. $x + y = 6$ $(2, 4)$
   $3x - y = 2$

8. $8x - 6y = -8$ $(-4, -4)$
   $-5x + 4y = 4$

9. $2x - 3y = 1$ $(4, -3)$
   $x + 2y = -2$

10. $3y - 5x = 8$ $(-1, 1)$
    $4y - 4x = 0$

11. $16x - 10y = 9$ $\left(\dfrac{1}{4}, -\dfrac{1}{2}\right)$
    $2x + 3y = -1$

12. $9x + 15y = -3$ $\left(-\dfrac{2}{3}, \dfrac{1}{5}\right)$
    $3x + 10y = 0$

B *Match the system of equations to its graph.*

13. $3x + 5y = 9$
    $3x - 2y = -12$

14. $3x + 4y = -6$
    $3x - 2y = 12$

15. $2x - 3y = 6$
    $4x - 6y = -18$

16. $2x - 3y = 6$
    $4x - 6y = 12$

a.

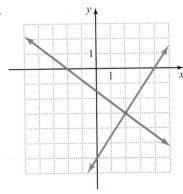

b.

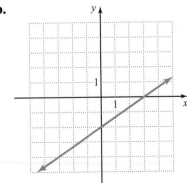

continued

c.

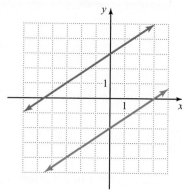

d.

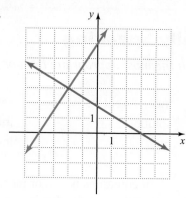

Solve by graphing.

17. $y = 2x$

$y = \dfrac{1}{3}x$

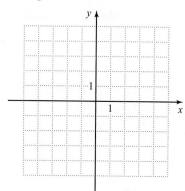

18. $y = -3x$

$y = \dfrac{1}{2}x$

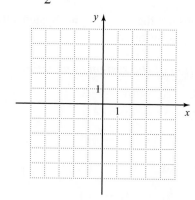

19. $y = -\dfrac{1}{2}x + 1$

$y = x - 5$

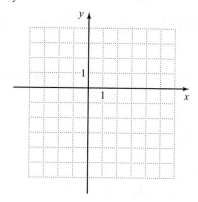

20. $y = 2x - 3$

$y = -\dfrac{1}{3}x + 4$

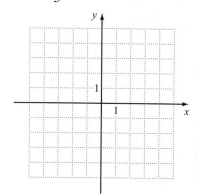

21. $x + y = -5$

$4x - 3y = 1$

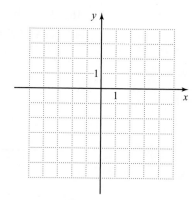

22. $x - y = -1$

$2x + 5y = -2$

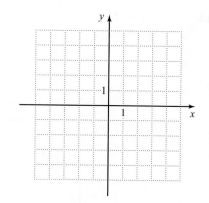

23. $x - y = -3$
  $y = 7 - 3x$

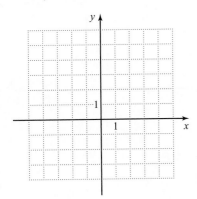

24. $x = 2y + 5$
  $y + x = 2$

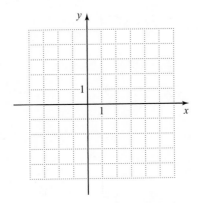

25. $4x = 12 - 3y$
  $-6y = 8x + 6$

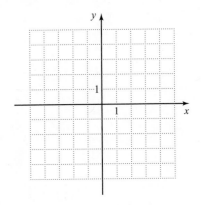

26. $x - 4y = -4$
  $-12y = 24 - 3x$

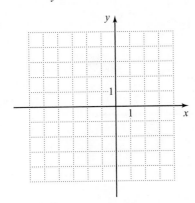

27. $y - 5 = -2$
  $3x = 6$

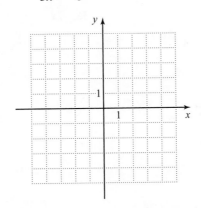

28. $x + 4 = 3$
  $5y = -20$

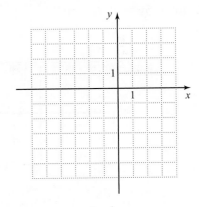

29. $3x - 6y = 12$
  $4y + 8 = 2x$

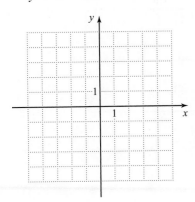

30. $6x - 12y = -24$
  $5x + 20 = 10y$

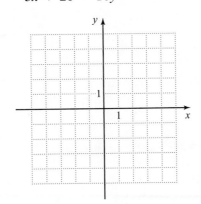

31. $-2x + 8y = 12$
  $3x + 2y = -4$

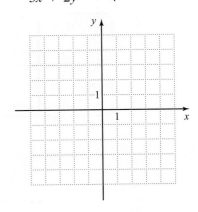

32. $4y + 3x = -15$
$2y - 5x = -1$

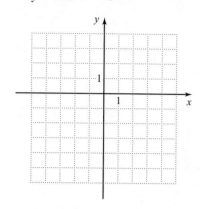

33. $0.5x - 1.5y = 1$
$2x - 6y - 4 = 0$

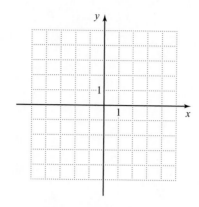

34. $8x + 16y + 48 = 0$
$1.6x + 3.2y = -9.6$

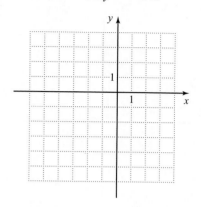

Solve the system using a grapher. Round solutions to the nearest thousandth.

35. $6x - 5y = 9$
$y = 3x - 2$

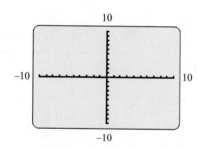

36. $7x + 3y = 10$
$y = 1 - 8x$

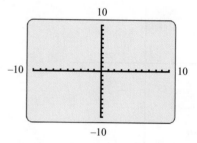

37. $3.7x + 2.1y = -4$
$-0.6x - 2y = 6.8$

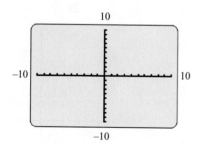

38. $5.3x - 7.2y = -3.9$
$-1.4x + 0.1y = -5$

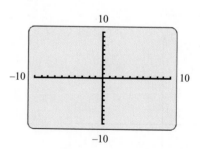

Mixed Practice

Solve.

39. Which system of equations matches the given graph?

a. $4x - 3y = 12$
 $\quad x + 2y = 2$

b. $4x + 3y = 12$
 $\quad -x + 2y = 2$

c. $4x + 3y = 12$
 $\quad x + 2y = 2$

d. $4x - 3y = 12$
 $\quad -x + 2y = 2$

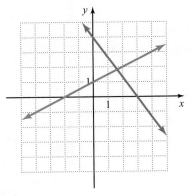

40. Determine whether the ordered pair $(-2, 1)$ is a solution of the following system:

$$3x - 5y = -11$$
$$-4x + 2y = -6$$

Solve by graphing.

41. $y = -2x + 4$

 $y = \dfrac{1}{3}x - 3$

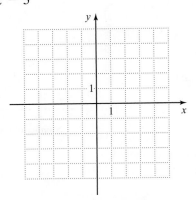

42. $x + 3y = 9$
 $\quad 2x = -6y - 12$

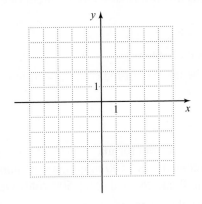

43. $2x - 3y = 6$
 $\quad 6y + 12 = 4x$

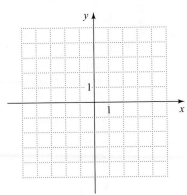

44. $2x + y = 3$
 $\quad -2y = 4 - x$

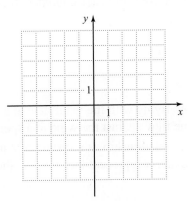

Applications

C *Solve.*

45. A state college received 12,000 applications for admission. Three times as many applications were from in-state residents as from out-of-state residents.

   a. Write the information as a system of equations, letting *x* represent the number of applications from in-state residents and *y* represent the number of applications from out-of-state residents.

   b. Graph the system.

   c. How many applications were from in-state residents? How many were from out-of-state residents?

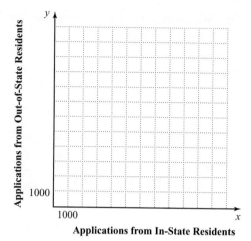

46. Last year, the number of male babies *m* was 50 fewer than the number of female babies *f* born at a local hospital. The records department reported a total of 750 births that year.

   a. Express this information as a system of equations.

   b. Graph the system.

   c. Find the number of male babies and female babies born at the hospital last year.

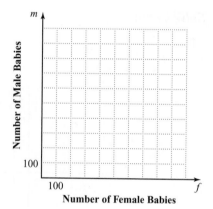

47. Two candidates ran for mayor. The number of votes *w* that the winning candidate received was 2500 more than the number of votes *l* received by the losing candidate. A total of 11,500 votes were cast in the election.

   a. Express the given information as a system of equations.

   b. Graph the system.

   c. How many votes did each candidate receive?

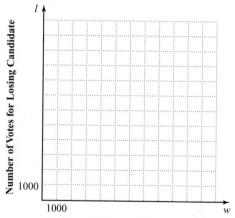

48. An airplane flying with the wind flew at a speed of 480 mph. On the return trip, the airplane flew against the wind at a speed of 400 mph.

   a. Write the relationship as a system of equations, letting *p* represent the speed of the airplane and *w* represent the speed of the wind.

   b. Graph the system.

   c. What was the speed of the plane in calm air and the speed of the wind?

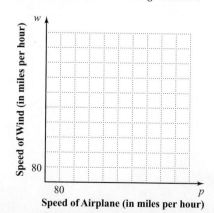

49. A medical supply manufacturer makes plastic sharps containers for needle disposal. The monthly cost to manufacture the sharps containers is $875 plus $1.50 per container. The manufacturer sells the containers for $4 each. How many containers must the company sell per month in order to break even?

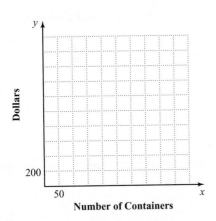

50. A shoe company sells hiking boots for $65 per pair. The company's fixed costs are $5000 and the variable costs are $15 per pair of hiking boots produced. How many pairs must the company sell in order to break even?

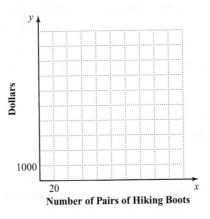

51. A student works two part-time jobs to help pay for college. Her job at a coffee shop pays $8 per hour and her job at an office pays $12 per hour. Because of her class schedule, she works only 20 hr per week. If she earns $216 per week, how many hours does she work at each job?

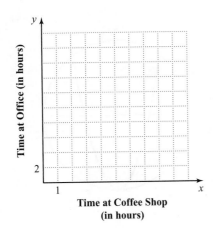

52. The owner of a car rental company added 30 new cars to his fleet. Each new compact car cost $9000 and each new sedan cost $15,000. If the owner spent $330,000 for the new cars, how many of each type of car did he purchase?

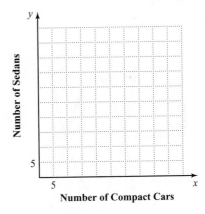

• Check your answers on page A-33.

MINDStretchers

Technology

1. Consider the equation $2x - 7 = 5x + 8$.

   a. Graph each side of the equation using a grapher, letting $Y1 = 2x - 7$ and $Y2 = 5x + 8$.

   b. What is the x-coordinate of the intersection of the graphs in part (a)?

   c. Solve the original equation. What do you notice about the solution of the equation and your answer in part (b)?

   d. Explain how you can solve a linear equation in one variable by graphing.

Mathematical Reasoning

2. Consider the following system of equations: $y = Ax + B$
$$y = Cx + D$$

   Under what circumstances will the system have exactly one solution? Under what circumstances will the system have no solution? Under what circumstances will the system have infinitely many solutions?

Groupwork

3. Not all systems of equations involve linear equations. Each of the following is a graph of a nonlinear system of equations. With a partner, explain how you can find the solution(s) of each system. Then, state the solution(s) of each system.

   a. $y = x^2 - 4$
   $y = -x^2 + 4$

   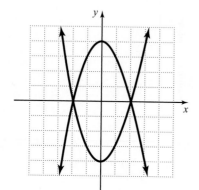

   b. $x^2 + y^2 = 25$
   $x^2 - y = 5$

   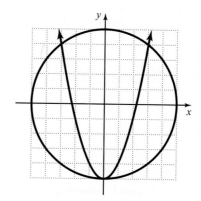

4.2 Solving Systems of Linear Equations Algebraically

In the previous section, we approximated solutions of systems of linear equations by graphing the equations. In this section, we use algebra to find exact solutions of systems of linear equations.

Here, we focus on two algebraic methods for solving systems of equations in two variables, namely the *substitution method* and the *elimination* (or *addition*) *method*. First, let's consider the substitution method.

A To solve a system of linear equations by substitution

B To solve a system of linear equations by elimination

C To solve applied problems involving systems of linear equations

Solving Systems by the Substitution Method

In applying the **substitution method**, we solve a system by first solving one linear equation in one variable. This method is especially suited to solving systems in which a variable is isolated.

EXAMPLE 1

Solve by the substitution method:

(1) $\quad 3x + 5y = -1$
(2) $\quad\quad\quad y = -3x - 5$

Solution Since equation (2) is solved for y in terms of x, we can substitute the expression $-3x - 5$ for y in equation (1), and then solve the resulting equation for x.

(1)
$$3x + 5y = -1$$
$$3x + 5(-3x - 5) = -1 \quad \text{Substitute } -3x - 5 \text{ for } y.$$
$$3x - 15x - 25 = -1$$
$$-12x - 25 = -1$$
$$-12x = 24$$
$$x = -2$$

Now that we have found the value of x, we can substitute it in either equation of the original system to find the corresponding value of y. Substituting -2 for x in equation (2), we get:

(2)
$$y = -3x - 5$$
$$y = -3(-2) - 5$$
$$y = 6 - 5$$
$$y = 1$$

Since $x = -2$ and $y = 1$, the solution of the system is $(-2, 1)$.

Check

We can check the solution by substituting -2 for x and 1 for y in both the original equations.

(1)
$$3x + 5y = -1$$
$$3(-2) + 5(1) \overset{?}{=} -1$$
$$-6 + 5 \overset{?}{=} -1$$
$$-1 = -1 \quad \text{True}$$

(2)
$$y = -3x - 5$$
$$1 \overset{?}{=} -3(-2) - 5$$
$$1 \overset{?}{=} 6 - 5$$
$$1 = 1 \quad \text{True}$$

The ordered pair $(-2, 1)$ is the solution of the system since it satisfies both equations. Note that throughout the remainder of this section, the check is left as an exercise.

PRACTICE 1

Solve by the substitution method:

(1) $\quad\quad\quad x = 2y - 4$
(2) $\quad -4x + y = 2$

In general, we use the following procedure to solve a system of linear equations in two variables by substitution:

> ### To Solve a System of Linear Equations by Substitution
>
> - In one of the equations, solve for either variable in terms of the other variable.
>
> - In the other equation, substitute the expression equal to the variable found in the previous step. Then, solve the resulting equation for the remaining variable.
>
> - Substitute the value found in the previous step into either of the original equations and solve for the other variable.
>
> - Check the proposed solution in both equations of the original system.

EXAMPLE 2

Solve for x and y by substitution:

$$\textbf{(1)} \quad x + 3y = 5$$
$$\textbf{(2)} \quad 2x - 4y = -5$$

Solution We first need to solve x or y in either of the equations. Let's solve for x in equation (1) since the coefficient of x is 1.

$$\textbf{(1)} \quad x + 3y = 5$$
$$x = -3y + 5$$

Next, substitute the expression $-3y + 5$ for x in equation (2), and then solve the resulting equation for y.

$$\textbf{(2)} \qquad 2x - 4y = -5$$
$$2(-3y + 5) - 4y = -5$$
$$-6y + 10 - 4y = -5$$
$$-10y + 10 = -5$$
$$-10y = -15$$
$$y = \frac{-15}{-10} = \frac{3}{2}$$

To find x, we substitute $\frac{3}{2}$ for y in equation (1).

$$\textbf{(1)} \qquad x + 3y = 5$$
$$x + 3\left(\frac{3}{2}\right) = 5$$
$$x + \frac{9}{2} = 5$$
$$x = \frac{10}{2} - \frac{9}{2} = \frac{1}{2}$$

So the solution is $\left(\frac{1}{2}, \frac{3}{2}\right)$. Note that neither variable is an integer, so that if we had solved this system by the graphing method, we would have had to estimate the coordinates of the point of intersection.

PRACTICE 2

Solve for x and y by substitution:

$$\textbf{(1)} \quad 2x - y = 1$$
$$\textbf{(2)} \quad 2x - 5y = -1$$

EXAMPLE 3

Solve by the substitution method: **(1)** $3x + y = -2$
                                      (2) $2 - y = 3x$

Solution   Solve for y in equation (1) since the coefficient of y is 1.

$$\textbf{(1)} \quad 3x + y = -2$$
$$y = -3x - 2$$

Substitute the expression $-3x - 2$ for y in equation (2), and then solve the resulting equation for x.

$$\textbf{(1)} \qquad 2 - y = 3x$$
$$2 - (-3x - 2) = 3x$$
$$2 + 3x + 2 = 3x$$
$$4 = 0 \qquad \text{False}$$

The statement $4 = 0$ is false. This means that there is no value of x that makes the last equation true. Therefore, the original system of equations has no solution.

PRACTICE 3

Solve by the substitution method:

(1) $-2x + y = 4$
(2) $4x = 7 + 2y$

EXAMPLE 4

Solve by substitution: **(1)** $6x - 3y = -12$
                          (2) $2y = 8 + 4x$

Solution   Let's solve for y in equation (2).

$$\textbf{(2)} \quad 2y = 8 + 4x$$
$$y = 2x + 4 \qquad \text{Divide each side by 2.}$$

Substitute the expression $2x + 4$ for y in equation (1), and then solve for x.

$$\textbf{(2)} \qquad 6x - 3y = -12$$
$$6x - 3(2x + 4) = -12$$
$$6x - 6x - 12 = -12$$
$$-12 = -12 \qquad \text{True}$$

The statement $-12 = -12$ is true. This means that every value of x makes the last equation true. Therefore, the original system of equations has infinitely many solutions, namely all ordered pairs that satisfy both equations.

PRACTICE 4

Solve by substitution:

(1) $3y = 9x + 6$
(2) $6x - 2y = -4$

How would you describe the graphs of each system given in Examples 1 through 4? Explain.

TIP   When solving a system of linear equations in two variables algebraically,
- if the result is a false statement, then the system has no solution.
- if the result is a true statement, then the system has infinitely many solutions.

Solving Systems by the Elimination Method

The second algebraic method that we consider for solving a system of linear equations in two variables is the **elimination** (or **addition**) **method**. As in the substitution method, the elimination method involves changing a given system of two equations in two variables to one equation in one variable. The elimination method is especially suited to solving systems in which coefficients of one variable are opposites.

The elimination method for solving systems is based on the following property of equality: If $a = b$ and $c = d$, then $a + c = b + d$. This property allows us to "add" equations.

EXAMPLE 5

Use the elimination method to solve:
$$\textbf{(1)} \quad 7x + y = 22$$
$$\textbf{(2)} \quad 5x - y = 14$$

Solution First, we decide which variable to eliminate. Since the coefficients of the y-terms in the two equations are opposites (1 and -1), we eliminate y if we add the equations.

$$\textbf{(1)} \quad 7x + y = 22$$
$$\textbf{(2)} \quad \underline{5x - y = 14}$$
$$12x \qquad = 36 \qquad \text{Add the equations.}$$
$$x = 3$$

To find y, we substitute 3 for x in either of the original equations. Let's choose equation (1).

$$\textbf{(1)} \qquad 7x + y = 22$$
$$7(3) + y = 22$$
$$21 + y = 22$$
$$y = 1$$

So the solution is $(3, 1)$.

PRACTICE 5

Use the elimination method to solve:

$$\textbf{(1)} \qquad 7x + 2y = 10$$
$$\textbf{(2)} \quad -7x + \ y = -16$$

EXAMPLE 6

Solve by the elimination method:
$$\textbf{(1)} \quad 2x - 6y = 42 - 2x$$
$$\textbf{(2)} \quad 2x - 3y = 23$$

Solution We begin by writing equation (1) in the general form $Ax + By = C$.

$$\textbf{(1)} \quad 2x - 6y = 42 - 2x \xrightarrow{\text{Write in general form.}} 4x - 6y = 42$$
$$\textbf{(2)} \quad 2x - 3y = 23 \qquad\qquad\qquad\qquad 2x - 3y = 23$$

Note that adding the two equations will not eliminate a variable. Instead, if we multiply each side of equation (2) by -2, the coefficients of y in the two equations will be opposites.

$$\textbf{(1)} \quad 4x - 6y = 42 \qquad\qquad\qquad 4x - 6y = 42$$
$$\textbf{(2)} \quad 2x - 3y = 23 \xrightarrow{\text{Multiply by } -2.} \underline{-4x + 6y = -46}$$
$$0 = -4 \quad \begin{array}{l}\text{Add the}\\\text{equations.}\end{array}$$

The statement $0 = -4$ is false for all values of the variables, so the original system has no solution.

PRACTICE 6

Use the elimination method to solve:

$$\textbf{(1)} \quad 3x - \ y = 0$$
$$\textbf{(2)} \quad 3x - 3y = 5 - 6x$$

EXAMPLE 7

Use the elimination method to solve:

(1) $\frac{3}{4}x + \frac{1}{3}y = -\frac{1}{2}$

(2) $\frac{1}{2}x - \frac{2}{3}y = -3$

Solution Note that if we multiply equation (1) by 2, we can eliminate the y-terms by adding.

(1) $\frac{3}{4}x + \frac{1}{3}y = -\frac{1}{2}$ $\xrightarrow{\text{Multiply by 2.}}$ $\frac{3}{2}x + \frac{2}{3}y = -1$

(2) $\frac{1}{2}x - \frac{2}{3}y = -3$ $\underline{\frac{1}{2}x - \frac{2}{3}y = -3}$

 $\frac{4}{2}x = -4$ Add the equations.

 $2x = -4$

 $x = -2$

To solve for y, we substitute -2 for x in equation (2).

(2) $\frac{1}{2}x - \frac{2}{3}y = -3$

$\frac{1}{2}(-2) - \frac{2}{3}y = -3$

$-1 - \frac{2}{3}y = -3$

$-\frac{2}{3}y = -2$

$y = 3$

So the solution is $(-2, 3)$.

EXAMPLE 8

Solve by elimination:

(1) $5x - 7y = 24$

(2) $3x - 5y = 16$

Solution This system is different from the previous examples because there is no single integer that we can multiply either equation by that will eliminate a variable when we add the equations. Instead, we must multiply *both* equations by integers that lead to a variable being eliminated. Multiplying equation (1) by 3 and equation (2) by -5 eliminates the x terms when the equations are added.

(1) $5x - 7y = 24$ $\xrightarrow{\text{Multiply by 3.}}$ $15x - 21y = 72$

(2) $3x - 5y = 16$ $\xrightarrow{\text{Multiply by } -5.}$ $\underline{-15x + 25y = -80}$ Add the equations.

 $4y = -8$

 $y = -2$

Solve by the elimination method:

(1) $\frac{3}{4}x + \frac{1}{2}y = 5$

(2) $-\frac{1}{4}x + \frac{1}{6}y = -\frac{7}{3}$

Solve by elimination:

(1) $4x + 3y = -13$

(2) $5x - 2y = -22$

EXAMPLE 8 (continued)

Now, let's substitute -2 for y in equation (2), and then solve for x.

$$
\begin{aligned}
(2) \qquad 3x - 5y &= 16 \\
3x - 5(-2) &= 16 \\
3x + 10 &= 16 \\
3x &= 6 \\
x &= 2
\end{aligned}
$$

So the solution is $(2, -2)$.

The preceding examples suggest the following procedure for solving systems of linear equations in two variables by elimination:

To Solve a System of Two Linear Equations by Elimination

- Write both equations in the general form $Ax + By = C$.

- Choose the variable that you want to eliminate.

- If necessary, multiply one or both equations by appropriate numbers so that the coefficients of the variable to be eliminated are opposites.

- Add the equations. Then, solve the resulting equation for the remaining variable.

- Substitute the value found in the previous step into either of the original equations and solve for the other variable.

- Check by substituting the values in both equations of the original system.

We can use our knowledge of the substitution method and the elimination method to solve some applied problems.

First, let's consider a problem that involves quantities of items sold and the total value of these items. Such problems are common in business.

EXAMPLE 9

On a particular airline route, a full-price coach ticket costs $310 and a discounted coach ticket costs $210. On one of these flights, there were 180 passengers, which resulted in total ticket sales of $45,800. How many of each type of ticket were sold?

Solution Let f represent the quantity of full-price coach tickets sold and d represent the quantity of discounted coach tickets sold. We can organize the given information in a table as follows:

	Kind of Ticket		
	Full-Price Coach	Discounted Coach	Total
Quantity Sold	f	d	180
Price Per Ticket	310	210	
Total Ticket Sales	$310f$	$210d$	45,800

PRACTICE 9

An outlet store is having a sale on women's shoes. Some shoes are selling for $20 a pair and others for $25 a pair. At the end of the day, the total receipts for the sale of 65 pairs of shoes were $1500. How many pairs of $20 shoes were sold?

We can use this table to write the following system:

(1) $\qquad f + d = 180$ ← Total quantity sold

(2) $\quad 310f + 210d = 45,800$ ← Total ticket sales

Let's solve the system using the substitution method. In applying the substitution method, we begin by solving equation (1) for f.

(1) $\quad f + d = 180 \xrightarrow{\text{Solve for } f.} f = 180 - d$

Next, we substitute the expression $180 - d$ for f in equation (2), and then solve for d.

$$
\begin{aligned}
\textbf{(2)} \qquad 310f + 210d &= 45,800 \\
310(180 - d) + 210d &= 45,800 \\
55,800 - 310d + 210d &= 45,800 \\
-100d &= -10,000 \\
d &= 100
\end{aligned}
$$

To find f, we substitute 100 for d in either of the original equations. Let's choose equation (1).

$$
\begin{aligned}
\textbf{(2)} \qquad f + d &= 180 \\
f + 100 &= 180 \\
f &= 80
\end{aligned}
$$

Since $f = 80$ and $d = 100$, there were 80 full-price tickets and 100 discounted tickets sold.

In Section 2.1, we solved *motion* problems using one linear equation. Here we consider motion problems using a system of two linear equations.

EXAMPLE 10

An earthquake sent shock waves through the earth. A primary wave travels at a rate of 5 mi/sec, and a secondary wave, which starts at the same time, travels at a rate of 3 mi/sec. A seismologist working in a monitoring station noted that the time between the primary and secondary waves of the earthquake was 16 sec.

a. Write two equations to represent this situation.

b. How far from the station did the earthquake occur?

Solution

a. Using the distance formula $d = rt$, we organize the given information in a table. Let d represent

	Rate	× Time	= Distance
Primary Wave	5	t	d
Secondary Wave	3	$t + 16$	d

the distance between the source of the earthquake and the seismic station and t represent the length of time that the waves traveled.

From the table, we write the following system:

(1) $\quad d = 5t$

(2) $\quad d = 3(t + 16)$

continued

PRACTICE 10

A Coast Guard cutter is sent information that a boat left the harbor 3 hr earlier traveling east at 20 mph. The Coast Guard cutter, traveling at 32 mph east, leaves the harbor to reach the boat.

EXAMPLE 10 (continued)

b. We can solve the system in part (a) to determine how far from the station the earthquake occurred. Since $d = 5t$, let's substitute $5t$ for d in equation (2). Then, we solve for t.

$$\begin{aligned} \textbf{(2)} \quad 5t &= 3(t + 16) \\ 5t &= 3t + 48 \\ 2t &= 48 \\ t &= 24 \end{aligned}$$

Since $t = 24$, we substitute 24 for t in equation (1) to find the value of d.

$$\begin{aligned} \textbf{(1)} \quad d &= 5t \\ d &= 5(24) \\ d &= 120 \end{aligned}$$

Since $d = 120$, the earthquake occurred 120 mi from the station.

a. Write a system of equations to represent this situation.

b. How long after the boat leaves the harbor will it take the Coast Guard cutter to reach the boat?

In Section 2.1, we solved *mixture* (or *solution*) problems using one linear equation. Here, we consider these kinds of problems using a system of two linear equations.

EXAMPLE 11

A biochemist has two alcohol solutions. One is a 20% alcohol solution and the other is a 50% alcohol solution. She needs 12 L of a solution that is 45% alcohol. How many liters of each of the original solutions should she mix?

Solution   First, we organize the given information in a table. Let x represent the number of liters of the 20% alcohol solution and y represent the number of liters of the 50% alcohol solution.

Action	Percent of Alcohol	Amount of Solution (L)	Amount of Alcohol (L)
Start with	20%	x	$0.20x$
Add	50%	y	$0.50y$
Finish with	45%	12	$0.45(12)$

Since the total amount of alcohol in the first two solutions must equal the amount of alcohol in the desired solution, we can use the table to write the following system of equations:

$$\begin{aligned} \textbf{(1)} \quad & x + y = 12 \\ \textbf{(2)} \quad & 0.20x + 0.50y = 0.45(12) \end{aligned}$$

Let's solve this system by elimination. To do this, we first multiply equation (1) by -20 and equation (2) by 100. Then by adding, we get:

$$\begin{aligned} \textbf{(1)} \quad -20x - 20y &= -240 \\ \textbf{(2)} \quad \underline{20x + 50y} &= \underline{540} \\ 30y &= 300 \\ y &= 10 \end{aligned}$$

Substituting 10 for y in equation (1) gives us:

$$\begin{aligned} \textbf{(1)} \quad x + y &= 12 \\ x + 10 &= 12 \\ x &= 2 \end{aligned}$$

Since $x = 2$ and $y = 10$, we conclude that 2 L of the 20% alcohol solution and 10 L of the 50% alcohol solution are needed to make 12 L of the 45% alcohol solution.

PRACTICE 11

In order to get a thicker and richer sauce, a chef combines a sauce that is 70% tomato paste with the original sauce that is 40% tomato paste. How much of each sauce should she mix if she wants 4 L of sauce that is 62.5% tomato paste?

Mathematically Speaking

Fill in each blank with the most appropriate term or phrase from the given list.

no solution	addition	solution
remaining variable	infinitely many solutions	reciprocals
substitution		both variables
opposites	one variable	

1. The elimination method for solving systems of equations is also called the _____ method.

2. The first step in the substitution method is to solve one of the equations for _____.

3. In the substitution method, after making the first substitution, solve the resulting equation for the _____.

4. When solving a system of linear equations algebraically, if the result is a true statement, then the system has _____.

5. When solving a system of linear equations algebraically, if the result is a false statement, then the system has _____.

6. The elimination method is useful when solving systems in which coefficients of one variable are _____.

A *Solve by substitution.*

7. $y = x - 3$
   $y = 2x + 1$

8. $y = 5 - x$
   $y = 3x - 7$

9. $y = \dfrac{1}{2}x + 3$
   $x = 4y + 6$

10. $y = 6x - 9$
    $x = \dfrac{1}{3}y - 1$

11. $3x + 2y = -8$
    $y = -4x + 11$

12. $5x - y = 2$
    $y = 2x - 5$

13. $x + y = 13$
    $x = -(y - 13)$

14. $-x + y = -2$
    $-y = 2 - x$

15. $7y = 21 - 14x$
    $-5y = 20x$

16. $9y = 18x + 36$
    $-6y = -30x$

17. $10a - b = 5$
    $10a + 3b = -7$

18. $p + 4q = 8$
    $3p - 4q = 16$

19. $x - 5y = -1$
    $7x - y = 10$

20. $8x - 3y = 1$
    $-12x - 6y = -5$

21. $2x + 12 = -2y$
    $x + y = 6$

22. $x - y = -5$
    $-3y - 15 = -3x$

23. $8s - 4t = 36$
    $3s + 5t = 20$

24. $7k - 2n = 15$
    $18k - 6n = 42$

25. $6x - 9y - 27 = 0$
    $10x - 12y - 29 = 0$

26. $3x + 4y + 16 = 0$
    $-9x - 8y - 30 = 0$

B *Solve by elimination.*

27. $x + y = 3$
$x - y = 5$

28. $y - x = 6$
$y + x = 4$

29. $-3x + 2y = -2$
$3x - 3y = 1$

30. $4x + 2y = 9$
$4x - 2y = 7$

31. $2x - 5y = 7$
$6x - 5y = -3$

32. $3x + 2y = -6$
$3x + 7y = 4$

33. $7c + 3d = 0$
$7c - 9d = 0$

34. $2x - 8y = 8$
$5x - 8y = 20$

35. $3x - 6y = -10$
$6x - 12y = -20$

36. $16x + 12y = -2$
$12x + 9y = -1.5$

37. $\dfrac{1}{8}x - \dfrac{3}{5}y = -7$
$\dfrac{1}{2}x + \dfrac{1}{5}y = -2$

38. $-\dfrac{1}{2}x + \dfrac{1}{3}y = 0$
$\dfrac{3}{2}x + 2y = -9$

39. $9x - 13 = 7y + 11x$
$4x + 8y = -17$

40. $5x + 10 = 16 - 2y$
$15x - 2y = 10x$

41. $2x + 3y = 6y - 4$
$-\dfrac{1}{2}(4x - 6y) = 7$

42. $\dfrac{1}{3}(9x - 18) = 5y$
$-(3x - 5y) = -6$

43. $3x + 4y = 1$
$2x - 3y = 12$

44. $2x - 3y = -3$
$-3x + 2y = 3$

45. $5x + 2y = 11$
$4x + 7y = -2$

46. $3x - 6y = -9$
$4x - 3y = 13$

47. $4u + 14v = -18$
$2u + 7v = 9$

48. $18x + 27y = 7.5$
$4x + 6y = 20$

49. $2.5a - 1.25b = -4$
$-3a - 0.2b = -2$

50. $1.6p - 3.2q = -6.4$
$-5p + 2.5q = 1.25$

Mixed Practice

Solve by substitution or elimination.

51. $2a + 3b = -3$
$2a + 7b = 9$

52. $y - x = 9$
$x = -(9 - y)$

53. $6x + 2y = 4$
$y = 1 - 2x$

54. $x - 2y = -3$
$2x - 4y = 3$

55. $x - y = 3$
$2y - 6 = 2x$

56. $2a + 5b = 11$
$3a + 4b = 6$

57. $\dfrac{1}{5}x - \dfrac{1}{4}y = 4$
$-\dfrac{2}{5}x + \dfrac{3}{4}y = -7$

58. $a + 4b = 2$
$5a - b = 3$

Applications

C *Solve.*

59. Uptown Towing Company charges $60 to tow a car plus $25 per day for vehicle storage. Downtown Towing Company charges $75 for towing plus $20 per day for vehicle storage.

  a. Write an equation that shows how much each company charges in terms of the number of days of storage.

  b. For how many days of storage will both companies charge the same?

60. Fifteen minutes after one car leaves a freeway rest stop, a second car leaves the rest stop traveling in the same direction. The first car travels at an average speed of 56 mph and the second travels at an average speed of 64 mph.

  a. Write two equations to represent this situation.

  b. How long will it take the second car to catch up to the first car?

61. A homeowner sections off part of her backyard for a vegetable garden. The width of the garden is 4 ft shorter than the length. The total perimeter of the garden is 52 ft. What are the dimensions of the garden?

62. A magazine publisher sells 845,760 more magazines through newsstands than through subscriptions. If the magazine has a total circulation of 1.52 million, how many magazines are sold through newsstands and how many are sold through subscriptions?

63. On a riverboat trip up the Wailua River in Kauai, a boat traveled against the current at an average speed of 18 mph. On the return trip down the river, the boat traveled with the current at an average speed of 22 mph.

  a. Express this information as a system of equations.

  b. Find the speed of the boat in still water and the speed of the current.

64. A couple budgets one-third of their monthly income for their rent and their car payment. The difference between the rent and the car payment is $802. The couple's monthly income is $4494.

  a. Write this information as a system of equations.

  b. What is the couple's monthly rent and car payment?

65. A coffee shop sells 12-oz and 20-oz cups of coffee. On a particular day, the shop sold a total of 508 cups of coffee. If the shop sold 3 times as many 20-oz cups of coffee as 12-oz cups of coffee, how many cups of each size did the coffee shop sell that day?

66. A local performing-arts group holds its annual winter arts festival in December. The group charges $20 for adult admission tickets and $12 for children's admission tickets. The group raised $13,344 on the sale of 824 tickets. How many of each type of ticket were sold?

67. A public storage facility rents small and large storage lockers. A small storage locker has 200 ft^2 of storage space and a large storage locker has 800 ft^2 of storage space. The facility has 54 storage lockers and a total of 22,800 ft^2 of storage space. How many large storage lockers does the facility have?

68. A discount linen store sells all items for either $15 or $30. In March, the sales totaled $12,570. If the store sold 563 items that month, how many $15 items were sold?

69. Twice the amount of money that was invested in a low-risk fund was invested in a high-risk fund. After one year, the low-risk fund earned 6% and the high-risk fund lost 9%. The investments had a net loss of $210. How much was invested in each fund?

70. A student takes out two loans for tuition totaling $8000. One of the loans charged 8% simple interest and the other charged $7\frac{1}{4}$% simple interest. After one year, the interest owed on the loans was $621.25. How much did the student borrow at each rate?

71. A chemist needs 10 L of a 40% alcohol solution. How many liters of a 65% alcohol solution and a 25% alcohol solution must she combine to get the desired solution?

72. The owner of a natural foods store made 20 lb of a dried fruit and granola blend that is 50% dried fruit by mixing a dried fruit and granola blend that is 20% dried fruit with a blend that is 60% dried fruit. How many pounds of each type were used to get the desired blend?

• Check your answers on page A-34.

MINDStretchers

Writing

1. A system of linear equations has the solutions $(1, 3)$ and $(4, -2)$.
   Explain how you can find another solution of this system.

Critical Thinking

2. Find A and B so that the system

$$Ax - 2By = 6$$
$$3Ax - By = -12$$

has the solution $(2, 3)$.

Mathematical Reasoning

3. A student claims that there is no solution to the following problem: One landscaping company charges a daily flat fee of $30 plus an hourly rate of $12 per hour for service. Another landscaping company charges a daily flat fee of $25 plus an hourly rate of $10 for service. For what number of hours will both companies charge the same amount? Is the student correct? Explain.

4.3 Solving Systems of Linear Equations in Three Variables

OBJECTIVES

A To determine whether an ordered triple is a solution of a system of linear equations

B To solve a system of linear equations in three variables

C To solve applied problems involving systems of linear equations in three variables

In Sections 4.1 and 4.2, we discussed situations that could be described by systems of linear equations in two variables. In this section, we look at more complex situations involving relationships that can be described by linear equations in three variables. For instance, consider the following problem from chemistry.

A molecule is the smallest physical unit of an element or compound. The molecule of water, pictured to the right, consists of 2 atoms of hydrogen and 1 atom of oxygen. The following table describes several molecules.

Water Molecule

Compound	Molecule	Atomic Structure	Approximate Molecular Weight
Water	H_2O	2 atoms of hydrogen and 1 atom of oxygen	18
Hydrogen peroxide	H_2O_2	2 atoms of hydrogen and 2 atoms of oxygen	34
Sulfuric acid	H_2SO_4	2 atoms of hydrogen, 1 atom of sulfur, and 4 atoms of oxygen	98

The molecular weight of water is twice the atomic weight of hydrogen h, added to the atomic weight of oxygen o. The molecular weight of hydrogen peroxide is twice the atomic weight of hydrogen plus twice the atomic weight of oxygen, and the atomic weight of sulfuric acid is twice the atomic weight of hydrogen plus the atomic weight of sulfur s, plus 4 times the atomic weight of oxygen. This leads to the following system of equations:

$$2h \quad + \quad o = 18$$
$$2h \quad + 2o = 34$$
$$2h + s + 4o = 98$$

If we can solve this system of equations, we will be able to find the atomic weights of the hydrogen, oxygen, and sulfur atoms.

Introduction to Systems of Linear Equations in Three Variables

A *solution* of a system of linear equations consists of the values for the variables that satisfy all three equations of the system.

DEFINITION

A **solution** of a system of linear equations in three variables is an **ordered triple** of numbers that makes all three equations in the system true.

EXAMPLE 1

Consider the following system of equations:

$$2x + y + 3z = 1$$
$$x - y - 2z = 6$$
$$3x + 2y = 11$$

a. Is $(3, 1, -2)$ a solution of the system?

b. Is $(0, 4, -1)$ a solution of the system?

Solution

a. To determine if the ordered triple $(3, 1, -2)$ is a solution of this system, we substitute 3 for x, 1 for y, and -2 for z and check if all three equations are true.

			Simplifies to		
$2x + y + 3z = 1$	$\rightarrow$	$2(3) + 1 + 3(-2) \stackrel{?}{=} 1$	$\xrightarrow{\text{Simplifies to}}$	$1 = 1$	True
$x - y - 2z = 6$	$\rightarrow$	$3 - 1 - 2(-2) \stackrel{?}{=} 6$	$\xrightarrow{\text{Simplifies to}}$	$6 = 6$	True
$3x + 2y = 11$	$\rightarrow$	$3(3) + 2(1) \stackrel{?}{=} 11$	$\xrightarrow{\text{Simplifies to}}$	$11 = 11$	True

The ordered triple $(3, 1, -2)$ is a solution of the system because it satisfies all three equations.

b. To determine if the ordered triple $(0, 4, -1)$ is a solution of this system, we substitute 0 for x, 4 for y, and -1 for z and check if all three equations are true.

			Simplifies to		
$2x + y + 3z = 1$	$\rightarrow$	$2(0) + 4 + 3(-1) \stackrel{?}{=} 1$	$\xrightarrow{\text{Simplifies to}}$	$1 = 1$	True
$x - y - 2z = 6$	$\rightarrow$	$0 - 4 - 2(-1) \stackrel{?}{=} 6$	$\xrightarrow{\text{Simplifies to}}$	$-2 = 6$	False
$3x + 2y = 11$	$\rightarrow$	$3(0) + 2(4) \stackrel{?}{=} 11$	$\xrightarrow{\text{Simplifies to}}$	$8 = 11$	False

The ordered triple $(0, 4, -1)$ is not a solution of the system because it does not satisfy all three equations.

PRACTICE 1

Determine whether the ordered triple is a solution of the following system.

$$x + 3y - 2z = 9$$
$$-2x + 4z = 0$$
$$3x - 5y + z = -29$$

a. $(6, 1, 3)$

b. $(-4, 3, -2)$

We can graph a linear equation in three variables, such as $2x + y + 3z = 1$. Although we are not considering such graphs in detail, we give a brief discussion of them since they can help us to visualize the possible number of solutions a system of three linear equations can have.

To begin with, an ordered triple of numbers (x, y, z) can be thought of as the coordinates of a point in space. The coordinates are defined with respect to three axes. The point $(2, 4, 3)$, for instance, is shown here.

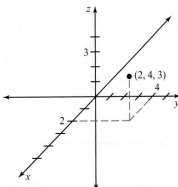

If we were to graph in this coordinate system any linear equation in three variables, we would find that the graph is a plane. So a system of three such equations has as its graph the points where three planes in space intersect. These planes can intersect in a variety of ways. The following graphs illustrate some of these ways:

One solution: three planes that intersect at one common point

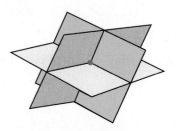

No solution: three planes that do not intersect

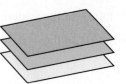

Infinitely many solutions: three planes that intersect at infinitely many points along a line

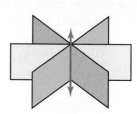

We see from these graphs that a system of linear equations in three variables can have one solution, no solution, or infinitely many solutions.

Solving Systems of Three Linear Equations by the Elimination Method

Just as we used the elimination method to solve systems of two linear equations, so we can use the elimination method to solve systems of three linear equations. Applying this method, we eliminate one of the variables in order to get a system of two equations in two variables. Then, we solve this system of two equations using the elimination method, as we did in Section 4.2.

EXAMPLE 2

Solve by the elimination method:

$$\text{(1)} \quad x - y - 2z = 4$$
$$\text{(2)} \quad -x + 2y + z = 1$$
$$\text{(3)} \quad -x + y - 3z = 11$$

Solution   First, we decide which two equations to choose in order to eliminate one of the three variables. Since the coefficients of the x-terms in equations (1) and (2) are opposites, we can eliminate x if we add these equations, getting a linear equation in y and z. Let's call this equation in y and z equation (4).

$$\text{(1)} \quad x - y - 2z = 4$$
$$\text{(2)} \quad \underline{-x + 2y + z = 1}$$
$$\text{(4)} \qquad\quad y - z = 5 \qquad \text{Add the equations.}$$

Next, we use a different pair of equations and eliminate the *same* variable, x. Let's add equations (2) and (3) by first multiplying

PRACTICE 2

Solve by the elimination method:

$$\text{(1)} \quad x + y + z = 2$$
$$\text{(2)} \quad 2x - y + 5z = -5$$
$$\text{(3)} \quad -x + 2y + 2z = 1$$

equation (2) by -1. We get another linear equation in y and z. Let's call it equation (5).

(2) $\quad -x + 2y + z = 1 \quad \xrightarrow{\text{Multiply by} -1.} \quad x - 2y - z = -1$

(3) $\quad -x + y - 3z = 11 \qquad\qquad\qquad -x + y - 3z = 11$

$\qquad\qquad\qquad\qquad\qquad\qquad\qquad$ **(5)** $\qquad -y - 4z = 10 \quad$ Add the equations.

Now, we solve equations (4) and (5) for y and z using the elimination method again. Note that if we add the equations, we can eliminate the y-terms, and then solve for z.

$$
\begin{array}{rl}
\textbf{(4)} & y - z = 5 \\
\textbf{(5)} & -y - 4z = 10 \\
\hline
& -5z = 15 \quad \text{Add the equations.} \\
& z = -3
\end{array}
$$

To find y, we can substitute -3 for z in either equation (4) or (5). Let's use equation (4).

$$
\begin{array}{rl}
\textbf{(4)} & y - z = 5 \\
& y - (-3) = 5 \\
& y + 3 = 5 \\
& y = 2
\end{array}
$$

Finally, we substitute 2 for y and -3 for z in one of the equations of the *original* system, and then solve for x. Let's use equation (1).

$$
\begin{array}{rl}
\textbf{(1)} & x - y - 2z = 4 \\
& x - 2 - 2(-3) = 4 \\
& x - 2 + 6 = 4 \\
& x + 4 = 4 \\
& x = 0
\end{array}
$$

Since $x = 0$, $y = 2$, and $z = -3$, the ordered triple $(0, 2, -3)$ is the solution.

Check To determine if the ordered triple $(0, 2, -3)$ is the solution to the original system, we check to see if this ordered triple satisfies all three equations.

(1) $\quad x - y - 2z = 4$ $\qquad\qquad$ **(2)** $\quad -x + 2y + z = 1$

$\qquad 0 - 2 - 2(-3) \overset{?}{=} 4$ $\qquad\qquad\qquad -0 + 2(2) + (-3) \overset{?}{=} 1$

$\qquad\qquad -2 + 6 \overset{?}{=} 4$ $\qquad\qquad\qquad\qquad 4 + (-3) \overset{?}{=} 1$

$\qquad\qquad\qquad\quad 4 = 4 \quad$ True $\qquad\qquad\qquad\qquad\qquad 1 = 1 \quad$ True

(3) $\quad -x + y - 3z = 11$

$\qquad -0 + 2 - 3(-3) \overset{?}{=} 11$

$\qquad\qquad 2 + 9 \overset{?}{=} 11$

$\qquad\qquad\quad 11 = 11 \quad$ True

The ordered triple $(0, 2, -3)$ satisfies each of the original equations, so the solution is $(0, 2, -3)$. Throughout the remainder of this section, the check is left as an exercise.

Example 2 suggests the following procedure for solving systems of linear equations in three variables using the elimination method:

> ## To Solve a System of Linear Equations in Three Variables by Elimination
>
> - Write all equations in the general form $Ax + By + Cz = D$.
>
> - Choose a pair of equations and use them to eliminate a variable, getting an equation in two variables.
>
> - Next, use a different pair of equations to eliminate the *same* variable that was eliminated in the previous step, getting another equation in the same two variables.
>
> - Solve the system of equations formed by the equations found in the previous two steps, getting the values of two of the variables.
>
> - Substitute the values found in the previous step in any of the original equations to find the value of the third variable.
>
> - Check by substituting the values found in the previous two steps in all three equations of the original system.

EXAMPLE 3

Solve by elimination:

$$\begin{aligned}
\textbf{(1)} \quad 2x + 3y + 12z &= 4 \\
\textbf{(2)} \quad 4x - 6y + 6z &= 1 \\
\textbf{(3)} \quad x + y + z &= 1
\end{aligned}$$

Solution   Let's use equations (2) and (3) to eliminate y. First, we multiply equation (3) by 6. Then, we add the resulting equation to equation (2), getting an equation in x and z, which we call equation (4).

$$\begin{aligned}
\textbf{(2)} \quad 4x - 6y + 6z &= 1 \\
\textbf{(3)} \quad x + y + z &= 1
\end{aligned} \quad \xrightarrow{\text{Multiply by 6.}} \quad \begin{aligned}
4x - 6y + 6z &= 1 \\
6x + 6y + 6z &= 6 \\
\hline
\textbf{(4)} \quad 10x + 12z &= 7
\end{aligned} \quad \text{Add the equations.}$$

Next, we use a different pair of equations and eliminate the same variable y. Let's multiply equation (3) by -3. Then, we add the resulting equation to equation (1) to get another equation in x and z, which we call equation (5).

$$\begin{aligned}
\textbf{(1)} \quad 2x + 3y + 12z &= 4 \\
\textbf{(3)} \quad x + y + z &= 1
\end{aligned} \quad \xrightarrow{\text{Multiply by } -3} \quad \begin{aligned}
2x + 3y + 12z &= 4 \\
-3x - 3y - 3z &= -3 \\
\hline
\textbf{(5)} \quad -x + 9z &= 1
\end{aligned} \quad \text{Add the equations.}$$

Now, we solve equations (4) and (5) for x and z using elimination. If we multiply equation (5) by 10, we can eliminate the x-terms. Then, we can solve for z.

$$\begin{aligned}
\textbf{(4)} \quad 10x + 12z &= 7 \\
\textbf{(5)} \quad -x + 9z &= 1
\end{aligned} \quad \xrightarrow{\text{Multiply by 10.}} \quad \begin{aligned}
10x + 12z &= 7 \\
-10x + 90z &= 10 \\
\hline
102z &= 17
\end{aligned} \quad \text{Add the equations.}$$

$$z = \frac{17}{102} = \frac{1}{6}$$

PRACTICE 3

Solve by elimination:

$$\begin{aligned}
\textbf{(1)} \quad 2x + y + 2z &= 1 \\
\textbf{(2)} \quad x + 2y + z &= 2 \\
\textbf{(3)} \quad x - y - z &= 0
\end{aligned}$$

We can substitute $\frac{1}{6}$ for z in either equation (4) or equation (5) to find x.

Using equation (5), we get:

$$\textbf{(5)} \qquad -x + 9z = 1$$

$$-x + 9\left(\frac{1}{6}\right) = 1$$

$$-x + \frac{9}{6} = 1$$

$$-x = -\frac{3}{6}$$

$$x = \frac{1}{2}$$

Finally, we substitute $\frac{1}{2}$ for x and $\frac{1}{6}$ for z in one of the original equations and solve for y. Let's use equation (3).

$$\textbf{(3)} \qquad x + y + z = 1$$

$$\frac{1}{2} + y + \frac{1}{6} = 1$$

$$y + \frac{4}{6} = 1$$

$$y = \frac{2}{6}$$

$$y = \frac{1}{3}$$

So the solution is $x = \frac{1}{2}$, $y = \frac{1}{3}$, and $z = \frac{1}{6}$, that is, $\left(\frac{1}{2}, \frac{1}{3}, \frac{1}{6}\right)$.

TIP When solving a system of linear equations in three variables, make sure that you eliminate the *same* variable in two pairs of equations to get a system of equations in two variables.

Some linear systems of equations in three variables have missing terms. When this is the case, we can omit one elimination step. In Example 4, each of the equations has a missing variable.

EXAMPLE 4

Solve by the elimination method:

(1) $\qquad x - 2z = -5$

(2) $-2x + z = 4$

(3) $\qquad -y + 3z = 3$

Solution The variable y is already eliminated in equations (1) and (2). Let's use these equations to eliminate x.

(1) $\qquad x - 2z = -5 \xrightarrow{\text{Multiply by 2.}} \quad 2x - 4z = -10$

(2) $-2x + z = 4 \qquad\qquad\qquad \underline{-2x + z = 4}$

$\qquad\qquad\qquad\qquad\qquad\qquad -3z = -6 \qquad$ Add the equations.

$\qquad\qquad\qquad\qquad\qquad\qquad\qquad z = 2$

PRACTICE 4

Solve by the elimination method:

(1) $\quad 3x + 4z = 5$

(2) $\quad 2x - 5y = 8$

(3) $\quad 2y + 3z = 2$

EXAMPLE 4 (continued)

Now, we can substitute 2 for z in either equation (1) or equation (2) to find x. Using equation (1), we get:

$$\begin{array}{rl}
\textbf{(1)} & x - 2z = -5 \\
& x - 2(2) = -5 \\
& x - 4 = -5 \\
& x = -1
\end{array}$$

Finally, we substitute 2 for z in equation (3) to find y.

$$\begin{array}{rl}
\textbf{(3)} & -y + 3z = 3 \\
& -y + 3(2) = 3 \\
& -y + 6 = 3 \\
& -y = -3 \\
& y = 3
\end{array}$$

So the solution is $(-1, 3, 2)$.

Just as with systems of linear equations in two variables, a linear system with three variables may have no solution or infinitely many solutions.

EXAMPLE 5

Solve by elimination:

$$\begin{array}{rl}
\textbf{(1)} & x + 2y - 4z = 7 \\
\textbf{(2)} & x - y + z = 5 \\
\textbf{(3)} & 2x + y - 3z = 6
\end{array}$$

Solution Since the coefficients of y are opposites in equations (2) and (3), we can eliminate y by adding these equations.

$$\begin{array}{rl}
\textbf{(2)} & x - y + z = 5 \\
\textbf{(3)} & \underline{2x + y - 3z = 6} \\
\textbf{(4)} & 3x - 2z = 11 \qquad \text{Add the equations.}
\end{array}$$

Next, we eliminate y again by using equations (1) and (3).

$$\begin{array}{rl}
\textbf{(1)} & x + 2y - 4z = 7 \\
\textbf{(3)} & 2x + y - 3z = 6
\end{array}
\quad \xrightarrow{\text{Multiply by } -2.} \quad
\begin{array}{rl}
& x + 2y - 4z = 7 \\
& \underline{-4x - 2y + 6z = -12} \\
\textbf{(5)} & -3x + 2z = -5
\end{array}
\quad \text{Add the equations.}$$

Now, using equations (4) and (5), we get

$$\begin{array}{rl}
\textbf{(4)} & 3x - 2z = 11 \\
\textbf{(5)} & \underline{-3x + 2z = -5} \\
& 0 = 6 \qquad \text{Add the equations.}
\end{array}$$

Since the statement $0 = 6$ is false for all values of the variables, the original system has no solution common to all three equations in the system.

PRACTICE 5

Solve by elimination:

$$\begin{array}{rl}
\textbf{(1)} & x - 3y + 2z = 1 \\
\textbf{(2)} & x - 2y + 3z = 5 \\
\textbf{(3)} & 2x - 6y + 4z = 3
\end{array}$$

EXAMPLE 6

Use the elimination method to solve:

$$(1) \quad x + 2y - 4z = 7$$
$$(2) \quad 2x + y - 3z = 12$$
$$(3) \quad x - y + z = 5$$

Solution Let's add equations (2) and (3) to eliminate y.

$$(2) \quad 2x + y - 3z = 12$$
$$(3) \quad \underline{x - y + z = 5}$$
$$(4) \quad 3x - 2z = 17 \qquad \text{Add the equations.}$$

Next, we use equations (1) and (2) to eliminate y again.

$$(1) \quad x + 2y - 4z = 7$$
$$(2) \quad 2x + y - 3z = 12$$

$\xrightarrow{\text{Multiply by} -2.}$

$$x + 2y - 4z = 7$$
$$\underline{-4x - 2y + 6z = -24}$$
$$(5) \quad -3x + 2z = -17$$

Add the equations.

Now, we add equations (4) and (5) to solve for x and z.

$$(4) \quad 3x - 2z = 17$$
$$(5) \quad \underline{-3x + 2z = -17}$$
$$0 = 0 \qquad \text{Add the equations.}$$

Since the statement $0 = 0$ is true for all values of the variables, the original system has infinitely many solutions, namely all ordered triples that satisfy the three equations.

How would you describe the graph of the original system in Example 6?

PRACTICE 6

Use the elimination method to solve:

$$(1) \quad x - y + z = 3$$
$$(2) \quad 2x - 2y - 2z = 6$$
$$(3) \quad -4x + 4y - 4z = -12$$

TIP Using algebra to solve a system of linear equations in three variables:
- If the result is a false statement, then the system has no solution.
- If the result is a true statement, then the system has infinitely many solutions.

Just as with systems of linear equations in two variables, systems of linear equations in three variables model many applied problems.

EXAMPLE 7

A company placed $2,000,000 in three different accounts. Part was placed in U.S. Treasury bonds paying 4.5% annually, twice as much was placed in short-term notes paying 5%, and the rest was placed in municipal bonds paying 4%. The total income after one year was $91,000. How much did the company place in each account?

Solution Let t represent the amount placed in U.S. Treasury bonds, s the amount placed in short-term notes, and m the amount placed in municipal bonds. From the given information, we get the following equations:

PRACTICE 7

A multimedia student saved $3200 working part-time as a Web designer. She invested her savings in a growth fund, an income fund, and a money market fund with return rates of 10%, 7%, and 5%, respectively. To maximize her return, she placed twice as much money in the growth fund as in the money market fund.

continued

EXAMPLE 7 (continued)

How should she invest the $3200 to get a return of $250 after one year?

	Amount in U.S. Treasury bonds	+	amount in short-term notes	+	amount in municipal bonds	equals	2,000,000.

Equation t $+$ s $+$ m $=$ 2,000,000

	Amount in short-term notes	equals	twice	the amount in treasury bonds.

Equation s $=$ 2 $\cdot$ t

Using the interest formula $I = Prt$, we can write a word sentence and an equation to express the total income earned after one year.

	Amount of income earned at 4.5%	+	amount of income earned at 5%	+	amount of income earned at 4%	equals	91,000.

Equation $0.045t$ $+$ $0.05s$ $+$ $0.04m$ $=$ 91,000

So the system of equations to solve is:

$$\textbf{(1)} \qquad t + s + m = 2{,}000{,}000$$
$$\textbf{(2)} \qquad s = 2t$$
$$\textbf{(3)} \quad 0.045t + 0.05s + 0.04m = 91{,}000$$

We can write equation (2) as $-2t + s = 0$. Since this equation does not contain the variable m, let's use equations (1) and (3) to eliminate m.

(1) $t + s + m = 2{,}000{,}000$ — Multiply by -40. →

(3) $0.045t + 0.05s + 0.04m = 91{,}000$ — Multiply by 1000. →

$$-40t - 40s - 40m = -80{,}000{,}000$$
$$45t + 50s + 40m = 91{,}000{,}000$$
$$\textbf{(4)} \qquad 5t + 10s = 11{,}000{,}000$$

Now, we use equations (2) and (4) to find t and s.

(2) $-2t + s = 0$ — Multiply by -10. → $20t - 10s = 0$

(4) $5t + 10s = 11{,}000{,}000$ $\qquad\qquad\quad 5t + 10s = 11{,}000{,}000$

$$25t = 11{,}000{,}000$$
$$t = 440{,}000$$

Substituting 440,000 for t in equation (2), we can solve for s.

$$\textbf{(2)} \qquad -2t + s = 0$$
$$-2(440{,}000) + s = 0$$
$$-880{,}000 + s = 0$$
$$s = 880{,}000$$

To find m, we substitute 440,000 for t and 880,000 for s in equation (1).

$$\textbf{(1)} \qquad t + s + m = 2{,}000{,}000$$
$$440{,}000 + 880{,}000 + m = 2{,}000{,}000$$
$$1{,}320{,}000 + m = 2{,}000{,}000$$
$$m = 680{,}000$$

So $440,000 was placed in U.S. Treasury bonds, $880,000 in short-term notes, and $680,000 in municipal bonds.

A *Determine whether the ordered triple is a solution of the system of equations.*

1. $x + y - z = -6$ $(2, -3, 5)$
$2x - y + z = 12$
$-3x - 2y - 4z = -20$

2. $6x + 3y + 2z = 4$ $\left(\dfrac{1}{3}, -2, 4\right)$
$-12x + 2y + 4z = 8$
$9x - 10y - 5z = 3$

3. $2x - 2y - 4z = 1$ $\left(-2, -\dfrac{1}{2}, -1\right)$
$5x - 8y + 3z = 9$
$-x + 6y + 5z = -8$

4. $4x - 2y + z = -1$ $(-1, 0, 3)$
$x - 3y - 5z = -16$
$3x + y + z = 6$

5. $\dfrac{3}{5}x - \dfrac{1}{4}y = 10$ $(20, 8, 7)$
$\dfrac{3}{20}x + 2z = 17$
$-9y + 10z = -2$

6. $\dfrac{1}{2}y - \dfrac{2}{3}z = 0$ $\left(\dfrac{1}{10}, 12, 9\right)$
$30x + \dfrac{4}{9}z = 7$
$10x + \dfrac{1}{4}y = 4$

B *Solve by the elimination method.*

7. $2x = 8$
$x - 4z = 12$
$3x - 2y + z = 0$

8. $x + 3y + z = 1$
$5y + 8z = 7$
$- 3z = 3$

9. $x + y + z = -1$
$2x - y - z = -5$
$x + 2y - z = 6$

10. $x - y - z = 3$
$-x + y + 3z = 3$
$-x - 3y + 2z = 8$

11. $3x - y + 5z = -20$
$2x + 4y - 2z = 15$
$-4x - 2y - z = -1$

12. $2x + 5y - z = 4$
$x + 3y + z = -2$
$-3x - 2y + 3z = 1$

13. $-2x + 6y - 3z = 0$
$x - y + z = 2$
$2x - 5y - z = 1$

14. $6x - 2y + 7z = 10$
$-3x - y + z = 0$
$x + y + z = 4$

15. $8y + 3z = 3$
$5x + 4y = 2$
$10x + 6z = 4$

16. $12x - 14z = -2$
$6y - 7z = 7$
$6x + 2y = 4$

17. $\dfrac{2}{3}x - \dfrac{1}{2}y = 1$
$\dfrac{1}{4}y + \dfrac{1}{3}z = 2$
$\dfrac{1}{3}x - \dfrac{1}{2}z = -5$

18. $\dfrac{2}{3}y + \dfrac{3}{5}z = 2$
$\dfrac{3}{4}x - \dfrac{1}{2}y = 0$
$\dfrac{1}{4}x + \dfrac{2}{5}z = 1$

19. $4x - y + 8z = -2$
$x + 2y + 4z = 3$
$2x - 5y = 1$

20. $6x + y - 5z = 8$
$3x - 2y + z = 4$
$x + y - 2z = 0$

21. $4x + 3y - 2z = -7$
$5x - 3y + 4z = 24$
$8x + 2y + z = 10$

22. $2x \quad\quad - 4z = 4$
$3x + 2y + 2z = 3$
$-x + 2y - 6z = -1$

23. $x + 4y - 2z = 6$
$2x + 3y + 2z = 7$
$x - y + 4z = 1$

24. $x + 7y + 5z = 14$
$3x + y - 15z = 44$
$2x + 4y - 5z = 29$

25. $7x - 6y + 3z = 13$
$2x + 3y + 2z = -2$
$6x - 5y + 7z = -11$

26. $2x + 3y + 5z = 0$
$5x + 2y + 3z = 12$
$3x + 5y + 2z = 18$

Mixed Practice

Solve.

27. Determine if the ordered triple $(1, -2, -1)$ is a solution of the system.

$$3x + 2y + z = -2$$
$$-x - 3y + 4z = 1$$
$$5x - 4y + 4z = 9$$

Solve by the elimination method.

28. $2x - y + z = 5$
$-2x + 2y + 2z = -2$
$2x - 3y - 3z = 1$

29. $x - 2y + z = 3$
$2x - y - z = 4$
$x + y - 2z = 1$

30. $2x - 3y \quad\quad = -5$
$-4x \quad\quad + 2z = -4$
$\quad - 3y - 2z = -4$

31. $-x + 4y - z = 1$
$x - y - 3z = 2$
$2x - 11y + 6z = -2$

32. $3x + y + 2z = 10$
$-6x + 2y - 3z = -14$
$3x - 3y + 5z = -4$

Applications

C *Solve.*

33. A caterer for a wedding reception charges $50 for a beef dinner, $42 for a chicken dinner, and $48 for a fish dinner. Twice as many beef dinners as fish dinners were ordered. In all, 192 dinners were ordered for the reception and the total catering bill for the dinners was $8988. How many of each type of dinner were ordered?

34. An apartment building has studio, one-bedroom, and two-bedroom apartments. The building has a total of 60 apartments. There are half as many studio apartments as one-bedroom apartments in the building. The rent for a studio apartment is $650, for a one-bedroom is $875, and for a two-bedroom is $1250. If all the apartments are occupied, the owner of the apartment building collects $53,400 in rent per month. How many of each type of apartment are in the building?

35. According to the U.S. Postal Service, a package delivered by parcel post must have a combined length and girth that does not exceed 130 in. A mailroom supervisor wants to send a package that has a combined length, width, and height of 65 in. The height of the package is 5 in. shorter than the width. Twice the length of the package is equal to 3 times the width. Does the package meet the postal service's parcel post restrictions? Explain. [*Hint:* Girth $= 2($width $+$ height$)$] (*Source:* usps.com)

36. On one of her sales routes, a sales representative drives from Austin to Dallas, from Dallas to Houston, and then from Houston back to Austin. The distance from Austin to Dallas is 33 mi more than the distance from Houston to Austin. The distance from Dallas to Houston is 85 mi less than twice the distance from Houston to Austin. If the sales representative drives a total of 596 mi on her route, determine the distance she drives between each city.

37. Each coin presently in circulation has particular specifications, such as weight, diameter, and thickness, set by the U.S. Mint. A nickel, a dime, and a quarter have a combined thickness of 5.05 mm. The combined thickness of four nickels, two quarters, and two dimes is 14 mm. A nickel is 0.6 mm thicker than a dime. Determine the thickness of each type of coin. (*Source:* U.S. Department of the Treasury, U.S. Mint)

38. The FDA recommends a daily fiber intake of 25 g for a 2000-calorie diet. Eating one apple, one banana, and one pear results in a total fiber intake of 11.5 g. Two apples, two bananas, and one pear result in a fiber intake of 17.9 g. A pear contains 2 g more fiber than a banana. How many grams of fiber does each fruit contain? (*Source:* U.S. Department of Agriculture, Nutrient Data Laboratory)

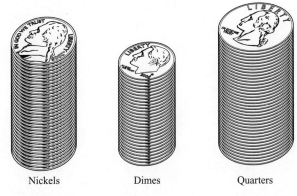

Nickels   Dimes   Quarters

39. The Masters is an annual golf tournament held at Augusta National Golf Club in Augusta, Georgia. The course has 18 holes consisting of par-3, par-4, and par-5 holes. If a golfer in the tournament gets a par score on every hole, then his total score is 72. There are six fewer par-5 holes than par-4 holes on the course. How many of each type of hole are on the course? (*Source:* David Owen, *The Making of the Masters*)

40. General admission tickets to an amusement park cost $39.99 each. The admission tickets for children 48 in. and under and for senior citizens cost $24.99 each. On Saturday, the amusement park had 7250 visitors. The park sold 823 more children's tickets than senior citizen tickets. If the amusement park collected $256,582.50 on the sale of admission tickets, how many of each type of admission ticket did the amusement park sell on Saturday?

41. A college student is traveling to Santo Domingo on spring break. He purchased 15 traveler's checks worth $280, in denominations of $10, $20, or $50. He bought 4 times as many $10 checks as $50 checks. How many checks in each denomination did he buy?

42. To diversify its stock portfolio, a small company invested $25,000 in three different stock funds: a growth and income fund, bonds, and a balanced fund. The company invested half the amount in bonds that it invested in the growth and income fund. The growth and income fund had a return of 10%, the bonds had a return of 4%, and the balanced fund had a loss of 3%. If the net return on the investments was $900, how much did the company invest in each fund?

• Check your answers on page A-34.

MIND*Stretchers*

Mathematical Reasoning

1. The method for solving a system of equations in three variables can be extended to four linear equations in four variables. Solve the following system:

$$w + x - y + z = -5$$
$$2w - 3x + 4y + z = 6$$
$$w - 2x - y + 3z = 0$$
$$x + 2y + 2z = 5$$

Writing

2. Write a word problem that can be solved using the following system of equations:

$$x + y + z = 100$$
$$5x + 3y + 2z = 365$$
$$x - y = 15$$

Critical Thinking

3. Find values of A, B, and C so that the following system has the solution $(4, 1, -3)$:

$$Ax + By + Cz = 13$$
$$Bx + Cy + Az = -18$$
$$Cx + Ay + Bz = 5$$

4.4 Solving Systems of Linear Equations by Using Matrices

In this section, we focus on a method of solving systems of two or three linear equations by using *matrices*. This procedure is based on the elimination method without writing the variables, that is, writing just the coefficients and the constants of the equations. This *matrix method* is often used on a graphing calculator or a computer when solving systems with many equations.

A To solve a system of two or three linear equations using matrices

B To solve applied problems involving systems of linear equations

Solving Systems by Using Matrices

A rectangular array of numbers such as

$$
\begin{array}{ccc}
\text{Column} & \text{Column} & \text{Column} \\
1 & 2 & 3 \\
\downarrow & \downarrow & \downarrow
\end{array}
$$

$$
\begin{array}{l}
\text{Row 1} \longrightarrow \\
\text{Row 2} \longrightarrow
\end{array}
\begin{bmatrix}
2 & 3 & 8 \\
1 & 5 & 3
\end{bmatrix}
$$

is called a *matrix*.

> **DEFINITIONS**
>
> A **matrix** (plural, **matrices**) is a rectangular array of numbers. The numbers of the matrix are called **elements** or **entries**. The **rows** of a matrix are horizontal, and the **columns** are vertical. A **square matrix** has the same number of rows as columns.

Here are other examples of matrices.

$$
\begin{array}{cc}
\text{Column 1} & \text{Column 2} \\
\downarrow & \downarrow
\end{array}
$$

$$
\begin{array}{l}
\text{Row 1} \longrightarrow \\
\text{Row 2} \longrightarrow
\end{array}
\begin{bmatrix}
-1 & 0 \\
0 & -1
\end{bmatrix}
$$

This square matrix has 2 rows and 2 columns. It is called a 2 × 2 (read "two by two") matrix.

$$
\begin{array}{cccc}
\text{Column} & \text{Column} & \text{Column} & \text{Column} \\
1 & 2 & 3 & 4 \\
\downarrow & \downarrow & \downarrow & \downarrow
\end{array}
$$

$$
\begin{array}{l}
\text{Row 1} \longrightarrow \\
\text{Row 2} \longrightarrow \\
\text{Row 3} \longrightarrow
\end{array}
\begin{bmatrix}
4 & 1 & 1 & 3 \\
-1 & 1 & -2 & -11 \\
1 & 2 & 2 & -1
\end{bmatrix}
$$

This matrix has 3 rows and 4 columns. It is called a 3 × 4 (read "three by four") matrix.

How many elements are there in an $n \times n$ matrix? In an $m \times n$ matrix? Explain.

Let's look at how a system of linear equations relates to a matrix.

System of Equations		**Augmented Matrix**
Equation (1)	$5x + y = 3$	Row (1)
Equation (2)	$-2x + 2y = 4$	Row (2)

$$
\begin{bmatrix}
5 & 1 & \big| & 3 \\
-2 & 2 & \big| & 4
\end{bmatrix}
$$

Note that coefficients of the variables and the constants in equations (1) and (2) correspond, respectively, to the elements in rows (1) and (2) of the **augmented matrix**. The augmented matrix is sometimes called the **corresponding matrix**. Note that in the augmented matrix there is a vertical bar that separates the coefficients on the left from the constants on the right.

We have already solved systems of equations by using multiples of one or more of the equations to eliminate variables. We can do the same thing to an augmented matrix by using **row operations**. When we perform row operations on a matrix, the result is an **equivalent matrix**, that is, a matrix corresponding to a system of equations equivalent to the original system. We can solve the system using row operations.

Matrix Row Operations

The following row operations produce an equivalent matrix:

- Interchanging any two rows

- Multiplying (or dividing) the elements of any row by the same nonzero number

- Multiplying (or dividing) the elements of any row by a nonzero number and adding the products to their corresponding elements in any other row

In solving a system of linear equations using matrices, we use these row operations to write a series of equivalent matrices to obtain a simplified matrix of the form

$$\begin{bmatrix} 1 & a & | & b \\ 0 & 1 & | & c \end{bmatrix}$$

for a system of linear equations in two variables, or

$$\begin{bmatrix} 1 & a & b & | & d \\ 0 & 1 & c & | & e \\ 0 & 0 & 1 & | & f \end{bmatrix}$$

for a system of linear equations in three variables.

Note that the first nonzero element of each row is 1 and that the elements below them are 0 in both of the simplified forms.

EXAMPLE 1

Use matrices to solve the system:

$$(1) \quad 2x - y = -4$$
$$(2) \quad x + 3y = 5$$

Solution First, we write the augmented matrix for the system.

System of Equations	Augmented Matrix

$$2x - y = -4 \qquad \begin{bmatrix} 2 & -1 & | & -4 \\ 1 & 3 & | & 5 \end{bmatrix}$$
$$x + 3y = 5$$

Next, we use matrix row operations to get a 1 in row (1), column (1). To do this, we can interchange row (1) and row (2), getting:

$$\left(\begin{bmatrix} 1 & 3 & | & 5 \\ 2 & -1 & | & -4 \end{bmatrix}\right.$$

PRACTICE 1

Use matrices to solve the system:

$$(1) \quad 2x + y = 3$$
$$(2) \quad x - 3y = 12$$

Now, we need 0 below the 1 in column (1); that is, we want to replace the 2 in row (2), column (1) with a 0. To do this, we can multiply row (1) by -2 and add the products to row (2).

$$\begin{bmatrix} 1 & 3 & | & 5 \\ 2 + (-2)(1) & -1 + (-2)(3) & | & -4 + (-2)(5) \end{bmatrix} = \begin{bmatrix} 1 & 3 & | & 5 \\ 0 & -7 & | & -14 \end{bmatrix}$$

Finally, to get a 1 in row (2), column (2), we can divide row (2) by -7.

$$\begin{bmatrix} 1 & 3 & | & 5 \\ 0 \div (-7) & -7 \div (-7) & | & -14 \div (-7) \end{bmatrix} = \begin{bmatrix} 1 & 3 & | & 5 \\ 0 & 1 & | & 2 \end{bmatrix}$$

This matrix is now in the desired form and corresponds to the system below:

$$x + 3y = 5$$
$$y = 2$$

From this system, we see that $y = 2$. To find x, we substitute 2 for y in the equation $x + 3y = 5$.

$$x + 3y = 5$$
$$x + 3(2) = 5$$
$$x + 6 = 5$$
$$x = -1$$

So the solution is $(-1, 2)$.

Example 1 suggests the following rule for solving linear systems by using matrices:

To Solve a System of Linear Equations by Using Matrices

- Write the augmented matrix for the system.

- Use matrix row operations to get a matrix of the form

$$\begin{bmatrix} 1 & a & | & b \\ 0 & 1 & | & c \end{bmatrix} \quad \text{or} \quad \begin{bmatrix} 1 & a & b & | & d \\ 0 & 1 & c & | & e \\ 0 & 0 & 1 & | & f \end{bmatrix}.$$

System of two equations   System of three equations

- Write the system of linear equations that corresponds to the matrix in the previous step and find the solution.

Now, let's apply the matrix method to solve linear systems with three variables.

EXAMPLE 2

Solve the following system using matrices:

(1) $x + 3y - 6z = 7$
(2) $2x - y + 2z = 0$
(3) $x + y + 2z = -1$

PRACTICE 2

Use matrices to solve the following system:

(1) $x - y + 2z = -2$
(2) $x + y - 4z = 5$
(3) $3x + 2y + 4z = 18$

EXAMPLE 2 (continued)

Solution   We begin by writing the augmented matrix for the system.

System of Equations   **Augmented Matrix**

$$
\begin{aligned}
x + 3y - 6z &= 7 \\
2x - y + 2z &= 0 \\
x + y + 2z &= -1
\end{aligned}
\qquad
\left[\begin{array}{rrr|r}
1 & 3 & -6 & 7 \\
2 & -1 & 2 & 0 \\
1 & 1 & 2 & -1
\end{array}\right]
$$

Since there is already a 1 in row (1), column (1), we start simplifying the matrix by getting 0's under the 1 in column (1). To do this, we can multiply row (3) by -2 and add it to row (2).

$$
\left[\begin{array}{ccc|c}
1 & 3 & -6 & 7 \\
2 + (-2)(1) & -1 + (-2)(1) & 2 + (-2)(2) & 0 + (-2)(-1) \\
1 & 1 & 2 & -1
\end{array}\right]
$$

$$
= \left[\begin{array}{rrr|r}
1 & 3 & -6 & 7 \\
0 & -3 & -2 & 2 \\
1 & 1 & 2 & -1
\end{array}\right]
$$

To get a 0 in row (3), column (1), we can multiply row (1) by -1 and add it to row (3).

$$
\left[\begin{array}{ccc|c}
1 & 3 & -6 & 7 \\
0 & -3 & -2 & 2 \\
1 + (-1)(1) & 1 + (-1)(3) & 2 + (-1)(-6) & -1 + (-1)(7)
\end{array}\right]
$$

$$
= \left[\begin{array}{rrr|r}
1 & 3 & -6 & 7 \\
0 & -3 & -2 & 2 \\
0 & -2 & 8 & -8
\end{array}\right]
$$

To get a 1 in row (2), column (2), we first multiply row (2) by -1.

$$
\left[\begin{array}{rrr|r}
1 & 3 & -6 & 7 \\
0 & 3 & 2 & -2 \\
0 & -2 & 8 & -8
\end{array}\right]
$$

Then, add row (3) to row (2).

$$
\left[\begin{array}{ccc|c}
1 & 3 & -6 & 7 \\
0 & 3 + (-2) & 2 + 8 & -2 + (-8) \\
0 & -2 & 8 & -8
\end{array}\right]
= \left[\begin{array}{rrr|r}
1 & 3 & -6 & 7 \\
0 & 1 & 10 & -10 \\
0 & -2 & 8 & -8
\end{array}\right]
$$

To get a 0 in row (3), column (2), we can multiply row (2) by 2 and add it to row (3).

$$
\left[\begin{array}{ccc|c}
1 & 3 & -6 & 7 \\
0 & 1 & 10 & -10 \\
0 + 2(0) & -2 + 2(1) & 8 + 2(10) & -8 + 2(-10)
\end{array}\right]
$$

$$
= \left[\begin{array}{rrr|r}
1 & 3 & -6 & 7 \\
0 & 1 & 10 & -10 \\
0 & 0 & 28 & -28
\end{array}\right]
$$

Next, we multiply row (3) by $\dfrac{1}{28}$ to get a 1 in row (3), column (3).

$$\begin{bmatrix} 1 & 3 & -6 & 7 \\ 0 & 1 & 10 & -10 \\ \frac{1}{28}(0) & \frac{1}{28}(0) & \frac{1}{28}(28) & \frac{1}{28}(-28) \end{bmatrix} = \begin{bmatrix} 1 & 3 & -6 & 7 \\ 0 & 1 & 10 & -10 \\ 0 & 0 & 1 & -1 \end{bmatrix}$$

The matrix is now in the desired form and corresponds to the following system:

$$x + 3y - 6z = 7$$
$$y + 10z = -10$$
$$z = -1$$

From this system, we see that $z = -1$. To find y, we substitute -1 for z in the equation $y + 10z = -10$.

$$y + 10z = -10$$
$$y + 10(-1) = -10$$
$$y - 10 = -10$$
$$y = 0$$

Finally, substituting 0 for y and -1 for z in the equation $x + 3y - 6z = 7$, we can solve for x.

$$x + 3y - 6z = 7$$
$$x + 3(0) - 6(-1) = 7$$
$$x + 0 + 6 = 7$$
$$x = 1$$

So the solution is $x = 1$, $y = 0$, and $z = -1$, or $(1, 0, -1)$.

EXAMPLE 3

Solve the system using matrices:

(1) $\quad\quad x - y = -5$
(2) $\quad 2x + 3y - z = 6$
(3) $\quad\quad\quad y + z = 2$

Solution Note that there is no z-term in equation (1) and no x-term in equation (3). When a term is missing in an equation, we use 0 in the augmented matrix.

System of Equations	**Augmented Matrix**

$$\begin{array}{c} x - y = -5 \\ 2x + 3y - z = 6 \\ y + z = 2 \end{array} \qquad \begin{bmatrix} 1 & -1 & 0 & -5 \\ 2 & 3 & -1 & 6 \\ 0 & 1 & 1 & 2 \end{bmatrix}$$

To get a 0 in row (2), column (1), we multiply row (1) by -2 and add it to row (2).

$$\begin{bmatrix} 1 & -1 & 0 & -5 \\ 2 + (-2)(1) & 3 + (-2)(-1) & -1 + (-2)(0) & 6 + (-2)(-5) \\ 0 & 1 & 1 & 2 \end{bmatrix}$$

$$= \begin{bmatrix} 1 & -1 & 0 & -5 \\ 0 & 5 & -1 & 16 \\ 0 & 1 & 1 & 2 \end{bmatrix}$$

PRACTICE 3

Use matrices to solve the system:

(1) $\quad x + y + z = -3$
(2) $\quad\quad\quad 3y + 2z = 0$
(3) $\quad 4x - y = -8$

EXAMPLE 3 (continued)

Interchanging row (2) and row (3), we get a 1 in row (2), column (2).

$$\begin{bmatrix} 1 & -1 & 0 & | & -5 \\ 0 & 1 & 1 & | & 2 \\ 0 & 5 & -1 & | & 16 \end{bmatrix}$$

To get 0 in row (3), column (2), we multiply row (2) by -5 and add it to row (3).

$$\begin{bmatrix} 1 & -1 & 0 & | & -5 \\ 0 & 1 & 1 & | & 2 \\ 0 + (-5)(0) & 5 + (-5)(1) & -1 + (-5)(1) & | & 16 + (-5)(2) \end{bmatrix}$$

$$= \begin{bmatrix} 1 & -1 & 0 & | & -5 \\ 0 & 1 & 1 & | & 2 \\ 0 & 0 & -6 & | & 6 \end{bmatrix}$$

Finally, we multiply row (3) by $-\dfrac{1}{6}$ to get a 1 in row (3), column (3).

$$\begin{bmatrix} 1 & -1 & 0 & | & -5 \\ 0 & 1 & 1 & | & 2 \\ (-\frac{1}{6})(0) & (-\frac{1}{6})(0) & (-\frac{1}{6})(-6) & | & (-\frac{1}{6})(6) \end{bmatrix} = \begin{bmatrix} 1 & -1 & 0 & | & -5 \\ 0 & 1 & 1 & | & 2 \\ 0 & 0 & 1 & | & -1 \end{bmatrix}$$

The matrix is now in the desired form and corresponds to the following system:

$$x - y \qquad = -5$$
$$y + z = 2$$
$$z = -1$$

From this system, we see that $z = -1$. To find y, we substitute -1 for z in the equation $y + z = 2$.

$$y + z = 2$$
$$y + (-1) = 2$$
$$y = 3$$

Substituting 3 for y in the equation $x - y = -5$, we can find x.

$$x - y = -5$$
$$x - 3 = -5$$
$$x = -2$$

So the solution is $(-2, 3, -1)$.

EXAMPLE 4

A plane flies 500 mph with a tailwind for 3 hr. Making the return trip against the same wind, it takes 4 hr. What is the plane's speed p in still air? What is the speed w of the wind?

Solution We know that the combined speed of the plane with the wind is 500 mph. So

$$p + w = 500$$

The combined speed of the plane against the wind is $p - w$. Since rate $\cdot$ time $=$ distance, the number of miles the plane flies with the wind in 3 hr is $(p + w) \cdot 3$, or $3(p + w)$. Similarly, the number of miles the plane flies against the wind in 4 hr is $(p - w)4$, or $4(p - w)$. Since the distance going and returning is the same, we can conclude that:

$$3(p + w) = 4(p - w)$$

Combining these two equations and simplifying the second one gives us the following system:

$$p + \ w = 500$$
$$-p + 7w = 0$$

This system is equivalent to the augmented matrix below:

$$\begin{bmatrix} 1 & 1 & | & 500 \\ -1 & 7 & | & 0 \end{bmatrix}$$

We need to get a 0 in row (2), column (1). To do this, we add row (1) to row (2), giving us:

$$\begin{bmatrix} 1 & 1 & | & 500 \\ -1+1 & 7+1 & | & 0+500 \end{bmatrix} = \begin{bmatrix} 1 & 1 & | & 500 \\ 0 & 8 & | & 500 \end{bmatrix}$$

To get a 1 in row (2), column (2), we multiply row (2) by $\dfrac{1}{8}$.

$$\begin{bmatrix} 1 & 1 & | & 500 \\ \frac{1}{8}(0) & \frac{1}{8}(8) & | & \frac{1}{8}(500) \end{bmatrix} = \begin{bmatrix} 1 & 1 & | & 500 \\ 0 & 1 & | & 62.5 \end{bmatrix}$$

This matrix is in the desired form and is equivalent to the following system of equations:

$$p + w = 500$$
$$w = 62.5$$

Substituting 62.5 for w in the equation $p + w = 500$, we can solve for p.

$$p + w = 500$$
$$p + 62.5 = 500$$
$$p = 437.5$$

So the speed of the wind is 62.5 mph, and the speed of the plane in still air is 437.5 mph.

PRACTICE 4

Traveling to his favorite spot 30 mi away, a fisherman in his boat takes 2 hr at top speed against the current. Returning at top speed with the current, the trip takes only 1 hr. Find both the top speed of the boat in still water and the speed of the current.

4.4 **Exercises**

FOR
EXTRA
HELP

MyMathLab

Math XL
PRACTICE

WATCH

READ

REVIEW

Mathematically Speaking

Fill in each blank with the most appropriate term or phrase from the given list.

constants	matrix	coefficients
rows	adding the	columns
augmented	products to	multiplying the result
square	equivalent	with
numeral	element	

1. A(n) _____ is a rectangular array of numbers, such as $\begin{bmatrix} -3 & 4 & 1 \\ 2 & 0 & 9 \end{bmatrix}$.

2. In the matrix $\begin{bmatrix} -3 & 4 & 1 \\ 2 & 0 & 9 \end{bmatrix}$, the entry -3 is also called a(n) _____.

3. The _____ of a matrix are horizontal.

4. A(n) _____ matrix has the same number of rows as columns.

5. In an augmented matrix, the _____ are on the left side of the vertical bar.

6. Performing row operations on a matrix results in a(n) _____ matrix.

7. One of the row operations is multiplying the elements of any row by a nonzero number and then _____ their corresponding elements in any other row.

8. When an equation is missing a term, we represent that term with a 0 in the _____ matrix.

A *Solve using matrices.*

9. $x - 7y = 12$
   $x - 4y = 6$

10. $6x + y = 7$
    $8x + y = 11$

11. $3x + 4y = -1$
    $2x + 3y = -1$

12. $5x - 2y = 6$
    $-4x + 3y = 5$

13. $9x - 15y = 12$
    $-3x + 5y = -4$

14. $7x - 4y = 3$
    $14x - 8y = -8$

15. $\frac{1}{2}x - \frac{1}{3}y = -7$
    $3x + 5y = 21$

16. $2x + 6y = 0$
    $\frac{2}{3}x - y = -3$

17. $x + 3y - 2z = 10$
    $3x - 2y - z = 9$
    $4x - y + 5z = 1$

18. $5x - 2y + 2z = -2$
    $-x + 3y + 6z = -20$
    $x + y - z = 8$

19. $6x - 4y - 3z = -4$
    $2x + 8y - 7z = -11$
    $-x + 2y + z = 2$

20. $2x + 9y - 6z = -4$
    $-3x - 15y + 18z = 4$
    $-4x - 11y + 8z = 11$

21. $5x + 7y + 3z = -48$
    $-8x + 2y + 4z = 24$
    $12x - 6y - 9z = -21$

22. $15x - 10y + 5z = -10$
    $6x + 4y - 2z = 4$
    $2x - 5y - 3z = 28$

23. $7x - y + 8z = -9$
    $3x + 3y - 10z = -5$
    $3.5x - 0.5y + 4z = -14$

24.
$$1.5x + 2y + 3z = 3$$
$$3x + 4y - 2z = 2$$
$$6x + 8y + 4z = 8$$

25.
$$y + 5z = -15$$
$$4x - 4y = -20$$
$$8x - 2z = 8$$

26.
$$7x + 9y = -3$$
$$-2x + 3z = 3$$
$$5y - z = 11$$

27.
$$2x + 3y = 7$$
$$-11y - 4z = -5$$
$$x + y - 6z = 0$$

28.
$$10x + 8y - 4z = 1$$
$$3x - 4y = -4$$
$$-x + 2y + 3z = 1$$

Mixed Practice

Solve using matrices.

29.
$$-2x - y = 5$$
$$x + 2y = 2$$

30.
$$2x - 8y = 10$$
$$-x + 4y = 5$$

31.
$$2x + 5y = -1$$
$$\frac{1}{4}x + 2y = 4$$

32.
$$x + 2y - z = -3$$
$$3x + 2y + 2z = 0$$
$$x - 2y - 3z = -1$$

33.
$$2x - y + z = 1$$
$$3x - 3y + 4z = 5$$
$$x - 2y + 3z = 4$$

34.
$$-2x + y = -5$$
$$-4y - 2z = 0$$
$$3x + 6z = -3$$

Applications

Ⓑ *Solve.*

◉ 35. As part of an experiment, a chemistry student is required to make 5 L of a 25% saline solution by combining a 15% saline solution with a 40% saline solution. How much of each solution does she need to make the required solution?

36. An electronics company has two manufacturing plants. Each month plant A produces 3 times as many plasma televisions as plant B. The combined monthly production of both manufacturing plants is 18,000 plasma televisions. How many plasma televisions does each plant produce per month?

37. A triathlon is a long-distance race that consists of three phases: a swim, a bike ride, and a run. An athlete who competed in a triathlon swam at a rate of 1.8 mph, biked at a rate of 16.8 mph, and ran at a rate of 6.55 mph. His combined time to complete the swim and bike ride was twice the time it took him to complete the run. The total length of the race is 140.6 mi. If it took him 12 hr to complete the triathlon, how long did it take him to complete each phase of the race?

38. An organization of college alumni holds its annual dinner and dance to raise money for student scholarships. Attendees can choose one of three types of ticket shown in the table. The total number of dinner tickets and dance tickets sold was half the number of combination tickets sold. If the organization raised $49,970 on the sale of 570 tickets, how many of each type of ticket were sold for the event?

Type of Ticket	Price
Dinner	$85
Dance	$30
Combination dinner and dance	$100

39. The following table shows some of the nutritional information for a cheeseburger, medium order of french fries, and a baked apple pie served at a local restaurant:

	Cheeseburger	Medium French Fries	Baked Apple Pie
Calories	330	450	260
Fat (in grams g)	14	22	13
Carbohydrates (in grams g)	6	57	34

How many of the cheeseburgers, medium orders of fries, and pies does a customer eat if he consumes 1960 calories, 90 g of fat, and 143 g of carbohydrates? How many of each item were consumed?

40. A beneficiary invested $50,000 of his inheritance in three accounts earning 6.25%, 4.5%, and 5% simple interest. The total amount invested in the accounts earning 4.5% and 5% interest was $3000 more than the amount he invested in the account earning 6.25% interest. After one year, the accounts earned a total of $2732.50 in interest. How much did he invest in each account?

• Check your answers on page A-34.

MINDStretchers

Research

1. A procedure for solving systems of equations, from which the matrix row operations were adapted, is called *Gauss–Jordan elimination*. Either in your college library or on the web, investigate how this procedure works, and compare it to the matrix row operations presented in this section. Write a brief summary of your findings.

Mathematical Reasoning

2. Are the matrices shown here equivalent? Explain.

$$\begin{bmatrix} 2 & -2 & 4 & | & -2 \\ 1 & 2 & 3 & | & 1 \\ -1 & 3 & -2 & | & 3 \end{bmatrix} \qquad \begin{bmatrix} 1 & 2 & 3 & | & 1 \\ 0 & -3 & -1 & | & -2 \\ 0 & 5 & 1 & | & 4 \end{bmatrix}$$

Writing

3. Suppose we are solving a system of linear equations using matrices.

a. Explain why, in terms of the system of equations, interchanging any two rows will produce an equivalent matrix.

b. Will changing any two columns produce an equivalent matrix? Explain.

Cultural Note

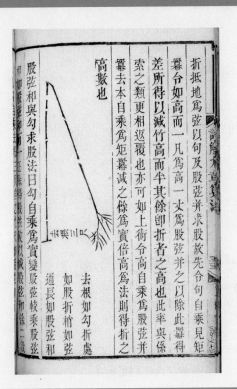

More than 2000 years ago, Chinese mathematicians, writing in the text *Nine Chapters on the Mathematical Art* (pictured here), used matrices to solve systems of linear equations. This text showed methods for solving everyday problems in engineering, surveying, trade, and taxation. It played a fundamental role in the development of mathematics in China.

(*Sources:* Jean-Claude Martzloff, *A History of Chinese Mathematics,* 1997; Li Yan and Du Shiran, *Chinese Mathematics, a Concise History,* 1987)

4.5 Solving Systems of Linear Inequalities

In Section 3.5, we graphed on a coordinate plane a single linear inequality in two variables, such as $3x - 2y > 4$. There, we saw that a solution of such an inequality is an ordered pair of real numbers that when substituted for the variables satisfy the inequality. We also saw that the graph of this inequality (the set of all ordered pair solutions) consists of a region of points in the coordinate plane whose coordinates satisfy the inequality.

In this section, we consider the graph of a *system of inequalities.*

DEFINITION

A **system** of inequalities consists of two or more inequalities considered simultaneously, that is, together.

For instance,

$$y \leq \quad 3x - 6$$
$$y > -4x + 2$$

is a system of linear inequalities. As with single inequalities, the *solutions* of a system of inequalities are ordered pairs.

DEFINITION

A **solution** of a system of inequalities in two variables is an ordered pair of numbers that makes both inequalities in the system true.

Solving Systems of Linear Inequalities by Graphing

We can solve a system of linear inequalities by graphing each inequality on the same coordinate plane. Every point in the region of overlap is a solution of both inequalities and is, therefore, a solution of the system.

EXAMPLE 1

Graph the solutions of the system:

$$y \leq 3x - 6$$
$$y > -4x + 2$$

Solution Let's begin by graphing each inequality. First, we graph $y \leq 3x - 6$. The boundary line is the graph of the corresponding equation $y = 3x - 6$. We draw this as a solid line since the inequality involves the $\leq$ symbol. Then, in either half-plane we choose a test point, for instance, $(3, 0)$. Since this point satisfies the inequality, the half-plane containing it is part of the graph.

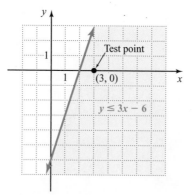

PRACTICE 1

Graph the solutions of the system:

$$y \geq x - 5$$
$$y < -3x - 2$$

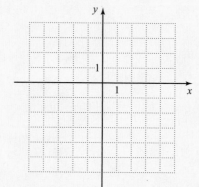

Next, we graph $y > -4x + 2$. The boundary line is the graph of the corresponding equation $y = -4x + 2$. We draw this as a broken line since the inequality involves the $>$ symbol. Then, in either half-plane we choose a test point, for instance, $(0, 4)$. Since this point satisfies the inequality, the half-plane containing it is part of the graph.

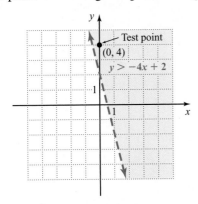

Finally, we draw each inequality on the same coordinate plane. Since a solution of the system of linear inequalities must satisfy each inequality, the solutions of the system are all the points that lie in the intersection of the shaded regions, that is, in the overlapping region of the two graphs. Points on part of the boundary line $y = 3x - 6$ are solutions, but points on the boundary line $y = -4x + 2$ are not.

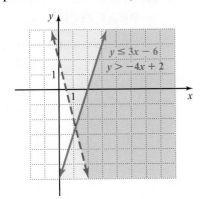

EXAMPLE 2

Solve the following system by graphing:

$$-x - y < 2$$
$$y - 2x > 1$$

Solution We begin by graphing both inequalities on the same coordinate plane. To graph each boundary line, we solve each inequality for y, getting:

$$
\begin{array}{ll}
-x - y < 2 & y - 2x > 1 \\
\quad -y < x + 2 & \quad\quad y > 2x + 1 \\
\quad\quad y > -x - 2 &
\end{array}
$$

PRACTICE 2

Solve the following system by graphing:

$$2x + y < 1$$
$$-y + 3x < 1$$

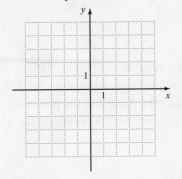

EXAMPLE 2 (continued)

Next, we graph each boundary line. Then, for each inequality, we shade the half-plane that contains its solutions.

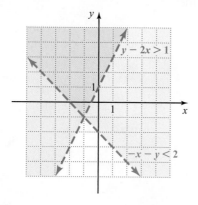

The solutions of the system are all the points that lie in the intersection of the shaded regions, that is, in the overlapping region of the two graphs. Note that none of the points on the boundary lines is a solution.

Now, let's look at solving a system consisting of three linear inequalities in two variables.

EXAMPLE 3

Graph the solutions of the system.

$$x + y < 8$$
$$x \geq 0$$
$$y \geq 0$$

Solution   We graph all three inequalities on the same coordinate plane. First, we graph the boundary lines. Then, for each inequality we shade the half-plane that contains its solutions.

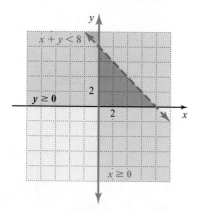

The solutions of the system are points that lie in the intersection of the three shaded regions, that is, the dark purple overlapping region of all three graphs. Note that the points on parts of the lines $x = 0$ and $y = 0$ are solutions, but points on the line $x + y = 8$ are not.

PRACTICE 3

Graph the solutions of the system.

$$y + 2x \geq 4$$
$$x > -3$$
$$y \geq 1$$

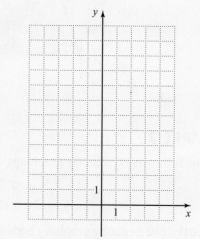

Just as with systems of linear equations, many applied problems can be modeled by systems of linear inequalities.

EXAMPLE 4

A company's employment test has two parts—a verbal subtest and a mathematics subtest. An applicant can earn a maximum combined score of 100 points. To qualify for a particular position, the applicant is required to have a math score of at least 40 and a verbal score of at least 25.

a. Write a system of inequalities to model the situation in which the applicant qualifies for a position.

b. Solve the system by graphing.

c. Describe the region in which the prospective employee qualifies for the position.

Solution

a. Let v represent the verbal score and m represent the math score. Since a prospective employee can earn a maximum combined score of 100 points, we write $v + m \leq 100$. If a successful applicant must score at least 40 on the math subtest and at least 25 on the verbal subtest, then we can write two inequalities, $m \geq 40$ and $v \geq 25$. So the system is

$$v + m \leq 100$$
$$m \geq 40$$
$$v \geq 25$$

b. Solve by graphing: $\quad v + m \leq 100$
$$m \geq 40$$
$$v \geq 25$$

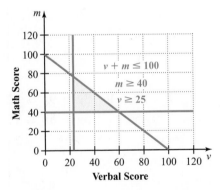

The solutions of the system are points that lie in the intersection of the three shaded regions, including the highlighted parts of the boundary lines.

c. The region in which the applicant qualifies for a position is the overlapping area bounded by the lines $m = 40$, $v + m = 100$, and $v = 25$.

PRACTICE 4

Suppose the president of the student government association formed a committee to raise funds for cancer research. The committee must have from 7 to 10 members. The number of first-time freshmen should be greater than the number of returning students.

a. Write a system of inequalities to model the problem.

b. Graph the solutions of the system.

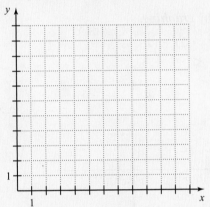

c. List the combinations of first-time freshmen and returning students that may participate in the committee. Explain why the list is finite.

Mathematically Speaking

Fill in each blank with the most appropriate term or phrase from the given list.

overlapping	triples	pairs
number line	a system of	unshaded
simultaneous	coordinate plane	

1. Two or more linear inequalities considered simultaneously, that is, together, are called _____ linear inequalities.

2. The solutions of a system of linear inequalities in two variables are ordered _____.

3. A system of linear inequalities in two variables can be solved by graphing each inequality on the same _____.

4. In the graph of a system of linear inequalities, every point in the _____ region is a solution of the system.

A *Solve by graphing.*

5. $y > 2x - 1$
   $y < -x + 3$

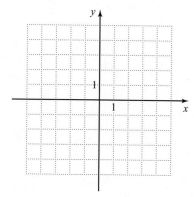

6. $y < 3x - 2$
   $y > -2x + 1$

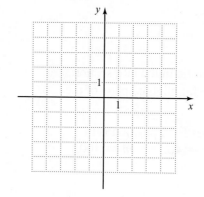

7. $y \leq \frac{1}{3}x + 3$
   $y < -\frac{1}{2}x + 1$

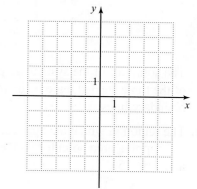

8. $y > -\frac{1}{4}x - 1$
   $y \geq \frac{3}{2}x$

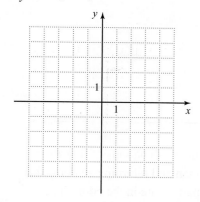

9. $4x + 2y \geq -6$
   $12x - 3y \geq -6$

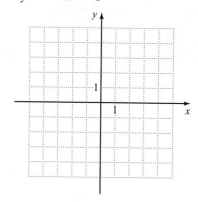

10. $3y - 9x \leq 12$
    $5y + 5x \leq 15$

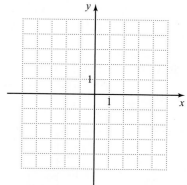

11. $3x - 2y \leq 8$
$\quad -x - 3y > 0$

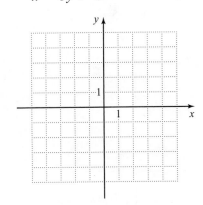

12. $-2x - 4y < 16$
$\quad x - 2y \geq 8$

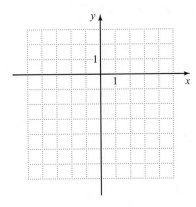

13. $4x + 2y \leq 4$
$\quad 2x + y > -3$

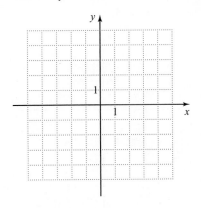

14. $x + 4y \leq 20$
$\quad 3x + 12y > -24$

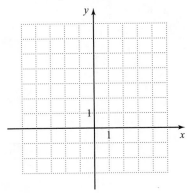

15. $3x - 9y < 18$
$\quad 1.5x + 0.5y > 1.5$

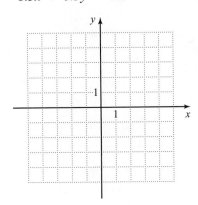

16. $1.2x + 0.4y < -0.4$
$\quad 5x - 15y > 10$

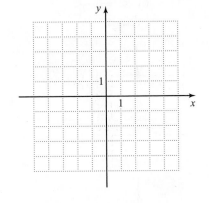

17. $y - x > 1$
$\quad x \leq 4$
$\quad y > 0$

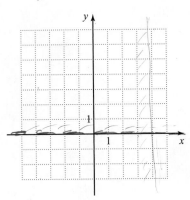

18. $x + y < 3$
$\quad x \leq 1$
$\quad y \geq -3$

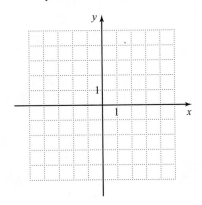

19. $-2x - y \geq 3$
$\quad x \geq -4$
$\quad y \geq -2$

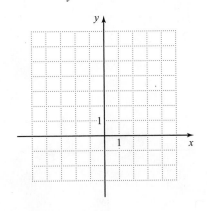

20. $y + 2x \geq -1$
$x \leq 2$
$y \leq 3$

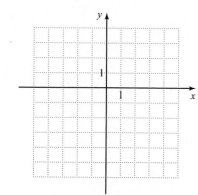

21. $3x - 2y < 2$
$x + 3y < 12$
$x > -2$

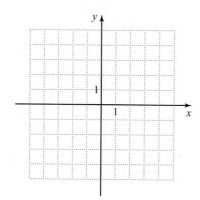

22. $4x + 8y > 8$
$2x - y > 3$
$x < 5$

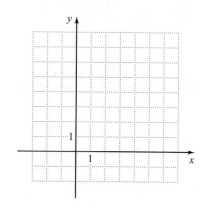

23. $12x - 4y \geq 16$
$3x - 6y < -6$
$y < 4$

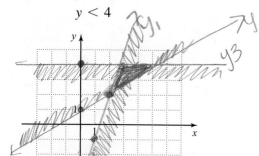

24. $10x + 5y > -5$
$9x - 3y \geq 0$
$y \geq -4$

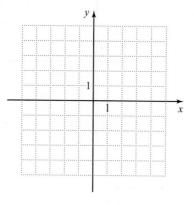

25. $-x + y > -2$
$x + y < 2$
$2x - y > 1$

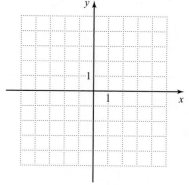

26. $y + x \leq -3$
$y - x \leq 1$
$2y - x > -2$

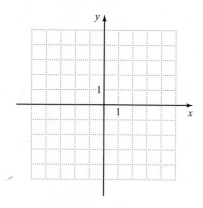

Mixed Practice

Solve by graphing.

27. $y \le -3x - 2$
$y < x + 2$

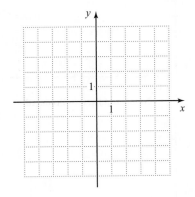

28. $-3x + 2y \le -6$
$\frac{1}{4}x + y > 1$

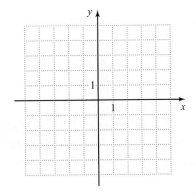

29. $6x + 3y < 9$
$-4x - 2y < 4$

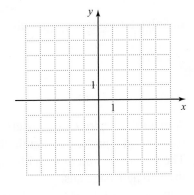

30. $x \ge 1$
$y > -3$
$x + y \ge 1$

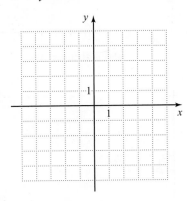

31. $x > -3$
$y < 4$
$-2x + y \ge 3$

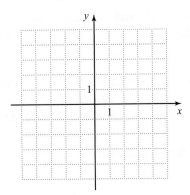

32. $y \ge -4$
$2x + y < -1$
$-3x + y \le 3$

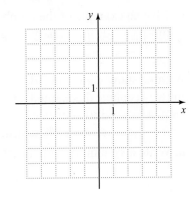

Applications

B *Solve.*

33. A furniture manufacturer produces dining tables and chairs. Each table requires 2 hr for assembly and 3 hr for finishing and painting. Each chair requires 1.5 hr for assembly and 1 hr for finishing and painting. The manufacturer allots at most 360 hours per month for assembly and at most 400 hours per month for finishing and painting.

a. Express this information as a system of inequalities.

b. Solve the system by graphing.

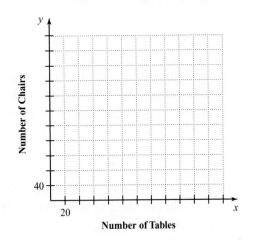

34. A factory produces two types of machine parts. Part A requires 5 hr for fabrication and $1\frac{1}{2}$ hr for testing. Part B requires 2 hr for fabrication and $\frac{1}{2}$ hr for testing. The factory manager allots no more than 150 hours per week for fabrication and no more than 40 hours per week for testing.

   a. Write a system of inequalities to represent the situation.

   b. Solve the system by graphing.

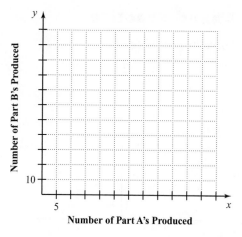

Number of Part A's Produced

35. A magazine charges advertisers $150 for a half-page ad and $225 for a full-page ad. In a particular month, the revenue from ads was less than $15,000. The magazine sold more half-page ads than full-page ads, including more than 20 half-page ads.

   a. Write a system of inequalities to model the information.

   b. Solve the system by graphing.

   c. Give an example of the number of half-page ads and the number of full-page ads that satisfy the given conditions.

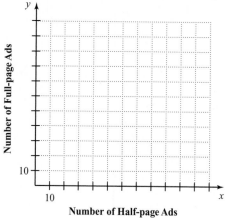

Number of Half-page Ads

36. A game show winner decided to invest no more than $8500 of his winnings in two accounts that earned a 6.5% and 5% return. He wanted the accounts to earn at least $300 after one year, and planned to invest less than $2500 in the account that earned a 5% return.

   a. Write a system of inequalities to model the information.

   b. Solve the system by graphing.

   c. Give an example of the amounts of money invested in the accounts that satisfy the given conditions.

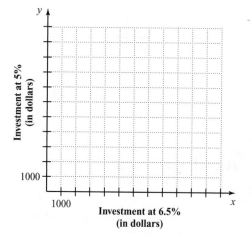

Investment at 6.5% (in dollars)

37. A part-time student works at two jobs to support herself. She can work at her retail job no more than 20 hr/wk and at her office job no more than 30 hr/wk. In order to make enough money, she must work a minimum of 35 hr/wk.

   a. Express this information as a system of inequalities.

   b. Graph the solutions of the system.

   c. If she works 15 hr at her retail job, how many hours must she work at her office job?

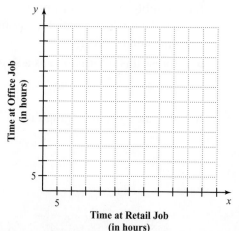

Time at Retail Job (in hours)

38. An investor has a maximum of $10,000 to buy shares of stock in a pharmaceutical company and in a technology company. The pharmaceutical stock costs $40 per share, and the technology stock costs $25 per share. He wants to buy no more than 200 shares of the pharmaceutical stock and at least 50 shares of the technology stock.

a. Write a system of inequalities to model this problem.

b. Graph the solutions of the system.

c. If he buys 100 shares of the technology stock, how many shares of the pharmaceutical stock can he buy?

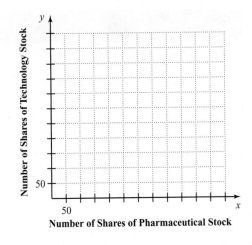

39. The owner of a garden shop decides to fence off part of his lot to build an outdoor nursery. The perimeter of the lot cannot exceed 400 ft. The length of the nursery is to be at least 25 ft longer than the width. The width must be at least 25 ft.

a. Express this information as a system of inequalities.

b. Solve the system by graphing.

c. What does the solution region represent?

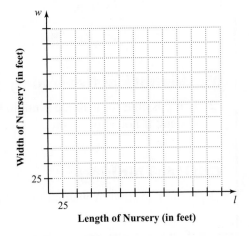

40. A donut shop charges $0.90 for a donut and $1.50 for a muffin. The manager of the shop would like to have daily sales of at least $465. The shop can make at most 30 dozen donuts and 20 dozen muffins per day.

a. Express this information as a system of inequalities.

b. Solve the system by graphing.

c. What does the solution region represent?

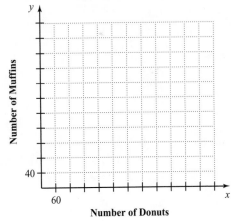

• Check your answers on page A-34.

MINDStretchers

Writing

1. Write a problem situation whose solutions lie in the region shown in the following graph.

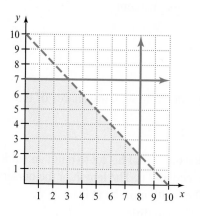

Mathematical Reasoning

2. Is it possible for a single point to be the only solution of a system of inequalities? If so, give an example of such a system. If not, explain why.

Critical Thinking

3. Write a system of inequalities whose solutions consist of all points on the line $y = 2x + 3$.

Key Concepts and Skills

CONCEPT   SKILL

Concept/Skill	Description	Example
[4.1] System of equations	Two or more equations considered simultaneously, that is, together.	$x + y = 6$ $y = -2x + 5$
[4.1] Solution of a system of equations in two variables	An ordered pair of numbers that makes both equations in the system true.	The ordered pair $(2, 3)$ is a solution of the system $x + y = 5$ $x - y = -1$
[4.1] Types of systems of two linear equations	• The system is *consistent* and the equations are *independent*. The graphs intersect at exactly one point. The system has one solution. • The system is *inconsistent* and the equations are *independent*. The graphs are parallel lines. The system has no solution. • The system is *consistent* and the equations are *dependent*. The graphs are the same line. The system has infinitely many solutions.	
[4.2] To solve a system of linear equations by substitution	• In one of the equations, solve for either variable in terms of the other variable. • In the other equation, substitute the expression equal to the variable found in the previous step. Then, solve the resulting equation for the remaining variable. • Substitute the value found in the previous step in either of the original equations and solve for the other variable. • Check the proposed solution in both equations of the original system.	**(1)** $\quad x + y = 6$ **(2)** $\quad 2x + y = 3$ Solve equation (1) for y: **(1)** $\quad y = -x + 6$ Substitute $-x + 6$ for y in equation (2): **(2)** $\quad 2x + (-x + 6) = 3$ $\quad\quad\quad\quad\quad x + 6 = 3$ $\quad\quad\quad\quad\quad\quad\quad x = -3$ Substitute -3 for x in equation (1): **(1)** $\quad x + y = 6$ $\quad\quad -3 + y = 6$ $\quad\quad\quad\quad\quad y = 9$ So the solution is $(-3, 9)$. **Check** **(1)** $\quad x + y = 6$ $\quad -3 + 9 \overset{?}{=} 6$ $\quad\quad\quad\quad 6 = 6 \quad$ True **(2)** $\quad\quad 2x + y = 3$ $\quad 2(-3) + 9 \overset{?}{=} 3$ $\quad\quad\quad\quad\quad 3 = 3 \quad$ True

continued

Concept/Skill	Description	Example
[4.2] To solve a system of linear equations in two variables by elimination	• Write both equations in the general form $Ax + By = C$. • Choose the variable that you want to eliminate. • If necessary, multiply one or both equations by the appropriate number(s) so that the coefficients of the variables to be eliminated are opposites. • Add the equations. Then, solve the resulting equation for the remaining variable. • Substitute the value found in the previous step in either of the original equations and solve for the other variable. • Check by substituting the values in both equations of the original system.	**(1)** $2x + y = 6$ **(2)** $x + y = 5$ To eliminate y, multiply equation (2) by -1. Then, add the equations. $$2x + y = 6$$ $$\underline{-x - y = -5}$$ $$x = 1$$ Solve for y by substituting 1 for x in equation (2). **(2)** $x + y = 5$ $1 + y = 5$ $y = 4$ So the solution is $(1, 4)$. **Check** **(1)** $2x + y = 6$ $2(1) + 4 \stackrel{?}{=} 6$ $6 = 6$ True **(2)** $x + y = 5$ $1 + 4 \stackrel{?}{=} 5$ $5 = 5$ True
[4.3] Solution of a system of linear equations in three variables	An ordered triple of numbers that makes all three equations in the system true.	The ordered triple $(1, 2, 0)$ is a solution of the following system: $x - 2y + 3z = -3$ $4x + 5y - z = 14$ $2x - y + 6z = 0$
[4.3] To solve a system of linear equations in three variables by elimination	• Write all equations in the general form $Ax + By + Cz = D$. • Choose a pair of equations and use them to eliminate a variable, getting an equation in two variables. • Next, use a different pair of equations to eliminate the same variable as in the previous step, getting another equation in the same two variables. • Solve the system of equations formed by the equations found in the previous two steps, getting the values of two of the variables. • Substitute the values found in the previous step in any of the original equations to find the value of the third variable. • Check by substituting the values in all three equations of the original system.	**(1)** $x + y + z = 3$ **(2)** $2x + y + z = 4$ **(3)** $x + 2y - z = -4$ Add equations (1) and (3) to eliminate the z-terms. **(1)** $x + y + z = 3$ **(3)** $\underline{x + 2y - z = -4}$ **(4)** $2x + 3y = -1$ Add equations (2) and (3) to eliminate the z-terms. **(2)** $2x + y + z = 4$ **(3)** $\underline{x + 2y - z = -4}$ **(5)** $3x + 3y = 0$ Solve the system. **(4)** $2x + 3y = -1$ **(5)** $3x + 3y = 0$

CONCEPT   SKILL

Concept/Skill	Description	Example
		To eliminate the y-terms, multiply equation (4) by -1 and add the equations.
		(4) $\quad -2x - 3y = 1$
		(5) $\quad \underline{3x + 3y = 0}$
		$\qquad\qquad\quad x = 1$
		Substitute 1 for x in equation (5).
		(5) $\quad 3x + 3y = 0$
		$\quad 3(1) + 3y = 0$
		$\quad\quad 3 + 3y = 0$
		$\qquad\quad 3y = -3$
		$\qquad\qquad y = -1$
		Substitute 1 for x and -1 for y in equation (1).
		(1) $\qquad x + y + z = 3$
		$\quad 1 + (-1) + z = 3$
		$\qquad\qquad\quad z = 3$
		So the solution is $(1, -1, 3)$.
		Check
		(1) $\qquad x + y + z = 3$
		$\quad 1 + (-1) + 3 \overset{?}{=} 3$
		$\qquad\qquad 3 = 3 \quad$ True
		(2) $\qquad 2x + y + z = 4$
		$\quad 2(1) + (-1) + 3 \overset{?}{=} 4$
		$\qquad\qquad\quad 4 = 4 \quad$ True
		(3) $\qquad x + 2y - z = -4$
		$\quad 1 + 2(-1) - 3 \overset{?}{=} -4$
		$\qquad\qquad -4 = -4 \quad$ True
[4.4] **Matrix**	A rectangular array of numbers. The numbers of a matrix are called **elements** or **entries**. The **rows** of a matrix are horizontal, and the **columns** are vertical. A **square matrix** has the same number of rows and columns.	$\begin{array}{ccc} \text{Column} & \text{Column} & \text{Column} \\ 1 & 2 & 3 \\ \downarrow & \downarrow & \downarrow \end{array}$ $\begin{array}{l} \text{Row 1} \rightarrow \\ \text{Row 2} \rightarrow \end{array} \begin{bmatrix} 2 & 3 & 8 \\ 1 & 5 & 3 \end{bmatrix}$
[4.4] **Equivalent matrices**	Matrices that correspond to equivalent systems of equations.	$\begin{bmatrix} 3 & -2 & \vert & 5 \\ 2 & 1 & \vert & 1 \end{bmatrix}$ corresponds to the system $\begin{array}{l} 3x - 2y = 5 \\ 2x + y = 1 \end{array}$ and $\begin{bmatrix} 2 & 1 & \vert & 1 \\ 3 & -2 & \vert & 5 \end{bmatrix}$ corresponds to the system $\begin{array}{l} 2x + y = 1 \\ 3x - 2y = 5. \end{array}$ Since the systems are equivalent, so are the matrices.

continued

CONCEPT   SKILL

Concept/Skill	Description	Example
[4.4] **Matrix row operations**	The following row operations produce an equivalent matrix: • Interchanging any two rows • Multiplying (or dividing) the elements of any row by the same nonzero number • Multiplying (or dividing) the elements of any row by a nonzero number and adding the products to their corresponding elements in any other row	Interchanging row (1) and row (2): $\begin{bmatrix} 1 & 2 & \vert & 4 \\ 1 & -1 & \vert & 1 \end{bmatrix}$ is equivalent to $\begin{bmatrix} 1 & -1 & \vert & 1 \\ 1 & 2 & \vert & 4 \end{bmatrix}$. Multiplying row (2) by -1: $\begin{bmatrix} 1 & 2 & \vert & 4 \\ 1 & -1 & \vert & 1 \end{bmatrix}$ is equivalent to $\begin{bmatrix} 1 & 2 & \vert & 4 \\ -1 & 1 & \vert & -1 \end{bmatrix}$. Multiplying row (1) by 3 and adding to row (2): $\begin{bmatrix} 1 & 2 & \vert & 4 \\ 1 & -1 & \vert & 1 \end{bmatrix}$ is equivalent to $\begin{bmatrix} 1 & 2 & \vert & 4 \\ 4 & 5 & \vert & 13 \end{bmatrix}$
[4.4] **To solve a system of linear equations by using matrices**	• Write the augmented matrix for the system. • Use matrix row operations to get a matrix of the form $\begin{bmatrix} 1 & a & \vert & b \\ 0 & 1 & \vert & c \end{bmatrix}$ or $\begin{bmatrix} 1 & a & b & \vert & d \\ 0 & 1 & c & \vert & e \\ 0 & 0 & 1 & \vert & f \end{bmatrix}$ System of two equations System of three equations • Write the system of linear equations that corresponds to the matrix in the previous step, and find the solution.	$x + 2y = 4$ $x - y = 1$ Writing the augmented matrix: $\begin{bmatrix} 1 & 2 & \vert & 4 \\ 1 & -1 & \vert & 1 \end{bmatrix}$ Multiplying row (1) by -1 and adding it to row (2): $\begin{bmatrix} 1 & 2 & \vert & 4 \\ 0 & -3 & \vert & -3 \end{bmatrix}$ Multiplying row (2) by $-\frac{1}{3}$: $\begin{bmatrix} 1 & 2 & \vert & 4 \\ 0 & 1 & \vert & 1 \end{bmatrix}$ The system that corresponds to the simplified matrix is $x + 2y = 4$ $y = 1$ Substituting 1 for y in $x + 2y = 4$: $x + 2(1) = 4$ $x = 2$ So the solution is $(2, 1)$.
[4.5] **System of linear inequalities**	Two or more inequalities considered simultaneously, that is, together.	$y \geq x + 3$ $y < 3x - 2$
[4.5] **Solution of a system of inequalities in two variables**	An ordered pair of numbers that makes both inequalities in the system true.	The ordered pair $(-2, 3)$ is a solution of the following system of inequalities: $2x + 3y > 1$ $x - y < 3$

Say Why
Fill in each blank.

1. A system with dependent equations _____ has/does not have infinitely many solutions because _____ _____.

2. The ordered pair $(3, -2)$ _____ is/is not a solution to the system $\begin{array}{l} x + 2y = -1 \\ 3x - y = -11 \end{array}$ because _____ _____.

3. To solve the system $\begin{array}{l} -5x + 2y = 3 \\ y = 4x + 1 \end{array}$ using the substitution method, an expression for y rather than for x _____ would/would not be substituted because _____ _____.

4. To solve the system $\begin{array}{l} x + y = -2 \\ 3x + y = 7 \end{array}$ using the elimination method, the two equations _____ would/would not be added because _____ .

5. The substitution method and the elimination method for solving a system of linear equations _____ are/are not similar because _____ _____ _____.

6. $\begin{bmatrix} 1 & 0 & -3 & -2 \\ 2 & 1 & 5 & -1 \\ -3 & 1 & 0 & 2 \end{bmatrix}$ _____ is/is not a 4 × 3 matrix because _____ _____ .

[4.1] *Determine whether the ordered pair is a solution of the system of equations.*

7. $\begin{array}{l} 4x - 3y = -2 \quad (1, 2) \\ -x + 6y = 11 \end{array}$

8. $\begin{array}{l} 3x + 2y = -10 \quad \left(-\dfrac{2}{3}, -4\right) \\ 9x - 5y = 14 \end{array}$

Solve by graphing.

9. $\begin{array}{l} y = x + 4 \\ y = 1 - 2x \end{array}$

10. $\begin{array}{l} 5y - 15 = 0 \\ 4x + 3y = 9 \end{array}$

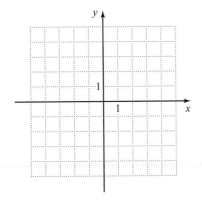

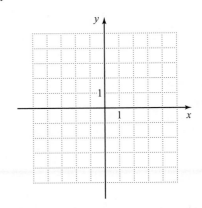

11. $2x - 4y = -8$
$-x + 2y = -4$

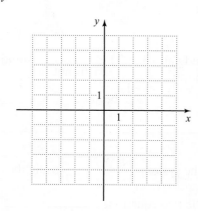

12. $6x = 16 - 4y$
$y = -\dfrac{3}{2}x + 4$

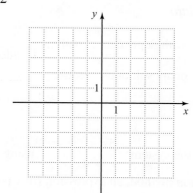

[4.2] *Solve by substitution.*

13. $y = 3x + 7$
$y = 7x - 9$

14. $y = 8x - 10$
$x = \dfrac{1}{2}y - 1$

15. $4x + 2y = 10$
$y = -(2x + 5)$

16. $5a - 3b = -16$
$6a = 2b$

Solve by elimination.

17. $x - 4y = 11$
$3x + 4y = 17$

18. $2p + 5q = -8$
$2p - 3q = 8$

19. $3s - 2t = 9$
$4s - 6t = 2$

20. $1.5x + 2.5y = -3$
$6x + 10y = -12$

[4.3] *Determine whether the ordered triple is a solution of the system of equations.*

21. $5x + 2y + 2z = -4$ $(-2, 3, 0)$
$2x - 3y + z = -13$
$3x + y + 4z = -3$

22. $9x + 8y + 4z = -7$ $\left(-\dfrac{1}{3}, -\dfrac{1}{4}, -\dfrac{1}{2}\right)$
$6x - 4y - 2z = -5$
$2y - 3z = 1$

Solve.

23. $x - y + z = 1$
$3x + y - 2z = -3$
$x + 4y - z = 2$

24. $4x + 3y - 6z = -4$
$x + y + z = 1$
$2x + 9y = -2$

25. $3y - 4z = 4$
$-2x + 3y = 8$
$3x - 6z = 5$

26. $\dfrac{3}{4}x - \dfrac{2}{3}y = -5$
$\dfrac{1}{2}x + y - \dfrac{1}{4}z = 0$
$\dfrac{1}{3}y + 3z = 13$

[4.4] *Solve each system using matrices.*

27. $3x - 5y = 6$
$5x - 11y = 18$

28. $4x - 9y = -7$
$\dfrac{2}{3}x - y + 2z = 0$
$8x + 2y + 15z = 11$

[4.5] *Graph each system of inequalities.*

29. $6x - 4y \leq -12$
$4x + 2y > 2$

30. $y \leq -x - 2$
$y \geq \dfrac{1}{2}x + 1$
$y \leq 3$

Mixed Applications

Solve.

31. After making the monthly mortgage payment, a family has $2185 left of their monthly income. The mortgage is 24% of the monthly income. What is the family's monthly income? How much is paid for the mortgage?

32. A college has a total of 9500 students. There are 1254 more female students than male students. How many male and female students attend the college?

33. A city council approves a plan to build a new park. The park is to have a soccer field whose perimeter is 400 yd. The width of the field is 40 yd less than the length. What is the area of the soccer field?

34. An athlete swims at a rate of 2.1 mph going with the current. Going against the current, she swims at a rate of 1.1 mph. Assuming that the athlete swims at a constant rate, find her rate in still water and the rate of the current.

35. Parker Plumbing charges $100 for a house call plus an additional $40 per hour for service. Plumbing Solutions charges $75 for a house call plus an additional $50 per hour for service. In how many hours will both plumbing services charge the same amount for a house call?

36. A sporting goods manufacturer produces bicycle helmets at a monthly cost of $1200 plus $1.20 per helmet. The manufacturer sells the helmets for $6 each. What is the manufacturer's break-even point for production?

37. A company merges two departments. As a result, 60% of the employees in one department and 30% of the employees in the other department are laid off. If 50% of the original 180 employees were laid off, how many employees were in each department before the merger?

38. Ten minutes after his sister leaves for school, a boy notices that she left her lunch on the kitchen table. To catch up with her, he rides his skateboard. The boy can skateboard at an average rate of 6 mph and his sister walks at a rate of 3 mph. How long will it take the boy to catch up to his sister?

39. Ticket prices at a movie theater are $5.50 for any movies before 5:00 P.M. and $8 for any movies after 5:00 P.M. On a particular day, the theater sold 4 times as many tickets after 5:00 P.M. as it sold before 5:00 P.M. If the ticket receipts that day totaled $31,725, how many of each type of ticket did the movie theater sell?

40. A store has a sale on all winter boots and outerwear. The store offers 50% off on each pair of boots and 30% off on each item of outerwear purchased. A customer purchases boots and outerwear and pays $285 after the discount. If the customer saved $165, what was the original price of the merchandise purchased?

41. A basketball team scored a total of 112 points in a game on a combination of 1-point, 2-point, and 3-point baskets. The number of 2-point baskets made was 10 more than half the sum of the number of 1-point baskets and 3-point baskets made. The team made a total of 55 baskets. How many of each type of basket did the team make?

42. A student at a college can take 1-credit, 2-credit, and 4-credit courses. Over two semesters, a student took a total of 10 courses and earned 32 credits toward graduation. The number of 2-credit courses he took was two-thirds the number of 4-credit courses he took. How many of each type of course did the student take over the two semesters?

43. A convenience store sells 16-oz, 24-oz, and 32-oz fountain drinks. A 16-oz drink costs $0.99, a 24-oz drink costs $1.09, and a 32-oz drink costs $1.19. On a particular day, the store collects $515.45 on the sale of 465 fountain drinks. The store sold 25 more 32-oz than 24-oz fountain drinks. How many of each size fountain drink did the store sell that day?

44. At a hospital, the median salary of a registered nurse is $23,420 more than the median salary of a licensed practical nurse. A medical assistant makes $10,730 less than a licensed practical nurse. The combined median salary of a licensed practical nurse and a medical assistant is $4880 more than the median salary of a registered nurse. Determine the median salary for each job.

45. In a discussion of income tax reform, one plan proposed was for a flat tax to be paid at a rate of 20% on earnings above $10 thousand. A second plan was for a rate of 30% on earnings in excess of $15 thousand.

   a. Express each proposal as a linear equation.

   b. Graph both equations for incomes over $10 thousand.

   c. How much has to be earned for each person to pay the same amount of taxes under each plan?

   d. Under which plan would a person earning $50 thousand pay more taxes?

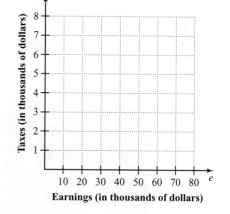

46. A family invested no more than $12,000 in two funds earning 6% and 8% simple interest. The amount of interest earned on the investment at 8% was greater than the amount of interest earned on the investment at 6%. They invested at least $4500 in the account earning 8% simple interest.

   a. Write a system of inequalities to model this problem.

   b. Solve the system by graphing.

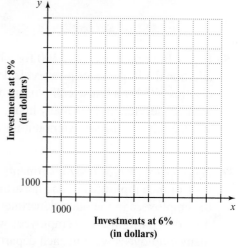

• Check your answers on page A-35.

CHAPTER 4   Posttest

FOR
EXTRA
HELP

CHAPTER
Test Prep
VIDEOS

The Chapter Test Prep Videos with test solutions are available on DVD,
in MyMathLab, and on You Tube ⁎ (search "AkstIntermediate
Alg" and click on "Channels").

To see if you have mastered the topics in this chapter, take this test.

1. Is $\left(\dfrac{1}{4}, -2\right)$ a solution of the following system?

$$8x + 2y = 0$$

$$3x - \frac{1}{4}y = -1$$

Solve by graphing.

2.
$$y = -2x + 3$$
$$3x - 2y = 8$$

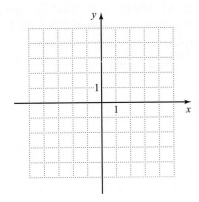

3.
$$5x = 20$$
$$-4x - 6y = 8$$

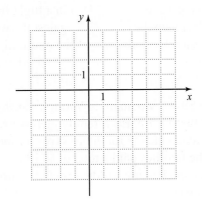

4.
$$1.5x - 3y = 6$$
$$-3x + 6y = -6$$

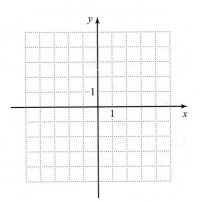

Solve by substitution.

5. $x = 6y - 11$

$$y = \frac{1}{3}x + 3$$

6.
$$5p - q = 1$$
$$7p - 2q = -1$$

7.
$$8x = 4y - 24$$
$$1.2x - 0.6y = -3.6$$

Solve by elimination.

8. $2x - 2y = 5$
$5x + 2y = 9$

9. $4x + 5y - 20 = 0$
$6x - 3y + 12 = 0$

10. $\dfrac{1}{4}u - 3v = -4$

$\dfrac{2}{3}u - 8v = -11$

11. Is $\left(-2, -1, \dfrac{1}{2}\right)$ a solution of the following system?

$$3x + 4y - 8z = -14$$
$$2x - 3y + 6z = 2$$
$$x + 2y - 2z = -5$$

Solve.

12. $-2x - y + 3z = 5$
$4x - 3y + z = 29$
$3x + 2y - 4z = -9$

13.
$$1.4y - 0.9z = -1.3$$
$$1.5x + 3.5y = 2$$
$$-0.1x - 0.2y + 0.3z = 0.8$$

14. Solve using matrices.

$$2x - 4y + 5z = 7$$
$$x - 3y + 2z = 5$$
$$3x - y + z = -6$$

15. Graph the solutions of the system of inequalities.

$$6x - 4y \geq -16$$
$$x + 2y > -2$$
$$x \leq 3$$

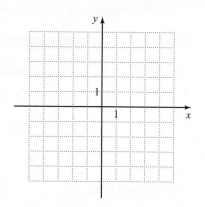

Solve.

16. A regulation field in Canadian football has a perimeter of 350 yd. The width of the field is 55 yd less than the length. What are the dimensions of a regulation football field? (*Source:* Canadian Football League)

17. On a flight from Boston to Chicago, an airplane flying against the wind flew at an average speed of 468 mph. On the return flight, the airplane flew with the wind at an average speed of 486 mph. What was the rate of the airplane in still air?

18. Four families went on a one-day ski trip to Stowe Mountain in Vermont during the prime season. There were a total of 19 people on the ski trip. Each family member rented either skis or a snowboard for the day. The daily rental charge for either skis or a snowboard was $43 for adults and $34 for children. The families paid a total of $718 for rental equipment for the day. How many adults and how many children rented equipment?

19. The nutritional information for packaged foods can be found on their nutrition labels. The table shows some of the nutritional information for one serving of yogurt (4 oz), a plain bagel, and orange juice (8 oz).

	Yogurt	Plain Bagel	Orange Juice
Calories	100	290	110
Protein (in grams)	4	11	2
Calcium (in milligrams)	150	100	20

How many servings of each food are needed to get 820 calories, 25 g of protein, and 460 mg of calcium?

20. Two students agree to split the driving time on a cross-country road trip. One student drives at an average rate of 64 mph and the other drives at an average speed of 60 mph. They would like to drive at least 300 miles per day, and they plan to spend no more than 12 hours driving per day. The student that drives faster agrees to drive more hours.

a. Write a system of inequalities to model the problem.

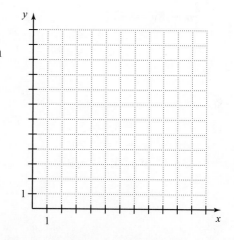

b. Graph the solutions of the system.

c. Find two solutions of the system.

• Check your answers on page A-36.

Cumulative Review Exercises

To help you review, solve these problems.

1. Find the value: $|5 - 5^2| - |(-2)(3)|$

2. Evaluate the expression for each value of the variable:

x	-1	-0.5	0	0.5	1
$0.3x$					

3. Calculate, writing the result in scientific notation:
$(2.5 \times 10^6) \div (5 \times 10^{-3})$

4. Solve: $9x - \dfrac{1}{2}(12 - 4x) = 6x + 14$

5. Solve: $8 - |x - 5| = 3$

6. Solve and graph: $-1 < 5 - 2x \le 7$

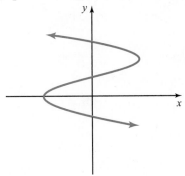

7. Graph: $4x - 6y = 12$

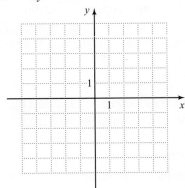

8. Write the slope-intercept form of the equation of a line that is perpendicular to the line $12x + 3y = -9$ and that passes through the point $(4, -1)$.

9. Evaluate $g(x) = x^2 - 5x - 7$ for $g(-2)$.

10. Does the graph below represent a function?

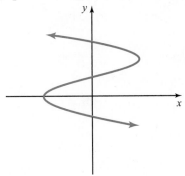

11. Solve by graphing: $\begin{aligned} y &= -x + 2 \\ y &= 2x + 5 \end{aligned}$

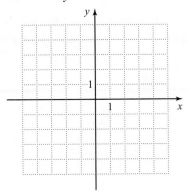

12. Solve by substitution: $\begin{aligned} 4x + 2y &= 6 \\ 2x + y &= -3 \end{aligned}$

13. Solve by elimination:
$$3x - 2y = 10$$
$$2x + 3y = -2$$

14. Solve by graphing:
$$-3x + y \leq -3$$
$$x + 2y > 8$$

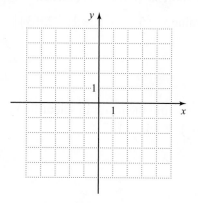

15. The expression $2000(1.1)^t$ represents the value of a $2000 investment that earns 10% interest per year, compounded annually for t years. What is the value of this investment at the end of 2 years?

16. An auto repair shop charges $35 per hour for labor. A customer's repair bill amounts to $322. If parts cost $112, how many hours was the customer charged for labor?

17. The electronic scale at a grocery store has an error of ± 0.05 lb. The actual weight of the tomatoes a customer purchases is 2.43 lb. If the tomatoes sell for $1.49 per pound, what is the range of prices that a customer could be charged for the purchase?

18. One second after an object is launched upward, its velocity is 88 ft/sec. Three seconds later, its velocity is 24 ft/sec. The velocity v and the time t are modeled by a linear equation.

a. Write an equation in slope-intercept form that expresses the object's velocity in terms of time.

b. Graph the equation on the coordinate plane.

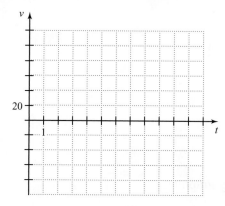

c. Find the t- and v-intercepts of the graph. Explain their significance within the context of the situation.

19. A college club with 24 members is going on a hiking trip. Each of the 5 faculty advisors will drive a van or a car. A van can seat 7 people, and a car 5 people. How many of each type of vehicle could carry all 29 people to the hiking area in one trip?

a. Write a system of inequalities to model the situation.

b. Solve the system by graphing.

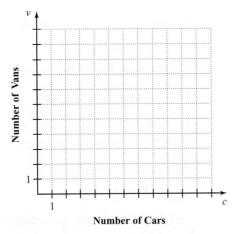

c. Give an example of a possible number of vans and cars that would satisfy the conditions.

20. Last season, a college student was the top basketball player on her team. She scored 506 points making 284 shots, which included three-point field goals, two-point field goals, and one-point free throws. She made 40 more two-point field goals than one-point free throws. Find the number of each type of shot that she scored.

• Check your answers on page A-37.

Polynomials

Making Predictions with Algebra

Have you ever wondered how almanacs predict future trends? Although no one can know the future, it's possible to make useful predictions based on the best available data.

Statisticians make projections from known data by means of a technique called *linear regression.* With this technique, data relating to two variables of interest—say, the number of terms that you have been attending college and the number of credits that you have earned—are plotted on a coordinate plane to form a *scatter diagram,* such as the one shown to the right above. If the points happen to lie on a straight line, we draw that line and extend it to project the data. If the points are not on a line—usually the case—we use the line that most closely passes through these points.

To determine which line to choose for making predictions, we calculate the distance between the y-value of each plotted point and the y-value of the point on the line directly above or below it. The distances are squared, resulting in expressions of the form $(Y - y)^2$, and then are added for all the plotted points: $(Y_1 - y_1)^2 + (Y_2 - y_2)^2 + (Y_3 - y_3)^2 + (Y_4 - y_4)^2$. The line that minimizes this sum is *the line of best fit,* as shown in the bottom graph. (*Source:* Mario Triola, *Elementary Statistics,* Pearson Education, 2010)

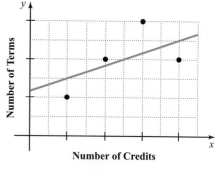

Number of Credits

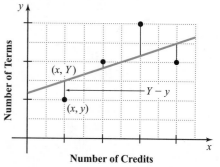

Number of Credits

1. For the polynomial $6x^4 - 2x + \dfrac{x^3}{4} - x^6 + 3x^2,$

   a. list the terms _____.

   b. list the coefficients _____.

   c. write the polynomial in descending order _____.

   d. identify the degree _____.

   e. identify the leading term and leading coefficient _____.

2. Add $(5n^3 - n^2 + 8n - 12)$ and $(4n^2 + 9 - 7n^3)$.

3. Subtract $(1 + 9x - x^2 - 10x^4)$ from $(8x^4 - 3x^2 + 2x)$.

Perform the indicated operation.

4. $\dfrac{1}{4}pq^2(8p^2 - 4pq + 16q^2)$

5. $(3x - y)(2x + 5y)$

6. $(6n - 1)^2$

7. $(8a + 3b)(8a - 3b)$

8. $(u + 2)(u^2 - 2u + 4)$

9. $(6x^4 - x^3 + x - 9) \div (2x + 1)$

Factor.

10. $x^2 - 4x - 32$

11. $3x^2 + 14x - 24$

12. $50p^2q - 40pq^2 + 8q^3$

13. $36a^2b^2 - 121$

14. $9x^3 - 9x^2y - 4x + 4y$

Solve.

15. $12t^2 - 18t = 0$

16. $(x - 4)(x - 6) = 3$

17. A toy maker's profit (in dollars) from selling x dolls is given by the polynomial $0.25x^2 - 20x - 125$. Find the profit from selling 100 dolls.

18. Suppose the parking lot at a shopping mall is expanded. The increase in the length of the parking area is double the increase in its width, as shown in the diagram. Write the area of the new parking lot as a polynomial in x.

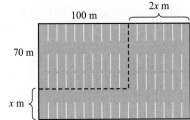

19. As the temperature of a lightbulb's filament changes from T_1 to T_2, the energy that the filament radiates changes by the quantity $aT_1^4 - aT_2^4$. Factor this expression.

20. An initial amount of $64,000 was invested in a high-risk growth fund. After two years, the investment was worth $81,000. To find the average rate of return r on the investment, a broker uses the equation $64,000(1 + r)^2 = 81,000$. What was the rate of return on the investment? Express this rate as a percent.

• Check your answers on page A-37.

5.1 Addition and Subtraction of Polynomials

OBJECTIVES

A To identify a polynomial

B To classify a polynomial

C To write a polynomial in descending order

D To simplify a polynomial

E To evaluate a polynomial

F To add or subtract polynomials

G To solve applied problems involving polynomials

What Polynomials Are and Why They Are Important

In Chapter 1, we discussed algebraic expressions in general. This chapter focuses on a special kind of algebraic expression called a *polynomial*. Many phenomena in science and business can be described by polynomials, as we will see.

After introducing polynomials, we consider ways of classifying polynomials according to a variety of properties. Then, we move on to a discussion of how to simplify and evaluate polynomials. Next, we cover operations on polynomials. The final sections of the chapter deal with factoring polynomials and applying this skill to solving equations.

Introduction to Polynomials

Polynomials are formed by adding and subtracting *monomials*.

> **DEFINITION**
>
> A **monomial** is an expression that is the product of a real number and variables raised to nonnegative integer powers.

Examples of monomials are:

$$8n^3, \frac{4}{5}x^2, -2p^2q^5, 12$$

Recall that in the expression $8n^3$, 8 is called the *coefficient*. Monomials (or terms) serve as building blocks for the larger set of *polynomials*.

> **DEFINITION**
>
> A **polynomial** is an algebraic expression with one or more monomials added or subtracted.

Examples of polynomials	Examples of algebraic expressions that are *not* polynomials
$4x^2 - x + 6$	$5x^{-2} - 3$
$3t^2 - 5$	$\dfrac{n^2}{4} - 2 + \dfrac{1}{n}$
$-6.3x^2$	$5\sqrt{x}$
$\dfrac{x - 1}{3}$	$\dfrac{x - 1}{x^2 + 3}$
$5pqr - 2p^2 - 3q^2 + 5$	

EXAMPLE 1

Determine whether each of the following expressions is a polynomial.

a. $6y^2 - y + 5$ **b.** $\dfrac{3}{2x + 1}$

Solution

a. The expression $6y^2 - y + 5$ can be written as the sum of the monomials $6y^2$, $-y$, and $5x^0$. Therefore, it is a polynomial.

b. The expression $\dfrac{3}{2x + 1}$ cannot be expressed as the sum of monomials. So it is not a polynomial.

PRACTICE 1

Indicate if each of the following expressions is a polynomial.

a. $3x - 4x^3 - 8$

b. $7x^{-1} + 3x - 1$

EXAMPLE 2

Consider the polynomial $6x^4 - x^3 + 7$.

a. Identify the terms of the polynomial.

b. For each term, identify its coefficient.

Solution

a. The terms are $6x^4$, $-x^3$, and 7.

b. The coefficients of the terms are 6, -1, and 7, respectively. Note that we can write a constant such as 7 as $7x^0$.

PRACTICE 2

Consider the polynomial

$$-x^3 + 9x^2 - 3x + 13.$$

a. Identify the terms of the polynomial.

b. For each term, identify its coefficients.

Classification of Polynomials

There are several ways of classifying polynomials.

Number of Variables   One way is according to the number of variables in the polynomial. A polynomial such as $3x^2 - 6x + 9$ is said to be *in one variable,* namely x. The polynomial $t^5 + 11t^4 - 7t^3 - t^2 + 10t - 50$ is also in one variable, t. On the other hand, the polynomial $3x^4y - 5x^2y^3 + 9$ is *in two variables*, namely x and y.

Number of Terms   Another way to classify polynomials is according to the number of terms in the polynomial. As we have already seen, a polynomial with just *one* term is called a *monomial.* A polynomial with *two* terms, such as $3t^2 - 5$, is called a *binomial.* A polynomial with *three* terms, for example $4x^2 - x + 6$, is called a *trinomial.* And polynomials with *four or more* terms, for example $13x^4 + 2x^3 - 5x^2 + 3x - 8$, are simply called *polynomials.*

Degree   A third way of classifying a polynomial is according to its degree. Let's first consider a monomial in one variable. The **degree of a monomial** in one variable is the power of the variable in the monomial.

$$5x^4 \longleftarrow \text{Of the fourth degree, or of degree 4}$$

$$-4n \longleftarrow \text{Of the first degree, or of degree 1}$$

Since a constant, such as 6, can be written as $\underline{6x^0}$, the degree of a nonzero constant is 0. For a monomial in more than one variable, its degree is the *sum of the powers* of the variables. For instance, the degree of $5x^2y^4$ is $2 + 4$, or 6. The **degree of a polynomial** is the highest degree of any of its terms. For instance, the degree of $5x^3 + 9x^2 - 2x - 6$ is 3, since 3 is the highest degree of any term.

EXAMPLE 3

Classify each polynomial by the number of terms. Then, identify the degree of each polynomial.

a. $p^2 - 10p + 16$ **b.** $7x - 3x^2 + 21 + 2x^3$

c. -18 **d.** $m^2n^2 - 3mn$

Solution

a. $p^2 - 10p + 16$ is a trinomial of degree 2.

b. $7x - 3x^2 + 21 + 2x^3$ is a polynomial of degree 3.

c. $-18 = -18x^0$ is a monomial of degree 0.

d. The degree of the term m^2n^2 is 4, and the degree of the term $-3mn$ is 2. Since 4 is the highest degree of any term of the polynomial, the degree of the polynomial is 4. So $m^2n^2 - 3mn$ is a binomial of degree 4.

PRACTICE 3

Classify each polynomial by the number of terms. Then, indicate the degree of each polynomial.

a. $n + 2$

b. -10

c. $9 - 4x - x^3 + 3x^5 + x^2$

d. $4xy^4 - 7x^2y + 1$

DEFINITIONS

The **leading term** of a polynomial is the term in the polynomial with the highest degree. The coefficient of that term is called the **leading coefficient**. The term of degree 0 is called the **constant term**.

In the polynomial $5t^2 - 8t + 6$, the leading term is $5t^2$, the leading coefficient is 5, and the constant term is 6.

The terms of a polynomial are usually arranged in *descending order of degree*. That is, we write the leading term on the left, then the term of the next highest degree, and so forth. For instance, the polynomial $5x^3 + 6x^2 - 2x - 4$ is written in descending order. Sometimes, however, the terms of a polynomial are written in ascending order of degree, as in $-4 - 2x + 6x^2 + 5x^3$.

EXAMPLE 4

Rearrange the following polynomials in descending order. Then, identify the leading term and leading coefficient.

a. $5x^2 + 2x^4 - x^5 + x^3 - 7 + 4x$ **b.** $9x^2 - 6 + 8x^3$

Solution

a. We write the polynomial so that the term with the highest exponent of x is on the left, the next highest exponent comes second, and so on.

$$5x^2 + 2x^4 - x^5 + x^3 - 7 + 4x = -x^5 + 2x^4 + x^3 + 5x^2 + 4x^1 - 7x^0$$
$$= -x^5 + 2x^4 + x^3 + 5x^2 + 4x - 7$$

The leading term of the polynomial is $-x^5$ and the leading coefficient is -1.

b. We write the polynomial so that the exponents decrease from left to right.

$$9x^2 - 6 + 8x^3 = 8x^3 + 9x^2 - 6$$

The leading term is $8x^3$ and the leading coefficient is 8.

PRACTICE 4

Write each polynomial in descending order and identify the leading term and leading coefficient.

a. $4x - 5x^4 + 7x^2 + 11$

b. $12x - 3x^3 + 9x^5$

Terms of a polynomial with coefficient 0 are usually not written. The unwritten term is said to be a *missing term*. For instance, in Example 4(b), note that $0x$ is a missing term. So we can write the polynomial $8x^3 + 9x^2 - 6$ as follows:

$$8x^3 + 9x^2 + 0x - 6.$$

The concept of using missing terms as placeholders is important in the division of polynomials, as we will see in Section 5.3.

When the terms of a polynomial in more than one variable are arranged in descending order, it is commonly done with respect to one of the variables.

EXAMPLE 5	PRACTICE 5
Arrange the polynomial $2xy + x^2 + y^2$ in descending powers of y.	Consider the polynomial $4p^2 + 4q^2 - 8pq$. Rewrite the terms in descending powers of q.
Solution	
Descending powers of y: $y^2 + 2xy + x^2$.	

Simplifying and Evaluating Polynomials

Recall that in Section 1.5, we discussed, in general, how to simplify an algebraic expression by combining like terms. Let's apply this technique to a polynomial.

EXAMPLE 6	PRACTICE 6
Write in descending order and simplify: $2x + 6x^2 - 3 - 4x^3 + 8 - 5x^2 - 2x$	Write in descending order and combine like terms:
Solution	$3x^3 - 9x + 10 + 8x^2 - 5x^3 + 9x + 2$

$$2x + 6x^2 - 3 - 4x^3 + 8 - 5x^2 - 2x$$
$$= -4x^3 + 6x^2 - 5x^2 + 2x - 2x - 3 + 8$$
$$= -4x^3 + x^2 + 0x + 5 \qquad \text{Combine like terms.}$$
$$= -4x^3 + x^2 + 5 \qquad \text{Group like terms in descending order.}$$

Polynomials, like other algebraic expressions, are evaluated by replacing each variable with the given number and then carrying out the computation.

EXAMPLE 7	PRACTICE 7
Evaluate $3x^2 - 8x - 10$ for the given values of the variable.	Evaluate $2x^2 - 3x + 1$ for the given values of x.
a. $x = 2$ **b.** $x = -2$	**a.** $x = 4$
Solution	**b.** $x = -4$
a. Substituting 2 for x, we get:	

$$3x^2 - 8x - 10 = 3(2)^2 - 8(2) - 10 \qquad \text{Substitute 2 for } x.$$
$$= 3(4) - 16 - 10$$
$$= -14$$

EXAMPLE 7 (continued)

b. Substituting -2 for x gives us:

$$3x^2 - 8x - 10 = 3(-2)^2 - 8(-2) - 10 \quad \text{Substitute } -2 \text{ for } x.$$
$$= 3(4) - (-16) - 10$$
$$= 18$$

EXAMPLE 8

The number of hybrid automobile sales (in thousands) of Honda Civics is approximated by the polynomial $-4.4x^2 + 8.1x + 30.7$, where x represents the number of years after 2006. To the nearest whole number, estimate how many thousands of hybrid Honda Civics were sold in 2009. (*Source:* afdc.energy.gov)

Solution

Since 2009 is three years after 2006, we need to substitute 3 for x in the polynomial.

$$-4.4x^2 + 8.1x + 30.7 = -4.4(3)^2 + 8.1(3) + 30.7 = 15.4$$

So 15 thousand hybrid Honda Civics were sold in 2009.

PRACTICE 8

The gross domestic product (GDP) in the United States is the total value of all goods and services produced in the country. For each year between 2005 and 2009, the GDP (in billions of dollars) can be approximated by the expression $-149.3x^2 + 1076.9x + 12{,}390.6$, where x is the number of years after 2005. To the nearest trillion dollars, estimate the U.S. GDP in 2009. (*Source:* www.gpoaccess.gov)

Adding and Subtracting Polynomials

Now, we consider operations on polynomials. In this section, we focus on addition and subtraction.

Polynomials can be added and subtracted either horizontally or vertically. No matter which format we choose, however, the key is to combine like terms.

EXAMPLE 9

Find the sum: $(4x^3 + 2x^2 - 7) + (-3x^3 + 6x - 12)$

Solution To add polynomials horizontally, we remove the parentheses and combine like terms.

First polynomial   Second polynomial
$$(4x^3 + 2x^2 - 7) + (-3x^3 + 6x - 12)$$
$$= 4x^3 + 2x^2 - 7 - 3x^3 + 6x - 12$$
$$= x^3 + 2x^2 + 6x - 19 \quad \text{Combine like terms.}$$

PRACTICE 9

Add $6x - 3$ and $9x^2 - 3x - 40$.

In general, a polynomial is considered to be in simplest form when all terms are simplified, like terms are combined, and parentheses are removed. Usually, the terms are then reordered in descending powers of the variable.

To add polynomials vertically, we position like terms of the polynomials in the same column. We usually begin by making sure that each polynomial is written in descending powers of the variable, leaving a space for each missing term in the polynomials.

EXAMPLE 10

Add vertically: $8x^2 + 2x - 1, 2x + 9,$ and $x^2 + 18$

Solution

First, we check that each polynomial is written in descending powers of x. Next, we rewrite the polynomials vertically, with like terms positioned in the same column:

First-degree terms

Second-degree terms ⟶ Zero-degree (constant) terms

$$8x^2 + 2x - 1$$
$$2x + 9$$
$$x^2 \quad\;\; + 18$$
$$\overline{9x^2 + 4x + 26}\quad \text{Add within columns.}$$

So the sum in simplest form is $9x^2 + 4x + 26$.

PRACTICE 10

Find the sum of $2n^2 - 12, 9n + 6,$ and $3n^2 - n + 4,$ using a vertical format.

Recall that when a minus sign precedes terms in parentheses, we remove the parentheses and change the sign of each term. To subtract polynomials horizontally, we change the signs of the terms *in the polynomial being subtracted* and then add.

EXAMPLE 11

Subtract: $(4x^2 + 3x - 8) - (-x^2 + 3x - 5)$

Solution   Since the polynomial $-x^2 + 3x - 5$ is preceded by a minus sign in the expression $(4x^2 + 3x - 8) - (-x^2 + 3x - 5)$, we begin by removing parentheses and changing the sign of each term of the polynomial $-x^2 + 3x - 5$. Then, we combine like terms.

$$(4x^2 + 3x - 8) - (-x^2 + 3x - 5) = 4x^2 + 3x - 8 + x^2 - 3x + 5$$
$$= 5x^2 - 3$$

PRACTICE 11

Find the difference:
$(4x - 1) - (3x^2 + 10x - 1)$

To subtract polynomials vertically—a skill that comes up in dividing polynomials—we position the expressions so that like terms are in the same column. Then, we change the signs of all the terms in the polynomial being subtracted, and add.

EXAMPLE 12

Subtract $(6x^2 + 2xy - 9y^2)$ from $(7y^2 - 10xy)$ using a vertical format.

Solution

$$\begin{array}{r} -10xy + 7y^2 \\ - (6x^2 + 2xy - 9y^2) \\ \hline \end{array}$$ Position like terms in the same column.

$$\begin{array}{r} -10xy + 7y^2 \\ - 6x^2 - 2xy + 9y^2 \\ \hline \end{array}$$ Change the signs of the terms in the polynomial being subtracted.

$$\begin{array}{r} -10xy + 7y^2 \\ - 6x^2 - 2xy + 9y^2 \\ \hline -6x^2 - 12xy + 16y^2 \end{array}$$ Add.

PRACTICE 12

Subtract $(p^2 - 11pq + 7q^2)$ from $(3p^2 + 5pq - 8q^2)$ using a vertical format.

Some expressions involve both the addition and subtraction of polynomials.

EXAMPLE 13

Simplify: $(3a^2 - 4ab) - (6ab + 2b^2) + (a^2 + 13ab - b^2)$

Solution

$(3a^2 - 4ab) - \underbrace{(6ab + 2b^2)}_{\substack{\text{Polynomial} \\ \text{preceded by a} \\ \text{minus sign:} \\ \textit{change} \text{ signs of} \\ \text{terms.}}} + \underbrace{(a^2 + 13ab - b^2)}_{\substack{\text{Polynomial} \\ \text{preceded by a} \\ \text{plus sign:} \\ \textit{keep} \text{ signs of} \\ \text{terms.}}} = 3a^2 - 4ab - 6ab$

$\qquad\qquad\qquad\qquad\qquad\qquad\qquad\qquad\quad - 2b^2 + a^2 + 13ab - b^2$

$\qquad\qquad\qquad\qquad\qquad\qquad\qquad\qquad = 4a^2 + 3ab - 3b^2$

PRACTICE 13

Simplify: $(8xy - 2y^2) + (4x^2 + 10xy - y^2) - (2x^2 - 5y^2)$

A *polynomial function* is a function in which the rule that defines the function is a polynomial. For instance, the function $f(x) = 3x^2 - x + 5$ is a polynomial function. We can use our knowledge of adding and subtracting polynomials to add and subtract polynomial functions.

EXAMPLE 14

Given $f(x) = 15x - 1$ and $g(x) = 5x^2 - 4x + 6$, find:
a. $f(x) + g(x)$ **b.** $f(x) - g(x)$

Solution

a. $f(x) + g(x) = (15x - 1) + (5x^2 - 4x + 6)$
$\qquad\qquad\qquad = 15x - 1 + 5x^2 - 4x + 6$
$\qquad\qquad\qquad = 5x^2 + 11x + 5$

b. $f(x) - g(x) = (15x - 1) - (5x^2 - 4x + 6)$
$\qquad\qquad\qquad = 15x - 1 - 5x^2 + 4x - 6$
$\qquad\qquad\qquad = -5x^2 + 19x - 7$

PRACTICE 14

Given $f(x) = 4 - 2x - x^2$ and $g(x) = 3x^2 + 9x$, find:

a. $f(x) + g(x)$

b. $g(x) - f(x)$

EXAMPLE 15

The total U.S. imports of petroleum (in thousands of barrels per day) in a given year is approximated by the polynomial $-37x^3 + 23x^2 - 13x + 13{,}718$, where x represents the number of years since 2005. The corresponding total of U.S. exports of petroleum is $-9x^3 + 84x^2 + 28x + 1175$. Express the difference between the imports and the exports in a given year as a polynomial in x. (*Source:* eia.gov)

Solution
The difference between the imports and the exports can be represented by the following expression:

$(-37x^3 + 23x^2 - 13x + 13{,}718) - (-9x^3 + 84x^2 + 28x + 1175)$.

We subtract the second polynomial from the first:

$$\begin{array}{r} -37x^3 + 23x^2 - 13x + 13{,}718 \\ - (-9x^3 + 84x^2 + 28x + 1175) \\ \hline \end{array}$$

Changing the signs in the second polynomial and then adding, we get:

$$\begin{array}{r} -37x^3 + 23x^2 - 13x + 13{,}718 \\ 9x^3 - 84x^2 - 28x - 1175 \\ \hline -28x^3 - 61x^2 - 41x + 12{,}543 \end{array}$$

So the difference between the total U.S. imports and exports of petroleum is approximately $(-28x^3 - 61x^2 - 41x + 12{,}543)$ thousand barrels.

PRACTICE 15

For a given year, the number (in thousands) of male commissioned officers in the U.S. Department of Defense is approximated by $327x^3 - 1651x^2 + 3x + 178{,}279$, where x represents the number of years since 2004. The corresponding polynomial for female commissioned officers is $149x^3 - 745x^2 + 310x + 34{,}022$. Find a polynomial that represents how many thousands more male commissioned officers than female commissioned officers there were for a given year. (*Source:* prhome.defense.gov)

Mathematically Speaking

Fill in each blank with the most appropriate term or phrase from the given list.

polynomial	monomial	multiplied or divided	highest
missing term	terms		leading term
leading coefficient	constant term	variables	lowest
added or subtracted	of degree three	with three terms	

1. A(n) _____ is an expression that is the product of a real number and variables raised to nonnegative integer powers.

2. A polynomial is an algebraic expression with one or more monomials _____.

3. A polynomial such as $4x^2y - xy^2$ is in two _____, namely x and y.

4. Trinomials are polynomials _____.

5. The degree of a polynomial is the _____ degree of any of its terms.

6. The _____ of a polynomial is the term in the polynomial with the highest degree.

7. The term of degree 0 is called the _____.

8. We can write the polynomial $5x^3 + 4x - 3$ as $5x^2 + 0x^2 + 4x - 3$, and the term $0x^2$ is called a(n) _____.

A *Determine whether each of the following expressions is a polynomial. If so, identify its terms and coefficients.*

9. $5x^3 - x^2 - 6x + 7$

10. $-y^4 + 9y - 2$

11. $\dfrac{4}{n} - n + 3n^2 - n^3$

12. $4n^3 + 5n - n^{-1}$

13. $2x + \dfrac{x^2}{7} - x^3 + 5x^4$

14. $7x^5 - 2x^4 + \dfrac{x}{6} - 10$

15. $3a^2 - 8ab + 5b^2$

16. $4p^2q^2 - 9$

17. $\dfrac{p^2 - pq}{10}$

18. $4n - \dfrac{n^2}{3}$

B *Classify each of the following polynomials by the number of terms it has. Then, identify the degree of the polynomial.*

19. $2n^3 + 5n^2 - n + 3$

20. $p^6 - p^4 - 2p^3 + 6p^2 - 15$

21. $16s^2 - t^2$

22. $7x^3y - x^4y^3$

23. $-8n$

24. $5a^4b$

25. $x^5 - 4x^3 + 7x$

26. $x^4 + 9x^2 - 10$

C *Write each polynomial in descending order. Then, identify the leading term and leading coefficient.*

27. $2 - 3x^2 + 9x$

28. $4x + x^3 - 6x^2$

29. $5x^5 - 2x^3 + 8x^4 - x + 10$

30. $x^2 - 7 + x^4 + 2x$

31. $8x^4 - x^6 + 5x - 11 - x^3 + 4x^2$

32. $3x^2 - 6x^5 + 4x^3 - 2x + 10x^9 - 7x^6$

33. Write $y^3 - 4x^2y - x^3 + 5xy^2$ in descending powers of x.

34. Write $4ab - 5a^4b^2 - 7a^2b^2 + a^3b^4$ in descending powers of a.

35. Rewrite $p^5q^4 + 3p^2q^2 + 2p^3q - pq^3 + 4p^4$ in descending powers of p.

36. Rewrite $7v^3 - u^2v + uv^2 + 2u^5 - 6v$ in descending powers of u.

D *Simplify, and then write in descending order.*

37. $-5x^2 - 8 + 3x^2 + 10x - 1$

38. $6x^2 + 12x + 15 - 9x^2 - 5x$

39. $4n^3 - 7n + 9n - n^3 + 2 - 6n^2 - 2 + 7n^2$

40. $13 - 8y^3 + y^4 + 4y - 2y^2 - y^4 + 8y^2 - 14 - 4y$

41. $\frac{1}{2}x^2 + \frac{3}{8} - 6x^4 + \frac{3}{4} - \frac{2}{3}x^2$

42. $-9 - \frac{1}{3}x - \frac{1}{4}x^2 + \frac{1}{6}x - \frac{1}{4}x^2$

E *Evaluate each polynomial for the given values of the variable.*

43. $2x^2 - 7x + 6$
   a. $x = 2$
   b. $x = -3$

44. $3x^2 + 11x - 4$
   a. $x = 1$
   b. $x = -4$

45. $8x^3 - 4x^2 - 2x + 7$
   a. $x = \frac{3}{2}$
   b. $x = -\frac{1}{2}$

46. $-27x^3 - 9x^2 + 3x + 10$
   a. $x = -\frac{1}{3}$
   b. $x = \frac{4}{3}$

F *Add or subtract horizontally.*

47. $(3x + 9) + (2x^2 - 4x + 1)$

48. $(7x^2 - 6x) + (-4x^2 + x - 8)$

49. $(a^3 - 5a^2 + 3) - (5a^2 + a - 3)$

50. $(2n^2 - 3n + 1) - (3n^4 - 2n^2 - 6)$

51. Add $3x^4 - 7x^3 + x^2 + 8x - 10, 8x^3 - 1$ and $10 - 9x - 7x^2 + x^4$.

52. Add $11x^3 + 9x^2 - 13x - 4, 16 - 8x + x^2$ and $5 + 6x - 9x^2 - 15x^3$.

53. Subtract $13p^5 + p^4 + 2p^3 - 3p^2 - 11p - 1$ from $12p^5 - 2p^3 + 4p^2 - 11p$.

54. Subtract $5a^4 - 9a^3 - a^2 + 6a - 2$ from $10a^4 + 9a^3 - a - 7$.

55. Add $x^2 + 5xy + 6y^2, 9x^2y - 2xy^2 - 3xy + x^2 - y^2$ and $2x^2y - 12xy^2$.

56. Add $4p^2 - 9q^2, 5pq - 11p^2 + 7q^2 - p^2q$ and $-8pq + p^2q + 7p^2$.

Add or subtract vertically.

57. Add $4y - 3y^2 - y^3$ and $6y^3 + 8y^2 - 3y + 2$.

58. Add $5x^3 + 2x^2 - 1$ and $3 - 7x + x^2 - 4x^3$.

59. Subtract $p^2 - 10p + 21$ from $3p^4 - 2p^2 - 5$.

60. Subtract $9x^3 - 4x + 8$ from $-6x^3 + 3x^2 - 11x$.

61. Add $12 - 3x - 3x^2$, $8x^2 - 4x - 13$ and $7x - 5x^2$.

62. Add $4n^5 - 5n^4 + 2n$, $7n^3 - n^2$ and $2n^2 - 7n^3 + 3n^4$.

63. Subtract $9y^3 - 11xy^2 + 2x^3$ from $-x^3 + 10x^2y - 14xy^2 + 16y^3$.

64. Subtract $-4a^2 - ab + 1$ from $4a^3 - a^2 - 3ab + 2$.

Simplify.

◉ 65. $(3x^2 - 7x + 5) - (5x - 8x^2) + (2 + 6x - 9x^2)$

66. $(4x^3 - 3x^2) + (x^2 + 2x - 10) - (x^3 + x - 15)$

67. $(12n^3 + 16) - (11n^3 - 9n^2 + 3n + 8) - (-10n - 13)$

68. $(7a^4 + 7a^2 - 14) - (-2a^4 - 6a^3) - (8a^2 - 9)$

Given the functions $f(x)$ and $g(x)$, find $f(x) + g(x)$ and $f(x) - g(x)$.

69. $f(x) = 2x^2 - 8$ and $g(x) = 2x^2 + 7x + 5$

70. $f(x) = x^3 + x + 1$ and $g(x) = x + 1$

71. $f(x) = -5x^4 + 6x^2 - 3$ and $g(x) = 4x^4 - 3x^3 + 2x^2 - x$

72. $f(x) = -3x^2 + 8x - 2$ and $g(x) = 3x^2 - 2x + 1$

Mixed Practice

Solve.

73. Subtract $5a^4 - a^3 - 3a^2 + 3$ from $3a^4 - 3a^3 - a + 4$.

74. Identify the terms and coefficients of the polynomial
$12x^4 - \dfrac{3}{4}x^3 - x + 5$.

75. Add $6 + 4n^2 - 2n$, $3n + 4n^2$, and $-2n^2 - n + 4$ vertically.

76. Given the functions $f(x)$ and $g(x)$, find $f(x) + g(x)$ and $f(x) - g(x)$, where $f(x) = x^2 - x + 2$ and $g(x) = x^3 - 3x^2 - 3$.

77. Simplify $6x^2 - 7x + x^3 - 2 - x + 4x^3 - x^2 + 9$, and write it in descending order.

78. Evaluate the polynomial $16x^3 - 4x^2 - 2x + 3$ for the given values of x.

    a. $x = -1$ **b.** $x = \dfrac{1}{2}$

79. Determine whether the expression $\dfrac{3n^2}{5} - 5 + 2n^4$ is a polynomial.

80. Classify the polynomial $2a^5 - 9a^3 + 6$ by the number of terms it has, and identify its degree.

81. Simplify
$(-2a^2 - 3a + 4) - (6a - 5a^2) - (a^2 + 5a^3 - 7)$.

82. Write the polynomial $x^2 - 2x^3 + 5x$ in descending order. Then, identify the leading term and leading coefficient.

Applications

C *Solve.*

83. The height above ground of a penny t sec after it is dropped from the observation deck of the Space Needle in Seattle, Washington, is modeled by the polynomial $-16t^2 + 520$. Use the polynomial to complete the table.

Time t (in seconds)	0	1	2	3
Height (in feet)				

84. A box manufacturer makes custom gift boxes for a client. Each gift box is to have a square base and a height that is 6 in. greater than the length x of a side of the square base, as shown to the right. The volume of each box is given by the polynomial $x^3 + 6x^2$, where x is the length of the square base in inches. Use the polynomial to complete the table.

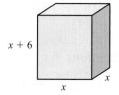

$x + 6$

x

x

Length x of a Side of the Square Base (in inches)	1	2	3	4
Volume (in cubic inches)				

85. A factory owner determines that her monthly revenue (in dollars) can be approximated by the polynomial $18n + 3.5n^2 - 0.005n^3$, where n is the number of units sold each month. In May, 100 units were sold. What was the factory's revenue in May?

86. An initial amount of $1000 is invested in an account earning compound interest. The polynomial $1000r^3 + 3000r^2 + 3000r$ represents the amount of interest earned after three compounding periods, where r is the rate of return (in decimal form) for each period. If the rate of return for each period is 2%, how much interest is earned after three compounding periods?

87. The estimated number of students (in thousands) enrolled in elementary schools can be modeled by the polynomial $-56x^3 + 382x^2 - 675x + 32,438$, where x is the number of years after 2005. The corresponding polynomial for students enrolled in secondary schools is $-41x^2 - 77x + 17,332$. Find a polynomial that represents the total number of students enrolled in elementary and secondary schools. (*Source:* census.gov)

88. Consumer credit is categorized as either revolving credit, such as credit cards, or nonrevolving credit, such as automobile loans. The consumer credit outstanding (in billions of dollars) for revolving credit can be approximated by the polynomial $1.3x^4 - 22.1x^3 + 71.7x^2 - 9.5x + 828.6$, where x represents the number of years after 2005. The corresponding polynomial for nonrevolving credit is $-4.8x^3 + 13.9x^2 + 51.6x + 1459.8$. Find a polynomial that approximates the total consumer credit outstanding. (*Source:* federalreserve.gov)

89. The total number (in thousands) of takeoffs and landings at the Hartsfield-Jackson International Airport in Atlanta can be modeled by the polynomial $7x^3 - 37x^2 + 48x + 976$, where x is the number of years after 2006. The corresponding polynomial for the McCarran International Airport in Las Vegas is $-14x^2 + 8x + 619$. Find a polynomial that represents how many more takeoffs and landings occurred at Atlanta Airport than at Las Vegas Airport in a given year. (*Source:* airports.org)

90. The U.S. grain supply (in millions of metric tons) from domestic production can be approximated by the polynomial $18x^3 - 57x^2 + 16x + 386$, where x is the number of years after 2005. The corresponding polynomial for the supply from imports is $2x^3 - 22x^2 + 51x + 49$. Write a polynomial that gives the total grain supply from production and imports for a given year. (*Source:* usda.mannlib.cornell.edu)

91. A shoe manufacturer estimates that the monthly cost to produce x pairs of shoes is given by the function $C(x) = 12x + 10,000$. The monthly revenue from selling x pairs of shoes is given by the function $R(x) = 40x - 0.005x^2$. Find a function $P(x)$ that represents the manufacturer's monthly profit from selling x pairs of shoes.

92. The U.S. annual production of biodiesel (in millions of gallons) is represented by the polynomial $-35x^3 + 160x^2 + 23x + 93$, where x is the number of years after 2005. The corresponding polynomial for annual consumption is $12x^3 - 100x^2 + 278x + 87$. Find a polynomial that represents the difference between the annual production and annual consumption of biodiesel for a given year. (*Source:* afdc.energy.gov)

• Check your answers on page A-37.

MINDStretchers

Mathematical Reasoning

1. Graphs of polynomial functions are said to be *continuous*, meaning they can be drawn without a break. Consider the following graphs. Which of these graphs could not represent a polynomial function? Explain.

a.
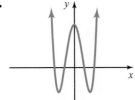

b.
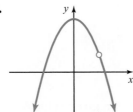

c.
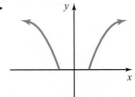

d.
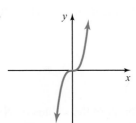

Critical Thinking

2. Indicate whether each statement is always true, sometimes true, or never true. Justify your answers.

a. The sum (or difference) of two polynomials is a polynomial.

b. A polynomial can have nonconstant terms raised to negative integer powers.

continued

c. The sum of *n*th-degree polynomials is an *n*th-degree polynomial.

d. The sum (or difference) of two polynomials has a degree that is higher than the degree of either of the original polynomials.

e. The sum (or difference) of two binomials is a binomial.

Investigation

3. Consider the graphs of the functions $f(x) = x^2 - 4$, $g(x) = -x^2 + 2x + 3$, and $h(x) = 2x - 1$ shown below.

$f(x) = x^2 - 4$ $g(x) = -x^2 + 2x + 3$ $h(x) = 2x - 1$

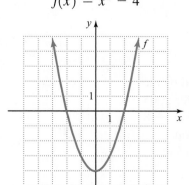 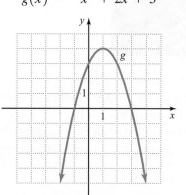 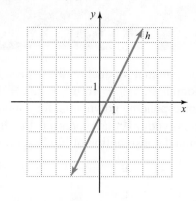

a. Use the graphs to complete the table.

x	−2	−1	0	1	2
$f(x)$					
$g(x)$					
$h(x)$					

b. Use the table in part (a) to complete the following table:

x	−2	−1	0	1	2
$f(x) + g(x)$					

c. Compare the values of $f(x) + g(x)$ to the values of $h(x)$ for a given value of x. What do you notice?

d. What conclusion can you draw about the relationship between $f(x)$, $g(x)$, and $h(x)$? Confirm your answer algebraically.

5.2 Multiplication of Polynomials

OBJECTIVES

A To multiply polynomials

B To multiply binomials using FOIL

C To use a special product formula to square a binomial or to multiply the sum and difference of two binomials

D To solve applied problems involving the multiplication of polynomials

In this section, we discuss how to multiply polynomials. We start with finding the product of two monomials.

Multiplying Monomials, Binomials, and Other Polynomials

Since a monomial is a single term that generally contains a variable raised to a power, we use the product rule of exponents to multiply monomials.

EXAMPLE 1

Multiply.

a. $-5x^2(2x)$ **b.** $(-3x^3y)(-4x^2y^4)(xy)$ **c.** $(-2p^2r)^3$

Solution

a. $(-5x^2)(2x) = (-5 \cdot 2)(x^2 \cdot x)$ Use the commutative and associative properties of multiplication.

$= -10(x^{2+1})$ Use the product rule of exponents.

$= -10x^3$

b. $(-3x^3y)(-4x^2y^4)(xy) = (-3)(-4)(1)(x^3 \cdot x^2 \cdot x)(y \cdot y^4 \cdot y)$

$= 12x^{3+2+1}y^{1+4+1}$

$= 12x^6y^6$

Note that the variables in a product are generally written in alphabetical order.

c. $(-2p^2r)^3 = (-2p^2r)(-2p^2r)(-2p^2r)$

$= [(-2)(-2)(-2)](p^2 \cdot p^2 \cdot p^2)(r \cdot r \cdot r)$

$= -8p^6r^3$

PRACTICE 1

Find each product.

a. $(-4x^2)$ and $(-6x^3)$

b. $(7ab)(3a^2b^3)(-2a)$

c. $(-10x^3y^2)^2$

Note that in Example 1(c), we could have used the rule for raising a product to a power. Can you explain how?

To multiply a polynomial by a monomial, we use the distributive property. Consider, for example, the product $2x(x - 4)$. To find this product, we distribute the monomial $2x$ over the binomial $(x - 4)$ as follows:

$$2x(x - 4) = 2x \cdot x + 2x \cdot (-4)$$ Use the distributive property.

$$= 2x^2 - 8x$$

EXAMPLE 2

Multiply. Simplify, if possible.

a. $(-8x + 1)(-3x^2)$ **b.** $2p^2q(4p^3 - 3pq + q^3)$

c. $4x^2(2x + 5) + x^2(5x^2 - 6x + 1)$

Solution

a. $(-8x + 1)(-3x^2) = (-3x^2)(-8x + 1)$ Use the commutative property of multiplication.

$$= (-3x^2)(-8x) + (-3x^2)(1)$$ Use the distributive property.

$$= 24x^3 - 3x^2$$ Simplify.

b. $2p^2q(4p^3 - 3pq + q^3) = 2p^2q(4p^3) + 2p^2q(-3pq) + 2p^2q(q^3)$

$$= 8p^5q - 6p^3q^2 + 2p^2q^4$$

c. $4x^2(2x + 5) + x^2(5x^2 - 6x + 1)$

$$= 4x^2(2x) + 4x^2(5) + x^2(5x^2) + x^2(-6x) + x^2(1)$$

$$= 8x^3 + 20x^2 + 5x^4 - 6x^3 + x^2$$

$$= 5x^4 + 2x^3 + 21x^2$$

Note that the simplified polynomial is written in descending order.

PRACTICE 2

Find each product. Simplify, if possible.

a. $(4x^2 - 9)(-2x)$

b. $-3a^3b^2(-6a^3b^5 - ab^2 + 10b)$

c. $4y^3(-3y^2 + 5y + 4) - 2y^2(y^2 + 2y - 1)$

To multiply binomials, we can also use the distributive property. Consider the product $(x + 2)(x - 7)$. We apply the distributive property, thinking of the first binomial factor as a single number, which we distribute over the second binomial factor.

$$(x + 2)(x - 7) = (x + 2)(x) + (x + 2)(-7)$$ Use the distributive property.

$$= x(x + 2) + (-7)(x + 2)$$ Use the commutative property.

$$= x^2 + 2x - 7x - 14$$ Use the distributive property.

$$= x^2 - 5x - 14$$ Combine like terms.

EXAMPLE 3

Find the product: $(2x + 5)(8x - 1)$

Solution

$(2x + 5)(8x - 1) = (2x + 5)(8x) + (2x + 5)(-1)$ Use the distributive property.

$$= (8x)(2x + 5) + (-1)(2x + 5)$$ Use the commutative property.

$$= 16x^2 + 40x - 2x - 5$$ Use the distributive property.

$$= 16x^2 + 38x - 5$$ Combine like terms.

PRACTICE 3

Multiply: $(4n - 3)(2n + 7)$

Another way of multiplying two binomials, based on the distributive property, is the *FOIL method*. FOIL stands for **F**irst, **O**uter, **I**nner, and **L**ast.

To Multiply Two Binomials Using the FOIL Method

Consider $(a + b)(c + d)$.

- Multiply the two first terms in the binomials.

$$(a + b)(c + d) \quad \text{Product } ac$$

F

- Multiply the two outer terms.

$$(a + b)(c + d) \quad \text{Product } ad$$

O

- Multiply the two inner terms.

$$(a + b)(c + d) \quad \text{Product } bc$$

I

- Multiply the two last terms.

$$(a + b)(c + d) \quad \text{Product } bd$$

L

The product of the two binomials is the sum of these four products:

$$(a + b)(c + d) = ac + ad + bc + bd$$

EXAMPLE 4

Multiply.

a. $(2x - 3)(x - 1)$ **b.** $(4x + 7)^2$ **c.** $(5a + b)(6a - b)$

Solution

a. We will use the FOIL method to find this product.

First   Last

$$(2x - 3) \quad (x - 1)$$

Inner

Outer

F $\quad (2x)(x) = 2x^2$
O $\quad (2x)(-1) = -2x$
I $\quad (-3)(x) = -3x$
L $\quad (-3)(-1) = 3$

So the product $(2x - 3)(x - 1)$ is the sum of these four products:

$$2x^2 - 2x - 3x + 3$$

which simplifies to $2x^2 - 5x + 3$. Notice here that the middle term $-5x$ is the sum of the outer and inner products, which are like terms.

$$\underbrace{(-2x) + (-3x) = (-5x)}_{\text{The middle term}}$$

PRACTICE 4

Simplify.

a. $(x + 1)(5x - 2)$

b. $(3x - 4)^2$

c. $(2m - n)(3m + n)$

EXAMPLE 4 (continued)

b. $(4x + 7)^2 = (4x + 7)(4x + 7)$

$$\begin{array}{cccc} \text{F} & \text{O} & \text{I} & \text{L} \end{array}$$
$$= (4x)(4x) + (4x)(7) + (7)(4x) + (7)(7)$$
$$= \quad 16x^2 \quad + \quad 28x \quad + \quad 28x \quad + \quad 49$$
$$= 16x^2 + 56x + 49$$

c. $(5a + b)(6a - b) = (5a)(6a) + (5a)(-b) + (b)(6a) + (b)(-b)$
$$= \quad 30a^2 \quad - \quad 5ab \quad + \quad 6ab \quad - \quad b^2$$
$$= 30a^2 + ab - b^2$$

To multiply polynomials with any number of terms, say the trinomial $(x^2 + 3x - 1)$ and the binomial $(x + 2)$, we again use the distributive property.

$$
\begin{aligned}
(x^2 + 3x - 1)(x + 2) &= (x^2 + 3x - 1)(x) + (x^2 + 3x - 1)(2) & \text{Use the distributive property.} \\
&= x(x^2 + 3x - 1) + 2(x^2 + 3x - 1) & \text{Use the commutative property.} \\
&= (x^3 + 3x^2 - x) + (2x^2 + 6x - 2) & \text{Use the distributive property.} \\
&= x^3 + 5x^2 + 5x - 2 & \text{Combine like terms.}
\end{aligned}
$$

To find the product of polynomials, we can also write the problem in a vertical format. When using this format, we position the terms of the partial products so that each column contains like terms. Then, we add down each column as illustrated in the next example.

EXAMPLE 5

Multiply.

a. $(3x^3 - x + 1)(x + 2)$ **b.** $(4x^2 + x - 6)(5x^2 - x - 1)$

Solution

a. We begin by rewriting the problem in a vertical format.

— Use $+ 0x^2$ for the missing x^2 term.

$$
\begin{array}{r}
3x^3 + 0x^2 - x + 1 \\
x + 2 \\
\hline
6x^3 + \quad\quad - 2x + 2 \\
3x^4 \quad\quad - x^2 + x \\
\hline
3x^4 + 6x^3 - x^2 - x + 2
\end{array}
$$

Multiply $(3x^3 + 0x^2 - x + 1)$ by 2.
Multiply $(3x^3 + 0x^2 - x + 1)$ by x.
Add.

b.

$$
\begin{array}{r}
4x^2 + x - 6 \\
5x^2 - x - 1 \\
\hline
- 4x^2 - x + 6 \\
- 4x^3 - x^2 + 6x \\
20x^4 + 5x^3 - 30x^2 \\
\hline
20x^4 + x^3 - 35x^2 + 5x + 6
\end{array}
$$

PRACTICE 5

Find the product of the following:

a. $5x^3 + x^2 - 2$ and $x + 8$

b. $(x^2 + 3x - 2)(3x^2 - x + 1)$

Can you explain in Example 5(a) the advantage of writing $0x^2$ in the polynomial $3x^3 - x + 1$?

EXAMPLE 6

Multiply: $(a - b)^3$

Solution We first multiply $(a - b)$ by $(a - b)$.

$$
\begin{array}{r}
a - b \\
a - b \\
\hline
-\ ab + b^2 \\
a^2 -\ ab \\
\hline
a^2 - 2ab + b^2
\end{array}
$$

Next, we multiply $(a^2 - 2ab + b^2)$ by $(a - b)$.

$$
\begin{array}{r}
a^2 - 2ab + b^2 \\
a - b \\
\hline
-\ a^2b\ + 2ab^2 - b^3 \\
a^3 - 2a^2b\ +\ ab^2 \\
\hline
a^3 - 3a^2b\ + 3ab^2 - b^3
\end{array}
$$

So we can conclude that $(a - b)^3 = a^3 - 3a^2b + 3ab^2 - b^3$.

PRACTICE 6

Simplify: $(p + 2q)^3$

Special Products

Special product formulas allow us to use shortcuts to find the products of binomials that have certain forms. The first formula we consider applies to the *square of a binomial*, such as $(x + 3)^2$. Using FOIL to find the product, we get:

$$
\begin{aligned}
(x + 3)^2 &= (x + 3)(x + 3) \\
&= x^2 + 3x + 3x + 9 \\
&= x^2 + 6x + 9
\end{aligned}
$$

Note that the expression $x^2 + 6x + 9$ is equal to $(x)^2 + 2(x)(3) + (3)^2$. So $(x + 3)^2$ is equal to the square of x plus twice the product of x and 3 plus the square of 3. Similarly, we can show that $(x - 3)^2 = (x)^2 - 2(x)(3) + (3)^2 = x^2 - 6x + 9$.

The Square of a Binomial

$$(a + b)^2 = a^2 + 2ab + b^2$$
$$(a - b)^2 = a^2 - 2ab + b^2$$

Let's compare the formula for the square of a sum of two terms $(a + b)^2$ to the formula for the square of a difference of two terms $(a - b)^2$. We see that the middle term for the square of a sum is *positive*, whereas the middle term for the square of a difference is *negative*.

EXAMPLE 7

Multiply.

a. $(3x + 1)^2$ **b.** $(7 - 2n)^2$

c. $(4p - 3q)^2$ **d.** $(x^3 + 5y^2)^2$

Solution

a. Here, we are squaring the sum of two terms, so we use the formula for the square of a sum.

First term   Second term   The square of the first term   The square of the second term

$$(3x + 1)^2 = (3x)^2 + \underbrace{2(3x)(1)}_{\substack{\text{Twice the product of} \\ \text{the two terms}}} + (1)^2$$

$$= 9x^2 + 6x + 1$$

b. Here, we are squaring the difference of two terms, so we use the formula for the square of a difference.

$$(7 - 2n)^2 = (7)^2 - 2(7)(2n) + (2n)^2$$
$$= 49 - 28n + 4n^2, \text{ or } 4n^2 - 28n + 49$$

c. $(4p - 3q)^2 = (4p)^2 - 2(4p)(3q) + (3q)^2$
$$= 16p^2 - 24pq + 9q^2$$

d. $(x^3 + 5y^2)^2 = (x^3)^2 + 2(x^3)(5y^2) + (5y^2)^2$
$$= x^6 + 10x^3y^2 + 25y^4$$

PRACTICE 7

Multiply.

a. $(4p + 7)^2$

b. $(10 - 2x)^2$

c. $(5x + 3y)^2$

d. $(6m^3 - n^4)^2$

Now, let's look at a formula for multiplying the sum of two terms by the difference of the same two terms. For instance, consider the product $(x + 3)(x - 3)$. Using the FOIL method, we get:

$$(x + 3)(x - 3) = x^2 - 3x + 3x - 9$$
$$= x^2 - 9$$

From the FOIL method we see that the outer and inner products are opposites, resulting in a middle term of $0x$. So the product $(x + 3)(x - 3)$ is equal to the square of the first term x minus the square of the second term 3, that is, the difference of their squares. This observation leads to the following formula for the *product of the sum and difference of the same two terms*:

The Product of the Sum and Difference of Two Terms

$$(a + b)(a - b) = a^2 - b^2$$

EXAMPLE 8

Multiply.

a. $(5n - 1)(5n + 1)$ **b.** $(2m + n)(2m - n)$

c. $(4a^2 - 3b^2)(4a^2 + 3b^2)$

PRACTICE 8

Multiply.

a. $(2s + 9)(2s - 9)$

b. $(4a + 7b)(4a - 7b)$

c. $(8x^4 - y^2)(8x^4 + y^2)$

Solution

a. Switching the order of the two factors, we apply the formula for finding the product of the sum and the difference of two terms.

First term	Second term	The square of the first term	The square of the second term
↓	↓	↓	↓

$$(5n + 1)(5n - 1) = (5n)^2 - (1)^2$$
$$= 25n^2 - 1$$

b. $(2m + n)(2m - n) = (2m)^2 - (n)^2 = 4m^2 - n^2$

c. $(4a^2 - 3b^2)(4a^2 + 3b^2) = (4a^2)^2 - (3b^2)^2 = 16a^4 - 9b^4$

In the next example, we use our knowledge of multiplying polynomials to multiply two polynomial functions.

EXAMPLE 9

Given that $f(x) = 2x^2 - x + 1$ and $g(x) = 3x + 4$, find $f(x) \cdot g(x)$.

Solution $f(x) \cdot g(x) = (2x^2 - x + 1) \cdot (3x + 4)$
$$= 6x^3 - 3x^2 + 3x + 8x^2 - 4x + 4$$
$$= 6x^3 + 5x^2 - x + 4$$

PRACTICE 9

Suppose $f(x) = -5x + 1$ and $g(x) = x^2 - 4x + 2$. Find $f(x) \cdot g(x)$.

We can also multiply polynomials to evaluate polynomial functions, as shown in the next example:

EXAMPLE 10

Given that $f(x) = x^2 + 2x - 1$, find:

a. $f(x + 3)$ **b.** $f(x + 3) - f(x)$ **c.** $f(x + h) - f(x)$

Solution

a. $f(x + 3) = (x + 3)^2 + 2(x + 3) - 1$
$$= (x^2 + 6x + 9) + 2(x + 3) - 1$$
$$= x^2 + 6x + 9 + 2x + 6 - 1$$
$$= x^2 + 8x + 14$$

b. From part (a), we know that $f(x + 3) = x^2 + 8x + 14$.
$$f(x + 3) - f(x) = (x^2 + 8x + 14) - (x^2 + 2x - 1)$$
$$= x^2 + 8x + 14 - x^2 - 2x + 1$$
$$= 6x + 15$$

c. $f(x + h) - f(x) = [(x + h)^2 + 2(x + h) - 1]$
$$- [x^2 + 2x - 1]$$
$$= [x^2 + 2xh + h^2 + 2x + 2h - 1]$$
$$- [x^2 + 2x - 1]$$
$$= x^2 + 2xh + h^2 + 2x + 2h - 1$$
$$- x^2 - 2x + 1$$
$$= 2xh + h^2 + 2h$$

PRACTICE 10

Given that $f(x) = 4x^2 - 1$, find:

a. $f(x + 2)$

b. $f(x + 2) - f(x)$

c. $f(x + h) - f(x)$

EXAMPLE 11

Newspapers come in different sizes. For instance, a *broadsheet* is typically 29.5 in. by 23.5 in. and a *tabloid* is 16.9 in. by 11.0 in. *USA Today*, the most widely circulated print newspaper in the country, has a height 10 in. more than its width w. Express as a polynomial the area of this newspaper's front page.
(*Sources:* papersizes.org and wikipedia.org)

Solution The front page of a newspaper is rectangular in shape, so its area is the product of its width and its height: $w(w + 10) = w^2 + 10w$. The area of the front page can be represented by $(w^2 + 10w)$ in^2.

PRACTICE 11

A company's total revenue R for a particular item is given by the following equation:

$$R = px$$

Here, p is the price of the item and x is the number of items sold. Write a polynomial in p for the revenue if

$$x = -\frac{1}{4}p + 100.$$

Mathematically Speaking

Fill in each blank with the most appropriate term or phrase from the given list.

distributive property	three	square of a binomial
monomials	binomials	two
negative	FOIL method	
product rule of exponents	positive	

1. To multiply monomials, we use the _____.

2. One way to multiply two _____ is to use the FOIL method.

3. In the formula for the square of a sum of two terms, $(a + b)^2$, the coefficient of the middle term is _____.

4. In the formula for the square of the difference of two terms, $(a - b)^2$, the coefficient of the middle term is _____.

5. The formula $(a + b)^2 = a^2 + 2ab + b^2$ is called the _____.

6. The product of the sum and difference of the same two terms, $(a + b)(a - b)$, has _____ terms.

A *Find the product.*

7. $(6n^3)(-5n^3)$

8. $(-9x^4)(7x^2)$

9. $\left(-\dfrac{2}{3}rt^2\right)(-9r^3t)$

10. $\left(-\dfrac{3}{5}xy\right)(-5x^2y)$

11. $(10x^5)(-x^3)(-2x^4)$

12. $(-8n^2)(-n^6)(-3n)$

13. $(-3pr) \cdot 2p^3q \cdot q^2$

14. $2a^2 \cdot (-6ab^4) \cdot b^3c$

15. $(-4ab^5)^3$

16. $(-3p^2q^3)^4$

17. $(-2x)(5 - 4x)$

18. $(5y + 7)(3y)$

19. $4n^2(6n - 1)$

20. $(-2y^3)(5y - 6)$

21. $(-3x)(x^2 - 4x + 5)$

22. $6x(1 - 2x - x^2)$

23. $\dfrac{1}{2}n^3(12m^2 + 8n)$

24. $-\dfrac{1}{3}t^4(9r^2 - 15t^2)$

25. $(7x^5 - 4x^3 + 1)(4x^2y^4)$

26. $(10p^2 + 7p + 9)(3p^2q)$

27. $(-2a^2b^3)(-8a^3 + 3a^2b - 4ab^2 + b^3)$

28. $(-3cd^2)(5c^4 - c^3d + 2cd^2 - 3d^3)$

Simplify.

29. $2y(1 + 4y - 4y^2) + 3y^2(y - 2)$

30. $4a^2(a^2 - 3a - 2) + 5a(a^2 - 2a^3)$

31. $6x^4(2x - 1) - 4x^3(2x^2 - 5) + 3(2x^5 + x^4 - 8x^3)$

32. $9(3n^3 - 2n^2) + 2n(-10n^2 + 6n) - 5n^2(n + 3)$

33. $\dfrac{1}{4}p^2(8p^2 - 12pq + 4q^2) - q^2(3q^2 - 5pq + p^2)$

34. $2x^3(4x^2 + 6xy - 9y^3) - \dfrac{1}{5}y^3(20y^2 - 45x^3)$

B *Multiply.*

35. $(x + 2)(x + 4)$ 36. $(x + 5)(x + 7)$ 37. $(n - 6)(n - 3)$ 38. $(y - 8)(y - 2)$

39. $(5 - a)(a + 7)$ 40. $(x + 9)(4 - x)$ 41. $(y + 7)^2$ 42. $(a - 5)^2$

43. $(2x - 1)(x + 4)$ 44. $(3x - 8)(x + 6)$ 45. $(3 - 2x)(2 + 3x)$ 46. $(5 + 4y)(4 - 2y)$

47. $(5x + 3)(6x + 5)$ 48. $(2x - 9)(2x - 7)$ 49. $(4x - 9)^2$ 50. $(8p + 3)^2$

51. $(a - b)(2a + 3b)$ 52. $(x + y)(4x - 7y)$ 53. $(9p + 10q)(2p - q)$ 54. $(5s - 3t)(2s + t)$

55. $(7x - 11y)(8x - 7y)$ 56. $(6p + 13q)(3p + 4q)$ 57. $\left(\dfrac{1}{2}x + 6y\right)\left(\dfrac{2}{3}x + 8y\right)$ 58. $\left(\dfrac{1}{4}x - 12y\right)\left(\dfrac{1}{2}x - 4y\right)$

59. $(n - 1)(n^2 + 3n + 2)$ 60. $(t - 4)(t^2 + t - 2)$ 61. $(3x + 2)(x^2 - 5x + 6)$ 62. $(2x + 3)(x^2 - 3x - 7)$

63. $(5x^2 - 10x + 25)\left(\dfrac{1}{5}x^2 - 2\right)$ 64. $(9x^2 - 3x + 6)\left(\dfrac{1}{3}x^2 + 1\right)$

65. $(y - 7x^2)(2x^3 + x^2 - 6)$ 66. $(x - 2y^2)(4 + 4x - x^3)$

67. $(x^2 - 3x - 4)(x^2 + 3x + 4)$ 68. $(a^2 + 5a - 6)(a^2 - 5a - 6)$

69. $(x^3 + 4y^4)(3x^2 + y)$ 70. $(2x - 7y^3)(2x^2 - y^3)$

71. $(4p^2 - 3q^2)(3p^2 + 4q^2)$ 72. $(3a^3 + 5b^3)(5a^3 - 3b^3)$

C *Use a special product formula to simplify.*

73. $(t - 10)(t + 10)$ 74. $(x + 7)(x - 7)$ 75. $(x + 8)^2$ 76. $(n - 2)^2$

77. $(4n - 3)(4n + 3)$ 78. $(6t - 1)(6t + 1)$ 79. $(2n - 5)^2$ 80. $(3t - 1)^2$

81. $\left(b - \dfrac{1}{3}a\right)^2$ 82. $\left(q + \dfrac{1}{2}p\right)^2$

83. $(5x + y)(5x - y)$ 84. $(x - 8y)(x + 8y)$ 85. $(11y - 9x)(9x + 11y)$ 86. $(10a + 3b)(3b - 10a)$

87. $(4p + 7q)^2$ 88. $(6x + 5y)^2$ 89. $(2x^2 - 9y^4)^2$ 90. $(8s^3 - 3t^2)^2$

91. $(p^4 - 8q^2)(p^4 + 8q^2)$ 92. $(10x^3 + y^3)(10x^3 - y^3)$

93. $[(x + 4) - y][(x + 4) + y]$

94. $[(p - 7) + q][(p - 7) - q]$

95. $[(1 - a) - b]^2$

96. $[(3 - x) + y]^2$

Multiply.

97. $(x - 3)(4x + 1)(4x - 1)$

98. $(3x + 2)(3x - 2)(x + 5)$

99. $(2n - 1)^3$

100. $(5y + 2)^3$

101. $(6a + b)(a - 6b)(a + b)$

102. $(x - 4y)(4x + y)(x - y)$

103. $(x + 2y)^3$

104. $(2a - b)^3$

105. $(2x + 3y)(2x - 3y)(4x^2 + 9y^2)$

106. $(4p^2 + q^4)(2p - q^2)(2p + q^2)$

Given $f(x)$ and $g(x)$, find $f(x) \cdot g(x)$.

107. $f(x) = 3x^2 - 1$ and $g(x) = 5x + 2$

108. $f(x) = 2x^2 + 3x$ and $g(x) = x - 4$

109. $f(x) = x^2 + 7x - 3$ and $g(x) = x^2 - 6x$

110. $f(x) = -x^2 + 8$ and $g(x) = x^2 - x - 9$

Evaluate the function for the given expressions.

111. $f(x) = x^2 - 2x + 5$
    a. $f(n - 1)$
    b. $f(2n + 3)$

112. $f(x) = x^2 + 3x - 7$
    a. $f(a + 2)$
    b. $f(3a - 2)$

113. $f(x) = x^2 - x - 9$
    a. $f(x - 2)$
    b. $f(x - 2) - f(x)$
    c. $f(x + h) - f(x)$

114. $f(x) = x^2 + 5x - 4$
    a. $f(x + 4)$
    b. $f(x + 4) - f(x)$
    c. $f(x + h) - f(x)$

Mixed Practice

Solve.

115. Simplify $5a(3a^3 - 2a) - 3a^2(-a - 2) + 2a(2a^2 + a)$.

116. Given that $f(x) = x^2 - 3x + 2$, find $f(2a - 3)$.

117. If $f(x) = 4x^2 + x$ and $g(x) = 2x - 5$, find $f(x) \cdot g(x)$.

Find the product.

118. $(2r^2t^2)(-5rt^3)(-r)$

119. $\frac{1}{2}x(4x^2 - 2x + 8)$

120. $(2m - 3)(4m + 7)$

121. $(x - 2)(x^2 - 3x + 2)$

Use a special product formula to multiply.

122. $(a - 7b)(a + 7b)$

123. $(4x - 5y)^2$

124. $(2x - y)(2x + y)(x - y)$

Ⓓ Applications

Solve.

125. The work output of a heat engine is usually produced by an expanding gas. If the volume of the gas changes from V_1 to V_2 at constant pressure p, then the work output can be computed using the expression $p(V_2 - V_1)$. Write this expression without parentheses.

126. A computer retailer's total revenue for selling x laptops at a price of p dollars per laptop is given by the equation $R = xp$. Write a polynomial in x for the revenue if $p = 1200 - \dfrac{1}{2}x$.

127. A homeowner plans to extend the length and width of her existing patio by x ft, as shown in the following figure:

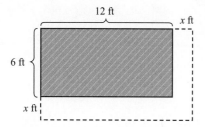

a. Write a polynomial for the area of the new patio.

b. By how many square feet will the size of the patio increase?

128. To make room for additional office cubicles, an office manager decides to reduce the floor space of each existing cubicle by decreasing its length and width by n ft.

a. If the original office cubicles measure 8 ft by 10 ft, find a polynomial that represents the new area of each cubicle.

b. By how many square feet will the floor space in each cubicle decrease?

129. A machine is used to cut out sheets of tin into square ceiling panels measuring x in. on a side. The machine has an allowable error of 0.1 in.

a. Express in terms of x the longest and shortest possible true dimensions of a panel cut out by the machine.

b. What is the difference in area between the largest possible panel and the smallest possible panel in terms of x?

130. To improve the flow of traffic, the radius of a traffic circle that had been r ft was increased by 10 ft.

a. Find the area of the new traffic circle (in square feet) as a polynomial in standard form.

b. By how much did enlarging the traffic circle increase its area?

131. An open box is to be made from an 18-inch by 12-inch sheet of cardboard by cutting squares of equal area from each corner and folding up the sides (see the diagram below). Write a polynomial expression for the volume of the box.

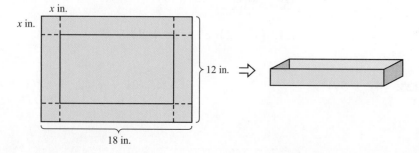

132. A circular swimming pool is surrounded by a deck as shown in the figure to the right. The outer border of the deck forms a square. Write a polynomial that represents the area of the deck.

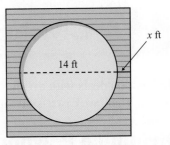

• Check your answers on page A-38.

MINDStretchers

Patterns

1. Find each of the polynomial products.
$$(x - 1)(x + 1)$$
$$(x - 1)(x^2 + x + 1)$$
$$(x - 1)(x^3 + x^2 + x + 1)$$
$$(x - 1)(x^4 + x^3 + x^2 + x + 1)$$

Describe the pattern you observe in these products. Then, use the pattern to find the product of
$$(x - 1)(x^n + x^{n-1} + x^{n-2} + \cdots + x + 1),$$
where n is a positive integer.

Mathematical Reasoning

2. What is the degree of the product of two polynomials if one of the polynomials is of degree m and the other is of degree n? Justify your answer.

Groupwork

3. Geometric models can be used to visualize the product of two binomials. Consider each of the following figures. Together with a partner, use the figures to write two different expressions for the area of the large outer rectangle. Then, show algebraically that the two expressions are equal.

a.

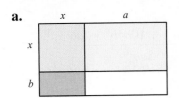

b.

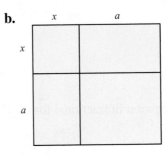

c.

5.3 Division of Polynomials

This section deals with dividing polynomials. We begin with dividing a polynomial by a monomial. Then, we discuss dividing one polynomial by another polynomial.

A To divide a polynomial by a monomial

Dividing a Polynomial by a Monomial

B To divide a polynomial by a polynomial

When we divide a polynomial by a monomial, we rely on our knowledge of dividing monomials. Let's consider the quotient $15x^6 \div 3x^2$.

C To solve applied problems involving the division of polynomials

$$15x^6 \div 3x^2 = \frac{15x^6}{3x^2}$$

$$= \frac{15}{3} \cdot \frac{x^6}{x^2}$$

$$= 5x^{6-2} \quad \text{Divide the coefficients, and use}$$
$$\quad\quad\quad\quad \text{the quotient rule of exponents.}$$
$$= 5x^4$$

Now, let's consider the quotient of a polynomial and a monomial. The key here is to write the quotient as the sum of fractions, and then to simplify each fraction.

EXAMPLE 1

Find the quotient.

a. $\dfrac{10x^7}{5x^2}$

b. $(8x^2 + 6x) \div (2x)$

c. $\dfrac{12x^4 - 6x^3 + 9x^2}{3x^2}$

d. $\dfrac{8p^4q - 12pq^3 + 3q}{-4p}$

Solution

a. $\dfrac{10x^7}{5x^2} = \dfrac{10}{5} \cdot \dfrac{x^7}{x^2}$

$\quad\quad = 2x^{7-2}$

$\quad\quad = 2x^5$

b. We write the given expression in fractional form.

$$(8x^2 + 6x) \div (2x) = \frac{\overbrace{8x^2 + 6x}^{\text{Polynomial}}}{\underset{\uparrow}{2x}}$$

$$\text{Monomial}$$

We can write $\dfrac{8x^2 + 6x}{2x}$ as the sum of two fractions.

$$\frac{8x^2 + 6x}{2x} = \frac{8x^2}{2x} + \frac{6x}{2x}$$

$$= 4x + 3$$

PRACTICE 1

Divide.

a. $12x^3 \div 3x$

b. $(7x^3 - 14x^2) \div (-7x)$

c. $\dfrac{6x^8 + 10x^5 - 4x^3}{2x^3}$

d. $\dfrac{-5a^7b^6 + 10a^2b^4 - ab}{5ab^3}$

c. $\dfrac{12x^4 - 6x^3 + 9x^2}{3x^2} = \dfrac{12x^4}{3x^2} - \dfrac{6x^3}{3x^2} + \dfrac{9x^2}{3x^2}$

$$= 4x^2 - 2x + 3$$

d. $\dfrac{8p^4q - 12pq^3 + 3q}{-4p} = \dfrac{8p^4q}{-4p} - \dfrac{12pq^3}{-4p} + \dfrac{3q}{-4p}$

$$= -2p^3q + 3q^3 - \dfrac{3q}{4p}$$

Dividing a Polynomial by a Polynomial

Now, we look at dividing one polynomial by another. The process of dividing polynomials is similar to the long division of whole numbers, as shown in the following examples:

$$
\begin{array}{r}
14 \longleftarrow \text{Quotient} \\
\text{Divisor} \longrightarrow 15\overline{)\ 217} \longleftarrow \text{Dividend} \\
-15 \\
\hline
67 \\
-60 \\
\hline
7 \longleftarrow \text{Remainder}
\end{array}
$$

So $217 \div 15$ is $14\frac{7}{15}$.

Recall that in dividing whole numbers, when there is a remainder, we write it as the numerator of a fraction with the divisor as the denominator. To divide two polynomials, we set up the problem just as if we were dividing two whole numbers, being careful to correctly distinguish between the dividend and the divisor. For example, consider the following quotient:

$$(x^2 - 6x - 27) \div (x + 3)$$

$$
\begin{array}{r}
x \\
x + 3\overline{)x^2 - 6x - 27}
\end{array}
$$
Divide the first term in the dividend by the first term in the divisor: $x^2 \div x = x$. Place x in the quotient above the x-term in the dividend.

$$
\begin{array}{r}
x \\
x + 3\overline{)x^2 - 6x - 27} \\
\underline{x^2 + 3x}
\end{array}
$$
Multiply x in the quotient by the divisor: $x(x + 3) = x^2 + 3x$.

$$
\begin{array}{r}
x \\
x + 3\overline{)x^2 - 6x - 27} \\
x^2 + 3x \\
\hline
-9x
\end{array}
$$
Subtract $(x^2 + 3x)$ from $(x^2 - 6x)$ by changing the sign of each term in $(x^2 + 3x)$ and then adding to get $-9x$:
$(x^2 - 6x) - (x^2 + 3x) = x^2 - 6x - x^2 - 3x = -9x.$

$$
\begin{array}{r}
x \\
x + 3\overline{)x^2 - 6x - 27} \\
\underline{x^2 + 3x} \\
-9x - 27
\end{array}
$$
Bring down the next term in the dividend, namely -27.

$$
\begin{array}{r}
x - 9 \\
x + 3\overline{)x^2 - 6x - 27} \\
\underline{x^2 + 3x} \\
-9x - 27
\end{array}
$$
Divide x into $-9x$: $-9x \div x = -9$.
Place -9 in the quotient above the constant term in the dividend.

$$\begin{array}{r} x - 9 \\ x + 3 \overline{)x^2 - 6x - 27} \\ \underline{x^2 + 3x} \\ -9x - 27 \\ \underline{-9x - 27} \end{array}$$

Multiply -9 in the quotient by the divisor: $-9(x + 3) = -9x + 27$.

$$\begin{array}{r} x - 9 \\ x + 3 \overline{)x^2 - 6x - 27} \\ \underline{x^2 + 3x} \\ -9x - 27 \\ \underline{-9x - 27} \\ 0 \end{array}$$

Subtract $(-9x - 27)$ from $(-9x - 27)$:
$(-9x - 27) - (-9x - 27) = -9x - 27 + 9x + 27 = 0.$

$0 \longleftarrow$ Remainder

So $(x^2 - 6x - 27) \div (x + 3)$ is $x - 9$.

> **TIP** When dividing one polynomial by another, put *both* polynomials in descending order.

Note that we can check a problem involving division of polynomials in the same way that we check division of whole numbers:

$$\text{Divisor} \cdot \text{Quotient} + \text{Remainder} = \text{Dividend}$$

EXAMPLE 2

Divide and check.

a. $3x - 1 \overline{)6x^2 + 13x - 5}$ **b.** $(3x^2 - 8x - 40) \div (x - 5)$

c. $\dfrac{9x - 6 + 6x^2}{1 - 2x}$

Solution

a.
$$\begin{array}{r} 2x \\ 3x - 1 \overline{)6x^2 + 13x - 5} \\ \underline{6x^2 - 2x} \\ 15x - 5 \end{array}$$

Divide $6x^2$ by $3x$, getting $2x$ in the quotient.
Multiply $2x$ by $(3x - 1)$, getting $6x^2 - 2x$.
Subtract $(6x^2 - 2x)$ from $(6x^2 + 13x)$,
getting $15x$, and then bring down -5.

$$\begin{array}{r} 2x + 5 \\ 3x - 1 \overline{)6x^2 + 13x - 5} \\ \underline{6x^2 - 2x} \\ 15x - 5 \\ \underline{15x - 5} \\ 0 \end{array}$$

Divide $15x$ by $3x$, getting 5 in the quotient.
Multiply 5 by $(3x - 1)$, getting $15x - 5$.
Subtract $(15x - 5)$ from $(15x - 5)$,
getting 0.

So $(6x^2 + 13x - 5) \div (3x - 1) = 2x + 5$. Note that each term in the quotient is above a term in the dividend of the same degree.

b.
$$\begin{array}{r} 3x + 7 \\ x - 5 \overline{)3x^2 - 8x - 40} \\ \underline{3x^2 - 15x} \\ 7x - 40 \\ \underline{7x - 35} \\ -5 \end{array}$$

$-5 \longleftarrow$ Remainder

PRACTICE 2

Find the quotient and check.

a. $(3x^2 - 19x - 14) \div (3x + 2)$

b. $(2x^2 - 11x + 5) \div (x - 6)$

c. $\dfrac{8n^2 - 3 + 10n}{1 - 4n}$

The remainder cannot be divided by the divisor, and so we write the answer as $3x + 7 + \dfrac{-5}{x - 5}$.

c. Before finding the quotient $\dfrac{9x - 6 + 6x^2}{1 - 2x}$, we place the terms in both the divisor and the dividend in descending order. Here, we need to rearrange the terms in both the dividend and the divisor.

$$
\begin{array}{r}
-3x - 6 \\
-2x + 1 \overline{)6x^2 + 9x - 6} \\
\underline{6x^2 - 3x} \\
12x - 6 \\
\underline{12x - 6} \\
0
\end{array}
$$

So $(6x^2 + 9x - 6) \div (1 - 2x) = -3x - 6$.

When dividing polynomials, repeat the division until the degree of the remainder is less than the degree of the divisor. The dividend or the divisor may have *missing terms*, as shown in the following example:

EXAMPLE 3

Divide: $x^2 + 1\overline{)4x^3 + x^2 + 2}$

Solution Since there is no x-term in either the dividend or the divisor, we can insert $0x$ as a placeholder for each missing term.

$$
\begin{array}{r}
4x + 1 \\
x^2 + 0x + 1 \overline{)4x^3 + x^2 + 0x + 2} \\
\underline{4x^3 + 0x^2 + 4x} \\
x^2 - 4x + 2 \\
\underline{-\,x^2 + 0x + 1} \\
-4x + 1 \quad \longleftarrow \text{Remainder}
\end{array}
$$

So $\dfrac{4x^3 + x^2 + 2}{x^2 + 1} = 4x + 1 + \dfrac{-4x + 1}{x^2 + 1}$.

Note that in Example 3, the division process ends because the degree of the remainder $-4x + 1$ is 1, which is less than 2, the degree of the divisor $x^2 + 1$.

EXAMPLE 4

If $f(x) = x^2 - 4x + 2$ and $g(x) = x - 2$, find $\dfrac{f(x)}{g(x)}$.

Solution

$$
\dfrac{f(x)}{g(x)} = \dfrac{x^2 - 4x + 2}{x - 2} = x - 2 + \dfrac{-2}{x - 2}
$$

PRACTICE 3

Divide: $\dfrac{2n^3 + 5n - 1}{n^2 - 3}$

PRACTICE 4

Suppose that $f(x) = 4x^2 + 5$ and $g(x) = 2x - 1$. Find $\dfrac{f(x)}{g(x)}$.

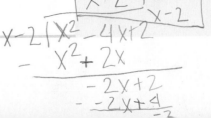

EXAMPLE 5

If $10 is invested at an interest rate of r percent per year (in decimal form), compounded annually, the value of the investment S in dollars at the end of the nth year is given by the following equation:

$$S = 10(1 + r)^n$$

How many times greater is the future value of the investment after 2 yr as compared with the value after 1 yr?

Solution   After 2 yr, the value of the investment is given by $10(1 + r)^2$, which can be written $10(1 + 2r + r^2)$ or $10r^2 + 20r + 10$. After only 1 yr, its value is $10(1 + r)$, or $10r + 10$. So the ratio of the value after 2 yr to the value after 1 yr can be represented by $\dfrac{10r^2 + 20r + 10}{10r + 10}$. To find this quotient, we divide:

$$
\begin{array}{r}
r + 1 \\
10r + 10{\overline{\smash{\big)}\,10r^2 + 20r + 10}} \\
\underline{10r^2 + 10r} \\
10r + 10 \\
\underline{10r + 10} \\
0
\end{array}
$$

Therefore, the value of the investment after 2 yr is $(r + 1)$ times the value after 1 yr.

PRACTICE 5

A *geometric series* is a sum of terms where each term after the first is formed by multiplying the previous term by the same quantity. For example, the series $5 + 5r + 5r^2$ has three terms, where the first term is 5 and each of the other terms is r times the previous term. Check by dividing that the sum of the first three terms can be calculated from the formula $\dfrac{5 - 5r^3}{1 - r}$.

Can you explain how we could have predicted the result in Example 5?

Mathematically Speaking

Fill in each blank with the most appropriate term or phrase from the given list.

divisor	remainder	missing terms
leading terms	dividend	quotient

1. In the division problem $(-8x^2 + 26x - 19) \div (2x - 3) = (-4x + 7) + \dfrac{2}{2x - 3}$, $(-8x^2 + 26x - 19)$ is called the _____.

2. In the division problem $(x^2 + 5x - 1) \div (x - 1)$, $(x - 1)$ is called the _____.

3. When dividing polynomials with _____, placeholders are inserted.

4. In polynomial division, when the degree of the _____ is less than the degree of the divisor, the division process ends.

Ⓐ *Find the quotient.*

5. $\dfrac{6x^4}{2x^3}$

6. $\dfrac{10n^6}{5n^2}$

7. $\dfrac{20y^5}{-4y^3}$

8. $\dfrac{-32x^3}{8x^2}$

9. $\dfrac{-9x^5y^2}{-3xy}$

10. $\dfrac{42a^2b^9}{6a^2b^4}$

11. $(6x^3 + 9x^2) \div (3x)$

12. $(15x^4 - 25x) \div (5x)$

13. $\dfrac{24n^3 - 10n^2 + 4n}{-4}$

14. $\dfrac{16y^3 - 8y^2 - 12}{-8}$

15. $(54a^5 - 6a^4 + 36a^3) \div (6a^3)$

16. $(4x^6 + 10x^4 - 2x^3) \div (-2x^3)$

17. $\dfrac{16t^4 + 10t^3 - 18t^2 - 8t}{-2t^2}$

18. $\dfrac{30a^4 + 12a^3 - 24a^2 - 18a}{6a^2}$

19. $\dfrac{8p^2 - 16pq + 28q^2}{4q^2}$

20. $\dfrac{3x^2 - 9xy - 72y^2}{9y^2}$

21. $\dfrac{9x^3y + 6x^2y^2 - 12xy^3}{-3xy}$

22. $\dfrac{5p^4q - 20p^2q^2 - 25pq^3}{-5pq}$

23. $\dfrac{8a^4b^3 - 4a^3b^2 + a^2b - 12ab^2}{4a^2b^2}$

24. $\dfrac{10x^3y^2 - 5x^2y^3 + 20x^2y^4}{10x^2y^3}$

Ⓑ *Divide.*

25. $(x^2 - 4x + 4) \div (x - 2)$

26. $(x^2 + x - 12) \div (x + 4)$

27. $\dfrac{n^3 + 5n^2 - n - 5}{n + 1}$

28. $\dfrac{x^3 - 2x^2 - 9x + 18}{x - 3}$

29. $\dfrac{11 - 39x - 8x^2 + 3x^3}{5 - x}$

30. $\dfrac{-5n^2 - 3n - 17 + 2n^3}{2 - n}$

31. $\dfrac{5x^2 - 11x - 12}{5x + 4}$

32. $\dfrac{2x^2 + x - 28}{2x - 7}$

33. $(12 + 5n^2 - 3n^3) \div (3n + 4)$

34. $(13t^2 - 6t^3 - 8t + 4) \div (6t - 1)$ **35.** $(4x^4 - 7x^2 + x - 3) \div (2x + 3)$ **36.** $(9n^3 - 3n^2 + 4) \div (3n + 2)$

37. $(13a^2 - 5a - 7 + 6a^3) \div (1 - 3a)$

38. $(-8x^4 + 20x + 2x^3 - 6x^2 + 8) \div (5 - 4x)$

39. $\dfrac{4y^5 - 7y^4 + 7y^3 - 7y^2 + 7y - 3}{4y - 3}$

40. $\dfrac{-5x^5 - x^4 + x^3 - x^2 + x - 6}{5x + 6}$

41. $\dfrac{8x^3 + 1}{2x + 1}$

42. $\dfrac{27x^3 - 1}{3x - 1}$

43. $\dfrac{x^4 - x^3 + 5x^2 - 3x + 6}{x^2 + 3}$

44. $\dfrac{x^4 + 4x^3 - 3x^2 - 8x + 2}{x^2 - 2}$

45. $\dfrac{2x^3 - x^2 + 5}{x^2 - 4}$

46. $\dfrac{3n^3 - 2n^2 + 10n - 4}{n^2 + 5}$

47. $\dfrac{p^4 - 16}{p^2 + 4}$

48. $\dfrac{a^4 - 1}{a^2 - 1}$

49. $\dfrac{n^3 + 2}{n^2 - 2n + 3}$

50. $\dfrac{x^3 + 10x}{x^2 + 3x + 1}$

Given $f(x)$ and $g(x)$, find $\dfrac{f(x)}{g(x)}$.

51. $f(x) = x^2 - 5x - 24$ and $g(x) = x - 8$

52. $f(x) = x^2 + 13x + 42$ and $g(x) = x + 7$

53. $f(x) = 4x^3 - 9x + 11$ and $g(x) = 2x + 3$

54. $f(x) = 9x^3 - 4x + 5$ and $g(x) = 3x - 2$

Mixed Practice

Solve.

55. Given $f(x) = x^2 - 3x - 28$ and $g(x) = x + 4$, find $\dfrac{f(x)}{g(x)}$.

56. Simplify $\dfrac{48a^7}{16a^3}$.

Find the quotient.

57. $(15y^4 - 45y^3 + 27y^2) \div (-3y^3)$

58. $\dfrac{12x^2 + 20xy - 16xy^2}{4y^2}$

Divide.

59. $\dfrac{-2x^2 - 15x + 5 + 2x^3}{3 - x}$

60. $(8x^2 + 14x - 15) \div (4x - 3)$

61. $(10n^3 + 11n^2 + 9) \div (2n + 3)$

62. $\dfrac{a^4 - 2a^3 - a^2 - 6a - 12}{a^2 + 3}$

Applications

C *Solve.*

63. A long-distance phone service charges $7.95 per month plus $0.05 per minute for long-distance phone calls. The average cost per minute of a phone call (in dollars) is given by the expression $\dfrac{7.95 + 0.05x}{x}$, where x is the total number of monthly calling minutes.

a. Use division to rewrite this expression.

b. Use the expression from part (a) to find the average cost per minute, to the nearest cent, if 5 hr of long-distance calls are made in a month.

64. A travel agency's average profit (in dollars) per ski-vacation package sold is given by the expression $\dfrac{1500 + 228x - 4x^2}{x}$, where x is the number of packages sold.

a. Use division to rewrite this expression.

b. Use the expression from part (a) to find the travel agency's average profit per package if 50 ski-vacation packages are sold.

65. A company gives its employees an annual salary increase of p percent. An employee is offered a starting salary of $30,000. His cumulative earnings after 3 yr is given by the expression $\dfrac{30,000(1 - r^3)}{1 - r}$, where r represents $(100 + p)$ percent expressed in decimal form.

a. Use long division to simplify this expression.

b. Use the expression from part (a) to determine the cumulative amount the employee will earn over 3 yr if the annual increase is 4%.

66. A factory manufactures cardboard boxes by cutting x-inch by x-inch squares from the corners of a rectangular piece of cardboard. The volume of a box made from a particular piece of cardboard with length 72 in. can be calculated using the expression $4x^3 - 240x^2 + 3456x$.

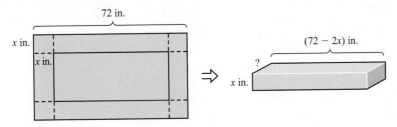

a. Use long division to find the width of the box in terms of x.

b. What is the area of the piece of cardboard that was used to make this box?

• Check your answers on page A-38.

MINDStretchers

Patterns

1. Evaluate each polynomial for the given value of x. Then, divide the polynomial by $x - a$, where a is the given value of x.

   a. $x^2 + 5x + 6$ $x = 2$
   b. $x^2 - 7x - 4$ $x = -3$
   c. $2x^2 + 2x - 1$ $x = -1$
   d. $x^3 + x^2 - 3x + 8$ $x = 1$
   e. $3x^3 + 8x^2 - 8$ $x = -2$

      i. What do you notice about the value of the polynomial for $x = a$ and the numerator of the remainder when the polynomial is divided by $x - a$?

      ii. What does this pattern suggest about the remainder of a polynomial when the polynomial is divided by $x - a$?

      iii. Using the pattern observed above, find the remainder when $5x^4 - x^3 + 3x - 10$ is divided by $x - 1$. Divide to confirm your answer.

Critical Thinking

2. Find a constant k so that the remainder is -4 when $4x^3 - 8x^2 + x + k$ is divided by $2x - 3$.

Mathematical Reasoning

3. Divide $6x^{4n} - 5x^{3n} + 3x^{2n} + 7x^n - 3$ by $2x^n - 1$.

What Factoring Is and Why It Is Important

In Section 5.2, we discussed how to multiply two polynomial factors in order to find their product.

Multiplying

$$\underbrace{(2x + 3)}_{\text{Factor}}\underbrace{(3x - 4)}_{\text{Factor}} = \underbrace{6x^2 + x - 12}_{\text{Product}}$$

Now, let's reverse the process by beginning with a polynomial and expressing it as a product of factors. Rewriting a polynomial as a product is called *factoring* the polynomial.

Factoring

$$\underbrace{6x^2 + x - 12}_{\text{Polynomial}} = \underbrace{(2x + 3)}_{\text{Factor}}\underbrace{(3x - 4)}_{\text{Factor}}$$

Factoring polynomials helps us to solve various types of equations as well as to simplify algebraic expressions. In this section, we discuss two types of factoring—factoring out the greatest common factor and factoring by grouping.

Factoring Out the Greatest Common Factor

The *greatest common factor* (or *GCF*) of two or more integers is the largest integer that is a factor of each integer. For instance, 12 is the greatest common factor of 24 and 36. We can extend the concept of greatest common factor to monomials. For instance, to find the GCF of $-18x^2y^3$ and $15x^4$, we begin by writing the monomials in factored form.

$$-18x^2y^3 = -1 \cdot 2 \cdot 3 \cdot 3 \cdot x \cdot x \cdot y \cdot y \cdot y$$
$$15x^4 = 3 \cdot 5 \cdot x \cdot x \cdot x \cdot x$$

From the factorizations we see that each monomial has one factor of 3 and two factors of x in common. So the greatest common factor of $-18x^2y^3$ and $15x^4$ is $3x^2$. Note that 3 is the greatest common factor of the coefficients and that x^2 is the highest power of the variable that is common to both monomials.

> **DEFINITION**
>
> The **greatest common factor (GCF)** of two or more monomials is the product of the greatest common factor of the coefficients and the highest powers of the variables common to all of the monomials.

When the terms of a polynomial have common factors, we can use the distributive property in reverse to factor the polynomial. Consider, for instance, the polynomial $15x^3 + 10x^2$. The GCF of the terms $15x^3$ and $10x^2$ is $5x^2$. So we can write the polynomial in factored form by dividing out the GCF of the terms.

$$15x^3 + 10x^2 \longrightarrow 5x^2 \cdot 3x + 5x^2 \cdot 2 \longrightarrow \underset{\underset{\text{GCF}}{\uparrow}}{5x^2} \underbrace{(3x + 2)}_{\substack{\text{Binomial} \\ \text{factor}}}$$

So the factored form of $15x^3 + 10x^2$ is $5x^2(3x + 2)$.

Note that unless otherwise stated, we assume that all coefficients of expressions to be factored are integers so that we can find the GCF of these coefficients.

EXAMPLE 1

Factor out the GCF.

a. $6x^3 + 4x$ **b.** $20n^2 - 5n + 10$ **c.** $7y - 6$

Solution

a. $6x^3 + 4x = 2x(3x^2) + 2x(2)$ Factor out the GCF $2x$ from each term.
$\qquad\qquad = 2x(3x^2 + 2)$ Use the distributive property.

So the *factorization* of $6x^3 + 4x$ is $2x(3x^2 + 2)$.

b. $20n^2 - 5n + 10 = 5(4n^2) - 5(n) + 5(2)$ Factor out the GCF 5 from each term.

$\qquad\qquad\qquad = 5(4n^2 - n + 2)$

c. The polynomial $7y - 6$ is not factorable.

PRACTICE 1

Factor out the greatest common factor.

a. $12y^2 + 8y$

b. $9x^2 + 12x - 6$

c. $5a^2 + 2b$

$2x(3x^2 + 2)$

In the next example, we factor out the GCF of a polynomial expression in more than one variable.

EXAMPLE 2

Factor.

a. $-10a^2 - 20ab$ **b.** $8x^3y^2 + 12x^2y^3 - 24xy^4$

Solution

a. $-10a^2 - 20ab = 10a(-a) - 10a(2b)$ Factor out the GCF $10a$ from each term.

$\qquad\qquad\qquad = 10a(-a - 2b)$

b. $8x^3y^2 + 12x^2y^3 - 24xy^4 = 4xy^2(2x^2) + 4xy^2(3xy) - 4xy^2(6y^2)$
$\qquad\qquad\qquad\qquad = 4xy^2(2x^2 + 3xy - 6y^2)$

PRACTICE 2

Factor out the greatest common factor.

a. $-14a^2b - 21a$

b. $6a^4b - 36a^3b^2 + 12ab^3$

$-10(a)(a + 2b)$

Note that it follows from the definition of the GCF of a polynomial that every GCF has a positive coefficient. However, sometimes it is useful to factor out a common factor with a negative coefficient. For instance, instead of factoring out $10a$ from $-10a^2 - 20ab$ in Example 2(a), we could have factored out $-10a$. If we had done this, we would have gotten $(-10a)(a) + (-10a)(2b) = -10a(a + 2b)$.

Recall our discussion of solving literal equations in Section 2.2. Solving some literal equations requires factoring out the greatest common factor.

EXAMPLE 3

Solve $y + bm = mx + a$ for m in terms of y, b, x, and a.

Solution To solve for m, we bring all the terms involving m to the left side of the equation and the other terms to the right side of the equation.

$$y + bm = mx + a$$
$$bm - mx = a - y$$
$$m(b - x) = a - y$$
$$m = \frac{a - y}{b - x}$$

PRACTICE 3

Solve for l: $S = 2lw + 2lh + 2wh$

Factoring by Grouping

For some polynomials, the greatest common factor is not a monomial, but a binomial instead.

EXAMPLE 4

Factor: $7x(x - 3) + 2(x - 3)$

Solution The GCF of the polynomial is the binomial $(x - 3)$, which we can factor out.

$$7x(x - 3) + 2(x - 3) = (x - 3)(7x + 2)$$

So $(x - 3)(7x + 2)$ is the polynomial in factored form.

PRACTICE 4

Factor: $(a - 2)(6b) + (a - 2)7$

As we have seen in Example 4, the GCF of a polynomial may be a binomial expression. We can then factor out this binomial using the distributive property. However, in some algebraic expressions, when trying to factor out a common binomial term we must rewrite one of the binomials as the product of -1 and the opposite binomial. For example, we can rewrite the binomial $(2 - a)$ as $-1(a - 2)$, as shown in the next example.

EXAMPLE 5

Factor: $x(a - 2) + 4(2 - a)$

Solution The binomial factors $(a - 2)$ and $(2 - a)$ are opposites, so we factor out -1 from the binomial $(2 - a)$ and rewrite the original expression.

$$\begin{aligned}
x(a - 2) + 4(2 - a) &= x(a - 2) + 4[-1(a - 2)] && \text{Factor out } -1 \\
&&& \text{from } (2 - a). \\
&= x(a - 2) - 4(a - 2) && \text{Simplify.} \\
&= (a - 2)(x - 4) && \text{Use the} \\
&&& \text{distributive} \\
&&& \text{property.}
\end{aligned}$$

PRACTICE 5

Factor: $4p(n - 1) + 3(1 - n)$

When trying to factor a polynomial that has four terms, it may be possible to group pairs of terms in such a way that a common binomial factor can be found. This method is called **factoring by grouping**.

EXAMPLE 6

Factor by grouping.

a. $y^2 + 7y - xy - 7x$ **b.** $2h - 2k - h^4 + h^3k$

Solution

a. $y^2 + 7y - xy - 7x$

$$\begin{aligned}
&= (y^2 + 7y) + (-xy - 7x) && \text{Group the first two terms and the last} \\
&&& \text{two terms.} \\
&= (y^2 + 7y) - (xy + 7x) && \text{Factor out } -1 \text{ in the second group.} \\
&&& -xy - 7x = -(xy + 7x) \\
&= y(y + 7) - x(y + 7) && \text{Factor out the GCF from each group.} \\
&= (y + 7)(y - x) && \text{Write in factored form.}
\end{aligned}$$

PRACTICE 6

Factor by grouping.

a. $5a^2 + 15a - 2ab - 6b$

b. $10y - 10z - y^5 + y^4z$

EXAMPLE 6 (continued)

b. $2h - 2k - h^4 + h^3k$

$= (2h - 2k) + (-h^4 + h^3k)$ Group the first two terms and the last two terms.

$= (2h - 2k) - (h^4 - h^3k)$ $-h^4 + h^3k = -(h^4 - h^3k)$

$= 2(h - k) - h^3(h - k)$ Factor out the GCF from each group.

$= (h - k)(2 - h^3)$ Use the distributive property.

EXAMPLE 7

We learn in physics that the momentum of an object is the product of its mass and its velocity. So when an object with mass m increases in velocity from v_1 to v_2, its momentum increases by $mv_2 - mv_1$. Factor this expression.

Solution

$$mv_2 - mv_1 = m(v_2 - v_1)$$

PRACTICE 7

Recall that a polygon is a many-sided geometric figure. The number of diagonals of a polygon with n sides is given by the expression $\frac{1}{2}n^2 - \frac{3}{2}n$.

Write this expression in factored form.

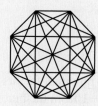

Mathematically Speaking

Fill in each blank with the most appropriate term or phrase from the given list.

using the GCF	negative	binomial
opposite binomial	sum	common multiples
common factors	grouping	highest
lowest	product	positive

1. To factor a polynomial is to write it as a _____.

2. The greatest common factor of two or more monomials is the product of the greatest common factor of the coefficients and the _____ powers of the variables common to all of the monomials.

3. The distributive property in reverse can be used to factor a binomial when its terms have _____.

4. Every GCF has a _____ coefficient.

5. When factoring out a common binomial term from the algebraic expression $3(x - 2) + x(2 - x)$, write one of the binomials as the product of -1 and the _____.

6. Four-term polynomials such as $16xy + 4y + 4x + 1$ may be factored by _____.

A *Factor out the greatest common factor.*

7. $2a^3 + 8a$

8. $5x^2 - 20x$

9. $32 - 40a + 24a^2$

10. $60n^4 - 45n^2 - 30$

11. $3x^4 + 7x^3 - 9x^2$

12. $12p^5 - 4p^4 - 24p$

13. $6n^3 - 4n^2 + 3$

14. $9x^2 - 28y^2$

15. $-15a^5 + 9a^4 - 18a^3$

16. $12p^3q + 21q^2$

17. $6x^2y^2 + 18xy^3 + 36y^4$

18. $14a^5 - 28a^4b - 35a^3b^2$

19. $-27p^6q^2 - 45p^5q^5 - 36p^4q^3 + 72p^3q^4$

20. $-48x^5y + 12x^3y^3 - 24x^3y^4 + 60x^2y^2$

21. $-a^3b^2 - 3a^2b^4c - abc^2$

22. $-2p^2q^2r^2 - 6pqr^3 - 18q^3r^2$

23. $3x(x - 5) + 2(x - 5)$

24. $4a(a - 1) + 3(a - 1)$

25. $(2x - 3y)(3x) - (2x - 3y)(2y)$

26. $5p(6p + q) - 2q(6p + q)$

27. $x(x + y)^2 - y(x + y)^2$

28. $3x(x - y)^2 + y(x - y)^2$

29. $x^2(x + 2y^2) - 3y(x + 2y^2) + (x + 2y^2)$

30. $7a^3(a - b^2) + 8b(a - b^2) - 5(a - b^2)$

Solve for the indicated variable.

31. $A = S + St$, for S

32. $p = l - rl$, for l

33. $A = \frac{1}{2}b_1h + \frac{1}{2}b_2h$, for h

34. $s = \frac{1}{2}v_1t + \frac{1}{2}v_2t$, for t

B *Factor by grouping.*

35. $x(y - 7) + 10(7 - y)$

36. $p(q - 1) + 5(1 - q)$

37. $4x(z - y) - (y - z)$

38. $8b(c - a) - (a - c)$

39. $(2x - y) + 2x(y - 2x)$

40. $(4s - r) - 4t(r - 4s)$

41. $ab + 5a + cb + 5c$

42. $xy + 7y + xz + 7z$

43. $x^2 - 9x + 2x - 18$

44. $x^2 - 3x + 4x - 12$

45. $9xy - 3y + 3x - 1$

46. $16ab + 4b + 4a + 1$

47. $10st - 6s - 25t + 15$

48. $8uv + 20v - 14u - 35$

49. $14uw + 15vw - 28ux - 30vx$

50. $6pr + 7ps - 12qr - 14qs$

51. $2a^2 - 4ab + ab - 2b^2$

52. $3p^2 - 9pq + pq - 3q^2$

53. $6x^2 + 3xy - 2xz - yz$

54. $20a^2 - 5ab + 8ac - 2bc$

55. $x^3 - xy + y^2 - x^2y$

56. $a^2 - ab - b^3 + ab^2$

57. $12c^3 + 4cd^2 + 9c^2d + 3d^3$

58. $18y^3 + 24xy^2 + 3x^2y + 4x^3$

59. $3ab - 8c + 4bc - 6a$

60. $2xy - 12z + 8x - 3yz$

61. $4p^2 - p^2r - q^3r + 4q^3$

62. $3x^2 - x^2y - yz^2 + 3z^2$

Mixed Practice

Solve for the indicated variable.

63. Solve $M = \frac{1}{2}pq + \frac{1}{2}pr$ for p.

Factor by grouping.

64. $6a^2 - 8ab + 3ac - 4bc$ **65.** $3x(y - z) - (z - y)$ **66.** $a^2 - 5a + 3a - 15$ **67.** $6rs - 14s + 15r - 35$

Factor out the greatest common factor.

68. $2a(2a - b) - 3b(2a - b)$

69. $-3xy^2z - 9x^2yz^2 + 6xyz$

70. $20x^4 - 10x^2 - 15x$

Applications

C *Solve.*

71. The height above the ground (in feet) of a stone t sec after it is dropped from a bridge 720 ft above ground is given by the polynomial $720 - 16t^2$.

   a. Factor the polynomial.

   b. Use the factored form in part (a) to find the height of the stone 5 sec after it is dropped.

72. A computer manufacturer estimates that the weekly cost to produce n computers is $(9000 + 450n)$ dollars.

   a. Factor this expression.

   b. Use the factored form in part (a) to determine the cost of producing 120 computers.

73. The area (in square meters) of an Olympic-size swimming pool is given by the expression $l^2 - 25l$, where l is the length of the pool (in meters).

   a. Factor the expression for the area.

   b. The width of an Olympic-size swimming pool is 25 m. What is the length? Explain how you know.

74. A lottery winner invests P dollars in an account at interest rate r (in decimal form), compounded annually. The cumulative amount of interest earned over 3 yr is given by the expression $3Pr + 3Pr^2 + Pr^3$.

   a. Factor the expression.

   b. After 3 yr the account earned a total of $6305 in interest. If the interest rate was 5%, how much was initially invested?

75. After 2 yr, the total amount of money in an account that pays interest rate r (in decimal form), compounded annually, is given by $P + Pr + (P + Pr)r$. Factor to show that the given expression can be written as $P(1 + r)^2$.

76. A department store has a sale advertising a certain percentage off on all items. A customer buys one item that regularly sells for x dollars and another item that regularly sells for y dollars. The total purchase price of the two items is given by $Px + Tx + Py + Ty$, where P is the percent (in decimal form) of the regular price that the customer paid and T is the sales tax rate (in decimal form) applied before the discount. Factor this expression.

77. After t yr, an investment of $\$P$ growing at an annual simple interest rate r is worth $\$A$. The formula $A = P + Prt$ expresses A in terms of the other variables.

   a. Solve this formula for P.

   b. An investment earning 5% simple interest is worth $2400 after 4 yr. Use the result from part (a) to determine the amount of money initially invested.

78. The surface area S of a box with no top is given by the formula $S = lw + 2lh + 2wh$, where w, l, and h represent the width, length and height of the box, respectively.

   a. Solve this formula for l.

   b. Use the result in part (a) to determine the length of a topless box constructed from $9\frac{1}{2}$ ft^2 of cardboard if the width of the box is 2 ft and the height is $\frac{1}{2}$ ft.

• Check your answers on page A-38.

MINDStretchers

Critical Thinking

1. Consider the monomials ax^n and bx^m. What must be true about the coefficients and the exponents of these two monomials if their GCF is bx^n?

Mathematical Reasoning

2. Factor the polynomial $24x^{n-1}y^{n+1} - 36x^ny^{n+2} + 48x^{n-2}y^{n+3}$.

Historical

3. Euclid, a Greek mathematician, developed a method that can be used to find the greatest common factor of two positive integers m and n, where $m \geq n$. This method, appropriately called the Euclidean algorithm, works by finding and using the remainders of a series of divisions. The first step is to divide m by n and find the remainder r_1. Next, divide n by r_1 and find the remainder r_2. Then, divide r_1 by r_2 and find the remainder. Continue the process of dividing the remainder of the previous step by the new remainder until a remainder of 0 is reached. The last nonzero remainder found is the GCF. For example, the GCF of 352 and 224 is found as follows:

$m \div n \longrightarrow \quad 352 \div 224 = 1$ remainder $128 \longleftarrow r_1$

$n \div r_1 \longrightarrow \quad 224 \div 128 = 1$ remainder $96 \longleftarrow r_2$

$r_1 \div r_2 \longrightarrow \quad 128 \div 96 = 1$ remainder $32 \longleftarrow r_3$

$r_2 \div r_3 \longrightarrow \quad 96 \div 32 = 3$ remainder $\mathbf{0} \longleftarrow$ *Remainder is 0.*

The last nonzero remainder is 32, which is therefore the GCF of 352 and 224.

Use the Euclidean algorithm to find the GCF of each pair of numbers.

a. 208 and 96

b. 456 and 312

c. 1116 and 828

5.5 Factoring Trinomials

In this section, we discuss factoring trinomials. We first consider how to factor trinomials of the form $x^2 + bx + c$, where the leading coefficient is equal to 1. Then, we move on to the more complex case of factoring trinomials of the form $ax^2 + bx + c$, where the leading coefficient a is not 1.

A To factor a trinomial of the form $x^2 + bx + c$

B To factor a trinomial of the form $ax^2 + bx + c$, $a \neq 1$

C To solve applied problems involving factoring

Factoring $x^2 + bx + c$

Recall from Section 5.2 the FOIL method of multiplying two binomials.

$$
\begin{array}{cccc}
\text{F} & \text{O} & \text{I} & \text{L}
\end{array}
$$
$$(x + 4)(x + 3) = x^2 + 3x + 4x + 12$$
$$= x^2 + 7x + 12$$

Note that when multiplying these two binomials, we get a trinomial whose leading coefficient is 1.

The factorization of the trinomial $x^2 + bx + c$ is the product of two binomials of the form $(x + ?)(x + ?)$, where each question mark represents a number. To factor a trinomial, we apply the FOIL method in reverse. That is, we look for two binomials whose product is the trinomial. This method of factoring, sometimes called the **trial-and-error method**, involves listing all the possible combinations of factors to find the correct factorization. First, we consider a trinomial in which the constant term is positive.

EXAMPLE 1

Factor: $x^2 + 8x + 15$

Solution Applying the FOIL method in reverse, we know that the first term of each factor is x, that is, the factorization is $(x + ?)(x + ?)$. Since both the constant term and the coefficient of the x-term are positive, we need to find two positive integers whose product is 15 and whose sum is 8. The following table shows how to test pairs of factors of 15, the constant term of the trinomial:

Factors of 15	Possible Binomial Factors	Sum of Outer and Inner Products
1, 15	$(x + 1)(x + 15)$	$15x + x = 16x$
3, 5	$(x + 3)(x + 5)$	$5x + 3x = 8x \leftarrow$ Correct middle term

So $x^2 + 8x + 15 = (x + 3)(x + 5)$, or $(x + 5)(x + 3)$.

Check

We can confirm that the factors are correct by multiplying.

$$(x + 3)(x + 5) = x^2 + \underbrace{5x + 3x}_{} + 15 = x^2 + \underset{\uparrow}{8x} + 15$$

Sum of the outer and inner products   Middle term

PRACTICE 1

Factor: $x^2 + 3x + 2$

EXAMPLE 2

Factor: $x^2 - 5x + 6$

Solution In this trinomial, the constant term is positive and the coefficient of the x-term is negative. So we need to find two negative integers whose product is 6 and whose sum is -5.

Factors of 6	Possible Binomial Factors	Sum of Outer and Inner Products	
$-1, -6$	$(x - 1)(x - 6)$	$-6x - x = -7x$	
$-2, -3$	$(x - 2)(x - 3)$	$-3x - 2x = -5x \leftarrow$	Correct middle term

So $x^2 - 5x + 6 = (x - 2)(x - 3)$.

PRACTICE 2

Factor: $x^2 - 10x + 9$

TIP When the constant term c of the trinomial $x^2 + bx + c$ is positive, the constant terms of the binomial have the same sign.
- The constant terms are both positive when b, the coefficient of the x-term, is positive.
- The constant terms are both negative when b is negative.

Next, we consider a trinomial in which the constant term is negative.

EXAMPLE 3

Factor: $n^2 + 2n - 8$

Solution In this trinomial, the constant term is negative and the coefficient of the n-term is positive. We must find two integers whose product is -8 and whose sum is 2. One integer will be positive and the other negative.

Factors of -8	Possible Binomial Factors	Sum of Outer and Inner Products	
$1, -8$	$(n + 1)(n - 8)$	$-8n + n = -7n$	
$-1, 8$	$(n - 1)(n + 8)$	$8n + (-n) = 7n$	
$2, -4$	$(n + 2)(n - 4)$	$-4n + 2n = -2n$	
$-2, 4$	$(n - 2)(n + 4)$	$4n + (-2n) = 2n \leftarrow$	Correct middle term

So $n^2 + 2n - 8 = (n - 2)(n + 4)$.

PRACTICE 3

Factor: $p^2 - 3p - 28$

TIP When the constant term c of the trinomial $x^2 + bx + c$ is negative, then one binomial factor has a positive constant term and the other has a negative constant term.
- The positive constant term has the larger absolute value when b is positive.
- The negative constant term has the larger absolute value when b is negative.

It is important to note that not every trinomial can be expressed as the product of binomial factors with coefficients that are integers. For instance, the trinomial $x^2 + 5x + 1$ cannot be expressed as the product of two binomials with integer coefficients since there is no pair of integers whose product is 1 and whose sum is 5. Polynomials that are not factorable are called **prime polynomials**.

Now, let's consider a trinomial of the form $x^2 + bxy + cy^2$. This trinomial contains more than one variable. In applying the trial-and-error method to factor these trinomials, we consider binomial factors of the form $(x + ?y)(x + ?y)$, where the question marks represent two numbers whose product is c, the coefficient of the y^2-term. The sum of the factors of c is b, the coefficient of the xy-term.

EXAMPLE 4

Factor: $x^2 - 5xy - 24y^2$

Solution Since the y^2-term is negative and the xy-term is negative, we look for two integers, one positive and the other negative, whose product is -24. Since the sum -5 is negative, the negative factor must have the larger absolute value.

Factors of -24	Possible Binomial Factors	Sum of Outer and Inner Products	
1, −24	$(x + y)(x - 24y)$	$-24xy + xy = -23xy$	
2, −12	$(x + 2y)(x - 12y)$	$-12xy + 2xy = -10xy$	Correct
3, −8	$(x + 3y)(x - 8y)$	$-8xy + 3xy = -5xy$	← middle
4, −6	$(x + 4y)(x - 6y)$	$-6xy + 4xy = -2xy$	term

So $x^2 - 5xy - 24y^2 = (x + 3y)(x - 8y)$.

PRACTICE 4

Factor: $p^2 + 3pq - 40q^2$

Note that in any of the previous examples, once we find the correct middle term, we can stop testing possible factors. In Example 4, we could have stopped the process of testing possible factors after finding the correct middle term.

Some polynomials have a common factor. When factoring such a polynomial, we begin by looking for its GCF, as discussed in the previous section. The remaining factor sometimes can still be factored. A polynomial is said to be *factored completely* when it is expressed as the product of a monomial and one or more prime polynomials. In the remainder of this text, "to factor" means to factor completely.

EXAMPLE 5

Factor.

a. $2x^2 + 14x - 36$ **b.** $-x^2 + 12x - 20$

Solution

a. $2x^2 + 14x - 36 = 2(x^2 + 7x - 18)$ Factor out the GCF 2 from each term.

$\qquad\qquad\qquad\quad = 2(x - 2)(x + 9)$ Factor $x^2 + 7x - 18$.

So $2x^2 + 14x - 36 = 2(x - 2)(x + 9)$. Note that after factoring out the GCF 2, neither of the remaining binomial factors of the trinomial has a common factor.

b. $-x^2 + 12x - 20 = -1(x^2 - 12x + 20)$ Factor out -1 so that the leading coefficient is 1.

$\qquad\qquad\qquad\quad = -1(x - 2)(x - 10)$ Factor $x^2 - 12x + 20$.

So $-x^2 + 12x - 20 = -(x - 2)(x - 10)$.

PRACTICE 5

Write in factored form.

a. $y^3 - 7y^2 - 8y$

b. $-n^2 + 3n + 10$

Some trinomials that are not quadratic can be rewritten as quadratic expressions by making an appropriate substitution. Consider the following example:

EXAMPLE 6

Factor: $x^4 + 6x^2 + 5$

Solution   Note that in the expression $x^4 + 6x^2 + 5$, both variable terms have even exponents, namely 4 and 2. This suggests rewriting the expression as $(x^2)^2 + 6x^2 + 5$. If we then give a name to the quantity x^2, say u, we can substitute u for x^2 and factor the resulting expression in terms of u.

$$(x^2)^2 + 6x^2 + 5 = u^2 + 6u + 5 \qquad \text{Substitute } u \text{ for } x^2.$$
$$= (u + 1)(u + 5) \qquad \text{Factor } u^2 + 6u + 5.$$

Once we have factored, we substitute x^2 for u, giving us:

$$(u + 1)(u + 5) = (x^2 + 1)(x^2 + 5)$$

So we can conclude that $x^4 + 6x^2 + 5 = (x^2 + 1)(x^2 + 5)$.

PRACTICE 6

Factor: $n^4 + 8n^2 + 15$

How would you factor the given polynomial in Example 6 without using substitution?

Factoring Trinomials Whose Leading Coefficient Is Not 1

Next, we consider how to factor a trinomial where the leading coefficient is not equal to 1. Let's begin by examining the product of two binomials.

$$\begin{array}{cccccc} & & \text{F} & \text{O} & \text{I} & \text{L} \\ (3x + 2)(x + 1) = & (3x)(x) & + & (3x)(1) & + & (2)(x) & + & (2)(1) \\ = & 3x^2 & & + 3x & & + 2x & & + 2 \\ = & 3x^2 & & + 5x & & + 2 \end{array}$$

Now, suppose that we reverse the process, starting with the product $3x^2 + 5x + 2$ and then finding its binomial factors. First, we check for common factors. Since there are no common factors other than 1 and -1, we use the **trial-and-error** process, applying the FOIL method in reverse. In this situation, however, we list and test all the possible factors of both the leading coefficient 3 and the constant term 2. The goal is to find a combination of factors where the sum of the outer and inner products is the middle term of the trinomial, $5x$. In other words, we are looking for four integers which meet the following condition:

$$3x^2 + 5x + 2 = (?x + ?)(?x + ?)$$

Note that we know from the preceding discussion that the correct factorization is $(3x + 2)(x + 1)$.

EXAMPLE 7

Factor: $6x^2 + 7x - 5$

Solution   First, we note that there are no common factors. Next, we look for combinations of the factors of 6, the coefficient of the leading term, and of -5, the constant term, that will give us a middle term whose coefficient is 7.

PRACTICE 7

Factor: $4x^2 + 16x + 7$

Factors of 6	Factors of -5	Possible Binomial Factors	Middle Term
6, 1	1, −5	$(6x + 1)(x - 5)$	$-30x + x = -29x$
		$(6x - 5)(x + 1)$	$6x - 5x = x$
	−1, 5	$(6x - 1)(x + 5)$	$30x - x = 29x$
		$(6x + 5)(x - 1)$	$-6x + 5x = -x$
3, 2	1, −5	$(3x + 1)(2x - 5)$	$-15x + 2x = -13x$
		$(3x - 5)(2x + 1)$	$3x - 10x = -7x$
	−1, 5	$(3x - 1)(2x + 5)$	$15x - 2x = 13x$
		$(3x + 5)(2x - 1)$	$-3x + 10x = 7x$ ← Correct middle term

So $6x^2 + 7x - 5 = (3x + 5)(2x - 1)$.

Check

We can verify our factorization by multiplying the factors, as follows:

$$(3x + 5)(2x - 1) = 6x^2 - 3x + 10x - 5 = 6x^2 + 7x - 5$$

So our factorization is correct.

Consider the following binomial factors from Example 7:

The signs of the constant terms are reversed. $(3x - 5)(2x + 1) = 6x^2 - 7x - 5$ The sign of the middle term changes.
$(3x + 5)(2x - 1) = 6x^2 + 7x - 5$

Note that reversing the signs of the constants in the binomial factors changes the sign of the middle term.

Now, we consider factoring a trinomial of the form $ax^2 + bxy + cy^2$. This trinomial contains more than one variable and has binomial factors that are of the form $(?x + ?y)(?x + ?y)$.

EXAMPLE 8

Factor: $15x^2 + 35xy + 10y^2$

Solution Since 5 is the GCF of the trinomial, let's first factor it out.

$$15x^2 + 35xy + 10y^2 = 5(3x^2 + 7xy + 2y^2)$$

Next, we factor $3x^2 + 7xy + 2y^2$. We look for a combination of factors of 3 and 2 that will give us the middle term with a coefficient of 7.

Factors of 3	Factors of 2	Possible Binomial Factors	Middle Term
3, 1	1, 2	$(3x + y)(x + 2y)$	$6xy + xy = 7xy$ ← Correct middle term
		$(3x + 2y)(x + y)$	$3xy + 2xy = 5xy$

So $15x^2 + 35xy + 10y^2 = 5(3x + y)(x + 2y)$.

PRACTICE 8

Factor: $16c^2 - 8cd - 24d^2$

Now, let's consider an alternative procedure for factoring the trinomial $ax^2 + bx + c$ based on grouping. This method, which the next example illustrates, is sometimes called the **ac method**.

EXAMPLE 9

Factor: $2x^2 - 5x - 3$

Solution   First, we check that the terms of $2x^2 - 5x - 3$ have no common factor. Next, instead of listing the factors of 2 and -3 as in the trial-and-error method, we begin by forming their product.

$$ac = (2)(-3) = -6$$

Then, we look for two factors of the number ac that add up to b, that is, to -5. The numbers -6 and 1 satisfy these conditions, since $(-6)(1) = -6$ and $(-6) + 1 = -5$. Using these factors to split up the middle term in the original trinomial, we rewrite the trinomial.

$$
\begin{aligned}
2x^2 - 5x - 3 &= 2x^2 - 6x + x - 3 &&\text{Split up the middle term.}\\
&= (2x^2 - 6x) + (x - 3) &&\text{Group the first two terms}\\
& &&\text{and the last two terms.}\\
&= 2x(x - 3) + 1(x - 3) &&\text{Factor out the GCF from}\\
& &&\text{each group.}\\
&= (x - 3)(2x + 1) &&\text{Factor out } (x - 3).
\end{aligned}
$$

So $2x^2 - 5x - 3 = (x - 3)(2x + 1)$. As usual, we can check that this factorization is correct by multiplication.

PRACTICE 9

Factor: $3x^2 - 2x - 8$

EXAMPLE 10

An object is thrown straight upward with an initial velocity of 24 ft/sec from a height of 40 ft. The height of the object above the ground in feet after t sec is represented by the expression $-16t^2 + 24t + 40$. Factor this expression.

Solution   We factor the expression $-16t^2 + 24t + 40$.

$$
\begin{aligned}
-16t^2 + 24t + 40 &= -8(2t^2 - 3t - 5)\\
&= -8(2t - 5)(t + 1)
\end{aligned}
$$

So the factorization is $-8(2t - 5)(t + 1)$.

PRACTICE 10

After saving $50 the first month, for each successive month a thrifty family saves $10 more than the previous month. After x mo, the family has saved $\left[50x + 5x(x - 1)\right]$ dollars. Write this expression in factored form.

In Section 5.7, we will see that writing quadratic expressions in factored form can be the key to solving equations involving those expressions.

Mathematically Speaking

Fill in each blank with the most appropriate term or phrase from the given list.

number b whose sum is ac	monomial and one or more prime polynomials	factors of c
multiplication	trial-and-error	number ac whose sum is b
factors of b	product	pair of binomials
FOIL		sum

1. Listing all the possible combinations of factors to find the correct factorization of a trinomial is sometimes called the _____ method.

2. To factor $x^2 + bx + c$ using the trial-and-error method, write the trinomial as $(x + ?)(x + ?)$, where the question marks are two factors of c whose _____ is b.

3. To factor a trinomial of the form $x^2 + bxy + cy^2$, try to write it in the form $(x + ?y)(x + ?y)$, where the question marks represent _____.

4. A polynomial is always factored completely when it is expressed as the product of a _____.

5. In the ac method for factoring a trinomial of the form $ax^2 + bxy + cy^2$, look for two factors of the _____.

6. Any factorization can be checked by using _____.

A *Fill in the missing factor.*

7. $x^2 - 4x + 3 = (x - 1)(\quad)$

8. $x^2 + 3x - 10 = (x - 2)(\quad)$

9. $x^2 + 12x + 35 = (x + 5)(\quad)$

10. $x^2 - x - 20 = (x + 4)(\quad)$

11. $x^2 + 2xy - 8y^2 = (x - 2y)(\quad)$

12. $x^2 - 15xy + 56y^2 = (x - 7y)(\quad)$

Factor, if possible.

13. $x^2 + 7x + 12$

14. $x^2 + 11x + 18$

15. $n^2 - 12n + 35$

16. $y^2 - 17y + 30$

17. $t^2 + 2t - 48$

18. $a^2 + 5a - 14$

19. $x^2 - 3x - 54$

20. $n^2 - n - 42$

21. $18 - 7y - y^2$

22. $20 + 8p - p^2$

23. $x^2 + 12x + 36$

24. $x^2 - 6x + 9$

25. $p^2 - 8p - 12$

26. $n^2 + 6n - 8$

27. $x^2 - 5xy + 6y^2$

28. $s^2 + 9st + 20t^2$

29. $a^2 - 2ab - 15b^2$

30. $p^2 - pq - 2q^2$

31. $x^2 - 3xy + 4y^2$

32. $a^2 + ab + 12b^2$

33. $c^2 - cd - 30d^2$

34. $x^2 - 8xy - 48y^2$

35. $5p^2 + 20p + 20$

36. $3x^2 + 18x + 27$

37. $-y^2 + 4y + 21$

38. $-a^2 + 14a + 32$

39. $2c^3 - 4c^2 - 96c$

40. $4t^4 + 4t^3 - 48t^2$

41. $-p^2q^2 - 8pq^2 + 20q^2$

42. $-xy^2 - 13xy - 40x$

43. $-10m^2n^2 + 12m^2n^3 - 2m^2n^4$

44. $24ab^2 - 6a^2b^2 - 3a^3b^2$

45. $a^2b^2 + ab - 72$

46. $x^2y^2 - 4xy - 45$

47. $x^4 - 7x^2 - 30$

48. $x^4 + 7x^2 - 60$

49. $y^6 + 9y^3 - 52$

50. $a^8 - 6a^4 - 72$

51. $64n - 12n^5 - n^9$

52. $28x^2 + 3x^4 - x^6$

53. $p^6q^4 - 2p^3q^2 - 63$

54. $x^4y^2 + 11x^2y + 30$

B *Fill in the missing factor.*

55. $2x^2 + 7x + 3 = ($ $)(x + 3)$

56. $3x^2 - 10x + 8 = ($ $)(x - 2)$

57. $4x^2 + 8x - 5 = (2x - 1)($ $)$

58. $6x^2 - x - 15 = (2x + 3)($ $)$

59. $12x^2 - 35xy + 18y^2 = (3x - 2y)($ $)$

60. $8x^2 + 30xy + 25y^2 = (2x + 5y)($ $)$

Factor, if possible.

61. $3x^2 + 11x + 6$

62. $2y^2 + 9y + 9$

63. $2a^2 - 13a + 18$

64. $3n^2 - 14n + 16$

65. $4t^2 + 4t - 15$

66. $4x^2 + 5x - 21$

67. $6n^2 - n - 12$

68. $5y^2 + 6y - 8$

69. $20 - 4m - 3m^2$

70. $18 + 5r - 2r^2$

71. $5a^2 - 14a - 8$

72. $6t^2 - 17t + 14$

73. $4p^2 + 12p + 9$

74. $9n^2 - 6n + 1$

75. $2x^2 - 17xy + 30y^2$

76. $8a^2 - 22ab + 15b^2$

77. $8p^2 - 2pq - 21q^2$

78. $3a^2 + ab - 24b^2$

79. $6c^2 + 5cd - 6d^2$

80. $10x^2 - 21xy - 10y^2$

81. $12a^2 + 17ab + 2b^2$

82. $8p^2 + 22pq + 15q^2$

83. $15n^2 - 50n - 40$

84. $12b^2 + 54b - 30$

85. $16x^2y^2 + 16xy^2 + 4y^2$

86. $18p^2q^2 - 24p^2q + 8p^2$

87. $-8a^2 - 14ab + 72b^2$

88. $-21c^2 + 70cd - 49d^2$

89. $30ab^4 - 3a^2b^3 - 9a^3b^2$

90. $20u^2v^4 - 15u^3v^3 - 5u^4v^2$

91. $12x^2y^2 + 40xy + 25$

92. $15p^2q^2 + 32pq + 16$

93. $5n^4 + 26n^2 - 24$

94. $7p^4 + 16p^2 - 15$

95. $4a^4 - 13a^2b^3 + 3b^6$

96. $6x^6 - 43x^3y^2 + 42y^4$

97. $6x^4y + 23x^3y^3 + 20x^2y^5$

98. $18u^5v - 45u^3v^3 + 18uv^5$

Mixed Practice

Fill in the missing factor.

99. $a^2 - 5a - 24 = (\quad)(a + 3)$

100. $12x^2 + 7x - 45 = (4x + 9)(\quad)$

Factor, if possible.

101. $m^4 - 15m^2 + 56$

102. $y^2 - 15y + 36$

103. $10a^4b^2 - 5a^3b^3 - 5a^2b^4$

104. $x^2 + 4x - 10$

105. $3b^3 - 3b^2 - 36b$

106. $-2a^2 - a + 21$

107. $4p^2 + 28pq + 49q^2$

108. $8x^2 - 12x - 20$

Applications

Solve.

109. A city parks department increased the size of a rectangular ice-skating rink in its largest park by adding x ft to the length and width. The new rink has an area given by $(x^2 + 140x + 4000)$ ft^2.

   a. Factor the expression.
   b. What were the dimensions of the original ice-skating rink?

110. On November 26, 2010, the temperature (in degrees Fahrenheit) in Orlando, Florida, from 11:00 A.M. to 5:00 P.M. could be modeled by the polynomial $-x^2 + 3x + 70$, where x represents the number of hours past 11:00 A.M. (*Source:* forecast.weather.gov)

   a. Factor this polynomial.
   b. What was the temperature in Orlando that day at noon?

111. The height (in feet) above the ground of an object t sec after it is thrown downward from a height of 192 ft with an initial velocity of 64 ft/sec is given by the polynomial $192 - 64t - 16t^2$. Factor this polynomial.

112. The owner of a company determines that her profit (in dollars) from selling x items is given by the polynomial $x^2 + 8x - 768$. Factor this polynomial.

113. Show that the polynomial $4n^2 - 4n - 15$, where n is an integer, represents the product of two integers that differ by 8.

114. A stone is tossed upward with an initial velocity of 24 ft/sec from a bridge 280 ft above a river. The height of the stone above the river (in feet) t sec after it is tossed upward is given by the polynomial $-16t^2 + 24t + 280$. Factor this polynomial.

• Check your answers on page A-39.

MINDStretchers

Writing

1. When factoring a trinomial by the trial-and-error method, why do we list the pairs of numbers with the correct *product* instead of the pairs of numbers with the correct *sum*?

Critical Thinking

2. Find two values for k, one positive and one negative, such that the trinomial can be factored as the product of two binomials.
   a. $x^2 + kx + 18$
   b. $x^2 - 8x + k$
   c. $4x^2 + kx - 6$
   d. $kx^2 - 3x - 1$

Groupwork

3. Algebra tiles can be used to create geometric models of polynomials. For example, the polynomial $2x^2 + 5x + 3$ can be represented geometrically as follows:

Copy or trace several of each of the following algebra tiles onto a separate piece of paper. Then, cut out each tile.

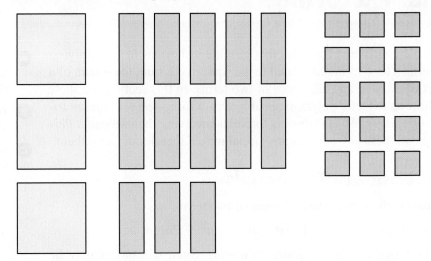

Working with a partner, position the tiles to form a rectangle that models each of the following polynomials and use the model to write the factors of the polynomial. Verify your answer algebraically. With your partner, discuss how algebra tiles can be used to factor polynomials.

a. $x^2 + 6x + 8$

b. $x^2 + 8x + 15$

c. $2x^2 + 7x + 5$

d. $3x^2 + 13x + 12$

OBJECTIVES

A To factor a perfect square trinomial or a difference of squares

B To factor a sum or difference of cubes

C To solve applied problems involving factoring

Recall that in Section 5.2, we considered formulas that provide a shortcut for multiplying binomials of a particular form. These formulas related to the square of a sum, the square of a difference, and the product of the sum and difference of two terms. In this section, we show that these formulas also allow us to factor polynomials of a certain form—*perfect square trinomials* and the *difference of squares*. We also discuss formulas involving the *sum and difference of cubes*. Recognizing that these polynomials are special makes it easier to factor them.

Factoring Perfect Square Trinomials

Recall the formulas for squaring the sum or the difference of two terms.

$$(a + b)^2 = a^2 + 2ab + b^2 \quad \text{and} \quad (a - b)^2 = a^2 - 2ab + b^2$$

Each of these products is called a *perfect square trinomial*. Such trinomials may be factored by reversing the multiplication process.

Factoring a Perfect Square Trinomial

$$a^2 + 2ab + b^2 = (a + b)^2$$
$$a^2 - 2ab + b^2 = (a - b)^2$$

The first formula shows us how to factor a trinomial that happens to be the sum of the squares of two terms *plus* twice their product. In this formula, the terms are a and b, the sum of their squares is $a^2 + b^2$, and $+2ab$ is twice their product. The formula says that the factorization of a trinomial of this form is the square of the *sum* of the two terms, that is, $(a + b)^2$.

The second formula applies when we want to factor a trinomial that is the sum of the squares of two terms *minus* twice their product. In this formula, the terms are again a and b, the sum of their squares is $a^2 + b^2$, and $-2ab$ is minus twice their product. According to this formula, the factorization of such a trinomial is the square of the *difference* of the two terms, namely $(a - b)^2$.

Note that the formulas $a^2 + 2ab + b^2 = (a + b)^2$ and $a^2 - 2ab + b^2 = (a - b)^2$ apply only when the trinomial that we are trying to factor is a perfect square.

EXAMPLE 1

Determine whether each expression is a perfect square trinomial.

a. $x^2 - 6x + 9$ **b.** $x^2 + 7x + 1$ **c.** $4x^2 + 20xy + 25y^2$

Solution

a. In $x^2 - 6x + 9$, x^2 and 9 are perfect squares and correspond to a^2 and b^2, respectively, in the formula $a^2 - 2ab + b^2$. So a corresponds to x, and b corresponds to 3. The middle term $-6x$, or $-2 \cdot x \cdot 3$, corresponds to $-2ab$.

$$x^2 - 6x + 9 = x^2 - \underbrace{2 \cdot x \cdot 3}_{} + (3)^2$$
$$ \uparrow \qquad \uparrow \qquad \uparrow$$
$$ a^2 \quad - \quad 2ab \quad + \quad b^2$$

Therefore, the polynomial is a perfect square trinomial.

PRACTICE 1

Indicate whether each trinomial is a perfect square.

a. $x^2 + 4x + 4$

b. $-6t^2 - 2t + 1$

c. $x^2 - 8xy + 16y^2$

b. For $x^2 + 7x + 1$, x^2 and 1 are both perfect squares. In the first formula, $a^2 + 2ab + b^2$, a corresponds to x and b to 1. However, the middle term $7x$ is not twice the product of x and 1. So $x^2 + 7x + 1$ is not a perfect square trinomial.

c. In $4x^2 + 20xy + 25y^2$, $4x^2$ is equal to $(2x)^2$ and $25y^2$ equals $(5y)^2$. So in the formula $a^2 + 2ab + b^2$, a corresponds to $2x$ and b to $5y$. The middle term $20xy$ can be expressed as $2 \cdot 2x \cdot 5y$.

$$4x^2 + 20xy + 25y^2 = (2x)^2 + \underbrace{2(2x)(5y)}_{} + (5y)^2$$
$$\qquad\qquad\quad a^2 \;\; + \qquad 2ab \qquad + \quad b^2$$

So the trinomial is a perfect square.

Now, let's look at *factoring* perfect square trinomials.

EXAMPLE 2

Factor.

a. $x^2 + 10x + 25$ **b.** $x^2 + 1 - 2x$

c. $9x^2 - 12xy + 4y^2$

Solution

a. For the trinomial $x^2 + 10x + 25$ the first term x^2 and the last term 25, or 5^2, are perfect squares. The middle term $10x$, or $2 \cdot x \cdot 5$, is twice the product of x and 5. It follows that $x^2 + 10x + 25$ is a perfect square trinomial. Since its middle term is positive, we apply the formula for the square of a sum.

$$x^2 + 10x + 25 = x^2 + \underbrace{2 \cdot x \cdot 5}_{} + (5)^2 = (x + 5)^2$$
$$\qquad\qquad a^2 + \qquad 2ab \qquad + \quad b^2 \; = \; (a + b)^2$$

We can check our answer by multiplying out $(x + 5)^2$.

$(x + 5)^2 = (x + 5)(x + 5) = x^2 + 2 \cdot 5x + 25$, or $x^2 + 10x + 25$

b. Let's begin by rewriting the trinomial in descending order: $x^2 - 2x + 1$.

$$x^2 - 2x + 1 = x^2 - \underbrace{2 \cdot x \cdot 1}_{} + (1)^2 = (x - 1)^2$$
$$\qquad\qquad a^2 - \qquad 2ab \qquad + \quad b^2 \; = \; (a - b)^2$$

c. Here the trinomial contains more than one variable.

$$9x^2 - 12xy + 4y^2 = (3x)^2 - \underbrace{2 \cdot 3x \cdot 2y}_{} + (2y)^2 = (3x - 2y)^2$$
$$\qquad\qquad\quad a^2 \quad - \qquad 2ab \qquad + \quad b^2 \; = \; (a - b)^2$$

PRACTICE 2

Express as the square of a binomial.

a. $n^2 + 8n + 16$

b. $t^2 + 9 - 6t$

c. $16c^2 - 8cd + d^2$

Factoring the Difference of Squares

Recall that when finding the product of the sum and the difference of the same terms, we get:

$$(a + b)(a - b) = a^2 - b^2$$

The product is a binomial that is called a *difference of squares*. We can factor such binomials by reversing the multiplication process.

> **Factoring the Difference of Squares**
> $$a^2 - b^2 = (a + b)(a - b)$$

This formula is a shortcut for factoring a binomial equal to the square of one term minus the square of another term. The terms are a and b, and so their squares are a^2 and b^2. The formula says that the factorization of a binomial that is the difference of the squares of two terms is the sum of the two terms times the difference of the same two terms, that is, $(a + b)(a - b)$.

EXAMPLE 3

Indicate whether each binomial involves a difference of squares.

a. $x^2 - 25$ **b.** $3x^2 + 3y^2$ **c.** $9p^6 - q^2$

Solution

a. In $x^2 - 25$, both x^2 and 25 are perfect squares and correspond to a^2 and b^2, respectively, in the formula $a^2 - b^2$.

$$x^2 - 25 = x^2 - 5^2$$
$$\underset{a^2 - b^2}{\uparrow \qquad \uparrow}$$

Here a corresponds to x, and b corresponds to 5. So $x^2 - 25$ is a difference of squares.

b. $3x^2 + 3y^2 = 3(x^2 + y^2)$. The factor $x^2 + y^2$ is not a difference of squares but a sum of squares, which is not factorable. So the completely factored form of $3x^2 + 3y^2$ is $3(x^2 + y^2)$.

c. $9p^6 - q^2$ can be written as $(3p^3)^2 - q^2$ and so is a difference of squares.

PRACTICE 3

Determine whether each binomial involves a difference of squares.

a. $x^2 - 9$

b. $7x^4 + 7y^4$

c. $r^4 - 64s^6$

> **TIP** After factoring out the greatest common factor, a sum of squares is not factorable using real numbers.

EXAMPLE 4

Factor.

a. $x^2 - 49$ **b.** $4x^2 - y^2$ **c.** $36y^2 - 100$

Solution

a. $x^2 - 49$ is a difference of squares.

$$x^2 - 49 = x^2 - 7^2 = \underbrace{(x + 7)}\underbrace{(x - 7)}$$
$$\underset{a^2 - b^2 \ = \ (a + b) \ (a - b)}{\uparrow \qquad \uparrow \qquad \uparrow \qquad \uparrow}$$

PRACTICE 4

Factor.

a. $y^2 - 81$

b. $x^2 - 100y^2$

c. $8x^2 - 50$

Check We can check that $(x + 7)(x - 7)$ is the factorization by multiplying.

$$(x + 7)(x - 7) = x^2 \underbrace{- 7x + 7x}_{\text{Sum is 0.}} - 49 = x^2 - 49$$

b. $4x^2 - y^2 = (2x)^2 - y^2 = \underbrace{(2x + y)}\underbrace{(2x - y)}$

$$a^2 \ - \ b^2 \ = \quad (a + b) \quad (a - b)$$

c. $36y^2 - 100 = 4(9y^2 - 25)$ Factor out the GCF from each term.

$$= 4\left[(3y)^2 - 5^2\right]$$ Express as the difference of squares.

$$= 4(3y + 5)(3y - 5)$$

In some differences of squares, one or both of the expressions being squared are binomials or powers of the variables.

EXAMPLE 5	**PRACTICE 5**
Factor each difference of squares.	Write in factored form.
a. $9 - (a + b)^2$ **b.** $4x^4 - 9y^6$	**a.** $16 - (2x - y)^2$
Solution	
a. $9 - (a + b)^2 = 3^2 - (a + b)^2$	**b.** $64x^8 - 25y^2$
$\qquad = [3 + (a + b)][(3 - (a + b)]$	
$\qquad = (3 + a + b)(3 - a - b)$	
b. $4x^4 - 9y^6 = (2x^2)^2 - (3y^3)^2 = (2x^2 + 3y^3)(2x^2 - 3y^3)$	

The Sum and Difference of Cubes

Recall that the cube of a number is that number raised to the third power.

$$2^3 = 2 \cdot 2 \cdot 2 = 8$$
$$x^3 = x \cdot x \cdot x$$
$$(4y)^3 = 4 \cdot 4 \cdot 4 \cdot y \cdot y \cdot y = 64y^3$$

Note that the cube of a negative number is negative, for example,

$$(-3)^3 = (-3)(-3)(-3) = -27$$

Expressions such as $2^3, x^3, 64y^3$, and -27 are called *perfect cubes*.

An expression that is the *sum or the difference of two perfect cubes* can be factored using the following formulas:

Factoring the Sum or Difference of Cubes

$$a^3 + b^3 = (a + b)(a^2 - ab + b^2)$$
$$a^3 - b^3 = (a - b)(a^2 + ab + b^2)$$

The first formula shows us how to factor a binomial that is the sum of the cubes of two terms. In this formula, the terms are a and b and the sum of their cubes is $a^3 + b^3$. The formula says that the factorization of a binomial of this form is the product of the sum of the two terms $a + b$ and the square of the first term *minus* the product of the terms plus the square of the second term, that is, $a^2 - ab + b^2$.

We can apply the second formula to factor the difference of two cubes, say $a^3 - b^3$. The factorization of a binomial of this form is the product of the difference of the two terms $a - b$ and the square of the first term *plus* the product of the terms plus the square of the second term, that is, $a^2 + ab + b^2$.

TIP In these two formulas, the sign of b^3 on the left side of the equation is:

- the *same* as that of b in the binomial factor on the right.
- the *opposite* of that of the middle term of the trinomial on the right.

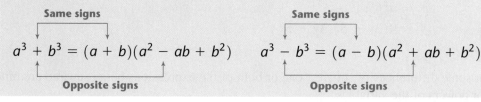

$$a^3 + b^3 = (a + b)(a^2 - ab + b^2) \qquad a^3 - b^3 = (a - b)(a^2 + ab + b^2)$$

EXAMPLE 6

Indicate whether each binomial is a sum or difference of cubes.

a. $x^3 + y^3$ **b.** $n^6 - 1$ **c.** $8n^3 + 125p$

Solution

a. In the expression, x^3 and y^3 are perfect cubes, so $x^3 + y^3$ is a sum of cubes.

$$\underset{a^3\ +\ b^3}{\underset{\uparrow\qquad\uparrow}{x^3\ +\ y^3}}$$

b. The expression $n^6 - 1$ can be written as $(n^2)^3 - (1)^3$, and so is a difference of cubes.

c. $8n^3 + 125p$ is not a sum of cubes because p is not a perfect cube.

PRACTICE 6

Determine whether each binomial is a sum or difference of cubes.

a. $p^3 - q^3$

b. $8 + x^6$

c. $x^4 + 27y^9$

Now, let's factor the sum or difference of cubes.

EXAMPLE 7

Factor.

a. $8 + y^3$ **b.** $p^6 - 64q^3$ **c.** $(r + s)^3 + (r - s)^3$

Solution

a. $8 + y^3$ is a sum of cubes.

$$8 + y^3 = 2^3 + y^3 = \underbrace{(2 + y)}\underbrace{(2^2 - 2y + y^2)} = (2 + y)(4 - 2y + y^2)$$

$$\underset{a^3\ +\ b^3}{} \ = \ \underset{(a + b)}{} \ \underset{(a^2 - ab + b^2)}{}$$

So the factorization of $8 + y^3$ is $(2 + y)(4 - 2y + y^2)$.

PRACTICE 7

Factor.

a. $125 - y^3$

b. $27m^3n^6 + 1$

c. $(x + 1)^3 - (x - 1)^3$

b. $p^6 - 64q^3$ is a difference of cubes.

$$p^6 - 64q^3 = (p^2)^3 - (4q)^3$$

$$\uparrow \qquad \uparrow$$
$$a^3 - \qquad b^3$$

$$= \underbrace{(p^2 - 4q)}\underbrace{\left[(p^2)^2 + p^2 \cdot (4q) + (4q)^2\right]}$$

$$\uparrow \qquad\qquad\qquad \uparrow$$
$$(a - b) \qquad\qquad (a^2 + ab + b^2)$$

$$= (p^2 - 4q)(p^4 + 4p^2q + 16q^2)$$

c. $(r + s)^3 + (r - s)^3$ is a sum of cubes. Here a corresponds to $(r + s)$ and b corresponds to $(r - s)$.

$$(r + s)^3 + (r - s)^3$$

$$= [(r + s) + (r - s)][(r + s)^2 - (r + s)(r - s) + (r - s)^2]$$

$$= [2r][(r^2 + 2rs + s^2) - (r^2 - s^2) + (r^2 - 2rs + s^2)]$$

$$= [2r][r^2 + 2rs + s^2 - r^2 + s^2 + r^2 - 2rs + s^2]$$

$$= (2r)(r^2 + 3s^2)$$

EXAMPLE 8

When the velocity of a rocket increases from v_1 to v_2, the force caused by air resistance increases by $kv_2^2 - kv_1^2$, for a constant k. Write this expression in factored form.

Solution The expression for the force is given by $kv_2^2 - kv_1^2$. Factoring this expression, we get

$$kv_2^2 - kv_1^2 = k(v_2^2 - v_1^2) = k(v_2 + v_1)(v_2 - v_1)$$

PRACTICE 8

A ball is dropped from the roof of a building 100 ft above the ground. The height of the ball (in feet) t sec after it is dropped is given by the expression $100 - 16t^2$. Write this expression in factored form.

Mathematically Speaking

Fill in each blank with the most appropriate term or phrase from the given list.

sum of the squares	square of the product	difference of the squares
prime	sum of cubes	perfect square
cube of a binomial	binomial	
difference of cubes	product	

1. A trinomial can be factored in the form $(a + b)^2$ if the trinomial is the _____ of two terms plus twice their product.

2. Trinomials of the form $a^2 + 2ab + b^2$ or the form $a^2 - 2ab + b^2$ are _____ trinomials.

3. If a binomial is the _____ of two terms, its factorization has the form $(a + b)(a - b)$.

4. The product $(a + b)(a^2 - ab + b^2)$ is the factorization of the _____.

A *Determine whether each polynomial is a perfect square trinomial, a difference of squares, or neither.*

5. $x^2 + 6x + 9$

6. $x^2 - 14x + 49$

7. $x^2 - 18$

8. $x^2 - 21$

9. $16x^2 - 40xy + 25y^2$

10. $64 + 48x^3 + 9x^6$

11. $4x^2 - 9y^2$

12. $25x^2 - 16y^2$

13. $9x^4 - 4x^2 + 1$

14. $49x^2 + 84xy - 36y^2$

15. $-y^4 + 16x^4$

16. $-y^4 + 25x^4$

17. $25x^6 + 10x^3y^2 - y^4$

18. $16x^4y^4 - 24x^2y^2 + 9$

Factor, if possible.

19. $x^2 - 12x + 36$

20. $x^2 - 16x + 64$

21. $n^2 + 20n + 100$

22. $y^2 + 18y + 81$

23. $4a^2 + 25 - 20a$

24. $4 + 49r^2 - 28r$

25. $64 + 48n - 9n^2$

26. $25 - 40x + 16x^2$

27. $x^2 - 16$

28. $t^2 - 36$

29. $121 - y^2$

30. $144 - a^2$

31. $9x^2 - 4$

32. $25n^2 - 1$

33. $0.16r^2 - 0.81$

34. $0.04p^2 - 0.09$

35. $64x^2 + 16xy - y^2$

36. $9a^2 - 12ab - 4b^2$

37. $72u^2 - 98v^2$

38. $3y^2 - 12x^2$

39. $\frac{1}{9}a^2 - 4b^2$

40. $49c^2 - \frac{1}{25}d^2$

41. $\frac{1}{4}u^2 - uv + v^2$

42. $\frac{1}{16}x^2 - 3xy + 36y^2$

43. $49 - 56ab + 16a^2b^2$

44. $9 + 60pq + 100p^2q^2$

45. $(2u - v)^2 - 64$

46. $9 - (3a + b)^2$

47. $49 - 4(x + y)^2$

48. $16(p - q)^2 - 1$

49. $p^6 - 22p^3 + 121$

50. $n^4 + 8n^2 + 16$

51. $9a^8 + 48a^4b + 64b^2$

52. $81p^6 - 18p^3q^2 + q^4$

53. $4a^4 - 225$

54. $36y^6 - 169$

55. $49x^6 - 144y^4$

56. $121u^8 - 64v^4$

57. $100p^4q^2 - 9r^2$

58. $36x^2y^2 - 49z^4$

59. $(3p + q)^2 - (2p - q)^2$

60. $(a - 2b)^2 - (a - 3b)^2$

B *Determine whether each polynomial is a sum of cubes, a difference of cubes, or neither.*

61. $x^3 - 64$

62. $125 + x^3$

63. $x^6 - 3y^3$

64. $8x^9 - 27y^6$

65. $y^{12} + 0.008x^3$

66. $1.25x^3 + 27y^3$

Factor, if possible.

67. $x^3 + 1$

68. $a^3 + 27$

69. $p^3 - 8$

70. $y^3 - 1$

71. $27t^2 - 4$

72. $8x^3 - 125$

73. $\frac{1}{8} - a^3$

74. $-\frac{1}{27} + t^3$

75. $125x^3 + y^3$

76. $u^3 + 216v^3$

77. $0.064b^3 - 0.027a^3$

78. $0.125p^3 - 0.001q^3$

79. $27p^3q^3 + 20$

80. $9x^3y^3 - 8$

81. $a^6 - 8$

82. $y^9 - 27$

83. $64x^9 + 27y^3$

84. $125a^3 - 8b^6$

85. $8p^{12} - q^9$

86. $216x^6 + 125y^6$

87. $-a^3b^3 - 64c^6$

88. $-27u^3v^6 - w^3$

89. $27 - (a + 1)^3$

90. $(x - 2)^3 + 1$

91. $(x - 2)^3 + (x + 2)^3$

92. $(p + 3)^3 - (3 - p)^3$

Factor.

93. $27x^4 - 18x^3 + 3x^2$

94. $5p^5 + 20p^4 + 20p^3$

95. $16x^3y - 100xy^3$

96. $36a^4b^2 - 16a^2b^4$

97. $54u^4v^3 + 128uv^6$

98. $375x^5y^2 - 24x^2y^8$

99. $32x^5 - 16x^3y^2 + 2xy^4$

100. $64a^6bc^2 + 16a^3b^4c^2 + b^7c^2$

101. $-x^8y^4 + 256y^4$

102. $-5x^2y^8 + 5x^2$

103. $243p^4 - 48q^4$

104. $10r^8 - 160t^4$

105. $x^2(y - z) + (z - y)$

106. $a^2(b - c) + 4(c - b)$

107. $u^3(v + w) + 8(v + w)$

108. $p^3(q - r) - 27(q - r)$

109. $4x^2y^4 - 9y^4 - 4x^2z^4 + 9z^4$

110. $a^4b^3 - b^3 - c^3 + a^4c^3$

111. $p^2 - 4pq + 4q^2 - 36$

112. $u^2 - 6uv + 9v^2 - 25$

113. $4c^2 - (a^2 + 10ab + 25b^2)$

114. $z^2 - (4x^2 + 12xy + 9y^2)$

Factor. Assume that all exponents are positive.

115. $x^{4n} - 16$

116. $y^{4n} - 1$

117. $t^{4a} - s^2$

118. $x^{2n} - y^2$

119. $25a^{2n} - 4b^{2m}$

120. $16x^{2a} - 9y^{2b}$

Mixed Practice

Solve.

121. Determine whether the polynomial $-8x^3 + 125y^6$ is a sum of cubes, a difference of cubes, or neither.

122. Determine whether the polynomial $4x^2 + 10xy + 25y^2$ is a perfect square trinomial, a difference of squares, or neither.

Factor, if possible.

123. $25(2a - b)^2 - 9$

124. $36 + 9x^2 - 36x$

125. $4a^4 - 24a^3 + 36a^2$

126. $x^{8a} - y^2$, where a is positive

127. $64a^2 - \dfrac{1}{36}b^2$

128. $8x^3y^3 - 125$

129. $32x^3y - 18xy^3$

130. $s^9 + 27t^{12}$

Applications

C *Solve.*

131. Find an expression, in factored form, for the area of the matte border surrounding the picture shown below.

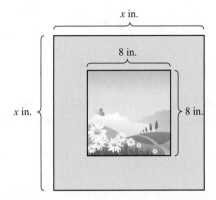

x in.

8 in.

x in.

8 in.

132. To prevent breakage during shipping, a company packages its snow globes in foam blocks that have a hole cut out of the middle, as shown. Write an expression, in factored form, that represents the volume of the foam block.

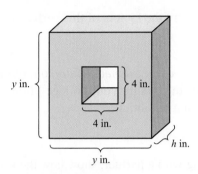

y in.

4 in.

4 in.

h in.

y in.

133. The height (in feet) of an object dropped from the top of a building 324 ft above the ground is given by the expression $324 - 16t^2$, where t is time (in seconds). Rewrite this expression in factored form.

134. The ring-shaped region shown is the region that lies between two concentric circles. The area of the ring is given by $\pi r_2^2 - \pi r_1^2$. Factor this expression.

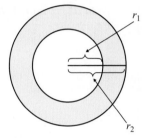

r_1

r_2

135. A couple invests \$10,000 in an account at an annual interest rate r, compounded twice a year. The amount in the account after 1 yr can be calculated using the expression $10{,}000 + 10{,}000r + 2500r^2$. Show that the expression can be written as $10{,}000\left(1 + \dfrac{r}{2}\right)^2$.

136. A 12-inch by 12-inch piece of cardboard is used to make an open box by cutting x-inch by x-inch squares from each corner and turning up the sides. The volume of the resulting box can be modeled by the polynomial $144x - 48x^2 + 4x^3$. Write the polynomial in factored form.

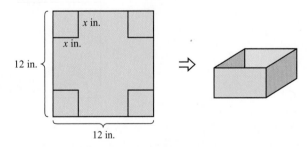

x in.

x in.

12 in.

12 in.

137. A furniture store sells two sizes of cube-shaped stackable storage cabinets. The dimensions of the interior storage space of the smaller cube is x-inch by x-inch by x-inch. The interior storage space of the larger cube is y-inch by y-inch by y-inch. Write an expression, in factored form, that shows the combined storage space of a large and a small cube.

138. If the radius of a spherical balloon increases from r_1 to r_2, then the increase in the volume of the balloon is given by the expression $\dfrac{4}{3}\pi r_2^3 - \dfrac{4}{3}\pi r_1^3$. Write this expression in factored form.

• Check your answers on page A-39.

MINDStretchers

Mathematical Reasoning

1. Suppose the polynomial $x^m - 2x^p y^q + y^n$ is a perfect square trinomial.

   a. What must be true of m and n? Explain.

   b. How are m and p related? How are n and q related?

   c. What are the factors of this perfect square trinomial?

Research

2. Biologists use a *Punnett square* to represent the way in which the genes from two parents might combine to produce an offspring. Using your college library or the web, investigate the relationship between a Punnett square and the square of a binomial.

Groupwork

3. Working with a partner, show how the geometric model can be used to verify the special factoring formula.

   a. $a^2 + 2ab + b^2 = (a + b)^2$

   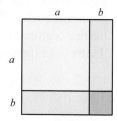

   b. $a^2 - b^2 = (a + b)(a - b)$

   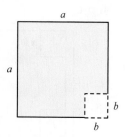

5.7 Solving Quadratic Equations by Factoring

This section deals with a kind of equation that we have not previously discussed, namely, *quadratic equations*. Such equations are used in physics, finance, and construction.

A To solve a quadratic equation by factoring

B To solve applied problems using quadratic equations

DEFINITION

A **second-degree** or **quadratic equation** is an equation that can be written in the form $ax^2 + bx + c = 0$, where a, b, and c are real numbers and $a \neq 0$.

Some examples of quadratic equations are:

$$x^2 - x - 6 = 0 \qquad 2n^2 = 8 \qquad (x + 1)^2 = 0$$

As in the case of a linear equation, a value is said to be a *solution* of a quadratic equation if substituting the value for the variable makes the equation a true statement.

Using Factoring to Solve Quadratic Equations

In this section, we consider quadratic equations that can be solved by factoring. In Chapter 8, we will consider additional approaches to solving quadratic equations.

The key to solving quadratic equations by factoring is to apply the *zero-product property*.

The Zero-Product Property

If $ab = 0$, then $a = 0$, $b = 0$, or both a and $b = 0$.

In words, this property states that if the product of two factors is zero, then either one or both of the factors must be zero.

Let's see how to use the zero-product property to solve a quadratic equation that is already in factored form.

EXAMPLE 1

Solve and check: $(2x + 1)(x - 6) = 0$

Solution Since $(2x + 1)(x - 6) = 0$, the zero-product property tells us that at least one of the factors must equal 0; that is, either $2x + 1 = 0$ or $x - 6 = 0$.

$$(2x + 1)(x - 6) = 0$$

$2x + 1 = 0$ or $x - 6 = 0$ Set each factor equal to 0.

$2x = -1$ $\qquad x = 6$ Solve each equation for x.

$$x = -\frac{1}{2}$$

PRACTICE 1

Solve: $(3x + 1)(x - 7) = 0$

EXAMPLE 1 (continued)

Check

To verify that $-\dfrac{1}{2}$ and 6 are the solutions, we substitute.

Substitute $-\dfrac{1}{2}$ for x.   Substitute 6 for x.

$$(2x + 1)(x - 6) = 0 \qquad\qquad (2x + 1)(x - 6) = 0$$

$$\left[2\left(-\frac{1}{2}\right) + 1\right]\left(-\frac{1}{2} - 6\right) \overset{?}{=} 0 \qquad [2(6) + 1](6 - 6) \overset{?}{=} 0$$

$$(-1 + 1)\left(-6\frac{1}{2}\right) = 0 \qquad\qquad\qquad (13)(0) \overset{?}{=} 0$$

$$(0)\left(-6\frac{1}{2}\right) \overset{?}{=} 0 \qquad\qquad\qquad\qquad\qquad 0 = 0 \quad \text{True}$$

$$0 = 0 \quad \text{True}$$

So the solutions of the equation $(2x + 1)(x - 6) = 0$ are $-\dfrac{1}{2}$ and 6.

When the quadratic expression in an equation is not written in factored form, we must factor it before solving.

EXAMPLE 2

Solve: $x^2 - 7x + 12 = 0$

Solution   Before we can use the zero-product property, we must first factor the quadratic expression on the left side of the equation.

$$x^2 - 7x + 12 = 0$$
$$(x - 3)(x - 4) = 0$$

Next, we set each factor equal to 0 and solve each equation for x.

$$x - 3 = 0 \quad \text{or} \quad x - 4 = 0$$
$$x = 3 \qquad\qquad x = 4$$

Check

Substitute 3 for x.   Substitute 4 for x.

$$x^2 - 7x + 12 = 0 \qquad\qquad x^2 - 7x + 12 = 0$$
$$3^2 - 7(3) + 12 \overset{?}{=} 0 \qquad 4^2 - 7(4) + 12 \overset{?}{=} 0$$
$$9 - 21 + 12 \overset{?}{=} 0 \qquad 16 - 28 + 12 \overset{?}{=} 0$$
$$0 = 0 \quad \text{True} \qquad\qquad\qquad 0 = 0 \quad \text{True}$$

So the solutions are 3 and 4.

PRACTICE 2

Solve: $x^2 + 9x + 20 = 0$

In order to use the zero-product property to solve a quadratic equation, we must first write the equation in *standard form*, $ax^2 + bx + c = 0$, and then factor.

EXAMPLE 3

Solve.

a. $2x^2 + x = 3$

b. $3x(x + 2) = 24$

Solution

a.
$$2x^2 + x = 3$$
$$2x^2 + x - 3 = 0 \qquad \text{Write the equation in standard form by adding } -3 \text{ to each side.}$$
$$(2x + 3)(x - 1) = 0 \qquad \text{Factor the left side of the equation.}$$
$$2x + 3 = 0 \quad \text{or} \quad x - 1 = 0 \qquad \text{Set each factor equal to 0.}$$
$$2x = -3 \qquad\qquad x = 1 \qquad \text{Solve for } x.$$
$$x = -\frac{3}{2}$$

Check

Substitute $-\frac{3}{2}$ for x. 　　　　　Substitute 1 for x.

$$2x^2 + x = 3 \qquad\qquad 2x^2 + x = 3$$
$$2\left(-\frac{3}{2}\right)^2 + \left(-\frac{3}{2}\right) \stackrel{?}{=} 3 \qquad\qquad 2(1)^2 + 1 \stackrel{?}{=} 3$$
$$2\left(\frac{9}{4}\right) - \frac{3}{2} \stackrel{?}{=} 3 \qquad\qquad 2 + 1 \stackrel{?}{=} 3$$
$$3 = 3 \quad \text{True} \qquad\qquad 3 = 3 \quad \text{True}$$

So the solutions are $-\frac{3}{2}$ and 1.

b.
$$3x(x + 2) = 24$$
$$3x^2 + 6x = 24 \qquad \text{Multiply.}$$
$$3x^2 + 6x - 24 = 0 \qquad \text{Write in standard form by adding } -24 \text{ to each side.}$$
$$3(x^2 + 2x - 8) = 0 \qquad \text{Factor out the GCF.}$$
$$3(x - 2)(x + 4) = 0 \qquad \text{Factor the left side of the equation.}$$

Now, we set factors containing variables equal to 0, and then solve for x.

$$x - 2 = 0 \quad \text{or} \quad x + 4 = 0 \qquad \text{Set each factor equal to 0.}$$
$$x = 2 \qquad\qquad x = -4 \qquad \text{Solve for } x.$$

Check

Substitute 2 for x. 　　　　Substitute -4 for x.

$$3x(x + 2) = 24 \qquad\qquad 3x(x + 2) = 24$$
$$3(2)(2 + 2) \stackrel{?}{=} 24 \qquad\qquad 3(-4)(-4 + 2) \stackrel{?}{=} 24$$
$$24 = 24 \quad \text{True} \qquad\qquad 24 = 24 \quad \text{True}$$

So the solutions are 2 and -4.

PRACTICE 3

Solve.

a. $3y^2 - 4y = 4$

b. $2x(x - 2) = 30$

These examples lead us to the following strategy (or rule) for solving a quadratic equation by factoring.

To Solve a Quadratic Equation by Factoring

- If necessary, rewrite the equation in standard form, with 0 on one side.

- Factor the other side.

- Use the zero-product property to get two simple linear equations.

- Solve the linear equations.

- Check by substituting the solutions in the original quadratic equation.

- State the solutions to the original equation.

EXAMPLE 4

A swimmer jumps from a diving board 8 ft above a pool. After t sec, the swimmer's height h above the water (in feet) is given by the expression $-16t^2 + 8t + 8$. In how many seconds will the swimmer hit the water?

Solution   We need to solve the equation $-16t^2 + 8t + 8 = 0$. Factoring the left side of the equation, we get:

$$-16t^2 + 8t + 8 = 0$$
$$-8(2t^2 - t - 1) = 0$$
$$-8(2t + 1)(t - 1) = 0$$

Setting each factor containing a variable equal to 0, gives us:

$$2t + 1 = 0 \quad \text{or} \quad t - 1 = 0$$
$$t = -\frac{1}{2} \quad \text{or} \quad t = 1$$

Since the answer must be positive, we can conclude that the swimmer will hit the water after 1 sec.

PRACTICE 4

A young couple invested $16,000 in a high-risk growth fund. After 2 yr, their investment was worth $25,000. Their broker used the equation

$$16{,}000(1 + r)^2 = 25{,}000$$

to find the average annual rate of return r. What is this rate?

We can also apply what we know about solving quadratic equations to problems involving the *Pythagorean theorem*. This theorem relates the lengths of the three sides of a right triangle.

A right triangle has one 90° angle. The side opposite the 90° angle is called the *hypotenuse*. The other sides are called *legs*. The *Pythagorean theorem* states that for every right triangle, the sum of the squares of the lengths of the legs equals the square of the length of the hypotenuse:

$$a^2 + b^2 = c^2$$

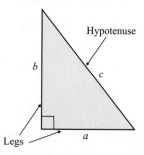

EXAMPLE 5

The mainsail on the sailboat shown below is a right triangle in which the hypotenuse is called the *leech*. The length of the leech is 13 ft, and the height of the mainsail is 7 ft greater than the length of its base. If sailcloth costs $15/ft^2, what is the cost of a new mainsail?

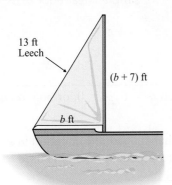

13 ft
Leech

$(b + 7)$ ft

b ft

Solution   We know that the length of the leech is 13 ft, and the height of the mainsail is 7 ft greater than the length of the base. To find the area of the sail, we need to first find the base b using the Pythagorean theorem:

$$b^2 + (b + 7)^2 = 13^2$$
$$b^2 + b^2 + 14b + 49 = 169$$
$$2b^2 + 14b - 120 = 0$$
$$b^2 + 7b - 60 = 0$$
$$(b + 12)(b - 5) = 0$$
$$b + 12 = 0 \quad \text{or} \quad b - 5 = 0$$
$$b = -12 \qquad \qquad b = 5$$

Since b represents a length and a length cannot be negative, we choose the positive solution 5. So the base of the mainsail is 5 ft long.

To find the cost of the sail, we must first compute its area. The formula for the area of a triangle is $\frac{1}{2}$ its base times its height. In this case, the base is 5 ft and the height is 12 ft. So the area of the sail is $\frac{1}{2} \cdot 5 \cdot 12 = 30$, or 30 ft^2. Since the cloth costs $15/ft^2, the total cost of the cloth is $15 \cdot 30$, or $450.

PRACTICE 5

Two cars leave an intersection, one traveling west and the other south. After some time, the faster car is 1 mi farther away from the intersection than the slower car. At that time, the two cars are 5 mi apart. How far did each car travel?

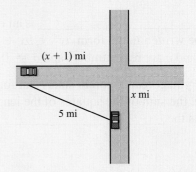

$(x + 1)$ mi

x mi

5 mi

Mathematically Speaking

Fill in each blank with the most appropriate term or phrase from the given list.

square root of the length of the hypotenuse	quadratic equation	square of the length of the hypotenuse
one simple linear equation	identity property	two simple linear equations
binomial equation	zero-product property	

1. A(n) _____ is an equation that can be written in the form $ax^2 + bx + c = 0$, where a, b, and c are real numbers and $a \neq 0$.

2. The _____ states that if $ab = 0$, then $a = 0$, $b = 0$, or both a and $b = 0$.

3. The Pythagorean theorem states that for every right triangle, the sum of the squares of the lengths of the legs equals the _____.

4. When a quadratic equation is solved by factoring, the zero-product property is used to get _____.

A *Solve.*

5. $(x + 3)(x - 4) = 0$

6. $(x - 1)(x + 7) = 0$

7. $(4n - 3)(n - 2) = 0$

8. $(5a - 2)(a - 9) = 0$

9. $(2x + 1)(2x + 3) = 0$

10. $(3y + 1)(2y + 5) = 0$

11. $(3 - p)^2 = 0$

12. $(4 - n)^2 = 0$

13. $x^2 + 5x = 0$

14. $r^2 + r = 0$

15. $6n^2 - 3n = 0$

16. $20y^2 - 5y = 0$

17. $x^2 + 8x + 12 = 0$

18. $y^2 + 10y + 24 = 0$

19. $a^2 - 3a + 2 = 0$

20. $p^2 - 6p + 5 = 0$

21. $36 + 5t - t^2 = 0$

22. $30 - 7x - x^2 = 0$

23. $2x^2 + 5x - 7 = 0$

24. $3n^2 - 14n + 8 = 0$

25. $0 = 4r^2 - 20r + 25$

26. $0 = 16a^2 + 8a + 1$

27. $20x^2 - 45 = 0$

28. $48a^2 - 3 = 0$

29. $6x^2 + 28x + 30 = 0$

30. $10y^2 + 35y - 20 = 0$

31. $r^2 = 49$

32. $v^2 = 100$

33. $9x = 15x^2$

34. $8a^2 = 16a$

35. $p^2 - 32 = -4p$

36. $n^2 - 18 = 3n$

37. $19t + 36 = 6t^2$

38. $12 - 16a = -5a^2$

39. $x^2 + 3 = 10x - 2x^2$

40. $7p^2 + 5 = 3p^2 + 9p$

41. $y(y - 2) = 8$

42. $r(r + 11) = -30$

43. $n(3n + 17) + 20 = 0$

44. $x(2x - 1) - 15 = 0$

45. $3a(2a - 3) = 2a - 4$

46. $2y(4y + 5) = 4y - 1$

47. $(v - 6)(v + 4) = -9$

48. $(p + 8)(p - 5) = 14$

49. $(2x - 1)(2x + 1) = 3x$

50. $(3x + 4)(3x - 4) = -10x$

51. $(5t - 3)(t - 2) = 3t^2 - 4t + 2$

52. $(3n + 1)(4n + 1) = 7n^2 - 6n - 5$

53. $x(x - 4) + 3(4 - x) = 0$

54. $y(y + 2) - 5(y + 2) = 0$

55. $2r(r + 1) + 3(r - 6) = -6$

56. $a(5 - 3a) - 4(a + 1) = -8$

57. $3(x^2 - 6) + 7x = x(x + 2)$

58. $6(n^2 + 2) - 11n = n(n + 5)$

Given $f(x)$ and $g(x)$, find all values of x such that $f(x) = g(x)$.

59. $f(x) = x^2 - 10x + 28; g(x) = 4$

60. $f(x) = x^2 + 3x - 7; g(x) = 3$

61. $f(x) = 3x^2 + 4x; g(x) = 9x + 2$

62. $f(x) = 4x^2 - x; g(x) = 1 - x$

63. $f(x) = 5x^2 - 6x + 4; g(x) = x^2 - 13x + 1$

64. $f(x) = 2x^2 + 11x - 16; g(x) = 9 + x - x^2$

Mixed Practice

Solve.

65. $v(v - 6) = -8$

66. $(a + 7)(3a - 2) = 0$

67. Given $f(x) = 2x^2 + 3x$ and $g(x) = -6x + 5$, find all values of x such that $f(x) = g(x)$.

68. $5x^2 - 11x - 12 = 0$

69. $p^2 - 40 = 3p$

70. $(n - 3)(n + 2) = 14$

Applications

B *Solve.*

71. In a league with t teams, $\dfrac{t(t - 1)}{2}$ games must be played in a season if every team must play every other team once. For a league season to consist of 28 games, how many teams must be in the league?

72. A financial analyst estimates that a company's revenue for selling x units of a product is given by the equation $R = 100x - \dfrac{1}{4}x^2$. How many units must the company sell in order to generate a revenue of $10,000?

73. The school committee approves funds to construct a soccer field for the high school on a 5000-yd^2 plot of land. If the plan requires the length of the soccer field to be 50 yd longer than the width w, what are the dimensions of the soccer field?

74. A gardener decides to build a stone walkway of uniform width around her flower bed. The flower bed measures 10 ft by 12 ft. If she wants the area of the bed and the walkway together to be 224 ft^2, how wide should she make the walkway?

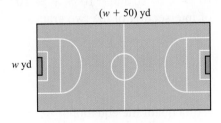

(w + 50) yd

w yd

75. The client of an investment broker invested $8000 in a high-risk fund. After 2 yr, the investment grew by $1680. To determine the average annual rate of return r on the investment, the broker used the equation $8000[(1 + r)^2 - 1] = 1680$. Find this rate of return on the investment expressed as a percent.

76. An object is thrown straight upward from a height of 200 ft with an initial velocity of 64 ft/sec. Its height h above the ground after t sec is given by the equation $h = -16t^2 + 64t + 200$. In how many seconds will the object be 120 ft above the ground?

77. A painter uses a 15-foot ladder to reach a window that is 12 ft above the ground. How far from the house should he place the ladder in order to reach the window?

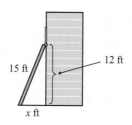

78. A train leaves the station traveling north. At the same time, a second train leaves the station traveling east. After some time, the train traveling north is 20 mi from the station and the train traveling east is 15 mi from the station. How far apart are the two trains at that time?

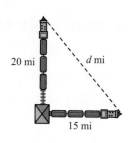

79. The width of a 15-inch laptop computer screen is 3 in. shorter than the length. What are the dimensions of the screen?

80. A 65-foot wire is attached to the top of an antenna and stretches to the ground d ft from the base of the antenna. The height of the antenna is 10 ft greater than twice the distance from the base of the antenna to the wire. How tall is the antenna?

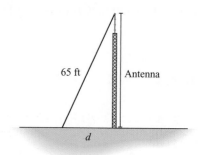

• Check your answers on page A-39.

MINDStretchers

Technology

1. Consider the quadratic equation $x^2 - x - 2 = 4$.

  a. Solve the equation algebraically.

  b. On a graphing calculator, enter the expression on the left side of the equation in Y1 and the expression on the right side in Y2 and graph the equations in the standard viewing window ($-10 \le x \le 10$ and $-10 \le y \le 10$). Use the **TRACE** or **INTERSECT** feature to find the point (or points) of intersection of the two graphs.

  c. Compare your answers in parts (a) and (b). What do you notice?

  d. Repeat the procedure for the equation $2x^2 - 3x - 6 = -1$.

  e. Explain how you can use a graphing calculator to verify the solutions of a quadratic equation.

Writing

2. Explain why the equation $(x + a)(x + b) = c$, where $c \ne 0$, cannot be solved by setting each factor equal to c.

Critical Thinking

3. The zero-product property is true for more than two factors. Use this fact and the techniques of this section to find the solutions of each of the following polynomial equations:

  a. $x^3 + 3x^2 - 18x = 0$
  b. $x^4 - 5x^2 + 4 = 0$
  c. $4x^3 + 12x^2 - x - 3 = 0$

Cultural Note

The nineteenth-century monk Gregor Mendel, pictured to the right, laid the foundation for the science of genetics in his experiments with garden peas. Mendel carried out a series of controlled crossbreedings using pea plants, some tall and some short. He recorded the number of offspring with each trait in the crosses. On the basis of these findings, Mendel proposed a series of formal rules that describes the behavior of hereditary factors characteristic of the pea plants—rules that are equally applicable to human genetic traits. At the beginning of the twentieth century, geneticists extended Mendel's rules of inheritance to a law that predicts the distribution of genes in various populations (the Hardy-Weinberg Law). This law is based on the *multiplication of binomials.* There are many other examples in the history of mathematics where a mathematical topic, such as the multiplication of binomials, was studied long before any important application of the topic was known.

(*Source:* Neil Campbell and Jane Reece, *Biology.* Benjamin Cummings, 2002)

Key Concepts and Skills

Concept/Skill	Description	Example
[5.1] **Monomial**	An expression that is the product of a real number and variables raised to nonnegative integer powers.	$2x^3$ $-6a^2b^4$
[5.1] **Polynomial**	A polynomial is an algebraic expression with one or more monomials added or subtracted.	$-7x \longleftarrow$ Monomial $3x + 5 \longleftarrow$ Binomial $\left.\rule{0pt}{2.5em}\right\}$ Polynomials $x^2 - 6x + 5 \longleftarrow$ Trinomial $x^3 + 4x^2 - x + 15$
[5.1] **Degree of a monomial**	For a monomial in one variable, the power of the variable. For a monomial in more than one variable, the sum of the powers of the variables.	$4x^2$ is of degree 2. $-3x^4y$ is of degree 5.
[5.1] **Degree of a polynomial**	The highest degree of any of its terms.	$4x^3 + 5x^2 - x + 10$ is of degree 3. $5x^2y^4 - xy + 10y$ is of degree 6.
[5.1] **Leading term of a polynomial**	The term in the polynomial with the highest degree. The coefficient of that term is called the **leading coefficient**. The term of degree 0 is called the **constant term**.	$3x^4 - 5x^2 + 6$ $3x^4 \longleftarrow$ Leading term $3 \longleftarrow$ Leading coefficient $6 \longleftarrow$ Constant term
[5.1] **To add polynomials**	• Add the like terms.	Add horizontally: $(4x^2 + 3x - 9) + (8x^2 + 7)$ $(4x^2 + 3x - 9) + (8x^2 + 7)$ $= 4x^2 + 3x - 9 + 8x^2 + 7$ $= 12x^2 + 3x - 2$ Add vertically: $(4x^2 + 3x - 9) + (8x^2 + 7)$ $\quad 4x^2 + 3x - 9$ $\underline{\quad 8x^2 \qquad + 7}$ $\quad 12x^2 + 3x - 2$
[5.1] **To subtract polynomials**	• Change the sign of each term of the polynomial being subtracted. • Add.	Subtract horizontally: $(3x^2 - 5x + 2) - (-x^2 + 4x - 6)$ $(3x^2 - 5x + 2) - (-x^2 + 4x - 6)$ $= 3x^2 - 5x + 2 + x^2 - 4x + 6$ $= 4x^2 - 9x + 8$ Subtract vertically: $(3x^2 - 5x + 2) - (-x^2 + 4x - 6)$ $\quad 3x^2 - 5x + 2 \qquad \quad 3x^2 - 5x + 2$ $\underline{-(-x^2 + 4x - 6)} \quad \underline{+ \; x^2 - 4x + 6}$ $\qquad\qquad\qquad\qquad\qquad 4x^2 - 9x + 8$
[5.2] **To multiply monomials**	• Multiply the coefficients. • Multiply the variables, using the product rule of exponents.	$(-4a^2b)(2a^3b^2)$ $= (-4 \cdot 2)(a^2 \cdot a^3)(b \cdot b^2)$ $= -8a^5b^3$

CONCEPT   SKILL

Concept/Skill	Description	Example
[5.2] **To multiply a monomial and a polynomial**	• Use the distributive property.	$2x^3y^4(2x^5y^2 + 5y^3)$ $= 2x^3y^4 \cdot 2x^5y^2 + 2x^3y^4 \cdot 5y^3$ $= 4x^8y^6 + 10x^3y^7$

[5.2] **To multiply two binomials using the FOIL method**	Consider $(a + b)(c + d)$.
	 • Multiply the two *first* terms in the binomials.
	 $(a + b)(c + d)$ Product ac
	       F
	 • Multiply the two *outer* terms.
	 $(a + b)(c + d)$ Product ad
	       O
	 • Multiply the two *inner* terms
	 $(a + b)(c + d)$ Product bc
	       I
	 • Multiply the two *last* terms.
	 $(a + b)(c + d)$ Product bd
	       L
	 The product of the two binomials is the sum of these four products:
	 $(a + b)(c + d) = ac + ad + bc + bd$
		Last
		 First
		 $(4x - 1)(2x + 1)$
		 Inner
		 Outer
		
		 F $(4x)(2x) = 8x^2$
		 O $(4x)(1) = 4x$
		 I $(-1)(2x) = -2x$
		 L $(-1)(1) = -1$
		
		 The product $(4x - 1)(2x + 1)$ is the sum of these four products:
		 $8x^2 + 4x - 2x - 1 = 8x^2 + 2x - 1$

[5.2] **The square of a binomial**	$(a + b)^2 = a^2 + 2ab + b^2$ $(a - b)^2 = a^2 - 2ab + b^2$	$(x + 4)^2 = x^2 + 2(x)(4) + (4)^2$ $= x^2 + 8x + 16$ $(2y - 5)^2 = (2y)^2 - 2(2y)(5) + (5)^2$ $= 4y^2 - 20y + 25$
[5.2] **The product of the sum and difference of two terms**	$(a + b)(a - b) = a^2 - b^2$	$(2x + y)(2x - y) = (2x)^2 - (y)^2$ $= 4x^2 - y^2$
[5.3] **To divide a polynomial by a polynomial**	• Arrange each term of the dividend and divisor in descending order. • Divide the first term of the dividend by the first term of the divisor. The result is the first term of the quotient. • Multiply the first term of the quotient by the divisor and place the product under the dividend. • Subtract the product, found in the previous step, from the dividend. • Bring down the next term to form a new dividend. • Repeat the process until the degree of the remainder is less than the degree of the divisor.	$(2x^3 - 6x - 11) \div (x - 1)$ $\begin{array}{r} 2x^2 + 2x - 4 \\ x - 1 \overline{)\ 2x^3 + 0x^2 - 6x - 11} \\ \underline{-\ 2x^3 + 2x^2} \\ 2x^2 - 6x \\ \underline{-\ 2x^2 + 2x} \\ -4x - 11 \\ \underline{4x - 4} \\ -15 \end{array}$ $2x^2 + 2x - 4 + \dfrac{-15}{x - 1}$

continued

Concept/Skill	Description	Example
[5.4] GCF of two or more monomials	The product of the greatest common factor (GCF) of the coefficients and the highest powers of the variables common to all of the monomials.	The GCF of $8x^3y$, $20xz$, and $32x^2z^2$ is $4x$.
[5.4] To factor by grouping	• Group pairs of terms and factor out a GCF in each group. • Factor out the common binomial factor.	$2xy - 4x + 3y - 6$ $= 2x(y - 2) + 3(y - 2)$ $= (y - 2)(2x + 3)$
[5.5] Prime polynomial	A polynomial that is not factorable.	$x^2 + 5x + 1$
[5.5] To factor a trinomial of the form $ax^2 + bx + c$, where $a = 1$	• List and test the factors of c to find two numbers for $(x + ?)(x + ?)$ whose sum is b.	$x^2 - 6x + 5 = (x - 5)(x - 1)$ because Product of factors $(-5)(-1) = 5$ $(-5) + (-1) = -6$ Sum of factors
[5.5] To factor a trinomial $ax^2 + bx + c$, where $a \neq 1$ (trial-and-error method)	• List and test all possible factors of a and c to find four numbers for $(?x + ?)(?x + ?)$ so that the product of the leading coefficients of the binomial factors is a, the product of the constant terms of the binomial factors is c, and the coefficients of the inner and outer products add up to b.	$2x^2 + 7x + 5 = (2x + 5)(x + 1)$ because $2 \cdot 1 = 2, 5 \cdot 1 = 5$, and $5x + 2x = 7x$.
[5.5] To factor a trinomial $ax^2 + bx + c$, where $a \neq 1$ (ac method)	• Form the product ac. • Find two factors of ac that add up to b. • Use these factors to split up the middle term in the original trinomial. • Group the first two terms and the last two terms. • Factor out the common factor from each group. • Factor out the common binomial factor.	For $3x^2 - x - 10$, $ac = 3 \cdot (-10) = -30$. $5 \cdot (-6) = -30$ $5 + (-6) = -1$ $3x^2 - x - 10$ $= 3x^2 + 5x - 6x - 10$ $= (3x^2 + 5x) + (-6x - 10)$ $= x(3x + 5) - 2(3x + 5)$ $= (3x + 5)(x - 2)$
[5.6] Factoring a perfect square trinomial	$a^2 + 2ab + b^2 = (a + b)^2$ $a^2 - 2ab + b^2 = (a - b)^2$	$x^2 + 10x + 25 = (x + 5)^2$ $9x^2 - 6xy + y^2 = (3x - y)^2$
[5.6] Factoring the difference of squares	$a^2 - b^2 = (a + b)(a - b)$	$x^2 - 36 = (x + 6)(x - 6)$ $4p^6 - 9q^4 = (2p^3 + 3q^2)(2p^3 - 3q^2)$
[5.6] Factoring the sum or difference of cubes	$a^3 + b^3 = (a + b)(a^2 - ab + b^2)$ $a^3 - b^3 = (a - b)(a^2 + ab + b^2)$	$x^3 + 27 = (x + 3)(x^2 - 3 \cdot x + 3^2)$ $\quad = (x + 3)(x^2 - 3x + 9)$ $8x^3 - y^3$ $= (2x)^3 - y^3$ $= (2x - y)[(2x)^2 + 2x \cdot y + y^2]$ $= (2x - y)(4x^2 + 2xy + y^2)$

CONCEPT   SKILL

Concept/Skill	Description	Example
[5.7] **Second-degree or quadratic equation**	An equation that can be written in the form $ax^2 + bx + c = 0$, where a, b, and c are real numbers and $a \neq 0$.	$x^2 - 2x + 8 = 0$ $6n^2 + 13n = 5$
[5.7] **The zero-product property**	If $ab = 0$, then $a = 0$, $b = 0$, or both a and $b = 0$.	If $2x(3x + 1) = 0$, then $2x = 0$ or $3x + 1 = 0$.
[5.7] **To solve a quadratic equation by factoring**	• If necessary, rewrite the equation in standard form, with 0 on one side. • Factor the other side. • Use the zero-product property to get two simple linear equations. • Solve the linear equations. • Check by substituting the solutions in the original quadratic equation. • State the solutions to the original equation.	$$x^2 = 5x + 6$$ $$x^2 - 5x - 6 = 0$$ $$(x - 6)(x + 1) = 0$$ $$x - 6 = 0 \quad \text{or} \quad x + 1 = 0$$ $$x = 6 \qquad\qquad x = -1$$ **Check** Substitute 6 for x. $$x^2 = 5x + 6$$ $$6^2 \stackrel{?}{=} 5(6) + 6$$ $$36 = 36 \quad \text{True}$$ Substitute -1 for x. $$x^2 = 5x + 6$$ $$(-1)^2 \stackrel{?}{=} 5(-1) + 6$$ $$1 = 1 \quad \text{True}$$ So the solutions are -1 and 6.

Say Why

Fill in each blank.

1. The expression $-\dfrac{1}{2}x^{-3}$ _____ a monomial
   is/is not

   because _____
   _____.

2. The expression $\dfrac{3x^2 - 2x + 5}{4}$ _____ a
   is/is not

   polynomial because _____
   _____.

3. The degree of the monomial a^2b^5 _____ 10
   is/is not

   because _____
   _____.

4. The leading coefficient of the polynomial

   $6y^3 + 2y^4 + 5y$ _____ 2 because
   is/is not

   _____
   _____.

5. The GCF of $8x^3y$ and $12x^2y^2$ _____ $4x^2y$ because
   is/is not

   _____
   _____
   _____.

6. $3x^2 = 7 - x$ _____ an example of a second-
   is/is not

   degree equation because _____

   _____
   _____.

[5.1] *Determine whether each expression is a polynomial.*

7. $\dfrac{t^3}{4} - 3t^2 + 2t - 8$

8. $4x^2 - 5x^{-1}$

Identify the terms and coefficients of the polynomial.

9. $3n - 1$

10. $x^5 - 4x^3 - x^2 + \dfrac{x}{2}$

Classify the polynomial as a monomial, binomial, trinomial, or other polynomial. Then, identify the degree of the polynomial.

11. $7a^3 - 6a^2$

12. $-2xy$

13. $9p^4 - 3p^3 + 8p - 11$

14. $4x^2 - 5x^5 + 1$

Rewrite each polynomial in descending order. Then, identify the leading term and the leading coefficient.

15. $5y^2 - 3y - y^6 + y^3 - 4y^4 + 10$

16. $7 - 2x + 9x^2 - \dfrac{x^3}{5}$

Simplify. Then, write the polynomial in descending order of powers.

17. $1 - 6n^2 + 3n - n^2 + 2n^3 + 7 + 8n$

18. $x^2 - \dfrac{1}{4}x^4 - \dfrac{1}{2}x + 2x^4 - x + 3x^2$

Evaluate the polynomial for the given values of the variable.

19. $8x^2 - 16x - 3$
  a. $x = 3$
  b. $x = -\dfrac{3}{4}$

20. $-x^3 + 6x^2 + 5x - 9$
  a. $x = 2$
  b. $x = -2$

Perform the indicated operations.

21. $(9y^2 + 2y - 13) + (17 + y - 10y^2 - 3y^3)$

22. $(12n^5 + 3n^4 - 11n^2 - 6n) - (5n^5 - 2n + 3n^4 - 16n^2 + 1)$

23. Subtract $2a^3 - 10a^2 + \dfrac{1}{3}a - 3$ from $a^3 - \dfrac{2}{3}a - 15a^2 + 9$.

24. Add $4x^2 - 7x - 11, 9x - x^2 + 2x^3$, and $8x + 4$.

Simplify.

25. $(7x^2 + 1) - (3x^2 - 5x + 2) - (2 + x + 4x^2)$

26. $(8n^3 - 1) + (2n^3 + 3n^2 - 4n + 6) - (5n^3 - 3n^2)$

[5.2] *Find the product.*

27. $(-3x^2)(2 - 4x)$

28. $\left(\dfrac{1}{3}ab^3\right)(9a^2 - 18ab - 3b^2)$

Simplify.

29. $2x^2(2x - 4) - x(3x^2 - 8x - 5)$

30. $6b(6a^3 - a^2b^2 + b) - 4a^2(5ab + b^3) - b^2$

Multiply.

31. $(n - 6)(n + 10)$

32. $(3t + 1)(t + 2)$

33. $(2x - 7)(2x - 5)$

34. $(4x + 3y)(3x - 4y)$

35. $(3p^2 - pq - 2q^2)(p^2 + q)$

36. $(u^2 - 4v^2)(u^4 + 4u^2v^2 + 16v^4)$

37. $(x^2 - 3x + 6)(2x^2 + x - 3)$

38. $(3a^2 - 7a + 1)(2a^3 + a^2 - 3a - 4)$

39. $(5x - 3)(5x + 3)$

40. $(3p^2 + q)(q - 3p^2)$

41. $(4n - 7)^2$

42. $(9a + 2b)^2$

43. $(2a + 3)^3$

44. $(4x - y)(x + y)(4x + y)$

[5.3] *Divide.*

45. $(32a^4 - 16a^2 + 24a - 4) \div (8a)$

46. $\dfrac{9p^3q^4 - 3p^2q^3 + 6pq^2 - q}{-3p^2q}$

47. $(16xy^3z - 8xy^2z^2 + 12x^2y^2z) \div (4xyz)$

48. $(2x^2 + 9x - 35) \div (2x - 5)$

49. $(5x^3 - x^2 + 10x + 2) \div (x^2 + 2)$

50. $(125x^3 - 8) \div (5x - 2)$

51. $(4x^4 - 18x^3 + 9x^2 - 10) \div (x - 4)$

52. $\dfrac{9x^3 + 2x - 1}{3x - 1}$

[5.4] *Factor.*

53. $36x^5y^3 - 27x^4y - 9x^3y^2 + 18x^2y^4$

54. $2p(p - q) - (q - p)$

55. $x^2 + 14y + 7x + 2xy$

56. $2a^2 + 6b - 4a - 3ab$

57. $ab - 5a - 8b + 40$

58. $3u^2 + 9u^2v - 2w^2 - 6vw^2$

Solve the equation for the indicated variable.

59. $S = \dfrac{1}{2}an + \dfrac{1}{2}bn$, for n

60. $T = kx_1 + kx_2$, for k

[5.5] *Factor, if possible.*

61. $x^2 - 9x + 20$

62. $5 - 5p - p^2$

63. $a^2 + 10ab + 21b^2$

64. $-4p^4q^2 + 4p^3q^3 + 48p^2q^4$

65. $n^5 + 14n^3 - 32n$

66. $4a^2 - 28a + 49$

67. $3p^2 - 38pq - 24q^2$

68. $12u^2 - 10uv - 12v^2$

69. $-30x^3y - 42x^2y - 12xy$

70. $8p^6 - 2p^3q^2 - q^4$

[5.6] *Factor, if possible.*

71. $4n^2 - 44n + 121$

72. $9a^2 + 30ab + 25b^2$

73. $100y^2 - 49x^2$

74. $3u^3 + 81$

75. $64c^3 - 27d^3$

76. $12x^6y^4 - 12x^3y^2 + 3$

77. $(x + y)^2 - z^2$

78. $1 + (3a + 1)^3$

79. $32u^{2n} - 2v^{2m}$

80. $x^2(x - 1) + 9(1 - x)$

[5.7] *Solve.*

81. $(x + 1)(5x - 4) = 0$

82. $n^2 - 64 = 0$

83. $5y^2 = 15y$

84. $8 - 2t - 3t^2 = 0$

85. $12k^2 - 60k = -75$

86. $5r^2 + 18 = 21r - r^2$

87. $3a(3a + 4) = -4$

88. $(3x - 4)(3x + 5) = 5x^2 - 13$

Mixed Applications

Solve.

89. The revenue that an outerwear manufacturer takes in is given by the equation $R = xp$, where x represents the number of jackets sold and p is the price (in dollars) for each jacket. If the manufacturer determines that $p = 250 - \frac{1}{4}x$, write a polynomial for the revenue in terms of x.

90. A company produces and sells computer CD-ROM discs. The company's weekly cost (in dollars) to produce x discs is given by the function $C(x) = 0.1x + 1500$. The revenue for selling x discs (in dollars) is given by the function $R(x) = 0.25x$. Find a function $P(x)$ that represents the company's weekly profit for selling x discs. (*Hint:* $P(x) = R(x) - C(x)$)

91. A company manufactures beach balls. The radius of a small beach ball is r_1 in. and the radius of a large beach ball is r_2 in. The difference in the surface areas of the two beach balls is given by the expression $4\pi r_2^2 - 4\pi r_1^2$. Write this expression in factored form.

92. The in-state tuition at a college increases by the same percent every year. A full-time student pays \$2000 her first year. The total amount she pays for the second year of college can be calculated using the expression $\dfrac{2000(1 - r^2)}{1 - r}$. Simplify this expression.

93. The owner of a store determines that his daily revenue R for selling DVD movies (in dollars) is given by the equation $R = 200x - 2x^2$, where x is the number of DVDs sold per day. How many DVDs must be sold per day to generate revenue of \$5000?

94. Two airplanes depart from the same airport at the same time. One airplane flies due west, and the other flies due north. After some time, the faster plane is 150 mi farther from the airport than the slower airplane. If the airplanes are 750 mi apart at that time, how far did each airplane travel?

95. The owner of a kennel expanded the rectangular dog run used for exercising the dogs by adding x ft to the length and width of the existing dog run. The area of the new dog run is given by the polynomial $x^2 + 58x + 720$.

96. A 30-inch by 40-inch piece of cardboard is used to make open boxes by cutting x-inch by x-inch squares from each corner. If specifications require the area of the base of the box to be 336 inch^2, what size square should be cut from each corner?

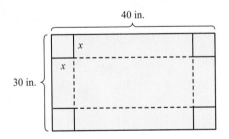

a. Factor the polynomial.

b. What was the area of the original dog run?

97. In order to install cable television in a home, a cable technician runs cable wire diagonally across a rectangular attic floor (see diagram). How much cable wire did the technician use in the attic?

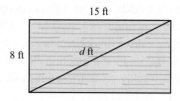

98. As part of a home remodeling project an architect plans to make an office out of one of the rooms in the house by increasing the width and decreasing the length the same number of feet, as shown in the following diagram. Write a polynomial expression that represents the area of the office.

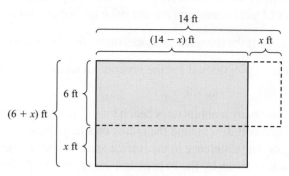

99. An x-inch by x-inch picture is surrounded by a matte border y in. wide. The total area of the picture including the matte border is $(x^2 + 4xy + 4y^2)$ in.^{2}. Factor the expression for the area.

100. A department store offers a 15% discount on every item in the store. A customer purchases a shirt that costs p dollars and a skirt and a pair of jeans that cost q dollars each. The expression $0.85p + 1.7q$ represents the total purchase price with the discount.

   a. Write the expression in factored form.

   b. Use the factored form of the expression in part (a) to find the total purchase price if the original price of the shirt was $30 and the original price of the skirt and jeans was $45 each.

   c. How much did the customer save in all with the 15% discount?

• Check your answers on page A-39.

CHAPTER 5   Posttest

 FOR EXTRA HELP

 CHAPTER Test Prep VIDEOS

The Chapter Test Prep Videos with test solutions are available on DVD, in MyMathLab, and on YouTube™ (search "AkstIntermediate Alg" and click on "Channels").

To see if you have mastered the topics in this chapter, take this test.

1. For the polynomial $1 + 5n^3 - 4n^5 - n^4 - \dfrac{n}{3}$,

   a. list the terms _____

   b. list the coefficients _____

   c. write the polynomial in descending order

   _____

   d. identify the degree _____

   e. identify the leading term and leading coefficient

   _____

2. Add $(2x^4 + 6x^3 - x^2 - 3x + 8)$ and $(3x^4 - 4x + x^2 - 5x^3)$.

3. Subtract $(4y - 9y^3 + 7y^2 - 11)$ from $(8y^2 - 10y^3 + y - 14)$.

Perform the indicated operation.

4. $\left(-\dfrac{2}{3}x^2y^3\right)(12x^3y + 9xy - 6y - 15y^2)$

5. $(2a - 7b)(7a + 2b)$

6. $(5 - 3p)^2$

7. $(-9v + 4u)(4u + 9v)$

8. $(1 - 3x - x^2)(2x^2 + 5x - 2)$

9. $\dfrac{9x^4 - 10x^2 + 1}{3x - 1}$

Factor.

10. $8x^2 - 6x - 35$

11. $-6a^3b^3 - 36a^2b^2 - 48ab$

12. $81p^2 - 100q^2$

13. $64 - n^3$

14. $x^3 + x^2y - 4yz^2 - 4xz^2$

Solve.

15. $4r^2 - 16 = 0$

16. $(x + 3)(2x - 5) = 6$

17. The average size of a farm (in acres) in the United States can be modeled by the polynomial $2x^3 - 11x^2 + 5x + 443$, where x is the number of years since 2005. Approximate the average size of a farm in the year 2006. (*Source:* nass.vsda.gov)

18. The amount in an account earning interest compounded annually is given by the expression $P(1 + r)^t$, where P is the principal amount invested, r is the rate of return per year (in decimal form), and t is the number of years that the money has been invested.

   a. Suppose $1000 is invested in an account. Write a polynomial that shows the amount in the account after 2 yr and after 3 yr.

   b. What is the difference between the amount of money in the account after 3 yr as compared to 2 yr? Express your answer as a polynomial in r.

19. The distance (in feet) that an accelerating car travels in the t sec after it begins to accelerate can be represented by $20t + 5t^2$. If this distance is 160 ft, for how many seconds was the car accelerating?

20. How far from the base of the building shown should the ladder be placed in order to reach a point on the building 16 ft above the ground?

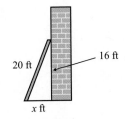

20 ft

16 ft

x ft

• Check your answers on page A-40.

Cumulative Review Exercises

To help you review, solve these problems.

1. Rewrite the following expression without grouping symbols: $-(x - 3.4y)$

2. Simplify: $(6a^4c)(2a^2b^{-1}c)^{-3}$

3. Solve: $5x - (3x - 7) = \dfrac{1}{2}(12 - 4x) + 11$

4. Solve the following formula for y_1: $m = \dfrac{y_2 - y_1}{x_2 - x_1}$

5. Solve: $|2x + 3| \le 9$

6. Graph: $6x - 4y = -4$

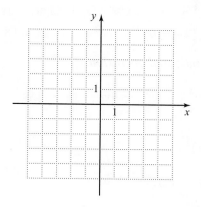

7. Graph: $2x - y \ge 1$

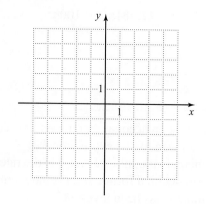

8. Solve the system:
$$2x - y + 3z = -3$$
$$4x + 5y - 2z = 16$$
$$x - 2y - z = -2$$

9. Solve the following system:
$$4x \quad\quad - z = 16$$
$$6y + 3z = -12$$
$$5x - 5y \quad\quad = 15$$

10. Solve by graphing:
$$y < 1 - x$$
$$y \ge \dfrac{1}{3}x - 3$$

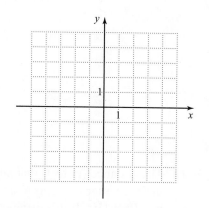

11. Multiply: $\left(x + \dfrac{2}{3}y\right)^2$

12. Divide: $(x^3 - x^2 - 6x + 8) \div (x - 2)$

13. Factor, if possible: $a^2b^2 + 3ab^2 - 18b^2$

14. Factor, if possible: $8x^3y^3 + 27$

15. An electric company charges a monthly service fee of $4 plus $0.08 per kilowatt-hour of electricity used. Write an equation that relates a customer's monthly bill, b, to the number of kilowatt-hours, h, of electricity used per month.

16. A couple has $15,000 to invest in two accounts earning simple interest. If one account earns 5% interest per year and the other earns 6% interest per year, how much should they invest in each account in order to earn at least $875 after 1 yr?

17. The population of a city increases by approximately 2000 people per year. Five years ago, its population was 100,000.

   a. Express the population y of the city in terms of x, the number of years from now.

   b. What was the city's population last year?

18. A discount store sells all used paperback books for $2.50 each and all used hardcover books for $6 each. Last week, the bookstore sold 548 books. If the sales totaled $2007, how many of each type of book did the store sell?

19. The price (in dollars) of a Disneyland one-day adult admission ticket can be modeled by the function $H(x) = -0.05x^3 + 0.4x^2 + 4x + 47$, where x is the number of years since 2003.
   (*Source:* ocresort.ocregister.com)

   a. Use functional notation to express the price of an admissions ticket in 2004.

   b. Evaluate $H(9)$

20. On November 26, 2010, the temperature (in degrees Fahrenheit) in Anaheim, California from 11:00 A.M. to 6 P.M. could be modeled by the polynomial $-x^2 + 4x + 60$, where x represents the number of hours past 11:00 A.M. According to this model, at what time was the temperature 55°F? (*Source:* forecast.weather.gov)

• Check your answers on page A-40.

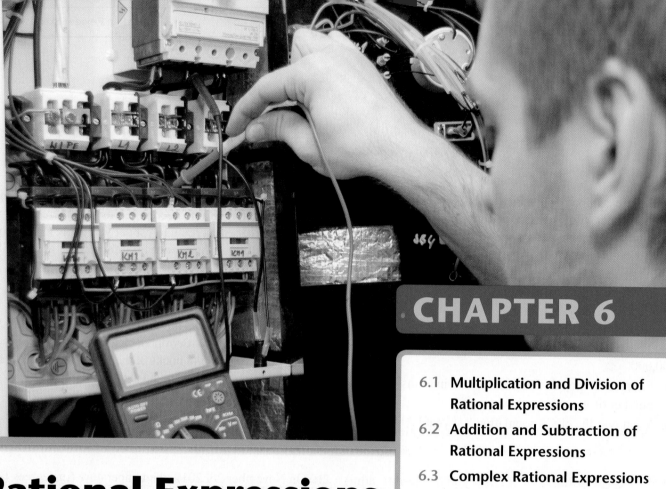

Rational Expressions and Equations

Rational Equations and Electrical Circuits

Electricians work with parallel circuits when wiring houses and other buildings. A parallel circuit provides more than one path for the electricity to follow. Electrical devices that are plugged into the circuit act as resistors, that is, as devices that use electricity to do work. Some examples include lamps, heaters, and toasters. If you turn off your lamp, all the other electrical devices will continue to work because the electricity flows through the other branches of the parallel circuit.

Relying on electricity, we need to understand that as more branches are added to a parallel circuit (that is, as more devices are plugged in and turned on), the total resistance in the circuit decreases. This reduced resistance causes more current to flow through the wiring in the wall. Consequently, the circuit could become overloaded, creating a fire hazard. So it is dangerous to plug too many devices into the same circuit.

Electricians calculate the total resistance of a circuit with resistors connected in parallel by using the following formula:

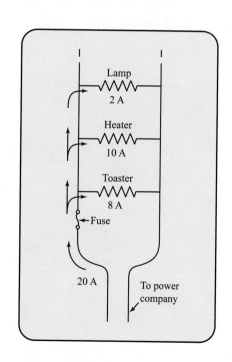

$$\frac{1}{R_T} = \frac{1}{R_1} + \frac{1}{R_2} + \cdots + \frac{1}{R_n},$$

that is, the reciprocal of the total resistance equals the sum of the reciprocals of the individual resistances. (*Sources:* Peter J. Nolan, *Fundamentals of College Physics*: Wm. C. Brown, 1993; W. Thomas Griffith, *The Physics of Everyday Phenomena*: Wm. C. Brown, 1992)

1. Identify all the values of the variable for which each expression is undefined.

   a. $\dfrac{x}{x^2 - 3x}$

   b. $\dfrac{n - 1}{n^2 + 9n + 14}$

Simplify.

2. $\dfrac{6a^3 - 9ab + a^2b^2}{3a^2b}$

3. $\dfrac{2p - 1}{1 - 4p^2}$

4. $\dfrac{y^2 - 2y - 35}{2y^2 + 13y + 15}$

Perform the indicated operation.

5. $\dfrac{8x^2y}{4x - 16} \cdot \dfrac{4 - x}{3x^3y - x^2y^2}$

6. $\dfrac{t^2 + 2t + 1}{t^2 - 2t} \cdot \dfrac{t^2 - 11t + 18}{t^2 - 8t - 9}$

7. $\dfrac{p^2 - 9q^2}{pq + 3q^2} \div (p^2 + 5pq - 24q^2)$

8. $\dfrac{4n^2 - 13n - 12}{n^3 - 64} \div \dfrac{8n^2 + 6n - 9}{n^2 + 4n + 16}$

9. $\dfrac{2}{9y^2} + \dfrac{1}{3y^2 - 6y}$

10. $\dfrac{r - 3}{r^2 - 1} - \dfrac{r + 1}{1 - r}$

11. $\dfrac{5}{x^2 - x - 12} + \dfrac{x}{x^2 + x - 6}$

12. $\dfrac{2p + 3}{4p^2 - 12p + 9} - \dfrac{p}{2p^2 - p - 3}$

13. Simplify each expression.

    a. $\dfrac{\dfrac{3}{x^2} - \dfrac{1}{3y^2}}{\dfrac{3}{x} + \dfrac{1}{y}}$

    b. $\dfrac{\dfrac{x^2 - 25}{x + 6}}{\dfrac{5 - x}{x^2 + 2x - 24}}$

Solve.

14. $\dfrac{3}{x + 2} - \dfrac{1}{x + 8} = \dfrac{12}{x^2 + 10x + 16}$

15. $\dfrac{2}{n + 1} + \dfrac{2}{n + 4} = -1$

16. $\dfrac{4}{y} = \dfrac{y}{y - 1}$

17. Find the constant of variation and the variation equation in which y varies inversely as x and $y = 2.4$ when $x = 25$.

18. A car traveled 150 mi in the same time it took a train to travel 120 mi. If the average speed of the train was 12 mph less than the average speed of the car, find the average speed of the car and the average speed of the train.

19. A pharmacy technician assists the pharmacist by preparing prescribed medications. An experienced pharmacy technician can fill the same number of prescriptions in 6 hr that an inexperienced technician fills in 9 hr. If the two technicians work together, how long will it take them to fill the prescriptions?

20. Hooke's Law states that the distance d a spring is stretched varies directly with the force F applied to the spring. If a force of 12 lb stretches a spring 3 in., how far will the spring stretch when a force of 30 lb is applied?

• Check your answers on page A-40.

6.1 Multiplication and Division of Rational Expressions

What Rational Expressions Are and Why They Are Important

In this chapter, we move beyond our previous discussion of polynomials to consider a broader category of algebraic expressions, namely, *rational expressions*. Rational expressions, sometimes called *algebraic fractions*, are useful in many disciplines, including the physical sciences, the social sciences, medicine, and business.

For example, the work of photographers who study optics can involve rational expressions. In taking pictures, photographers compute a camera's "*f*-stop," the ratio of the focal length (the distance L from the lens to the film) to the aperture (the diameter d of the lens). The *f*-stop can be represented by the rational expression $\dfrac{L}{d}$.

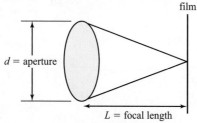

d = aperture

film

L = focal length

(Graphic adapted from Leslie Strobel, John Compton, Ira Current, and Richard Zakia, *Basic Photographic Materials and Processes*: Focal Press, 2000)

Now, we discuss how to simplify rational expressions, as well as how to perform operations on rational expressions. We also focus on how to solve equations involving rational expressions and problems involving direct and inverse variations.

Introduction to Rational Expressions

Rational expressions in algebra are similar to rational numbers. In arithmetic, a rational number such as $\dfrac{5}{8}$ is the quotient of two integers, whereas in algebra a rational expression such as $\dfrac{L}{d}$ is the quotient of two polynomials.

> **DEFINITION**
>
> A **rational expression** $\dfrac{P}{Q}$ is an algebraic expression that can be written as the quotient of two polynomials, P and Q, where $Q \neq 0$.

Other examples of rational expressions include the following:

$$\frac{-x^2}{2xy} \qquad \frac{3x+1}{x-5} \qquad \frac{5x}{x^2-x-6}$$

Since we can write a rational expression in terms of division, we must be sure that its denominator does not equal 0. When a variable is replaced with a value that makes the denominator 0, the rational expression is undefined for that value.

Consider, for instance, the rational expression $\dfrac{5}{x+3}$. This expression is undefined when the denominator is equal to 0, that is, when $x + 3 = 0$. Solving the equation gives us $x = -3$. When -3 is substituted for x in the rational expression, we get:

$$\frac{5}{x+3} = \frac{5}{-3+3} = \frac{5}{0} \leftarrow \text{Undefined}$$

We can also see why this expression is undefined at this x-value by considering the graph of $y = \dfrac{5}{x+3}$. Note that the graph never intersects the line $x = -3$. This line, shown on the coordinate plane as dashed, is called an *asymptote*. No point on the graph of $y = \dfrac{5}{x+3}$ has x-value -3; that is, the graph is not defined at $x = -3$.

So we also see graphically that the expression $\dfrac{5}{x+3}$ is undefined when x is equal to -3.

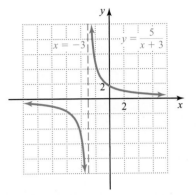

EXAMPLE 1

Find all numbers for which each rational expression is undefined.

a. $\dfrac{3x+1}{x-5}$ **b.** $\dfrac{5x}{x^2-x-6}$

Solution To find the values for which the rational expressions are undefined, we set each denominator equal to 0 and then solve for x.

a. $\dfrac{3x+1}{x-5}$ $\quad x - 5 = 0$

$x = 5$

So $\dfrac{3x+1}{x-5}$ is undefined when x is equal to 5.

b. $\dfrac{5x}{x^2-x-6}$ $\quad x^2 - x - 6 = 0$

$(x-3)(x+2) = 0$

$x - 3 = 0 \quad \text{or} \quad x + 2 = 0$

$x = 3 \qquad\qquad x = -2$

So $\dfrac{5x}{x^2-x-6}$ is undefined when x is equal to 3 or -2.

PRACTICE 1

Find the values of the variable for which each rational expression is undefined.

a. $\dfrac{2m-7}{m+4}$

b. $\dfrac{5x}{x^2-36}$

Note that throughout the remainder of this book, *we will assume that no variable in a rational expression represents a value that makes the denominator* 0.

Simplifying Rational Expressions

A rational expression (like a rational number) is considered to be in *simplified form* (or in *lowest terms*) when its numerator and denominator have no common factor other than 1.

To simplify a rational expression, we use the following rule for finding *equivalent rational expressions,* which are rational expressions that have the same value where they are both defined:

To Find an Equivalent Rational Expression

If P, Q, and R are polynomials, then

$$\frac{P}{Q} = \frac{PR}{QR},$$

where $Q \neq 0$ and $R \neq 0$.

In words, this rule states that to find an equivalent rational expression, we can multiply or divide the numerator and denominator of a rational expression by the same nonzero polynomial.

$$\frac{3}{x} = \frac{3 \cdot x^2}{x \cdot x^2} \quad \text{and} \quad \frac{x^2}{x - 1} = \frac{x^2(x + 5)}{(x - 1)(x + 5)}$$

We can also use this rule to find an equivalent expression by dividing the numerator and denominator by the same nonzero polynomial.

$$\frac{x^2 - 9}{x^2 + 6x + 9} = \frac{(x + 3)(x - 3)}{(x + 3)(x + 3)} \qquad \text{Factor the numerator and the denominator.}$$

$$= \frac{\cancel{(x + 3)}(x - 3)}{\cancel{(x + 3)}(x + 3)} \qquad \text{Divide out the common factor } x + 3.$$

$$= \frac{x - 3}{x + 3}$$

So $\dfrac{x^2 - 9}{x^2 + 6x + 9}$ is equivalent to $\dfrac{x - 3}{x + 3}$.

The preceding example leads us to the following rule for simplifying rational expressions:

To Simplify a Rational Expression

* Factor the numerator and denominator.

* Divide out any common factors.

For the remainder of this book, *we generally simplify any answer that is a rational expression.*

EXAMPLE 2

Write in simplest form.

a. $\dfrac{-28x^2y^5}{4xy^6}$ b. $\dfrac{6x^3}{18x^4 - 6x^2}$

Solution

a. $\dfrac{-28x^2y^5}{4xy^6} = \dfrac{4xy^5(-7x)}{4xy^5(y)}$ Factor the numerator and denominator.

$\qquad = \dfrac{\cancel{4xy^5}(-7x)}{\cancel{4xy^5}(y)}$ Divide out the common factor $4xy^5$.

$\qquad = -\dfrac{7x}{y}$

PRACTICE 2

Write in simplest form.

a. $\dfrac{-72m^2n}{-12m^3n^2}$

b. $\dfrac{5a^4}{25a^2 - 15a^3}$

b. $\dfrac{6x^3}{18x^4 - 6x^2} = \dfrac{6x^2(x)}{6x^2(3x^2 - 1)}$ Factor the numerator and denominator.

$\qquad = \dfrac{\cancel{6x^2}(x)}{\cancel{6x^2}(3x^2 - 1)}$ Divide out the common factor $6x^2$.

$\qquad = \dfrac{x}{3x^2 - 1}$

EXAMPLE 3

Simplify, if possible.

a. $\dfrac{16 + 4x}{2x + 8}$ **b.** $\dfrac{x - 3}{3 - x}$ **c.** $\dfrac{x + 2}{x - 5}$

Solution

a. $\dfrac{16 + 4x}{2x + 8} = \dfrac{4(4 + x)}{2(x + 4)}$ Factor the numerator and the denominator.

$\qquad = \dfrac{4(x + 4)}{2(x + 4)}$ Use the commutative property to rewrite $4 + x$ as $x + 4$.

$\qquad = \dfrac{\overset{2}{\cancel{4}}\cancel{(x + 4)}}{\cancel{2}\cancel{(x + 4)}}$ Divide out the common factors 2 and $x + 4$.

$\qquad = 2$

b. $\dfrac{x - 3}{3 - x} = \dfrac{x - 3}{-(x - 3)}$ Factor out -1 in the denominator.

$\qquad = \dfrac{\cancel{x - 3}}{-\cancel{(x - 3)}}$ Divide out the common factor $x - 3$.

$\qquad = \dfrac{1}{-1}$ Simplify.

$\qquad = -1$

c. $\dfrac{x + 2}{x - 5}$ The numerator $x + 2$ and the denominator $x - 5$ have no common factor other than 1. So this expression cannot be simplified.

PRACTICE 3

Simplify, if possible.

a. $\dfrac{5x + 15}{9 + 3x}$

b. $\dfrac{5 - x}{x - 5}$

c. $\dfrac{x + 1}{x - 1}$

Note that when the terms in the numerator and denominator of a rational expression differ only in sign, as in Example 3(b), they are opposites of each other; such expressions always simplify to -1.

TIP   When simplifying a rational expression, do not divide out common terms in a sum or difference in the numerator and denominator of the expression. For example, do not divide out the x's in the expression $\dfrac{x + 2}{x - 5}$.

EXAMPLE 4

Simplify.

a. $\dfrac{x^2 - x - 42}{x^2 + 8x + 12}$ **b.** $\dfrac{xy - 5y + 3x - 15}{y + 3}$

Solution

a. $\dfrac{x^2 - x - 42}{x^2 + 8x + 12} = \dfrac{(x + 6)(x - 7)}{(x + 2)(x + 6)}$ Factor the numerator and the denominator.

$= \dfrac{\cancel{(x + 6)}(x - 7)}{(x + 2)\cancel{(x + 6)}}$ Divide out the common factor $x + 6$.

$= \dfrac{x - 7}{x + 2}$

b. $\dfrac{xy - 5y + 3x - 15}{y + 3} = \dfrac{y(x - 5) + 3(x - 5)}{y + 3}$ Factor the numerator by grouping.

$= \dfrac{(x - 5)(y + 3)}{y + 3}$ Write the numerator in factored form.

$= \dfrac{(x - 5)\cancel{(y + 3)}}{\cancel{y + 3}}$ Divide out the common factor $y + 3$.

$= x - 5$

PRACTICE 4

Write in simplest form.

a. $\dfrac{n^2 + 11n + 18}{n^2 + 7n + 10}$

b. $\dfrac{y - x}{y^2 - xy + y - x}$

EXAMPLE 5

An expression important in the study of nuclear energy is $\dfrac{mu^2 - mv^2}{mu - mv}$, where m represents mass and u and v represent velocities. Write this expression in simplified form.

Solution

$$\dfrac{mu^2 - mv^2}{mu - mv} = \dfrac{m(u^2 - v^2)}{m(u - v)}$$

$$= \dfrac{m(u + v)(u - v)}{m(u - v)}$$

$$= \dfrac{\cancel{m}(u + v)\cancel{(u - v)}}{\cancel{m}\cancel{(u - v)}}$$

$$= u + v$$

PRACTICE 5

A lighting fixture globe with radius r is packed in a cubic box with side $2r$. The surface area of the globe is $4\pi r^2$, and the surface area of the box is $24r^2$. Write the ratio of the surface area of the globe to the surface area of the box in simplified form.

Multiplying Rational Expressions

Operations on rational expressions are similar to those on rational numbers. Just as we did with rational numbers, to multiply rational expressions we multiply their numerators to get the numerator of the product and multiply their denominators to get the denominator of the product. For instance,

$$\dfrac{3}{x} \cdot \dfrac{x - 1}{4} = \dfrac{3(x - 1)}{4x}.$$

When multiplying rational expressions, it is helpful to factor the numerators and denominators, divide out any common factors, and then multiply the remaining numerators and denominators.

$$\frac{3x}{x^2 + 5x} \cdot \frac{3x + 15}{6x} = \frac{3\cancel{x}}{\cancel{x}(x\cancel{+5})} \cdot \frac{\cancel{3}(x\cancel{+5})}{\underset{2}{\cancel{6}x}} = \frac{3}{2x}$$

This example suggests the following rule:

To Multiply Rational Expressions

- Factor the numerators and denominators.

- Divide the numerators and denominators by all common factors.

- Multiply the remaining factors in the numerators and the remaining factors in the denominators.

EXAMPLE 6

Multiply.

a. $\dfrac{2x}{x^2 - 6x} \cdot \dfrac{5x - 30}{3x}$

b. $\dfrac{3x^2 - 10xy - 8y^2}{9x + 12y} \cdot \dfrac{3x + 12y}{4x^2 - 64y^2}$

c. $\dfrac{n^2 + 4n - 21}{n^2 - 12n + 32} \cdot \dfrac{2n^2 - 4n - 16}{6 + n - n^2}$

Solution

a. $\dfrac{2x}{x^2 - 6x} \cdot \dfrac{5x - 30}{3x} = \dfrac{2x}{x(x - 6)} \cdot \dfrac{5(x - 6)}{3x}$ Factor the numerator and denominator.

$= \dfrac{2\cancel{x}}{\cancel{x}(x\cancel{-6})} \cdot \dfrac{5(x\cancel{-6})}{3x}$ Divide out common factors.

$= \dfrac{10}{3x}$ Multiply the remaining factors in the numerator and in the denominator.

b. $\dfrac{3x^2 - 10xy - 8y^2}{9x + 12y} \cdot \dfrac{3x + 12y}{4x^2 - 64y^2}$

$= \dfrac{(3x + 2y)(x - 4y)}{3(3x + 4y)} \cdot \dfrac{3(x + 4y)}{4(x + 4y)(x - 4y)}$

$= \dfrac{(3x + 2y)(x\cancel{-4y})}{3(3x + 4y)} \cdot \dfrac{\cancel{3}(x\cancel{+4y})}{4(x\cancel{+4y})(x\cancel{-4y})}$

$= \dfrac{3x + 2y}{4(3x + 4y)}$

PRACTICE 6

Find the product.

a. $\dfrac{7x - 14}{-20x^2} \cdot \dfrac{40x}{3x^2 - 6x}$

b. $\dfrac{5x^2 - 5y^2}{10x - 15y} \cdot \dfrac{2x^2 - 3xy}{x^2 + 2xy + y^2}$

c. $\dfrac{a^2 - a - 2}{4a^2 + 16a - 20} \cdot \dfrac{a^2 + 11a + 24}{16 - 6a - a^2}$

EXAMPLE 6 (continued)

c. $\dfrac{n^2 + 4n - 21}{n^2 - 12n + 32} \cdot \dfrac{2n^2 - 4n - 16}{6 + n - n^2}$

$$= \dfrac{(n - 3)(n + 7)}{(n - 4)(n - 8)} \cdot \dfrac{2(n + 2)(n - 4)}{(3 - n)(2 + n)}$$

$$= \dfrac{(n - 3)(n + 7)}{(n - 4)(n - 8)} \cdot \dfrac{2(n + 2)(n - 4)}{(3 - n)(n + 2)}$$

$$= \dfrac{(n - 3)(n + 7)}{(n - 8)} \cdot \dfrac{2}{-(n - 3)}$$

$$= \dfrac{(n - 3)(n + 7)}{(n - 8)} \cdot \dfrac{2}{-(n - 3)}$$

$$= \dfrac{2(n + 7)}{-(n - 8)}$$

$$= -\dfrac{2(n + 7)}{n - 8}$$

Dividing Rational Expressions

Recall that the reciprocal (or multiplicative inverse) of a rational number is formed by interchanging its numerator and denominator. For instance, the reciprocal of $\dfrac{2}{3}$ is $\dfrac{3}{2}$. Similarly, if $\dfrac{P}{Q}$ is a rational expression and $P \neq 0$, then its reciprocal is $\dfrac{Q}{P}$. For example, the reciprocal of $\dfrac{x + 2}{x - 3}$ is $\dfrac{x - 3}{x + 2}$, where $x \neq -2$.

Just as with dividing rational numbers, to divide rational expressions we take the reciprocal of the divisor and change the operation to multiplication, as shown below:

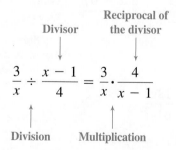

In general, when dividing rational expressions, we use the following procedure:

To Divide Rational Expressions

• Take the reciprocal of the divisor, and change the operation to multiplication.

• Follow the rule for multiplying rational expressions.

EXAMPLE 7

Find the quotient.

a. $\dfrac{6x^3}{8x - 40} \div \dfrac{30x^4}{3x - 15}$

b. $\dfrac{x - 1}{x^2 + 4x + 4} \div \dfrac{1 - x^2}{x^2 + x - 2}$

c. $\dfrac{n^2 + 4n + 3}{5n + 10} \div (n^2 - n - 12)$

Solution

a. $\dfrac{6x^3}{8x - 40} \div \dfrac{30x^4}{3x - 15} = \dfrac{6x^3}{8x - 40} \cdot \dfrac{3x - 15}{30x^4}$

Take the reciprocal of the divisor and change the operation to multiplication.

$= \dfrac{6x^3}{8(x - 5)} \cdot \dfrac{3(x - 5)}{30x^4}$

Factor the numerator and denominator.

$= \dfrac{6x^3}{8\cancel{(x - 5)}} \cdot \dfrac{3\cancel{(x - 5)}}{30x^4}$
$\qquad\qquad {}_{5}\;\;{}_{x}$

Divide out common factors.

$= \dfrac{3}{40x}$

Multiply.

b. $\dfrac{x - 1}{x^2 + 4x + 4} \div \dfrac{1 - x^2}{x^2 + x - 2}$

Take the reciprocal of the divisor and change the operation to multiplication.

$= \dfrac{x - 1}{x^2 + 4x + 4} \cdot \dfrac{x^2 + x - 2}{1 - x^2}$

$= \dfrac{x - 1}{(x + 2)(x + 2)} \cdot \dfrac{(x + 2)(x - 1)}{(1 + x)(1 - x)}$

Factor the numerators and denominators.

$= \dfrac{x - 1}{(x + 2)(x + 2)} \cdot \dfrac{(x + 2)(x - 1)}{-(1 + x)(x - 1)}$

Factor out -1 from $(1 - x)$ in the denominator.

$= \dfrac{x - 1}{\cancel{(x + 2)}(x + 2)} \cdot \dfrac{\cancel{(x + 2)}\cancel{(x - 1)}}{-(1 + x)\cancel{(x - 1)}}$

Divide out common factors.

$= \dfrac{x - 1}{-(x + 2)(x + 1)}$

Multiply.

$= -\dfrac{x - 1}{(x + 2)(x + 1)}$

Simplify.

c. $\dfrac{n^2 + 4n + 3}{5n + 10} \div (n^2 - n - 12)$

$= \dfrac{n^2 + 4n + 3}{5n + 10} \div \dfrac{n^2 - n - 12}{1}$

$= \dfrac{n^2 + 4n + 3}{5n + 10} \cdot \dfrac{1}{n^2 - n - 12}$

$= \dfrac{(n + 3)(n + 1)}{5(n + 2)} \cdot \dfrac{1}{(n + 3)(n - 4)}$

$= \dfrac{\cancel{(n + 3)}(n + 1)}{5(n + 2)} \cdot \dfrac{1}{\cancel{(n + 3)}(n - 4)} = \dfrac{n + 1}{5(n + 2)(n - 4)}$

PRACTICE 7

Divide.

a. $\dfrac{6a - 14}{25c^2} \div \dfrac{9a - 21}{5c}$

b. $\dfrac{x^2 - 4}{x^2 + 3x - 10} \div \dfrac{x + 2}{x^2 + 2x - 15}$

c. $\dfrac{t^2 + 6t - 27}{t^2 + 5t - 36} \div (4t - 12)$

EXAMPLE 8

Perform the indicated operations.

$$\frac{x - y}{4x^2 - 9y^2} \cdot \frac{3y + 2x}{x + y} \div \frac{x - y}{x^2 + 2xy + y^2}$$

Solution

$$\frac{x - y}{4x^2 - 9y^2} \cdot \frac{3y + 2x}{x + y} \div \frac{x - y}{x^2 + 2xy + y^2}$$

$$= \frac{x - y}{4x^2 - 9y^2} \cdot \frac{3y + 2x}{x + y} \cdot \frac{x^2 + 2xy + y^2}{x - y}$$

$$= \frac{x - y}{(2x + 3y)(2x - 3y)} \cdot \frac{3y + 2x}{x + y} \cdot \frac{(x + y)(x + y)}{x - y}$$

$$= \frac{\cancel{x - y}}{(\cancel{2x + 3y})(2x - 3y)} \cdot \frac{\cancel{3y + 2x}}{\cancel{x + y}} \cdot \frac{\cancel{(x + y)}(x + y)}{\cancel{x - y}}$$

$$= \frac{x + y}{2x - 3y}$$

PRACTICE 8

Perform the indicated operations.

$$\frac{2r^2 + 3rs + s^2}{r^2 - 49s^2} \div \frac{2r + s}{r - 7s} \cdot \frac{r + 7s}{2r - 3s}$$

In Chapter 5, we discussed polynomial functions such as $P(x) = x + 1$. Now, we consider rational functions. A **rational function** is a function in which the rule that defines the function is a rational expression. A rational function is of the form $f(x) = \frac{P(x)}{Q(x)}$, where $P(x)$ and $Q(x)$ are polynomials and $Q(x) \neq 0$.

EXAMPLE 9

If $f(x) = \frac{8x}{x + 1}$ and $g(x) = \frac{6x}{x - 2}$, find:

a. $f(x) \cdot g(x)$

b. $f(x) \div g(x)$

Solution

a. $f(x) \cdot g(x) = \dfrac{8x}{x + 1} \cdot \dfrac{6x}{x - 2}$ Substitute the expressions for $f(x)$ and $g(x)$.

$$= \frac{48x^2}{(x + 1)(x - 2)}$$

b. $f(x) \div g(x) = \dfrac{8x}{x + 1} \div \dfrac{6x}{x - 2}$ Substitute the expressions for $f(x)$ and $g(x)$.

$$= \frac{\overset{4}{\cancel{8x}}}{x + 1} \cdot \frac{x - 2}{\underset{3}{\cancel{6x}}}$$

$$= \frac{4(x - 2)}{3(x + 1)}$$

PRACTICE 9

If $g(t) = \dfrac{t + 3}{t - 3}$ and $h(t) = \dfrac{t + 3}{t}$, find:

a. $g(t) \cdot h(t)$

b. $g(t) \div h(t)$

EXAMPLE 10

Electricians use the formula

$$P = \frac{V^2}{R + r} \cdot \frac{r}{R + r}$$

when studying the resistance in a heating element. Here, P represents power, V voltage, and R and r resistances. Multiply the expression on the right side of the equation.

Solution

$$P = \frac{V^2}{R + r} \cdot \frac{r}{R + r}$$

$$= \frac{V^2 r}{(R + r)(R + r)}$$

$$= \frac{V^2 r}{(R + r)^2}$$

PRACTICE 10

Suppose the probability that an event occurs is $\frac{a}{b}$ and the probability that an independent (or unrelated) event occurs is $\frac{c}{d}$. The probability that neither event will occur can be represented by the expression $\left(\frac{b - a}{b}\right)\left(\frac{d - c}{d}\right)$. Write this product as a single rational expression.

Mathematically Speaking

Fill in each blank with the most appropriate term or phrase from the given list.

rational number	rational expression	multiply	factoring
divide	denominator	no common factor	has an asymptote
equivalent	simplified rational expression	common factors	intersects the *y*-axis
multiplying		equivalent rational expression	
numerator	opposites		

1. A(n) _____ $\dfrac{P}{Q}$ is an algebraic expression that can be written as the quotient of two polynomials, *P* and *Q*, where $Q \neq 0$.

2. When the _____ of a rational expression is equal to 0, the expression is undefined.

3. The graph of $y = \dfrac{4}{x-2}$ _____ at $x = 2$.

4. A rational expression is simplified when its numerator and denominator have _____ other than 1.

5. Multiplying or dividing the numerator and denominator of a rational expression by the same nonzero polynomial always results in a(n) _____.

6. A rational expression can be simplified by _____ the numerator and denominator and then dividing out any common factors.

7. If the terms in the numerator and denominator of a rational expression are _____, the expression simplifies to −1.

8. Helpful first steps in multiplying rational expressions are to factor the numerators and denominators and then to _____ them by all common factors.

A *Identify the values for which the given rational expression is undefined.*

9. $\dfrac{5}{x}$

10. $-\dfrac{8}{3t}$

11. $\dfrac{2n-1}{4}$

12. $\dfrac{1-7a}{2}$

13. $\dfrac{4x+5}{6x-3}$

14. $\dfrac{3n-2}{2n+8}$

15. $\dfrac{n+6}{n^2-8n+12}$

16. $\dfrac{5-3t}{t^2+3t-40}$

17. $\dfrac{x^2+10}{x^2-9}$

18. $\dfrac{6x}{25-x^2}$

19. $\dfrac{x^2-2x+1}{2x^2+x-3}$

20. $\dfrac{n^2+5n+6}{3n^2+10n+8}$

B *Determine whether each pair of rational expressions is equivalent.*

21. $\dfrac{2x}{3x^2-4x}$ and $\dfrac{2}{3x-4}$

22. $\dfrac{4t^2+5t}{2t^3-t^2}$ and $\dfrac{4t+5}{2t^2-t}$

23. $\dfrac{n^2+1}{n^2-1}$ and $\dfrac{n+1}{n-1}$

24. $\dfrac{x^2-4}{x-2}$ and $x+2$

25. $\dfrac{2x^2-7x+6}{16-6x-x^2}$ and $\dfrac{2x-3}{x+8}$

26. $\dfrac{n^2+10n+24}{n^2-3n-28}$ and $\dfrac{n+6}{n-7}$

Simplify, if possible.

27. $\dfrac{24a^3b}{3ab^3}$

28. $\dfrac{-18p^2q^3}{8p^3q}$

29. $\dfrac{6x^3 - 4x}{2x^2}$

30. $\dfrac{14n^4 + 7n^2}{-7n^3}$

31. $\dfrac{-12p^2q}{8p + 4q}$

32. $\dfrac{36x^3}{5x - 6xy}$

33. $\dfrac{-7 + a}{a - 7}$

34. $\dfrac{n - 1}{1 - n}$

35. $\dfrac{5 - 2y}{5y - 2}$

36. $\dfrac{4 - x}{4x - 1}$

37. $\dfrac{9n^2 - 3}{4 - 12n^2}$

38. $\dfrac{10 - 5y}{6y^2 - 12y}$

39. $\dfrac{x^2 - 16}{4x + 16}$

40. $\dfrac{2a + 3}{4a^2 - 9}$

41. $\dfrac{n^2 - 6n + 9}{n^2 + n - 12}$

42. $\dfrac{2p^2 + 14p + 20}{p^2 + 10p + 25}$

43. $\dfrac{r^2 - 8r + 15}{60 + 3r - 3r^2}$

44. $\dfrac{18 - 7t - t^2}{t^2 - 10t + 16}$

45. $\dfrac{2x^2 + 21xy + 27y^2}{3x^2 + 22xy - 45y^2}$

46. $\dfrac{4p^2 - 8pq - 5q^2}{2p^2 - 11pq - 6q^2}$

47. $\dfrac{2ab + 14b - 6a - 42}{2a^2b^2 - 6a^2b}$

48. $\dfrac{3u^2v^2 + 12uv^2}{3u^2 + 12u + 3uv + 12v}$

49. $\dfrac{64 - y^3}{16 - y^2}$

50. $\dfrac{n^2 + 2}{n^3 + 8}$

C *Multiply. Express answers in lowest terms.*

51. $\dfrac{6x^4y}{3y} \cdot \dfrac{9y^2}{18x^3y}$

52. $\dfrac{24ab}{15a^3} \cdot \dfrac{20a^3}{16a^2b^2}$

53. $\dfrac{8y - 4}{6y^2 - 12y} \cdot \dfrac{3y}{2y - 1}$

54. $\dfrac{3n + 1}{12n^3} \cdot \dfrac{9n^3 - 18n^2}{6n - 12}$

55. $\dfrac{p^2 + p - 30}{p^3 - 2p^2} \cdot \dfrac{2p^2 - 4p}{p^2 + 12p + 36}$

56. $\dfrac{5a^2 - 15a}{3a + 6} \cdot \dfrac{a^2 + 4a + 4}{a^2 - 5a + 6}$

57. $\dfrac{x^2 - 2xy}{4y^2 - x^2} \cdot \dfrac{x^2 + xy - 2y^2}{x^2 + y^2}$

58. $\dfrac{a^2 - b^2}{3b^2 + 2ab - a^2} \cdot \dfrac{a^2 + 3ab}{b^2 - ab}$

59. $\dfrac{r^2 - 7r + 12}{r^2 + r - 20} \cdot \dfrac{r^2 + 10r + 25}{24 - 5r - r^2}$

60. $\dfrac{x^2 - 4x - 45}{x^2 - 3x - 54} \cdot \dfrac{48 + 2x - x^2}{x^2 - 10x + 16}$

61. $\dfrac{2n^2 + 11n + 12}{n^2 + 3n - 4} \cdot \dfrac{n^2 - 9n + 8}{6n^2 + 5n - 6}$

62. $\dfrac{r^2 - r - 30}{3r^2 + 16r + 5} \cdot \dfrac{9r^2 + 6r + 1}{r^2 - 4r - 12}$

63. $\dfrac{x^2 - 4}{x^2 - 10x + 16} \cdot \dfrac{8 - x}{x^3 + 8}$

64. $\dfrac{9 - n}{n^3 - 64} \cdot \dfrac{n^2 + 5n - 36}{n^2 - 81}$

65. $\dfrac{4x^2 - 49y^2}{4x^3 + 28x^2y + 49xy^2} \cdot \dfrac{12xy + 6x^2y}{2x^2 + 4x + 7xy + 14y}$

66. $\dfrac{12b - 3ab + 4a - a^2}{20a^2 - 5a^3} \cdot \dfrac{2a^3b^2 + 12a^2b^3 + 18ab^4}{4a^2 - 36b^2}$

D *Divide. Express answers in lowest terms.*

67. $\dfrac{32x^4}{9xy^2} \div \dfrac{24x^2y^4}{15y^3}$

68. $\dfrac{6p^2q^2}{27q^4} \div \dfrac{21p^2}{54p^5q^3}$

69. $\dfrac{4t - 10}{12t^2} \div \dfrac{2t - 5}{6t^2 + 9t}$

70. $\dfrac{3n^2 + 7n}{8n^4 - 6n^3} \div \dfrac{12n + 28}{16n}$

71. $\dfrac{a^2 - 2a}{3a^3 + 9a^2} \div \dfrac{a^2 - 4a + 4}{a^2 + a - 6}$

72. $\dfrac{4x^3 + 24x^2}{x^2 + 4x - 12} \div \dfrac{5x^2 + 20x}{x^2 + 2x - 8}$

73. $\dfrac{p^2 - 16q^2}{p^2q - 16pq^2 + 64q^3} \div \dfrac{8q - 2p}{p^2 - 4pq - 32q^2}$

74. $\dfrac{3a^2 - 13ab + 4b^2}{3a^3b - 4a^2b^2 + ab^3} \div \dfrac{7b^2 - 6ab - a^2}{21a^3b^2 + 3a^4b}$

75. $\dfrac{3x^2 + 11x - 4}{x^2 + 13x + 36} \div \dfrac{6x^2 + 7x - 3}{2x^2 + 15x - 27}$

76. $\dfrac{4a^2 + 8a - 5}{2a^2 - 9a + 4} \div \dfrac{20 + 3a - 9a^2}{3a^2 - 8a - 16}$

77. $(3t^2 + 12t + 48) \div \dfrac{t^3 - 64}{2t^2 - 32}$

78. $\dfrac{4p^2 - 12p + 36}{27 - 3p^2} \div (p^3 + 27)$

79. $\dfrac{4p^2 - q^2}{q^3 - 8p^3} \div \dfrac{2p^2 - 6pq + pq - 3q^2}{4p^2 - 3pq - 27q^2}$

80. $\dfrac{2xy + 10y^2 + x^2 + 5xy}{6y^2 + xy - x^2} \div \dfrac{x^3 + 125y^3}{x^2 - 9y^2}$

Perform the indicated operations.

81. $\dfrac{x^2 - 25}{x - 4} \div \dfrac{x^2 - 2x - 15}{x^2 - 10x + 24} \cdot \dfrac{x + 3}{x^2 + 10x + 25}$

82. $\dfrac{y - 3}{y^2 - 8y + 16} \cdot \dfrac{y^2 - 16}{y + 4} \div \dfrac{y^2 + 3y - 18}{y^2 + 11y + 30}$

Given $f(x)$ and $g(x)$, find $f(x) \cdot g(x)$ and $f(x) \div g(x)$.

83. $f(x) = \dfrac{x - 4}{x^2 + x}$ and $g(x) = \dfrac{2x}{x + 1}$

84. $f(x) = \dfrac{x^3 - 3x^2}{x + 5}$ and $g(x) = \dfrac{4x^2}{x - 3}$

85. $f(x) = \dfrac{x^2 - 7x + 12}{x + 3}$ and $g(x) = \dfrac{9 - x^2}{x - 4}$

86. $f(x) = \dfrac{x + 6}{4 - x^2}$ and $g(x) = \dfrac{2 - x}{x^2 + 8x + 12}$

Mixed Practice

Solve.

87. Determine whether the rational expressions $\dfrac{n^2 + 7n}{3n^3 + 2n}$ and $\dfrac{n + 7}{3n^2 + 2}$ are equivalent.

88. Given $f(x) = \dfrac{x^2 - 4}{x - 5}$ and $g(x) = \dfrac{x^2 - 7x + 10}{x + 2}$, find $f(x) \cdot g(x)$ and $f(x) \div g(x)$.

89. Identify the values for which $\dfrac{n + 4}{n^2 + 6n - 40}$ is undefined.

Simplify, if possible.

90. $\dfrac{p^2 + 7p + 12}{p^2 - 3p - 18}$

91. $\dfrac{-b^2 + 4b + 5}{b^2 - 2b - 3}$

92. $\dfrac{r^2 - 2}{r^2 + 8}$

93. $\dfrac{15a^4 - 6a^2}{3a^3}$

Divide. Express the answer in lowest terms.

94. $\dfrac{4y^2 + 3y}{2y^4 - 4y^3} \div \dfrac{20y + 15}{20y}$

Applications

E *Solve.*

95. An environmental agency determines that it will cost approximately $\dfrac{125p}{100 - p}$ million dollars to clean $p\%$ of the chemical pollutants in a river. Under what circumstance is this expression undefined?

96. The force that one charge exerts on another charge is given by the expression $\dfrac{kq_1q_2}{d^2}$, where q_1 and q_2 are the magnitudes of the charges, d is the distance between them, and k is a constant. Under what circumstance is this force undefined?

97. A geometric progression is a sequence of terms in which each term is r times the preceding term. The sum of the first three terms of such a sequence is given by the expression $\dfrac{a - ar^3}{1 - r}$ (where a is the first term). Simplify the expression.

98. The surface area of a cylindrical can is given by the expression $2\pi rh + 2\pi r^2$, where r is the radius of the circular base and h is the height. The expression for the volume of this can is $\pi r^2 h$. Write the ratio of the surface area of the can to its volume in simplest form.

99. In studying electrical circuits, a physics student can use the expression

$$\left(\dfrac{V}{R}\right)^2 \cdot R,$$

where V is the potential difference in volts and R is the resistance in ohms, to find the power needed to maintain an electric current. Simplify this expression.

100. In order for an investment to be worth A dollars after 2 yr, the amount of money that must be invested in an account at interest rate r, compounded annually, is given by the following expression:

$$A\left(\dfrac{1}{1 + r}\right)\left(\dfrac{1}{1 + r}\right)$$

Write this product as a single rational expression.

101. Inflation is a sustained increase in prices of goods and services. When the rate of inflation i grows, the purchasing power of the dollar generally decreases. Economists use the expression

$$\frac{1}{100 + i} \div \frac{1}{i}$$

to calculate the amount by which the value of a dollar decreases. Write this quotient as a single rational expression. (*Source:* wikipedia.org)

102. The distance between two cities, such as Chicago and Milwaukee, on the same longitude can be found using the expression

$$(v - u) \div \frac{360}{C},$$

where u and v are angles and C is the circumference of Earth. Write this quotient as a single rational expression. (*Source:* infoplease.com)

• Check your answers on page A-40.

MINDStretchers

Mathematical Reasoning

1. Find a rational expression that is undefined at $x = 0$ and $x = -1$ and has a value of 3 at $x = 6$.

Writing

2. When multiplying rational expressions, we often factor the numerators and denominators and divide out any common factors before we multiply the numerators and denominators. Alternatively, we can multiply the numerators and denominators first, and then simplify. Discuss the pros and cons of each approach.

Technology

3. With a grapher, graph each of the following functions. Then, use the results to identify the domain and range of each function. (*Hint:* Recall that $y = f(x)$.)

a. $f(x) = \dfrac{x - 1}{x + 1}$

b. $f(x) = \dfrac{3x}{x^2 - 4}$

c. $f(x) = \dfrac{2}{x^2 + 1}$

6.2 Addition and Subtraction of Rational Expressions

OBJECTIVES

Adding and Subtracting Rational Expressions with the Same Denominator

In arithmetic, to add or subtract fractions with the same denominator we add or subtract the numerators, and keep the same denominator. In algebra, we add or subtract rational expressions with the same denominator in a similar manner. For instance,

A	To add or subtract rational expressions with the same denominator
B	To find the least common denominator (LCD) of two or more rational expressions
C	To add or subtract rational expressions with different denominators
D	To solve applied problems involving the addition or subtraction of rational expressions

Add
numerators.

$$\frac{4}{x} + \frac{3}{x} = \frac{\overbrace{4+3}}{x} = \frac{7}{x}$$

Keep same
denominator.

Subtract
numerators.

$$\frac{4}{x} - \frac{3}{x} = \frac{\overbrace{4-3}}{x} = \frac{1}{x}$$

Keep same
denominator.

To Add (or Subtract) Rational Expressions with the Same Denominator

- Add (or subtract) the numerators, and keep the same denominator.

- Simplify, if possible.

EXAMPLE 1

Add.

a. $\dfrac{9x}{x-y} + \dfrac{x}{x-y}$

b. $\dfrac{15y}{3y+1} + \dfrac{5}{3y+1}$

Solution

a.
$$\frac{9x}{x-y} + \frac{x}{x-y} = \frac{9x+x}{x-y}$$
Add the numerators, and keep the same denominator.

$$= \frac{10x}{x-y}$$
Simplify.

b.
$$\frac{15y}{3y+1} + \frac{5}{3y+1} = \frac{15y+5}{3y+1}$$
Add the numerators, and keep the same denominator.

$$= \frac{5(3y+1)}{3y+1}$$
Factor the numerator, and divide out the common factor.

$$= 5$$
Simplify.

PRACTICE 1

Find the sum.

a. $\dfrac{3y}{x+y} + \dfrac{2y}{x+y}$

b. $\dfrac{5x+1}{2x-3} + \dfrac{x-10}{2x-3}$

EXAMPLE 2

Subtract.

a. $\dfrac{7x + 2}{x + 9} - \dfrac{8x + 3}{x + 9}$

b. $\dfrac{7y}{x^2 + y^2} - \dfrac{6y - x}{x^2 + y^2}$

Solution

a.
$$\dfrac{7x + 2}{x + 9} - \dfrac{8x + 3}{x + 9} = \dfrac{(7x + 2) - (8x + 3)}{x + 9}$$ Subtract the numerators.

$$= \dfrac{7x + 2 - 8x - 3}{x + 9}$$ Use the distributive property in the numerator.

$$= \dfrac{-x - 1}{x + 9}$$ Combine like terms.

$$= \dfrac{-(x + 1)}{x + 9}$$ Factor out -1 in the numerator.

$$= -\dfrac{x + 1}{x + 9}$$

b.
$$\dfrac{7y}{x^2 + y^2} - \dfrac{6y - x}{x^2 + y^2} = \dfrac{7y - (6y - x)}{x^2 + y^2}$$

$$= \dfrac{7y - 6y + x}{x^2 + y^2}$$

$$= \dfrac{x + y}{x^2 + y^2}$$

PRACTICE 2

Find the difference.

a. $\dfrac{r - 5s}{s + 4r} - \dfrac{r + 11s}{s + 4r}$

b. $\dfrac{3n - 1}{n^2 + 5} - \dfrac{2 - n}{n^2 + 5}$

The Least Common Denominator of Rational Expressions

Recall that adding or subtracting unlike fractions involves finding the least common denominator (LCD). Similarly, to add or subtract rational expressions with different denominators, it is often helpful to find the LCD.

To find the LCD of rational expressions, we begin by factoring their denominators. For instance, consider the rational expressions $\dfrac{4x}{3x^2 - 3x}$ and $\dfrac{6}{x^2 - 2x + 1}$, which we can write as:

$$\underbrace{\dfrac{4x}{3x(x - 1)}}_{} \quad \text{and} \quad \underbrace{\dfrac{6}{(x - 1)(x - 1)}}_{}$$

Denominators in factored form

The LCD of the rational expressions is the product of the different factors in the denominators, where the power of each factor is the greatest number of times that it occurs in any single denominator. So the LCD of $\dfrac{4x}{3x(x - 1)}$ and $\dfrac{6}{(x - 1)(x - 1)}$ is

$$\text{LCD} = 3x(x - 1)(x - 1) = 3x(x - 1)^2.$$

To Find the LCD of Rational Expressions

- Factor each denominator completely.

- Multiply the factors found in the previous step, using for the power of each factor the greatest number of times that it occurs in any of the denominators. The product of all these factors is the LCD.

EXAMPLE 3

Find the LCD of each group of rational expressions.

a. $\dfrac{3}{7p^2q}$ and $\dfrac{9}{5pq^2}$

b. $\dfrac{2}{x+y}$ and $\dfrac{2}{x-y}$

c. $\dfrac{x}{x^2-1}$, $\dfrac{2}{x^2+x-2}$, and $\dfrac{3x}{x^2+2x+1}$

Solution

a. We begin by factoring the denominators of $\dfrac{3}{7p^2q}$ and $\dfrac{9}{5pq^2}$.

Factor $7p^2q$: $\ 7 \cdot p^2 \cdot q$
Factor $5pq^2$: $\ 5 \cdot p \cdot q^2$

The factors p and q each appear at most twice in any single denominator. The factors 7 and 5 each appear at most once in any denominator. The LCD is the product of 7, 5, p^2, and q^2, that is,

$$\text{LCD} = 7 \cdot 5 \cdot p^2 \cdot q^2 = 35p^2q^2$$

b. $\dfrac{2}{x+y}$ and $\dfrac{2}{x-y}$

Factor $x + y$: $\ x + y$
Factor $x - y$: $\ x - y$

No factor is repeated more than once in any denominator. The LCD is the product of the factors.

$$\text{LCD} = (x+y)(x-y)$$

c. $\dfrac{x}{x^2-1}$, $\dfrac{2}{x^2+x-2}$, and $\dfrac{3x}{x^2+2x+1}$

Factor $x^2 - 1$: $(x+1)(x-1)$
Factor $x^2 + x - 2$: $(x-1)(x+2)$
Factor $x^2 + 2x + 1$: $(x+1)(x+1) = (x+1)^2$
$\text{LCD} = (x-1)(x+2)(x+1)^2$

PRACTICE 3

Find the LCD.

a. $\dfrac{5}{6x^2y}$ and $\dfrac{3}{4xy^3}$

b. $\dfrac{m}{m-n}$ and $\dfrac{m}{m+n}$

c. $\dfrac{2y}{x^2-2xy+y^2}$, $\dfrac{x}{x^2-y^2}$, and $\dfrac{x-y}{x^2+2xy+y^2}$

Adding and Subtracting Rational Expressions with Different Denominators

In order to add or subtract rational expressions with different denominators, we need to change each expression to an equivalent rational expression whose denominator is the LCD.

EXAMPLE 4

Write the following rational expressions in terms of their LCD.

a. $\dfrac{5}{x-2}$ and $\dfrac{4x}{3x+1}$

b. $\dfrac{a}{a^2-1}$ and $\dfrac{4a}{2a^2-a-1}$

Solution

a. To find the LCD, we begin by factoring the denominators of the expressions.

Factor $x-2$: $x-2$

Factor $3x+1$: $3x+1$

The LCD of $\dfrac{5}{x-2}$ and $\dfrac{4x}{3x+1}$ is $(x-2)(3x+1)$.

We multiply the numerator and denominator of $\dfrac{5}{x-2}$ by $(3x+1)$ so that the denominator also becomes the LCD.

$$\frac{5}{x-2} = \frac{5(3x+1)}{(x-2)(3x+1)}$$

We multiply the numerator and denominator of $\dfrac{4x}{3x+1}$ by $(x-2)$ so that the denominator becomes the LCD.

$$\frac{4x}{3x+1} = \frac{4x(x-2)}{(3x+1)(x-2)}$$

b. Begin by factoring the denominators to find the LCD.

Factor a^2-1: $(a+1)(a-1)$

Factor $2a^2-a-1$: $(2a+1)(a-1)$

The LCD of $\dfrac{a}{a^2-1}$ and $\dfrac{4a}{2a^2-a-1}$ is $(a+1)(a-1)(2a+1)$.

Now, we write each rational expression in terms of the LCD.

$$\frac{a}{a^2-1} = \frac{a(2a+1)}{(a+1)(a-1)(2a+1)} \qquad \text{Multiply the numerator and denominator by } (2a+1).$$

So $\dfrac{a}{a^2-1}$ is equivalent to $\dfrac{a(2a+1)}{(a+1)(a-1)(2a+1)}$.

$$\frac{4a}{2a^2-a-1} = \frac{4a(a+1)}{(2a+1)(a-1)(a+1)} \qquad \text{Multiply the numerator and denominator by } (a+1).$$

So $\dfrac{4a}{2a^2-a-1}$ is equivalent to $\dfrac{4a(a+1)}{(2a+1)(a-1)(a+1)}$.

PRACTICE 4

Write the following rational expressions in terms of their LCD.

a. $\dfrac{3n}{6n+8}$ and $\dfrac{5}{n-4}$

b. $\dfrac{7}{3x-1}$ and $\dfrac{2x+3}{3x^2+11x-4}$

Now, let's look at how to add and subtract rational expressions with different denominators using the concept of equivalent rational expressions.

To Add (or Subtract) Rational Expressions with Different Denominators

- Find the LCD of the rational expressions.
- Write each rational expression with a common denominator, using the LCD.
- Add (or subtract) the numerators, keeping the denominator.
- Simplify, if possible.

EXAMPLE 5

Add.

a. $\dfrac{4}{y - 2} + \dfrac{y}{y - 3}$ **b.** $\dfrac{2}{n - 6} + \dfrac{n - 4}{n^2 + n - 42}$

Solution

a. The LCD of the rational expressions is $(y - 2)(y - 3)$.

$$\dfrac{4}{y - 2} + \dfrac{y}{y - 3} = \dfrac{4(y - 3)}{(y - 2)(y - 3)} + \dfrac{y(y - 2)}{(y - 3)(y - 2)}$$ Write in terms of the LCD.

$$= \dfrac{4y - 12}{(y - 2)(y - 3)} + \dfrac{y^2 - 2y}{(y - 2)(y - 3)}$$ Simplify the numerators.

$$= \dfrac{(4y - 12) + (y^2 - 2y)}{(y - 2)(y - 3)}$$ Add the numerators.

$$= \dfrac{y^2 + 2y - 12}{(y - 2)(y - 3)}$$ Combine like terms in the numerator.

Note that the expression in the numerator cannot be factored and that the numerator and denominator have no factors in common. Since there are no common factors in the numerator and denominator, the rational expression is in simplest form.

b. $\dfrac{2}{n - 6} + \dfrac{n - 4}{n^2 + n - 42} = \dfrac{2}{n - 6} + \dfrac{n - 4}{(n - 6)(n + 7)}$ Factor the denominators.

$$= \dfrac{2(n + 7)}{(n - 6)(n + 7)} + \dfrac{n - 4}{(n - 6)(n + 7)}$$ Write in terms of the LCD $(n - 6)(n + 7)$.

$$= \dfrac{2n + 14}{(n - 6)(n + 7)} + \dfrac{n - 4}{(n - 6)(n + 7)}$$ Simplify the left numerator.

$$= \dfrac{2n + 14 + n - 4}{(n - 6)(n + 7)}$$ Add the numerators.

$$= \dfrac{3n + 10}{(n - 6)(n + 7)}$$ Combine like terms in the numerator.

Since there are no common factors in the numerator and denominator, the rational expression is in simplest form.

PRACTICE 5

Find the sum.

a. $\dfrac{x}{x - 1} + \dfrac{3}{x + 1}$

b. $\dfrac{t + 5}{t^2 - 9t + 20} + \dfrac{6}{t - 4}$

EXAMPLE 6

Subtract.

a. $\dfrac{2m + 5}{m - 1} - \dfrac{m}{1 - m}$

b. $\dfrac{4n + 1}{n^2 + 3n + 2} - \dfrac{n - 4}{n^2 + 5n + 6}$

Solution

a. The LCD of the rational expressions can be either $m - 1$ or $1 - m$, since $1 - m = -(m - 1)$. Let's use $m - 1$ as the LCD. So we factor out -1 in the denominator of the expression $\dfrac{m}{1 - m}$.

$$\frac{2m + 5}{m - 1} - \frac{m}{1 - m} = \frac{2m + 5}{m - 1} - \frac{m}{-(m - 1)} \qquad \text{Write } 1 - m \text{ as } -(m - 1).$$

$$= \frac{2m + 5}{m - 1} - \left(\frac{-m}{m - 1}\right) \qquad \text{Write } \frac{m}{-(m - 1)} \text{ as } \frac{-m}{m - 1}.$$

$$= \frac{2m + 5 - (-m)}{m - 1} \qquad \text{Subtract the numerators.}$$

$$= \frac{2m + 5 + m}{m - 1} \qquad \text{Simplify the numerator.}$$

$$= \frac{3m + 5}{m - 1} \qquad \text{Combine like terms in the numerator.}$$

b. $\dfrac{4n + 1}{n^2 + 3n + 2} - \dfrac{n - 4}{n^2 + 5n + 6} = \dfrac{4n + 1}{(n + 2)(n + 1)} - \dfrac{n - 4}{(n + 3)(n + 2)}$

The LCD is $(n + 2)(n + 1)(n + 3)$.

$$= \frac{(4n + 1)(n + 3)}{(n + 2)(n + 1)(n + 3)} - \frac{(n - 4)(n + 1)}{(n + 3)(n + 2)(n + 1)}$$

$$= \frac{4n^2 + 13n + 3}{(n + 3)(n + 2)(n + 1)} - \frac{n^2 - 3n - 4}{(n + 3)(n + 2)(n + 1)}$$

$$= \frac{(4n^2 + 13n + 3) - (n^2 - 3n - 4)}{(n + 3)(n + 2)(n + 1)}$$

$$= \frac{4n^2 + 13n + 3 - n^2 + 3n + 4}{(n + 3)(n + 2)(n + 1)}$$

$$= \frac{3n^2 + 16n + 7}{(n + 3)(n + 2)(n + 1)}$$

PRACTICE 6

Find the difference.

a. $\dfrac{2}{x - 3} - \dfrac{x - 1}{3 - x}$

b. $\dfrac{7y + 1}{y^2 + 4y + 3} - \dfrac{7y - 2}{y^2 + 2y + 1}$

EXAMPLE 7

Perform the indicated operations.

$$\frac{6}{x^2 + 2x - 15} + \frac{3}{x^2 - 3x} - \frac{5}{x^2 - 9}$$

Solution

$$\frac{6}{x^2 + 2x - 15} + \frac{3}{x^2 - 3x} - \frac{5}{x^2 - 9}$$

$$= \frac{6}{(x + 5)(x - 3)} + \frac{3}{x(x - 3)} - \frac{5}{(x + 3)(x - 3)}$$

$$= \frac{6x(x + 3)}{(x + 5)(x - 3)x(x + 3)} + \frac{3(x + 5)(x + 3)}{x(x - 3)(x + 5)(x + 3)}$$

$$- \frac{5x(x + 5)}{(x + 3)(x - 3)x(x + 5)}$$

$$= \frac{6x^2 + 18x}{x(x + 5)(x + 3)(x - 3)} + \frac{3x^2 + 24x + 45}{x(x + 5)(x + 3)(x - 3)}$$

$$- \frac{5x^2 + 25x}{x(x + 5)(x + 3)(x - 3)}$$

$$= \frac{(6x^2 + 18x) + (3x^2 + 24x + 45) - (5x^2 + 25x)}{x(x + 5)(x + 3)(x - 3)}$$

$$= \frac{6x^2 + 18x + 3x^2 + 24x + 45 - 5x^2 - 25x}{x(x + 5)(x + 3)(x - 3)}$$

$$= \frac{4x^2 + 17x + 45}{x(x + 5)(x + 3)(x - 3)}$$

PRACTICE 7

Perform the indicated operations.

$$\frac{4}{2x^2 + x - 6} - \frac{2}{2x^2 - 3x} + \frac{1}{x^2 - 4}$$

We can apply our knowledge of adding and subtracting rational expressions to adding and subtracting rational functions.

EXAMPLE 8

Consider the functions $f(x) = \frac{x}{x - 1}$ and $g(x) = \frac{7}{x^2 - x}$. Find:

a. $f(x) + g(x)$ **b.** $f(x) - g(x)$

Solution

a. $f(x) + g(x) = \frac{x}{x - 1} + \frac{7}{x^2 - x}$

$$= \frac{x}{x - 1} + \frac{7}{x(x - 1)} \qquad \text{The LCD is } x(x - 1).$$

$$= \frac{x \cdot x}{x(x - 1)} + \frac{7}{x(x - 1)}$$

$$= \frac{x^2}{x(x - 1)} + \frac{7}{x(x - 1)}$$

$$= \frac{x^2 + 7}{x(x - 1)}$$

PRACTICE 8

Given that $f(x) = \frac{2x}{x^2 - 25}$ and $g(x) = \frac{3}{x + 5}$, find:

a. $f(x) + g(x)$ **b.** $f(x) - g(x)$

EXAMPLE 8 (continued)

b. Since we know the LCD of the rational expressions, we can simply subtract the equivalent forms to find the difference.

$$f(x) - g(x) = \frac{x}{x - 1} - \frac{7}{x^2 - x}$$

$$= \frac{x^2}{x(x - 1)} - \frac{7}{x(x - 1)}$$

$$= \frac{x^2 - 7}{x(x - 1)}$$

EXAMPLE 9

A local bank pays an interest rate r, compounded annually, on all account balances. If a customer wanted the balance in his account to be \$2000 at the end of one year, he would need to have a current balance of $\dfrac{2000}{1 + r}$ dollars. However, if he were willing to wait two years for the balance to reach \$2000, then his current balance would need to be only $\dfrac{2000}{(1 + r)^2}$ dollars.

a. Express the difference between the quantities $\dfrac{2000}{1 + r}$ and $\dfrac{2000}{(1 + r)^2}$ as a single rational expression.

b. If the local bank's interest rate is 3%, evaluate the expression found in part (a), rounded to the nearest dollar.

Solution

a. The LCD of the two quantities $\dfrac{2000}{1 + r}$ and $\dfrac{2000}{(1 + r)^2}$ is $(1 + r)^2$.

The difference between the quantities is found as follows:

$$\frac{2000}{1 + r} - \frac{2000}{(1 + r)^2} = \frac{2000(1 + r)}{(1 + r)^2} - \frac{2000}{(1 + r)^2}$$

$$= \frac{2000 + 2000r}{(1 + r)^2} - \frac{2000}{(1 + r)^2}$$

$$= \frac{2000 + 2000r - 2000}{(1 + r)^2}$$

$$= \frac{2000r}{(1 + r)^2}$$

So the difference between the two quantities is $\dfrac{2000r}{(1 + r)^2}$ dollars.

b. We are asked to evaluate the expression $\dfrac{2000r}{(1 + r)^2}$ when r is equal to 3%, or 0.03.

$$\frac{2000r}{(1 + r)^2} = \frac{2000(0.03)}{(1 + 0.03)^2} = \frac{60}{1.0609} \approx 56.56$$

So the expression is equal to \$57, to the nearest dollar.

PRACTICE 9

To maintain a checking account, a bank charges \$4 per month and \$0.05 per check. For n checks, the cost per check is $\left(\dfrac{4}{n} + 0.05\right)$ dollars.

a. How much less is the cost per check for $(n + 1)$ checks than for n checks? Express your answer as a single rational expression.

b. Evaluate the expression found in part (a) for 10 checks, rounded to the nearest cent.

Mathematically Speaking

Fill in each blank with the most appropriate term or phrase from the given list.

different denominators	least	greatest
find the product of the denominators	same denominator	factor each denominator completely

1. To add (or subtract) rational expressions with the same denominator, add (or subtract) the numerators and keep the _____.

2. To find the LCD of rational expressions, first _____.

3. To add (or subtract) rational expressions with _____, first find the LCD of the rational expressions.

4. The LCD of rational expressions is the product of the different factors in the denominators, where the power of each factor is the _____ number of times that it occurs in any single denominator.

A *Perform the indicated operation. Simplify, if possible.*

5. $\dfrac{n - 10}{4n^2} + \dfrac{3n + 2}{4n^2}$

6. $\dfrac{a + 7b}{5a^2b} + \dfrac{4a + 3b}{5a^2b}$

7. $\dfrac{4n - 15}{n - 6} - \dfrac{2n - 3}{n - 6}$

8. $\dfrac{3r + 4}{r + 3} - \dfrac{5r + 10}{r + 3}$

9. $\dfrac{3y - 2x}{x^2 - y^2} - \dfrac{4y - 3x}{x^2 - y^2}$

10. $\dfrac{8v - u^2}{2u^2 - v^2} - \dfrac{12v - u^2}{2u^2 - v^2}$

11. $\dfrac{y^2 - 3}{y^2 - y - 12} + \dfrac{7 + y - y^2}{y^2 - y - 12}$

12. $\dfrac{11p - 10}{p^2 + p - 20} + \dfrac{3p^2 + 2p}{p^2 + p - 20}$

B *Find the LCD of each group of rational expressions. Then, write each expression in terms of the LCD.*

13. $\dfrac{3}{8x^2y^3}$ and $\dfrac{1}{6x^3y^2}$

14. $\dfrac{2}{5a^3b^2}$ and $\dfrac{3}{4a^2b^4}$

15. $\dfrac{n}{n - 2}$ and $\dfrac{7}{n - 1}$

16. $\dfrac{1}{y + 4}$ and $\dfrac{3y}{y - 5}$

17. $\dfrac{3}{2t^2 + 12t}$ and $\dfrac{4}{3t^3 + 18t^2}$

18. $\dfrac{2}{5x^2 - 10x^3}$ and $\dfrac{6}{x^3 - 2x^4}$

19. $\dfrac{2p - q}{p^2 - 9q^2}$ and $\dfrac{3}{p - 3q}$

20. $\dfrac{x - y}{x^2 - 16y^2}$ and $\dfrac{4}{4y + x}$

21. $\dfrac{2n}{n^2 - 6n + 8}$ and $\dfrac{3n}{n^2 + 4n - 12}$

22. $\dfrac{4r}{r^2 - 6r + 9}$ and $\dfrac{r}{r^2 - 11r + 24}$

23. $\dfrac{1}{3n^2 + 5n - 2}$, $\dfrac{n}{n^2 - n - 6}$, and $\dfrac{n - 4}{3n^2 - 10n + 3}$

24. $\dfrac{4t}{t^2 - 3t - 18}$, $\dfrac{3}{2t^2 + 7t + 3}$, and $\dfrac{t - 1}{2t^2 - 11t - 6}$

C *Perform the indicated operations.*

25. $\dfrac{1}{8x^2y} + \dfrac{1}{12xy^2}$

26. $\dfrac{4}{9a^2} + \dfrac{2}{15ab}$

27. $\dfrac{3}{p} - 4p$

28. $2 - \dfrac{1}{3x^2}$

29. $\dfrac{n}{n + 4} + \dfrac{2}{n + 1}$

30. $\dfrac{3}{p + 5} + \dfrac{p}{p - 2}$

31. $\dfrac{5}{x - 2y} + \dfrac{2}{2x - y}$

32. $\dfrac{3}{p - 5q} + \dfrac{1}{5p + q}$

33. $\dfrac{7a}{a - 3} + \dfrac{5a + 6}{3 - a}$

34. $\dfrac{1 - 3t}{1 - t} - \dfrac{2t}{t - 1}$

35. $\dfrac{1}{4 - 2r} + \dfrac{7}{3r^2 - 6r}$

36. $\dfrac{3}{15y^2 - 5y^3} + \dfrac{2}{3y - 9}$

37. $\dfrac{4t - 1}{t^2 - 16} + \dfrac{2}{4 - t}$

38. $\dfrac{3}{x - 6} + \dfrac{x + 8}{36 - x^2}$

39. $\dfrac{n^2 - 4}{n^2 - 10n + 21} - \dfrac{4}{n - 7}$

40. $\dfrac{5}{a + 4} - \dfrac{1 - a^2}{a^2 + 2a - 8}$

41. $\dfrac{6}{u^2 + 11uv + 30v^2} + \dfrac{3}{u^2 + 2uv - 24v^2}$

42. $\dfrac{2}{p^2 - 2pq - 35q^2} + \dfrac{2}{p^2 - 8pq + 7q^2}$

43. $\dfrac{4}{x^2 - 2x - 3} - \dfrac{x}{x^2 + 3x + 2}$

44. $\dfrac{n}{n^2 - 4} - \dfrac{1}{n^2 + 6n + 8}$

45. $\dfrac{-2x}{2x^2 + 5x - 12} + \dfrac{3x}{3x^2 + 13x + 4}$

46. $\dfrac{-4t}{5 - 12t + 4t^2} + \dfrac{2t}{2t^2 - 13t + 6}$

47. $\dfrac{4a + 3}{a^2 + 2a - 8} - \dfrac{3a - 2}{a^2 - 4a + 4}$

48. $\dfrac{1 - 2n}{n^2 + 6n + 9} - \dfrac{1 - 5n}{n^2 - 3n - 18}$

49. $\dfrac{1 - x}{6x^4 - 9x^3 - 15x^2} + \dfrac{1}{2x^3 - 2x}$

50. $\dfrac{r + 4}{4r^3 - 16r} + \dfrac{2}{8 - 2r - 3r^2}$

51. $\dfrac{4}{v^2 - 9} - \dfrac{2}{v + 1} + \dfrac{v - 6}{v^2 - 2v - 3}$

52. $\dfrac{p - 10}{p^2 - 2p - 15} + \dfrac{3}{p^2 - 25} - \dfrac{2}{p + 3}$

53. $\dfrac{3}{2x + 8} - \dfrac{4x}{4x^2 - 4x + 1} - \dfrac{x + 7}{4 - 7x - 2x^2}$

54. $\dfrac{3x}{3x^2 - 7x + 4} - \dfrac{x + 1}{16 - 9x^2} - \dfrac{2x}{3x^2 + x - 4}$

Given $f(x)$ and $g(x)$, find $f(x) + g(x)$ and $f(x) - g(x)$.

55. $f(x) = \dfrac{8}{x^2 + 10x + 24}$ and $g(x) = \dfrac{2}{x^2 - 5x - 36}$

56. $f(x) = \dfrac{7}{x^2 - 11x + 18}$ and $g(x) = \dfrac{4}{x^2 - 3x - 54}$

57. $f(x) = \dfrac{x + 4}{3x^2 - 12x}$ and $g(x) = -\dfrac{2}{x^2}$

58. $f(x) = -\dfrac{3}{x^2}$ and $g(x) = \dfrac{x + 3}{4x^2 - 32x}$

Mixed Practice

Solve.

59. Given $f(x) = \dfrac{1 + x}{2x^2 + 6x}$ and $g(x) = -\dfrac{3}{x^2}$, find $f(x) + g(x)$ and $f(x) - g(x)$.

60. Find the LCD of $\dfrac{2r}{r^2 - 9r + 18}$ and $\dfrac{5r}{r^2 + r - 12}$. Then, write each expression in terms of the LCD.

Perform the indicated operations.

61. $\dfrac{y}{y - 1} - \dfrac{y}{3} + \dfrac{4}{y + 2}$

62. $\dfrac{2}{t + 2} - \dfrac{3 - t^2}{t^2 - 2t - 8}$

63. $\dfrac{3}{x - 3y} + \dfrac{7}{3x - y}$

64. $\dfrac{5n - 2}{n^2 - 2n - 8} + \dfrac{3n + 1}{n^2 - 8n + 16}$

65. $\dfrac{4a - 5b}{a^2 - b^2} - \dfrac{3a - 6b}{a^2 - b^2}$

66. $\dfrac{3x}{3x^2 + 5x - 2} - \dfrac{4x}{4x^2 + 9x + 2}$

Applications

D *Solve.*

67. A cylindrical water tank is drained for a short period of time. The change in the water level can be found using the expression $\dfrac{V_1}{\pi r^2} - \dfrac{V_2}{\pi r^2}$, where V_1 and V_2 represent the original and new volume of the water in the tank, respectively, and r is the radius of the tank. Write the change in water level as a single rational expression.

68. A homeowner decides to increase the area of her patio by increasing the width and keeping the length the same. The change in the width of the patio is $\left(\dfrac{A_2}{l} - \dfrac{A_1}{l}\right)$ ft, where A_1 is the original area of the patio, A_2 is the new area of the patio, and l is the length of the patio. Write the change in width as a single rational expression.

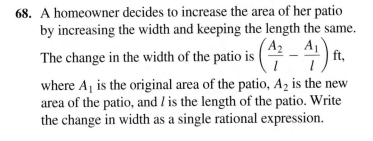

69. To determine the percent growth in sales from the previous year, the owner of a company uses the expression $100\left(\dfrac{S_1}{S_0} - 1\right)$, where S_1 represents the current year's sales and S_0 represents last year's sales. Write this expression as a single rational expression.

70. If $\dfrac{a}{b}$ represents the probability of an event, then the probability that the event will not occur is $1 - \dfrac{a}{b}$. Write $1 - \dfrac{a}{b}$ as a single rational expression.

71. A capacitor is a device that stores electric charge and energy. The reciprocal of the equivalent capacitance of two capacitors joined in series (see figure below) is given by the sum $\dfrac{1}{C_1} + \dfrac{1}{C_2}$. Write this sum as a single rational expression.

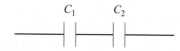

72. The reciprocal of the focal length of a combination of two thin lenses in contact is given by $\dfrac{1}{f_1} + \dfrac{1}{f_2}$, where f_1 and f_2 are focal lengths of the individual lenses. Write this expression as a single rational expression.

73. An airplane flies m mi with a wind whose speed is w mph. On the return flight, the airplane flies against the same wind. The expression $\dfrac{m}{s + w} + \dfrac{m}{s - w}$, where s is the speed of the airplane in still air, represents the total time in hours that it takes to make the round-trip flight. Write a single rational expression representing the total time.

74. On a math test, the fraction $\dfrac{a}{b}$ of the students in the class scored in the 90's, and $p\%$ scored in the 80's. Write as a single rational expression the fraction of students who scored either in the 80's or in the 90's.

75. A girl bicycled the 4 mi to her friend's house at an average rate of r mph. On the return trip, she bicycled at twice the rate. How long did the whole trip take? Write the answer as a single rational expression.

76. The distance between a nurse's home and the hospital where she works is 18 mi. On a particular day, she drove home at an average speed of s mph, which was 8 mph slower than the average speed she had driven to work. If she drove the same route each way, how much longer did it take her to get home that day? Write the answer as a single rational expression.

• Check your answers on page A-41.

MINDStretchers

Mathematical Reasoning

1. Consider the rational expressions $\dfrac{P}{Q}$ and $\dfrac{R}{S}$, where P, Q, R, and S are polynomials. Under what circumstances will the LCD of the rational expressions be equal to the product of the individual denominators QS? Give examples to support your answer.

Groupwork

2. Working with a partner, write each of the following expressions as a single rational expression.

   a. $2x^{-3} + (3x)^{-1}$

   b. $(2x - 1)^{-2} - (2x - 1)^{-1}$

   c. $(x^2 - 9)^{-1} + 2(x - 3)^{-1}$

   d. $\left(\dfrac{x + 1}{3}\right)^{-1} - \left(\dfrac{x - 4}{2}\right)^{-1}$

Investigation

3. Consider the functions $f(x) = \dfrac{1}{x - 2}$, $g(x) = \dfrac{x - 1}{x + 1}$, and $h(x) = \dfrac{x^2 - 2x + 3}{x^2 - x - 2}$.

   a. Suppose $x = 3$. Evaluate $f(3)$, $g(3)$, and $h(3)$.

   b. Find $f(3) + g(3)$. Compare your answer to the value $h(3)$. What do you notice?

   c. Repeat parts (a) and (b) for several other values of x.

   d. What do your answers in parts (b) and (c) suggest about $f(x) + g(x)$ and $h(x)$? Verify your conclusion.

OBJECTIVES

In this section, we discuss complex rational expressions and how to simplify them. A rational expression whose numerator, denominator, or both contain one or more rational expressions is called a *complex rational expression* or a *complex algebraic fraction*. For example, the expression below is a complex rational expression.

A   To simplify a complex rational expression

B   To solve applied problems involving complex rational expressions

$$\dfrac{1 + \dfrac{1}{x}}{1 - \dfrac{1}{x}} \left.\begin{matrix}\\ \\ \end{matrix}\right.$$

$\left.\begin{matrix}\\\end{matrix}\right\}$ Rational expression in the numerator

← Main fraction line

$\left.\begin{matrix}\\\end{matrix}\right\}$ Rational expression in the denominator

Here are some additional examples of complex rational expressions:

$$\dfrac{\dfrac{n}{2}}{\dfrac{n}{4}} \qquad \dfrac{\dfrac{1}{y}}{x - y \over 5} \qquad \dfrac{\dfrac{1}{x^2} - \dfrac{1}{y^2}}{\dfrac{3}{x} + \dfrac{3}{y}}$$

A complex rational expression is simplified when it is in the form $\dfrac{P}{Q}$, where P and Q are polynomials that have no common factors other than 1. We consider two methods of simplifying a complex rational expression: the *division method* and the *LCD method*.

First, let's look at the division method. In this method, we simplify complex rational expressions by writing both the numerator and denominator as single rational expressions. Then, we divide.

To Simplify a Complex Rational Expression: The Division Method

- Write both the numerator and denominator as single rational expressions in simplified form.

- Write the expression as the numerator divided by the denominator.

- Divide.

- Simplify, if possible.

EXAMPLE 1

Simplify: $\dfrac{\dfrac{5x^2}{x + 1}}{\dfrac{3x}{x + 1}}$

Solution   Both the numerator and the denominator of the complex rational expression are already expressed as single rational expressions in simplified form. So we can write the expression as the numerator divided by the denominator.

PRACTICE 1

Simplify: $\dfrac{\dfrac{n - 1}{n}}{\dfrac{n - 1}{n^2}}$

$$\frac{\dfrac{5x^2}{x+1}}{\dfrac{3x}{x+1}} = \frac{5x^2}{x+1} \div \frac{3x}{x+1}$$

Rewrite the complex fraction as the numerator divided by the denominator.

$$= \frac{5x^2}{x+1} \cdot \frac{x+1}{3x}$$

Take the reciprocal of the divisor and change the operation to multiplication.

$$= \frac{\overset{x}{5x^2}}{\cancel{x+1}} \cdot \frac{\cancel{x+1}}{3\cancel{x}}$$

Divide out common factors.

$$= \frac{5x}{3}$$

Simplify.

EXAMPLE 2

Simplify: $\dfrac{3}{\dfrac{4}{x^2}+\dfrac{1}{x}}$

Solution   First, we express the denominator as a single rational expression. The LCD of the rational expressions in the denominator is x^2.

$$\frac{3}{\dfrac{4}{x^2}+\dfrac{1}{x}} = \frac{3}{\dfrac{4}{x^2}+\dfrac{1 \cdot x}{x \cdot x}}$$

Write in terms of the LCD. Add the rational expressions in the denominator.

$$= \frac{3}{\dfrac{4+x}{x^2}}$$

$$= 3 \div \frac{4+x}{x^2}$$

Write as the numerator divided by the denominator.

$$= 3 \cdot \frac{x^2}{4+x}$$

Multiply by the reciprocal.

$$= \frac{3x^2}{4+x}$$

Simplify.

EXAMPLE 3

Simplify: $\dfrac{\dfrac{1}{x^2}-\dfrac{1}{y^2}}{\dfrac{3}{x}+\dfrac{3}{y}}$

Solution   Let's write the numerator and the denominator so that each is a single rational expression.

PRACTICE 2

Simplify: $\dfrac{\dfrac{1}{y}+\dfrac{3}{y^2}}{2y}$

PRACTICE 3

Simplify: $\dfrac{\dfrac{4}{m}+\dfrac{4}{n}}{\dfrac{1}{m^2}-\dfrac{1}{n^2}}$

EXAMPLE 3 (continued)

$$\frac{\dfrac{1}{x^2} - \dfrac{1}{y^2}}{\dfrac{3}{x} + \dfrac{3}{y}} = \frac{\dfrac{1}{x^2} \cdot \dfrac{y^2}{y^2} - \dfrac{1}{y^2} \cdot \dfrac{x^2}{x^2}}{\dfrac{3}{x} \cdot \dfrac{y}{y} + \dfrac{3}{y} \cdot \dfrac{x}{x}}$$

$$= \frac{\dfrac{y^2}{x^2 y^2} - \dfrac{x^2}{x^2 y^2}}{\dfrac{3y}{xy} + \dfrac{3x}{xy}}$$

$$= \frac{\dfrac{y^2 - x^2}{x^2 y^2}}{\dfrac{3y + 3x}{xy}}$$

$$= \frac{y^2 - x^2}{x^2 y^2} \div \frac{3y + 3x}{xy}$$

$$= \frac{y^2 - x^2}{x^2 y^2} \cdot \frac{xy}{3y + 3x}$$

$$= \frac{(y - x)(y + x)}{x^2 y^2} \cdot \frac{xy}{3(y + x)}$$

$$= \frac{(y - x)(y + x)}{\underset{x\ y}{x^2 y^2}} \cdot \frac{xy}{3(y + x)}$$

$$= \frac{y - x}{3xy}$$

If the complex fraction in Example 3 had been expressed as $\dfrac{x^{-2} - y^{-2}}{3x^{-1} + 3y^{-1}}$, would we get the same answer when simplifying?

Now, let's consider the LCD method for simplifying a complex rational expression. This method uses the LCD of all the rational expressions that appear within the complex rational expression.

> ### To Simplify a Complex Rational Expression: The LCD Method
>
> - Find the LCD of all rational expressions within the complex rational expression.
> - Multiply the numerator and denominator of the complex rational expression by this LCD.
> - Simplify, if possible.

EXAMPLE 4

Simplify: $\dfrac{\dfrac{(x+2)^2}{2x^3}}{\dfrac{x+2}{3x}}$

Solution Multiply the numerator and denominator by the LCD $6x^3$.

$$\dfrac{\dfrac{(x+2)^2}{2x^3}}{\dfrac{x+2}{3x}} = \dfrac{\dfrac{(x+2)^2}{2x^3} \cdot 6x^3}{\dfrac{x+2}{3x} \cdot 6x^3}$$

$$= \dfrac{3(x+2)^2}{2x^2(x+2)} \qquad \text{Simplify the numerator and the denominator.}$$

$$= \dfrac{3(\overset{(x+2)}{\cancel{x+2}})^2}{2x^2\cancel{(x+2)}} \qquad \text{Divide out the common factor.}$$

$$= \dfrac{3(x+2)}{2x^2}$$

PRACTICE 4

Simplify: $\dfrac{\dfrac{y^2-9}{5y}}{\dfrac{y+3}{6y^2}}$

EXAMPLE 5

Simplify: $\dfrac{2+\dfrac{1}{x}}{4-\dfrac{1}{x^2}}$

Solution

$$\dfrac{2+\dfrac{1}{x}}{4-\dfrac{1}{x^2}} = \dfrac{\left(2+\dfrac{1}{x}\right)\cdot x^2}{\left(4-\dfrac{1}{x^2}\right)\cdot x^2} \qquad \begin{array}{l}\text{Multiply the numerator and denominator}\\ \text{by the LCD } x^2.\end{array}$$

$$= \dfrac{2\cdot x^2 + \dfrac{1}{x}\cdot x^2}{4\cdot x^2 - \dfrac{1}{x^2}\cdot x^2} \qquad \text{Use the distributive property.}$$

$$= \dfrac{2x^2+x}{4x^2-1} \qquad \text{Simplify.}$$

$$= \dfrac{x(2x+1)}{(2x+1)(2x-1)} \qquad \text{Factor the numerator and denominator.}$$

$$= \dfrac{x\cancel{(2x+1)}}{\cancel{(2x+1)}(2x-1)} \qquad \text{Divide out the common factor.}$$

$$= \dfrac{x}{2x-1}$$

PRACTICE 5

Simplify: $\dfrac{3-\dfrac{2}{n}}{9-\dfrac{4}{n^2}}$

EXAMPLE 6

Simplify: $\dfrac{\dfrac{2}{x} - \dfrac{2}{y}}{\dfrac{2}{x^3} - \dfrac{2}{y^3}}$

Solution Multiply the numerator and denominator by the LCD x^3y^3.

$$\dfrac{\dfrac{2}{x} - \dfrac{2}{y}}{\dfrac{2}{x^3} - \dfrac{2}{y^3}} = \dfrac{\left(\dfrac{2}{x} - \dfrac{2}{y}\right) \cdot x^3y^3}{\left(\dfrac{2}{x^3} - \dfrac{2}{y^3}\right) \cdot x^3y^3}$$

$$= \dfrac{\dfrac{2}{x} \cdot x^3y^3 - \dfrac{2}{y} \cdot x^3y^3}{\dfrac{2}{x^3} \cdot x^3y^3 - \dfrac{2}{y^3} \cdot x^3y^3}$$

$$= \dfrac{2x^2y^3 - 2x^3y^2}{2y^3 - 2x^3}$$

$$= \dfrac{2x^2y^2(y - x)}{2(y^3 - x^3)}$$

$$= \dfrac{2x^2y^2(y - x)}{2(y - x)(y^2 + xy + x^2)}$$

$$= \dfrac{x^2y^2}{y^2 + xy + x^2}$$

PRACTICE 6

Simplify: $\dfrac{\dfrac{1}{p^3} + \dfrac{1}{q^3}}{\dfrac{1}{p} + \dfrac{1}{q}}$

EXAMPLE 7

The *harmonic mean* is a kind of average used in the study of motion, music, and geometry. The harmonic mean of three numbers x, y, and z is given by the expression $\dfrac{3}{\dfrac{1}{x} + \dfrac{1}{y} + \dfrac{1}{z}}$. Simplify this complex rational expression.

Solution $\dfrac{3}{\dfrac{1}{x} + \dfrac{1}{y} + \dfrac{1}{z}} = \dfrac{3 \cdot xyz}{\left(\dfrac{1}{x} + \dfrac{1}{y} + \dfrac{1}{z}\right) \cdot xyz}$

$$= \dfrac{3 \cdot xyz}{\dfrac{1}{x} \cdot xyz + \dfrac{1}{y} \cdot xyz + \dfrac{1}{z} \cdot xyz}$$

$$= \dfrac{3xyz}{yz + xz + xy}$$

So the harmonic mean of x, y, and z is $\dfrac{3xyz}{yz + xz + xy}$.

PRACTICE 7

A resistor is an electrical device such as a lightbulb that offers resistance to the flow of electricity. When four bulbs with resistances R_1, R_2, R_3, and R_4 are connected in a certain circuit, the combined resistance is given by the following expression:

$$\dfrac{1}{\dfrac{1}{R_1} + \dfrac{1}{R_2} + \dfrac{1}{R_3} + \dfrac{1}{R_4}}$$

Simplify this expression for the combined resistance.

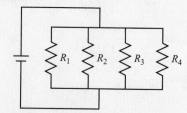

Mathematically Speaking

Fill in each blank with the most appropriate term or phrase from the given list.

reciprocal	LCD method
algebraic	have no common factors other than 1
are factored	division method
rational	

1. A rational expression whose numerator, denominator, or both contain one or more _____ expressions is called a complex rational expression or a complex algebraic fraction.

2. A complex rational expression is simplified when it is in the form $\dfrac{P}{Q}$, where P and Q are polynomials that _____.

3. To simplify a complex rational expression using the _____, first write both the numerator and the denominator as single rational expressions in simplified form.

4. To simplify a complex rational expression using the _____, first find the LCD of all rational expressions within the complex rational expression.

A *Simplify.*

5. $\dfrac{\dfrac{12}{3}}{5n}$

6. $\dfrac{\dfrac{6}{y}}{20}$

7. $\dfrac{\dfrac{6}{x}}{4x^2}$

8. $\dfrac{3x}{\dfrac{9}{x^3}}$

9. $\dfrac{\dfrac{a+2}{2a^2}}{\dfrac{a+2}{4a}}$

10. $\dfrac{\dfrac{n-4}{6n}}{\dfrac{n-4}{8n^2}}$

11. $\dfrac{\dfrac{10r^2}{3r-2}}{\dfrac{25r^3}{3r-2}}$

12. $\dfrac{\dfrac{21y}{2y+1}}{\dfrac{14y^4}{2y+1}}$

13. $\dfrac{\dfrac{s^2-16}{4s^3}}{\dfrac{4-s}{3s}}$

14. $\dfrac{\dfrac{p-5}{6p^2}}{\dfrac{25-p^2}{9p^3}}$

15. $\dfrac{\dfrac{4n-8}{5n+15}}{\dfrac{2n-4}{n-5}}$

16. $\dfrac{\dfrac{9-3a}{4a+12}}{\dfrac{a-3}{6a-24}}$

17. $\dfrac{\dfrac{3}{n}}{1-\dfrac{1}{n}}$

18. $\dfrac{\dfrac{4}{t}}{\dfrac{2}{t}-3}$

19. $\dfrac{1+\dfrac{2}{t}}{1-\dfrac{4}{t^2}}$

20. $\dfrac{\dfrac{9}{x^2}-1}{\dfrac{3}{x}-1}$

21. $\dfrac{\dfrac{p-12}{p}+p}{p+4}$

22. $\dfrac{2r-1}{\dfrac{3r-2}{r}+2r}$

23. $\dfrac{\dfrac{4}{y}-\dfrac{y}{x^2}}{\dfrac{1}{x}-\dfrac{2}{y}}$

24. $\dfrac{\dfrac{5}{p}-\dfrac{1}{q}}{\dfrac{1}{5q^2}-\dfrac{5}{p^2}}$

25. $\dfrac{\dfrac{3}{4n^3} - \dfrac{1}{2n^2}}{\dfrac{4}{3n^2} + \dfrac{1}{2n}}$

26. $\dfrac{\dfrac{2}{9a} + \dfrac{5}{6a^2}}{\dfrac{3}{8a^3} + \dfrac{3}{2a^4}}$

27. $\dfrac{2 - \dfrac{11}{x} + \dfrac{12}{x^2}}{1 - \dfrac{2}{x} - \dfrac{8}{x^2}}$

28. $\dfrac{1 + \dfrac{6}{t} + \dfrac{9}{t^2}}{3 + \dfrac{5}{t} - \dfrac{12}{t^2}}$

29. $\dfrac{\dfrac{a}{b^2} - \dfrac{b}{a^2}}{\dfrac{3}{a^3} - \dfrac{3}{b^3}}$

30. $\dfrac{\dfrac{p^2}{4q} + \dfrac{q^2}{4p}}{\dfrac{q}{2p^2} + \dfrac{p}{2q^2}}$

31. $\dfrac{\dfrac{6}{x - y} + \dfrac{9}{x + y}}{\dfrac{9}{x - y} - \dfrac{6}{x + y}}$

32. $\dfrac{\dfrac{2}{a - b} + \dfrac{4}{a + b}}{\dfrac{4}{a + b} - \dfrac{2}{a - b}}$

33. $\dfrac{\dfrac{2}{x^2 - 3x + 2} - \dfrac{2}{2 - x}}{\dfrac{4}{x^2 - 4} + \dfrac{4}{x^2 + x - 2}}$

34. $\dfrac{\dfrac{3}{x^2 + 6x + 9} + \dfrac{3}{x + 3}}{\dfrac{6}{x^2 - 9} + \dfrac{6}{3 - x}}$

35. $\dfrac{\dfrac{2n^2 - 3n - 20}{10 - 11n - 6n^2}}{\dfrac{n^2 - 8n + 16}{3n^2 + 10n - 8}}$

36. $\dfrac{\dfrac{4p^2 - 12p + 9}{2p^2 + 7p - 15}}{\dfrac{2p^2 - 15p + 18}{p^2 - p - 30}}$

37. $\dfrac{x^{-1} + x^{-2}}{3x^{-1}}$

38. $\dfrac{n^{-3} - n^{-2}}{4n^{-1}}$

39. $\dfrac{a^{-3}}{a^{-1} + b^{-1}}$

40. $\dfrac{p^{-1} - q^{-1}}{6q^{-2}}$

Mixed Practice

Simplify.

41. $\dfrac{\dfrac{6}{r}}{2 - \dfrac{3}{r}}$

42. $\dfrac{\dfrac{1}{3a^3} + \dfrac{3}{4a^2}}{\dfrac{1}{2a^4} + \dfrac{2}{9a}}$

43. $\dfrac{\dfrac{t^2 - 9}{3t^3}}{\dfrac{t - 3}{2t}}$

44. $\dfrac{\dfrac{1}{3x^2} - \dfrac{3}{y^2}}{\dfrac{1}{x} - \dfrac{3}{y}}$

45. $\dfrac{\dfrac{2x - 4}{3x + 12}}{\dfrac{2 - x}{9x + 45}}$

46. $\dfrac{4 + \dfrac{1}{x} - \dfrac{5}{x^2}}{1 - \dfrac{4}{x} + \dfrac{3}{x^2}}$

Applications

B *Solve.*

47. The acceleration of an object when a force of F is applied to it can be represented by the expression

$$\dfrac{F}{\dfrac{w}{g}},$$

where w is the weight of the object and g is the acceleration due to gravity. Simplify this expression.

48. The ratio of the surface area of a sphere to its volume is given by the expression

$$\dfrac{4\pi r^2}{\dfrac{4\pi r^3}{3}},$$

where r is the radius of the sphere. Simplify this expression.

49. The change in the pitch of a car horn as it moves away from an observer can be described by the *Doppler effect*. As the source of a sound moves away, the observer hears the sound with a frequency given by the expression

$$\frac{f}{1 + \dfrac{v}{S}},$$

where f is the actual frequency, v is the speed of the source of the sound, and S is the speed of sound. Simplify this expression.

51. A commuter travels to work, driving a distance d at speed r. At the end of the day, she returns home, driving the same distance at speed R. Her average speed for the round-trip can be represented by the expression

$$\frac{2d}{\dfrac{d}{r} + \dfrac{d}{R}}.$$

a. Simplify this expression.

b. Use the simplified expression found in part (a) to determine the commuter's average speed if her work is 20 mi from home and she drives at 40 mph to work and at 60 mph on the way home.

53. The total current flowing through a particular circuit is given by the expression

$$\frac{V}{\dfrac{1}{R} + \dfrac{1}{R - 2}},$$

where V represents voltage and R represents resistance. Simplify this expression.

50. In the study of probability, the *odds in favor* of an event occurring are given by the expression

$$\frac{\dfrac{m}{n}}{1 - \dfrac{m}{n}},$$

where $\dfrac{m}{n}$ represents the probability of the event. Write this expression in simplest form.

52. On the first leg of a trip, a family travels x mi for t hr. On the second leg of the trip, the family travels y mi in the same number of hours. The expression

$$\frac{\dfrac{x + y}{t}}{2}$$

represents the average speed the family travels for the entire trip.

a. Simplify this expression.

b. Suppose the family travels 243 mi in $4\frac{1}{2}$ hr on the first leg of the trip, and then 279 mi in the same amount of time on the second leg of the trip. Use the expression found in part (a) to determine their average speed for the entire trip.

54. The time it takes for two landscapers to plant trees in a client's garden is given by the expression

$$\frac{1}{\dfrac{1}{t_1} + \dfrac{1}{t_2}},$$

where t_1 and t_2 represent the time it takes the first and second landscaper, respectively, to complete the job working alone. Simplify this expression.

• Check your answers on page A-41.

MINDStretchers

Investigation

1. Consider the rational expression $\dfrac{1}{x}$.

   a. Complete the table.

x	$\dfrac{1}{1000}$	$\dfrac{1}{100}$	$\dfrac{1}{10}$	$\dfrac{1}{5}$	$\dfrac{1}{2}$	1	2	5	10	100	1000
$\dfrac{1}{x}$											

   b. What happens to the value of $\dfrac{1}{x}$ as x gets larger? What happens as x gets smaller?

Groupwork

2. Work with a partner to simplify each of the following *continued fractions*:

   a. $\dfrac{1}{1 + \dfrac{1}{1 + \dfrac{1}{1 + \dfrac{1}{x}}}}$

   b. $\dfrac{1}{1 + \dfrac{1}{1 + \dfrac{1}{1 + \dfrac{1}{x + 1}}}}$

Critical Thinking

3. Consider the function $f(x) = \dfrac{2}{x - 1}$.

   Evaluate $\dfrac{f(x + h) - f(x)}{h}$.

6.4 Solving Rational Equations

OBJECTIVES

Ⓐ To solve an equation involving a rational expression

Ⓑ To solve applied problems involving rational equations

What Rational Equations Are and Why They Are Important

In previous sections of this chapter, we discussed rational expressions. Now, we consider *rational* or *fractional equations*, that is, equations that contain one or more rational expressions.

Rational equations can be used to model many situations, including those involving rates of work, motion, and medicine. The formula $C = \dfrac{aA}{a + 12}$, used by pediatricians, is an example of a rational equation. In this equation, C represents a child's dosage of a medicine, a stands for the child's age, and A is an adult dosage of this medicine. Here are other examples of rational equations:

$$\frac{x}{3} - \frac{1}{2} = \frac{x}{6} \qquad\qquad \frac{2}{t - 1} + \frac{1}{t + 1} = \frac{5}{t^2 - 1}$$

Solving Rational Equations

In solving rational equations, it is important not to confuse a rational expression with a rational equation. Expressions contain no equal sign, whereas equations do. Consider these examples:

$$\frac{x}{8} + \frac{3}{4} \qquad\qquad \frac{x}{8} + \frac{3}{4} = \frac{x}{2}$$

Rational expression $\qquad\qquad$ Rational equation

Let's look at how to solve the equation $\dfrac{x}{8} + \dfrac{3}{4} = \dfrac{x}{2}$. The rational expressions in this equation are $\dfrac{x}{8}, \dfrac{3}{4}$, and $\dfrac{x}{2}$. The denominators of these expressions are 8, 4, and 2, so the LCD is 8.

$$\frac{x}{8} + \frac{3}{4} = \frac{x}{2}$$

$$\left(\frac{x}{8} + \frac{3}{4}\right) \cdot 8 = \frac{x}{2} \cdot 8 \qquad \text{Multiply each side of the equation by the LCD.}$$

$$\frac{x}{8} \cdot 8 + \frac{3}{4} \cdot 8 = \frac{x}{2} \cdot 8 \qquad \text{Use the distributive property.}$$

$$x + 6 = 4x \qquad \text{Simplify.}$$

$$6 = 3x$$

$$2 = x, \text{ or } x = 2$$

To verify the solution, we check to see if substituting 2 for x in the original equation results in a true statement.

$$\frac{x}{8} + \frac{3}{4} = \frac{x}{2}$$

$$\frac{2}{8} + \frac{3}{4} \overset{?}{=} \frac{2}{2} \qquad \text{Substitute 2 for } x \text{ in the original equation.}$$

$$\frac{8}{8} \overset{?}{=} \frac{2}{2}$$

$$1 = 1 \qquad \text{True}$$

So the solution of the equation is 2.

The following rule gives a general procedure for solving rational equations:

> ## To Solve a Rational Equation
> - Find the LCD of all the denominators.
> - Multiply each side of the equation by the LCD.
> - Solve the resulting equation.
> - Check your solution(s) in the original equation.

EXAMPLE 1

Solve and check: $\dfrac{4}{3x} - \dfrac{x+2}{x} = 1$

Solution The LCD of the denominators in the equation is $3x$.

$$\frac{4}{3x} - \frac{x+2}{x} = 1$$

$$3x \cdot \left(\frac{4}{3x} - \frac{x+2}{x}\right) = 3x \cdot 1 \qquad \text{Multiply each side of the equation by the LCD.}$$

$$3x \cdot \frac{4}{3x} - 3x \cdot \left(\frac{x+2}{x}\right) = 3x \cdot 1 \qquad \text{Use the distributive property.}$$

$$4 - 3(x+2) = 3x \qquad \text{Simplify.}$$

$$4 - 3x - 6 = 3x$$

$$-2 - 3x = 3x$$

$$-2 = 6x$$

$$\frac{-2}{6} = x$$

$$x = -\frac{1}{3}$$

Check

$$\frac{4}{3x} - \frac{x+2}{x} = 1$$

$$\frac{4}{3\left(-\dfrac{1}{3}\right)} - \frac{-\dfrac{1}{3} + 2}{-\dfrac{1}{3}} \overset{?}{=} 1 \qquad \text{Substitute } -\frac{1}{3} \text{ for } x \text{ in the original equation.}$$

$$\frac{4}{-1} - \frac{\dfrac{5}{3}}{-\dfrac{1}{3}} \overset{?}{=} 1 \quad \longrightarrow \quad -4 - (-5) \overset{?}{=} 1$$

$$1 = 1 \quad \text{True}$$

So the solution is $-\dfrac{1}{3}$.

PRACTICE 1

Solve and check: $\dfrac{1}{2y} - \dfrac{y+1}{y} = 2$

When multiplying each side of a rational equation by a variable expression, the resulting equation may have a solution that does not satisfy the original equation. If such a number makes a denominator in the original equation 0, then the rational expression with that denominator is undefined. These numbers, called *extraneous solutions*, are *not* solutions of the original equation. So in solving a rational equation, it is particularly important to check for extraneous solutions.

EXAMPLE 2

Solve and check:

$$\frac{x}{x + 4} = 3 - \frac{4}{x + 4}$$

Solution The LCD of the denominators in the equation is $x + 4$.

$$\frac{x}{x + 4} = 3 - \frac{4}{x + 4}$$

$$(x + 4) \cdot \frac{x}{x + 4} = (x + 4) \cdot \left(3 - \frac{4}{x + 4}\right) \qquad \text{Multiply each side of the equation by the LCD.}$$

$$(x + 4) \cdot \frac{x}{x + 4} = (x + 4) \cdot 3 - (x + 4) \cdot \frac{4}{x + 4}$$

$$x = 3(x + 4) - 4$$

$$x = 3x + 12 - 4$$

$$x = 3x + 8$$

$$-2x = 8$$

$$x = -4$$

Check

$$\frac{x}{x + 4} = 3 - \frac{4}{x + 4}$$

$$\frac{-4}{-4 + 4} \stackrel{?}{=} 3 - \frac{4}{-4 + 4} \qquad \text{Substitute } -4 \text{ for } x \text{ in the original equation.}$$

$$\frac{-4}{0} \stackrel{?}{=} 3 - \frac{4}{0} \qquad \text{Undefined}$$

Since substituting -4 for x in the original equation makes the rational expressions undefined, we say -4 is an extraneous solution and conclude that the original equation has no solution.

PRACTICE 2

Solve and check:

$$\frac{y}{y - 3} = \frac{3}{y - 3} - 1$$

Without solving, explain how we would know that -4 is not a solution of the equation in Example 2.

EXAMPLE 3

Solve and check: $\dfrac{x}{x+2} - \dfrac{2}{2-x} = \dfrac{x+6}{x^2-4}$

Solution

$$\dfrac{x}{x+2} - \dfrac{2}{2-x} = \dfrac{x+6}{x^2-4}$$

$$\dfrac{x}{x+2} - \dfrac{2}{2-x} = \dfrac{x+6}{(x+2)(x-2)}$$ Factor x^2-4.

$$\dfrac{x}{x+2} - \dfrac{2}{-(x-2)} = \dfrac{x+6}{(x+2)(x-2)}$$ Write $(2-x)$ as $-(x-2)$.

$$\dfrac{x}{x+2} + \dfrac{2}{x-2} = \dfrac{x+6}{(x+2)(x-2)}$$ Simplify.

$$\dfrac{x}{x+2}\cdot(x+2)(x-2) + \dfrac{2}{x-2}\cdot(x+2)(x-2) = \dfrac{x+6}{(x+2)(x-2)}\cdot(x+2)(x-2)$$ Multiply each side of the equation by the LCD $(x+2)(x-2)$.

$$\dfrac{x}{x+2}\cdot(x+2)(x-2) + \dfrac{2}{x-2}\cdot(x+2)(x-2) = \dfrac{x+6}{(x+2)(x-2)}\cdot(x+2)(x-2)$$

$$x(x-2) + 2(x+2) = x+6$$

$$x^2 - 2x + 2x + 4 = x+6$$ Distribute.

$$x^2 - x - 2 = 0$$

$$(x+1)(x-2) = 0$$ Factor.

$$x+1 = 0 \quad \text{or} \quad x-2 = 0$$ Set each factor equal to 0.

$$x = -1 \qquad\qquad x = 2$$

Check

Substitute -1 for x.

$$\dfrac{x}{x+2} - \dfrac{2}{2-x} = \dfrac{x+6}{x^2-4}$$

$$\dfrac{-1}{-1+2} - \dfrac{2}{2-(-1)} \stackrel{?}{=} \dfrac{-1+6}{(-1)^2-4}$$

$$\dfrac{-1}{1} - \dfrac{2}{3} \stackrel{?}{=} \dfrac{5}{-3}$$

$$\dfrac{-3-2}{3} \stackrel{?}{=} \dfrac{5}{-3}$$

$$-\dfrac{5}{3} = -\dfrac{5}{3} \quad \text{True}$$

Substitute 2 for x.

$$\dfrac{x}{x+2} - \dfrac{2}{2-x} = \dfrac{x+6}{x^2-4}$$

$$\dfrac{2}{2+2} - \dfrac{2}{2-2} \stackrel{?}{=} \dfrac{2+6}{(2)^2-4}$$

$$\dfrac{2}{4} - \dfrac{2}{0} = \dfrac{8}{0} \quad \text{Undefined}$$

Note that when -1 is substituted for x in the original equation, we get a true statement. On the other hand, when 2 is substituted, two of the denominators of the rational equation are equal to 0. So -1 is a solution, whereas 2 is an extraneous solution.

PRACTICE 3

Solve and check: $\dfrac{m}{m+3} - \dfrac{2}{3-m} = \dfrac{2m+6}{m^2-9}$

EXAMPLE 4

Solve and check: $\dfrac{x}{x+6} = \dfrac{1}{x+2}$

Solution

$$\dfrac{x}{x+6} = \dfrac{1}{x+2}$$

$$\dfrac{x}{x+6} \cdot (x+6)(x+2) = \dfrac{1}{x+2} \cdot (x+6)(x+2)$$

Multiply each side of the equation by the LCD $(x+6)(x+2)$.

$$\dfrac{x}{x+6} \cdot (x+6)(x+2) = \dfrac{1}{x+2} \cdot (x+6)(x+2)$$

$$x(x+2) = x+6$$

$$x^2 + 2x = x+6$$

$$x^2 + x - 6 = 0$$

$$(x+3)(x-2) = 0$$

$$x+3 = 0 \quad \text{or} \quad x-2 = 0$$

$$x = -3 \qquad\qquad x = 2$$

The check, which we leave as an exercise, confirms that both -3 and 2 are solutions.

PRACTICE 4

Solve and check: $\dfrac{n-3}{3} = \dfrac{3}{n+5}$

Equations of the type shown in Example 4 commonly arise from ratio and proportion problems. A *ratio* is a comparison of two numbers or two quantities expressed as a quotient. Since we can write rational expressions as quotients, they can also be written as ratios. A *proportion* is a statement that two ratios are equal.

When two rational expressions, $\dfrac{a}{b}$ and $\dfrac{c}{d}$, are equal to one another, we can solve the equation either by applying the LCD method or by using an alternative approach called the **cross-product method.**

Let's solve Example 4 using the cross-product method. With this method, we set the cross products equal to each other.

$$\dfrac{x}{x+6} = \dfrac{1}{x+2}$$

Cross products

$$x \cdot (x+2) = 1 \cdot (x+6) \qquad \text{Set the cross products equal.}$$

$$x^2 + 2x = x+6$$

$$x^2 + x - 6 = 0$$

$$(x+3)(x-2) = 0$$

$$x+3 = 0 \quad \text{or} \quad x-2 = 0$$

$$x = -3 \qquad\qquad x = 2$$

Note that we get the same solutions for Example 4 whether multiplying by the LCD or setting the cross products equal to each other. In general, when two rational expressions are equal to one another, we can solve the equation by either the LCD method or the cross-product method.

EXAMPLE 5

Solve using the cross-product method: $\dfrac{6}{2x-5} = \dfrac{2}{x+3}$

Solution

$$\frac{6}{2x-5} = \frac{2}{x+3}$$

$$6(x+3) = 2(2x-5) \quad \text{Set the cross products equal to each other.}$$

$$6x + 18 = 4x - 10$$

$$2x = -28$$

$$x = -14$$

The check, which we leave as an exercise, confirms that -14 is the solution.

PRACTICE 5

Solve using the cross-product method: $\dfrac{5}{t-2} = \dfrac{3}{2t+3}$

EXAMPLE 6

If $f(x) = x - \dfrac{4}{x}$, find all values of x for which $f(x) = 3$.

Solution Since $f(x) = x - \dfrac{4}{x}$, we want to find all values of x for which $3 = x - \dfrac{4}{x}$.

$$3 = x - \frac{4}{x}$$

$$3 \cdot x = \left(x - \frac{4}{x}\right) \cdot x \quad \begin{array}{l}\text{Multiply each side of the equation}\\\text{by the LCD } x.\end{array}$$

$$3x = x^2 - 4 \quad \text{Use the distributive property.}$$

$$0 = x^2 - 3x - 4$$

$$x^2 - 3x - 4 = 0$$

$$(x-4)(x+1) = 0$$

$$x - 4 = 0 \quad \text{or} \quad x + 1 = 0$$

$$x = 4 \qquad\qquad x = -1$$

Checking confirms that the values of x for which $f(x) = 3$ are 4 and -1.

PRACTICE 6

Given that $g(x) = \dfrac{x(x+2)}{x+6}$, find all values of x for which $g(x) = 1$.

Rational equations are used to solve many types of applied problems. We will discuss several of these. First, let's consider a problem involving *rates*. A rate is the ratio of unlike quantities, that is, quantities with different units.

EXAMPLE 7

A dripping faucet wastes about 15 gal of water daily. At this rate, how much water is wasted in 4 hr?

Solution The rate of the dripping faucet is 15 gal/day. We want to find the number of gallons of water wasted in 4 hr. We write a proportion consisting of two equal rates that have the same units written in the same order. Let x represent the number of gallons of water wasted in 4 hr. (Recall that 1 day = 24 hr.) So we write the following proportion and cross multiply.

PRACTICE 7

A certain iceberg in Antarctica moves about 2 ft/yr. At this rate, how long will the iceberg take to move $1\frac{1}{2}$ ft?

$$\text{Gallons} \longrightarrow \frac{15}{24} = \frac{x}{4} \longleftarrow \text{Gallons}$$
$$\text{Hours} \longrightarrow \frac{15}{24} = \frac{x}{4} \longleftarrow \text{Hours}$$
$$24 \cdot x = 4 \cdot 15$$
$$24x = 60$$
$$x = 2.5$$

So 2.5 gal of water are wasted in 4 hr.

Can you write a different proportion to solve Example 7? Explain.

Another application of rational equations is motion problems. Recall that in these problems, an object moves at a uniform or average rate r for time t and travels a distance d. These three quantities are related by the following formula:

$$d = r \cdot t$$

If we solve this equation for t, we get $\frac{d}{r} = t$, a different form of the formula that we apply in the next example where two times of travel are equal.

EXAMPLE 8

A canoe travels 15 mph in still water. Find the speed of the river's current if the canoe traveled 20 mi down the river in the same time it took to travel 10 mi up the river.

Solution   Let r represent the speed of the river's current. When traveling downriver, the speed of the river's current is added to that of the canoe. So the rate of the canoe going downriver is $15 + r$. When traveling upriver, the speed of the river's current is subtracted from that of the canoe. The rate of the canoe going upriver is, therefore, $15 - r$.

Applying the formula $\frac{d}{r} = t$, we can set up the following table:

	d	$\div$	r	$=$	t
Downriver	20		$15 + r$		$\frac{20}{15 + r}$
Upriver	10		$15 - r$		$\frac{10}{15 - r}$

Since the two times are equal, we write the following proportion, and then cross multiply to solve for r:

$$\frac{20}{15 + r} = \frac{10}{15 - r}$$
$$20(15 - r) = 10(15 + r)$$
$$300 - 20r = 150 + 10r$$
$$150 = 30r$$
$$r = 5$$

So the speed of the river's current is 5 mph.

PRACTICE 8

A local bus travels 9 mph more slowly than the express bus. The express bus travels 90 mi in the same time it takes the local bus to travel 75 mi. Find the speed of the local bus.

We can also use the formula $\dfrac{d}{r} = t$ when the total time of travel for two objects is known. Consider the next example.

EXAMPLE 9

A commuter drove 24 mi to work, and then returned home following the same route. Since there was less traffic going to work than on the return trip, she traveled 10 mph faster going to work than coming back home. If her round-trip to and from work took 2 hr, at what speed did the commuter return home?

Solution   We are looking for the commuter's speed returning home. So we let r represent that speed. Since she traveled 10 mph faster going to work than returning home, her speed going to work must have been $r + 10$.

Applying the formula $\dfrac{d}{r} = t$, we can set up the following table:

	d	$\div$	r	$=$	t
Going to Work	24		$r + 10$		$\dfrac{24}{r + 10}$
Returning Home	24		r		$\dfrac{24}{r}$

Since the round-trip took a total of 2 hr, we get the following rational equation:

$$\frac{24}{r + 10} + \frac{24}{r} = 2$$

To solve this equation, we multiply both sides of the equation by the LCD, $r(r + 10)$.

$$r(r + 10) \cdot \frac{24}{r+10} + r(r + 10) \cdot \frac{24}{r} = r(r + 10) \cdot 2$$

$$24r + 24(r + 10) = 2r(r + 10)$$

$$24r + 24r + 240 = 2r^2 + 20r$$

$$-2r^2 + 28r + 240 = 0$$

$$-2(r^2 - 14r - 120) = 0$$

$$-2(r - 20)(r + 6) = 0$$

$$r - 20 = 0 \quad \text{or} \quad r + 6 = 0$$

$$r = 20 \qquad\qquad r = -6$$

We reject the speed -6 because speed cannot be negative. To check the other solution, we can substitute 20 for r in the original equation. This confirms that the commuter returned home at the rate of 20 mph.

PRACTICE 9

A cargo ship travels 7 km/hr in still water. It travels 45 km up the river and the same distance down the river in a total time of 14 hr. Find the speed of the river's current.

Rational equations also play an important role in applications known as *work problems*. Like motion problems, work problems involve rate and time. The key to solving a work problem is to determine the rate of work, that is, the fraction of a task that is completed in one unit of time. In these problems, we typically want to compute how long it will take to complete the task.

EXAMPLE 10

One construction crew can pave a road in 24 hr, and a second crew can do the same job in 18 hr. How long would it take for the two crews to do the job working together?

Solution   Since this is a work problem, we can use the following equation to determine how long it would take the two crews to do the job together:

$$(\text{Rate of work}) \cdot (\text{Time worked}) = \text{Part of the task completed}$$

Using this equation, we can set up a table. Let t represent the time in hours it would take the two crews to complete the job working together.

Since the first crew needs 24 hr to pave the road, they complete $\dfrac{1}{24}$ of

the job in an hour. Similarly, the second road crew completes $\dfrac{1}{18}$ of the job in an hour.

	Rate of Work	$\cdot$ Time Worked	=	Part of the Task Completed
First Crew	$\dfrac{1}{24}$	t		$\dfrac{1}{24} \cdot t$
Second Crew	$\dfrac{1}{18}$	t		$\dfrac{1}{18} \cdot t$

Since the sum of the parts of the task completed must equal 1, the whole task, we have:

$$\frac{1}{24} \cdot t + \frac{1}{18} \cdot t = 1$$

$$\frac{t}{24} + \frac{t}{18} = 1$$

$$\overset{3}{\cancel{72}} \cdot \frac{t}{24} + \overset{4}{\cancel{72}} \cdot \frac{t}{18} = 72 \cdot 1$$

$$3t + 4t = 72$$

$$7t = 72$$

$$t = 10\tfrac{2}{7}$$

So it would take the two crews working together $10\tfrac{2}{7}$ hr to complete the job. We leave the check as an exercise.

Some rational equations are formulas or literal equations that relate two or more variables. Let's look at the following example.

PRACTICE 10

One printing press can print an order of college admission brochures in 12 hr. Another press can complete the same job in 18 hr. How long would it take if the college used both presses?

EXAMPLE 11

To produce x units, it costs a factory $3000 in fixed costs and $20 per unit. The cost per unit C can be represented by $C = \dfrac{20x + 3000}{x}$.

a. Solve the equation for x in terms of C.

b. Using the equation found in part (a), determine how many units must be produced so that the cost per unit is $25.

Solution

a. Solving for x in terms of C, we get:

$$C = \frac{20x + 3000}{x}$$
$$Cx = 20x + 3000$$
$$Cx - 20x = 3000$$
$$x(C - 20) = 3000$$
$$x = \frac{3000}{C - 20}$$

b. To determine how many units must be produced to make the cost per unit $25, we substitute 25 for C in the equation found in part (a).

$$x = \frac{3000}{C - 20} = \frac{3000}{\mathbf{25} - 20} = \frac{3000}{5} = 600$$

So 600 units must be produced for the cost per unit to be $25.

PRACTICE 11

Recall from geometry that a trapezoid is a four-sided geometric figure, two of whose sides (the bases) are parallel. Suppose we draw the two diagonals of a trapezoid and also the line parallel to the bases that passes through the point of intersection of the diagonals (see the figure below).

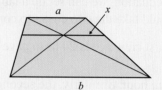

It can be shown that the three lengths x, a, and b are related by the following equation:

$$\frac{2}{x} = \frac{1}{a} + \frac{1}{b}$$

a. Solve the equation for x in terms of a and b.

b. Using the equation found in part (a), determine the value of x if the lengths of the two bases are 6 in. and 4 in.

Mathematically Speaking

Fill in each blank with the most appropriate term or phrase from the given list.

GCF	rate	equal
division	proportion	extraneous
cross-product	LCD	ratio

1. To solve a rational equation, first find the _____ of all the rational expressions.

2. In solving a rational equation, always check for _____ solutions, which make a denominator in the original equation equal to 0.

3. A(n) _____ is a comparison of two numbers or two quantities expressed as a quotient.

4. A(n) _____ is a statement that two ratios are equal.

5. If two rational expressions are equal to each other, the resulting equation can be solved by using either the LCD method or the _____ method.

6. A(n) _____ is the ratio of unlike quantities, that is, quantities with different units.

A *Solve and check.*

7. $\dfrac{x}{10} + \dfrac{3}{5} = \dfrac{x}{20}$

8. $\dfrac{5}{8} - \dfrac{y}{4} = \dfrac{y}{16}$

9. $\dfrac{r}{8} + \dfrac{r-4}{12} = \dfrac{r}{24}$

10. $\dfrac{n-2}{2} - \dfrac{n}{6} = \dfrac{4n}{9}$

11. $\dfrac{x+5}{x} - \dfrac{2}{3} = 0$

12. $\dfrac{4}{5} - \dfrac{t-2}{t} = 0$

13. $\dfrac{7}{r+20} = \dfrac{2}{r}$

14. $\dfrac{9}{x-6} = \dfrac{6}{x}$

15. $\dfrac{2}{p} - \dfrac{3p+1}{4p} = -1$

16. $\dfrac{2y+3}{5y} + \dfrac{1}{y} = 2$

17. $\dfrac{8}{a-1} - 4 = \dfrac{2a}{a-1}$

18. $2 + \dfrac{t}{t+3} = \dfrac{6}{t+3}$

19. $\dfrac{1}{x+5} = \dfrac{5}{x+5}$

20. $\dfrac{4}{n-8} = \dfrac{7}{n-8}$

21. $\dfrac{2}{n} + n = 3$

22. $y - \dfrac{6}{y} = -1$

23. $\dfrac{a}{4} = \dfrac{9}{a}$

24. $\dfrac{8}{y} = \dfrac{y}{2}$

25. $\dfrac{t}{t-1} - \dfrac{t^2}{t-1} = 5$

26. $\dfrac{x^2}{x+2} - \dfrac{6x}{x+2} = -2$

27. $4 + \dfrac{7}{p} = \dfrac{15}{p^2}$

28. $3 - \dfrac{12}{x^2} = \dfrac{5}{x}$

29. $\dfrac{r+7}{2} + \dfrac{4}{r} = \dfrac{1}{2}$

30. $\dfrac{1}{3} - \dfrac{x-1}{x} = \dfrac{x}{3}$

31. $\dfrac{1}{x} + \dfrac{2}{x+10} = \dfrac{x}{x+10}$

32. $\dfrac{7}{n} + \dfrac{n}{n-3} = \dfrac{3}{n-3}$

33. $\dfrac{6}{n+2} = \dfrac{10}{n-2}$

34. $\dfrac{5}{y-1} = \dfrac{9}{y+3}$

35. $\dfrac{1}{a+1} = \dfrac{4}{a+8}$

36. $\dfrac{2}{t-1} = \dfrac{6}{t-9}$

37. $\dfrac{3-7r}{r^2-9} - \dfrac{r}{3-r} = \dfrac{10}{r+3}$

38. $\dfrac{p}{p+4} + \dfrac{12}{p-4} = \dfrac{8-6p}{p^2-16}$

39. $\dfrac{x}{x-1} - 1 = -\dfrac{7}{x-5}$

40. $\dfrac{4}{a+2} + \dfrac{2a}{a-3} = 2$

41. $\dfrac{x}{3} = \dfrac{6}{x-7}$

42. $\dfrac{9}{y+6} = \dfrac{y}{8}$

43. $\dfrac{t-4}{8} = \dfrac{-3}{t+6}$

44. $\dfrac{p+1}{-2} = \dfrac{2}{p-4}$

45. $\dfrac{x}{x+5} - \dfrac{3}{x+4} = \dfrac{7x+1}{x^2+9x+20}$

46. $\dfrac{5-8x}{x^2-x-42} + \dfrac{x}{x-7} = \dfrac{-1}{x+6}$

47. $\dfrac{n-1}{n^2-2n} - \dfrac{n}{4n-8} = 0$

48. $\dfrac{x-4}{x^2+3x} + \dfrac{x}{2x+6} = 0$

49. $-\dfrac{8}{y^2-1} = -\dfrac{y+5}{y^2+y}$

50. $\dfrac{-6}{r^2-4} = \dfrac{r-1}{r^2+2r}$

51. $\dfrac{2y}{y^2-9} + \dfrac{y-1}{6-2y} = 1$

52. $\dfrac{n+2}{1-n} + 1 = \dfrac{1-4n}{2n^2-2}$

Solve each equation for the indicated variable.

53. $P = \dfrac{W}{t}$ for t

54. $I = \dfrac{P}{A}$ for A

55. $\dfrac{I_2}{I_1} = \dfrac{a^2}{b^2}$ for I_2

56. $\dfrac{p_1V_1}{T_1} = \dfrac{p_2V_2}{T_2}$ for V_2

57. $\dfrac{1}{R} = \dfrac{1}{R_1} + \dfrac{1}{R_2}$ for R_1

58. $\dfrac{1}{f} = \dfrac{1}{f_1} + \dfrac{1}{f_2}$ for f_2

59. $V = \dfrac{m_1v_1 + m_2v_2}{m_1 + m_2}$ for m_1

60. $a = \dfrac{v_2 - v_1}{t_2 - t_1}$ for t_2

Solve.

61. If $f(x) = \dfrac{x+9}{x-1}$, find all values of x for which $f(x) = \dfrac{3}{2}$.

62. If $g(x) = \dfrac{x}{3x-4}$, find all values of x for which $g(x) = \dfrac{1}{4}$.

63. If $g(x) = \dfrac{2}{x-1} - \dfrac{4}{x+5}$, find all values of x for which $g(x) = -2$.

64. If $f(x) = \dfrac{6x}{x+2} + x$, find all values of x for which $f(x) = 8$.

65. If $f(x) = \dfrac{x+3}{4}$ and $g(x) = \dfrac{x}{x-6}$, find all values of x for which $f(x) = g(x)$.

66. If $f(x) = \dfrac{1}{x^2-1}$ and $g(x) = \dfrac{3}{x+1}$, find all values of x for which $f(x) = g(x)$.

Mixed Practice

Multiply. Express each answer in lowest terms.

67. $\dfrac{x^2-x-6}{x^3-5x^2} \cdot \dfrac{4x^2-20x}{x^2+4x+4}$

68. $\dfrac{2a^2-ab-b^2}{a^2+b^2} \cdot \dfrac{2a^2-ab}{4a^2-b^2}$

Solve and check.

69. $\dfrac{8}{x+3} = \dfrac{2}{x-3}$

70. $1 - \dfrac{21}{x^2} = \dfrac{4}{x}$

71. $\dfrac{a-6}{a^2+5a} + \dfrac{a}{3a+15} = 0$

72. $\dfrac{2}{m+2} + 5 = \dfrac{2m}{m+2}$

73. $\dfrac{y-5}{6} - \dfrac{y}{4} = \dfrac{y}{3}$

74. $\dfrac{a+1}{a+6} = \dfrac{a}{a+6}$

Solve.

75. If $g(x) = \dfrac{-3x}{x+3} + x$, find all values of x for which
$g(x) = 4$.

Solve the equation for the indicated variable.

76. $\dfrac{1}{a} + \dfrac{1}{b} = \dfrac{2}{h}$ for h.

B Applications

Solve.

77. The human resources department of a company surveyed employees to determine employee satisfaction with the company's benefits package. Of the 150 employees surveyed, 124 said that they were satisfied. At this rate, approximately how many of the company's 1128 employees are satisfied with their benefits package?

78. The EPA mileage estimates for highway driving indicate that a new sedan averages 33 mpg. If the car has a gas tank with a capacity of 17.1 gal, how many miles can be driven on a full tank of gas? Round the answer to the nearest mile.

79. The speed of an airplane in still air is 420 mph. Find the speed of the wind if the plane traveled 1125 mi with the wind in the same time it took to travel 975 mi against the wind.

80. On a road trip, a group of students traveled 210 mi before stopping at a rest area. After leaving the rest area, they traveled an additional 255 mi at an average speed that was 4 mph faster than their speed before stopping. If the entire trip took 8 hr of driving time, what was the average speed during each part of the trip?

81. For the cardio portion of a workout routine, a personal trainer recommends that a client run 5 mi on the treadmill followed by a 1 mi cool-down walk. The client runs twice as fast as she walks on the treadmill. If she spends 70 min on the treadmill, at what rate did she run during her cardio workout?

82. During a major snowstorm, a state highway department reduced the speed limit on a major highway by 25 mph. A car traveled 24 mi at the reduced speed limit during the snowstorm in the same amount of time that it took to drive 39 mi at the regular posted speed limit. What is the regular posted speed limit on the highway?

83. The current of a river is 1 mph. A crew team rows 12 mi upriver and 12 mi downriver. If it takes the team 1 hr and 16 min round-trip, what is the rate at which the team rows in still water?

84. A train makes a trip at 70 mph, and a bus makes the same trip at 50 mph. If the train takes 2 hr less than the bus to make this trip, how far did they each travel?

85. One grounds crew can prepare a college football field in 9 hr. Another grounds crew can prepare the same field in 6 hr. Working together, how long will it take them to prepare the field?

86. An older machine at a beverage bottling plant can fill the same number of bottles in 4 hr that a newer machine can fill in half the time. How many hours will it take both machines working together to fill the same number of bottles?

87. At a warehouse, it takes two employees $1\frac{1}{3}$ hr to load a truck when they work together. If it takes one employee 3 hr to load the truck working alone, how long does it take the other employee working alone?

88. The inlet pipe of a water tank can fill the tank in $1\frac{1}{2}$ hr. The outlet pipe can empty the tank in 1 hr. If both pipes are open, how long will it take to empty the tank?

89. The focal length f of a thin lens is given by the *lensmaker's equation*

$$\frac{1}{f} = (n - 1)\left(\frac{1}{r_1} - \frac{1}{r_2}\right),$$

where n is the index of refraction of the lens material and r_1 and r_2 are the radii of the curvature of the lens surfaces.

a. Solve the formula for n.

b. Find the index of refraction of a glass lens if the focal length is 30 cm, $r_1 = 2.5$ cm, and $r_2 = 3$ cm.

90. The formula $P = \dfrac{A}{1 + r}$ can be used to determine the amount P (in dollars) that must be invested in an account at interest rate r (in decimal form), compounded annually, in order to have A dollars at the end of one year.

a. Solve this formula for r in terms of A and P.

b. What is the interest rate (expressed as a percent) on an account if \$1500 was initially invested and the account had \$1597.50 after one year?

• Check your answers on page A-41.

MINDStretchers

Critical Thinking

1. Consider each of the following rational equations. Without graphing the equations, identify any x- and y-intercepts of their graphs.

a. $y = \dfrac{1}{x}$

b. $y = \dfrac{x + 3}{x - 2}$

c. $y = \dfrac{4}{x + 6}$

Mathematical Reasoning

2. Solve the following system of equations:

$$\frac{4}{x} + \frac{3}{y} = 2$$

$$\frac{5}{x} + \frac{2}{y} = -1$$

Groupwork

3. Working with a partner, solve each of the following equations:

a. $\dfrac{x}{1 + \dfrac{1}{x + 1}} = x - 3$

b. $\dfrac{2 - \dfrac{1}{x}}{4 - \dfrac{1}{x^2}} = 1$

6.5 Variation

What Variation Is and Why It Is Important

OBJECTIVES

A To determine whether the variation between variables is direct or inverse.

B To write and solve an equation expressing direct variation, inverse variation, joint variation, or combined variation

C To solve applied problems involving variation

In this section, we consider a type of equation or formula that relates one variable to one or more other variables by means of multiplication, division, or both operations. We call this type of equation a *variation equation*. Variation equations or formulas show how one quantity changes in relation to other quantities. Such quantities can vary *directly, inversely*, or *jointly*. For example, for a fixed speed, distance traveled *varies directly* as time traveled. For a fixed distance, time traveled *varies inversely* as speed. And in general, distance traveled *varies jointly* as time traveled and speed.

Now, let's look at how to write equations expressing direct variation, inverse variation, joint variation, and also combined variation, and how to solve applied problems involving these types of variation.

Direct Variation

Let's first discuss *direct variation*. Consider the circumference of a circle, which is given by the formula $C = \pi d$, where d is the diameter of the circle. In this formula, the circumference is always π times the diameter; that is, it is always a constant multiple of d. In such a case, we say that C *varies directly* as d, or that C is *directly proportional* to d.

DEFINITION

If a relationship between two variables is described by an equation of the form $y = kx$, where k is a positive constant, we say that we have **direct variation**, that y **varies directly** as x, or that y is **directly proportional** to x. The number k is called the **constant of variation** or the **constant of proportionality**, and $y = kx$ is called a **direct variation equation**.

In the direct variation equation $y = kx$, the relationship between x and y is linear. Recall from Section 3.3 that the graph of this equation is a straight line with slope k that passes through the origin.

For example, the graph of the direct variation equation $C = \pi d$ is a line that passes through the origin with slope π. In the graph to the right, we see that as d increases, C increases. Similarly, as d decreases, C decreases. Note that for $C = \pi d$, the number π is the constant of proportionality.

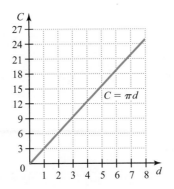

In general, for the direct variation equation $y = kx$, as x increases, y increases. Similarly, as x decreases, y decreases.

EXAMPLE 1

Suppose y varies directly as x, and $y = 25$ when $x = 15$. Find the constant of variation and the direct variation equation.

Solution We know that y varies directly as x, and so we write $y = kx$. If $y = 25$ when $x = 15$, we get:

$$y = kx$$

$$25 = k(15) \qquad \text{Substitute 25 for } y \text{ and 15 for } x.$$

$$\frac{25}{15} = k$$

$$k = \frac{5}{3}$$

So the constant of variation is $\frac{5}{3}$, and the direct variation equation is $y = \frac{5}{3}x$.

PRACTICE 1

Find the constant of variation and the direct variation equation in which y varies directly as x and $y = 16$ when $x = 20$.

Let's consider an application involving direct variation.

EXAMPLE 2

In chemistry, Charles's Law deals with the properties of a sample of gas. The law states that if the pressure of the gas remains constant, then its volume V is directly proportional to the Kelvin temperature T. If the gas in a hot-air balloon occupies 100 cm^3 at 200 K, what volume would it occupy at a temperature of 150 K while the pressure of the gas remains constant?

Solution Using Charles's Law, we know that the volume V of a sample of gas is directly proportional to the temperature, T. So we write

$$V = kT,$$

where k is the constant of variation. We are given that the sample of gas occupies 100 cm^3 at 200 K; that is, V equals 100 cm^3 when T equals 200 K. Substituting this information in the variation equation, we can find the value of k.

$$V = kT$$

$$100 = k(200)$$

$$\frac{100}{200} = k$$

$$k = \frac{1}{2}$$

Substituting $\frac{1}{2}$ for k in the direct variation equation, we get:

$$V = \frac{1}{2}T$$

PRACTICE 2

In a clinical class, a student nurse was told that the amount A of the drug gentamicine administered to patients is directly proportional to the patient's weight M (in kilograms). Suppose 160 mg of gentamicine is administered to a patient whose weight is 40 kg. How much of the drug should be administered to a patient whose weight is 55 kg?

Now, we find the volume when the temperature is 150 K.

$$V = \frac{1}{2}T$$

$$V = \frac{1}{2}(150)$$

$$V = 75$$

So when the temperature is 150 K, the gas will occupy 75 cm^3.

Inverse Variation

In a direct variation relationship, we saw that as x increases, y increases, and as x decreases, y decreases. Here, we discuss another kind of variation called *inverse variation*. An example of inverse variation can be found by looking at the distance formula, $d = rt$. Solving the formula for t, we get:

$$t = \frac{d}{r}$$

If the distance d is constant, we say that t *varies inversely* as r or that t is *inversely proportional* to r.

> **DEFINITION**
>
> If a relationship between two variables is described by an equation in the form $y = \frac{k}{x}$, where k is a positive constant, we say that we have **inverse variation**, that y **varies inversely** as x, or y is **inversely proportional** to x. The number k is called the **constant of variation** or the **constant of proportionality**, and $y = \frac{k}{x}$ is called an **inverse variation equation**.

In the inverse variation equation $y = \frac{k}{x}$, the relationship between x and y is nonlinear.

Similarly, in the inverse variation equation $t = \frac{d}{r}$, where d is a distance traveled, r is rate, and t is time traveled, the relationship between r and t is nonlinear. For instance, suppose $d = 8$ mi. From the nonlinear graph $t = \frac{8}{r}$, we see that as r increases, t decreases.

r	$t = \dfrac{8}{r}$	(r, t)
1	$t = \dfrac{8}{1} = 8$	$(1, 8)$
2	$t = \dfrac{8}{2} = 4$	$(2, 4)$
4	$t = \dfrac{8}{4} = 2$	$(4, 2)$
8	$t = \dfrac{8}{8} = 1$	$(8, 1)$

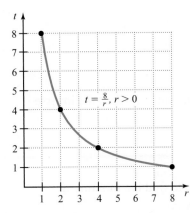

$$t = \tfrac{8}{r}, r > 0$$

Note that for $t = \frac{8}{r}$, the constant of variation is 8.

In general, for the inverse variation equation $y = \frac{k}{x}$, as x increases, y decreases. Similarly, as x decreases, y increases.

EXAMPLE 3

Find the constant of variation and the inverse variation equation in which y varies inversely as x if $y = 3.5$ when $x = 6$.

Solution   We know that y varies inversely as x, so we write $y = \dfrac{k}{x}$. Since $y = 3.5$ when $x = 6$, we get:

$$y = \frac{k}{x}$$

$$3.5 = \frac{k}{6} \qquad \text{Substitute 3.5 for } y \text{ and 6 for } x.$$

$$k = 21.0, \text{ or } 21$$

So the constant of variation is 21, and the inverse variation equation is $y = \dfrac{21}{x}$.

PRACTICE 3

Suppose y varies inversely as x and $y = 1.6$ when $x = 30$. Find the constant of variation and the inverse variation equation.

Now, let's consider an applied problem that can be modeled by inverse variation.

EXAMPLE 4

The marketing division of a large company found that the demand d for one of its products varies inversely with the price p of the product. The monthly demand is 60,000 units when the price of the product is $9. Find the monthly demand if the company reduces the price to $6.

Solution   We are given that the demand d for a product varies inversely with the price p of the product. So we write

$$d = \frac{k}{p},$$

where k is the constant of variation. We can use this inverse variation equation to find k when d equals 60,000 units and p equals $9. Substituting 60,000 for d and 9 for p, we get:

$$d = \frac{k}{p}$$

$$60,000 = \frac{k}{9}$$

$$540,000 = k$$

So the value of k is 540,000. Substituting this value for k in the inverse variation equation gives us

$$d = \frac{540,000}{p}.$$

Now, let's use the inverse variation equation to find the monthly demand for the product when the price is $6.

$$d = \frac{540,000}{6} = 90,000$$

So the monthly demand is 90,000 units when the price is reduced to $6.

PRACTICE 4

In photography, the f-stop for a particular lens is inversely proportional to the aperture (the maximum diameter of the opening of the lens). For a certain lens, an f-stop of 2 corresponds to an aperture diameter of 25 mm. Find the f-stop of this lens when the aperture diameter is 12.5 mm.

Joint Variation

In some relationships, one quantity may vary as the product of two or more other quantities. For example, in the interest formula $I = prt$, if the principal p is constant, then the amount of interest earned I varies directly as the product of the interest rate r and the time t. This kind of variation is called *joint variation*. In this case, we say the interest varies jointly as the rate and the time.

> **DEFINITION**
>
> If a relationship among three variables is described by an equation of the form $y = kxz$, where k is a positive constant, we say we have **joint variation**, that y **varies jointly** as x and z, or that y is **jointly proportional** to x and z. The number k is called the **constant of variation** or the **constant of proportionality**, and $y = kxz$ is called a **joint variation equation**.

EXAMPLE 5

Suppose that y varies jointly as x and z and that $y = 240$ when $x = 8$ and $z = 12$. Find the constant of variation and the joint variation equation.

Solution   We know that y varies jointly as x and z, so we write $y = kxz$. Since $y = 240$ when $x = 8$ and $z = 12$, we get:

$$y = kxz$$

$$240 = k(8)(12) \qquad \text{Substitute 240 for } y, 8 \text{ for } x, \text{ and 12 for } z.$$

$$240 = 96k$$

$$\frac{240}{96} = k$$

$$k = \frac{5}{2}$$

So the constant of variation is $\frac{5}{2}$, and the joint variation equation is $y = \frac{5}{2}xz$.

PRACTICE 5

Find the constant of variation and the joint variation equation in which y varies jointly as x and z and $y = 150$ when $x = 15$ and $z = 20$.

In Example 5, we say y varies jointly as x and z. In this situation, "and" does not mean "to add." Can you explain why?

Consider the following application of joint variation.

EXAMPLE 6

The volume V of a rectangular structure of a given height is jointly proportional to its length l and its width w. An architect plans to construct two rectangular buildings with the same height in an office complex. One building is 100 ft long and 20 ft wide, and has a volume of 96,000 ft³. Find the volume of the other building with the same height that is 90 ft long and 25 ft wide.

Solution   We are given that the volume V of a rectangular structure of a given height is jointly proportional to its length l and width w. Since the height is constant, we write:

$$V = klw,$$

PRACTICE 6

For a given period of time t, the simple interest earned I varies jointly as the interest rate r and the principal p. Suppose one employee invests \$2000 at 3.5% for a given period of time earning \$140 interest. For the same period of time, how much money must another employee invest at 4% to earn \$180 interest?

EXAMPLE 6 (continued)

where k is the constant of variation. So we use this relationship to find k when the volume is 96,000 ft^3, the length is 100 ft, and the width is 20 ft.

$$V = klw$$
$$96,000 = k(100)(20)$$
$$96,000 = 2000k$$
$$\frac{96,000}{2000} = k$$
$$k = 48$$

Substituting 48 for k in the joint variation equation, we get:

$$V = 48lw$$

Now, we find the volume of the other building given that the length is 90 ft and the width is 25 ft.

$$V = 48lw$$
$$V = 48(90)(25)$$
$$V = 108,000$$

So the other building has a volume of 108,000 ft^3.

Combined Variation

Direct variation, inverse variation, and joint variation may be combined. A given relationship may be a combination of two or all three kinds of variation. Let's look at some examples of *combined variation*.

EXAMPLE 7

Suppose y varies directly as the square of x and inversely as z.

a. Write the variation equation.

b. If $y = 60$ when $x = 2$ and $z = 10$, find the constant of variation using the equation found in part (a).

c. Using part (b), find y for $x = 6$ and $z = 15$.

Solution

a. We are given that y varies directly as the square of x and inversely as z. So we can write the equation as:

$$y = \frac{kx^2}{z}$$

b. We use the equation $y = \frac{kx^2}{z}$ to find k when $y = 60$, $x = 2$, and $z = 10$.

$$y = \frac{kx^2}{z}$$
$$60 = \frac{k(2)^2}{10} \qquad \text{Substitute 60 for } y, \text{ 2 for } x, \text{ and 10 for } z.$$
$$600 = 4k$$
$$k = 150$$

PRACTICE 7

Suppose w varies jointly as x and y and inversely as the square of z.

a. Write the equation for the variation.

b. If $w = 8$ when $x = 2$, $y = 6$, and $z = 6$, find the constant of variation using the equation found in part (a).

c. Using part (b), find w when $x = 2$, $y = 3$, and $z = 4$.

c. Now, we find y for $x = 6$ and $z = 15$.

$$y = \frac{150x^2}{z}$$

$$y = \frac{150(6)^2}{15} \quad \text{Substitute 6 for } x \text{ and 15 for } z.$$

$$y = \frac{150(36)}{15}$$

$$y = 360$$

So for this relationship, $y = 360$ for $x = 6$ and $z = 15$.

Many relationships, particularly in science, can be modeled by combined variation equations. Consider the following example:

EXAMPLE 8

The centrifugal force C of a body moving in a circle varies jointly as the radius r of the circular path and the body's mass m and inversely as the square of the time t it takes to complete one revolution. A 9-gram body moving in a circle radius 100 cm at a rate of one revolution in 2 sec has a centrifugal force of 9000 dynes, where a dyne is a unit of force. What is the centrifugal force of a 12-gram body moving in a circle with a radius of 100 cm at a rate of one revolution in 4 sec?

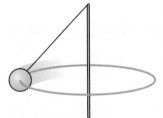

Solution We are given that the centrifugal force C of a body moving in a circle varies jointly with the radius r of the circular path and the body's mass m and inversely with the square of the time t it takes to make a complete revolution. So we write the combined variation equation.

$$C = \frac{krm}{t^2}$$

Using this equation, we can find the constant of variation k given that $C = 9000$ dynes, $r = 100$ cm, $m = 9$ g, and $t = 2$ sec.

$$C = \frac{krm}{t^2}$$

$$9000 = \frac{k(100)(9)}{(2)^2}$$

$$9000 = \frac{900k}{4}$$

$$36{,}000 = 900k$$

$$k = 40$$

Therefore, the equation is $C = \frac{40rm}{t^2}$.

Now, we find the centrifugal force of the 12-gram body moving in a circle with radius 100 cm at a rate of one revolution in 4 sec.

$$C = \frac{40rm}{t^2}$$

$$= \frac{40(100)(12)}{4^2} = 3000$$

So the centrifugal force is 3000 dynes.

PRACTICE 8

The *body mass index* (BMI) is used by physicians to determine a patient's total body fat. The BMI varies directly as a person's weight (in pounds) and inversely as the square of the person's height (in inches). A patient who weighs 165 lb and is 70 in. tall has a BMI of approximately 23. To the nearest whole number, find the approximate BMI of a patient who weighs 120 lb with a height of 65 in.

Mathematically Speaking

Fill in each blank with the most appropriate term or phrase from the given list.

combined variation	directly	joint variation
constant term	inverse variation	inversely
inversely	constant of variation	proportional
directly proportional	direct variation	

The following statements concern the relationship between two variables described by an equation of the form $y = kx$, where k is a positive constant.

1. We say that y varies _____ as x.

2. The number k is called the _____.

3. The relationship is a(n) _____.

The following statements concern the relationship between two variables described by an equation of the form $y = \dfrac{k}{x}$, where k is a positive constant.

4. We say that y is _____ to x.

5. The relationship is a(n) _____.

The following statement concerns the relationship among three variables described by an equation of the form $y = kxz$, where k is a positive constant.

6. The relationship is a(n) _____.

A *For each pair of variables given, indicate whether the second variable increases or decreases if the first variable increases and whether the variation between the variables is direct or inverse.*

7. The speed of a runner and the time it takes the runner to complete a 5-km race

8. The average thickness of DVD containers and the number of containers that fit on a shelf

9. The distance between two towns on a map and the actual distance between them

10. The side of a square and its perimeter

11. The cost of a plane ticket in dollars and its cost in euros

12. The real-estate tax paid on a house and the estimated value of the house

13. The number of beneficiaries named in a will and the average bequest each beneficiary receives

14. The length and width of a rectangle with area 20 ft^2

B *Use the given information to find the constant of variation and the variation equation.*

15. y varies directly as x; $y = 48$ when $x = 16$

16. y varies directly as x; $y = 35$ when $x = 7$

17. y varies directly as x; $y = 6$ when $x = 36$

18. y varies directly as x; $y = 15$ when $x = 40$

19. y varies directly as x; $y = 3$ when $x = \dfrac{1}{3}$

20. y varies directly as x; $y = 8$ when $x = \dfrac{1}{4}$

21. y varies directly as x; $y = 0.9$ when $x = 0.6$

22. y varies directly as x; $y = 0.5$ when $x = 0.8$

23. y varies inversely as x; $y = 13$ when $x = 3$

24. y varies inversely as x; $y = 25$ when $x = 4$

25. y varies inversely as x; $y = 1.8$ when $x = 15$

26. y varies inversely as x; $y = 2.1$ when $x = 20$

27. y varies inversely as x; $y = 0.7$ when $x = 0.4$

28. y varies inversely as x; $y = 0.1$ when $x = 1.9$

29. y varies inversely as x; $y = 27$ when $x = \dfrac{2}{3}$

30. y varies inversely as x; $y = 54$ when $x = \dfrac{1}{6}$

31. y varies jointly as x and z; $y = 160$ when $x = 10$ and $z = 4$

32. y varies jointly as x and z; $y = 216$ when $x = 18$ and $z = 6$

33. y varies jointly as x and z; $y = 360$ when $x = 25$ and $z = 12$

34. y varies jointly as x and z; $y = 120$ when $x = 16$ and $z = 15$

35. y varies jointly as x and z; $y = 63$ when $x = 4.2$ and $z = 5$

36. y varies jointly as x and z; $y = 90$ when $x = 2$ and $z = 0.9$

37. y varies jointly as x and z; $y = 4.5$ when $x = 0.6$ and $z = 0.3$

38. y varies jointly as x and z; $y = 5.6$ when $x = 0.5$ and $z = 0.7$

39. y varies directly as x and inversely as the square of z; $y = 20$ when $x = 4$ and $z = 5$

40. y varies directly as the cube of z and inversely as x; $y = 60$ when $x = 16$ and $z = 2$

41. y varies inversely as x and the square of z; $y = 100$ when $x = 20$ and $z = 0.5$

42. y varies inversely as the square of x and the square of z; $y = 250$ when $x = 0.2$ and $z = 10$

43. y varies jointly as x and w and inversely as the square of z; $y = 130$ when $x = 13$, $w = 16$, and $z = 0.4$

44. y varies jointly as x and the square of z and inversely as w; $y = 600$ when $x = 8$, $z = 5$, and $w = 0.5$

Mixed Practice

Use the given information to find the constant of variation and the variation equation.

45. y varies jointly as x and w, and inversely as the cube of z; $y = 15$ when $x = \dfrac{1}{2}$, $w = 8$, and $z = 2$.

46. y varies inversely as x; $y = 4$ when $x = 8$

47. y varies directly as x; $y = \dfrac{3}{7}$ when $x = \dfrac{9}{14}$

48. y varies jointly as x and z; $y = 9$ when $x = 0.6$ and $z = 3$

49. y varies directly as x; $y = 2$ when $x = \dfrac{1}{6}$

50. y varies inversely as x; $y = 36$ when $x = \dfrac{2}{3}$

For each pair of variables given, indicate whether the second variable increases or decreases if the first variable increases and whether the variation between the variables is direct or inverse.

51. The number of slices in a pizza and the amount of pizza in each slice

52. The number of credit hours taken and the total cost of the credit hours

53. The area of a wall and the amount of paint needed to cover the wall

54. On a fixed budget, the daily rate at a hotel and the number of days stayed at the hotel

Applications

Solve.

55. The amount A of state income tax that taxpayers pay in Illinois in a certain year is directly proportional to their gross income i. A person pays $1080 in income tax on a gross annual income of $36,000. (*Source:* Federation of Tax Administrators)

  a. Write a variation equation to represent the situation. Interpret the meaning of the constant of variation.

  b. How much state income tax will a person pay on a gross annual salary of $26,500?

56. A tire company's revenue R in dollars is directly proportional to the number of tires n it sells. In a particular month, the company generated revenue of $136,710 on the sale of 3255 tires.

  a. Write a variation equation to represent the situation and interpret the meaning of the constant of variation.

  b. What is the company's monthly revenue if 4000 tires are sold?

57. The length of a sound wave w (in meters) is inversely proportional to the frequency f of the sound (in hertz, or Hz). Bottlenose dolphins emit clicking sounds at different frequencies for communicating, orienting themselves to their surroundings, avoiding obstacles, and finding food. If the wavelength of a click that has a frequency of 300 Hz is 5.1 m, find the wavelength of a click that has a frequency of 500 Hz.

58. Physicists use Boyle's Law, which states that at constant temperature the volume V of a gas is inversely proportional to its pressure P. If the volume of a gas is 0.5 m³ when the pressure is 850 kilopascals (kP), find the volume of the gas when the pressure is 680 kP.

59. Kinetic energy is energy associated with motion. The kinetic energy E of an object is jointly proportional to its mass m and the square of its velocity v. A 0.142-kilogram baseball has 113.6 joules (J) of kinetic energy when its velocity is 40 m/sec. What is the kinetic energy of the same baseball when its velocity is 20 m/sec?

60. When an object is dropped, the distance it falls varies directly as the square of the time t the object has fallen. A sandbag falls 64 ft in the 2 sec after it is dropped from a hot-air balloon. How many feet did it fall 4 sec after it was dropped?

61. The illumination I from a light source varies inversely as the square of the distance d from the light source. If the illumination on a table from a lightbulb 2 m above the table is 21.6 lumens per square meter (lm/m^2), find the illumination when the lightbulb is 4 m above the table.

62. The resistance of a wire varies directly as its length L and inversely as its cross-sectional area A. If 2000 ft of wire with a cross-sectional area of 0.008 in² has a resistance of 1.9 ohms, what is the resistance of 5000 ft of the same wire?

• Check your answers on page A-41.

MINDStretchers

Writing

1. Use variation terminology to write a verbal statement for each equation.

   a. $V = \pi r^2 h$

   b. $F = \dfrac{mv^2}{r}$

Groupwork

2. Working with a partner, consider the direct variation equation $y = kx$ and the inverse variation equation $y = \dfrac{k}{x}$. Describe the effect on y if x changes as follows:

   a. x is doubled

   b. x is reduced by one-half

   c. x is multiplied by a factor of $\dfrac{a}{b}$

Mathematical Reasoning

3. Show that if y varies inversely as x and x varies inversely as z, then y varies directly as z.

Cultural Note

Hot air balloons were developed in France in the late 1700s. The first one, unmanned but with a sheep, a rooster, and a duck for passengers, was launched from Versailles in 1783. These balloons caught the attention of two French scientists, Jacques Charles and Joseph-Louis Gay-Lussac. Working separately, they investigated the relationships between gases and temperature. Charles is credited with discovering the direct relationship between temperature and volume of a gas ($V = kT$). Gay-Lussac is credited with finding the direct relationship between temperature and pressure ($P = kT$). These two laws were later combined to show the joint variation of pressure and volume to temperature ($PV = kT$).

(*Source:* Hugh Salzburg, *From Caveman to Chemist*, American Chemical Society, 1991)

Key Concepts and Skills

CONCEPT · SKILL

Concept/Skill	Description	Example
[6.1] **Rational expression**	An algebraic expression $\dfrac{P}{Q}$ that can be written as the quotient of two polynomials, P and Q, where $Q \neq 0$.	$\dfrac{2x + 3}{x - 2}$ where $x - 2 \neq 0$
[6.1] **To find an equivalent rational expression**	• If P, Q, and R are polynomials, then $\dfrac{P}{Q} = \dfrac{PR}{QR}$, where $Q \neq 0$ and $R \neq 0$.	$\dfrac{n - 3}{n} \cdot \dfrac{2n}{2n} = \dfrac{2n^2 - 6n}{2n^2}$
[6.1] **To simplify a rational expression**	• Factor the numerator and denominator. • Divide out any common factors.	$\dfrac{x^2 - 16}{x^2 + 13x + 36}$ $= \dfrac{(x + 4)(x - 4)}{(x + 4)(x + 9)} = \dfrac{x - 4}{x + 9}$
[6.1] **To multiply rational expressions**	• Factor the numerators and denominators. • Divide the numerators and denominators by all common factors. • Multiply the remaining factors in the numerators and the remaining factors in the denominators.	$\dfrac{4x}{x^2 - 5x} \cdot \dfrac{6x - 30}{5x}$ $= \dfrac{4x}{x(x - 5)} \cdot \dfrac{6(x - 5)}{5x}$ $= \dfrac{4x}{x(x - 5)} \cdot \dfrac{6(x - 5)}{5x}$ $= \dfrac{24}{5x}$
[6.1] **To divide rational expressions**	• Take the reciprocal of the divisor, and change the operation to multiplication. • Follow the rule for multiplying rational expressions.	$\dfrac{x - 2}{x^2 - 5x + 6} \div \dfrac{x - 1}{x^2 - 9}$ $= \dfrac{x - 2}{x^2 - 5x + 6} \cdot \dfrac{x^2 - 9}{x - 1}$ $= \dfrac{x - 2}{(x - 2)(x - 3)} \cdot \dfrac{(x + 3)(x - 3)}{x - 1}$ $= \dfrac{x + 3}{x - 1}$
[6.2] **To add (or subtract) rational expressions with the same denominator**	• Add (or subtract) the numerators, and keep the same denominator. • Simplify, if possible.	$\dfrac{3x}{x + 5} + \dfrac{15}{x + 5} = \dfrac{3x + 15}{x + 5}$ $= \dfrac{3(x + 5)}{x + 5}$ $= 3$ $\dfrac{x - 1}{x(x + 1)} - \dfrac{2x}{x(x + 1)} = \dfrac{x - 1 - 2x}{x(x + 1)}$ $= \dfrac{-x - 1}{x(x + 1)}$ $= \dfrac{-(x + 1)}{x(x + 1)}$ $= -\dfrac{1}{x}$

$$\boxed{\text{CONCEPT}} \quad \boxed{\text{SKILL}}$$

Concept/Skill	Description	Example
[6.2] **To find the LCD of rational expressions**	• Factor each denominator completely. • Multiply the factors found in the previous step, using for the power of each factor the greatest number of times that it occurs in any of the denominators. The product of all these factors is the LCD.	$\dfrac{1}{x^2 - 4x + 4}$ and $\dfrac{x}{x^2 - 4}$ Factor $x^2 - 4x + 4$: $(x - 2)(x - 2)$ Factor $x^2 - 4$: $(x + 2)(x - 2)$ LCD $= (x - 2)(x - 2)(x + 2)$, or $(x - 2)^2(x + 2)$
[6.2] **To add (or subtract) rational expressions with different denominators**	• Find the LCD of the rational expressions. • Write each rational expression with a common denominator, using the LCD. • Add (or subtract) the numerators, keeping the denominator. • Simplify, if possible.	$\dfrac{4}{y - 1} + \dfrac{y}{y + 3}$ $= \dfrac{4(y + 3)}{(y - 1)(y + 3)} + \dfrac{y(y - 1)}{(y + 3)(y - 1)}$ $= \dfrac{(4y + 12) + (y^2 - y)}{(y - 1)(y + 3)}$ $= \dfrac{4y + 12 + y^2 - y}{(y - 1)(y + 3)}$ $= \dfrac{y^2 + 3y + 12}{(y - 1)(y + 3)}$ $\dfrac{n}{n^2 - 3n + 2} - \dfrac{5n + 2}{n^2 - 5n + 6}$ $= \dfrac{n}{(n - 1)(n - 2)} - \dfrac{5n + 2}{(n - 2)(n - 3)}$ $= \dfrac{n(n - 3)}{(n - 1)(n - 2)(n - 3)}$ $\qquad - \dfrac{(5n + 2)(n - 1)}{(n - 2)(n - 3)(n - 1)}$ $= \dfrac{(n^2 - 3n) - (5n^2 - 3n - 2)}{(n - 1)(n - 2)(n - 3)}$ $= \dfrac{n^2 - 3n - 5n^2 + 3n + 2}{(n - 1)(n - 2)(n - 3)}$ $= \dfrac{-4n^2 + 2}{(n - 1)(n - 2)(n - 3)}$
[6.3] **Complex rational expression**	A rational expression whose numerator, denominator, or both contain one or more rational expressions.	$\dfrac{1 + \dfrac{1}{x}}{1 - \dfrac{1}{x}}$
[6.3] **To simplify a complex rational expression: the division method**	• Write both the numerator and denominator as single rational expressions in simplified form. • Write the expression as the numerator divided by the denominator. • Divide. • Simplify, if possible.	$\dfrac{\dfrac{1}{x} - \dfrac{1}{y}}{\dfrac{1}{x} + \dfrac{1}{y}} = \dfrac{\dfrac{1}{x} \cdot \dfrac{y}{y} - \dfrac{1}{y} \cdot \dfrac{x}{x}}{\dfrac{1}{x} \cdot \dfrac{y}{y} + \dfrac{1}{y} \cdot \dfrac{x}{x}} = \dfrac{\dfrac{y - x}{xy}}{\dfrac{y + x}{xy}}$ $= \dfrac{y - x}{xy} \div \dfrac{y + x}{xy}$ $= \dfrac{y - x}{xy} \cdot \dfrac{xy}{y + x}$ $= \dfrac{y - x}{y + x}$

continued

Concept/Skill	Description	Example
[6.3] **To simplify a complex rational expression: the LCD method**	• Find the LCD of all the rational expressions within the complex rational expression. • Multiply the numerator and denominator of the complex rational expression by this LCD. • Simplify, if possible.	$$\dfrac{\dfrac{x}{4} + \dfrac{3}{x^2}}{\dfrac{x-5}{2x}} = \dfrac{\left(\dfrac{x}{4} + \dfrac{3}{x^2}\right)(4x^2)}{\dfrac{x-5}{2x}(4x^2)}$$ $$= \dfrac{\left(\dfrac{x}{4}\right)(4x^2) + \dfrac{3}{x^2}(4x^2)}{\dfrac{x-5}{2x}(4x^2)}$$ $$= \dfrac{x^3 + 12}{(x-5)(2x)}$$
[6.4] **To solve a rational equation**	• Find the LCD of all the denominators. • Multiply each side of the equation by the LCD. • Solve the resulting equation. • Check your solution(s) in the original equation.	$$\dfrac{1}{x+2} - \dfrac{2}{x-1} = \dfrac{-7}{x^2+x-2}$$ $$(x+2)(x-1)\left(\dfrac{1}{x+2}\right)$$ $$- (x+2)(x-1)\left(\dfrac{2}{x-1}\right)$$ $$= \dfrac{-7}{(x+2)(x-1)}(x+2)(x-1)$$ $$(x-1) - 2(x+2) = -7$$ $$x - 1 - 2x - 4 = -7$$ $$-x - 5 = -7$$ $$-x = -2$$ $$x = 2$$ **Check** $$\dfrac{1}{x+2} - \dfrac{2}{x-1} = \dfrac{-7}{x^2+x-2}$$ $$\dfrac{1}{2+2} - \dfrac{2}{2-1} \overset{?}{=} \dfrac{-7}{2^2+2-2}$$ $$\dfrac{1}{4} - \dfrac{2}{1} \overset{?}{=} -\dfrac{7}{4}$$ $$-\dfrac{7}{4} = -\dfrac{7}{4} \quad \text{True}$$
[6.5] **Direct variation**	If a relationship between two variables is described by an equation in the form $y = kx$, where k is a positive constant, we say that we have **direct variation**, that y **varies directly** as x, or that y is **directly proportional** to x. The number k is called the **constant of variation** or the **constant of proportionality**, and $y = kx$ is called a **direct variation equation**.	$y = 2x$ y varies directly as x; 2 is the constant of variation.

Concept/Skill	Description	Example
[6.5] **Inverse variation**	If a relationship between two variables is described by an equation in the form $y = \dfrac{k}{x}$, where k is a positive constant, we say that we have **inverse variation**, that y **varies inversely** as x, or that y is **inversely proportional** to x. The number k is called the **constant of variation** or the **constant of proportionality**, and $y = \dfrac{k}{x}$ is called an **inverse variation equation**.	$y = \dfrac{4}{x}$ y varies inversely as x; 4 is the constant of variation.
[6.5] **Joint variation**	If a relationship among three variables is described by an equation of the form $y = kxz$, where k is a positive constant, we say that we have **joint variation**, that y **varies jointly** as x and z, or that y is **jointly proportional** to x and z. The number k is called the **constant of variation** or the **constant of proportionality**, and $y = kxz$ is called a **joint variation equation**.	$y = 5xz$ y varies jointly as x and z; 5 is the constant of variation.

Say Why
Fill in each blank.

1. The expression $\dfrac{x + 3}{x - 2}$ _____ undefined at $x = 2$
   is/is not
   because _____
   _____ .

2. The expression $\dfrac{3x^2 - 2x - 5}{3x - 5}$ _____ equivalent to
   is/is not
   $x + 1$ because _____
   _____ .

3. $\dfrac{\frac{1}{x}}{6}$ _____ a complex rational expression because
   is/is not
   _____
   _____ .

4. An extraneous solution to a rational equation
   _____ check because _____
   will/will not
   _____ .

5. The circumference C of a circle _____
   does/does not
   vary directly as the circle's radius r because
   _____ .

6. If two quantities vary inversely and the first increases,
   then the second _____ increase because
   does/does not
   _____ .

[6.1] *Identify the values for which the given rational function is undefined.*

7. $\dfrac{x}{2x - 1}$

8. $\dfrac{n^2 - n + 2}{n^2 - 6n + 8}$

Determine whether the rational expressions are equivalent.

9. $\dfrac{1 - t}{5 - t}$ and $\dfrac{t - 1}{t - 5}$

10. $\dfrac{4p^4 + 6p}{2p^3 - 3p^2}$ and $\dfrac{2p^2 + 2}{1 - p}$

Simplify, if possible.

11. $\dfrac{6x^3y^2}{9x^2y - 12xy^2}$

12. $\dfrac{2 - 6n}{12n - 4}$

13. $\dfrac{a^2 - 4b^2}{5a + 10b}$

14. $\dfrac{r^3 + 1}{r^2 + 1}$

15. $\dfrac{x^2 + 10x + 16}{x^2 - 5x - 14}$

16. $\dfrac{4t^2 - 4t + 1}{8 - 13t - 6t^2}$

Perform the indicated operation.

17. $\dfrac{16u^4v}{12v^3} \cdot \dfrac{9uv^4}{2u^2}$

18. $\dfrac{20pq}{8q^3} \div \dfrac{25p^5}{p^3q}$

19. $\dfrac{9ab - 18a}{6a^2} \div \dfrac{b^3 - 2b^2}{7a^2b - ab}$

20. $\dfrac{y^2 - 5y}{y^3 - 4y^2} \cdot \dfrac{y - 4}{15 - 3y}$

21. $\dfrac{x^2 - 4}{x^2 - 3x - 10} \cdot \dfrac{2x^2 - 50}{x^3 - 2x^2}$

22. $\dfrac{3 - 2r - r^2}{r^2 - 9} \div \dfrac{r^3 - 1}{r^2 + 7r - 30}$

23. $\dfrac{2x^2 - 13x + 20}{x^2 - 8x + 16} \div (4x^2 - 4x - 15)$

24. $\dfrac{3n^2 + 7n + 4}{n^2 - 5n - 6} \cdot \dfrac{n^2 + 3n - 54}{9n^2 + 24n + 16}$

[6.2] *Find the LCD of each pair of rational expressions. Then, write each expression in terms of the LCD.*

25. $\dfrac{x - 1}{x^3 + x^2}$ and $\dfrac{2}{x^2 + 2x + 1}$

26. $\dfrac{x + 5}{x^2 - 16x + 48}$ and $\dfrac{2x - 3}{x^2 - 10x - 24}$

Perform the indicated operation.

27. $\dfrac{8}{3x^2y} + \dfrac{4}{3x^2y}$

28. $\dfrac{2n + 9}{n + 6} - \dfrac{n + 3}{n + 6}$

29. $\dfrac{6}{p - 5} - \dfrac{8}{p - 4}$

30. $\dfrac{a - 4}{8a^2 - 16a} + \dfrac{2}{4a^3 - 8a^2}$

31. $\dfrac{1}{t - 3} - \dfrac{t^2 - 6t}{18 - 3t - t^2}$

32. $\dfrac{5y - 1}{y^2 - 4} + \dfrac{2}{2 - y}$

33. $\dfrac{2x}{x^2 + 2x - 3} + \dfrac{1}{x^2 - 2x + 1}$

34. $\dfrac{u + 1}{2u^2 + 11u + 12} - \dfrac{1 - u}{u^2 - 5u - 36}$

[6.3] *Simplify.*

35. $\dfrac{\dfrac{36x^3}{x - 4}}{\dfrac{18x}{2x - 8}}$

36. $\dfrac{\dfrac{1}{x^2} - 9}{\dfrac{1}{x} - 3}$

37. $\dfrac{1 - \dfrac{4}{p} - \dfrac{5}{p^2}}{1 + \dfrac{2}{p} + \dfrac{1}{p^2}}$

38. $\dfrac{\dfrac{3}{x + 3} + \dfrac{6}{x - 1}}{\dfrac{3}{x - 1} - \dfrac{6}{x + 3}}$

39. $\dfrac{\dfrac{r^2 - 8r + 12}{2r^2 - 13r + 6}}{\dfrac{r^2 - 4r + 4}{6r^2 + 5r - 4}}$

40. $\dfrac{2x^{-2} + x^{-1}}{x^{-3}}$

[6.4] *Solve.*

41. $\dfrac{y + 1}{3} - \dfrac{y}{8} = \dfrac{2y - 1}{24}$

42. $\dfrac{a - 12}{4a} + \dfrac{3}{a} = 1$

43. $\dfrac{t + 7}{t - 1} + 6 = \dfrac{2t}{1 - t}$

44. $\dfrac{5}{x + 3} = \dfrac{4}{x - 2}$

45. $\dfrac{n + 1}{n^2} - \dfrac{3}{n} + 1 = 0$

46. $\dfrac{10}{r^2 + 2r} + \dfrac{r}{r + 2} = \dfrac{5}{r}$

47. $\dfrac{x}{x + 6} = \dfrac{1}{x - 4}$

48. $\dfrac{p - 5}{p^2 + 2p + 1} - \dfrac{7}{p + 1} = \dfrac{p}{p + 1}$

Solve each equation for the indicated variable.

49. $F = \dfrac{mv^2}{r}$ for m

50. $R = \dfrac{R_1 R_2}{R_1 + R_2}$ for R_2

[6.5] *Use the given information to find the constant of variation and the variation equation.*

51. y varies directly as x; $y = 1.6$ when $x = 4$

52. y varies inversely as x; $y = \dfrac{1}{2}$ when $x = 3$

53. y varies jointly as x and z; $y = 144$ when $x = 4$ and $z = 6$

54. y varies directly as the square of x and inversely as the square of z; $y = 2$ when $x = 5$ and $z = 10$

Mixed Applications

Solve.

55. The owner of a small company determines that the average cost for producing x units of a product is given by the following expression:

$$\frac{1500 + 2x}{x}$$

For what value(s) of x is the expression undefined?

56. The maximum efficiency of a heat engine is given by the expression

$$1 - \frac{T_1}{T_2},$$

where T_1 and T_2 represent the Kelvin temperatures. Write this expression as a single rational expression.

57. In the study of electricity, the total current in a circuit can be calculated using the expression

$$V\left(\frac{1}{R_1} + \frac{1}{R_2}\right),$$

where V is voltage and R_1 and R_2 represent resistances. Write this expression as a single rational expression.

58. The focal length of a lens is given by the expression

$$\frac{1}{\dfrac{1}{p} + \dfrac{1}{q}},$$

where p is the distance between the lens and the object and q is the distance between the image and the lens. Simplify this expression.

59. A student's average speed on the 294-mile drive home was one-and-a-half times her speed on the way back to school. If the round-trip drive took 11 hr 40 min, find her average speed each way.

60. For a particular project, an office manager knows that it takes one data-entry clerk 16 hr and another clerk 12 hr to complete the project working alone. If the office manager has the clerks work on the project together, how long will it take them to complete the project?

61. When an object is dropped, its velocity v (in meters per second) is directly proportional to the time t (in seconds) the object has fallen. If the velocity of an object after 2 sec is 19.6 m/sec, what is its velocity after 5 sec?

62. The accommodation (or focusing ability) A of a person's eye is inversely proportional to the distance d that a person can bring an object to the eye and still see the object. If the accommodation of a person who can see an object clearly within 12.5 cm is 8 diopters, what is the accommodation of a person who can see an object clearly within 10 cm?

● Check your answers on page A-41.

CHAPTER 6   Posttest

FOR EXTRA HELP

CHAPTER Test Prep VIDEOS

The Chapter Test Prep Videos with test solutions are available on DVD, in MyMathLab, and on YouTube* (search "AkstIntermediate Alg" and click on "Channels").

To see if you have mastered the topics in this chapter, take this test.

1. Identify all the values of the variable for which each expression is undefined.

    a. $\dfrac{x^2}{4x - 2x^2}$ **b.** $\dfrac{3n + 5}{n^2 - 8n + 16}$

Simplify.

2. $\dfrac{8y^2 - 4xy^3 + 6x^2y}{12xy^2}$ **3.** $\dfrac{2 - 3a}{9a^2 - 4}$ **4.** $\dfrac{3n^2 - 8n + 4}{n^2 + 6n - 16}$

Perform the indicated operation.

5. $\dfrac{-6pq^2}{12q - 2p} \cdot \dfrac{p^2 - 36q^2}{9p^2q^3 - 3pq^2}$ **6.** $\dfrac{y^2 + 3y}{y^2 + 4y + 4} \cdot \dfrac{y^2 - 7y - 18}{y^2 - 6y - 27}$

7. $(a^2 - 9ab + 20b^2) \div \dfrac{a^2 - 16b^2}{a^2 + 4ab}$ **8.** $\dfrac{x^2 - 3xy + 9y^2}{6x^2 + 13xy - 5y^2} \div \dfrac{x^3 + 27y^3}{2x^2 + 11xy + 15y^2}$

9. $\dfrac{1}{6r^3} + \dfrac{1}{4r^3 - 12r^2}$ **10.** $\dfrac{1}{2t - 1} - \dfrac{3t - 11}{1 - 4t^2}$

11. $\dfrac{n}{n^2 - 6n + 8} + \dfrac{3}{n^2 - n - 2}$ **12.** $\dfrac{3y - 2}{9y^2 + 12y + 4} - \dfrac{y}{3y^2 - 7y - 6}$

13. Simplify each expression.

    a. $\dfrac{\dfrac{1}{x^2} - 4}{\dfrac{3}{x} + 6}$ **b.** $\dfrac{\dfrac{8}{a^2} - \dfrac{6}{a} + 1}{\dfrac{4}{a^2} - \dfrac{5}{a} + 1}$

Solve.

14. $\dfrac{3}{n - 9} + \dfrac{1}{n + 2} = \dfrac{n^2}{n^2 - 7n - 18}$ **15.** $\dfrac{x - 1}{x + 2} - 1 = \dfrac{2}{x - 5}$

16. $\dfrac{2}{y + 7} = \dfrac{y + 4}{2}$ **17.** Find the constant of variation and the variation equation in which y varies jointly as x and z and $y = 18.2$ when $x = 1.4$ and $z = 65$.

18. The speed of a boat in still water is 40 km/hr. If the boat travels 174 km downriver in the same amount of time it travels 146 km upriver, find the speed of the river's current.

19. One gardener can mow a lawn in 6 hr. If he works with a second gardener, they can mow the lawn in 2 hr 40 min. How long would it take the second gardener working alone to mow the lawn?

20. The measure of each exterior angle of a regular polygon varies inversely as the number of sides of the polygon. If the measure of each exterior angle of a regular octagon is 45°, find the measure of each exterior angle of a regular dodecagon.

Octagon   Dodecagon

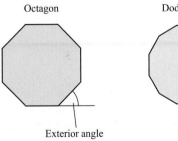

Exterior angle

• Check your answers on page A-42.

Cumulative Review Exercises

To help you review, solve the following problems.

1. Simplify: $(2xy^2)^2(4x^3y)^{-2}$

2. Solve: $8 - 3(1 - 3n) = 12 - 7n$

3. Find the slope of the line that passes through the points $(-1, 5)$ and $(2, 4)$.

4. Graph the equation $6x = 3y - 3$ on the coordinate plane.

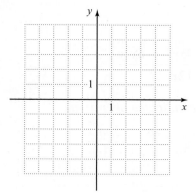

5. Graph the function and identify its domain and range: $f(x) = -2 + |x|$

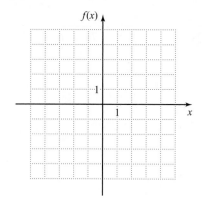

6. Solve by graphing:
$$-x + 2y = 4$$
$$-x + 2y = -1$$

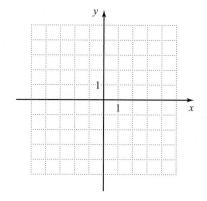

7. Solve the following system of equations.
$$2x - y + z = 3$$
$$x + 3y - z = -10$$
$$x + 2y + 4z = 7$$

8. Multiply: $-(5 - x^2)(x^2 - x + 1)$

9. Factor by grouping: $6xy - 10y + 9x - 15$

10. Factor, if possible: $2x^2 - 15xy + 28y^2$

11. Factor, if possible: $9x^2 - 30xy^3 + 25y^6$

12. Find the quotient: $\dfrac{4x^2 - 12x + 36}{27 - 3x^2} \div (x^3 + 27)$

13. Simplify: $\dfrac{\dfrac{1}{y} - \dfrac{3}{x}}{\dfrac{9}{x} - \dfrac{x}{y^2}}$

14. Solve and check: $3 + \dfrac{13}{x} = \dfrac{10}{x^2}$

15. The Hoover Dam is 726 ft high. If a tourist drops a coin from 3 ft above the top of the dam into the water below, the height of the coin after t sec is given by the expression $729 - 16t^2$. Factor this expression. (*Source:* desertusa.com)

16. Two trains leave a station at the same time, one heading north and the other heading west. After half an hour, the northbound train is 21 mi from the station and the trains are 35 mi apart. What is the speed of the westbound train?

17. A soft-drink manufacturer produces 12-ounce cans of cola. The manufacturer allows for a tolerance of ± 0.5 oz per case. If there are 24 cans per case, find the maximum and minimum amounts of cola in a case.

18. A stockbroker invests 3 times the amount of money in a low-risk fund as in a high-risk fund. After one year, the low-risk fund showed a profit of 9% and the high-risk fund showed a loss of 4%. If the total profit on the investments was $2760, how much did she invest in each fund?

19. The total expenditures (in millions of dollars) on dental services from 2004 to 2008 in the United States can be modeled by the polynomial $-27x^3 + 245x^2 + 4380x + 81{,}500$, where x represents the number of years since 2004. The binomial $3x + 293$ models the U.S. population (in millions) during the same years. (*Source:* cms.gov)

   a. Write a rational expression that models the total expenditures on dental services per person.

   b. Find to the nearest hundred dollars the amount of expenditures on dental services per person in 2005.

20. In baseball, the *range factor* is a statistic that measures how many plays a fielder made per game at a specific field position. The statistic is given by the expression $\dfrac{P + A}{\dfrac{D}{9}}$, where P is the number of putouts the fielder made, A is the number of assists he made, and D is the number of innings he played at the position. Simplify this expression.

(*Source:* baseball-almanac.com)

• Check your answers on page A-42.

Radical Expressions and Equations

Radicals and Rockets

Escape velocity is the minimum initial velocity that an object must achieve to be free of the gravitational bonds of a planet. A rocket taking off from the surface of our planet has to reach this velocity to enter orbit and deploy satellites or to fly to the Moon.

In general, the escape velocity from a planet depends on the planet's mass m, its radius r, and the universal gravitational constant G. Physicists have shown that this velocity can be modeled by the radical expression

$$\sqrt{\frac{2Gm}{r}}.$$

The value of this expression for Earth turns out to be approximately 11 km (or 7 mi) per sec.

(*Sources:* Isaac Asimov, *Asimov on Astronomy*, Doubleday and Company, 1974; Michael Zeilik, *Astronomy—The Evolving Universe*, Wiley, 1997)

CHAPTER 7 PRETEST

To see if you have already mastered the topics in this chapter, take this test. (Assume that all variables represent nonnegative real numbers.)

1. Evaluate:

   a. $2\sqrt{36}$ **b.** $\sqrt[3]{-64}$

2. Simplify: $\sqrt{100u^2v^4}$

3. Rewrite in radical notation, and then simplify:
   $(81x^4)^{3/4}$

4. Simplify:

   a. $\dfrac{8p^{2/3}}{(36p^{4/3})^{1/2}}$ **b.** $\sqrt[6]{x^2y^4}$

5. Multiply: $\sqrt{6a} \cdot \sqrt{7b}$

6. Divide: $\dfrac{\sqrt[3]{18r^2}}{\sqrt[3]{3r}}$

7. Simplify: $\sqrt{147x^5y^4}$

Perform the indicated operation.

8. $\sqrt{\dfrac{2p}{25q^8}}$

9. $3\sqrt{12} - 5\sqrt{3} + \sqrt{108}$

10. $(\sqrt{x} - 8)(2\sqrt{x} + 1)$

11. Simplify: $\dfrac{\sqrt{2x}}{\sqrt{27y}}$

12. Rationalize the denominator:
    $\dfrac{\sqrt{n}}{\sqrt{n} + \sqrt{3}}$

Solve.

13. $\sqrt{x + 4} - 9 = -3$

14. $\sqrt{x^2 - 2} = \sqrt{9x - 10}$

15. $\sqrt{2x} - \sqrt{x + 7} = -1$

16. Multiply: $(5 - \sqrt{-9})(4 - \sqrt{-25})$

17. Divide: $\dfrac{1 + 2i}{1 - 2i}$

18. If an object is dropped from a height of h ft above the ground, the time (in seconds) it takes for the object to be d ft above the ground is given by the following expression:

    $$\sqrt{\dfrac{h - d}{16}}$$

    Simplify this expression.

19. How long is the diagonal of the basketball court shown below?

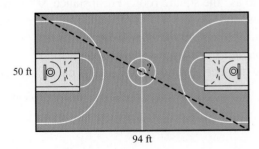

50 ft

94 ft

20. The rate at which firefighters spray water on a fire affects their ability to put out the fire. For a hose with a nozzle 2 in. in diameter, the flow rate f (in gallons per minute) is modeled by the formula $f = 120\sqrt{p}$, where p is the nozzle pressure in pounds per square inch. Solve this formula for p.

• Check your answers on page A-42.

7.1 Radical Expressions and Rational Exponents

What Radical Expressions Are and Why They Are Important

So far we have considered two types of algebraic expressions, namely polynomials and rational expressions. In this chapter, we extend the discussion to a third type called **radical expressions**. Radical expressions, such as $10 - \sqrt{2}$ or $\sqrt[3]{abc}$, are algebraic expressions that contain *radicals*. These expressions are used to solve a wide variety of problems in many fields, such as geometry and the physical sciences. For instance, physicists use the radical expression $\sqrt{\dfrac{h}{16}}$ to model the number of seconds that it takes an object to fall h ft.

This chapter deals with the meaning of radical expressions, ways of writing, simplifying or operating on them, and methods of solving equations involving them. In addition, we consider a set of numbers not previously discussed—the complex numbers.

Square Roots

Let's begin our discussion of radicals with the definition of a particular type of radical, namely the square root. Recall that we have already encountered this concept when we discussed irrational numbers in Section 1.1.

> **DEFINITION**
>
> The number b is a **square root** of a if $b^2 = a$, for any real numbers a and b and for a nonnegative.

Every positive number has two square roots, one positive and the other negative. For example, the square roots of 4 are $+2$ and -2 because $(+2)^2 = 4$ and $(-2)^2 = 4$. The symbol $\sqrt{}$, called the **radical sign**, stands for the positive or **principal square root** but is commonly referred to as "the square root." For instance $\sqrt{9}$ is read "the square root of 9" or "radical 9" and represents $+3$, or 3. By contrast, the negative square root of 9, namely -3, is represented by $-\sqrt{9}$. *Throughout the remainder of this text, when we speak of the square root of a number, we mean its principal square root.*

The number under a radical sign is called the **radicand**.

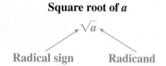

Square root of *a*

Radical sign   Radicand

Since the square of any real number is nonnegative, the square root of a negative number is not a real number. For instance, $\sqrt{-3}$ is not a real number.

A nonnegative rational number is said to be a **perfect square** if it is the square of another rational number. For instance, $\dfrac{1}{9}$ is a perfect square since $\dfrac{1}{9} = \left(\dfrac{1}{3}\right)^2$. On the other hand, 8 is not a perfect square since it is not the square of any rational number. Perfect squares play a special role in finding the value of square roots.

Now, let's look at evaluating square roots.

EXAMPLE 1

Find the value of the following radical expressions:

a. $\sqrt{49}$ **b.** $\sqrt{\dfrac{1}{9}}$

c. $-2\sqrt{16}$ **d.** $\sqrt{-4}$

Solution

a. $\sqrt{49} = 7$ since $7^2 = 49$.

b. $\sqrt{\dfrac{1}{9}} = \dfrac{1}{3}$ since $\left(\dfrac{1}{3}\right)^2 = \dfrac{1}{9}$.

c. $-2\sqrt{16} = -2 \cdot 4 = -8$.

d. $\sqrt{-4}$ is not a real number.

PRACTICE 1

Evaluate the following square roots:

a. $\sqrt{25}$

b. $\sqrt{\dfrac{9}{100}}$

c. $-4\sqrt{81}$

d. $\sqrt{-1}$

The square root of a number that is not a perfect square is an *irrational number*. Recall from Section 1.1 that irrational numbers have decimal representations that neither terminate nor repeat. For example, $\sqrt{7}$ and $\sqrt{10}$ are irrational numbers.

$$\sqrt{7} = 2.645751311\ldots, \text{ which is approximately } 2.646.$$
$$\sqrt{10} = 3.16227766\ldots, \text{ which is approximately } 3.162.$$

We can use a calculator to find the decimal approximation of these square roots rounded to a specific place value.

EXAMPLE 2

Evaluate $\sqrt{3}$, rounded to the nearest thousandth.

Solution Since the radicand 3 is not a perfect square, its square root is an irrational number. Using a calculator, we see that $\sqrt{3} = 1.7320508\ldots$. Rounding to the nearest thousandth, we get 1.732.

PRACTICE 2

Find the value of $\sqrt{6}$, rounded to the nearest thousandth.

In simplifying radicals, we use the fact that the operations of squaring and taking a square root undo each other; that is, they are inverse operations in the same way that multiplying by a number and dividing by the same number are inverse operations. The following two properties of radicals result from this relationship:

Squaring a Square Root

For any nonnegative real number a,

$$(\sqrt{a})^2 = a.$$

In words, this property states that when we take the square root of a nonnegative number and then square the result, we get the original number. For example, $(\sqrt{8})^2 = 8$.

> ### Taking the Square Root of a Square
> For any nonnegative real number a,
> $$\sqrt{a^2} = a.$$

In words, this property states that when we square a nonnegative number and then take the square root, the result is the original number. For example, $\sqrt{8^2} = 8$.

Some radicands contain variables. Consider, for instance, the expression $\sqrt{x}$. For negative values of x, the radicand is negative, and so the radical is not a real number. *For purposes of simplicity, we will assume throughout the remainder of this text that, unless otherwise stated, all variables in radicands are nonnegative.*

Some radicals have perfect square radicands containing variables, such as $\sqrt{x^6}$, $\sqrt{36y^8}$, and $\sqrt{49a^6b^2}$. Using the properties of exponents and the properties of square roots just stated, we can find their square roots.

$$\sqrt{x^6} = \sqrt{(x^3)^2} = x^3$$
$$\sqrt{36y^8} = \sqrt{(6y^4)^2} = 6y^4$$
$$\sqrt{49a^6b^2} = \sqrt{(7a^3b)^2} = 7a^3b$$

Note that the exponent of the variable(s) in any perfect square is always an even number. What relationship do you observe between the coefficients and the exponents of the radicands and those of the corresponding square roots?

EXAMPLE 3

Simplify.

a. $\sqrt{25n^4}$ **b.** $-\sqrt{9a^8b^{10}}$

Solution

a. $\sqrt{25n^4} = \sqrt{(5n^2)^2} = 5n^2$

b. $-\sqrt{9a^8b^{10}} = -\sqrt{(3a^4b^5)^2} = -3a^4b^5$

PRACTICE 3

Simplify.

a. $-\sqrt{4y^2}$

b. $\sqrt{36x^6y^6}$

Cube Roots

Not every radical is a square root. For instance, we can take the *cube root* of a number.

> **DEFINITION**
> The number b is the **cube root** of a if $b^3 = a$, for any real numbers a and b.
> The cube root of a is written $\sqrt[3]{a}$, where 3 is called the **index** of the radical.

For example, the cube root of 8, written $\sqrt[3]{8}$, is 2 since $2^3 = 8$. The cube root of -8, written $\sqrt[3]{-8}$, is -2, since $(-2)^3 = -8$.

As with square roots, the operations of cubing and taking a cube root undo each other and so are inverse operations. The following two properties stem from this relationship:

> ### Cubing a Cube Root
> For any real number a,
> $$(\sqrt[3]{a})^3 = a.$$

In words, this property states that when we take the cube root of a number and then cube the result, we get the original number. For example, $(\sqrt[3]{5})^3 = 5$.

Taking the Cube Root of a Cube

For any real number a,

$$\sqrt[3]{a^3} = a.$$

In words, this property states that when we cube a number and then take the cube root of the result, we get the original number. For example, $\sqrt[3]{5^3} = 5$.

A rational number is said to be a **perfect cube** if it is the cube of another rational number. For example, $\dfrac{1}{27}$ is a perfect cube since $\dfrac{1}{27} = \left(\dfrac{1}{3}\right)^3$. On the other hand, 16 is not a perfect cube since it is not the cube of any rational number.

EXAMPLE 4

Find the cube root.

a. $\sqrt[3]{64}$ b. $\sqrt[3]{-1}$ c. $\sqrt[3]{\dfrac{8}{27}}$ d. $\sqrt[3]{-27x^{12}}$

Solution

a. $\sqrt[3]{64} = \sqrt[3]{4^3} = 4$

b. $\sqrt[3]{-1} = \sqrt[3]{(-1)^3} = -1$

c. $\sqrt[3]{\dfrac{8}{27}} = \sqrt[3]{\left(\dfrac{2}{3}\right)^3} = \dfrac{2}{3}$

d. $\sqrt[3]{-27x^{12}} = \sqrt[3]{(-3x^4)^3} = -3x^4$

PRACTICE 4

Find the cube root.

a. $\sqrt[3]{216}$

b. $\sqrt[3]{-27}$

c. $\sqrt[3]{\dfrac{1}{125}}$

d. $\sqrt[3]{-64x^6}$

---

*n*th Roots

Just as we can raise a real number to powers other than 2 or 3, so we can find roots of a number other than the square root and the cube root.

DEFINITION

The number b is the ***n*th root** of a if $b^n = a$, for any real number a and for any positive integer n greater than 1. The *n*th root of a is written $\sqrt[n]{a}$, where n is called the **index** of the radical.

For instance, the fourth root of 16, written $\sqrt[4]{16}$, is 2 since $2^4 = 16$.

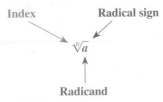

Index   Radical sign

$\sqrt[n]{a}$

Radicand

Note that for a square root, the index is understood to be 2 but is usually not written.

The nth root of any nonnegative real number is a real number. However, for negative radicands, the root is not a real number for every index:

- If the radicand is negative and the index is *even*, the radical expression is *not* a real number.

- If the radicand is negative and the index is *odd*, then the radical expression is a real number.

The next property of radicals involves taking the nth root of an nth power.

Taking the nth Root of an nth Power

For any real number a,

$$\sqrt[n]{a^n} = |a| \text{ for any even positive integer } n, \text{ and}$$

$$\sqrt[n]{a^n} = a \text{ for any odd positive integer } n \text{ greater than } 1.$$

In words, this property states that the nth root of the nth power of a equals the absolute value of a when n is even and equals a when n is odd. So, for example, $\sqrt{(-5)^2} = |-5| = 5$ and $\sqrt[3]{(-5)^3} = -5$. Also, $\sqrt{(-x)^2} = |x|$. Because we assume all variable radicands are positive, we can drop the absolute value sign when taking the root of a variable expression even when the index is even.

EXAMPLE 5

Write without a radical sign.

a. $\sqrt[4]{81}$ **b.** $\sqrt[5]{-243}$ **c.** $\sqrt[5]{32x^{10}}$

Solution

a. $\sqrt[4]{81} = \sqrt[4]{3^4} = 3$ **b.** $\sqrt[5]{-243} = \sqrt[5]{(-3)^5} = -3$

c. $\sqrt[5]{32x^{10}} = \sqrt[5]{(2x^2)^5} = 2x^2$

PRACTICE 5

Express without a radical sign.

a. $\sqrt[6]{64}$

b. $\sqrt[5]{-32}$

c. $\sqrt[4]{256y^8}$

The following table lists some perfect powers. Committing these powers to memory can simplify working with radicals.

Number	Perfect Squares	Perfect Cubes	Perfect Fourth Powers	Perfect Fifth Powers
1	1	1	1	1
2	4	8	16	32
3	9	27	81	243
4	16	64	256	
5	25	125	625	
6	36	216		
7	49			
8	64			
9	81			
10	100			
11	121			
12	144			

Rational Exponents

So far in this textbook, we have discussed only integral exponents. However, we now extend this discussion to include **rational exponents**, that is, exponents that are rational numbers. The concept of rational exponents gives us an alternative way to write radical expressions. Some examples of expressions written with rational exponents are:

$$8^{1/3} \qquad -32^{4/5} \qquad x^{-3/4}$$

Let's first consider rational exponents with numerator 1, as in $8^{1/3}$. The question is, how can we attach a meaning to such an expression so that the laws of exponents hold? Cubing this expression and applying the power rule of exponents will provide an answer.

$$(8^{(1/3)})^3 = 8^{(1/3)\cdot3} = 8^{3/3} = 8^1 = 8$$

Since the cube of $8^{1/3}$ is 8, $8^{1/3}$ must be the same as the cube root of 8, $\sqrt[3]{8}$. This suggests the following definition:

DEFINITION OF $a^{1/n}$

For any positive integer n greater than 1, and a real number a for which $\sqrt[n]{a}$ is a real number,

$$a^{1/n} = \sqrt[n]{a}.$$

In words, this definition states that a number raised to the power $\dfrac{1}{n}$ is the nth root of that number.

Note that the denominator of the rational exponent is identical to the index of the radical.

EXAMPLE 6

Write using radical notation. Then, simplify if possible.

a. $49^{1/2}$ **b.** $64^{1/3}$ **c.** $-2x^{1/2}$ **d.** $(81x^8)^{1/4}$

Solution

a. $49^{1/2} = \sqrt{49} = \sqrt{7^2} = 7$

b. $64^{1/3} = \sqrt[3]{64} = \sqrt[3]{4^3} = 4$

c. $-2x^{1/2} = -2(x^{1/2}) = -2\sqrt{x}$

d. $(81x^8)^{1/4} = \sqrt[4]{81x^8} = \sqrt[4]{(3x^2)^4} = 3x^2$

PRACTICE 6

Express as a radical expression. Then, simplify.

a. $36^{1/2}$

b. $27^{1/3}$

c. $-n^{1/4}$

d. $(-125y^9)^{1/3}$

Now, let's consider the meaning that we would give to *any* positive rational exponent so that the laws of exponents hold. Consider, for instance, the expression $8^{2/3}$. Applying the laws of exponents gives us the following:

$$
\begin{aligned}
8^{2/3} &= (8^{1/3})^2 \\
&= (\sqrt[3]{8})^2 \\
&= (2)^2 \\
&= 4
\end{aligned}
\qquad \text{or equivalently} \qquad
\begin{aligned}
8^{2/3} &= (8^2)^{1/3} \\
&= \sqrt[3]{8^2} \\
&= \sqrt[3]{64} \\
&= 4
\end{aligned}
$$

This example suggests the following definition:

DEFINITION OF $a^{m/n}$

For any positive integers m and n such that n is greater than 1, where $\dfrac{m}{n}$ is in simplest form and a is a real number for which $\sqrt[n]{a}$ is a real number,

$$a^{m/n} = \sqrt[n]{a^m}, \qquad \text{or equivalently} \qquad a^{m/n} = (\sqrt[n]{a})^m.$$

As previously mentioned, the denominator n of the rational exponent is the index of the radical. Also note that the numerator m is the power to which the base a (or $\sqrt[n]{a}$) is raised.

EXAMPLE 7

Rewrite using radical notation. Then, simplify if possible.

a. $16^{5/4}$ **b.** $(-8)^{2/3}$ **c.** $\left(\frac{1}{4}x^2\right)^{3/2}$

Solution

a. $16^{5/4} = (\sqrt[4]{16})^5 = (\sqrt[4]{2^4})^5 = 2^5 = 32$

b. $(-8)^{2/3} = (\sqrt[3]{-8})^2 = (-2)^2 = 4$

c. $\left(\frac{1}{4}x^2\right)^{3/2} = \left(\sqrt{\frac{1}{4}x^2}\right)^3$

$= \left(\sqrt{\left(\frac{1}{2}x\right)^2}\right)^3$

$= \left(\frac{1}{2}x\right)^3$

$= \frac{1}{8}x^3$

PRACTICE 7

Express in radical notation. Then, simplify.

a. $81^{3/4}$

b. $(-64)^{2/3}$

c. $\left(\frac{4}{9}y^4\right)^{5/2}$

Would we have gotten the same answer to Example 7(b) if we had first squared -8 and then taken the cube root of the result?

> **TIP** In evaluating $a^{m/n}$ for a given a, m, and n, it is usually easier to take the nth root before raising the base to the mth power, rather than the other way around.

Recall from Section 1.4 that to raise a number to a negative exponent, we raise the number to the corresponding positive exponent and then find the multiplicative inverse, that is, $a^{-n} = \dfrac{1}{a^n}$. The following definition extends the meaning of negative exponents to include negative rational exponents:

DEFINITION OF $a^{-m/n}$

For any positive integers m and n such that n is greater than 1, where $\dfrac{m}{n}$ is in simplest form and a is a real number for which $\sqrt[n]{a}$ is a nonzero real number,

$$a^{-m/n} = \frac{1}{a^{m/n}}.$$

For instance,

$$8^{-2/3} = \frac{1}{8^{2/3}}$$

$$= \frac{1}{(\sqrt[3]{8})^2}$$

$$= \frac{1}{2^2} = \frac{1}{4}$$

EXAMPLE 8

Simplify. **a.** $16^{-3/4}$ **b.** $(-8x^3)^{-2/3}$

Solution

a. $16^{-3/4} = \dfrac{1}{16^{3/4}}$

$\qquad\quad = \dfrac{1}{(\sqrt[4]{16})^3}$

$\qquad\quad = \dfrac{1}{(\sqrt[4]{2^4})^3}$

$\qquad\quad = \dfrac{1}{2^3} = \dfrac{1}{8}$

b. $(-8x^3)^{-2/3} = \dfrac{1}{(-8x^3)^{2/3}}$

$\qquad\qquad\qquad = \dfrac{1}{(\sqrt[3]{-8x^3})^2}$

$\qquad\qquad\qquad = \dfrac{1}{(-2x)^2} = \dfrac{1}{4x^2}$

PRACTICE 8

Simplify using the laws of exponents.

a. $-81^{-1/4}$

b. $(27a^6)^{-4/3}$

Expressions containing rational exponents, like those containing integral exponents, can be simplified using the laws of exponents. Review the laws of exponents in the following table:

Raising a number to a negative exponent	$x^{-a} = \dfrac{1}{x^a}$, for x nonzero
The product rule of exponents	$x^a \cdot x^b = x^{a+b}$
The quotient rule of exponents	$\dfrac{x^a}{x^b} = x^{a-b}$, for x nonzero
The power rule of exponents	$(x^a)^b = x^{ab}$
Raising a product to a power	$(xy)^a = x^a \cdot y^a$
Raising a quotient to a power	$\left(\dfrac{x}{y}\right)^a = \dfrac{x^a}{y^a}$, for y nonzero

EXAMPLE 9

Simplify, using the laws of exponents. Then, write the answer in radical notation and simplify, if possible.

a. $64^{2/3}64^{1/6}$ **b.** $\dfrac{y^{2/3}}{y^{1/3}}$ **c.** $(x^{3/8})^2$ **d.** $\left(\dfrac{8x^3}{y^6}\right)^{1/3}$

Solution

a. $64^{2/3}64^{1/6} = 64^{2/3+1/6}$ Use the product rule of exponents.

$\qquad\qquad\quad = 64^{5/6}$ Simplify.

$\qquad\qquad\quad = (\sqrt[6]{64})^5$ Write in radical notation.

$\qquad\qquad\quad = 2^5$ Simplify.

$\qquad\qquad\quad = 32$ Simplify.

PRACTICE 9

Use the laws of exponents to simplify. Then, write the answer in radical form and simplify, if possible.

a. $81^{1/4}81^{1/2}$

b. $\dfrac{n^{3/5}}{n^{2/5}}$

c. $(r^{1/6})^3$

d. $\left(\dfrac{x^4}{4y^2}\right)^{1/2}$

EXAMPLE 9 (continued)

b. $\dfrac{y^{2/3}}{y^{1/3}} = y^{2/3-1/3}$ Use the quotient rule of exponents.

$\qquad = y^{1/3}$ Simplify.

$\qquad = \sqrt[3]{y}$ Write in radical notation.

c. $(x^{3/8})^2 = x^{(3/8)\cdot 2}$ Use the power rule of exponents.

$\qquad = x^{3/4}$ Simplify.

$\qquad = \sqrt[4]{x^3}$ Write in radical notation.

d. $\left(\dfrac{8x^3}{y^6}\right)^{1/3} = \dfrac{(8x^3)^{1/3}}{(y^6)^{1/3}}$ Use the rule for raising a quotient to a power.

$\qquad = \dfrac{\sqrt[3]{8x^3}}{\sqrt[3]{y^6}}$ Write in radical notation.

$\qquad = \dfrac{2x}{y^2}$ Simplify.

EXAMPLE 10

The following formula relates the intensity of sound I to decibel level D:

$$I = 10^{D/10}$$

Write this formula using radical notation.

Solution

$$I = 10^{D/10} = \sqrt[10]{10^D}$$

So we can write the formula as $I = \sqrt[10]{10^D}$.

PRACTICE 10

In our solar system, the *period* of a planet, that is, the time that it takes to make a complete orbit around the Sun, can be approximated by the expression $2\pi\sqrt{\dfrac{d^3}{Gm}}$, where d is the average distance of the planet from the Sun, G is a constant, and m is the mass of the planet, in appropriate units. Write this radical expression as an exponential expression in simplest form.

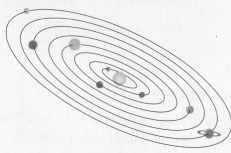

Mathematically Speaking

Fill in each blank with the most appropriate term or phrase from the given list.

is not	square root	raised to the power
irrational	is	radicand
multiplied by	cube root	principal
square	index	rational

1. The number b is a(n) _____ of a if $b^2 = a$, for any real numbers a and b and for a nonnegative.

2. The symbol $\sqrt{}$ stands for the positive or, _____, square root.

3. The number under the radical sign is called the _____.

4. The square root of a negative number _____ a real number.

5. The square root of a number that is not a perfect square is a(n) _____ number.

6. The number b is the _____ of a if $b^3 = a$, for any real numbers a and b.

7. The cube root of a is written $\sqrt[3]{a}$, where 3 is called the _____ of the radical.

8. A number _____ $\dfrac{1}{n}$ is the nth root of that number.

Ⓐ *Evaluate, if possible.*

9. $\sqrt{64}$

10. $\sqrt{49}$

11. $-\sqrt{100}$

12. $-\sqrt{144}$

13. $\sqrt{-36}$

14. $\sqrt{-9}$

15. $2\sqrt{16}$

16. $3\sqrt{25}$

17. $\sqrt[3]{27}$

18. $\sqrt[3]{125}$

19. $5\sqrt[3]{-8}$

20. $-6\sqrt[3]{-27}$

21. $\sqrt[4]{256}$

22. $\sqrt[5]{243}$

23. $8\sqrt[5]{-1}$

24. $7\sqrt[4]{81}$

25. $\sqrt{\dfrac{9}{16}}$

26. $\sqrt{\dfrac{1}{4}}$

27. $\sqrt[3]{-\dfrac{8}{125}}$

28. $-\sqrt[3]{\dfrac{27}{64}}$

29. $\sqrt{0.04}$

30. $\sqrt{1.21}$

🔲 *Use a calculator to approximate the root to the nearest thousandth.*

31. $\sqrt{21}$

32. $\sqrt{17}$

33. $\sqrt{46}$

34. $\sqrt{59}$

35. $\sqrt{14.25}$

36. $\sqrt{0.006}$

37. $\sqrt[3]{112}$

38. $\sqrt[3]{142}$

39. $\sqrt[5]{150}$

40. $\sqrt[4]{200}$

Ⓑ *Simplify.*

41. $\sqrt{x^8}$

42. $\sqrt{y^{10}}$

43. $\sqrt{16a^6}$

44. $\sqrt{49r^2}$

45. $9\sqrt{p^8q^4}$

46. $3\sqrt{u^8v^6}$

47. $\dfrac{1}{3}\sqrt{36x^{10}y^2}$

48. $-\dfrac{1}{4}\sqrt{64a^4b^{12}}$

49. $\sqrt[3]{-125u^9}$

50. $\sqrt[3]{-27r^6}$

51. $2\sqrt[3]{216u^3v^{12}}$

52. $7\sqrt[3]{27x^9y^9}$

53. $\sqrt[4]{16t^{12}}$

54. $\sqrt[4]{256n^{20}}$

55. $\sqrt[5]{p^5q^{15}}$

56. $\sqrt[5]{a^{10}b^{20}}$

C *Write using radical notation. Then, simplify, if possible.*

57. $16^{1/2}$

58. $81^{1/2}$

59. $-16^{1/2}$

60. $-81^{1/2}$

61. $(-64)^{1/3}$

62. $(-125)^{1/3}$

63. $6x^{1/4}$

64. $-y^{1/5}$

65. $(36a^2)^{1/2}$

66. $(25n^4)^{1/2}$

67. $(-216u^6)^{1/3}$

68. $(-32y^5)^{1/5}$

69. $27^{4/3}$

70. $4^{5/2}$

71. $-16^{3/2}$

72. $-64^{4/3}$

73. $(-27y^3)^{2/3}$

74. $(-32x^{10})^{3/5}$

75. $-81^{-3/4}$

76. $64^{-2/3}$

77. $\left(\dfrac{x^{10}}{4}\right)^{-1/2}$

78. $\left(\dfrac{8}{y^9}\right)^{-1/3}$

Write using radical notation. Simplify, if possible.

79. $16 \cdot 16^{1/2}$

80. $8^{1/3} \cdot 8$

81. $\dfrac{6^{3/5}}{6^{2/5}}$

82. $\dfrac{3^{4/3}}{3^{1/3}}$

83. $\left(\dfrac{1}{2}\right)^{3/2} \cdot \left(\dfrac{1}{2}\right)^{-1/2}$

84. $\left(\dfrac{1}{16}\right)^{-1/4} \cdot \left(\dfrac{1}{16}\right)^{3/4}$

85. $(9^{3/4})^{2/3}$

86. $(81^{1/6})^{3/2}$

87. $4n^{2/5} \cdot n^{1/5}$

88. $a^{5/6} \cdot 6a^{-1/2}$

89. $(y^{-4})^{-1/8}$

90. $(x^{-4/3})^3$

91. $(4x^2)^{-1/2}$

92. $(27y^6)^{1/3}$

93. $\dfrac{5r^{3/4}}{r^{1/2}}$

94. $\dfrac{p^{5/6}}{7p^{2/3}}$

95. $\left(\dfrac{a^2}{b^6}\right)^{1/3}$

96. $\left(\dfrac{u^3}{v^4}\right)^{1/4}$

97. $3x(16x^8)^{1/2}$

98. $-2n^2(64n^9)^{2/3}$

99. $\dfrac{(2p^{1/6})^6}{16p^3}$

100. $\dfrac{12x^3}{(3x^{3/4})^4}$

Mixed Practice

Simplify.

101. $\dfrac{1}{4}\sqrt{64a^6b^4}$

102. $\sqrt[3]{-8u^{12}v^3}$

Write using radical notation. Simplify, if possible.

103. $n^{-1/6} \cdot 3n^{2/3}$

104. $\left(\dfrac{x^2}{y^3}\right)^{1/3}$

Use a calculator to approximate the root to the nearest thousandth.

105. $\sqrt{0.009}$

106. $\sqrt[3]{36}$

Write using radical notation. Then, simplify, if possible.

107. $(81x^6)^{1/2}$

108. $(-125u^3)^{2/3}$

Evaluate, if possible.

109. $\sqrt{\dfrac{25}{121}}$

110. $2\sqrt[3]{-125}$

Applications

D *Solve.*

111. If an object is dropped, the time (in seconds) it takes the object to fall s ft is given by the expression $\frac{1}{4}\sqrt{s}$. Find the time it takes a stone dropped from a height of 100 ft to reach the ground.

112. The geometric mean is a statistic used in business and economics. The geometric mean of three numbers is given by the expression $\sqrt[3]{p}$, where p is the product of the three numbers. Find the geometric mean of 9, 3, and 8.

113. The length of the side of a square with area A can be computed using the expression $\sqrt{A}$. The area of the square picture frame, including the 1-inch wood border, is 25 in^2.

a. Find x, the length of the side of the picture frame.

b. What size photograph fits in the frame?

114. The length of one side of a cube with volume V can be computed using the expression $\sqrt[3]{V}$. A box manufacturer makes special-order cube-shaped boxes for a shipping company.

a. If the shipping company requires a box to have a volume of 1728 in^3, what is the length of each side of the box that the manufacturer must make?

b. If the shipping company requests a box that is one-eighth the volume of the box described in part (a), by what factor does the length of each side change?

115. The manager of an office uses the expression $8000(0.5)^{t/3}$ to calculate the value of a piece of office equipment t yr after it was purchased new for $8000.

a. Write this expression in radical form.

b. Find the value of the equipment 6 yr after it was purchased.

116. The number of Earth days it takes a planet in the solar system to revolve once around the Sun can be approximated by the expression $0.4(D)^{3/2}$, where D is the average distance (in millions of miles) of the planet from the Sun.

a. Write this expression in radical form.

b. The planet Mercury is an average distance of 36 million miles from the Sun. To the nearest day, how many Earth days does it take Mercury to revolve once around the Sun?

• Check your answers on page A-42.

MINDStretchers

Mathematical Reasoning

1. Consider the equation $y = \sqrt{x}$. Explain why the graph of this equation lies completely in Quadrant I.

Research

2. Before the introduction of calculators, a variety of methods were used to find the approximate value of square roots with radicands that are not perfect squares. Either in your college library or on the web, identify two such methods and write a brief summary of your findings.

Groupwork

3. Work with a partner to simplify each of the following radical expressions. Assume all variables and radicands are nonnegative real numbers.

a. $\sqrt{x^2 + 2xy + y^2}$

b. $(9a^2 + 12ab + 4b^2)^{-1/2}$

c. $\dfrac{x + y}{(x + y)^{1/3}}$

d. $\dfrac{(a^2 - 10a + 25)^{1/3}}{(a^2 - 10a + 25)^{1/4}}$

7.2 Simplifying Radical Expressions

Just as we rewrite rational expressions in lowest terms, so we rewrite radical expressions *in simplified form*. Simplified radical expressions are usually easier to work with and also to recognize when they are equal.

Rewriting radical expressions in terms of rational exponents and then using the laws of exponents often helps us to simplify these expressions. Once they are simplified, we can convert them back to radical notation.

EXAMPLE 1

Simplify each radical expression, if possible, by using rational exponents. Then, write in radical notation.

a. $\sqrt[8]{x^4}$ **b.** $\sqrt[4]{49}$ **c.** $\sqrt{x} \cdot \sqrt[4]{x}$ **d.** $\dfrac{\sqrt{x}}{\sqrt[3]{x}}$

Solution

a. $\sqrt[8]{x^4} = x^{4/8} = x^{1/2} = \sqrt{x}$

b. $\sqrt[4]{49} = 49^{1/4} = (7^2)^{1/4} = 7^{2/4} = 7^{1/2} = \sqrt{7}$

c. $\sqrt{x} \cdot \sqrt[4]{x} = x^{1/2} \cdot x^{1/4} = x^{1/2+1/4} = x^{3/4} = \sqrt[4]{x^3}$

d. $\dfrac{\sqrt{x}}{\sqrt[3]{x}} = \dfrac{x^{1/2}}{x^{1/3}} = x^{1/2-1/3} = x^{1/6} = \sqrt[6]{x}$

PRACTICE 1

If possible, simplify by using rational exponents. Then, write your answer as a radical expression.

a. $\sqrt[8]{y^2}$

b. $\sqrt[6]{25}$

c. $\sqrt{n} \cdot \sqrt[5]{n}$

d. $\dfrac{\sqrt{t}}{\sqrt[4]{t}}$

EXAMPLE 2

Simplify, if possible. Then, write using radical notation.

a. $\sqrt{\sqrt[3]{y}}$ **b.** $\sqrt[6]{r^2 s^4}$ **c.** $\sqrt[3]{3} \cdot \sqrt{2}$

Solution

a. $\sqrt{\sqrt[3]{y}} = (y^{1/3})^{1/2} = y^{1/6} = \sqrt[6]{y}$

b. $\sqrt[6]{r^2 s^4} = (r^2 s^4)^{1/6} = (r^2)^{1/6}(s^4)^{1/6} = r^{2/6} s^{4/6} = r^{1/3} s^{2/3}$

$= (rs^2)^{1/3} = \sqrt[3]{rs^2}$

c. The expression $\sqrt[3]{3} \cdot \sqrt{2} = 3^{1/3} \cdot 2^{1/2}$ cannot be simplified because the bases are not the same.

PRACTICE 2

Simplify, if possible. Then, write the answer as a radical expression.

a. $\sqrt[3]{\sqrt[4]{n}}$

b. $\sqrt[8]{a^2 b^6}$

c. $\sqrt[5]{5} \cdot \sqrt{7}$

Now, let's consider the product of two radicals with the same index, namely $\sqrt[3]{2} \cdot \sqrt[3]{5}$. Using rational exponents and the laws of exponents, we can simplify this expression as follows:

$$\sqrt[3]{2} \cdot \sqrt[3]{5} = 2^{1/3} \cdot 5^{1/3}$$
$$= (2 \cdot 5)^{1/3}$$
$$= \sqrt[3]{2 \cdot 5}$$
$$= \sqrt[3]{10}$$

So we see that $\sqrt[3]{2} \cdot \sqrt[3]{5} = \sqrt[3]{2 \cdot 5}$. This result leads us to the *product rule of radicals*, which we use to multiply radical expressions and to simplify them.

The Product Rule of Radicals

If $\sqrt[n]{a}$ and $\sqrt[n]{b}$ are real numbers, then

$$\sqrt[n]{a} \cdot \sqrt[n]{b} = \sqrt[n]{ab}.$$

In words, this rule states that to multiply radicals with the *same index*, multiply the radicands.

EXAMPLE 3

Multiply.

a. $\sqrt{3} \cdot \sqrt{2}$ **b.** $\sqrt[3]{5} \cdot \sqrt[3]{x}$ **c.** $\sqrt[4]{6n^2} \cdot \sqrt[4]{2n}$ **d.** $\sqrt{\dfrac{x}{3}} \cdot \sqrt{\dfrac{y}{2}}$

Solution

a. $\sqrt{3} \cdot \sqrt{2} = \sqrt{3 \cdot 2} = \sqrt{6}$

b. $\sqrt[3]{5} \cdot \sqrt[3]{x} = \sqrt[3]{5 \cdot x} = \sqrt[3]{5x}$

c. $\sqrt[4]{6n^2} \cdot \sqrt[4]{2n} = \sqrt[4]{6n^2 \cdot 2n} = \sqrt[4]{12n^3}$

d. $\sqrt{\dfrac{x}{3}} \cdot \sqrt{\dfrac{y}{2}} = \sqrt{\dfrac{x}{3} \cdot \dfrac{y}{2}} = \sqrt{\dfrac{xy}{6}}$

PRACTICE 3

Find the product.

a. $\sqrt{7} \cdot \sqrt{3}$

b. $\sqrt[5]{2} \cdot \sqrt[5]{y^3}$

c. $\sqrt[3]{4p} \cdot \sqrt[3]{7p}$

d. $\sqrt{\dfrac{3}{v}} \cdot \sqrt{\dfrac{5}{u}}$

We can use the product rule of radicals not only to multiply radicals but also to simplify them. When simplifying, we typically use the rule "in reverse," that is, in the form $\sqrt[n]{ab} = \sqrt[n]{a} \cdot \sqrt[n]{b}$.

A square root is not considered simplified if a perfect square factors the radicand. For instance, the radical $\sqrt{18}$ is not simplified, since the perfect square 9 is a factor of 18. Likewise, a cube root is not considered simplified if the radicand has a perfect cube factor. For example, the expression $\sqrt[3]{24}$ is not simplified since the perfect cube 8 is a factor of 24. In general, the expression $\sqrt[n]{a}$ is *not* simplified if the radicand a has a factor that is a perfect nth power.

EXAMPLE 4

Simplify.

a. $\sqrt{32}$ **b.** $-4\sqrt{75}$ **c.** $\sqrt[3]{54}$ **d.** $\sqrt[4]{80}$

Solution

a. $\sqrt{32} = \sqrt{16 \cdot 2}$ Factor out the perfect square 16.

$\qquad = \sqrt{16} \cdot \sqrt{2}$ Use the product rule of radicals.

$\qquad = 4\sqrt{2}$ Take the square root of the perfect square.

b. $-4\sqrt{75} = -4\sqrt{25 \cdot 3}$

$\qquad\quad = -4\sqrt{25} \cdot \sqrt{3}$

$\qquad\quad = -4 \cdot 5 \cdot \sqrt{3}$

$\qquad\quad = -20\sqrt{3}$

c. $\sqrt[3]{54} = \sqrt[3]{27 \cdot 2} = \sqrt[3]{27} \cdot \sqrt[3]{2} = 3\sqrt[3]{2}$

d. $\sqrt[4]{80} = \sqrt[4]{16 \cdot 5} = \sqrt[4]{16} \cdot \sqrt[4]{5} = 2\sqrt[4]{5}$

PRACTICE 4

Write in simplified form.

a. $\sqrt{72}$

b. $-7\sqrt{18}$

c. $\sqrt[3]{56}$

d. $\sqrt[4]{48}$

Some radical expressions contain radicands that have more than one factor that is a perfect nth power. For instance, the radicand in Example 4(a) has two perfect square factors, namely, 4 and 16. To simplify these radicals, we typically factor out the largest perfect nth-power factor.

EXAMPLE 5

Simplify.

a. $\sqrt{x^5}$ **b.** $\sqrt{12n^3}$ **c.** $\sqrt[3]{40x^3y^4}$

Solution

a. $\sqrt{x^5} = \sqrt{x^4 \cdot x}$
$= \sqrt{x^4} \cdot \sqrt{x}$
$= x^2\sqrt{x}$

b. $\sqrt{12n^3} = \sqrt{4n^2 \cdot 3n}$
$= \sqrt{4n^2} \cdot \sqrt{3n}$
$= 2n\sqrt{3n}$

c. $\sqrt[3]{40x^3y^4} = \sqrt[3]{8x^3y^3 \cdot 5y}$
$= \sqrt[3]{8x^3y^3} \cdot \sqrt[3]{5y}$
$= 2xy\sqrt[3]{5y}$

PRACTICE 5

Simplify.

a. $\sqrt{y^7}$

b. $\sqrt{18x^5}$

c. $\sqrt[3]{81a^4b^6}$

Some radical expressions consist of the quotient of radicals. For instance, consider the expression $\dfrac{\sqrt{2}}{\sqrt{5}}$. Using rational exponents and the laws of exponents, we can write this expression as follows:

$$\frac{\sqrt{2}}{\sqrt{5}} = \frac{2^{1/2}}{5^{1/2}}$$

$$= \left(\frac{2}{5}\right)^{1/2}$$

$$= \sqrt{\frac{2}{5}}$$

So $\dfrac{\sqrt{2}}{\sqrt{5}} = \sqrt{\dfrac{2}{5}}$. This result leads us to the *quotient rule of radicals*, which can be applied to dividing radicals and simplifying them.

The Quotient Rule of Radicals

For any integer $n > 1$ and any real numbers a and b for which $\sqrt[n]{a}$ and $\sqrt[n]{b}$ are real numbers and b is nonzero,

$$\frac{\sqrt[n]{a}}{\sqrt[n]{b}} = \sqrt[n]{\frac{a}{b}}.$$

In words, this rule states that to divide radicals, divide the radicands. Note that in the quotient rule, the index is the same in all three radicals.

We can use this rule to divide two radicals.

EXAMPLE 6

Divide. Simplify, if possible.

a. $\dfrac{\sqrt{15x}}{\sqrt{3}}$
b. $\dfrac{\sqrt{50}}{\sqrt{2}}$
c. $\dfrac{\sqrt[3]{18a^2b}}{\sqrt[3]{3ab}}$

Solution

a. $\dfrac{\sqrt{15x}}{\sqrt{3}} = \sqrt{\dfrac{15x}{3}} = \sqrt{5x}$

b. $\dfrac{\sqrt{50}}{\sqrt{2}} = \sqrt{\dfrac{50}{2}} = \sqrt{25} = 5$

c. $\dfrac{\sqrt[3]{18a^2b}}{\sqrt[3]{3ab}} = \sqrt[3]{\dfrac{18a^2b}{3ab}} = \sqrt[3]{6a}$

PRACTICE 6

Find the quotient. Simplify, if possible.

a. $\dfrac{\sqrt{42n}}{\sqrt{6}}$

b. $\dfrac{\sqrt{72}}{\sqrt{2}}$

c. $\dfrac{\sqrt[3]{10x^2y^2}}{\sqrt[3]{5x^2y}}$

A radical with a radicand in the form of a fraction is not considered to be simplified. To rewrite such radicals in simplest terms, we apply the quotient rule of radicals "in reverse":

$$\sqrt[n]{\dfrac{a}{b}} = \dfrac{\sqrt[n]{a}}{\sqrt[n]{b}}.$$

EXAMPLE 7

Simplify.
a. $\sqrt{\dfrac{9}{49}}$
b. $\sqrt{\dfrac{x}{4}}$
c. $\sqrt[3]{\dfrac{8x}{27y^6}}$
d. $\sqrt{\dfrac{5a^4b^{10}}{49c^2}}$

Solution

a. $\sqrt{\dfrac{9}{49}} = \dfrac{\sqrt{9}}{\sqrt{49}} = \dfrac{3}{7}$

b. $\sqrt{\dfrac{x}{4}} = \dfrac{\sqrt{x}}{\sqrt{4}} = \dfrac{\sqrt{x}}{2}$

c. $\sqrt[3]{\dfrac{8x}{27y^6}} = \dfrac{\sqrt[3]{8x}}{\sqrt[3]{27y^6}} = \dfrac{2\sqrt[3]{x}}{3y^2}$

d. $\sqrt{\dfrac{5a^4b^{10}}{49c^2}} = \dfrac{\sqrt{5a^4b^{10}}}{\sqrt{49c^2}} = \dfrac{a^2b^5\sqrt{5}}{7c}$

PRACTICE 7

Express in simplest terms.

a. $\sqrt{\dfrac{16}{25}}$
b. $\sqrt{\dfrac{n}{64}}$

c. $\sqrt[3]{\dfrac{64x}{125y^9}}$

d. $\sqrt{\dfrac{7p^6q^2}{36r^4}}$

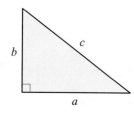

In the final examples of this section, we consider the problem of finding the distance between two points on a coordinate plane. To solve this problem, we apply our knowledge not only of radicals but also of the Pythagorean theorem from geometry.

Recall that the Pythagorean theorem states that for any right triangle, the sum of the squares of the lengths of the two legs equals the square of the length of the hypotenuse: $a^2 + b^2 = c^2$.

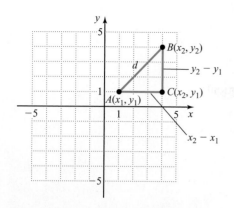

Now, consider any two points on a coordinate plane, $A(x_1, y_1)$ and $B(x_2, y_2)$. The distance between these two points d is the same as the length of line segment $\overline{AB}$. Note from the diagram on the left that this line segment is the hypotenuse of a right triangle whose third vertex is $C(x_2, y_1)$. Applying the Pythagorean theorem to this triangle gives us $d^2 = (x_2 - x_1)^2 + (y_2 - y_1)^2$. This relationship can be written as

$$d = \sqrt{(x_2 - x_1)^2 + (y_2 - y_1)^2},$$

which is known as **the distance formula**. In words, this formula states that the distance between two points on the coordinate plane is the square root of the sum of the squares of the difference of the x-values and the difference of the y-values.

EXAMPLE 8

Find the distance between the points $(3, 6)$ and $(-2, 1)$ on the coordinate plane.

Solution   Let $(3, 6)$ stand for (x_1, y_1) and $(-2, 1)$ for (x_2, y_2). Substituting these values in the distance formula, we get:

$$\begin{aligned}
d &= \sqrt{(x_2 - x_1)^2 + (y_2 - y_1)^2} \\
&= \sqrt{(-2 - 3)^2 + (1 - 6)^2} \\
&= \sqrt{(-5)^2 + (-5)^2} \\
&= \sqrt{25 + 25} \\
&= \sqrt{50} \\
&= 5\sqrt{2}
\end{aligned}$$

So the distance between $(3, 6)$ and $(-2, 1)$ on the coordinate plane is $5\sqrt{2}$ units.

PRACTICE 8

Find the distance between the points $(5, -7)$ and $(3, -1)$ on the coordinate plane.

We can use the distance formula to solve applied problems.

EXAMPLE 9

A hiker walked 3 mi north and then 3 mi east.

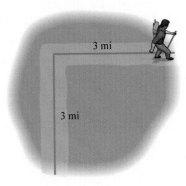

3 mi

3 mi

a. Using the hiker's starting point as the origin of a coordinate plane, find the coordinates of the starting and ending points.

b. How far is the hiker's destination from his starting point? Express the answer as a radical in simplified form.

Solution

a. The hiker's starting point is the origin of a coordinate system, with coordinates $(0, 0)$. So $x_1 = 0$ and $y_1 = 0$. The ending point for the hiker is $(3, 3)$, so $x_2 = 3$ and $y_2 = 3$.

b. To find the distance between these two points, we apply the distance formula.

$$d = \sqrt{(x_2 - x_1)^2 + (y_2 - y_1)^2}$$
$$d = \sqrt{(3 - 0)^2 + (3 - 0)^2} = \sqrt{9 + 9} = \sqrt{18} = \sqrt{9 \cdot 2} = 3\sqrt{2}$$

So the hiker was $3\sqrt{2}$ mi (or about 4.2 mi) from the starting point.

PRACTICE 9

One truck driver traveled west for 20 mi and then north for 50 mi. A second driver traveled 10 mi east and then 60 mi north. Both drivers had started from the same point.

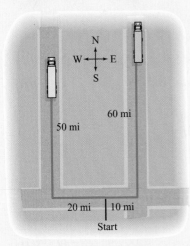

N
W ←→ E
S

60 mi

50 mi

20 mi | 10 mi

Start

a. Using their starting point as the origin of a coordinate plane, find the coordinates of the two destinations.

b. How far apart were the drivers at the end of their travels? Express the answer as a radical in simplified form.

Mathematically Speaking

Fill in each blank with the most appropriate term or phrase from the given list.

radicands	the same index	distance
square of the distance	perfect nth power	
multiple of n	different indices	

1. The product rule of radicals states that to multiply radicals with _____, multiply the radicands.

2. In general, the expression $\sqrt[n]{a}$ is not simplified if the radicand a has a factor that is a(n) _____.

3. The quotient rule of radicals states that to divide radicals with the same index, divide the _____.

4. The _____ between two points (x_1, y_1) and (x_2, y_2) on a coordinate plane is equal to $\sqrt{(x_2 - x_1)^2 + (y_2 - y_1)^2}$.

A *Use rational exponents to simplify the expression, if possible. Then, write the answer in radical form.*

5. $\sqrt[6]{n^2}$

6. $\sqrt[9]{a^6}$

7. $\sqrt[8]{16}$

8. $\sqrt[6]{64}$

9. $\sqrt[3]{x} \cdot \sqrt[6]{x}$

10. $\sqrt[4]{t} \cdot \sqrt[8]{t}$

11. $\sqrt[3]{p} \cdot \sqrt[4]{q}$

12. $\sqrt{a} \cdot \sqrt[5]{b}$

13. $\dfrac{\sqrt[3]{x}}{\sqrt[4]{x}}$

14. $\dfrac{\sqrt{n}}{\sqrt[6]{n}}$

15. $\sqrt{\sqrt{y}}$

16. $\sqrt[3]{\sqrt{x}}$

17. $\sqrt[4]{x^8 y^2}$

18. $\sqrt[6]{u^4 v^{12}}$

19. $\sqrt[6]{x^4} \cdot \sqrt[3]{x}$

20. $\sqrt[8]{n^4} \cdot \sqrt[4]{n^2}$

21. $\dfrac{\sqrt[3]{y^2}}{\sqrt[9]{y^3}}$

22. $\dfrac{\sqrt[4]{a^3}}{\sqrt[8]{a^6}}$

23. $\sqrt[4]{p^2} \cdot \sqrt{q}$

24. $\sqrt[6]{a^2} \cdot \sqrt[3]{b}$

Multiply.

25. $\sqrt{6} \cdot \sqrt{5}$

26. $\sqrt{7} \cdot \sqrt{2}$

27. $\sqrt{3x} \cdot \sqrt{2y}$

28. $\sqrt{5a} \cdot \sqrt{3b}$

29. $\sqrt[3]{4a^2} \cdot \sqrt[3]{9b}$

30. $\sqrt[3]{10x} \cdot \sqrt[3]{2xy}$

31. $\sqrt[4]{3n} \cdot \sqrt[4]{7n^2}$

32. $\sqrt[5]{6y^2} \cdot \sqrt[5]{4y^2}$

33. $\sqrt{\dfrac{x}{2}} \cdot \sqrt{\dfrac{6}{y}}$

34. $\sqrt{\dfrac{10}{p}} \cdot \sqrt{\dfrac{q}{5}}$

Simplify.

35. $\sqrt{24}$

36. $\sqrt{63}$

37. $-3\sqrt{80}$

38. $5\sqrt{98}$

39. $\sqrt[3]{81}$

40. $\sqrt[3]{72}$

41. $\sqrt[4]{96}$

42. $\sqrt[4]{162}$

43. $\sqrt[5]{64}$

44. $\sqrt[5]{160}$

45. $6\sqrt{x^7}$

46. $-2\sqrt{n^3}$

47. $\sqrt{20y}$

48. $\sqrt{72a}$

49. $\sqrt{200r^4}$

50. $\sqrt{216t^6}$

51. $\sqrt{54x^5 y^7}$

52. $\sqrt{125a^7 b^9}$

53. $\sqrt[3]{32n^5}$

54. $\sqrt[3]{108x^8}$

55. $\sqrt[5]{64n^8}$

56. $\sqrt[4]{243y^9}$

57. $\sqrt[3]{72x^7 y^9}$

58. $\sqrt[3]{250p^2 q^8}$

59. $\sqrt[4]{64x^5 y^{10}}$

60. $\sqrt[4]{81p^9 q^6}$

Divide.

61. $\dfrac{\sqrt{90}}{\sqrt{10}}$

62. $\dfrac{\sqrt{28}}{\sqrt{7}}$

63. $\dfrac{\sqrt{30n}}{\sqrt{6}}$

64. $\dfrac{\sqrt{65y}}{\sqrt{13}}$

65. $\dfrac{\sqrt{12x^3y}}{\sqrt{3x}}$

66. $\dfrac{\sqrt{54pq}}{\sqrt{6q}}$

67. $\dfrac{\sqrt[3]{16u^2}}{\sqrt[3]{4u}}$

68. $\dfrac{\sqrt[3]{45n^2}}{\sqrt[3]{5n^2}}$

69. $\dfrac{\sqrt[4]{24a^3b^2}}{\sqrt[4]{4a^2b}}$

70. $\dfrac{\sqrt[5]{36p^4q^3}}{\sqrt[5]{3p^2q^3}}$

Simplify.

71. $\sqrt{\dfrac{25}{16}}$

72. $\sqrt{\dfrac{49}{64}}$

73. $\sqrt{\dfrac{7}{81}}$

74. $\sqrt{\dfrac{3}{100}}$

75. $\sqrt{\dfrac{2}{n^6}}$

76. $\sqrt{\dfrac{3}{x^8}}$

77. $\sqrt{\dfrac{7a}{121}}$

78. $\sqrt{\dfrac{5n}{144}}$

79. $\sqrt{\dfrac{a}{9b^4}}$

80. $\sqrt{\dfrac{5p}{36q^6}}$

81. $\sqrt{\dfrac{9u}{25v^2}}$

82. $\sqrt{\dfrac{4x}{81y^4}}$

83. $\sqrt[3]{\dfrac{27a^2}{64}}$

84. $\sqrt[3]{\dfrac{3x^3}{8}}$

85. $\sqrt[3]{\dfrac{9a}{8b^6c^9}}$

86. $\sqrt[3]{\dfrac{4u^2}{125v^3w^{12}}}$

87. $\sqrt[4]{\dfrac{2a^4b^3}{81c^8}}$

88. $\sqrt[5]{\dfrac{p^2}{32q^5r^{10}}}$

89. $\sqrt{\dfrac{13a^4b}{9c^6d^2}}$

90. $\sqrt{\dfrac{10pq^2}{49r^8s^4}}$

Find the distance between the two points on the coordinate plane.

91. $(10, 3)$ and $(4, -3)$

92. $(-7, 8)$ and $(2, 5)$

93. $(12, 15)$ and $(6, 12)$

94. $(9, 2)$ and $(-1, -8)$

95. $(-4, -3)$ and $(-8, 0)$

96. $(-4, 3)$ and $(-8, 0)$

Mixed Practice

Perform the indicated operation.

97. Multiply: $\sqrt[4]{5x^2} \cdot \sqrt[4]{6xy^3}$

98. Divide: $\dfrac{\sqrt[3]{63x^2}}{\sqrt[3]{7x}}$

Use rational exponents to simplify the expression. Then, write the answer in radical form.

99. $\sqrt[3]{x^2} \cdot \sqrt[3]{x^4}$

100. $\dfrac{\sqrt[4]{n^3}}{\sqrt[6]{n^3}}$

Simplify.

101. $\sqrt{50r^6s^7}$

102. $\sqrt[4]{64}$

103. $\sqrt[3]{\dfrac{25x}{64y^3z^9}}$

104. $\sqrt{\dfrac{36a}{81b^2}}$

105. $\sqrt[3]{128u^4v^6}$

106. Find the distance between the points $(-1, 3)$ and $(4, -2)$ on the coordinate plane.

Applications

B *Solve.*

107. The velocity (in feet per second) of a fluid flowing out of a tank through a small opening d ft below the surface is given by the expression $\sqrt{64d}$. Simplify the expression.

108. The time (in seconds) it takes an object that is dropped to fall h ft is given by the expression $\sqrt{\dfrac{2h}{32}}$. Simplify this expression.

109. In IMAX theaters, movies are projected in high resolution, allowing the audience to sit close to the screen. Typically, all rows are within one screen height away. A standard IMAX screen has a diagonal of 27 m and a width of 22 m. Find the height of the screen to the nearest meter. (*Source:* wikipedia.org)

110. The owner of a factory finds that the price p (in dollars) of an item is given by the equation $p = 0.75\sqrt{x - 8}$, where x is the daily demand. Find the price of the item when the daily demand is 80 units.

111. Find the length of string d let out for the kite shown in the figure to the right. Express the answer as both a radical in simplest form and as a decimal rounded to the nearest tenth of a foot.

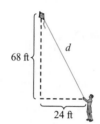

112. Two airplanes depart the same city at the same time. One flies east at a rate of 450 mph and the other flies south at a rate of 400 mph. How far apart are the two planes 1 hr after departure? Express the answer as both a radical in simplest form and as a decimal rounded to the nearest mile.

113. A sales rep drives from her home to a college 20 mi north. She then drives 48 mi east to another college.

   a. Using the origin of a coordinate plane as the sales rep's starting point, find the coordinates of the location of the second college.

   b. How far from her home is the second college?

114. To get to a friend's apartment from his apartment, a student walks 4 blocks west and 6 blocks south.

   a. Using the origin of a coordinate plane as the starting point, find the coordinates of the location of the friend's apartment.

   b. How far from the student's apartment is the friend's apartment?

• Check your answers on page A-42.

MINDStretchers

Writing

1. Explain what it means for the expression $\sqrt[n]{a}$ to be in simplest form.

Technology

2. Use a grapher to display the graphs of $y = \sqrt{\dfrac{1}{x^2 + 1}}$ and $y = \dfrac{1}{\sqrt{x^2 + 1}}$. Compare the two graphs. What conclusion can you draw from this comparison?

Groupwork

3. Working with a partner, simplify the following expressions:

   a. $\sqrt{x^3 + x^2 - x - 1}$

   b. $\dfrac{\sqrt{x^3 + 2x^2 - 9x - 18}}{\sqrt{x^2 + 5x + 6}}$

7.3 Addition and Subtraction of Radical Expressions

In contrast to other kinds of numbers such as fractions and decimals, sums and differences of many radicals cannot be simplified. For instance, there is no way to combine $\sqrt{5}$ and $\sqrt{2}$, and so we cannot simplify the expression $\sqrt{5} + \sqrt{2}$. Similarly, we cannot simplify the expression $\sqrt[5]{3} - 1$. However, we can approximate these expressions using a calculator.

Other radicals can be combined and simplified. In this section, we discuss how to add and subtract such radicals.

A To add or subtract radical expressions

B To solve applied problems involving the addition or subtraction of radical expressions

Combining Like Radicals

When adding or subtracting **like radicals**, we simplify the result. Radicals are said to be *like* if they have *the same index and also the same radicand*. For instance,

Same index

$7\sqrt[3]{5}$ and $4\sqrt[3]{5}$ are like radicals.

Same radicand

Different index

$\sqrt[3]{3}$ and $\sqrt[4]{3}$ are unlike radicals.

Same radicand

We use the distributive property to add or subtract like radicals, just as we do for adding or subtracting like terms.

Adding Like Terms

$$5x + 4x = (5 + 4)x = 9x$$

Like terms

Adding Like Radicals

$$5\sqrt{3} + 4\sqrt{3} = (5 + 4)\sqrt{3} = 9\sqrt{3}$$

Like radicals

Subtracting Like Terms

$$9n^3 - 5n^3 = (9 - 5)n^3 = 4n^3$$

Like terms

Subtracting Like Radicals

$$9\sqrt[3]{2} - 5\sqrt[3]{2} = (9 - 5)\sqrt[3]{2} = 4\sqrt[3]{2}$$

Like radicals

As these examples illustrate, we add or subtract like radicals by combining their coefficients, and then multiply this result by the radical.

EXAMPLE 1

Combine.

a. $7\sqrt[3]{10} + 4\sqrt[3]{10}$ **b.** $8\sqrt{x} - 3\sqrt{x} - 2\sqrt{x}$

c. $5\sqrt{2y + 1} - \sqrt{2y + 1}$ **d.** $\sqrt[3]{x} + 3\sqrt{x}$

Solution

a. $7\sqrt[3]{10} + 4\sqrt[3]{10} = (7 + 4)\sqrt[3]{10}$ Combine the coefficients using the distributive property.

$= 11\sqrt[3]{10}$ Simplify.

b. $8\sqrt{x} - 3\sqrt{x} - 2\sqrt{x} = (8 - 3 - 2)\sqrt{x} = 3\sqrt{x}$

c. $5\sqrt{2y + 1} - \sqrt{2y + 1} = (5 - 1)\sqrt{2y + 1} = 4\sqrt{2y + 1}$

d. $\sqrt[3]{x} + 3\sqrt{x}$

The terms cannot be combined because they are not like radicals.

PRACTICE 1

Add or subtract, as indicated.

a. $-\sqrt[4]{9} + 3\sqrt[4]{9}$

b. $6\sqrt{p} + \sqrt{p} - 2\sqrt{p}$

c. $-5\sqrt{t^2 - 4} + \sqrt{t^2 - 4}$

d. $5\sqrt[3]{a} - 2\sqrt[3]{b}$

> **TIP**   When adding or subtracting radicals, do not combine their radicands. For example, $\sqrt{5} + \sqrt{5} \neq \sqrt{10}$.

Combining Unlike Radicals

Some *unlike* radicals become *like* when they are simplified, and so can be combined.

EXAMPLE 2

Add or subtract, as indicated.

a. $\sqrt{48} + \sqrt{12}$

b. $5\sqrt{72} - \sqrt{50} + \sqrt{32}$

c. $\sqrt[3]{54} + 2\sqrt[3]{16} - 7\sqrt[3]{2}$

d. $\sqrt{45} - \sqrt{49} + \sqrt{20}$

Solution

a. $\sqrt{48} + \sqrt{12} = \sqrt{16 \cdot 3} + \sqrt{4 \cdot 3}$ Factor out perfect squares.

$\qquad = \sqrt{16}\sqrt{3} + \sqrt{4}\sqrt{3}$ Use the product rule of radicals.

$\qquad = 4\sqrt{3} + 2\sqrt{3}$ Take the square root of a perfect square.

$\qquad = (4 + 2)\sqrt{3}$ Use the distributive property.

$\qquad = 6\sqrt{3}$ Simplify.

b. $5\sqrt{72} - \sqrt{50} + \sqrt{32}$

$\qquad = 5\sqrt{36 \cdot 2} - \sqrt{25 \cdot 2} + \sqrt{16 \cdot 2}$ Factor out perfect squares.

$\qquad = 5 \cdot 6\sqrt{2} - 5\sqrt{2} + 4\sqrt{2}$ Use the product rule, and take the square root.

$\qquad = 30\sqrt{2} - 5\sqrt{2} + 4\sqrt{2}$ Simplify.

$\qquad = 29\sqrt{2}$ Combine like radicals.

c. $\sqrt[3]{54} + 2\sqrt[3]{16} - 7\sqrt[3]{2} = \sqrt[3]{27 \cdot 2} + 2\sqrt[3]{8 \cdot 2} - 7\sqrt[3]{2}$

$\qquad\qquad = 3\sqrt[3]{2} + 2 \cdot 2\sqrt[3]{2} - 7\sqrt[3]{2}$

$\qquad\qquad = (3 + 4 - 7)\sqrt[3]{2}$

$\qquad\qquad = 0$

d. $\sqrt{45} - \sqrt{49} + \sqrt{20} = \sqrt{9 \cdot 5} - \sqrt{49} + \sqrt{4 \cdot 5}$

$\qquad\qquad = 3\sqrt{5} - 7 + 2\sqrt{5}$

$\qquad\qquad = 5\sqrt{5} - 7$

PRACTICE 2

Combine.

a. $\sqrt{75} + \sqrt{27}$

b. $\sqrt{96} - 3\sqrt{54} - 7\sqrt{24}$

c. $9\sqrt[4]{2} - 4\sqrt[4]{162} - \sqrt[4]{32}$

d. $\sqrt{64} - \sqrt{28} + \sqrt{63}$

The radicands of radicals to be combined may contain one or more variables. For every radical, we assume that its value is a real number.

EXAMPLE 3

Simplify.

a. $\sqrt{9x} - 2\sqrt{x}$

b. $b\sqrt{a^2b} - 2a\sqrt{b}$

c. $\sqrt[3]{54t^4} + \sqrt[3]{-128t}$

Solution

a. $\sqrt{9x} - 2\sqrt{x} = 3\sqrt{x} - 2\sqrt{x}$

$\qquad\qquad = (3 - 2)\sqrt{x}$

$\qquad\qquad = \sqrt{x}$

b. $b\sqrt{a^2b} - 2a\sqrt{b} = b \cdot a\sqrt{b} - 2a\sqrt{b}$

$\qquad\qquad\quad = ab\sqrt{b} - 2a\sqrt{b}$

$\qquad\qquad\quad = (ab - 2a)\sqrt{b}$

c. $\sqrt[3]{54t^4} + \sqrt[3]{-128t} = \sqrt[3]{27 \cdot 2 \cdot t^3 \cdot t^1} + \sqrt[3]{(-64) \cdot 2 \cdot t^1}$

$\qquad\qquad\qquad\quad = 3t\sqrt[3]{2t} + (-4)\sqrt[3]{2t}$

$\qquad\qquad\qquad\quad = (3t - 4)\sqrt[3]{2t}$

PRACTICE 3

Combine.

a. $4\sqrt{25n} + \sqrt{n}$

b. $\sqrt[3]{27ab^3} - 11\sqrt[3]{a}$

c. $\sqrt{50x} - \sqrt{32x^3}$

EXAMPLE 4

If $f(x) = 10x\sqrt[4]{16x^3}$ and $g(x) = 2x\sqrt[4]{x^3}$, find:

a. $f(x) + g(x)$ $\qquad$ **b.** $f(x) - g(x)$

Solution

a. $f(x) + g(x) = 10x\sqrt[4]{16x^3} + 2x\sqrt[4]{x^3}$

$\qquad\qquad\quad = 10x \cdot 2\sqrt[4]{x^3} + 2x\sqrt[4]{x^3}$

$\qquad\qquad\quad = 20x\sqrt[4]{x^3} + 2x\sqrt[4]{x^3}$

$\qquad\qquad\quad = 22x\sqrt[4]{x^3}$

b. $f(x) - g(x) = 10x\sqrt[4]{16x^3} - 2x\sqrt[4]{x^3}$

$\qquad\qquad\quad = 20x\sqrt[4]{x^3} - 2x\sqrt[4]{x^3}$

$\qquad\qquad\quad = 18x\sqrt[4]{x^3}$

PRACTICE 4

If $p(x) = x\sqrt[3]{81x}$ and $q(x) = \sqrt[3]{24x^4}$, find:

a. $p(x) + q(x)$

b. $p(x) - q(x)$

As reviewed in Section 7.2, the Pythagorean theorem, which asserts that for every right triangle, the sum of the squares of the lengths of the two legs, a and b, equals the square of the length of the hypotenuse c. This relationship is usually expressed as $a^2 + b^2 = c^2$. However, we can also can write this equation as

$$c = \sqrt{a^2 + b^2}$$

EXAMPLE 5

The map to the right shows the route that a college recruiter takes in calling on colleges each week. She starts at home (A), visits colleges in towns B, C, and D in that order, and then returns home. ($ACBD$ is a rectangle.)

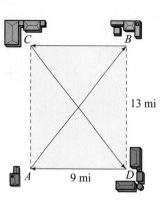

a. How long is the road from A to B? Express your answer as a radical.

b. What is the length of the recruiter's weekly trip?

Solution

a. Let x represent the length of $\overline{AB}$. The line segments $\overline{BD}$ and $\overline{DA}$ form the other two sides of a right triangle. The lengths of the sides are 13 mi and 9 mi. Using the Pythagorean theorem, we get:

$$x = \sqrt{(BD)^2 + (DA)^2}$$
$$x = \sqrt{13^2 + 9^2}$$
$$x = \sqrt{250}$$
$$x = \sqrt{25 \cdot 10}$$
$$x = 5\sqrt{10}$$

So the road from A to B is $5\sqrt{10}$ mi long.

b. The total length of the recruiter's trip is the sum of the lengths of the components of the trip:

$$\text{Trip length} = AB + BC + CD + DA$$
$$= 5\sqrt{10} + 9 + 5\sqrt{10} + 9$$
$$= 18 + 10\sqrt{10}$$

So the length of the recruiter's trip is $(18 + 10\sqrt{10})$ mi, or approximately 49.6 mi.

PRACTICE 5

The size of a television set is commonly given by the length of its diagonal. For the two sets pictured, find:

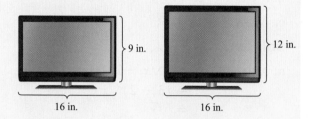

a. the lengths of their diagonals

b. the difference of the diagonal lengths

Mathematically Speaking

Fill in each blank with the most appropriate term or phrase from the given list.

unlike	simplified	reduced
indices	coefficients	like
area of a triangle formula	Pythagorean theorem	

1. Radicals that have the same index and the same radicand are called _____ radicals.

2. Like radicals can be added by combining their _____ and then multiplying this sum by the radical.

3. Some unlike radicals can become like radicals when they are _____.

4. The _____ can be expressed as $c = \sqrt{a^2 + b^2}$.

Ⓐ *Combine, if possible.*

5. $4\sqrt{3} + \sqrt{3}$

6. $2\sqrt{7} + 5\sqrt{7}$

7. $3\sqrt[3]{2} - 6\sqrt[3]{2}$

8. $11\sqrt[4]{6} - 9\sqrt[4]{6}$

9. $4\sqrt{2} - 2\sqrt{7}$

10. $10\sqrt[3]{5} + 6\sqrt{5}$

11. $-8\sqrt{y} - 12\sqrt{y}$

12. $-5\sqrt{x} - \sqrt{x}$

13. $7\sqrt{x} + 4\sqrt[3]{x}$

14. $2\sqrt{r} + 10\sqrt[3]{r}$

15. $-11\sqrt{n} + 2\sqrt{n} + 9\sqrt{n}$

16. $8\sqrt{a} - 13\sqrt{a} + 3\sqrt{a}$

17. $12\sqrt[4]{p} - 5\sqrt[4]{p} + \sqrt[4]{p}$

18. $-11\sqrt[3]{x} - \sqrt[3]{x} + 15\sqrt[3]{x}$

19. $9y\sqrt{3x} + 4y\sqrt{3x}$

20. $19a\sqrt{5b} - 10a\sqrt{5b}$

21. $-7\sqrt{2r - 1} + 3\sqrt{2r - 1}$

22. $8\sqrt{3y - 2} + 2\sqrt{3y - 2}$

23. $\sqrt{x^2 - 9} + \sqrt{x^2 - 9}$

24. $6\sqrt{a^2 - b^2} - 7\sqrt{a^2 - b^2}$

Simplify, if possible.

25. $\sqrt{6} + \sqrt{24}$

26. $\sqrt{45} + 4\sqrt{5}$

27. $\sqrt{72} - \sqrt{8}$

28. $\sqrt{12} - \sqrt{75}$

29. $3\sqrt{18} - 4\sqrt{2} + \sqrt{50}$

30. $-6\sqrt{28} + 2\sqrt{63} - 7\sqrt{7}$

31. $-5\sqrt[3]{24} + \sqrt[3]{-81} + 9\sqrt[3]{3}$

32. $7\sqrt[4]{4} - 3\sqrt[4]{64} - \sqrt[4]{324}$

33. $\sqrt{54x^3} - x\sqrt{150x}$

34. $y\sqrt{32y} - \sqrt{98y^3}$

35. $\dfrac{1}{4}\sqrt{128n^5} + \sqrt{242n}$

36. $\sqrt{490t} + \dfrac{1}{2}\sqrt{40t^3}$

37. $-4y\sqrt{xy^3} + 7x\sqrt{x^3y}$

38. $6a\sqrt{ab^5} - 9b\sqrt{a^3b}$

39. $\sqrt[3]{-125p^9} + p\sqrt[3]{-8p^6}$

40. $\sqrt[3]{-27r^6} + 4r\sqrt[3]{64r^3}$

41. $5\sqrt[4]{48ab^3} - 2\sqrt[4]{3a^5b^3}$

42. $10\sqrt[5]{2x^2y^7} + 4\sqrt[5]{64x^2y^2}$

43. $2\sqrt{125} + \dfrac{1}{6}\sqrt{144} - \dfrac{3}{4}\sqrt{80}$

44. $\dfrac{1}{2}\sqrt{128} - \sqrt{98} + \dfrac{1}{3}\sqrt{225}$

45. $\dfrac{3}{5}\sqrt[3]{-125} - \dfrac{1}{4}\sqrt[3]{48} + \dfrac{1}{2}\sqrt[3]{162}$

46. $-\dfrac{3}{2}\sqrt[3]{192} + 6\sqrt[3]{24} + \dfrac{2}{3}\sqrt[3]{-216}$

Given $f(x)$ and $g(x)$, find $f(x) + g(x)$ and $f(x) - g(x)$.

47. $f(x) = 5x\sqrt{20x}$ and $g(x) = 3\sqrt{5x^3}$

48. $f(x) = -\sqrt{108x^4}$ and $g(x) = x\sqrt{147x^2}$

49. $f(x) = 2x\sqrt[4]{64x}$ and $g(x) = -\sqrt[4]{4x^5}$

50. $f(x) = 3\sqrt[3]{6x^8}$ and $g(x) = -5x^2\sqrt[3]{162x^2}$

Mixed Practice

Simplify, if possible.

51. $-8\sqrt{24} - 3\sqrt{6} + 2\sqrt{150}$

52. $\sqrt[3]{16ab^4} - \sqrt[3]{54ab}$

53. $-9\sqrt{3b} - \sqrt{3b} + 5\sqrt{3b}$

54. $-6\sqrt{r} + 5\sqrt[3]{r}$

55. $\dfrac{1}{3}\sqrt{405k^3} - \sqrt{180k}$

56. Given $f(x) = -2x\sqrt{63x^2}$ and $g(x) = -9\sqrt{7x^4}$, find $f(x) + g(x)$ and $f(x) - g(x)$.

Applications

B *Solve.*

57. When an object is dropped from a given height, its velocity (in feet per second) when it has fallen h ft is given by the expression $\sqrt{64h}$. Find the difference in the velocity of an object when it has fallen 80 ft and when it has fallen 20 ft. Express the answer both as a radical in simplest form and as a decimal rounded to the nearest tenth.

58. The distance (in miles) a person can see to the horizon from an altitude of a ft is given by the expression $\sqrt{1.5a}$. How much farther can a person see from the window of an airplane flying at an altitude of 32,000 ft than from the window of an airplane flying at an altitude of 18,000 ft? Express the answer both as a radical in simplest form and as a decimal rounded to the nearest tenth.

59. The areas of two types of square floor tiles sold at a home improvement store are shown below. How much longer is the side of the larger tile? Express the answer as a radical in simplest form and as a decimal to the nearest tenth.

Area = 72 in²

Area = 128 in²

60. The Great Pyramid at Giza has a square base with an area of 52,900 m². What is the perimeter of the base?

61. The length of the ramp shown in the following figure can be determined using the expression $\sqrt{b^2 + h^2}$, where b is the length of the base of the ramp and h is the height of the ramp.

a. Find the length of the ramp if the height of the ramp is 2 ft and the base is 4 ft.

b. If the base of the ramp is increased by 2 ft, by approximately how many feet does the length of the ramp increase? (Round your answer to the nearest tenth.)

62. A guy wire of length l attached to the top of an antenna whose height is h is anchored to the ground a distance d from the center of the base of the antenna, as shown in the diagram to the right.

a. If the height of an antenna is 160 m, how far from the base's center will a 200-meter guy wire be anchored to the ground?

b. If the same length of wire is used on an antenna that is 175 m high, by approximately how many meters does the distance from the center of the antenna's base to the point where the guy wire is anchored decrease? (Round your answer to the nearest tenth.)

• Check your answers on page A-43.

MIND*Stretchers*

Critical Thinking

1. Find two radical expressions whose sum is $4\sqrt{2}$ and whose difference is $7\sqrt{2}$.

Groupwork

2. Working with a partner, consider the functions $f(x) = \sqrt{9x + 18}$, $g(x) = \sqrt{4x + 8}$, and $h(x) = 5\sqrt{x + 2}$.
   a. Evaluate $f(2)$, $g(2)$, and $h(2)$.
   b. Find the sum of $f(2)$ and $g(2)$, and compare it to the value of $h(2)$. What do you notice?

   c. Repeat parts (a) and (b) for several other values of x.
   d. From the result in parts (b) and (c), what conclusion is suggested about the relationship between $f(x)$, $g(x)$, and $h(x)$?
   e. Justify your answer to part (d) algebraically.

Mathematical Reasoning

3. For most values of a and b, $\sqrt{a} + \sqrt{b} \neq \sqrt{a + b}$. Are there any real numbers a and b such that $\sqrt{a} + \sqrt{b} = \sqrt{a + b}$? Justify your answer.

7.4 Multiplication and Division of Radical Expressions

The previous section covered the addition and subtraction of radical expressions. In this section, we discuss the multiplication and division of radical expressions.

A To multiply radical expressions

B To divide radical expressions

C To rationalize the denominator in a radical expression

D To solve applied problems involving the multiplication or division of radical expressions

Multiplying Radical Expressions

In considering how to find the product of radical expressions, let's begin with the simplest case—multiplying two radicals. The key is to apply the product rule of radicals discussed in Section 7.2.

$$\sqrt[n]{a} \cdot \sqrt[n]{b} = \sqrt[n]{ab}$$

EXAMPLE 1

Multiply and simplify.

a. $\sqrt{8n^3} \cdot \sqrt{3n}$ **b.** $(-5\sqrt[3]{7})(4\sqrt[3]{2})$

Solution

a. Here we use the product rule of radicals to multiply.

$$\sqrt{8n^3} \cdot \sqrt{3n} = \sqrt{8n^3 \cdot 3n}$$
$$= \sqrt{24n^4}$$
$$= \sqrt{4n^4 \cdot 6} \quad \text{Factor out a perfect square.}$$
$$= 2n^2\sqrt{6}$$

b. $(-5\sqrt[3]{7})(4\sqrt[3]{2}) = -5 \cdot 4 \cdot \sqrt[3]{7 \cdot 2}$
$$= -20\sqrt[3]{14}$$

PRACTICE 1

Find the product.

a. $\sqrt{6y} \cdot \sqrt{12y^7}$

b. $(-2\sqrt[3]{3})(7\sqrt[3]{5})$

Next, we consider how to multiply radical expressions that may contain more than one term. Just as with the multiplication of polynomials, the key here is to apply the distributive property.

EXAMPLE 2

Find the product and simplify.

a. $\sqrt{7}(3\sqrt{7} + 4)$ **b.** $\sqrt[3]{3}(5\sqrt[3]{2} - 1)$
c. $(\sqrt{3} - 1)(\sqrt{3} + 2)$ **d.** $(5\sqrt{n} - 1)(4\sqrt{n} + 3)$

Solution

a. $\sqrt{7}(3\sqrt{7} + 4) = \sqrt{7} \cdot 3\sqrt{7} + \sqrt{7} \cdot 4$ Use the distributive property.
$$= 21 + 4\sqrt{7} \quad \text{Simplify.}$$

b. $\sqrt[3]{3}(5\sqrt[3]{2} - 1) = \sqrt[3]{3} \cdot 5\sqrt[3]{2} - \sqrt[3]{3} \cdot 1$ Use the distributive property.
$$= 5\sqrt[3]{6} - \sqrt[3]{3} \quad \text{Use the product rule of radicals.}$$

PRACTICE 2

Multiply and simplify.

a. $\sqrt{5}(2\sqrt{5} - 4)$

b. $\sqrt[4]{5}(\sqrt[4]{7} + 6\sqrt[4]{2})$

c. $(\sqrt{5} + 4)(\sqrt{5} - 3)$

d. $(3\sqrt{x} - 1)(2\sqrt{x} + 1)$

First   Last

c. $(\sqrt{3} - 1)(\sqrt{3} + 2) = (\sqrt{3} - 1)(\sqrt{3} + 2)$

Inner

Outer

$$= (\sqrt{3})(\sqrt{3}) + 2\sqrt{3} - 1\sqrt{3} + (-1)(2) \qquad \text{Use the FOIL method.}$$
$$= 3 + 2\sqrt{3} - \sqrt{3} - 2 \qquad \text{Simplify.}$$
$$= \sqrt{3} + 1 \qquad \text{Combine like terms.}$$

First   Last

d. $(5\sqrt{n} - 1)(4\sqrt{n} + 3) = (5\sqrt{n} - 1)(4\sqrt{n} + 3)$

Inner

Outer

$$= (5\sqrt{n})(4\sqrt{n}) + (5\sqrt{n})3 + (-1)(4\sqrt{n}) + (-1)(3) \qquad \text{Use the FOIL method.}$$
$$= 5 \cdot 4 \cdot \sqrt{n^2} + 5 \cdot 3\sqrt{n} - 4\sqrt{n} - 3 \qquad \text{Multiply.}$$
$$= 20n + 15\sqrt{n} - 4\sqrt{n} - 3 \qquad \text{Simplify.}$$
$$= 20n + 11\sqrt{n} - 3 \qquad \text{Combine like terms.}$$

Recall that when multiplying the sum and difference of the same two terms, we can use the following formula:

$$(a + b)(a - b) = a^2 - b^2$$

One or more of the factors in this product may contain a radical, as in the next example.

EXAMPLE 3

Find the product.

a. $(\sqrt{7} + \sqrt{3})(\sqrt{7} - \sqrt{3})$ **b.** $(\sqrt{6n} + 1)(\sqrt{6n} - 1)$

c. $(2\sqrt{x} + 3\sqrt{y})(2\sqrt{x} - 3\sqrt{y})$

Solution

a. $(\sqrt{7} + \sqrt{3})(\sqrt{7} - \sqrt{3}) = (\sqrt{7})^2 - (\sqrt{3})^2$
$$= 7 - 3$$
$$= 4$$

b. $(\sqrt{6n} + 1)(\sqrt{6n} - 1) = (\sqrt{6n})^2 - 1^2 = 6n - 1$

c. $(2\sqrt{x} + 3\sqrt{y})(2\sqrt{x} - 3\sqrt{y}) = (2\sqrt{x})^2 - (3\sqrt{y})^2 = 4x - 9y$

PRACTICE 3

Multiply.

a. $(\sqrt{2} - \sqrt{5})(\sqrt{2} + \sqrt{5})$

b. $(\sqrt{2x} + 3)(\sqrt{2x} - 3)$

c. $(\sqrt{p} - 4\sqrt{q})(\sqrt{p} + 4\sqrt{q})$

When squaring binomials that contain radical terms, we can apply the following formulas for squaring a binomial:

$$(a + b)^2 = a^2 + 2ab + b^2$$
$$(a - b)^2 = a^2 - 2ab + b^2$$

EXAMPLE 4

Simplify.

a. $(\sqrt{5} + 3x)^2$ **b.** $(\sqrt{x+1} - 1)^2$

Solution

a. $(\sqrt{5} + 3x)^2 = (\sqrt{5})^2 + 2(\sqrt{5})(3x) + (3x)^2$

$\qquad\qquad\quad = 5 + 6\sqrt{5}x + 9x^2$

$\qquad\qquad\quad = 9x^2 + 6\sqrt{5}x + 5$

b. $(\sqrt{x+1} - 1)^2 = (\sqrt{x+1})^2 - 2(1)(\sqrt{x+1}) + 1$

$\qquad\qquad\qquad\quad = x + 1 - 2\sqrt{x+1} + 1$

$\qquad\qquad\qquad\quad = x - 2\sqrt{x+1} + 2$

PRACTICE 4

Simplify.

a. $(\sqrt{3} + 2b)^2$

b. $(\sqrt{n+1} - 3)^2$

EXAMPLE 5

In the right triangle shown, the length of hypotenuse AB is $\dfrac{2\sqrt{3}}{3}$ times the length of side AC. Find the length of AB expressed in simplest terms.

Solution

$$AB = \frac{2\sqrt{3}}{3} \cdot AC$$

$$= \frac{2\sqrt{3}}{3} \cdot \sqrt{5}$$

$$= \frac{2\sqrt{3}}{3} \cdot \frac{\sqrt{5}}{1}$$

$$= \frac{2\sqrt{3} \cdot \sqrt{5}}{3}$$

$$= \frac{2\sqrt{15}}{3}$$

So the length of AB is $\dfrac{2\sqrt{15}}{3}$.

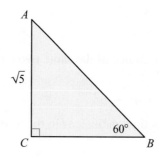

PRACTICE 5

Find the area of the rectangle shown, expressed as a simplified radical.

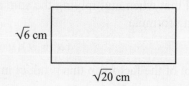

$\sqrt{6}$ cm

$\sqrt{20}$ cm

Dividing Radical Expressions

Now, let's consider division. To divide radical expressions, we use the quotient rule of radicals discussed in Section 7.2:

$$\frac{\sqrt[n]{a}}{\sqrt[n]{b}} = \sqrt[n]{\frac{a}{b}}$$

When dividing nth roots, this rule allows us to bring the radicands under a single radical sign. We can then divide the radicands. This approach is particularly useful if their quotient happens to be a perfect power.

EXAMPLE 6

Divide and simplify, if possible.

a. $\dfrac{3\sqrt{50}}{\sqrt{2}}$ **b.** $\dfrac{-x\sqrt[3]{24x^8}}{\sqrt[3]{6x^3}}$

Solution

a. $\dfrac{3\sqrt{50}}{\sqrt{2}} = 3\sqrt{\dfrac{50}{2}}$ Use the quotient rule of radicals.

$= 3\sqrt{25}$

$= 3 \cdot 5$

$= 15$

b. $\dfrac{-x\sqrt[3]{24x^8}}{\sqrt[3]{6x^3}} = -x\sqrt[3]{\dfrac{24x^8}{6x^3}}$

$= -x\sqrt[3]{4x^5}$

$= -x \cdot x\sqrt[3]{4x^2}$

$= -x^2\sqrt[3]{4x^2}$

PRACTICE 6

Find the quotient. Simplify, if possible.

a. $\dfrac{5\sqrt{12}}{\sqrt{3}}$

b. $\dfrac{\sqrt[3]{18p^7}}{-5p\sqrt[3]{2p^2}}$

A radical such as $\sqrt{\frac{1}{3}}$ is not considered to be simplified, since its radicand is in the form of a fraction. To simplify, we can express the single radical as the quotient of two radicals by applying the quotient rule of radicals "in reverse," which we can write as follows:

$$\sqrt[n]{\dfrac{a}{b}} = \dfrac{\sqrt[n]{a}}{\sqrt[n]{b}}$$

Radicands with perfect powers in the denominator particularly lend themselves to this approach.

EXAMPLE 7

Simplify.

a. $\sqrt{\dfrac{4p^{10}}{q^6}}$ **b.** $\sqrt[3]{\dfrac{5x^4}{8}}$

Solution

a. $\sqrt{\dfrac{4p^{10}}{q^6}} = \dfrac{\sqrt{4p^{10}}}{\sqrt{q^6}} = \dfrac{2p^5}{q^3}$ **b.** $\sqrt[3]{\dfrac{5x^4}{8}} = \dfrac{\sqrt[3]{5x^4}}{\sqrt[3]{8}} = \dfrac{x\sqrt[3]{5x}}{2}$

PRACTICE 7

Simplify.

a. $\sqrt{\dfrac{y^2}{9x^8}}$

b. $\sqrt[3]{\dfrac{2n^5}{27}}$

Rationalizing the Denominator

Some radical expressions are written as fractions with radicals in their denominators:

$$\dfrac{1}{\sqrt{5}} \qquad \dfrac{2x}{\sqrt[3]{x}} \qquad \dfrac{z}{\sqrt{1-z^2}}$$

This type of expression, which can be difficult to evaluate without a calculator, is not considered to be simplified. However, we can always **rationalize the denominator** of such an expression, that is, rewrite the expression in an equivalent form that contains no radical in its denominator. To do this, we multiply the numerator and denominator by a factor that will make the radicand in the denominator a perfect power.

EXAMPLE 8

Rationalize the denominator.

a. $\dfrac{2}{\sqrt{3}}$ **b.** $\dfrac{\sqrt{x}}{\sqrt{20}}$ **c.** $\dfrac{\sqrt{49y^6}}{\sqrt{12}}$ **d.** $\dfrac{\sqrt[3]{4a^2b}}{\sqrt[3]{9c}}$

Solution

a. $\dfrac{2}{\sqrt{3}} = \dfrac{2}{\sqrt{3}} \cdot \dfrac{\sqrt{3}}{\sqrt{3}}$ **Multiply the numerator and denominator by $\sqrt{3}$.**

$\qquad = \dfrac{2\sqrt{3}}{\sqrt{3^2}}$

$\qquad = \dfrac{2\sqrt{3}}{3}$

b. $\dfrac{\sqrt{x}}{\sqrt{20}} = \dfrac{\sqrt{x}}{\sqrt{20}} \cdot \dfrac{\sqrt{20}}{\sqrt{20}} = \dfrac{\sqrt{20x}}{\sqrt{20^2}} = \dfrac{2\sqrt{5x}}{20} = \dfrac{\sqrt{5x}}{10}$

c. $\dfrac{\sqrt{49y^6}}{\sqrt{12}} = \dfrac{7y^3}{\sqrt{12}}$

$\qquad = \dfrac{7y^3}{\sqrt{12}} \cdot \dfrac{\sqrt{12}}{\sqrt{12}}$

$\qquad = \dfrac{7y^3\sqrt{12}}{\sqrt{12^2}}$

$\qquad = \dfrac{7y^3(2\sqrt{3})}{12}$

$\qquad = \dfrac{7y^3\sqrt{3}}{6}$

d. $\dfrac{\sqrt[3]{4a^2b}}{\sqrt[3]{9c}} = \dfrac{\sqrt[3]{4a^2b}}{\sqrt[3]{3^2c}} \cdot \dfrac{\sqrt[3]{3c^2}}{\sqrt[3]{3c^2}} = \dfrac{\sqrt[3]{12a^2bc^2}}{\sqrt[3]{(3c)^3}} = \dfrac{\sqrt[3]{12a^2bc^2}}{3c}$

Note that we multiplied the numerator and the denominator by $\sqrt[3]{3c^2}$ to make the radicand in the denominator a perfect cube.

PRACTICE 8

Simplify.

a. $\dfrac{4}{\sqrt{7}}$

b. $\dfrac{\sqrt{y}}{\sqrt{32}}$

c. $\dfrac{\sqrt{4p^4}}{\sqrt{18}}$

d. $\dfrac{\sqrt[3]{5pq}}{\sqrt[3]{2r^2}}$

When the radicand of a radical expression is a fraction, we can simplify the radical by first applying the quotient rule, and then rationalizing the denominator.

EXAMPLE 9

Simplify.   **a.** $\sqrt{\dfrac{2a}{5b}}$ **b.** $\sqrt[3]{\dfrac{m}{2}}$

Solution

a. $\sqrt{\dfrac{2a}{5b}} = \dfrac{\sqrt{2a}}{\sqrt{5b}}$

$\qquad = \dfrac{\sqrt{2a}}{\sqrt{5b}} \cdot \dfrac{\sqrt{5b}}{\sqrt{5b}}$

$\qquad = \dfrac{\sqrt{10ab}}{5b}$

b. $\sqrt[3]{\dfrac{m}{2}} = \dfrac{\sqrt[3]{m}}{\sqrt[3]{2}}$

$\qquad = \dfrac{\sqrt[3]{m}}{\sqrt[3]{2}} \cdot \dfrac{\sqrt[3]{2^2}}{\sqrt[3]{2^2}}$

$\qquad = \dfrac{\sqrt[3]{4m}}{\sqrt[3]{2^3}} = \dfrac{\sqrt[3]{4m}}{2}$

PRACTICE 9

Simplify.

a. $\sqrt{\dfrac{x}{3y}}$

b. $\sqrt[3]{\dfrac{5}{a}}$

TIP Remember that a radical expression is considered simplified if
- no radicand has a factor that is a perfect nth power,
- no radicand contains a fraction, and
- the expression contains no denominator with a radical.

Some radical expressions are in the form of fractions, with more than one term in the numerator. Any denominator that contains a radical can be rationalized.

EXAMPLE 10

Rationalize the denominator.

a. $\dfrac{4\sqrt{10} - \sqrt{6}}{\sqrt{2}}$

b. $\dfrac{\sqrt{y} - 2}{\sqrt{y}}$

Solution

a.
$$\frac{4\sqrt{10} - \sqrt{6}}{\sqrt{2}} = \frac{4\sqrt{10} - \sqrt{6}}{\sqrt{2}} \cdot \frac{\sqrt{2}}{\sqrt{2}}$$
$$= \frac{(4\sqrt{10} - \sqrt{6})\sqrt{2}}{2}$$
$$= \frac{4\sqrt{10} \cdot \sqrt{2} - \sqrt{6} \cdot \sqrt{2}}{2}$$
$$= \frac{4\sqrt{20} - \sqrt{12}}{2}$$
$$= \frac{4(2\sqrt{5}) - 2\sqrt{3}}{2}$$
$$= 2(2\sqrt{5}) - \sqrt{3}$$
$$= 4\sqrt{5} - \sqrt{3}$$

b.
$$\frac{\sqrt{y} - 2}{\sqrt{y}} = \frac{\sqrt{y} - 2}{\sqrt{y}} \cdot \frac{\sqrt{y}}{\sqrt{y}}$$
$$= \frac{(\sqrt{y} - 2)\sqrt{y}}{y}$$
$$= \frac{\sqrt{y} \cdot \sqrt{y} - 2 \cdot \sqrt{y}}{y}$$
$$= \frac{y - 2\sqrt{y}}{y}$$

PRACTICE 10

Express in simplified form.

a. $\dfrac{\sqrt{6} - 9}{\sqrt{3}}$

b. $\dfrac{5 + \sqrt{b}}{\sqrt{b}}$

Can you solve the problem in Example 10(a) another way?

So far, we have rationalized denominators consisting of a single radical term. But suppose that the radical expression in a denominator contains two terms. Here, we consider the case in which the denominator involves one or more square roots, as in $\dfrac{1}{\sqrt{x} + 5}$. The key to rationalizing such a denominator is to identify its *conjugate*. The expressions $a + b$ and

$a - b$ are called **conjugates** of one another. When multiplying a pair of conjugates, we can apply the formula for the difference of squares:

$$(a + b)(a - b) = a^2 - b^2$$

Recall that in Example 3(a), we multiplied two conjugates, $(\sqrt{7} + \sqrt{3})$ and $(\sqrt{7} - \sqrt{3})$, and found the product to be 4. Can you explain why there is no radical sign in this product?

The elimination of radical signs in the product of conjugates suggests a procedure for rationalizing a denominator with two terms, namely *multiplying both the numerator and the denominator by the conjugate of the denominator.*

EXAMPLE 11

Rationalize the denominator.

a. $\dfrac{6}{1 - \sqrt{3}}$ **b.** $\dfrac{x}{\sqrt{y} + \sqrt{z}}$

Solution

a. We multiply the numerator and the denominator by the conjugate of the denominator.

$$\frac{6}{1 - \sqrt{3}} = \frac{6}{1 - \sqrt{3}} \cdot \frac{1 + \sqrt{3}}{1 + \sqrt{3}}$$

$$= \frac{6(1 + \sqrt{3})}{(1 - \sqrt{3})(1 + \sqrt{3})}$$

$$= \frac{6 + 6\sqrt{3}}{1^2 - (\sqrt{3})^2}$$

$$= \frac{6 + 6\sqrt{3}}{1 - 3}$$

$$= \frac{2(3 + 3\sqrt{3})}{-2}$$

$$= -3 - 3\sqrt{3}$$

b. $\dfrac{x}{\sqrt{y} + \sqrt{z}} = \dfrac{x}{\sqrt{y} + \sqrt{z}} \cdot \dfrac{\sqrt{y} - \sqrt{z}}{\sqrt{y} - \sqrt{z}}$

$$= \frac{x(\sqrt{y} - \sqrt{z})}{(\sqrt{y} + \sqrt{z})(\sqrt{y} - \sqrt{z})}$$

$$= \frac{x\sqrt{y} - x\sqrt{z}}{y - z}$$

PRACTICE 11

Express in simplified form.

a. $\dfrac{4}{2 + \sqrt{2}}$

b. $\dfrac{a}{\sqrt{b} - \sqrt{c}}$

EXAMPLE 12

If $f(x) = 10x\sqrt[4]{16x^3}$ and $g(x) = 2x\sqrt[4]{x^3}$, find:

a. $f(x) \cdot g(x)$ **b.** $\dfrac{f(x)}{g(x)}$

Solution

a. $f(x) \cdot g(x) = 10x\sqrt[4]{16x^3} \cdot 2x\sqrt[4]{x^3}$

$= 20x^2\sqrt[4]{16x^6}$

$= 20x^2\sqrt[4]{16x^4 \cdot x^2}$ **Factor out a perfect fourth power.**

$= 20x^2 \cdot 2x\sqrt[4]{x^2}$

$= 40x^3\sqrt[4]{x^2}$

b. $\dfrac{f(x)}{g(x)} = \dfrac{10x\sqrt[4]{16x^3}}{2x\sqrt[4]{x^3}} = 5\sqrt[4]{\dfrac{16x^3}{x^3}} = 5\sqrt[4]{16} = 5 \cdot 2 = 10$

PRACTICE 12

If $p(x) = x\sqrt[3]{81x}$ and $q(x) = \sqrt[3]{24x^4}$, determine:

a. $p(x) \cdot q(x)$

b. $\dfrac{p(x)}{q(x)}$

EXAMPLE 13

The expression $\sqrt{64h}$ can be used to determine the velocity (in feet per second) of a free-falling object when the object has fallen h ft.

a. Find the velocity of a ball after it has fallen 5 ft and after it has fallen 10 ft.

b. What is the ratio of the greater velocity to the smaller velocity?

Solution

a. When the ball has fallen 5 ft, its velocity is:

$$\sqrt{64 \cdot 5} = 8\sqrt{5}, \text{ or } 8\sqrt{5} \text{ ft/sec}$$

When the ball has fallen 10 ft, the velocity is:

$$\sqrt{64 \cdot 10} = 8\sqrt{10}, \text{ or } 8\sqrt{10} \text{ ft/sec}$$

b. The ratio of the greater velocity to the smaller velocity is:

$$\frac{8\sqrt{10}}{8\sqrt{5}} = \frac{\sqrt{10}}{\sqrt{5}}$$

$$= \sqrt{\frac{10}{5}}$$

$$= \sqrt{2}$$

$$\approx 1.414$$

So the ratio of the velocities is approximately 1.414.

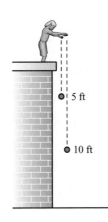

PRACTICE 13

The following diagram shows a sphere of volume V:

The radius of the sphere can be modeled by the expression $\dfrac{\sqrt[3]{3V}}{\sqrt[3]{4\pi}}$.

a. Rewrite this expression, rationalizing the denominator.

b. If the volume is doubled, is the radius doubled?

Mathematically Speaking

Fill in each blank with the most appropriate term or phrase from the given list.

numerator	denominator	squaring a binomial
perfect power	constant	multiply
product	rationalize	
commutative	distributive	
the sum and difference of the same two terms	quotient	

1. To multiply radical expressions containing more than one term, we use the _____ property.

2. When squaring binomials that contain radical terms, apply the formulas for _____.

3. To simplify a radical that has its radicand in the form of a fraction, we can use the _____ rule of radicals "in reverse."

4. To _____ the denominator means to rewrite an expression that has a radical in its denominator as an equivalent expression that does not have one.

5. A denominator can be rationalized by multiplying the numerator and denominator by a factor that makes the radicand in the denominator a(n) _____.

6. To rationalize a denominator with two terms, multiply both the numerator and the denominator by the conjugate of the _____.

A *Multiply and simplify.*

7. $\sqrt{12} \cdot \sqrt{8}$

8. $\sqrt{6} \cdot \sqrt{3}$

9. $(-4\sqrt{3})(\sqrt{7})$

10. $(\sqrt{11})(-6\sqrt{2})$

11. $(5\sqrt[3]{6})(3\sqrt[3]{9})$

12. $(2\sqrt[3]{4})(4\sqrt[3]{14})$

13. $(2\sqrt{10x})(7\sqrt{5x^5})$

14. $(4\sqrt{12y^5})(3\sqrt{6y^3})$

15. $(8\sqrt{ab^3})(-2\sqrt{a^3b})$

16. $(-6\sqrt{p^3q^5})(-5\sqrt{pq})$

17. $\sqrt[3]{12x^2y} \cdot \sqrt[3]{-16xy^4}$

18. $-\sqrt[4]{8x^3y^2} \cdot \sqrt[4]{50x^2y^6}$

19. $\sqrt{2}(\sqrt{8} - 4)$

20. $\sqrt{5}(\sqrt{20} + 3)$

21. $\sqrt{6}(2\sqrt{3} + \sqrt{12})$

22. $\sqrt{3}(\sqrt{8} - 5\sqrt{2})$

23. $-2\sqrt{3}(2\sqrt{5} - 6\sqrt{3})$

24. $-3\sqrt{2}(2\sqrt{7} - 4\sqrt{2})$

25. $\sqrt[3]{4}(5\sqrt[3]{12} + 2\sqrt[3]{3})$

26. $-\sqrt[4]{8}(\sqrt[4]{8} - 7\sqrt[4]{5})$

27. $\sqrt{x}(\sqrt{x^3} + \sqrt{2x})$

28. $\sqrt{y}(\sqrt{3y} - \sqrt{y})$

29. $\sqrt[4]{a^2}(3\sqrt[4]{2a^3} - \sqrt[4]{10a^2})$

30. $\sqrt[3]{2x^2}(7\sqrt[3]{4x} - \sqrt[3]{5})$

31. $(\sqrt{2} - 3)(\sqrt{2} + 4)$

32. $(\sqrt{5} + 1)(\sqrt{5} - 6)$

33. $(2 - 4\sqrt{3})(4 + 3\sqrt{3})$

34. $(-2\sqrt{5} + 1)(2\sqrt{5} - 1)$

35. $(\sqrt{8} + \sqrt{3})(\sqrt{2} + \sqrt{12})$

36. $(\sqrt{5} + \sqrt{10})(\sqrt{10} + \sqrt{20})$

37. $(2\sqrt{r} - 4)(8\sqrt{r} + 6)$

38. $(2\sqrt{p} + 7)(3\sqrt{p} - 5)$

39. $(\sqrt{6} + \sqrt{2})(\sqrt{6} - \sqrt{2})$

40. $(\sqrt{8} + \sqrt{3})(\sqrt{8} - \sqrt{3})$

41. $(\sqrt{x} - 8)(\sqrt{x} + 8)$

42. $(\sqrt{n} + 11)(\sqrt{n} - 11)$

43. $(\sqrt{5x} + \sqrt{y})(\sqrt{5x} - \sqrt{y})$

44. $(\sqrt{7q} - \sqrt{p})(\sqrt{7q} + \sqrt{p})$

45. $(\sqrt{x-1} - 5)(\sqrt{x-1} + 5)$

46. $(\sqrt{y+3} - 4)(\sqrt{y+3} + 4)$

47. $(3\sqrt{x} - 2)^2$

48. $(3 - 2\sqrt{a})^2$

49. $(\sqrt{6a} + 4)^2$

50. $(\sqrt{8a} + 2)^2$

51. $(1 - \sqrt{n+7})^2$

52. $(3 - \sqrt{x-6})^2$

B *Divide and simplify.*

53. $-\dfrac{\sqrt{32}}{4\sqrt{2}}$

54. $-\dfrac{\sqrt{108}}{2\sqrt{3}}$

55. $\dfrac{\sqrt{84x^3}}{\sqrt{7x}}$

56. $\dfrac{\sqrt{48a^4}}{\sqrt{3a^3}}$

57. $-\dfrac{n\sqrt{60n^7}}{2\sqrt{5n^3}}$

58. $-\dfrac{y\sqrt{108y^5}}{3\sqrt{6y}}$

59. $\dfrac{\sqrt[4]{128r^{10}}}{6r\sqrt[4]{4r^3}}$

60. $\dfrac{\sqrt[3]{256x^8}}{8x\sqrt[3]{2x^4}}$

61. $\sqrt{\dfrac{5a}{16}}$

62. $\sqrt{\dfrac{2x}{49}}$

63. $\sqrt{\dfrac{12x}{y^6}}$

64. $\sqrt{\dfrac{18u^2}{v^4}}$

65. $\sqrt[3]{\dfrac{54}{n^9}}$

66. $\sqrt[3]{\dfrac{r^5}{64}}$

67. $\sqrt{\dfrac{32a^5}{9b^8}}$

68. $\sqrt{\dfrac{75p^3}{81q^2}}$

C *Simplify.*

69. $\dfrac{4}{\sqrt{5}}$

70. $\dfrac{3}{\sqrt{6}}$

71. $\dfrac{\sqrt{y}}{\sqrt{40}}$

72. $\dfrac{\sqrt{u}}{\sqrt{63}}$

73. $\dfrac{\sqrt{3a}}{\sqrt{18}}$

74. $\dfrac{\sqrt{5p}}{\sqrt{24}}$

75. $\dfrac{\sqrt{9u}}{6\sqrt{v}}$

76. $\dfrac{\sqrt{36p}}{4\sqrt{q}}$

77. $\dfrac{\sqrt{25x^4}}{\sqrt{3y}}$

78. $\dfrac{\sqrt{64a^8}}{\sqrt{5b}}$

79. $\dfrac{\sqrt{10xy}}{\sqrt{9z}}$

80. $\dfrac{\sqrt{7pq}}{\sqrt{4r}}$

81. $\dfrac{\sqrt[3]{9x}}{\sqrt[3]{2y^2}}$

82. $\dfrac{\sqrt[3]{2a}}{\sqrt[3]{3b}}$

83. $\sqrt{\dfrac{11}{x}}$

84. $\sqrt{\dfrac{3}{n}}$

85. $\sqrt{\dfrac{7x}{3y}}$

86. $\sqrt{\dfrac{5p}{2q}}$

87. $\sqrt{\dfrac{5x^3}{48y^3}}$

88. $\sqrt{\dfrac{3u^4}{98v^5}}$

89. $\sqrt[3]{\dfrac{a^2b}{32c^4}}$

90. $\sqrt[3]{\dfrac{p^3q^2}{81r^2}}$

91. $\dfrac{2 - \sqrt{3}}{\sqrt{6}}$

92. $\dfrac{\sqrt{2} - 1}{\sqrt{10}}$

93. $\dfrac{\sqrt{a} - \sqrt{b}}{\sqrt{b}}$

94. $\dfrac{\sqrt{x} + \sqrt{y}}{\sqrt{x}}$

95. $\dfrac{\sqrt{5} + 10\sqrt{t}}{\sqrt{15t}}$

96. $\dfrac{4\sqrt{3n} - \sqrt{2}}{\sqrt{6n}}$

97. $\dfrac{\sqrt[3]{x} - 4}{\sqrt[3]{x^2}}$

98. $\dfrac{3 - \sqrt[3]{a^2}}{\sqrt[3]{a}}$

Rationalize the denominator.

99. $\dfrac{1}{2 + \sqrt{2}}$

100. $\dfrac{3}{\sqrt{5} + 1}$

101. $\dfrac{6}{\sqrt{2} - \sqrt{5}}$

102. $\dfrac{12}{\sqrt{3} - \sqrt{6}}$

103. $\dfrac{8}{2 + \sqrt{2x}}$

104. $\dfrac{9}{6 - \sqrt{3y}}$

105. $\dfrac{\sqrt{x}}{\sqrt{x} + y}$

106. $\dfrac{\sqrt{b}}{a - \sqrt{b}}$

107. $\dfrac{\sqrt{a}+3}{\sqrt{a}-\sqrt{2}}$

108. $\dfrac{\sqrt{n}-1}{\sqrt{n}-\sqrt{5}}$

109. $\dfrac{\sqrt{x}-\sqrt{y}}{\sqrt{x}+\sqrt{y}}$

110. $\dfrac{\sqrt{p}+\sqrt{q}}{\sqrt{p}-\sqrt{q}}$

111. $\dfrac{2\sqrt{a}+3\sqrt{b}}{3\sqrt{a}-2\sqrt{b}}$

112. $\dfrac{4\sqrt{x}+\sqrt{y}}{\sqrt{x}-4\sqrt{y}}$

For the given functions $f(x)$ and $g(x)$, find $f(x)\cdot g(x)$ and $\dfrac{f(x)}{g(x)}$.

113. $f(x)=3x\sqrt{2x}$ and $g(x)=\dfrac{1}{3}\sqrt{6x}$

114. $f(x)=4\sqrt[3]{9x^2}$ and $g(x)=x\sqrt[3]{12x^2}$

115. $f(x)=\sqrt{x}+1$ and $g(x)=\sqrt{x}-1$

116. $f(x)=\sqrt{x}+2$ and $g(x)=\sqrt{x}+4$

Mixed Practice

Simplify.

117. $(\sqrt{12}+\sqrt{3})(\sqrt{3}+\sqrt{6})$

118. $\dfrac{3\sqrt{2r}-\sqrt{5}}{\sqrt{10r}}$

119. $(-8\sqrt{ab^3})(7\sqrt{a^5b})$

120. $-\sqrt[3]{9}(4\sqrt[3]{18}-7\sqrt[3]{15})$

121. $\sqrt{\dfrac{75y}{z^4}}$

122. $(\sqrt{x+6}+3)(\sqrt{x+6}-3)$

123. $\sqrt{\dfrac{5x}{7y}}$

124. $\dfrac{\sqrt{81m^6}}{\sqrt{5n}}$

Solve.

125. Rationalize the denominator: $\dfrac{\sqrt{p}-4}{\sqrt{p}-\sqrt{3}}$

126. For the functions $f(x)=2x\sqrt{7x}$ and $g(x)=\dfrac{1}{2}\sqrt{14x}$, find $f(x)\cdot g(x)$ and $\dfrac{f(x)}{g(x)}$.

Applications

Ⓓ *Solve.*

127. The altitude of an equilateral triangle with side length a bisects the base of the triangle, as shown in the diagram to the right. (The area of a triangle equals one-half the product of its base and height.)

a. Write a formula for the area A of an equilateral triangle.

b. Use your formula from part (a) to find the area of a wall tile in the shape of an equilateral triangle with side length 8 in.

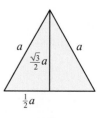

128. The length of the diagonal of a square with side length s is $s\sqrt{2}$. A circle is circumscribed about a square as shown in the figure to the right.

a. Express the area A of the circle in terms of s (The area of a circle with radius r is πr^2.).

b. Use your answer from part (a) to find the area of a circle circumscribed about a square with side length 4 in.

129. The period of a simple pendulum L ft long can be found using the expression $2\pi\sqrt{\dfrac{L}{32}}$. Simplify the expression by rationalizing the denominator.

130. The formula for the distance a person can see on any planet is given by the expression $\dfrac{\sqrt{rh}}{4\sqrt{165}}$, where r is the radius of the planet (in miles) and h is the height of the person (in feet). Simplify the expression by rationalizing the denominator.

131. The radius of a sphere with a surface area of A is given by $\sqrt{\dfrac{A}{4\pi}}$. Simplify the expression by rationalizing the denominator.

132. The annual rate of return r on an initial investment of P dollars whose value is V dollars after 3 yr is given by the following expression:

$$\frac{\sqrt[3]{V} - \sqrt[3]{P}}{\sqrt[3]{P}}$$

Rationalize the denominator.

• Check your answers on page A-43.

MINDStretchers

Critical Thinking

1. Some equations have solutions that contain radicals. Show that $3 - 2\sqrt{2}$ is a solution of the equation $x^2 - 6x + 1 = 0$.

Groupwork

2. Working with a partner, simplify each of the following expressions:

a. $\dfrac{\dfrac{x}{\sqrt{y}}}{\dfrac{y}{\sqrt{x}}}$

b. $\dfrac{\sqrt{x} - \dfrac{1}{\sqrt{x}}}{\sqrt{x} + \dfrac{1}{\sqrt{x}}}$

c. $\dfrac{\dfrac{1}{1 - \sqrt{x}} + \dfrac{1}{1 + \sqrt{x}}}{\dfrac{1}{1 + \sqrt{x}} - \dfrac{1}{1 - \sqrt{x}}}$

Technology

3. According to the product rule of radicals, $\sqrt{x + 2} \cdot \sqrt{x - 2} = \sqrt{x^2 - 4}$.

a. Using a grapher, graph $y = \sqrt{x + 2} \cdot \sqrt{x - 2}$ and $y = \sqrt{x^2 - 4}$ on separate coordinate planes.

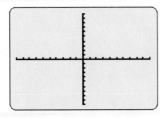 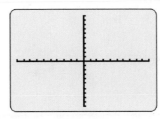

b. Explain why the graphs of the equations in part (a) are different.

7.5   Solving Radical Equations

Next, we consider equations containing radicals.

> **DEFINITION**
>
> A **radical equation** is an equation in which a variable appears in one or
> more radicands.

Some examples of radical equations are:

$$\sqrt{5x} = 18 \qquad \sqrt{3n} = \sqrt{2n-1} \qquad \sqrt[3]{y+2} - 6y = 5$$

Note that $x = \sqrt{5}$ is *not* a radical equation, since there is no variable under the radical sign.

In solving radical equations, we can use the *Power Property of Equality*.

> ## The Power Property of Equality
>
> For any real numbers a and b and any positive integer n, if $a = b$, then $a^n = b^n$.

To use this property to solve radical equations, we first isolate the radical, and then raise each side of the equation to the *same power* in order to produce a new equation that contains no radicals. When each side of the equation is raised to the same power, the solutions of the original equation are among the solutions of the new equation.

Consider, for instance, the equation $\sqrt{x} = 3$. If we square each side of the equation, that is, raise each side to the second power, we get:

$$\sqrt{x} = 3$$
$$(\sqrt{x})^2 = (3)^2 \qquad \text{Use the power property of equality (square).}$$
$$x = 9 \qquad \text{Simplify.}$$

When raising each side of the equation to the same power, the resulting equation may have one or more solutions that do not satisfy the original equation, that is, extraneous solutions. So when solving a radical equation, it is particularly important to check all possible solutions in the *original* equation.

For instance, we can check the solution of the equation in the preceding example by substituting 9 for x in the original equation.

$$\sqrt{x} = 3$$
$$\sqrt{9} \stackrel{?}{=} 3$$
$$3 = 3 \qquad \text{True}$$

So the solution of the equation is 9.

Let's consider some other examples of solving a radical equation.

EXAMPLE 1

Solve and check: $\sqrt{x} - 3 = 7$

Solution

$$\sqrt{x} - 3 = 7$$

$\sqrt{x} - 3 + 3 = 7 + 3$ Add 3 to each side of the equation.

$\sqrt{x} = 10$ Simplify.

$(\sqrt{x})^2 = (10)^2$ Use the power property of equality (square).

$x = 100$

Check

$$\sqrt{x} - 3 = 7$$

$\sqrt{100} - 3 \stackrel{?}{=} 7$ Substitute 100 for x.

$10 - 3 \stackrel{?}{=} 7$

$7 = 7$ True

So 100 is the solution.

PRACTICE 1

Solve and check: $\sqrt{y} + 2 = 10$

Some radical equations involve cube roots, or, more generally, radicals with an index greater than 2.

EXAMPLE 2

Solve and check: $\sqrt[3]{n - 5} = 2$

Solution   To solve this equation, we cube each side of the equation, that is, raise each side of the equation to the third power.

$$\sqrt[3]{n - 5} = 2$$

$(\sqrt[3]{n - 5})^3 = (2)^3$ Use the power property of equality (cube).

$n - 5 = 8$

$n = 13$

Check

$$\sqrt[3]{n - 5} = 2$$

$$\sqrt[3]{13 - 5} \stackrel{?}{=} 2$$

$$\sqrt[3]{8} \stackrel{?}{=} 2$$

$$2 = 2 \quad \text{True}$$

So 13 is the solution.

PRACTICE 2

Solve and check: $\sqrt[3]{2x + 7} = 3$

TIP   When solving a radical equation, raise each side of the equation to a power that is equal to the index of the radical in the equation.

EXAMPLE 3

Solve and check: $\sqrt{x+4} + 2 = 0$

Solution

$$\sqrt{x+4} + 2 = 0$$
$$\sqrt{x+4} = -2$$
$$(\sqrt{x+4})^2 = (-2)^2$$
$$x + 4 = 4$$
$$x = 0$$

Check

$$\sqrt{x+4} + 2 = 0$$
$$\sqrt{0+4} + 2 \overset{?}{=} 0 \qquad \text{Substitute 0 for } x.$$
$$\sqrt{4} + 2 \overset{?}{=} 0$$
$$2 + 2 \overset{?}{=} 0$$
$$4 = 0 \qquad \text{False}$$

Since our check fails, 0 is *not* a solution to the original equation. Note that 0 is an extraneous solution. Therefore, the original equation has no solution.

PRACTICE 3

Solve and check: $\sqrt{3t+1} + 4 = 0$

Some radical equations contain more than one radical. We solve by first isolating either of the radicals.

EXAMPLE 4

Solve and check: $\sqrt{2x+11} - \sqrt{4x+1} = 0$

Solution

$$\sqrt{2x+11} - \sqrt{4x+1} = 0$$

$$\sqrt{2x+11} = \sqrt{4x+1} \qquad \text{Isolate the radical on the left.}$$

$$(\sqrt{2x+11})^2 = (\sqrt{4x+1})^2 \qquad \text{Use the power property of equality (square).}$$

$$2x + 11 = 4x + 1 \qquad \text{Simplify.}$$

$$2x = 10$$

$$x = 5$$

Check

$$\sqrt{2x+11} - \sqrt{4x+1} = 0$$
$$\sqrt{2(5)+11} - \sqrt{4(5)+1} \overset{?}{=} 0 \qquad \text{Substitute 5 for } x.$$
$$\sqrt{21} - \sqrt{21} \overset{?}{=} 0$$
$$0 = 0 \qquad \text{True}$$

So 5 is the solution.

PRACTICE 4

Solve and check:
$$\sqrt{4n+5} - \sqrt{7n-4} = 0$$

In some radical equations, we must raise each side of the equation to a power more than once.

EXAMPLE 5

Solve and check: $\sqrt{2x} - \sqrt{x-2} = 2$

Solution

$$\sqrt{2x} - \sqrt{x-2} = 2$$

$\sqrt{2x} = \sqrt{x-2} + 2$	Isolate one of the radicals.
$(\sqrt{2x})^2 = (\sqrt{x-2} + 2)^2$	Use the power property of equality (square).
$2x = (x-2) + 2\cdot2\sqrt{x-2} + 2\cdot2$	Simplify.
$2x = x + 2 + 4\sqrt{x-2}$	Combine like terms.
$2x - (x+2) = x + 2 + 4\sqrt{x-2} - (x+2)$	Subtract $(x+2)$ from each side of the equation.
$x - 2 = 4\sqrt{x-2}$	Simplify.
$(x-2)^2 = (4\sqrt{x-2})^2$	Use the power property of equality (square).
$x^2 - 4x + 4 = 16(x-2)$	Simplify.
$x^2 - 4x + 4 = 16x - 32$	Use the distributive property.
$x^2 - 20x + 36 = 0$	Write in standard form.
$(x-18)(x-2) = 0$	Factor.

$$x - 18 = 0 \quad \text{or} \quad x - 2 = 0$$
$$x = 18 \qquad\qquad x = 2$$

Check

Substitute 18 for x.

$$\sqrt{2x} - \sqrt{x-2} = 2$$
$$\sqrt{2(18)} - \sqrt{18-2} \stackrel{?}{=} 2$$
$$\sqrt{36} - \sqrt{16} \stackrel{?}{=} 2$$
$$6 - 4 \stackrel{?}{=} 2$$
$$2 = 2 \qquad \text{True}$$

Substitute 2 for x.

$$\sqrt{2x} - \sqrt{x-2} = 2$$
$$\sqrt{2(2)} - \sqrt{2-2} \stackrel{?}{=} 2$$
$$\sqrt{4} - \sqrt{0} \stackrel{?}{=} 2$$
$$2 - 0 \stackrel{?}{=} 2$$
$$2 = 2 \qquad \text{True}$$

So both 18 and 2 are solutions.

PRACTICE 5

Solve and check: $\sqrt{2y+3} = \sqrt{y-2} + 2$

The preceding examples lead us to the following procedure for solving radical equations:

To Solve a Radical Equation

- Isolate a term with a radical.
- Raise each side of the equation to a power equal to the index of the radical.
- If the equation still contains a radical, repeat the preceding steps.
- Where possible, combine like terms.
- Solve the resulting equation.
- Check the possible solution(s) in the original equation.

EXAMPLE 6

Solve and check: $\sqrt[3]{x^2 - 5} + \sqrt[3]{2x - 3} = 0$

Solution

$$\sqrt[3]{x^2 - 5} + \sqrt[3]{2x - 3} = 0$$

$\sqrt[3]{x^2 - 5} = -\sqrt[3]{2x - 3}$ Isolate one of the radicals.

$(\sqrt[3]{x^2 - 5})^3 = (-\sqrt[3]{2x - 3})^3$ Use the power property of equality (cube).

$x^2 - 5 = -(2x - 3)$ Simplify.

$x^2 - 5 = -2x + 3$ Use the distributive property.

$x^2 + 2x - 8 = 0$ Write in standard form.

$(x - 2)(x + 4) = 0$ Factor.

$x - 2 = 0$ or $x + 4 = 0$

$x = 2$ $x = -4$

Check

Substitute 2 for x.   Substitute -4 for x.

$$\sqrt[3]{x^2 - 5} + \sqrt[3]{2x - 3} = 0 \qquad\qquad \sqrt[3]{x^2 - 5} + \sqrt[3]{2x - 3} = 0$$

$$\sqrt[3]{(2)^2 - 5} + \sqrt[3]{2(2) - 3} \stackrel{?}{=} 0 \qquad \sqrt[3]{(-4)^2 - 5} + \sqrt[3]{2(-4) - 3} \stackrel{?}{=} 0$$

$$\sqrt[3]{4 - 5} + \sqrt[3]{4 - 3} \stackrel{?}{=} 0 \qquad\qquad \sqrt[3]{16 - 5} + \sqrt[3]{-8 - 3} \stackrel{?}{=} 0$$

$$\sqrt[3]{-1} + \sqrt[3]{1} \stackrel{?}{=} 0 \qquad\qquad\qquad \sqrt[3]{11} + \sqrt[3]{-11} \stackrel{?}{=} 0$$

$$-1 + 1 \stackrel{?}{=} 0 \qquad\qquad\qquad\qquad \sqrt[3]{11} + \sqrt[3]{(-1)11} \stackrel{?}{=} 0$$

$$0 = 0 \quad \text{True} \qquad\qquad\qquad\qquad \sqrt[3]{11} - 1\sqrt[3]{11} \stackrel{?}{=} 0$$

$$0 = 0 \quad \text{True}$$

So both 2 and -4 are solutions.

PRACTICE 6

Solve and check: $\sqrt[3]{n^2 + 8} + \sqrt[3]{4 - 7n} = 0$

EXAMPLE 7

Solve and check: $\sqrt{x + 1} + 1 = -x$

Solution

$$\sqrt{x + 1} + 1 = -x$$

$\sqrt{x + 1} = -x - 1$ Isolate the radical.

$(\sqrt{x + 1})^2 = (-x - 1)^2$ Use the power property of equality (square).

$x + 1 = (-x)^2 - (-2x) + (-1)^2$

$x + 1 = x^2 + 2x + 1$

$x + 1 + (-x - 1) = x^2 + 2x + 1 + (-x - 1)$ Add $(-x - 1)$ to each side of the equation.

$0 = x^2 + x$

$x(x + 1) = 0$ Factor.

$x = 0$ or $x + 1 = 0$

$x = -1$

EXAMPLE 7 (continued)

Check

Substitute 0 for x.

$$\sqrt{x + 1} + 1 = -x$$
$$\sqrt{0 + 1} + 1 \overset{?}{=} 0$$
$$1 + 1 \overset{?}{=} 0$$
$$2 = 0 \quad \text{False}$$

Substitute -1 for x.

$$\sqrt{x + 1} + 1 = -x$$
$$\sqrt{-1 + 1} + 1 \overset{?}{=} -(-1)$$
$$0 + 1 \overset{?}{=} 1$$
$$1 = 1 \quad \text{True}$$

So the solution is -1.

PRACTICE 7

Solve and check: $\sqrt{y + 4} - y = 2$

Recall from Section 2.2 our discussion of solving a formula for one variable. Some of these formulas contain radicals.

EXAMPLE 8

The period of a pendulum is the time that it takes for a pendulum to swing back and forth. A pendulum L ft long will have a period of t sec, where

$$t = 2\pi\sqrt{\frac{L}{32}}$$

Rewrite this formula by solving for L.

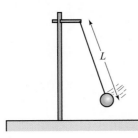

Solution

$$t = 2\pi\sqrt{\frac{L}{32}}$$

$$\frac{t}{2\pi} = \sqrt{\frac{L}{32}}$$

$$\left(\frac{t}{2\pi}\right)^2 = \left(\sqrt{\frac{L}{32}}\right)^2$$

$$\frac{t^2}{4\pi^2} = \frac{L}{32}$$

$$(32)\left(\frac{t^2}{4\pi^2}\right) = L$$

$$L = \frac{8t^2}{\pi^2}$$

PRACTICE 8

For an electrical appliance, such as a toaster, the formula

$$I = \sqrt{\frac{P}{R}}$$

shows the relationship between its resistance R, the amount of current I that it draws, and the power P that it consumes. Rewrite this formula by solving for P. (*Source*: Peter J. Nolan, *Fundamentals of College Physics*)

EXAMPLE 9

The formula

$$r = \sqrt{\frac{A}{4\pi}}$$

gives the radius of a sphere r in terms of the surface area A of the sphere.

a. Rewrite this formula by solving for A.

b. If the shape of Earth is approximately spherical with a radius of 3959 mi, find its surface area to the nearest 100 million mi^2. (Use 3.14 for π.) (*Source: The World Almanac and Book of Facts 2011*)

Solution

a. $r = \sqrt{\dfrac{A}{4\pi}}$

$r^2 = \dfrac{A}{4\pi}$

$A = 4\pi r^2$

b. $A = 4\pi r^2$

$= 4(3.14)(3959)^2 \approx 196,861,433.4$

We conclude that the surface area of Earth is approximately $200,000,000\ mi^2$.

PRACTICE 9

The average annual interest rate r earned in a 3-year investment can be represented by

$$r = \sqrt[3]{\frac{A}{P}} - 1,$$

where P is the initial investment and A is the value of the investment after 3 yr.

a. Rewrite this formula by solving for A.

b. If the interest rate was 0.03 (that is, 3%) on an initial investment of $10,000, find the value after 3 yr.

A *Solve and check.*

1. $\sqrt{3n} = 6$

2. $\sqrt{2a} = 10$

3. $\sqrt{x + 6} = 3$

4. $\sqrt{x - 7} = 4$

5. $\sqrt{5x - 6} = 2$

6. $\sqrt{4x + 1} = 5$

7. $\sqrt[3]{3y + 10} = -2$

8. $\sqrt[3]{12 - 5x} = 3$

9. $\sqrt{x} + 9 = 8$

10. $\sqrt{x} + 10 = 1$

11. $\sqrt{x} - 20 = -9$

12. $\sqrt{x} - 8 = 7$

13. $14 - \sqrt[3]{x} = 11$

14. $\sqrt[3]{n} + 15 = 13$

15. $\sqrt{6x} + 17 = 29$

16. $\sqrt{3n} - 11 = -2$

17. $\sqrt{2y - 1} - 8 = 5$

18. $\sqrt{5n - 4} + 7 = 16$

19. $14 - \sqrt{4a + 9} = 13$

20. $21 - \sqrt{3x - 15} = 24$

21. $\sqrt{5x - 1} - \sqrt{3x + 11} = 0$

22. $\sqrt{4n + 5} - \sqrt{2n + 7} = 0$

23. $2\sqrt{x - 3} = \sqrt{7x + 15}$

24. $\sqrt{6x - 11} = 3\sqrt{x - 7}$

25. $\sqrt[3]{3y - 19} = \sqrt[3]{6y + 26}$

26. $\sqrt[3]{2x + 9} = \sqrt[3]{3x + 14}$

27. $\sqrt{a^2 + 7} = \sqrt{5a + 1}$

28. $\sqrt{n^2 - 8} = \sqrt{2 - 3n}$

29. $\sqrt[3]{a^2 - 6} + \sqrt[3]{1 - 4a} = 0$

30. $\sqrt[3]{p^2 + 10} + \sqrt[3]{9p + 8} = 0$

31. $\sqrt{2x + 8} = -x$

32. $\sqrt{3y - 2} = y$

33. $2n - \sqrt{6 - 5n} = 0$

34. $-3a + \sqrt{6a - 1} = 0$

35. $x - 2 = \sqrt{4x - 11}$

36. $\sqrt{2x + 5} = x + 1$

37. $\sqrt{3t + 1} - 1 = 2t$

38. $\sqrt{5x + 9} + 3 = 2x$

39. $\sqrt{3n} + \sqrt{n - 2} = 4$

40. $\sqrt{x - 1} + \sqrt{2x} = 3$

41. $\sqrt{x - 2} + 1 = -\sqrt{x + 3}$

42. $4 - \sqrt{r + 6} = -\sqrt{r - 2}$

43. $\sqrt{x + 5} - 2 = \sqrt{x - 1}$

44. $\sqrt{14 - a} = \sqrt{a + 3} + 3$

45. $\sqrt{2y + 3} - \sqrt{3y + 7} = -1$

46. $\sqrt{2x - 1} + 1 = \sqrt{3x + 1}$

47. $\sqrt[3]{x^3 + 8} = x + 2$

48. $\sqrt[3]{8 - x^3} = 2 - x$

Mixed Practice

Solve.

49. $13 - \sqrt{4x - 7} = 18$

50. $\sqrt{7x + 4} - 2 = 3x$

51. $\sqrt{6x + 1} = 7$

52. $\sqrt[3]{p} + 13 = 9$

53. $\sqrt[3]{n^2 + 16} + \sqrt[3]{10n + 8} = 0$

54. $\sqrt{a^2 - 11} = \sqrt{7 - 3a}$

Applications

B *Solve.*

55. The distance d to the horizon for an object h mi above Earth's surface is given by the equation $d = \sqrt{8000h + h^2}$. How many miles above Earth's surface is a satellite if the distance to the horizon is 900 mi?

56. The diagonal of a rectangular box with a square base can be calculated using the formula $d = \sqrt{2x^2 + h^2}$, where x is the length of the side of the square base. The diagonal of the box shown in the figure to the right is 18 in. What is the height of the box?

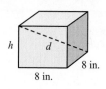

57. To calculate the minimum speed S (in miles per hour) that a car was traveling before skidding to a stop, traffic accident investigators use the formula $S = \sqrt{30fL}$, where f is the drag factor of the road surface and L is the length of a skid mark (in feet).

   a. Solve the formula for L in terms of S and f.

   b. Calculate the length of the skid marks for a car traveling at a speed of 30 mph that skids to a stop on a road surface with a drag factor of 0.5.

58. If an object is dropped, the velocity v (in feet per second) of the object after it has fallen d ft can be calculated using the formula $v = \sqrt{64d}$.

   a. Solve the formula for d in terms of v.

   b. Find the distance an object has fallen if its velocity is 80 ft/sec.

59. How many feet from the side of a house must a painter place a 15-foot ladder in order to reach a window 12 ft above the ground?

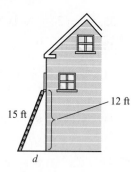

60. Two tugboats are towing a ship so that the tow ropes are perpendicular to each other. The tugboats exert forces of magnitude A and B on the ship. The formula for the magnitude of the combined force R is $R = \sqrt{A^2 + B^2}$. If the forces B and R have magnitude 6 tons and 10 tons respectively, find the magnitude of A.

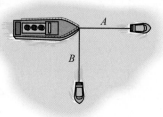

• Check your answers on page A-43.

MINDStretchers

Mathematical Reasoning

1. When raising each side of a radical equation to a power, is the resulting equation equivalent to the original equation? Explain.

Technology

2. Consider the equation $\sqrt{x - 1} = 3 - x$.

   a. Use a grapher to graph each side of the equation. Then, find the point of intersection of the two graphs.

   b. Solve the equation algebraically.

   c. What do you notice about the x-coordinate of the point of intersection in part (a) and your answer to part (b)?

   d. Explain how a grapher can be used to solve radical equations.

Critical Thinking

3. Solve each of the following equations:

   a. $\sqrt{\sqrt{x^2 - 9}} = 2$

   b. $\sqrt{\sqrt[3]{(2x + 1)^2}} = 2$

Cultural Note

In the 1600s, people were very interested in designing accurate clocks. The Dutch astronomer Christiaan Huygens invented the pendulum clock. Huygens chose to work with pendulums because they have an interesting property: the period T (the amount of time it takes for a pendulum to go back and forth once) is related to the length L of the pendulum and the force of gravity g by the formula:

$$T = 2\pi\sqrt{\frac{L}{g}}$$

Since the force of gravity is constant, the only thing that influences the period of a pendulum is its length. This principle has been used to design small wall clocks, cuckoo clocks, and grandfather clocks.

(*Source:* wikipedia.org)

7.6 Complex Numbers

In the preceding sections of this text, our discussions have centered on operations and expressions involving real numbers. Here, we extend the concept of number to include imaginary numbers and, more generally, complex numbers. The existence of these numbers other than real numbers may be difficult to accept. However, working with complex numbers has several advantages—they permit us to solve equations that have no real-number solutions and also to solve applied problems in physics, engineering, and graphics.

Imaginary and Complex Numbers

Recall that since the square of any real number is positive, there is no real number that is the square root of a negative number. Numbers represented by expressions such as $\sqrt{-1}$ and $\sqrt{-3}$ are not real numbers and so are called *imaginary numbers*. The imaginary number $\sqrt{-1}$, which is commonly represented by the letter i, will play a key role throughout this discussion.

> **DEFINITION**
>
> The imaginary number i is $\sqrt{-1}$; that is, $i^2 = -1$.

We can express any imaginary number in terms of i. For instance, consider the imaginary number $\sqrt{-9}$.

$$\sqrt{-9} = \sqrt{9(-1)}$$ Factor out -1 within the radicand.

$$= \sqrt{9} \cdot \sqrt{-1}$$ Use the product rule of radicals. It does not hold for imaginary numbers in general, but does hold for -1 times a positive number.

$$= 3i$$ Take the square root of the real number 9, and substitute i for $\sqrt{-1}$.

Let's see if this result is correct by checking that the square of $3i$ is the original radicand.

$$(3i)^2 = (3i)(3i) = 9i^2 = 9(-1) = -9$$

Since the square of $3i$ is -9, our result is correct. This example suggests the following general property of imaginary numbers that will allow us to rewrite any imaginary number in terms of i.

> ## The Square Root of a Negative Number
>
> For any positive real number a,
> $$\sqrt{-a} = i\sqrt{a}.$$

Note that in the product of i and a radical, i is usually written before the radical. Do you think that writing i first makes the expression easier to read?

EXAMPLE 1

Write each square root in terms of i.

a. $\sqrt{-25}$ **b.** $\sqrt{-6}$ **c.** $-7\sqrt{-\dfrac{4}{9}}$

Solution

a. $\sqrt{-25} = i\sqrt{25}$ **b.** $\sqrt{-6} = i\sqrt{6}$
$\quad\quad\quad\quad = i \cdot 5$
$\quad\quad\quad\quad = 5i$

c. $-7\sqrt{-\dfrac{4}{9}} = -7i\sqrt{\dfrac{4}{9}}$

$\quad\quad\quad\quad = -7i\left(\dfrac{2}{3}\right)$

$\quad\quad\quad\quad = -\dfrac{14}{3}i$

PRACTICE 1

Express in terms of i.

a. $\sqrt{-36}$

b. $\sqrt{-2}$

c. $-10\sqrt{-\dfrac{1}{4}}$

We can use imaginary numbers to define a broader set of numbers that are called *complex numbers*.

DEFINITION

A **complex number** is any number that can be written in the form $a + bi$, where a and b are real numbers and $i = \sqrt{-1}$. In the complex number $a + bi$, a is called the **real part** and b is called the **imaginary part**.

Note that for any complex number, both its real part and its imaginary part are real numbers.

Every real number can be expressed as a complex number. For instance, the real number 4 can be written as $4 + 0i$. Similarly, every imaginary number can be expressed as complex, so we can write $2i$ as $0 + 2i$.

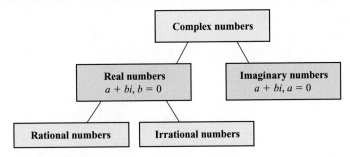

Consider the following examples of complex numbers:

Complex Number	Real Part a	Imaginary Part b
$2 + 3i$	2	3
$-1 - 5i$	-1	-5
$\dfrac{3}{4}\ \left(=\dfrac{3}{4}+0i\right)$	$\dfrac{3}{4}$	0
$i\ \ (=0+1i)$	0	1

Just as for real numbers, we can carry out the four basic arithmetic operations on complex numbers.

Adding and Subtracting Complex Numbers

The addition and subtraction of complex numbers is similar to the addition and subtraction of algebraic binomial expressions. In other words, to add complex numbers, we add their real parts and their imaginary parts separately. Similarly, to subtract complex numbers, we subtract their real parts and their imaginary parts separately.

EXAMPLE 2

Simplify.

a. $(2 - 5i) + (1 + 3i)$

b. $(-1 + 9i) - (2 - 4i)$

c. $(5 - 2\sqrt{-4}) + (-5 + 7\sqrt{-1})$

Solution

a.

Real parts   Imaginary parts

$(2 - 5i) + (1 + 3i) = \overbrace{(2 + 1)} + \overbrace{(-5 + 3)}i = 3 - 2i$

b. $(-1 + 9i) - (2 - 4i) = -1 + 9i - 2 + 4i$
$$= (-1 - 2) + (9 + 4)i$$
$$= -3 + 13i$$

c. $(5 - 2\sqrt{-4}) + (-5 + 7\sqrt{-1}) = (5 - 2\cdot 2i) + (-5 + 7i)$
$$= (5 - 4i) + (-5 + 7i)$$
$$= (5 - 5) + (-4 + 7)i$$
$$= 3i$$

PRACTICE 2

Combine.

a. $(6 + 5i) + (-2 - 6i)$

b. $(-4 + 3i) - (8 - i)$

c. $(1 + 5\sqrt{-9}) + (4 + \sqrt{-16})$

Multiplying Complex Numbers

Before considering how to find the product of complex numbers generally, let's look at some examples of multiplying by an imaginary number. When multiplying radicals with negative radicands, we must first express them in terms of i, since the product rule for radicals does not in general hold for complex numbers.

EXAMPLE 3

Multiply.

a. $\sqrt{-4} \cdot \sqrt{-9}$

b. $\sqrt{-2} \cdot \sqrt{-7}$

c. $-3i \cdot 6i$

d. $5i(3 - 10i)$

Solution

a. $\sqrt{-4} \cdot \sqrt{-9} = (2i)(3i)$
$$= 6i^2$$
$$= -6$$

b. $\sqrt{-2} \cdot \sqrt{-7} = (i\sqrt{2})(i\sqrt{7})$
$$= i^2\sqrt{14}$$
$$= -1 \cdot \sqrt{14}$$
$$= -\sqrt{14}$$

PRACTICE 3

Multiply.

a. $\sqrt{-36} \cdot \sqrt{-4}$

b. $\sqrt{-5} \cdot \sqrt{-3}$

c. $-4i \cdot 10i$

d. $-2i(5 + 6i)$

c. $-3i \cdot 6i = (-18)i^2$

$\qquad\qquad = (-18)(-1)$

$\qquad\qquad = 18$

d. $5i(3 - 10i) = 15i - 50i^2$ 　　　Use the distributive property.

$\qquad\qquad\quad = 15i - (-50)$

$\qquad\qquad\quad = 50 + 15i$

Next, we turn to finding the product of any complex numbers. The key is to multiply the complex numbers just as if they were algebraic binomials, using the FOIL method.

EXAMPLE 4

Multiply.

a. $(5 + 2i)(3 - i)$ 　　　　　　**b.** $(2 + 3i)(4 - 6i)$

Solution

a.

First　　Last

$(5 + 2i)(3 - i) = 15 - 5i + 6i - 2i^2$

Inner

Outer

$\qquad\qquad\qquad\quad = 15 - 5i + 6i + 2$

$\qquad\qquad\qquad\quad = 17 + i$

b. $(2 + 3i)(4 - 6i) = 8 - 12i + 12i - 18i^2$

$\qquad\qquad\qquad\qquad\quad = 8 - 12i + 12i + 18$

$\qquad\qquad\qquad\qquad\quad = 26$

PRACTICE 4

Multiply.

a. $(4 + 7i)(2 + i)$

b. $(3 - 2i)(1 - 3i)$

Complex Conjugates

The multiplication of *complex conjugates*, which play a special role in dividing complex numbers, requires special mention.

DEFINITION

The complex numbers $a + bi$ and $a - bi$ are called **complex conjugates**.

For instance, $5 + 2i$ and $5 - 2i$ are complex conjugates. So are $-8 - 3i$ and $-8 + 3i$.

EXAMPLE 5

Find the complex conjugate.

a. $3 + 2i$ 　　　　**b.** $-1 - 5i$ 　　　　**c.** $6i$

Solution

a. $3 - 2i$ 　　　　**b.** $-1 + 5i$

c. The number $6i$ can be written $0 + 6i$. The conjugate is $0 - 6i$, or $-6i$.

PRACTICE 5

Find the complex conjugate.

a. $-1 - 7i$

b. $8 + 9i$

c. $-3i$

Note that complex conjugates, such as $(3 + 6i)$ and $(3 - 6i)$, are binomials that are the sum and difference of the same two terms. Therefore, to multiply conjugates, we can use the special product formula for multiplying the sum and difference of two terms, $(a + b)(a - b) = a^2 - b^2$, as shown in the following example.

EXAMPLE 6

Find the product of $3 + 6i$ and $3 - 6i$.

Solution

$$(3 + 6i)(3 - 6i) = (3)^2 - (6i)^2$$
$$= 9 - 36i^2$$
$$= 9 + 36$$
$$= 45$$

Note that the product of these two complex numbers is a real number.

PRACTICE 6

Multiply $2 - 5i$ and its complex conjugate.

TIP In general, the product of a complex number $a + bi$ and its complex conjugate $a - bi$ is the real number $a^2 + b^2$.

Dividing Complex Numbers

Complex numbers are best divided by expressing the division problem as a fraction. We then calculate an equivalent fraction whose denominator is a real number. To do this, we multiply the numerator and the denominator of the fraction by the complex conjugate of the denominator. Consider the following example:

EXAMPLE 7

Divide.

a. $\dfrac{3}{2 + i}$ **b.** $\dfrac{4}{3i}$

Solution

a. To find the quotient of 3 and $2 + i$, we begin by multiplying both the numerator and the denominator by the complex conjugate of the denominator.

$$\dfrac{3}{2 + i} = \dfrac{3}{2 + i} \cdot \dfrac{2 - i}{2 - i} \qquad \text{Multiply both the numerator and denominator by } 2 - i, \text{ the conjugate of } 2 + i.$$

$$= \dfrac{3(2 - i)}{(2 + i)(2 - i)}$$

$$= \dfrac{6 - 3i}{4 + 1}$$

$$= \dfrac{6 - 3i}{5}$$

$$= \dfrac{6}{5} - \dfrac{3}{5}i \qquad \text{Write in } a + bi \text{ form.}$$

PRACTICE 7

Divide.

a. $\dfrac{7}{1 - 2i}$

b. $-\dfrac{2}{5i}$

b. $\dfrac{4}{3i} = \dfrac{4}{3i} \cdot \dfrac{-3i}{-3i}$

$= \dfrac{-12i}{-9i^2}$

$= \dfrac{-12i}{9}$

$= -\dfrac{4}{3}i$

EXAMPLE 8

Divide.

a. $\dfrac{1 - 3i}{1 + 2i}$

b. $\dfrac{5 + \sqrt{-25}}{3 + \sqrt{-9}}$

Solution

a. $\dfrac{1 - 3i}{1 + 2i} = \dfrac{1 - 3i}{1 + 2i} \cdot \dfrac{1 - 2i}{1 - 2i}$

$= \dfrac{(1 - 3i)(1 - 2i)}{(1 + 2i)(1 - 2i)}$

$= \dfrac{1 - 5i + 6i^2}{1 + 4}$

$= \dfrac{1 - 5i - 6}{5}$

$= \dfrac{-5 - 5i}{5}$

$= \dfrac{5(-1 - i)}{5}$

$= -1 - i$

b. $\dfrac{5 + \sqrt{-25}}{3 + \sqrt{-9}} = \dfrac{5 + 5i}{3 + 3i} = \dfrac{5(1 + i)}{3(1 + i)} = \dfrac{5}{3}$

PRACTICE 8

Divide.

a. $\dfrac{2 + i}{1 - 3i}$

b. $\dfrac{1 - \sqrt{-4}}{4 - \sqrt{-64}}$

Raising *i* to a Power

Finally, in our discussion of operations on complex numbers, let's consider powers of the number *i*. When raising *i* to consecutive whole-number exponents, an interesting pattern emerges.

$$i = \sqrt{-1} = i \qquad\qquad i^5 = i^4 \cdot i = i$$
$$i^2 = (\sqrt{-1})^2 = -1 \qquad i^6 = i^4 \cdot i^2 = 1(-1) = -1$$
$$i^3 = i^2 \cdot i = -i \qquad\qquad i^7 = i^4 \cdot i^3 = 1(-i) = -i$$
$$i^4 = i^2 \cdot i^2 = (-1)(-1) = 1 \qquad i^8 = i^4 \cdot i^4 = (1)(1) = 1$$

Can you explain why this pattern continues?

Generally, an expression involving complex numbers is not considered to be simplified if it contains a power of i greater than 1. So we can use the pattern of powers just noted to simplify such expressions. Because i raised to the fourth power is 1, we can readily evaluate i raised to higher integral powers by factoring out i^4 as many times as possible.

EXAMPLE 9

Evaluate i^{23}.

Solution $\qquad i^{23} = (i^4)^5 i^3 = 1^5 i^3 = i^3 = -i$

PRACTICE 9

Find the value of i^{30}.

An important application of complex numbers is the flow of alternating current in a circuit. This is the kind of current that flows through an appliance when we plug it into a wall outlet.

EXAMPLE 10

Ohm's law for alternating-current circuits states that $V = IZ$, where V is the voltage (in volts), I is the current (in amperes), and Z is the impedance (in ohms).

a. Rewrite Ohm's law by solving for Z.

b. If V can be represented by $11 + 10i$ and I by $2 + 3i$, find Z.

c. Check the answer in part (b) by substituting into Ohm's law.

Solution

a. Ohm's law states that $V = IZ$. We solve for Z by dividing both sides of the equation by I, giving us $Z = \dfrac{V}{I}$.

b. To find Z, we use the formula found in part (a) and substitute $11 + 10i$ for V and $2 + 3i$ for I:

$$Z = \frac{V}{I} = \frac{11 + 10i}{2 + 3i} \cdot \frac{2 - 3i}{2 - 3i} = \frac{22 - 33i + 20i + 30}{4 + 9} = 4 - i$$

So the impedance Z can be represented by $(4 - i)$ ohms.

c. The answer in part (b) can be checked as follows:

$$V = IZ = (2 + 3i)(4 - i) = 8 - 2i + 12i + 3 = 11 + 10i$$

The computed voltage, $11 + 10i$, agrees with the given information, confirming our answer in part (b).

PRACTICE 10

Use Ohm's law (stated in Example 10).

a. Rewrite Ohm's law by solving for I.

b. If V can be represented by $3 + 5i$ and Z by $1 - i$, find I.

c. Check the answer in part (b) by substituting into Ohm's law.

Mathematically Speaking

Fill in each blank with the most appropriate term or phrase from the given list.

complex number	irrational number	complex part
additive inverses	imaginary number	the square root property
real part	power	
multiple	complex conjugates	FOIL method

1. The _____ i is $\sqrt{-1}$; in other words, $i^2 = -1$.

2. A(n) _____ is any number that can be written in the form $a + bi$, where a and b are real numbers and $i = \sqrt{-1}$.

3. In the complex number $a + bi$, a is called the _____ and b is called the imaginary part.

4. To multiply complex numbers, use the _____.

5. The complex numbers $a + bi$ and $a - bi$ are called _____.

6. An expression involving complex numbers is generally not considered simplified if it contains a(n) _____ of i greater than 1.

A *Write in terms of i.*

7. $\sqrt{-4}$

8. $\sqrt{-49}$

9. $\sqrt{-\dfrac{1}{16}}$

10. $\sqrt{-\dfrac{9}{64}}$

11. $\sqrt{-3}$

12. $\sqrt{-10}$

13. $\sqrt{-18}$

14. $\sqrt{-24}$

15. $\sqrt{-500}$

16. $\sqrt{-72}$

17. $-\sqrt{-9}$

18. $-2\sqrt{-1}$

19. $6\sqrt{-\dfrac{5}{16}}$

20. $3\sqrt{-\dfrac{7}{36}}$

21. $-\dfrac{1}{4}\sqrt{-12}$

22. $-\dfrac{1}{2}\sqrt{-32}$

B *Perform the indicated operation.*

23. $(1 + 12i) + 8i$

24. $3i + (4 + 14i)$

25. $(3 - 15i) + (2 + 9i)$

26. $(6 + 11i) + (5 - 7i)$

27. $(7 - i) - (7 + 5i)$

28. $(10 - 4i) - (9 - 4i)$

29. $(-8 - 6i) - (-1 - 3i)$

30. $(2 + 13i) - (-3 - 2i)$

31. $16 - (18 + \sqrt{-4})$

32. $20 - (7 + \sqrt{-9})$

33. $(10 - 3\sqrt{-16}) + (2 - \sqrt{-25})$

34. $(4 + 2\sqrt{-4}) - (12 - 4\sqrt{-1})$

C *Multiply.*

35. $\sqrt{-25} \cdot \sqrt{-4}$

36. $\sqrt{-9} \cdot \sqrt{-16}$

37. $\sqrt{-3}(-\sqrt{-27})$

38. $-\sqrt{-6}(\sqrt{-24})$

39. $7i \cdot 9i$

40. $11i \cdot 5i$

41. $-2i(14i)$

42. $6i(-8i)$

43. $3i(1 - i)$

44. $2i(5 - 2i)$

45. $-i(12 + 7i)$

46. $-4i(4 + 8i)$

47. $\sqrt{-9}(7 + \sqrt{-16})$

48. $\sqrt{-1}(2 + \sqrt{-49})$

49. $-\sqrt{2}(\sqrt{8} - \sqrt{-18})$

50. $-\sqrt{3}(\sqrt{3} - \sqrt{-27})$

51. $(4 + 2i)(2 + 3i)$
52. $(3 + 5i)(1 + 4i)$
53. $(10 - i)(4 + 6i)$
54. $(8 + 3i)(1 - 2i)$

55. $(7i - 7)(3 - 5i)$
56. $(8 - i)(9i - 2)$
57. $(-4 - 2i)(2 - 4i)$
58. $(-3 - 6i)(6 - 3i)$

59. $(6 + 5i)(6 - 5i)$
60. $(7 + 3i)(7 - 3i)$
61. $(3 + 2i)^2$
62. $(4 + 3i)^2$

63. $(2 - 3i)^2$
64. $(3 - 4i)^2$

65. $(\sqrt{-1} + \sqrt{2})(\sqrt{-9} - \sqrt{8})$
66. $(\sqrt{3} - \sqrt{-4})(\sqrt{12} + \sqrt{-16})$

D *Find the complex conjugate of the complex number. Then, find the product of each complex number and its complex conjugate.*

67. $1 + 10i$
68. $5 + 2i$
69. $7 - i$
70. $4 - 3i$

71. $-9 + 6i$
72. $-4 + 12i$
73. $8i$
74. $5i$

75. $-11i$
76. $-9i$

E *Divide.*

77. $\dfrac{7}{4 + i}$
78. $\dfrac{-1}{2 + 3i}$
79. $\dfrac{-3}{1 - 5i}$
80. $\dfrac{3}{3 - i}$

81. $\dfrac{5}{4i}$
82. $\dfrac{9}{2i}$
83. $-\dfrac{2}{\sqrt{-49}}$
84. $-\dfrac{3}{\sqrt{-25}}$

85. $\dfrac{4 - 3i}{i}$
86. $\dfrac{6 + 2i}{2i}$
87. $\dfrac{3 + 5i}{1 + i}$
88. $\dfrac{2 - 4i}{3 - i}$

89. $\dfrac{6 + 3i}{2 - 2i}$
90. $\dfrac{4 - 5i}{5 + 3i}$
91. $\dfrac{2 - \sqrt{-16}}{5 - \sqrt{-100}}$
92. $\dfrac{6 + \sqrt{-9}}{4 + \sqrt{-4}}$

93. $\dfrac{8 - \sqrt{-36}}{6 + \sqrt{-64}}$
94. $\dfrac{9 + \sqrt{-9}}{3 - \sqrt{-81}}$

F *Evaluate.*

95. i^{18}
96. i^{20}
97. i^{35}
98. i^{41}

99. $i^{12} \cdot i^9$
100. $i^{15} \cdot i^{11}$
101. $\dfrac{i^{38}}{i^{19}}$
102. $\dfrac{i^{24}}{i^7}$

Mixed Practice

Solve.

103. Find the complex conjugate of $-8 + 5i$. Then, calculate the product of both numbers.

104. Evaluate: $i^{11} \cdot i^3$

Write in terms of i.

105. $4\sqrt{-\dfrac{3}{25}}$

106. $\sqrt{-72}$

Divide.

107. $\dfrac{2 - 7i}{1 - i}$

108. $\dfrac{-2}{2 - i}$

Simplify.

109. $-\sqrt{2}(\sqrt{32} - \sqrt{-50})$

110. $-6i(5 - 12i)$

111. $18 - (20 + \sqrt{-16})$

112. $(-7 + 8i)(3 - 6i)$

Applications

ⓒ *Solve.*

113. The total impedance in an alternating-current circuit connected in series is the sum of the individual impedances. If the impedance in one part of the circuit is $(3 + 9i)$ ohms and in another part is $(5 - 8i)$ ohms, what is the total impedance in the circuit?

114. The formula $V = IZ$ expresses the relationship between the voltage V (in volts), the current I (in amperes), and the impedance Z (in ohms) in an alternating-current circuit. If the current in a circuit is $(8 + 5i)$ amps and the impedance is $(9 + 3i)$ ohms, what is the voltage in the circuit?

115. Is the complex number $2 + 3i$ a solution of the equation $x^2 - 4x = -13$. Explain.

116. Some functions have the set of complex numbers for their domain. These functions play a role in *fractal geometry*, which is important in nature and art, including film animation. If the function $f(z) = 2z + 5 + 2i$, find $f(-3 + 6i)$.

• Check your answers on page A-43.

MINDStretchers

Mathematical Reasoning

1. Use the properties of exponents to show that the expression i^{4n+3} is equivalent to $-i$ for any whole number n.

Research

2. Casper Wessel was credited with the geometric interpretation of a complex number as a point in the *complex plane*. Either in your college library or on the web, investigate how to plot complex numbers. Write a summary of your findings and explain how to plot the complex number $2 - 3i$.

Critical Thinking

3. Some solutions of equations are complex numbers. Show that $1 + i$ and its conjugate are solutions of the equation $x^2 - 2x + 2 = 0$.

Key Concepts and Skills

CONCEPT   SKILL

Concept/Skill	Description	Example
[7.1] **Radical expression**	An algebraic expression that contains a radical.	$10 - \sqrt{2}$ $\sqrt[3]{abc}$
[7.1] **Perfect square**	A nonnegative rational number that is the square of another rational number.	$\dfrac{1}{9}$
[7.1] **Square root**	The number b is a square root of a if $b^2 = a$, for any real numbers a and b and for a nonnegative.	4 is a square root of 16 since $4^2 = 16$. -4 is a square root of 16 since $(-4)^2 = 16$.
[7.1] **Radicand**	The radicand is the number under $\sqrt{}$, the **radical sign**. This symbol stands for the positive or **principal square root**.	5 is the radicand of $\sqrt{5}$. The principal square root of 5 is $\sqrt{5}$.
[7.1] **Squaring a square root**	For any nonnegative real number a, $(\sqrt{a})^2 = a$.	$(\sqrt{6})^2 = 6$ $(\sqrt{x})^2 = x$
[7.1] **Taking the square root of a square**	For any nonnegative real number a, $\sqrt{a^2} = a$.	$\sqrt{6^2} = 6$ $\sqrt{x^2} = x$
[7.1] **Perfect cube**	A rational number that is the cube of another rational number.	$\dfrac{1}{27}$
[7.1] **Cube root**	The number b is the **cube root** of a if $b^3 = a$, for any real numbers a and b. The cube root of a is written $\sqrt[3]{a}$, where 3 is called the **index** of the radical.	2 is the cube root of 8 since $2^3 = 8$. -2 is the cube root of -8 since $(-2)^3 = -8$.
[7.1] **Cubing a cube root**	For any real number a, $(\sqrt[3]{a})^3 = a$.	$(\sqrt[3]{7})^3 = 7$ $(\sqrt[3]{n})^3 = n$
[7.1] **Taking the cube root of a cube**	For any real number a, $\sqrt[3]{a^3} = a$.	$\sqrt[3]{7^3} = 7$ $\sqrt[3]{n^3} = n$
[7.1] ***n*th root**	The number b is the ***n*th root** of a if $b^n = a$, for any real number a and for any positive integer n greater than 1. The *n*th root of a is written $\sqrt[n]{a}$, where n is called the **index** of the radical.	The fourth root of 81, written $\sqrt[4]{81}$, is 3 since $3^4 = 81$. The index of the radical is 4.
[7.1] **Taking the *n*th root of an *n*th power**	For any real number a, • $\sqrt[n]{a^n} = \lvert a \rvert$ for any even positive integer n, and • $\sqrt[n]{a^n} = a$ for any odd positive integer n greater than 1.	$\sqrt{(-5)^2} = \lvert -5 \rvert = 5$ $\sqrt[3]{(-5)^3} = -5$
[7.1] $a^{1/n}$	For any positive integer n greater than 1 and a real number a for which $\sqrt[n]{a}$ is a real number, $a^{1/n} = \sqrt[n]{a}$.	$5^{1/2} = \sqrt{5}$ $x^{1/3} = \sqrt[3]{x}$

Concept/Skill	Description	Example
[7.1] $a^{m/n}$	For any positive integers m and n such that n is greater than 1 where $\dfrac{m}{n}$ is in simplest form and a is a real number for which $\sqrt[n]{a}$ is a real number, $a^{m/n} = \sqrt[n]{a^m}$, or equivalently $a^{m/n} = (\sqrt[n]{a})^m$.	$5^{2/3} = (\sqrt[3]{5})^2$ or $5^{2/3} = \sqrt[3]{5^2}$ $n^{3/5} = (\sqrt[5]{n})^3$ or $n^{3/5} = \sqrt[5]{n^3}$
[7.1] $a^{-m/n}$	For any positive integers m and n such that n is greater than 1, where $\dfrac{m}{n}$ is in simplest form and a is a real number for which $\sqrt[n]{a}$ is a nonzero real number, $$a^{-m/n} = \frac{1}{a^{m/n}}.$$	$5^{-2/3} = \dfrac{1}{5^{2/3}}$ $n^{-3/5} = \dfrac{1}{n^{3/5}}$
[7.2] **The Product Rule of Radicals**	If $\sqrt[n]{a}$ and $\sqrt[n]{b}$ are real numbers, then $\sqrt[n]{a} \cdot \sqrt[n]{b} = \sqrt[n]{ab}$.	$\sqrt{2} \cdot \sqrt{3} = \sqrt{6}$ $\sqrt[4]{x} \cdot \sqrt[4]{y^2} = \sqrt[4]{xy^2}$
[7.2] **The Quotient Rule of Radicals**	For any integer $n > 1$ and any real numbers a and b for which $\sqrt[n]{a}$ and $\sqrt[n]{b}$ are real numbers and b is nonzero, $$\frac{\sqrt[n]{a}}{\sqrt[n]{b}} = \sqrt[n]{\frac{a}{b}}.$$	$\dfrac{\sqrt{2}}{\sqrt{3}} = \sqrt{\dfrac{2}{3}}$ $\dfrac{\sqrt[4]{x}}{\sqrt[4]{y^2}} = \sqrt[4]{\dfrac{x}{y^2}}$
[7.2] **The distance formula**	The distance d between two points, (x_1, y_1) and (x_2, y_2) on a coordinate plane is given by $$d = \sqrt{(x_2 - x_1)^2 + (y_2 - y_1)^2}.$$	The distance between $(3, 2)$ and $(5, 4)$ is: $\begin{aligned} d &= \sqrt{(x_2 - x_1)^2 + (y_2 - y_1)^2} \\ &= \sqrt{(5 - 3)^2 + (4 - 2)^2} \\ &= \sqrt{(2)^2 + (2)^2} \\ &= \sqrt{8} \\ &= 2\sqrt{2} \text{ units} \end{aligned}$
[7.3] **Like radicals**	Radicals that have the same radicand and the same index.	$2\sqrt[3]{5}$ and $3\sqrt[3]{5}$ are like radicals. $-\sqrt{3x}$ and $7\sqrt{3x}$ are like radicals.
[7.3] **To add (or subtract) radicals**	• Simplify the radicals if possible, and then add (or subtract) like radicals using the distributive property.	$4\sqrt{3} - 6\sqrt{3} = (4 - 6)\sqrt{3} = -2\sqrt{3}$ $\begin{aligned} \sqrt{8} + \sqrt{18} &= 2\sqrt{2} + 3\sqrt{2} \\ &= (2 + 3)\sqrt{2} \\ &= 5\sqrt{2} \end{aligned}$
[7.4] **To multiply radicals**	• Apply the product rule of radicals. • Simplify, if possible.	$\begin{aligned} (6\sqrt{2})(-2\sqrt{6}) &= -12\sqrt{12} \\ &= -12 \cdot 2\sqrt{3} \\ &= -24\sqrt{3} \end{aligned}$
[7.4] **To divide radicals**	• Apply the quotient rule of radicals. • Simplify, if possible.	$\begin{aligned} \dfrac{\sqrt{24x}}{\sqrt{6}} &= \sqrt{\dfrac{24x}{6}} \\ &= \sqrt{4x} \\ &= 2\sqrt{x} \end{aligned}$

continued

Concept/Skill	Description	Example
[7.4] To rationalize a denominator	• Multiply the numerator and denominator by a factor that will make the radicand in the denominator a perfect power.	$\dfrac{3}{\sqrt{5}} = \dfrac{3}{\sqrt{5}} \cdot \dfrac{\sqrt{5}}{\sqrt{5}} = \dfrac{3\sqrt{5}}{\sqrt{5^2}} = \dfrac{3\sqrt{5}}{5}$ $\sqrt[3]{\dfrac{2}{x}} = \dfrac{\sqrt[3]{2}}{\sqrt[3]{x}}$ $\quad = \dfrac{\sqrt[3]{2}}{\sqrt[3]{x}} \cdot \dfrac{\sqrt[3]{x^2}}{\sqrt[3]{x^2}}$ $\quad = \dfrac{\sqrt[3]{2x^2}}{\sqrt[3]{x^3}}$ $\quad = \dfrac{\sqrt[3]{2x^2}}{x}$
[7.4] Conjugates	The expressions $a + b$ and $a - b$ are conjugates.	$p + q$ and $p - q$
[7.4] To rationalize a denominator involving two terms	• Multiply both numerator and denominator by the conjugate of the denominator.	$\dfrac{3}{2 + \sqrt{5}} = \dfrac{3}{2 + \sqrt{5}} \cdot \dfrac{2 - \sqrt{5}}{2 - \sqrt{5}}$ $\quad = \dfrac{3(2 - \sqrt{5})}{(2 + \sqrt{5})(2 - \sqrt{5})}$ $\quad = \dfrac{6 - 3\sqrt{5}}{2^2 - (\sqrt{5})^2}$ $\quad = \dfrac{6 - 3\sqrt{5}}{4 - 5}$ $\quad = \dfrac{6 - 3\sqrt{5}}{-1}$ $\quad = -6 + 3\sqrt{5}$
[7.5] Radical equation	An equation in which a variable appears in one or more radicands.	$\sqrt{2x} = 6$ $\sqrt{x + 1} = \sqrt{2x - 3}$
[7.5] The power property of equality	For any real numbers a and b and any integer n, if $a = b$, then $a^n = b^n$.	If $\sqrt{x} = 3$, then $x = 9$.
[7.5] To solve a radical equation	• Isolate a term with a radical. • Raise each side of the equation to a power equal to the index of the radicand. • If the equation still contains a radical, repeat the preceding steps. • Where possible, combine like terms. • Solve the resulting equation. • Check the possible solution(s) in the original equation.	$\sqrt{3x + 4} + 7 = 12$ $\sqrt{3x + 4} = 5$ $(\sqrt{3x + 4})^2 = 5^2$ $3x + 4 = 25$ $3x = 21$ $x = 7$ **Check** $\sqrt{3x + 4} + 7 = 12$ $\sqrt{3(7) + 4} + 7 \overset{?}{=} 12$ $\sqrt{25} + 7 \overset{?}{=} 12$ $5 + 7 \overset{?}{=} 12$ $12 = 12 \quad$ True

CONCEPT   SKILL

Concept/Skill	Description	Example
[7.6] **Imaginary number**	A number that is the square root of a negative number.	$\sqrt{-5}$
[7.6] *i*	The imaginary number i is $\sqrt{-1}$; in other words, $i^2 = -1$.	$\sqrt{-1}$
[7.6] **Square root of a negative number**	For any positive real number a, $\sqrt{-a} = i\sqrt{a}$.	$\sqrt{-25} = 5i$ $\sqrt{-6} = i\sqrt{6}$
[7.6] **Complex number**	Any number that can be written in the form $a + bi$, where a and b are real numbers and $i = \sqrt{-1}$. In the complex number $a + bi$, a is called the **real part** and b is called the **imaginary part**.	$6 + 2i$, 3, and $-5i$ are complex numbers.
[7.6] **To add (or subtract) complex numbers**	• Add (or subtract) the real parts and imaginary parts separately.	$(2 + 4i) + (3 - 2i)$ $= (2 + 3) + (4 - 2)i$ $= 5 + 2i$ $(4 - i) - (7 - 6i)$ $= 4 - i - 7 + 6i$ $= (4 - 7) + (-1 + 6)i$ $= -3 + 5i$
[7.6] **To multiply complex numbers**	• Use the distributive property or the FOIL method. Then simplify, if possible.	$(2 + 3i)(1 - 2i) = 2 - 4i + 3i - 6i^2$ $= 2 - 4i + 3i + 6$ $= 8 - i$
[7.6] **Complex conjugates**	The complex numbers $a + bi$ and $a - bi$ are complex conjugates.	$6 + 2i$ and $6 - 2i$
[7.6] **To divide complex numbers**	• Multiply the numerator and denominator by the complex conjugate of the denominator.	$\dfrac{1 + 2i}{2 - i} = \dfrac{1 + 2i}{2 - i} \cdot \dfrac{2 + i}{2 + i}$ $= \dfrac{(1 + 2i)(2 + i)}{(2 - i)(2 + i)}$ $= \dfrac{2 + 5i + 2i^2}{4 + 1}$ $= \dfrac{2 + 5i - 2}{4 + 1}$ $= \dfrac{5i}{5}$ $= i$

Say Why
Fill in each blank.

1. The cube of $\sqrt[3]{5}$ _____ 5 because _____
   is/is not
   _____ .

2. The number $\dfrac{1}{64}$ _____ a perfect cube because
   is/is not
   _____ .

3. The sum of $8\sqrt[3]{7}$ and $5\sqrt[3]{7}$ _____ $13\sqrt[3]{7}$
   is/is not
   because _____
   _____ .

4. The expressions $x\sqrt[3]{2}$ and $y\sqrt[3]{2}$ _____ like
   are/are not
   radicals because _____
   _____ .

5. The expression $\sqrt{\dfrac{x}{3}}$ _____ considered to be
   is/is not
   simplified because _____
   _____ .

6. The expression $\dfrac{1}{\sqrt{x}}$ _____ considered to be
   is/is not
   simplified because _____
   _____ .

7. $\sqrt[3]{5} \cdot \sqrt[3]{5}$ _____ equal to $\sqrt[3]{25}$ because
   is/is not
   _____ .

8. The sum of a complex number and its conjugate
   _____ a real number because _____
   is/is not
   _____ .

To help you review this chapter, solve these problems.

[7.1] *Evaluate.*

9. $-6\sqrt{121}$

10. $2\sqrt[3]{-125}$

11. $\sqrt{\dfrac{1}{9}}$

12. $\sqrt{0.36}$

Simplify.

13. $\sqrt{81y^8}$

14. $-\sqrt{49a^6b^2}$

15. $\sqrt[3]{-216x^9}$

16. $\sqrt[5]{243p^{15}}$

Write using radical notation. Then, simplify, if possible.

17. $-64^{1/2}$

18. $7x^{1/3}$

19. $-(16n^4)^{3/4}$

20. $8^{-2/3}$

Simplify. Then, write the answer in radical notation.

21. $x^{1/4} \cdot x^{1/2}$

22. $\dfrac{r^{2/3}}{6r^{1/6}}$

23. $(25y^2)^{-1/2}$

24. $\dfrac{3a^{2/3}}{(6a^{1/6})^2}$

[7.2] *Use rational exponents to simplify. Then, write the answer in radical notation.*

25. $\sqrt[8]{x^2}$

26. $\sqrt[6]{n^4} \cdot \sqrt[3]{n}$

27. $\sqrt{\sqrt[4]{y^2}}$

28. $\sqrt[6]{p^3q^6}$

29. $\dfrac{\sqrt[3]{a^2}}{\sqrt{a}}$

30. $\sqrt[4]{x^2} \cdot \sqrt[10]{y^5}$

Multiply.

31. $\sqrt{10r} \cdot \sqrt{3s}$

32. $\sqrt[3]{4p} \cdot \sqrt[3]{7pq^2}$

Simplify.

33. $\sqrt{300n^3}$

34. $\sqrt{45x^5y^4}$

35. $\sqrt[3]{128t^7}$

36. $\sqrt[4]{96a^5b^{10}}$

Divide.

37. $\dfrac{\sqrt{35a}}{\sqrt{7}}$

38. $\dfrac{\sqrt[3]{12p^2q^2}}{\sqrt[3]{6pq^2}}$

Simplify.

39. $\sqrt{\dfrac{n}{25}}$

40. $\sqrt{\dfrac{6}{49y^4}}$

41. $\sqrt[3]{\dfrac{64u^2}{125v^9}}$

42. $\sqrt[4]{\dfrac{4p^4q^3}{81r^4s^8}}$

[7.3] *Perform the indicated operations.*

43. $9\sqrt{x} - 5\sqrt{x}$

44. $3\sqrt[3]{q^2} + 8\sqrt[3]{q^2}$

45. $\sqrt{48} + \sqrt{27}$

46. $-\sqrt{96} - 5\sqrt{6} + 3\sqrt{54}$

47. $6\sqrt[3]{56a^4} - \sqrt[3]{189a}$

48. $\dfrac{1}{3}\sqrt[3]{27p^5q} + 2p\sqrt[3]{p^2q}$

[7.4] *Multiply and simplify.*

49. $(-2\sqrt{3a})(3\sqrt{6a})$

50. $\sqrt{5}(4\sqrt{10} - 2\sqrt{5})$

51. $\sqrt[3]{2n}(\sqrt[3]{n^2} - \sqrt[3]{4})$

52. $(4\sqrt{t} - 5)(\sqrt{t} - 3)$

53. $(\sqrt{6} - \sqrt{x})(\sqrt{6} + \sqrt{x})$

54. $(\sqrt{2y} - 1)^2$

Divide and simplify.

55. $\dfrac{\sqrt{72n^3}}{4\sqrt{6}}$

56. $\sqrt{\dfrac{32a}{9b^4}}$

Simplify.

57. $\dfrac{1}{\sqrt{8}}$

58. $\dfrac{\sqrt{16x}}{2\sqrt{y}}$

59. $\sqrt{\dfrac{14p^2}{3q}}$

60. $\sqrt[3]{\dfrac{5v}{54u^5}}$

61. $\dfrac{\sqrt{10} - 3}{\sqrt{5}}$

62. $\dfrac{2\sqrt{6a} + \sqrt{2}}{\sqrt{2a}}$

Rationalize the denominator.

63. $\dfrac{4}{\sqrt{3} - 1}$

64. $\dfrac{\sqrt{x} + 2}{\sqrt{x} - \sqrt{5}}$

[7.5] *Solve and check.*

65. $\sqrt{x + 8} = 4$

66. $\sqrt{n} - 2 = 3$

67. $\sqrt{3n - 4} + 1 = -2$

68. $\sqrt{x^2 - 7} = \sqrt{5x + 7}$

69. $\sqrt[3]{2x - 3} = -2$

70. $\sqrt{x + 5} + 1 = \sqrt{3x + 4}$

[7.6] *Write in terms of i.*

71. $\sqrt{-36}$

72. $\sqrt{-125}$

Perform the indicated operation.

73. $(6 - 4i) + (2 + 9i)$

74. $(\sqrt{-4} - 3) - (\sqrt{-16} - 7)$

75. $\sqrt{-81} \cdot \sqrt{-1}$

76. $-2i(5i + 1)$

77. $(3 - 3i)(8 + 3i)$

78. $(5 - i)^2$

79. $\dfrac{-1}{4 - 4i}$

80. $\dfrac{3 - 4i}{6 - 2i}$

Evaluate.

81. i^{38}

82. i^{53}

Mixed Applications

Solve. Express answers in radical form.

83. When grown in a rich medium, the number of *E. coli* bacteria in a colony after t min can be calculated using the expression $N(2)^{t/20}$, where N is the number of bacteria present in the initial population.

  a. Write this expression in radical form.

  b. If 10 bacteria are present in the initial population, calculate the number of bacteria in the colony after 1 hr.

84. The velocity (in feet per second) needed for a spacecraft to escape the gravitational pull of Earth is given by the expression $\sqrt{64R}$, where R is the radius of the planet in feet. Simplify this expression.

85. When an object at rest is accelerated a m/sec^2, its velocity in meters per second after it has traveled d m is given by the expression $\sqrt{2ad}$.

  a. If a car at rest is accelerated 2 m/sec^2, find its velocity after it has traveled 50 m.

  b. How much faster would the object in part (a) be moving if the acceleration had been 4 m/sec^2?

86. The distance associated with the illumination I of an object is given by the expression

$$\sqrt{\dfrac{k}{I}}$$

where k is a constant. Simplify this expression by rationalizing the denominator.

87. An object with mass m is attached to a spring. The spring's period of oscillation is given by the expression

$$2\pi\sqrt{\dfrac{m}{k}},$$

where k is a constant associated with the spring. Simplify this expression by rationalizing the denominator.

88. To get to her office building, an administrative assistant drives 8 mi north and 4 mi east from her apartment. How far is the office building from her home?

89. A wire is to be attached to a telephone pole at a point 20 ft above the ground. If 24 ft of wire are used, how far from the pole does the wire need to be anchored to the ground?

90. The owner of a health foods store determines that the demand equation for selling a nutritional supplement is $p = 18 - 0.5\sqrt{x - 4}$, where p is the price (in dollars) and x is the number of bottles demanded per week. How many bottles are demanded per week if the price of the supplement is $13.50?

• Check your answers on page A-43.

CHAPTER 7 Posttest

FOR
EXTRA
HELP

CHAPTER
Test Prep VIDEOS

The Chapter Test Prep Videos with test solutions are available on DVD, in MyMathLab, and on You Tube * (search "AkstIntermediate Alg" and click on "Channels").

To see if you have mastered the topics in this chapter, take this test. (Assume that all variables represent nonnegative real numbers.)

1. Evaluate:
   a. $-3\sqrt{81}$ 　　　　 b. $\sqrt[3]{-216}$

2. Simplify: $\sqrt{144a^6b^2}$

3. Rewrite in radical notation and simplify: $(32x^{10})^{2/5}$

4. Simplify:
   a. $\dfrac{(16p^{1/3})^{3/2}}{8p^{1/3}}$ 　　　 b. $\sqrt[8]{x^6y^2}$

5. Multiply: $\sqrt[3]{5p^2} \cdot \sqrt[3]{4q}$

6. Divide: $\dfrac{\sqrt{56n}}{\sqrt{7n}}$

7. Simplify: $\sqrt{117x^3y^7}$

Perform the indicated operation.

8. $\sqrt{\dfrac{6u}{49v^6}}$

9. $-4\sqrt{24} + 2\sqrt{54} - 7\sqrt{6}$

10. $(4\sqrt{2} + 3)(2\sqrt{2} - 5)$

11. Simplify: $\dfrac{\sqrt{3a}}{\sqrt{50b}}$

12. Rationalize the denominator: $\dfrac{\sqrt{x}}{\sqrt{x} - \sqrt{y}}$

Solve.

13. $\sqrt{2x - 1} + 9 = 5$

14. $\sqrt{8 - 3x} = \sqrt{6 - x^2}$

15. $\sqrt{x + 3} + \sqrt{2x + 5} = 2$

16. Multiply: $(3 + \sqrt{-49})(1 - \sqrt{-16})$

17. Divide: $\dfrac{3 - 5i}{2 + 3i}$

18. The radius of a sphere with surface area S is given by the expression

$$\sqrt{\dfrac{S}{4\pi}}$$

   a. Simplify the expression.
   b. Find the radius of a beach ball if its surface area is 512 in². Express the answer in radical form.

19. The distance to the horizon of a satellite h mi above Earth's surface is given by the expression $\sqrt{8000h + h^2}$. What is the distance to the horizon from a satellite that is 200 mi above Earth?

20. The *demand equation* for a particular item manufactured at a factory is $p = 32 - \sqrt{x - 5}$, where p is the price of the item (in dollars) and x is the daily demand. What is the daily demand when the price is $20?

• Check your answers on page A-44.

Cumulative Review Exercises

To help you review, solve these problems.

1. Solve and graph: $2(3x - 4) > 9x + 7$

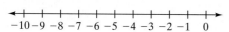

2. Solve: $-2 < \dfrac{1}{2}(-3x + 5) \leq 7$

3. Graph the equation $4x = 2y$ on the coordinate plane.

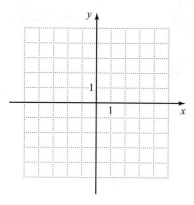

4. Graph: $3y - 2x + 9 \leq 0$

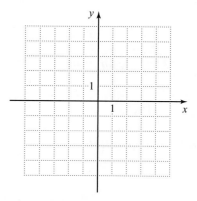

5. Solve the system of inequalities by graphing:

$$y < -\frac{1}{2}x + 3$$

$$y \geq x - 2$$

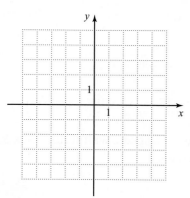

6. Subtract $(5x^2 - x - 6)$ from $(7x^2 - 4x - 9)$.

7. Factor completely: $x^3 - 2x^2 - 4x + 8$

8. Factor, if possible: $27x^6 - \dfrac{1}{8}y^3$

9. Solve: $3(x - 1)^2 = 12$

10. Add: $\dfrac{3}{x^2 - 2x + 1} + \dfrac{2}{4 - 3x - x^2}$

11. Simplify: $\dfrac{x^{-3} + 2x^{-2}}{3x^{-2}}$

12. Solve and check: $\dfrac{x}{6} = \dfrac{8}{x + 8}$

13. Simplify: $\sqrt[3]{54x^6y^4}$

14. Divide: $\dfrac{4 - 3i}{1 + i}$

15. To be acceptable, the width w of a machine part (in inches) must satisfy the inequality $|w - 3| \leq 0.002$. Find the least and greatest acceptable widths for this machine part.

16. A part-time employee at a grocery store has a weekly gross pay of $110 when he works 10 hr. If he works 25 hr, his gross pay is $275.

    a. Find a linear equation that relates the employee's gross pay p to the number of hours h he works per week.

    b. What does the slope of the line represent?

17. An investment of $2000 increases in value by $a\%$ the first year and by $(a + 1)\%$ the second year. What is the value of the investment at the end of the second year?

18. An object is thrown upward from a height of 250 ft above the ground with an initial velocity of 24 ft/sec. The height (in feet) above the ground of the object after t sec is given by the equation $h = -16t^2 + 24t + 250$. When will the object be 115 ft above the ground?

19. Suppose that the time t that it takes to complete a job varies inversely with the number of people n working. If it takes 2 painters 50 hr to paint a particular size house, how long will it take 8 painters to paint the house?

20. The volume of a regulation basketball is approximately 35 times that of a major league baseball. The radius r of a ball is given by the formula $r = \sqrt[3]{\dfrac{3V}{4\pi}}$, where V is the volume of the ball and $\pi \approx 3.14$. How many times a baseball's radius is a basketball's radius? (*Source:* wikipedia.org)

• Check your answers on page A-44.

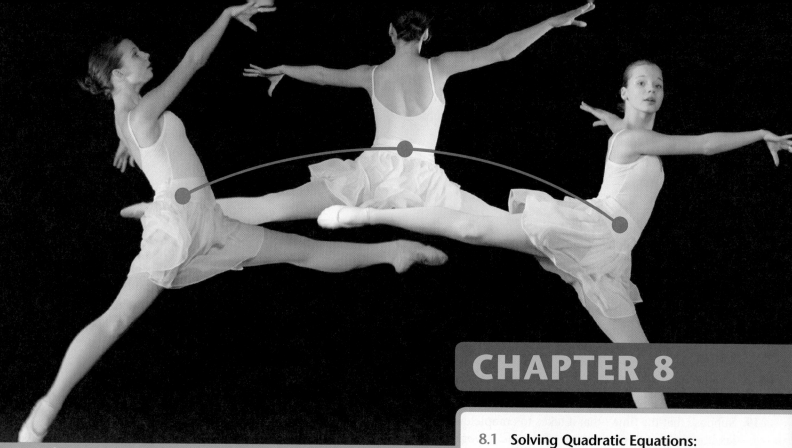

Quadratic Equations, Functions, and Inequalities

Dancing and Quadratic Functions

The motion of a dancer depends on a variety of physical factors. The forces on a dancer, for example, include gravity (pulling downward), support from the floor (pushing upward), and friction from the floor (exerted sideways).

For balance, the dancer's center of gravity must be directly above the area of contact with the floor. A smaller contact area with the floor makes balancing more difficult. When jumping, a dancer's center of gravity moves along a path that is parabolic, resembling the graph of a *quadratic function*. Great jumpers seem to ignore gravity as they float through the air in long, slow leaps. This illusion is created when the dancer raises or lowers her legs, changing the distance between her head and her center of gravity.

(*Source:* Kenneth Laws and Cynthia Harvey, *Physics, Dance, and the Pas de Deux*, Schirmer Books, 1994)

1. Solve by using the square root property:
   $3x^2 + 17 = 5$

2. Solve by completing the square:
   $4a^2 + 12a = 7$

3. Solve by using the quadratic formula:
   $2x^2 - 6x + 3 = 0$

4. Use the discriminant to determine the number and type of solutions of $3x^2 + 5x - 1 = 0$.

Solve.

5. $6(2n - 3)^2 = 48$

6. $x^2 - 4x + 12 = 0$

7. $3x^2 + 15x + 16 = x$

8. $2x^2 + 2x = 1 - 6x$

9. $5x^2 - 10x + 9 = 3$

10. $0.04x^2 - 0.12x + 0.09 = 0$

11. $x^4 - x^2 - 72 = 0$

12. Find a quadratic equation in x that has the solutions $-\dfrac{3}{2}$ and 4.

Identify the vertex, the equation of the axis of symmetry, and the x- and y-intercepts of the graph. Then, graph.

13. $f(x) = x^2 - 4x + 3$

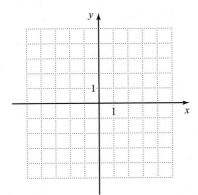

14. $f(x) = 12 - x - x^2$

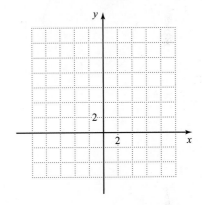

15. Graph the function $f(x) = \dfrac{1}{2}x^2 + x - 4$. Then, identify the domain and range.

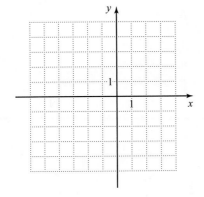

16. Solve. Then, graph the solution: $x^2 - 6x + 8 > 0$

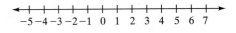

Solve.

17. A man tosses an object from the top of a tall building straight upward with an initial velocity of 20 ft/sec. The height h of the object above the point of release t sec after it is tossed upward is given by the equation $h = -16t^2 + 20t$. When is the object 50 ft below the point of release?

18. It takes a student on a boating trip 30 min less time to travel 6 mi downriver than to travel the same distance upriver. If the student travels at a speed of 9 mph in still water, find the speed of the current to the nearest mile per hour.

19. A gardener wants to make a circular planting area in her yard. The planting area will be surrounded by a cement border 1 ft in width. What radius will produce the maximum planting area if the total area including the border is to be 157 ft^2? Use $\pi \approx 3.14$ and round your answer to the nearest tenth.

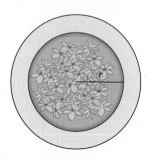

20. A health food company determines that its daily revenue R for selling x bottles of a new dietary supplement is modeled by the function $R(x) = 55x - 0.5x^2$. How many bottles must the company sell for revenue of at least $1200?

• Check your answers on page A-44.

8.1 Solving Quadratic Equations: The Square Root Property of Equality; Completing the Square

What Quadratic Equations Are and Why They Are Important

Recall from Section 5.7 that a *quadratic equation* (also called a second-degree equation) is an equation that can be written in the form $ax^2 + bx + c = 0$, where a, b, and c are real numbers and $a \neq 0$. In that section, we saw how quadratic equations can be used to model problems in physics and business. We solved such equations by factoring.

In this chapter, we consider three other methods of solving quadratic equations: the square root property of equality, completing the square, and, most importantly, the quadratic formula. Then, we discuss how to graph quadratic functions, and finally how to solve quadratic and rational inequalities.

The Square Root Property of Equality

Let's consider the equation $x^2 = 25$. We already solved the equivalent equation $x^2 - 25 = 0$ in Example 5 of Section 5.7 by factoring:

$$x^2 = 25$$
$$x^2 - 25 = 0$$
$$(x - 5)(x + 5) = 0$$
$$x - 5 = 0 \quad \text{or} \quad x + 5 = 0$$
$$x = 5 \qquad\qquad x = -5$$

So the solutions of $x^2 = 25$ are 5 and -5, which can be written as ± 5 (read "plus or minus 5"). Note that ± 5 can also be expressed as $\pm \sqrt{25}$. This suggests that we could also have solved the equation $x^2 = 25$ by taking the square root of each side of the equation.

> **The Square Root Property of Equality**
>
> If b is a real number and $x^2 = b$, then $x = \pm \sqrt{b}$; that is, $x = \sqrt{b}$ or $x = -\sqrt{b}$.

In words, this property states that if we take the square root of each side of an equation, the result is an equivalent equation.

We can apply the square root property of equality to solve quadratic equations of the form $x^2 = b$ or $ax^2 = b$. Note that for some of these equations, the solutions are not real numbers.

EXAMPLE 1

Solve by taking the square roots.

a. $x^2 = 36$

b. $z^2 + 4 = 0$

Solution

a. $x^2 = 36$

$\quad x = \pm \sqrt{36}$ Use the square root property of equality.

$\quad x = \pm 6$ Simplify.

PRACTICE 1

Use the square root property of equality to solve.

a. $n^2 = 49$

b. $y^2 + 9 = 0$

EXAMPLE 1 (continued)

Check

Substitute 6 for x.   Substitute -6 for x.

$$x^2 = 36 \qquad\qquad x^2 = 36$$
$$(6)^2 \stackrel{?}{=} 36 \qquad\qquad (-6)^2 \stackrel{?}{=} 36$$
$$36 = 36 \quad \text{True} \qquad 36 = 36 \quad \text{True}$$

So the solutions are 6 and -6.

b. $z^2 + 4 = 0$

$$z^2 = -4 \qquad \text{Add } -4 \text{ to each side of the equation.}$$
$$z = \pm\sqrt{-4} \qquad \text{Use the square root property of equality.}$$
$$z = \pm 2i \qquad \text{Simplify.}$$

Check

Substitute $2i$ for z.   Substitute $-2i$ for z.

$$z^2 + 4 = 0 \qquad\qquad z^2 + 4 = 0$$
$$(2i)^2 + 4 \stackrel{?}{=} 0 \qquad\qquad (-2i)^2 + 4 \stackrel{?}{=} 0$$
$$-4 + 4 \stackrel{?}{=} 0 \qquad\qquad -4 + 4 \stackrel{?}{=} 0$$
$$0 = 0 \quad \text{True} \qquad\qquad 0 = 0 \quad \text{True}$$

So the solutions are $2i$ and $-2i$.

EXAMPLE 2

Solve.

a. $3x^2 = 36$ **b.** $5x^2 + 40 = 0$

Solution

a. $3x^2 = 36$

$$x^2 = 12 \qquad \text{Divide each side of the equation by 3.}$$
$$x = \pm\sqrt{12} \qquad \text{Use the square root property of equality.}$$
$$x = \pm 2\sqrt{3} \qquad \text{Simplify.}$$

So the solutions are $2\sqrt{3}$ and $-2\sqrt{3}$.

b. $5x^2 + 40 = 0$

$$5x^2 = -40 \qquad \text{Subtract 40 from each side of the equation.}$$
$$x^2 = -8 \qquad \text{Divide each side of the equation by 5.}$$
$$x = \pm\sqrt{-8} \qquad \text{Use the square root property of equality.}$$
$$x = \pm 2i\sqrt{2} \qquad \text{Simplify.}$$

So the solutions are $2i\sqrt{2}$ and $-2i\sqrt{2}$.

PRACTICE 2

Solve.

a. $2y^2 = 16$

b. $3x^2 + 54 = 0$

We can also use the square root property of equality to solve quadratic equations of the form $(ax + b)^2 = d$ or $a(x + c)^2 = d$, that is, equations containing the square of a binomial.

EXAMPLE 3

Solve.

a. $(3x - 1)^2 = 18$

b. $(x + 5)^2 + 6 = 0$

c. $4(y + 1)^2 = 25$

Solution

a. $(3x - 1)^2 = 18$

$$3x - 1 = \pm\sqrt{18} \qquad \text{Use the square root property of equality.}$$

$$3x = 1 \pm 3\sqrt{2} \qquad \text{Add 1 to each side of the equation and simplify the radical.}$$

$$x = \frac{1 \pm 3\sqrt{2}}{3} \qquad \text{Divide each side of the equation by 3.}$$

So the solutions are $\dfrac{1 + 3\sqrt{2}}{3}$ and $\dfrac{1 - 3\sqrt{2}}{3}$.

b. $(x + 5)^2 + 6 = 0$

$$(x + 5)^2 = -6 \qquad \text{Subtract 6 from each side of the equation.}$$

$$x + 5 = \pm\sqrt{-6} \qquad \text{Use the square root property of equality.}$$

$$x = -5 \pm \sqrt{-6} \qquad \text{Subtract 5 from each side of the equation.}$$

$$x = -5 \pm i\sqrt{6} \qquad \text{Simplify the radical.}$$

So the solutions are $x = -5 + i\sqrt{6}$ and $x = -5 - i\sqrt{6}$.

c. $4(y + 1)^2 = 25$

$$(y + 1)^2 = \frac{25}{4} \qquad \text{Divide each side of the equation by 4.}$$

$$y + 1 = \pm\sqrt{\frac{25}{4}} \qquad \text{Use the square root property of equality.}$$

$$y = -1 \pm \frac{5}{2} \qquad \text{Subtract 1 from each side of the equation and simplify the radical.}$$

$$y = -1 + \frac{5}{2} \quad \text{or} \quad y = -1 - \frac{5}{2}$$

$$y = \frac{3}{2} \qquad\qquad y = -\frac{7}{2}$$

So the solutions are $\dfrac{3}{2}$ and $-\dfrac{7}{2}$.

PRACTICE 3

Solve.

a. $(2x + 1)^2 = 12$

b. $(n + 3)^2 + 2 = 0$

c. $9(y - 1)^2 = 49$

Some quadratic equations contain a perfect square trinomial. To solve these equations, we rewrite the trinomial as the square of a binomial, and then apply the square root property of equality, as shown in the next example.

EXAMPLE 4

Solve: $x^2 + 10x + 25 = 12$

Solution

$$x^2 + 10x + 25 = 12$$
$$(x + 5)^2 = 12$$
$$x + 5 = \pm\sqrt{12}$$
$$x = -5 \pm 2\sqrt{3}$$

So the solutions are $-5 + 2\sqrt{3}$ and $-5 - 2\sqrt{3}$.

PRACTICE 4

Solve: $y^2 - 6y + 9 = 24$

Completing the Square

Another method of solving quadratic equations is called **completing the square.** To apply this method, we must transform one side of an equation to a perfect square trinomial. For instance, consider the equation $x^2 + 6x = 1$. Note that the expression on the left side of the equation is *not* a perfect square. Let's recall our discussion of perfect square trinomials in Section 5.6.

Perfect square		The coefficient of the quadratic term is 1.		The coefficient of the linear term		The constant term
$(x + 1)^2$	$=$	x^2	$+$	$2x$	$+$	1
$(x + 2)^2$	$=$	x^2	$+$	$4x$	$+$	4
$(x + 3)^2$	$=$	x^2	$+$	$6x$	$+$	9
$(x + 4)^2$	$=$	x^2	$+$	$8x$	$+$	16

Note that in each of these perfect square trinomials, the coefficient of the quadratic term is 1 and the constant term is the square of one-half the coefficient of the linear term. For example, in $x^2 + \mathbf{8}x + \mathbf{16}$, we see that:

$$16 = \left(\frac{1}{2} \cdot 8\right)^2$$

More generally, when we square the binomial $x + c$ we get:

$$(x + c)^2 = x^2 + 2cx + c^2$$

where

$$c^2 = \left(\frac{1}{2} \cdot 2c\right)^2$$

The constant term   The coefficient of the linear term

To complete the square of the expression $x^2 + bx$, that is, to transform it to a perfect square trinomial, we find one-half of the coefficient of the linear term x, square this value, and then add it to the expression, giving us:

$$x^2 + bx + \left(\frac{1}{2}b\right)^2$$

EXAMPLE 5

Fill in the blank to make the trinomial a perfect square.

a. $y^2 - 12y + [\quad]$ **b.** $x^2 + 5x + [\quad]$

Solution

a. $y^2 - 12y + [\quad]$

The coefficient of the linear term y is -12.

$$\frac{1}{2}(-12) = -6 \qquad \text{Find one-half of the coefficient of the linear term.}$$

The constant term to be added to $y^2 - 12y$ is therefore $(-6)^2$, or 36. So we get $y^2 - 12y + 36$, which is equal to $(y - 6)^2$.

b. $x^2 + 5x + [\quad]$

First, we find one-half the coefficient of the linear term x:

$$\frac{1}{2} \cdot 5 = \frac{5}{2}$$

The constant term to be added to $x^2 + 5x$ is $\left(\frac{5}{2}\right)^2$, or $\frac{25}{4}$.

So we get $x^2 + 5x + \frac{25}{4}$, which is equal to $\left(x + \frac{5}{2}\right)^2$.

PRACTICE 5

Fill in the blank to make each trinomial a perfect square.

a. $x^2 + 6x + [\quad]$

b. $t^2 - t + [\quad]$

Now, let's apply the method of completing the square to solve a quadratic equation of the form $ax^2 + bx + c$, where $a \neq 1$. Consider, for instance, the equation $2x^2 + 12x = 8$. To complete the square, the coefficient of the quadratic term x^2 must be 1. So before completing the square in this equation, we must first divide each side of the equation by 2, the coefficient of x^2.

$$2x^2 + 12x = 8$$
$$x^2 + 6x = 4 \qquad \text{Divide each side of the equation by 2.}$$

Next, we take one-half of the coefficient of the linear term x: $\frac{1}{2}(6) = 3$. Then, we square the result: $3^2 = 9$. Now, we add 9 to the left side of the equation to complete the square. To maintain equality, we must also add 9 to the right side of the equation.

$$x^2 + 6x + 9 = 4 + 9$$

Finally, we solve the equation by rewriting $x^2 + 6x + 9$ as the square of a binomial, and then applying the square root property of equality.

$$(x + 3)^2 = 13$$
$$x + 3 = \pm\sqrt{13} \qquad \text{Use the square root property of equality.}$$
$$x = -3 \pm \sqrt{13} \qquad \text{Subtract 3 from each side of the equation.}$$

So the solutions are $-3 + \sqrt{13}$ and $-3 - \sqrt{13}$. Can you check that these solutions are correct?

This example suggests the following general method for solving a quadratic equation by completing the square:

To Solve a Quadratic Equation ($ax^2 + bx + c = 0$) by Completing the Square

- If $a \neq 1$, then divide the equation by a, the coefficient of the quadratic (or second-degree) term. If $a = 1$, then proceed to the next step.

- Move all terms with variables to one side of the equal sign and all constants to the other side.

- Take one-half the coefficient of the linear term, then square it and add this value to each side of the equation.

- Factor the side of the equation containing the variables, writing it as the square of a binomial.

- Use the square root property of equality.

- Solve the resulting equation.

EXAMPLE 6

Solve $x^2 - 5x - 2 = 0$ by completing the square.

Solution

$$x^2 - 5x - 2 = 0$$

$$x^2 - 5x = 2 \qquad \text{Add 2 to each side of the equation.}$$

$$x^2 - 5x + \frac{25}{4} = 2 + \frac{25}{4} \qquad \text{Take one-half of the coefficient of the linear term: } \frac{1}{2}(-5) = -\frac{5}{2}.$$

$$\text{Square it: } \left(-\frac{5}{2}\right)^2 = \frac{25}{4}. \text{ Add } \frac{25}{4} \text{ to each side of the equation.}$$

$$x^2 - 5x + \frac{25}{4} = \frac{33}{4} \qquad \text{Simplify the right side.}$$

$$\left(x - \frac{5}{2}\right)^2 = \frac{33}{4} \qquad \text{Write } x^2 - 5x + \frac{25}{4} \text{ as the square of a binomial.}$$

$$x - \frac{5}{2} = \pm\sqrt{\frac{33}{4}} \qquad \text{Use the square root property of equality.}$$

$$x = \frac{5}{2} \pm \sqrt{\frac{33}{4}} \qquad \text{Add } \frac{5}{2} \text{ to each side.}$$

$$x = \frac{5}{2} + \frac{\sqrt{33}}{2} \quad \text{or} \quad x = \frac{5}{2} - \frac{\sqrt{33}}{2}$$

$$x = \frac{5 + \sqrt{33}}{2} \qquad\qquad x = \frac{5 - \sqrt{33}}{2}$$

So the solutions are $\dfrac{5 + \sqrt{33}}{2}$ and $\dfrac{5 - \sqrt{33}}{2}$.

PRACTICE 6

Solve $t^2 + 3t = 1$ by completing the square.

EXAMPLE 7

Solve: $3x^2 + 2x + 5 = 0$

Solution

$3x^2 + 2x + 5 = 0$

$x^2 + \dfrac{2}{3}x + \dfrac{5}{3} = 0$ Divide each side of the equation by 3, the coefficient of the quadratic term.

$x^2 + \dfrac{2}{3}x = -\dfrac{5}{3}$ Add $-\dfrac{5}{3}$ to each side of the equation.

$x^2 + \dfrac{2}{3}x + \dfrac{1}{9} = -\dfrac{5}{3} + \dfrac{1}{9}$ Take one-half the coefficient of the linear term: $\dfrac{1}{2}\left(\dfrac{2}{3}\right) = \dfrac{1}{3}$.

Square it: $\left(\dfrac{1}{3}\right)^2 = \dfrac{1}{9}$. Add $\dfrac{1}{9}$ to each side of the equation.

$x^2 + \dfrac{2}{3}x + \dfrac{1}{9} = -\dfrac{14}{9}$ Simplify the right side.

$\left(x + \dfrac{1}{3}\right)^2 = -\dfrac{14}{9}$ Write $x^2 + \dfrac{2}{3}x + \dfrac{1}{9}$ as the square of a binomial.

$x + \dfrac{1}{3} = \pm\sqrt{-\dfrac{14}{9}}$ Use the square root property of equality.

$x = -\dfrac{1}{3} \pm \sqrt{-\dfrac{14}{9}}$ Add $-\dfrac{1}{3}$ to each side.

$x = -\dfrac{1}{3} + i\dfrac{\sqrt{14}}{3}$ or $x = -\dfrac{1}{3} - i\dfrac{\sqrt{14}}{3}$

$x = \dfrac{-1 + i\sqrt{14}}{3}$ $x = \dfrac{-1 - i\sqrt{14}}{3}$

So the solutions are $\dfrac{-1 + i\sqrt{14}}{3}$ and $\dfrac{-1 - i\sqrt{14}}{3}$.

PRACTICE 7

Solve for y: $2y^2 - y + 1 = 0$

EXAMPLE 8

If $f(x) = x^2 - 5x + 3$ and $g(x) = x + 2$, find all values of x for which $f(x) = g(x)$.

Solution

$$f(x) = g(x)$$
$$x^2 - 5x + 3 = x + 2$$
$$x^2 - 6x = -1$$
$$x^2 - 6x + 9 = -1 + 9$$
$$(x - 3)^2 = 8$$
$$x - 3 = \pm\sqrt{8}$$
$$x = 3 \pm \sqrt{8}$$
$$x = 3 \pm 2\sqrt{2}$$

So the solutions are $3 + 2\sqrt{2}$ and $3 - 2\sqrt{2}$.

PRACTICE 8

If $g(x) = x^2 + 6x$ and $h(x) = 2$, find all values of x for which $g(x)$ and $h(x)$ are equal.

EXAMPLE 9

The amount of pollution A (in parts per million) present in the air outside a factory on a recent weekday is modeled by the equation $A = -t^2 + 16t + 6$, where t is the number of hours after 7 A.M. Find the times, to the nearest hour, when there were 42 parts per million of pollution.

Solution   Substituting 42 for A, we solve the equation $42 = -t^2 + 16t + 6$:

$$42 = -t^2 + 16t + 6$$
$$t^2 - 16t - 6 = -42$$
$$t^2 - 16t = -36$$
$$t^2 - 16t + 64 = -36 + 64$$
$$(t - 8)^2 = 28$$
$$t - 8 = \pm\sqrt{28}$$
$$t = 8 \pm \sqrt{28}$$

The solutions to the equation are $8 + \sqrt{28}$ and $8 - \sqrt{28}$. Using a calculator, we can approximate the solutions to the nearest whole number by 3 and 13, respectively. Since t represents the number of hours after 7 A.M., the times with the given amount of pollution are about 10 A.M. and 8 P.M.

PRACTICE 9

Each day, a company produces x pieces of machinery. The company's daily profit P (in dollars) depends on the value of x and can be modeled by the equation $P = -x^2 + 200x - 4000$. For which value of x, to the nearest whole number, does the company make a daily profit of $6000?

In Example 9, which we solved using a calculator, why did we not express the answers as $8 \pm 2\sqrt{7}$?

A *Solve by using the square root property of equality.*

1. $x^2 = 16$

2. $x^2 = 81$

3. $y^2 = 24$

4. $n^2 = 63$

5. $a^2 + 25 = 0$

6. $p^2 + 64 = 0$

7. $4n^2 - 8 = 0$

8. $2x^2 - 6 = 0$

9. $\frac{1}{6}y^2 = 12$

10. $\frac{1}{3}a^2 = 32$

11. $3x^2 + 6 = 21$

12. $5y^2 + 19 = 54$

13. $8 - 9n^2 = 14$

14. $10 - 4p^2 = 21$

15. $(x - 1)^2 = 48$

16. $(x + 3)^2 = 18$

17. $(2n + 5)^2 = 75$

18. $(3t - 2)^2 = 108$

19. $(4a - 3)^2 + 9 = 1$

20. $(2p + 1)^2 + 30 = 6$

21. $16(x + 4)^2 = 81$

22. $25(y - 10)^2 = 36$

23. $x^2 - 6x + 9 = 80$

24. $n^2 + 8n + 16 = 50$

25. $4p^2 + 12p + 9 = 32$

26. $9x^2 - 12x + 4 = 27$

27. $(3n - 2)(3n + 2) = -52$

28. $(4t + 3)(4t - 3) = -49$

29. $2x - 1 = \dfrac{18}{2x - 1}$

30. $3n + 1 = \dfrac{20}{3n + 1}$

Solve the formula for the indicated variable.

31. Volume of a cylinder: $V = \pi r^2 h$, for r

32. Kinetic energy: $E = \dfrac{1}{2}mv^2$, for v

33. Centripetal force: $F = \dfrac{mv^2}{r}$, for v

34. Pythagorean theorem: $a^2 + b^2 = c^2$, for b

B *Fill in the blank to make each trinomial a perfect square.*

35. $x^2 - 12x + [\quad]$

36. $y^2 - 16y + [\quad]$

37. $n^2 + 7n + [\quad]$

38. $x^2 - 3x + [\quad]$

39. $t^2 - \dfrac{4}{3}t + [\quad]$

40. $a^2 + \dfrac{4}{5}a + [\quad]$

Solve by completing the square.

41. $x^2 - 8x = 0$

42. $p^2 + 12p = 0$

43. $n^2 - 3n = 4$

44. $t^2 - t = 2$

45. $x^2 + 4x - 2 = 0$

46. $x^2 - 6x + 6 = 0$

47. $a^2 + 7a = 3a - 4$

48. $y^2 - 4y = 4y - 16$

49. $x^2 - 9x + 4 = x - 25$

50. $p^2 + 2p + 35 = 10 - 6p$

51. $2n^2 - 8n = -24$

52. $3x^2 + 6x = -9$

53. $3x^2 - 12x - 84 = 0$

54. $5x^2 - 60x + 80 = 0$

55. $4a^2 + 20a - 12 = 0$

56. $2t^2 + 6t - 10 = 0$

57. $3y^2 - 9y + 15 = 0$ **58.** $4n^2 - 4n + 16 = 0$ **59.** $x^2 - \frac{4}{3}x - 4 = 0$ **60.** $x^2 - \frac{2}{5}x - 3 = 0$

61. $4y^2 + 11y + 6 = 0$ **62.** $2n^2 + 7n + 5 = 0$ **63.** $2p^2 + 7p = 6p - 8$ **64.** $3r^2 - r = r - 15$

Given $f(x)$ and $g(x)$, find all values of x for which $f(x) = g(x)$.

65. $f(x) = x^2 - 9$ and $g(x) = 4x - 6$

66. $f(x) = x^2 + 3x - 5$ and $g(x) = x$

67. $f(x) = 4x^2$ and $g(x) = x^2 - 6x + 6$

68. $f(x) = 2x^2 - 5x$ and $g(x) = -x + 14$

Mixed Practice

Solve.

69. The formula for the volume of a cone is $V = \frac{1}{3}\pi r^2 h$. Solve this formula for r.

70. Fill in the blank to make the trinomial a perfect square: $n^2 - 5n + [\ \ ]$

Solve by completing the square.

71. Given $f(x) = x^2 - 4x$ and $g(x) = 2x - 2$, find all values of x for which $f(x) = g(x)$.

72. $(4t - 3)^2 = 32$

73. $-3x^2 - 24x + 6 = 0$

74. $x^2 + 16x + 16 = 4x - 22$

Solve by using the square root property of equality or completing the square.

75. $x^2 - 10x + 25 = 45$

76. $\frac{1}{2}x^2 = 24$

Applications

C *Solve. Use a calculator, where appropriate.*

77. In a search-and-rescue mission, a team maps out a circular search area from the last known location of a group of hikers. If the search region is 78.5 mi², how far from the last known location, to the nearest mile, is the team searching? Use $\pi \approx 3.14$.

78. The projection televisions made by an electronics manufacturer have screen dimensions that are in the ratio of 3:4. What are the screen dimensions of a 50-inch projection television made by the company?

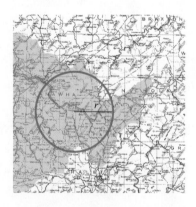

79. A student invests $1000 in an account earning r percent interest compounded annually. After two years, the amount A in the account is given by

$$A = 1000(1 + r)^2,$$

where r is in decimal form. What is the interest rate if the account has $1102.50 after two years?

80. The parks department of a city plans to build a circular wading pool in one of its parks. The wading pool is to be surrounded by a concrete ledge for sitting 2 ft in width. If the pool including the ledge has an area of 785 ft^2, what is the radius of the wading pool, to the nearest tenth of a foot? Use $\pi \approx 3.14$.

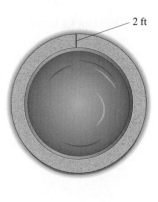

2 ft

81. The height h (in feet) relative to the point of release of an object t sec after it is thrown straight upward with an initial velocity of 32 ft/sec is given by the equation $h = 32t - 16t^2$.

a. How long after it is thrown upward will it take the object to return to the original height at which it was released?

b. How long after it is thrown upward will the object be 12 ft above the point of release?

82. A manufacturer of athletic shoes determines that its daily revenue R (in dollars) from selling x pairs of running shoes is modeled by the equation $R = 5x + 0.5x^2$.

a. How many pairs of running shoes must be sold per day to generate revenue of $1500?

b. If the answer in part (a) is doubled, does the daily revenue double? Explain.

83. A homeowner wants to build a rectangular patio using the house as one side, as shown in the figure below. He decides to make the length of the patio 10 ft longer than the width.

a. What are the dimensions of the patio if the homeowner wants to enclose an area of 144 ft^2?

b. If the fencing costs $14.95 per foot, how much will it cost to enclose the patio?

84. Supplies are dropped to marching troops from a military aircraft 3400 ft above the ground. The equation $h = -16t^2 + 3400$ models the height h (in feet) above the ground of the supplies t sec after they are dropped.

a. Approximately how long after being released will the supplies be 1700 ft above the ground?

b. Is the time it takes the supplies to reach the ground exactly twice the length of time found in the solution to part (a)? Explain.

85. Two trucks leave a truck stop at the same time. One truck goes north, driving at an average speed that is 16 mph slower than the truck driving east. After 15 min, the trucks are 20 mi apart. Find the average speed of each truck.

86. Two campus security officers meet at the campus center. One walks south and the other walks west. The officer walking south walks at an average rate that is 30 ft/min faster than the officer walking west. If they are 25 ft apart after 10 sec, find the rate at which each security officer is walking.

• Check your answers on page A-44.

MINDStretchers

Mathematical Reasoning

1. Consider the equation $x^3 = 1$. Can this equation be solved by taking the cube root of each side? Explain.

Writing

2. Consider the two methods for solving a quadratic equation presented in this section—the square root property of equality and completing the square. Discuss when it is better to use one method over the other to solve a quadratic equation.

Groupwork

3. Working with a partner, consider the equation $x^2 + 6x + c = 0$, where c is a constant.
   a. When does this equation have a single solution?
   b. For which values of c are the solutions of the equation real numbers?
   c. For which values of c are the solutions not real?

Cultural Note

As long ago as 2000 B.C., the Babylonians had developed techniques to solve quadratic equations. A major advance in algebra took place in the sixteenth century in connection with third-degree (cubic) and fourth-degree (quartic) equations. In 1545, Girolamo Cardano, responding to a challenge, published solutions to these equations. This triggered a lengthy dispute with another Italian mathematician, Niccolò Tartaglia, over credit for the discovery—one of a number of such disputes in the history of mathematics. Cardano was a multitalented Renaissance scholar. A prominent physician, he first described typhoid fever. An inveterate gambler, he published the earliest systematic treatise on the mathematics of probability. Also a talented inventor, Cardano's conviction that a code in one's mind is more secure than a physical key led to his inventing the combination lock.

(*Sources:* W. Dunham, *Journey Through Genius: The Great Theorems of Mathematics,* John Wiley and Sons, 1990; Peter L. Bernstein, *Against the Gods: The Remarkable Story of Risk,* John Wiley and Sons, 1996; Girolamo Cardano, *The Book of My Life,* J. M. Dent and Sons, 1931)

OBJECTIVES

Some quadratic equations cannot be solved either by factoring or by the square root property of equality. On the other hand, any quadratic equation in standard form, $ax^2 + bx + c = 0$, *can* be solved by completing the square.

A To solve a quadratic equation by using the quadratic formula

In fact, we can use the method of completing the square to derive a general formula, called the *quadratic formula*, to solve quadratic equations. In this section, we show how to use the quadratic formula to solve quadratic equations. We also discuss the *discriminant* of a quadratic equation—an expression that allows us to determine the number and type of solutions that the equation has without solving the equation.

B To determine the number and type of solutions to a quadratic equation by using the discriminant

C To solve applied problems involving the quadratic formula

Deriving the Quadratic Formula

Consider an arbitrary quadratic equation in standard form: $ax^2 + bx + c = 0$, where $a \neq 0$. We apply the method of completing the square to this equation to solve for x.

$$ax^2 + bx + c = 0$$

$$x^2 + \frac{b}{a}x + \frac{c}{a} = 0 \qquad \text{Divide each side of the equation by } a.$$

$$x^2 + \frac{b}{a}x = -\frac{c}{a} \qquad \text{Add } -\frac{c}{a} \text{ to each side.}$$

$$x^2 + \frac{b}{a}x + \frac{b^2}{4a^2} = -\frac{c}{a} + \frac{b^2}{4a^2} \qquad \text{Complete the square: } \left(\frac{1}{2} \cdot \frac{b}{a}\right)^2 = \frac{b^2}{4a^2}; \text{ add } \frac{b^2}{4a^2} \text{ to each side of the equation.}$$

$$\left(x + \frac{b}{2a}\right)^2 = \frac{b^2 - 4ac}{4a^2} \qquad \text{Write the left side as the square of a binomial, and write the right side as a single rational expression.}$$

$$x + \frac{b}{2a} = \pm\sqrt{\frac{b^2 - 4ac}{4a^2}} \qquad \text{Use the square root property of equality.}$$

$$x = -\frac{b}{2a} \pm \frac{\sqrt{b^2 - 4ac}}{2a} \qquad \text{Add } -\frac{b}{2a} \text{ to each side of the equation and simplify the radical.}$$

$$x = \frac{-b \pm \sqrt{b^2 - 4ac}}{2a} \qquad \text{Combine the rational expressions.}$$

So the solutions are $\dfrac{-b + \sqrt{b^2 - 4ac}}{2a}$ and $\dfrac{-b - \sqrt{b^2 - 4ac}}{2a}$. This result gives us the quadratic formula.

The Quadratic Formula

If $ax^2 + bx + c = 0$, where a, b, and c are real numbers and $a \neq 0$, then

$$x = \frac{-b \pm \sqrt{b^2 - 4ac}}{2a}.$$

We will be considering a number of examples of applying this formula to solve quadratic equations.

Solving Quadratic Equations by Using the Quadratic Formula

The quadratic formula allows us to solve any quadratic equation of the form $ax^2 + bx + c = 0$, where $a \neq 0$. For instance, consider the equation $2x^2 - x = 3$. To solve, we first write the equation in standard form:

$$2x^2 - x = 3$$
$$2x^2 - x - 3 = 0 \qquad \text{Add } -3 \text{ to each side of the equation.}$$

In this equation, $a = 2$, $b = -1$, and $c = -3$. So we substitute these values for a, b, and c in the quadratic formula, and then simplify.

$$x = \frac{-b \pm \sqrt{b^2 - 4ac}}{2a}$$

$$= \frac{-(-1) \pm \sqrt{(-1)^2 - 4(2)(-3)}}{2(2)} \qquad \text{Substitute 2 for } a, -1 \text{ for } b, \text{ and } -3 \text{ for } c.$$

$$= \frac{1 \pm \sqrt{1 + 24}}{4}$$

$$= \frac{1 \pm \sqrt{25}}{4}$$

$$= \frac{1 \pm 5}{4}$$

$$x = \frac{1 + 5}{4} \qquad \text{or} \qquad x = \frac{1 - 5}{4}$$

$$= \frac{6}{4} \qquad\qquad\qquad = \frac{-4}{4}$$

$$= \frac{3}{2} \qquad\qquad\qquad = -1$$

So the solutions are $\frac{3}{2}$ and -1.

Note that the equation $2x^2 - x = 3$ could also have been solved by factoring:

$$2x^2 - x - 3 = 0$$
$$(2x - 3)(x + 1) = 0$$
$$2x - 3 = 0 \quad \text{or} \quad x + 1 = 0$$
$$2x = 3 \qquad\qquad x = -1$$
$$x = \frac{3}{2}$$

We see that solving the equation by factoring yields the same solutions as those we found using the quadratic formula. Whenever the solutions to a quadratic equation are rational numbers, as in this example, the original equation can also be solved by factoring.

To Solve a Quadratic Equation ($ax^2 + bc + c = 0$) by Using the Quadratic Formula

- Write the equation in standard form, if necessary.

- Identify the coefficients a, b, and c.

- Substitute values for a, b, and c in the formula $x = \dfrac{-b \pm \sqrt{b^2 - 4ac}}{2a}$.

- Simplify.

EXAMPLE 1

Solve $x^2 + x - 3 = 0$ by using the quadratic formula.

Solution   The equation $x^2 + x - 3 = 0$ is already in standard form, where $a = 1$, $b = 1$, and $c = -3$. Substitute these values in the quadratic formula.

$$x = \frac{-b \pm \sqrt{b^2 - 4ac}}{2a}$$

$$x = \frac{-(1) \pm \sqrt{(1)^2 - 4(1)(-3)}}{2(1)}$$

$$= \frac{-1 \pm \sqrt{1 + 12}}{2}$$

$$= \frac{-1 \pm \sqrt{13}}{2}$$

$$x = \frac{-1 + \sqrt{13}}{2} \quad \text{or} \quad x = \frac{-1 - \sqrt{13}}{2}$$

So the solutions are $\dfrac{-1 + \sqrt{13}}{2}$ and $\dfrac{-1 - \sqrt{13}}{2}$.

PRACTICE 1

Solve: $y^2 + 3y - 5 = 0$

Can you explain how to check that the solutions in Example 1 are correct?

EXAMPLE 2

Use the quadratic formula to solve $3x^2 + 2x = -4$.

Solution   First, let's write $3x^2 + 2x = -4$ in standard form: $3x^2 + 2x + 4 = 0$. We see that $a = 3$, $b = 2$ and $c = 4$. Substitute these values in the formula.

$$x = \frac{-b \pm \sqrt{b^2 - 4ac}}{2a}$$

$$= \frac{-(2) \pm \sqrt{(2)^2 - 4(3)(4)}}{2(3)}$$

$$= \frac{-2 \pm \sqrt{4 - 48}}{6}$$

$$= \frac{-2 \pm \sqrt{-44}}{6}$$

$$= \frac{-2 \pm 2i\sqrt{11}}{6} = \frac{\overset{1}{2}(-1 \pm i\sqrt{11})}{\underset{3}{6}}$$

$$= \frac{-1 \pm i\sqrt{11}}{3}, \quad \text{or} \quad -\frac{1}{3} \pm \frac{i\sqrt{11}}{3}$$

So the solutions are $-\dfrac{1}{3} + \dfrac{i\sqrt{11}}{3}$ and $-\dfrac{1}{3} - i\dfrac{\sqrt{11}}{3}$.

PRACTICE 2

Solve: $6p^2 - 2p = -1$

In solving quadratic equations, we sometimes need to approximate the solutions. This is often the case when solving applied problems.

EXAMPLE 3

Solve $\dfrac{y^2}{3} + \dfrac{y}{5} = 1$. Round the solutions to the nearest thousandth.

Solution First, we multiply each side of the equation by the LCD, 15, to clear the equation of fractions.

$$\frac{y^2}{3} + \frac{y}{5} = 1$$

$$15 \cdot \frac{y^2}{3} + 15 \cdot \frac{y}{5} = 15 \cdot 1$$

$$5y^2 + 3y = 15$$

Then, we write the equation in standard form, $5y^2 + 3y - 15 = 0$, letting $a = 5, b = 3$, and $c = -15$. Substituting in the quadratic formula gives us:

$$y = \frac{-b \pm \sqrt{b^2 - 4ac}}{2a}$$

$$= \frac{-3 \pm \sqrt{(3)^2 - 4(5)(-15)}}{2(5)}$$

$$= \frac{-3 \pm \sqrt{9 + 300}}{10} = \frac{-3 \pm \sqrt{309}}{10}$$

So the solutions are:

$$\frac{-3 + \sqrt{309}}{10} \approx 1.458 \quad \text{and} \quad \frac{-3 - \sqrt{309}}{10} \approx -2.058$$

PRACTICE 3

Solve $\dfrac{n^2}{5} - \dfrac{n}{2} = 3$. Round the solutions to the nearest thousandth.

EXAMPLE 4

Solve: $0.04m^2 + 0.25 = 0.2m$

Solution In standard form, the equation becomes $0.04m^2 - 0.2m + 0.25 = 0$. To clear the equation of decimals, let's multiply through by 100:

$$0.04m^2 - 0.2m + 0.25 = 0$$

$$100(0.04)m^2 - 100(0.2)m + 100(0.25) = 100(0)$$

$$4m^2 - 20m + 25 = 0$$

So $a = 4, b = -20$, and $c = 25$. Substituting in the quadratic formula gives us:

$$m = \frac{-b \pm \sqrt{b^2 - 4ac}}{2a}$$

$$= \frac{-(-20) \pm \sqrt{(-20)^2 - 4(4)(25)}}{2(4)}$$

$$= \frac{20 \pm \sqrt{400 - 400}}{8}$$

$$= \frac{20 \pm \sqrt{0}}{8} = \frac{20}{8} = \frac{5}{2}$$

So there is only one solution, namely $\dfrac{5}{2}$.

PRACTICE 4

Use the quadratic formula to solve:
$0.01v^2 = -0.18v - 0.81$

Can you explain in terms of the quadratic formula when a quadratic equation will have a single solution?

The Discriminant

It is possible to determine the number and type of solutions to a quadratic equation without actually solving the equation. To do this, we compute the value of the radicand in the quadratic formula, that is, $b^2 - 4ac$. This quantity is called the **discriminant** of the equation.

Consider, for instance, the equation $x^2 - 5x + 3 = 0$. Here, $a = 1, b = -5$, and $c = 3$. Using the quadratic formula, we get:

$$x = \frac{-b \pm \sqrt{b^2 - 4ac}}{2a}$$

$$= \frac{-(-5) \pm \sqrt{(-5)^2 - 4(1)(3)}}{2(1)}$$

$$= \frac{5 \pm \sqrt{13}}{2}$$

Note that the value of this discriminant 13 is *positive* and that the quadratic equation has the *two real solutions* $\dfrac{5 + \sqrt{13}}{2}$ and $\dfrac{5 - \sqrt{13}}{2}$.

Next, let's consider the equation $x^2 - 3x + 5 = 0$. Here, $a = 1, b = -3$, and $c = 5$. Using the quadratic formula give us:

$$x = \frac{-b \pm \sqrt{b^2 - 4ac}}{2a}$$

$$= \frac{-(-3) \pm \sqrt{(-3)^2 - 4(1)(5)}}{2(1)}$$

$$= \frac{3 \pm \sqrt{-11}}{2}$$

Here, the value of the discriminant -11 is *negative* and the quadratic equation has *two complex solutions containing i,* $\dfrac{3 + i\sqrt{11}}{2}$ and $\dfrac{3 - i\sqrt{11}}{2}$.

Finally, we consider the equation $x^2 - 2x + 1 = 0$. Here, $a = 1, b = -2$, and $c = 1$. From the quadratic formula, we get:

$$x = \frac{-b \pm \sqrt{b^2 - 4ac}}{2a}$$

$$= \frac{-(-2) \pm \sqrt{(-2)^2 - 4(1)(1)}}{2(1)}$$

$$= \frac{2 \pm \sqrt{0}}{2}$$

$$= 1$$

The value of the discriminant in this case is *zero* and the quadratic equation has *one real solution*, namely 1.

The preceding examples suggest that the value of a quadratic equation's discriminant can tell us if the equation has one real solution, two real solutions, or two complex solutions. This is summarized in the following table:

Value of the Discriminant ($b^2 - 4ac$)	Number and Type of Solutions
Zero	One real solution
Positive	Two real solutions
Negative	Two complex solutions (containing i)

EXAMPLE 5

For each of the following equations, evaluate the discriminant. Then, use the discriminant to determine the number and type of solutions that the equation has.

a. $6x^2 + 11x - 7 = 0$ **b.** $5y^2 - y + 3 = 0$ **c.** $4x^2 = -12x - 9$

Solution

Equation	a	b	c	Discriminant ($b^2 - 4ac$)	Number of Solutions	Type of Solution
a. $6x^2 + 11x - 7 = 0$	6	11	−7	$(11)^2 - 4(6)(-7)$ $= 121 + 168 = 289$, which is *positive*	2	Real numbers
b. $5y^2 - y + 3 = 0$	5	−1	3	$(-1)^2 - 4(5)(3)$ $= 1 - 60 = -59$, which is *negative*	2	Complex numbers (containing i)
c. $4x^2 = -12x - 9$	4	12	9	$(12)^2 - 4(4)(9)$ $= 144 - 144 = 0$, which is *zero*	1	Real number

PRACTICE 5

Complete the following table:

Equation	a	b	c	Discriminant ($b^2 - 4ac$)	Number of Solutions	Type of Solution
a. $p^2 + 6p + 9 = 0$						
b. $x^2 + 6x - 9 = 0$						
c. $4n^2 = -3 - 2n$						

EXAMPLE 6

The total number in the United States of children adopted from other countries is modeled by the equation $y = -322x^2 + 3027x + 15,530$, where x represents the number of years since 1999. According to this model, in what year(s) did the number of adoptions reach approximately 20,680? (*Source:* adoption.state.gov)

Solution   We set y equal to 20,680, and then solve the resulting equation.

$$20,680 = -322x^2 + 3027x + 15,530$$
$$0 = -322x^2 + 3027x - 5150$$

So we write: $-322x^2 + 3027x - 5150 = 0$

Next, we substitute in the quadratic formula, letting $a = -322$, $b = 3027$, and $c = -5150$.

$$x = \frac{-b \pm \sqrt{b^2 - 4ac}}{2a}$$

$$= \frac{-3027 \pm \sqrt{(3027)^2 - 4(-322)(-5150)}}{2(-322)}$$

$$= \frac{-3027 \pm \sqrt{9,162,729 - 4(-322)(-5150)}}{2(-322)}$$

$$= \frac{-3027 \pm \sqrt{9,162,729 - 6,633,200}}{-644}$$

$$= \frac{-3027 \pm \sqrt{2,529,529}}{-644}$$

So the solutions are

$$\frac{-3027 + \sqrt{2,529,529}}{-644} \approx 2.2 \quad \text{and} \quad \frac{-3027 - \sqrt{2,529,529}}{-644} \approx 7.2.$$

Since both values for x are positive, we can conclude that $x \approx 2.2$ or 2 and $x \approx 7.1$ or 7. Therefore, the total number of adoptions in the United States of children from other countries was approximately 20,680 in $1999 + 2$, or 2001 and in $1999 + 7$, or 2006.

PRACTICE 6

The equation

$$y = -0.04x^2 + 2.09x + 21.45$$

represents the annual per capita consumption (in gallons) of bottled water from 2003 to 2007, where x represents the number of years since 2003. During this period of time, in what year was the annual per capita consumption of bottled water approximately 27.4 gallons? (*Source:* census. gov)

Solving a Quadratic Equation ($ax^2 + bx + c = 0$)

Method	When to Use
Factoring	Use when the polynomial can be factored.
The Square Root Property	Use when $b = 0$.
The Quadratic Formula	Use when the first two methods do not apply.
Completing the Square	Use only when specified. This method is easier to use when $a = 1$ and b is even.

Mathematically Speaking

Fill in each blank with the most appropriate term or phrase from the given list.

one	quadratic equation	negative
discriminant	numerator	quadratic formula
positive	no	

1. The _____ states that if $ax^2 + bx + c = 0$, where a, b, and c are real numbers and $a \neq 0$, then $x = \dfrac{-b \pm \sqrt{b^2 - 4ac}}{2a}$.

2. The _____ is the radicand in the quadratic formula.

3. If the discriminant of a quadratic equation equals zero, the equation has _____ real solution(s).

4. If a quadratic equation has two real solutions, its discriminant is _____.

A *Solve.*

5. $x^2 + 3x + 2 = 0$

6. $x^2 - 7x + 12 = 0$

7. $x^2 - 6x - 1 = 0$

8. $x^2 + 8x - 4 = 0$

9. $x^2 = x + 11$

10. $x^2 = 2 - 5x$

11. $x^2 - 4x + 13 = 8$

12. $x^2 + 2x + 11 = 6$

13. $3x^2 + 6x = 7$

14. $2x^2 - 10x = 1$

15. $6t^2 - 8t = 3t - 4$

16. $8n^2 + 5n = 7n + 3$

17. $2x^2 + 8x + 9 = 0$

18. $4x^2 - 10x + 7 = 0$

19. $1 - 5x^2 = 4x^2 + 6x$

20. $-3x^2 + 14x = 8x - 5$

21. $2y^2 - 9y + 10 = 1 + 3y - 2y^2$

22. $13p^2 + 8p - 2 = 4p^2 + 2p - 3$

23. $\dfrac{x^2}{4} - \dfrac{x}{2} = -3$

24. $\dfrac{x^2}{9} + \dfrac{x}{3} = -1$

25. $\dfrac{1}{2}x^2 + \dfrac{2}{3}x - \dfrac{5}{6} = 0$

26. $\dfrac{1}{8}x^2 - \dfrac{1}{4}x - \dfrac{1}{12} = 0$

27. $0.2x^2 + x + 0.8 = 0$

28. $0.5x^2 + 0.3x - 0.2 = 0$

29. $0.03x^2 - 0.12x + 0.24 = 0$

30. $0.02x^2 + 0.16x + 0.34 = 0$

31. $(x + 6)(x + 2) = 8$

32. $(x + 2)(x - 4) = 1$

33. $(2x - 3)^2 = 8(x + 1)$

34. $(3x + 1)^2 = 2(1 - 3x)$

Solve. Round the answers to the nearest thousandth.

35. $1.4x^2 - 2.7x - 0.1 = 0$

36. $0.6x^2 - 4.9x - 3.3 = 0$

37. $0.003x^2 + 0.23x + 1.124 = 0$

38. $2.04x^2 + 0.45x + 0.017 = 0$

B *Use the discriminant to determine the number and type of solutions for each equation.*

39. $x^2 + 2x + 4 = 0$

40. $7x^2 - x + 3 = 0$

41. $4x^2 - 12x = -9$

42. $x^2 - 8x + 16 = 0$

43. $3x^2 + 6x = -1$

44. $6x^2 = 2 - 5x$

45. $10x = 16 - 2x^2$

46. $8x^2 - 7x - 1 = 0$

47. $3x^2 + 10 = 0$

48. $2x^2 + 7 = 0$

Mixed Practice

Use the discriminant to determine the number and type of solutions for each equation.

49. $2x^2 + 8 = 7x$

50. $5x = 2 + 3x^2$

Solve.

51. $2x^2 - 4x + 2 = -3$

52. $x - 2x^2 - 3 = 2x^2 - 7x - 8$

53. $\dfrac{x^2}{4} - \dfrac{2x}{3} - \dfrac{1}{6} = 0$

54. $0.8x^2 - 3.7x - 0.5 = 0$
(Round to the nearest thousandth.)

Applications

C *Solve.*

55. The height h (in feet) of a stone t sec after it is thrown straight downward with an initial velocity of 20 ft/sec from a bridge 800 ft above the ground is given by the equation $h = -16t^2 - 20t + 800$. When is the stone 300 ft above the ground?

56. A software company's weekly profit P (in dollars) for selling x units of a new video game can be determined by the equation $P = -0.05x^2 + 48x - 100$. What is the smaller of the two numbers of units that must be sold in order to make a profit of $9000?

57. An open box is made from a 36-inch by 20-inch rectangular piece of cardboard by cutting squares of equal area from the corners, as shown in the following figure. If the design specifications require that the base of the box have an area of 465 in², what size squares should be cut from each corner?

58. A homeowner decides to increase the area of her rectangular garden by increasing the length and the width, as shown in the figure. What are the new dimensions of the garden if it is to have an area of 315 ft²?

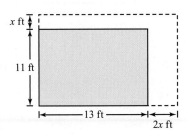

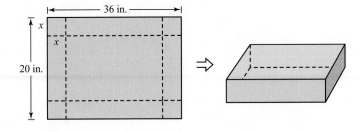

59. A travel company offers special pricing on a weekend getaway trip. For each person who buys a ticket, the price is reduced by \$5. The regular price of a ticket is \$300. To determine the revenue R (in dollars) from selling x tickets, the company uses the equation $R = x(300 - 5x)$. If the company wants to keep the reduced ticket price above \$200, how many tickets must it sell in order to generate revenue of \$3520?

60. One airplane flies east from an airport at an average speed of m mph. At the same time, a second airplane flies south from the same airport at an average speed that is 42 mph faster than the airplane traveling east. Thirty minutes after the two airplanes leave, they are 315 mi apart. Find the rate of each airplane to the nearest mile per hour.

61. In the United States, the number of reports R (in thousands) of consumer fraud for each year from 2005 to 2009 can be approximated by the equation $R = 6.3t^2 + 52.8t + 418$, where t is the number of years after 2005. In what year were there approximately 644 thousand reports of consumer fraud? (*Source:* ftc.gov)

62. The net revenue R (in billions of dollars) for Starbucks Corporation for the years from 2003 to 2009 can be approximated by the equation $R = -0.144x^2 + 2.067x + 3.069$, where x represents the number of years after 2003. In what year during this time period was the net revenue approximately \$7.8 billion? (*Source:* investor.starbucks.com)

• Check your answers on page A-44.

MINDStretchers

Critical Thinking

1. Solve each of the following equations:
  a. $2x^2 + \sqrt{3}x - 3 = 0$
  b. $x^2 - ix + 6 = 0$

Mathematical Reasoning

2. Consider the equation $ax^2 + bx + c = 0$. For what values of c will the equation have two real solutions? Two complex solutions (containing i)? Give examples to support your answers.

Writing

3. Do you think "discriminant" is a good name for the expression $b^2 - 4ac$? In a couple of sentences, explain why.

8.3 More on Quadratic Equations

We have discussed several methods of solving quadratic equations—factoring, the square root property, completing the square, and the quadratic formula. In this section, we use these methods to solve some additional types of problems relating to quadratic equations. The first of these problems involves equations that, while not quadratic, can be transformed to quadratic equations. In the second type of problem, we are given the solutions to an unknown quadratic equation and need to determine the equation.

Equations That Lead to Quadratic Equations

Recall from Section 6.4 that some equations, while not quadratic, lead to quadratic equations. For instance, in the following example, we solve a rational equation that leads to a quadratic equation.

EXAMPLE 1

Solve and check: $\dfrac{3}{x} - \dfrac{2}{2x - 1} = 1$

Solution First, we multiply each side of the equation by the LCD of the denominators, $x(2x - 1)$:

$$\frac{3}{x} - \frac{2}{2x - 1} = 1$$

$$\cancel{x}(2x - 1)\frac{3}{\cancel{x}} - x(\cancel{2x - 1})\frac{2}{\cancel{2x - 1}} = x(2x - 1)1$$
Multiply each side by the LCD, and divide out common factors.

$$3(2x - 1) - 2x = x(2x - 1)$$
Simplify.

$$6x - 3 - 2x = 2x^2 - x$$
Use the distributive property.

$$4x - 3 = 2x^2 - x$$
Combine like terms.

$$2x^2 - 5x + 3 = 0$$
Write in standard form.

Next, we factor:

$$(2x - 3)(x - 1) = 0$$

To solve for x, we set each factor equal to 0:

$$2x - 3 = 0 \quad \text{or} \quad x - 1 = 0$$
$$2x = 3 \qquad\qquad x = 1$$
$$x = \frac{3}{2}$$

PRACTICE 1

Solve and check: $1 = \dfrac{2}{x + 2} + x$

EXAMPLE 1 (continued)

Check

Substitute $\dfrac{3}{2}$ for x.

$$\frac{3}{x} - \frac{2}{2x-1} = 1$$

$$\frac{3}{\frac{3}{2}} - \frac{2}{2\left(\frac{3}{2}\right)-1} \overset{?}{=} 1$$

$$\frac{6}{3} - \frac{2}{3-1} \overset{?}{=} 1$$

$$2 - 1 \overset{?}{=} 1$$

$$1 = 1 \qquad \text{True}$$

Substitute 1 for x.

$$\frac{3}{x} - \frac{2}{2x-1} = 1$$

$$\frac{3}{1} - \frac{2}{2(1)-1} \overset{?}{=} 1$$

$$3 - 2 \overset{?}{=} 1$$

$$1 = 1 \qquad \text{True}$$

So the solutions are $\dfrac{3}{2}$ and 1.

Some equations are **quadratic in form**. That is, they can be rewritten in the form $au^2 + bu + c = 0$, where $a \neq 0$ and u represents an algebraic expression. For example, the equation $x^4 - 3x^2 - 4 = 0$ is quadratic in form since it can be written as $(x^2)^2 - 3(x^2) - 4 = 0$, or $u^2 - 3u - 4 = 0$, where u stands for x^2.

To solve an equation that is quadratic in form, we transform the equation to a quadratic equation by rewriting it in terms of a new variable, as shown in the following example:

EXAMPLE 2

Solve and check: $x^4 - 5x^2 - 6 = 0$

Solution   This equation is quadratic in form. Letting a new variable, say u, stand for x^2 gives us:

$$x^4 - 5x^2 - 6 = 0$$
$$(x^2)^2 - 5(x^2) - 6 = 0$$
$$(u)^2 - 5(u) - 6 = 0 \qquad \text{Substitute } u \text{ for } x^2.$$
$$u^2 - 5u - 6 = 0$$

Next, let's factor to solve for u:

$$(u - 6)(u + 1) = 0$$
$$u - 6 = 0 \quad \text{or} \quad u + 1 = 0$$
$$u = 6 \qquad\qquad u = -1$$

Having found the two possible values of u, we now solve for the original variable x. Since u represents x^2, we have:

$$x^2 = 6 \qquad \text{or} \quad x^2 = -1$$
$$x = \pm\sqrt{6} \qquad\qquad x = \pm\sqrt{-1}$$
$$x = \pm i$$

Check   We substitute these values for x in the *original* equation.

Substitute $\sqrt{6}$ for x.

$$x^4 - 5x^2 - 6 = 0$$
$$(\sqrt{6})^4 - 5(\sqrt{6})^2 - 6 \overset{?}{=} 0$$
$$6^2 - 5 \cdot 6 - 6 \overset{?}{=} 0$$
$$0 = 0 \qquad \text{True}$$

Substitute $-\sqrt{6}$ for x.

$$x^4 - 5x^2 - 6 = 0$$
$$(-\sqrt{6})^4 - 5(-\sqrt{6})^2 - 6 \overset{?}{=} 0$$
$$6^2 - 5 \cdot 6 - 6 \overset{?}{=} 0$$
$$0 = 0 \qquad \text{True}$$

PRACTICE 2

Solve and check: $n^4 - 9n^2 + 8 = 0$

Substitute i for x.

$$x^4 - 5x^2 - 6 = 0$$
$$(i)^4 - 5(i)^2 - 6 \stackrel{?}{=} 0$$
$$(-1)^2 - 5(-1) - 6 \stackrel{?}{=} 0$$
$$0 = 0 \quad \text{True}$$

Substitute $-i$ for x.

$$x^4 - 5x^2 - 6 = 0$$
$$(-i)^4 - 5(-i)^2 - 6 \stackrel{?}{=} 0$$
$$(-1)^2 - 5(-1) - 6 \stackrel{?}{=} 0$$
$$0 = 0 \quad \text{True}$$

All solutions check. So the original equation has four solutions, namely $\sqrt{6}$, $-\sqrt{6}$, i, and $-i$.

TIP

When introducing a new variable to substitute for a given expression, we can choose any letter to represent the new variable.

EXAMPLE 3

Solve and check: $y - 2\sqrt{y} = 8$

Solution   Recognizing that $(\sqrt{y})^2 = y$, we can write the given equation as follows:

$$(\sqrt{y})^2 - 2\sqrt{y} = 8$$

Substituting u for $\sqrt{y}$ gives us:

$$u^2 - 2u = 8$$
$$u^2 - 2u - 8 = 0$$

Now, we solve for u by factoring.

$$(u - 4)(u + 2) = 0$$
$$u - 4 = 0 \quad \text{or} \quad u + 2 = 0$$
$$u = 4 \qquad\qquad u = -2$$

Having found the possible values of u, we now solve for the original variable y. Since u represents $\sqrt{y}$, we get:

$$\sqrt{y} = 4 \qquad \text{or} \qquad \sqrt{y} = -2$$
$$(\sqrt{y})^2 = (4)^2 \qquad (\sqrt{y})^2 = (-2)^2$$
$$y = 16 \qquad\qquad y = 4$$

Check

Substitute 16 for y.

$$y - 2\sqrt{y} = 8$$
$$16 - 2\sqrt{16} \stackrel{?}{=} 8$$
$$8 = 8 \quad \text{True}$$

Substitute 4 for y.

$$y - 2\sqrt{y} = 8$$
$$4 - 2\sqrt{4} \stackrel{?}{=} 8$$
$$0 = 8 \quad \text{False}$$

So the only solution of the original equation is 16.

PRACTICE 3

Solve and check: $x - 3\sqrt{x} = 10$

EXAMPLE 4

Solve and check: $x^{1/2} - 4x^{1/4} + 3 = 0$

Solution   We can rewrite this equation as follows:

$$(x^{1/4})^2 - 4(x^{1/4}) + 3 = 0$$

Substituting u for $x^{1/4}$ and solving, we get:

$$u^2 - 4u + 3 = 0$$
$$(u - 3)(u - 1) = 0$$
$$u - 3 = 0 \quad \text{or} \quad u - 1 = 0$$
$$u = 3 \qquad\qquad u = 1$$

Having found the two possible values of u, we now solve for the original variable x. Since u represents $x^{1/4}$, it follows that:

$$
\begin{array}{ll}
x^{1/4} = 3 & \quad\text{or}\quad \quad x^{1/4} = 1 \\
\sqrt[4]{x} = 3 & \qquad\quad \sqrt[4]{x} = 1 \\
(\sqrt[4]{x})^4 = (3)^4 & \quad (\sqrt[4]{x})^4 = (1)^4 \\
x = 81 & \qquad\quad x = 1
\end{array}
$$

Check

Substitute 81 for x.   Substitute 1 for x.

$$
\begin{array}{ll}
x^{1/2} - 4x^{1/4} + 3 = 0 & \quad x^{1/2} - 4x^{1/4} + 3 = 0 \\
(81)^{1/2} - 4(81)^{1/4} + 3 \stackrel{?}{=} 0 & \quad (1)^{1/2} - 4(1)^{1/4} + 3 \stackrel{?}{=} 0 \\
9 - 4 \cdot 3 + 3 \stackrel{?}{=} 0 & \quad 1 - 4 \cdot 1 + 3 \stackrel{?}{=} 0 \\
9 - 12 + 3 \stackrel{?}{=} 0 & \quad 1 - 4 + 3 \stackrel{?}{=} 0 \\
0 = 0 \quad \text{True} & \quad 0 = 0 \quad \text{True}
\end{array}
$$

The solutions of the original equation are 81 and 1.

PRACTICE 4

Solve and check:
$$y^{2/3} - y^{1/3} - 20 = 0$$

Many applications can be modeled by equations that are not quadratic, but are quadratic in form, as in the next example.

EXAMPLE 5

A young businessman cycled to and from work, which is 2 mi from his apartment. Because of traffic, he cycled 3 mph faster going to work than coming home. It took him 10 min longer coming home than going to work. To the nearest tenth of a mile per hour, at what speed did he cycle from work?

Solution   Let r represent the cycling speed coming from work. Since the businessman cycled 3 mph faster going to work, we can represent this speed by $r + 3$. We know that the distance between home and work is 2 mi. Since the product of the rate and the time is the distance traveled, the cycling time is the quotient of the distance and the rate: $t = \dfrac{d}{r}$. Therefore, the time required to cycle *to* work

PRACTICE 5

It takes a painter 1 hr longer to paint an apartment than it takes his faster partner. If they take 5 hr working together to paint the apartment, how long would it take each of them to paint the apartment working alone?

is $\dfrac{2}{r+3}$ and the time required to cycle *from* work is $\dfrac{2}{r}$. We can summarize this information in a table.

	Rate	· Time =	Distance
To work	$r+3$	$\dfrac{2}{r+3}$	2
From work	r	$\dfrac{2}{r}$	2

We also know that the cycling time from work is 10 min longer than the cycling time to work. Since the cycling speeds are expressed in mph and there are 60 min in an hour, we rewrite the 10 min in terms of hours as follows:

$$10 \text{ min} = 10 \cdot \frac{1}{60} \text{ hr} = \frac{1}{6} \text{ hr}$$

So we conclude that the cycling time from work is $\dfrac{1}{6}$ hr longer than the cycling time to work, leading to the following equation:

$$\frac{2}{r} = \frac{2}{r+3} + \frac{1}{6}$$

To solve this equation, we multiply by the LCD of the denominators, $6r(r+3)$, and then divide out common factors.

$$6r(r+3)\frac{2}{r} = 6r(r+3)\frac{2}{r+3} + 6r(r+3)\frac{1}{6}$$

$$12(r+3) = 12r + r(r+3)$$
$$12r + 36 = 12r + r^2 + 3r$$
$$0 = r^2 + 3r - 36$$
$$r^2 + 3r - 36 = 0$$

Here, $a = 1$, $b = 3$, and $c = -36$. We substitute these values in the quadratic formula, and then simplify.

$$r = \frac{-b \pm \sqrt{b^2 - 4ac}}{2a}$$

$$= \frac{-(3) \pm \sqrt{(3)^2 - 4(1)(-36)}}{2(1)}$$

$$= \frac{-3 \pm \sqrt{9 + 144}}{2}$$

$$= \frac{-3 \pm \sqrt{153}}{2}$$

$$= \frac{-3 \pm 3\sqrt{17}}{2}$$

So the solutions to the equation are $\dfrac{-3 + 3\sqrt{17}}{2} \approx 4.685$ and $\dfrac{-3 - 3\sqrt{17}}{2} \approx -7.685$. Since the value of r in the context of our problem must be positive, we reject the negative solution and conclude that the cycling speed from work is approximately 4.7 mph.

Finding a Quadratic Equation That Has Given Solutions

In the typical problem involving a quadratic equation, we are given the equation, and are asked to find its solutions. But suppose that we know the solutions to a quadratic equation and want to find the original equation. For instance, let's say that -4 and 3 are solutions to a quadratic equation in x. Since $x = -4$, it follows that $x + 4 = 0$. Similarly, since $x = 3$, it follows that $x - 3 = 0$. Since both $x + 4$ and $x - 3$ are equal to 0, the product of these two factors must also be equal to 0.

$$(x + 4)(x - 3) = 0$$
$$x^2 + x - 12 = 0$$

We can conclude that $x^2 + x - 12 = 0$ might have been the original equation. Note that this answer is not unique, since multiplying each side of the equation by any constant results in a different quadratic equation with the same given solutions. For example, multiplying each side of the equation $x^2 + x - 12 = 0$ by 3 yields $3x^2 + 3x - 36 = 0$, which has the same solutions.

EXAMPLE 6

Find a quadratic equation that has solutions $x = 1$ and $x = -5$.

Solution

$$x = 1 \qquad\qquad x = -5$$
$$x - 1 = 0 \quad \text{Subtract 1} \qquad x + 5 = 0 \quad \text{Add 5}$$
$$\text{from each side.} \qquad\qquad \text{to each side.}$$

Setting the product of the two factors equal to 0 gives us:

$$(x - 1)(x + 5) = 0$$
$$x^2 + 4x - 5 = 0$$

This quadratic equation has solutions 1 and -5, as desired.

PRACTICE 6

Find a quadratic equation in y that has solutions 0 and -3.

EXAMPLE 7

Write a quadratic equation in t with integer coefficients and with solutions $\frac{1}{2}$ and $\frac{2}{3}$.

Solution

$$t = \frac{1}{2} \qquad\qquad t = \frac{2}{3}$$
$$t - \frac{1}{2} = 0 \quad \text{Subtract } \frac{1}{2} \text{ from} \qquad t - \frac{2}{3} = 0 \quad \text{Subtract } \frac{2}{3} \text{ from}$$
$$\text{each side.} \qquad\qquad \text{each side.}$$
$$2t - 1 = 0 \quad \text{Multiply each side} \qquad 3t - 2 = 0 \quad \text{Multiply each}$$
$$\text{by 2.} \qquad\qquad \text{side by 3.}$$

Setting the product of the two factors equal to 0, we get:

$$(2t - 1)(3t - 2) = 0$$
$$6t^2 - 7t + 2 = 0$$

This quadratic equation has the desired solutions.

PRACTICE 7

Find a quadratic equation in n with integer coefficients that has solutions $\frac{3}{4}$ and $\frac{1}{6}$.

Mathematical Reasoning

Fill in each blank with the most appropriate term or phrase from the given list.

product	in terms of a new variable	binomial
as a linear equation		quadratic
quadratic	sum	in form

1. The equation $\dfrac{4}{x-4} + \dfrac{3}{x-1} = 1$ leads to a
   _____ equation.

2. An equation is _____ if it can be rewritten in the form $au^2 + bu + c = 0$, where $a \neq 0$ and u represents an algebraic expression.

3. To solve an equation that is quadratic in form, rewrite it _____.

4. To find a quadratic equation whose solutions are 6 and -4, set the _____ of $x - 6$ and $x + 4$ equal to 0.

A *Solve and check.*

5. $2 - \dfrac{3}{x} + \dfrac{1}{x^2} = 0$

6. $3 - \dfrac{7}{x} = \dfrac{6}{x^2}$

7. $\dfrac{3}{p-1} + \dfrac{4}{p-4} = 1$

8. $\dfrac{3}{t+6} - \dfrac{1}{t-2} - 1 = 0$

9. $n^4 - 8n^2 + 12 = 0$

10. $a^4 + 2a^2 - 8 = 0$

11. $3x^4 + 11x^2 - 3 = -x^4$

12. $9y^4 - 5y^2 + 2 = 6$

13. $h - 4\sqrt{h} + 4 = 0$

14. $x - \sqrt{x} - 20 = 0$

15. $x^{1/2} + 4x^{1/4} - 32 = 0$

16. $x^{1/2} - 3x^{1/4} + 2 = 0$

17. $t^{2/3} + 3t^{1/3} + 2 = 0$

18. $y^{2/3} - y^{1/3} - 12 = 0$

19. $(n - 3)^2 - 5(n - 3) + 6 = 0$

20. $(p + 2)^2 + 4(p + 2) + 3 = 0$

21. $(2x + 4) + 2\sqrt{2x + 4} = 24$

22. $(3x - 2) - 6\sqrt{3x - 2} = -8$

B *Find a quadratic equation that has the given solutions.*

23. $x = 7, x = 2$

24. $x = 3, x = 6$

25. $y = -4, y = 9$

26. $n = 1, n = -10$

27. $t = \dfrac{2}{3}, t = -2$

28. $a = -\dfrac{1}{4}, a = -5$

29. $n = -\dfrac{3}{2}, n = -\dfrac{4}{3}$

30. $p = \dfrac{2}{5}, p = \dfrac{5}{2}$

31. $x = 4$

32. $y = -8$

33. $t = \sqrt{2}, t = -\sqrt{2}$

34. $x = \sqrt{5}, x = -\sqrt{5}$

35. $x = 3i, x = -3i$

36. $x = 6i, x = -6i$

Mixed Practice

Solve and check.

37. $y^{2/3} - y^{1/3} = 2$

38. $\dfrac{5}{x} + 2 = \dfrac{3}{x^2}$

39. $a^4 + 7a^2 - 12 = 6$

40. $b + 5\sqrt{b} - 24 = 0$

Find a quadratic equation that has the given solutions.

41. $a = -\dfrac{2}{5}, a = 3$

42. $x = -7$

Applications

C *Solve. Use a calculator, where appropriate.*

43. *Body mass index* (BMI) is an estimate of total body fat based on a person's height H (in meters) and weight W (in kilograms). The formula for this index is $BMI = \dfrac{W}{H^2}$. Find the approximate height, to the nearest tenth of a meter, of a person who weighs 52 kg and has a BMI of 20.

44. The formula $P = \dfrac{A}{(1 + r)^2}$ can be used to determine the amount P that must be invested in an account with an average interest rate r (in decimal form), compounded annually, in order to have A dollars at the end of two years. If an account with an initial investment of \$10,000 has \$11,290 after two years, find the interest rate to the nearest tenth of a percent.

45. Because of traffic, a commuter drives home at an average speed that is 8 mph slower than her average speed driving to work. If she lives 15 mi from work and the round-trip commute takes 1 hr, find the average speed she drives for each part of her commute to the nearest mile per hour.

46. A couple kayaks 9 mi upriver from their campsite to a waterfall. The trip downriver takes half an hour less than the trip upriver. If the couple kayaks at an average rate of 8 mph in still water, find the rate of the current. Round to the nearest tenth.

47. A swimming pool has a small and a large drain. If only the small drain is open, it takes 1.5 hr longer to empty the pool than if only the large drain is open. If both drains are open, the pool empties in 2 hr. To the nearest tenth of an hour, find the time it takes each drain to empty the pool if only one of the drains is open.

48. Working alone, it takes an inexperienced gardener 45 min longer than an experienced gardener to mow a client's lawn. If the two gardeners work together, they can mow the lawn in 30 min. Find the time it takes each gardener working alone to mow the lawn.

• Check your answers on page A-45.

MINDStretchers

Groupwork

1. Consider the solutions $\dfrac{-b + \sqrt{b^2 - 4ac}}{2a}$ and $\dfrac{-b - \sqrt{b^2 - 4ac}}{2a}$ of the quadratic equation $ax^2 + bx + c = 0$.

   Work with a partner to answer each of the following:

   a. Find the sum of the solutions. What do you notice?

   b. Find the product of the solutions. What do you notice?

   c. Explain how you could use the sum and product of the solutions of a quadratic equation to find its equation in standard form.

   d. Find a quadratic equation with solutions whose sum is $-\dfrac{2}{3}$ and whose product is $\dfrac{1}{2}$.

Investigation

2. Consider each of the following polynomial equations in one variable.

   a. $3x - 4 = 0$ **b.** $x^2 - 4x + 4 = 0$ **c.** $x^3 - 4x^2 + x - 4 = 0$

   d. $x^4 - 2x^2 - 3 = 0$ **e.** $x^5 - 2x^3 - x^2 + 2 = 0$ **f.** $x^6 - 7x^3 - 8 = 0$

       i. Use the techniques of this and previous sections to find *all* the solutions of each equation.

       ii. For each equation, compare the number of solutions to the degree of the polynomial. What do you notice?

Critical Thinking

3. Solve.

   a. $\dfrac{1}{(y + 1)^2} - 2\left(\dfrac{1}{y + 1}\right) = 3$ **b.** $2n^{-2} + 3n^{-1} + 1 = 0$ **c.** $x^{1/3} + 3x^{1/6} - 4 = 0$

8.4 Graphing Quadratic Functions

Recall our discussion of graphs of functions in Section 3.6. There we graphed the *quadratic function* $f(x) = x^2$. In this section, we focus on graphing more general **quadratic functions**, that is, functions of the form $f(x) = ax^2 + bx + c$.

The graph of a quadratic function is called a **parabola**. Parabolas model many phenomena in the real world. For instance, when a drinking fountain is turned on, the water's path is parabolic.

A To graph a quadratic function

B To identify the vertex, the axis of symmetry, and the intercepts of a parabola

C To determine whether the graph of a quadratic function opens upward or downward

D To determine the domain or range of a quadratic function

E To solve applied problems involving the graph of a quadratic function

Graphing Quadratic Functions

We can graph quadratic functions by first choosing several values of x, and then computing the corresponding values of $f(x)$ to make a table of values. Then, we plot the points with these coordinates on a coordinate plane and draw a smooth curve through the points. Recall that $y = f(x)$, so that the ordered pairs (x, y) and $(x, f(x))$ represent the same point.

Let's revisit the graph of the function $f(x) = x^2$. To graph this function, we first select some values for x, say $-2, -1, 0, 1$, and 2, and then find the corresponding values for $f(x)$. We put the results in a table.

x	$f(x) = x^2$	(x, y)
-2	$f(-2) = (-2)^2 = 4$	$(-2, 4)$
-1	$f(-1) = (-1)^2 = 1$	$(-1, 1)$
0	$f(0) = (0)^2 = 0$	$(0, 0)$
1	$f(1) = (1)^2 = 1$	$(1, 1)$
2	$f(2) = (2)^2 = 4$	$(2, 4)$

Next, we plot the points on a coordinate plane, and then draw a smooth curve passing through them. The result is the graph of $f(x) = x^2$, as shown below. Note that this parabola is U-shaped and that it opens upward.

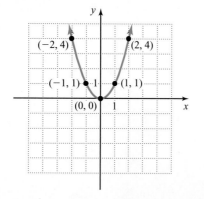

EXAMPLE 1

Graph $f(x) = 3x^2$.

Solution   First, we select several values for x, and then find the corresponding values for y.

x	$f(x) = 3x^2$	(x, y)
-2	$f(-2) = 3(-2)^2 = 3 \cdot 4 = 12$	$(-2, 12)$
-1	$f(-1) = 3(-1)^2 = 3 \cdot 1 = 3$	$(-1, 3)$
0	$f(0) = 3(0)^2 = 3 \cdot 0 = 0$	$(0, 0)$
1	$f(1) = 3(1)^2 = 3 \cdot 1 = 3$	$(1, 3)$
2	$f(2) = 3(2)^2 = 3 \cdot 4 = 12$	$(2, 12)$

Next, we plot the points on a coordinate plane, and then draw a smooth curve through them. The result is a parabola, which is the graph of $f(x) = 3x^2$.

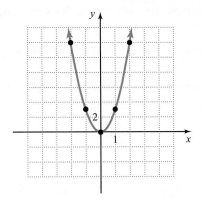

PRACTICE 1

Graph $f(x) = -3x^2$.

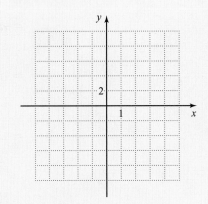

What are the differences and similarities between the graphs of $f(x) = 3x^2$ and $f(x) = -3x^2$?

The Vertex and the Axis of Symmetry

Parabolas have special characteristics that help us to sketch them. For instance, each parabola has either a highest or lowest point called a **vertex**. A graph of a quadratic equation also has an **axis of symmetry**, that is, a vertical line that passes through the vertex. If we were to fold the coordinate plane along this axis, the two parts of the parabola would coincide since each part is the mirror image of the other. That is, the parabola is symmetric with respect to its axis.

Consider the two graphs shown. The parabola on the left opens upward and has a lowest point with a **minimum** y-value occurring at the vertex. By contrast, the parabola on the right opens downward and has a highest point with a **maximum** y-value occurring at the vertex. Note that a parabola goes on indefinitely in two directions.

We can derive a formula for the coordinates of the vertex of a parabola given the corresponding quadratic function. Consider the general quadratic function $f(x) = ax^2 + bx + c$, and assume that a is

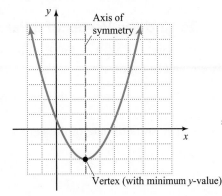

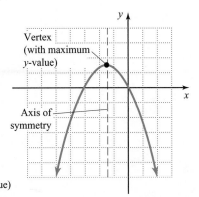

positive. Let's rewrite the right-hand side of this equation by completing the square for the variable terms, $ax^2 + bx$:

$$f(x) = (ax^2 + bx) + c$$

$$= a\left(x^2 + \frac{b}{a}x\right) + c \qquad \text{Factor out } a \text{ from } ax^2 + bx.$$

$$= a\left[x^2 + \frac{b}{a}x + \left(\frac{b}{2a}\right)^2\right] + c - a\left(\frac{b}{2a}\right)^2 \qquad \text{Complete the square for } x^2 + \frac{b}{a}x, \text{ and}$$
$$\text{subtract } a\left(\frac{b}{2a}\right)^2 \text{ from the same side.}$$

$$= a\left(x + \frac{b}{2a}\right)^2 + \left(c - \frac{b^2}{4a}\right) \qquad \text{Express } x^2 + \frac{b}{a}x + \left(\frac{b}{2a}\right)^2 \text{ as}$$
$$\text{the square of } x + \frac{b}{2a}.$$

$$= a\left(x + \frac{b}{2a}\right)^2 + \frac{4ac - b^2}{4a}$$

Note that the term $a\left(x + \frac{b}{2a}\right)^2$ involves the variable x. By contrast, the term $\dfrac{4ac - b^2}{4a}$ contains only constants and so does not depend on x. Since the expression $\left(x + \dfrac{b}{2a}\right)^2$ is a perfect square, its value is either positive or 0. Since a is positive, the term $a\left(x + \dfrac{b}{2a}\right)^2$ is nonnegative and has a minimum value at 0.

$$a\left(x + \frac{b}{2a}\right)^2 = 0$$

$$\left(x + \frac{b}{2a}\right)^2 = 0 \qquad \text{Divide each side of the equation by } a.$$

$$x + \frac{b}{2a} = 0 \qquad \text{Use the square root property of equality.}$$

$$x = -\frac{b}{2a} \qquad \text{Subtract } \frac{b}{2a} \text{ from each side of the equation.}$$

Since we assumed that a is positive, we conclude that the minimum value of $a\left(x + \dfrac{b}{2a}\right)^2$ occurs when $x = -\dfrac{b}{2a}$, and is therefore the x-value of the vertex. If we had assumed that a is negative, we could then show that the maximum value occurs when $x = -\dfrac{b}{2a}$, and so again is the x-value of the vertex.

To Find the Vertex of a Parabola

If a parabola is given by the function $f(x) = ax^2 + bx + c$, its vertex is found as follows:

- The x-coordinate of the vertex is $-\dfrac{b}{2a}$.

- The y-coordinate of the vertex is found by substituting $-\dfrac{b}{2a}$ for x in $f(x)$.

Using function notation, we can express the coordinates of the vertex as $\left(-\dfrac{b}{2a}, f\left(-\dfrac{b}{2a}\right)\right)$.

Knowing the coordinates of the vertex and also that the axis of symmetry is a vertical line that passes through the vertex leads us to the following:

The Equation of the Axis of Symmetry

For the graph of $f(x) = ax^2 + bx + c$, the equation of the axis of symmetry is

$$x = -\frac{b}{2a}.$$

Recall that in Example 1, the vertex of the parabola is the point $(0, 0)$ and the line of symmetry is the y-axis. This is true for the graph of any quadratic function of the form $f(x) = ax^2$.

Although we can always make a table of values to graph a quadratic function, we consider two other methods of graphing quadratic equations. One method involves graphing the vertex and several points on either side of the vertex. The other method involves graphing the x- and y-intercepts as well as the vertex.

EXAMPLE 2

Graph $f(x) = 6 - 3x^2$

Solution Writing the function as $f(x) = -3x^2 + 0x + 6$, we see that $a = -3$ and $b = 0$. Next, we find the coordinates of the vertex. The x-coordinate of the vertex is:

$$-\frac{b}{2a} = -\frac{0}{2(-3)} = 0$$

To find the y-coordinate of the vertex, we substitute 0 for x in $f(x)$:

$$\begin{aligned}
f(x) &= 6 - 3x^2 \\
f(0) &= 6 - 3(0)^2 \quad \text{Substitute 0 for } x \text{ in } f(x). \\
&= 6 - 0 \\
&= 6
\end{aligned}$$

So the vertex of the parabola is the point $(0, 6)$.
Next, we choose points on each side of the vertex.

x	$f(x) = 6 - 3x^2$	(x, y)	
-2	$f(-2) = 6 - 3(-2)^2 = -6$	$(-2, -6)$	
-1	$f(-1) = 6 - 3(-1)^2 = 3$	$(-1, 3)$	
0	$f(0) = 6 - 3(0)^2 = 6$	$(0, 6)$	← Vertex
1	$f(1) = 6 - 3(1)^2 = 3$	$(1, 3)$	
2	$f(2) = 6 - 3(2)^2 = -6$	$(2, -6)$	

PRACTICE 2

Graph $f(x) = -4x^2 + 1$

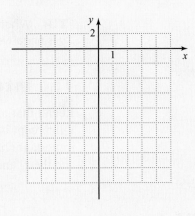

EXAMPLE 2 (continued)

Now, we plot the points and draw a smooth curve through them.

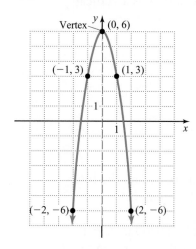

The result is a parabola, which is the graph of $f(x) = 6 - 3x^2$. Note that the y-axis, $x = 0$, is the axis of symmetry for this parabola.

In Example 1, we saw that for $f(x) = 3x^2$, a is positive and the parabola opens upward. On the other hand, for $f(x) = 6 - 3x^2$ in Example 2, a is negative and the parabola opens downward. Knowing the sign of a helps us to graph $f(x) = ax^2 + bx + c$.

> **TIP**   When a is positive, the parabola opens upward and has a minimum point. When a is negative, the parabola opens downward and has a maximum point.

The Intercepts

Recall that an x-intercept of a graph is a point where the graph crosses the x-axis, and the y-intercept is a point where the graph crosses the y-axis. The intercepts, like the vertex, are important characteristics of a parabola. We can graph a quadratic function using the vertex and the x- and y-intercepts.

EXAMPLE 3

Consider $y = x^2 - 8x + 12$.

a. Find the vertex and the axis of symmetry of the graph.

b. Determine if the graph opens upward or downward.

c. Find the x- and y-intercepts.

d. Sketch the graph.

Solution

a. First, let's find the vertex. Since $a = 1$ and $b = -8$, the x-coordinate of the vertex is:

$$-\frac{b}{2a} = -\frac{(-8)}{2(1)} = \frac{8}{2} = 4$$

PRACTICE 3

Consider $y = -x^2 + 4x + 5$.

a. Find the vertex and the axis of symmetry of the graph.

b. Determine if the graph opens upward or downward.

c. Find the x- and y-intercepts.

Substituting 4 for x, we get:

$$y = x^2 - 8x + 12$$
$$= (4)^2 - 8(4) + 12$$
$$= 16 - 32 + 12$$
$$= -4$$

So the vertex is the point $(4, -4)$. The equation of the axis of symmetry is $x = 4$, that is, the vertical line passing through the vertex.

b. The graph opens upward, since the leading coefficient 1 is positive.

c. Next, we find the x- and y-intercepts. To find the x-intercepts of a graph, we let $y = 0$ and solve for x.

$$y = x^2 - 8x + 12$$
$$0 = x^2 - 8x + 12$$
$$0 = (x - 6)(x - 2)$$
$$x - 6 = 0 \quad \text{or} \quad x - 2 = 0$$
$$x = 6 \qquad \qquad x = 2$$

So the x-intercepts are $(6, 0)$ and $(2, 0)$. Note that this parabola has two x-intercepts. To find the y-intercept, we let $x = 0$.

$$y = x^2 - 8x + 12$$
$$= (0)^2 - 8(0) + 12$$
$$= 12$$

The y-intercept is therefore $(0, 12)$.

d. To graph the parabola, let's first plot the vertex and the x- and y-intercepts on a coordinate plane. Then, we draw a smooth curve through the points. The result is the following parabola:

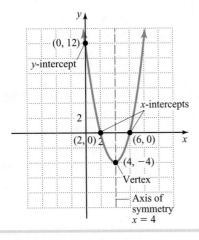

d. Sketch the graph.

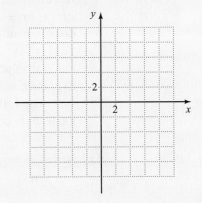

Note that in Example 3 the graph of $y = x^2 - 8x + 12$ has x-intercepts $(6, 0)$ and $(2, 0)$ and that the equation $x^2 - 8x + 12 = 0$ has solutions 6 and 2.

In general, the x-intercepts of the graph of $y = ax^2 + bx + c$ have as their x-coordinates the real solutions to the equation $ax^2 + bx + c = 0$.

The Discriminant

Recall from Section 8.2 that we can use the discriminant of the quadratic equation $ax^2 + bx + c = 0$ to determine the number and type of solutions of the equation. If the discriminant $b^2 - 4ac$ is positive, the equation will have two real solutions and the graph of $f(x) = ax^2 + bx + c$ will have two x-intercepts. If the discriminant is 0, there will be only one real solution and the graph will have only one x-intercept. And if the discriminant is negative, there will be no real solutions, and the graph will have no x-intercepts.

The following table shows the graph of $f(x) = ax^2 + bx + c$ for possible values of the leading coefficient a and the discriminant $b^2 - 4ac$:

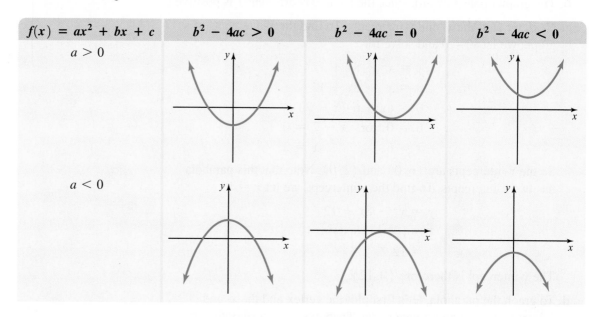

$f(x) = ax^2 + bx + c$	$b^2 - 4ac > 0$	$b^2 - 4ac = 0$	$b^2 - 4ac < 0$
$a > 0$			
$a < 0$			

Although the graph of $f(x) = ax^2 + bx + c$ may have two, one, or zero x-intercepts, it must have exactly one y-intercept. Can you explain why?

EXAMPLE 4

For each function, determine whether its graph opens upward or downward and whether there is a minimum or a maximum point. Then, find the number of x- and y-intercepts.

a. $f(x) = 2x^2 - x + 1$

b. $f(x) = x^2 - 6x + 9$

c. $f(x) = -2x^2 + x + 3$

Solution

a. For $f(x) = 2x^2 - x + 1$, the leading coefficient 2 is positive, so the graph opens upward and has a minimum point.

To find the number of x-intercepts, we evaluate the discriminant, $b^2 - 4ac$, where $a = 2$, $b = -1$, and $c = 1$.

$$b^2 - 4ac = (-1)^2 - 4(2)(1) = -7$$

Since the discriminant is negative, there is no x-intercept. There is one y-intercept.

PRACTICE 4

For each function, determine whether its graph opens upward or downward and whether there is a minimum or a maximum point. Then, find the number of x- and y-intercepts.

a. $f(x) = 4x^2 - 9x + 2$

b. $f(x) = -3x^2 + 5x - 7$

c. $f(x) = -4x^2 + 4x - 1$

b. For $f(x) = x^2 - 6x + 9$, the leading coefficient 1 is positive, so the graph opens upward and has a minimum point. The discriminant is:

$$b^2 - 4ac = (-6)^2 - 4(1)(9) = 0$$

Since the discriminant is 0, there is one x-intercept as well as one y-intercept.

c. For $f(x) = -2x^2 + x + 3$, the leading coefficient -2 is negative, so the graph opens downward and has a maximum point. The discriminant is:

$$b^2 - 4ac = (1)^2 - 4(-2)(3) = 25$$

Since the discriminant is positive, there are two x-intercepts and one y-intercept.

The Domain and Range of a Quadratic Function

Recall from Section 3.6 that the domain of the function $f(x) = x^2$ is the set of all real numbers. In fact, any function of the form $f(x) = ax^2 + bx + c$ is defined for all real values of x, so the domain is the set of all real numbers, or in interval notation, $(-\infty, \infty)$. To find the range of a quadratic function, we must identify the vertex of the parabola and determine whether it opens upward or downward.

EXAMPLE 5

Consider the function $f(x) = x^2 - 2x + 4$.

a. Find the vertex of the graph.

b. Determine whether the parabola opens upward or downward.

c. Sketch the parabola.

d. Determine the domain and range of $f(x)$.

Solution

a. Here, $a = 1, b = -2$, and $c = 4$.

For the vertex, $x = -\dfrac{b}{2a} = -\dfrac{(-2)}{2} = 1$. To find the corresponding y-value on the graph, we evaluate the function at $x = 1$.

$$f(1) = (1)^2 - 2(1) + 4 = 3$$

So the vertex is the point $(1, 3)$.

b. Since the leading coefficient 1 is positive, the parabola opens upward and the vertex is a minimum point.

c. The value of the discriminant is $b^2 - 4ac = (-2)^2 - 4(1)(4) = -12$. Since the discriminant is negative, we conclude that the graph has no x-intercepts. To find the y-intercept, we substitute 0 for x in $f(x)$.

$$f(0) = (0)^2 - 2(0) + 4 = 4$$

So the y-intercept is the point $(0, 4)$.

PRACTICE 5

Consider $f(x) = -x^2 - 3x - 4$.

a. Find the vertex of the graph.

b. Determine whether the parabola opens upward or downward.

c. Sketch the parabola.

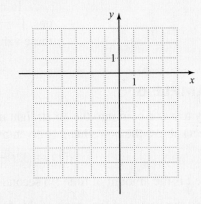

d. Determine the domain and range of $f(x)$.

EXAMPLE 5 (continued)

To sketch the parabola, let's plot the vertex, y-intercept, and a few additional points.

x	$f(x) = x^2 - 2x + 4$	(x, y)	
-2	$f(-2) = (-2)^2 - 2(-2) + 4 = 12$	$(-2, 12)$	
-1	$f(-1) = (-1)^2 - 2(-1) + 4 = 7$	$(-1, 7)$	
0	$f(0) = (0)^2 - 2(0) + 4 = 4$	$(0, 4)$	← y-intercept
1	$f(1) = (1)^2 - 2(1) + 4 = 3$	$(1, 3)$	← Vertex
2	$f(2) = (2)^2 - 2(2) + 4 = 4$	$(2, 4)$	

The result is the following parabola:

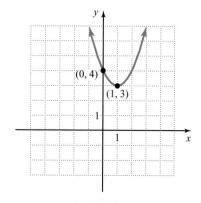

d. The domain is the set of all real numbers, that is, $(-\infty, \infty)$. From the graph, we see that the function takes on any y-value greater than or equal to 3, the y-value of the vertex. So the range of the function is the set of all real numbers greater than or equal to 3, that is, $[3, \infty)$.

In many real-life situations, we are interested in finding maximum and minimum values. For situations that can be described by a quadratic function, we can easily determine these values using our knowledge of the vertex of a parabola.

EXAMPLE 6

A boy tosses a coin into the air straight upward. The height of the coin (in feet) above the boy's hand is given by the function

$$f(t) = 32t - 16t^2,$$

where t is the amount of time (in seconds) after the toss.

a. Find the vertex of the graph of this function.

b. Determine if the vertex is a minimum or a maximum point.

c. Graph the function.

d. At what time will the coin reach its maximum height? What is that height?

PRACTICE 6

A rectangular garden with width w has a perimeter of 200 ft.

a. Find the length l of the garden expressed in terms of w.

b. Find the area A of the garden expressed as a function of w.

Solution

a. We can write the function as $f(t) = -16t^2 + 32t$. Here $a = -16$, $b = 32$, and $c = 0$. To find the coordinates of the vertex, we compute:

$$-\frac{b}{2a} = -\frac{32}{-32} = 1$$

Next, we compute the value of the function at $t = 1$.

$$f(1) = -16(1)^2 + 32(1) = 16$$

The vertex is the point $(1, 16)$.

b. Since the value of a, -16, is negative, the graph opens downward and has a maximum point.

c. Now, we graph the function for $t \geq 0$.

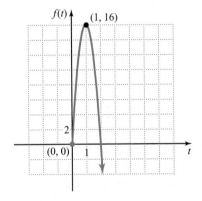

d. Since the vertex is the maximum point $(1, 16)$, the coin will reach its maximum height of 16 ft above the boy's hand 1 sec after the toss.

c. Graph the function.

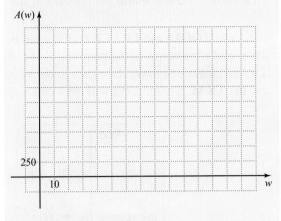

d. Determine the dimension of the garden that will maximize its area. What is that maximum area?

In Example 6(c), explain why we consider points in only the first and fourth quadrants when sketching the graph of the function.

Finding the Vertex and Intercepts of a Parabola Using a Grapher

Calculators with graphing capabilities and computers with graphing software allow us to graph quadratic functions. Recall that pressing the $\boxed{Y =}$ key opens a window in which we enter the function. For instance, if we want to graph the function $f(x) = x^2 - 4x + 3$, using either the $\boxed{\wedge}$ key or the $\boxed{x^2}$ key, we enter $x^2 - 4x + 3$ to the right of $\backslash Y1 =$. Pressing $\boxed{\text{GRAPH}}$ will display the graph of the function.

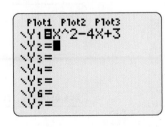

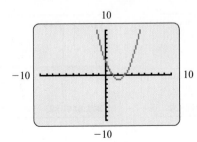

TIP When graphing a quadratic equation, set the viewing window so that it includes the vertex and the *x*- and *y*-intercepts of the parabola.

The **ZOOM** and **TRACE** features can be used to locate the coordinates of the vertex and the *x*- and *y*-intercepts of the parabola. So for the preceding graphed parabola we can use the **TRACE** feature to move the cursor to the vertex of the parabola. If we **ZOOM IN** and again use the **TRACE** feature, we will get a better approximation for the coordinates of the vertex.

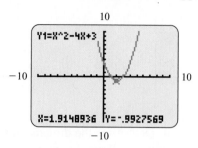 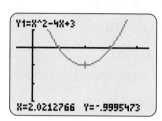

So the vertex is approximately $(2.02, -1.00)$.

TIP To get a more accurate approximation for the coordinates of the vertex, use the **ZOOM** and **TRACE** features several times.

We can also find the coordinates of the *x*- and *y*-intercepts using the **TRACE** and **ZOOM** features. To locate the *y*-intercept, use the **TRACE** feature to move the cursor until it rests on the point at which the graph crosses the *y*-axis. To locate the *x*-intercept(s), use the **TRACE** feature to move the cursor to the point at which the graph crosses the *x*-axis. As with the vertex, if we **ZOOM IN** and **TRACE** several times, we will get a better approximation of the coordinates.

To find the *y*-intercept:

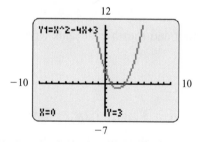

So the *y*-intercept is $(0, 3)$.

To find the *x*-intercepts:

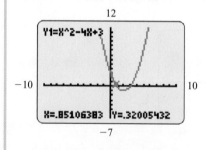

Trace to the left *x*-intercept

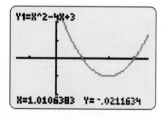

Zoom In and Trace

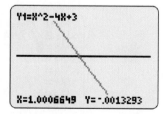

Zoom In and Trace three times

The coordinates of the left x-intercept are approximately $(1, 0)$. Carrying out the procedure for the right x-intercept, we find its coordinates to be approximately $(3, 0)$.

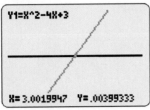

Zoom In and Trace three times

EXAMPLE 7

Consider the function $f(x) = -3x^2 - 3x + 5$.

a. Graph this equation using a grapher.

b. Use the **ZOOM** and **TRACE** features to find the coordinates of the vertex.

c. Use the **ZOOM** and **TRACE** features to find the coordinates of the x- and y-intercepts.

Solution

a. We press $\boxed{Y =}$ and enter $-3x^2 - 3x + 5$ to the right of $\backslash Y1 =$, using either the $\boxed{x^2}$ feature or $x \boxed{\wedge} 2$ to enter the quadratic term. Then, we press $\boxed{GRAPH}$. If necessary, we change the viewing window so that the vertex and the intercepts of the parabola are all displayed.

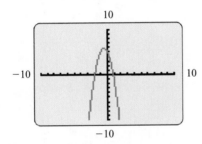

b. With the **TRACE** feature, we move the cursor until it rests on the parabola's vertex. To get a better approximation, we **ZOOM IN** and **TRACE** a few times.

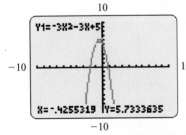

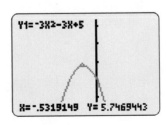

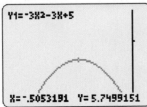

So the coordinates of the vertex are approximately $(-0.51, 5.75)$.

PRACTICE 7

Consider the equation $f(x) = 2x^2 + x - 4$.

a. Graph this equation using a graphing calculator or a computer.

b. Use the **TRACE** or **ZOOM** features to identify the coordinates of the vertex.

c. Use the **TRACE** and **ZOOM** features to find the x- and y-intercepts of the graph.

EXAMPLE 7 (continued)

c. Using the **TRACE** feature, we locate the coordinates of the y-intercept.

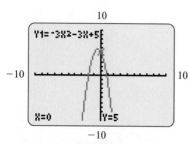

The y-intercept is $(0, 5)$.

Using the **ZOOM** and **TRACE** features, we find the x-intercepts of the parabola.

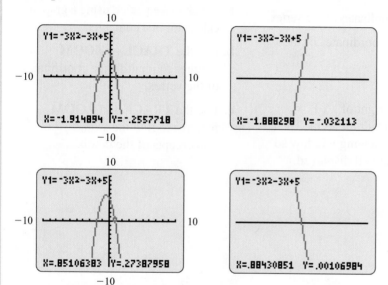

So the x-intercepts are approximately $(-1.9, 0)$ and $(0.9, 0)$.

Mathematically Speaking

Fill in each blank with the most appropriate term or phrase from the given list.

zero	minimum	vertex
maximum	downward	x-coordinate
axis of symmetry	quadratic	upward
y-coordinate	parabola	two
circle	squared	one

1. Functions of the form $f(x) = ax^2 + bx + c$ are called _____ functions.

2. The graph of a quadratic function is called a(n) _____.

3. The highest or lowest point of a parabola is called a(n) _____.

4. For a parabola given by the function $f(x) = ax^2 + bx + c$, the _____ of the vertex is $-\dfrac{b}{2a}$.

5. When a is positive, the parabola given by the function $f(x) = ax^2 + bx + c$ opens _____.

6. When a is negative, the parabola given by the function $f(x) = ax^2 + bx + c$ has a(n) _____ point.

7. If the discriminant of the quadratic equation given by the function $f(x) = ax^2 + bx + c$ is positive, the parabola has _____ x-intercept(s).

8. For a parabola given by the function $f(x) = ax^2 + bx + c$, the equation of the _____ is $x = -\dfrac{b}{2a}$.

Ⓐ *Complete the table. Then, plot the points, and sketch the graph of the parabola.*

9.

x	$y = f(x) = 2x^2$	(x, y)
-2		
-1		
0		
1		
2		

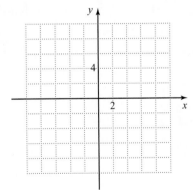

10.

x	$y = f(x) = -2x^2$	(x, y)
-2		
-1		
0		
1		
2		

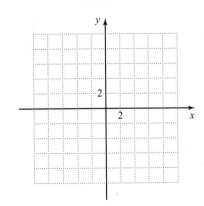

11.

x	$y = f(x) = \frac{1}{2}x^2$	(x,y)
-4		
-2		
0		
2		
4		

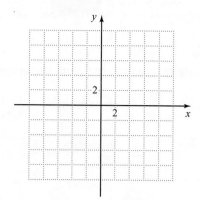

12.

x	$y = f(x) = \frac{1}{3}x^2$	(x,y)
-6		
-3		
0		
3		
6		

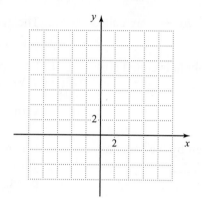

13.

x	$y = f(x) = 2 - x^2$	(x,y)
-3		
-2		
-1		
0		
1		
2		
3		

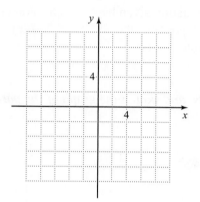

14.

x	$y = f(x) = x^2 - 3$	(x,y)
-3		
-2		
-1		
0		
1		
2		
3		

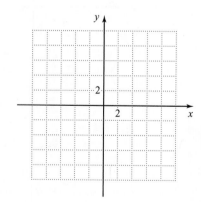

B *For each of the following functions, find the vertex, the axis of symmetry, and the x- and y-intercepts of the graph. Then, sketch the graph.*

15. $f(x) = x^2 - 8x$

Vertex:

Axis of symmetry:

x-intercept(s):

y-intercept:

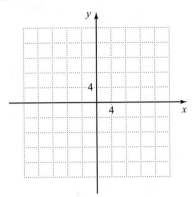

16. $f(x) = x^2 + 6x$

Vertex:

Axis of symmetry:

x-intercept(s):

y-intercept:

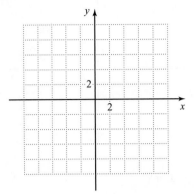

17. $f(x) = x^2 - 2x - 3$

Vertex:

Axis of symmetry:

x-intercept(s):

y-intercept:

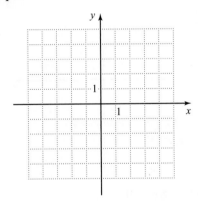

18. $f(x) = x^2 + 4x - 5$

Vertex:

Axis of symmetry:

x-intercept(s):

y-intercept:

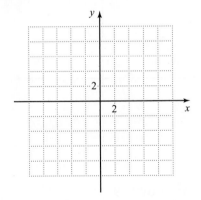

19. $f(x) = -x^2 + 4x + 12$

Vertex:

Axis of symmetry:

x-intercept(s):

y-intercept:

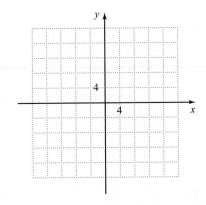

20. $f(x) = -x^2 - 2x + 8$

Vertex:

Axis of symmetry:

x-intercept(s):

y-intercept:

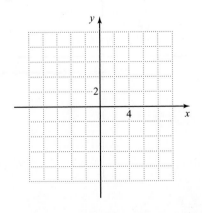

21. $f(x) = x^2 - 1$
Vertex:
Axis of symmetry:
x-intercept(s):
y-intercept:

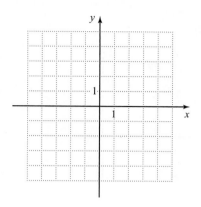

22. $f(x) = 4 - x^2$
Vertex:
Axis of symmetry:
x-intercept(s):
y-intercept:

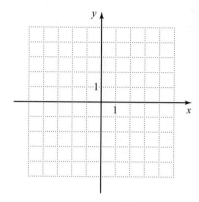

23. $f(x) = (x + 1)^2$
Vertex:
Axis of symmetry:
x-intercept(s):
y-intercept:

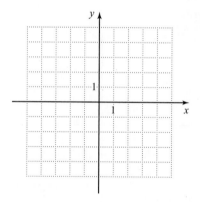

24. $f(x) = (x - 1)^2$
Vertex:
Axis of symmetry:
x-intercept(s):
y-intercept:

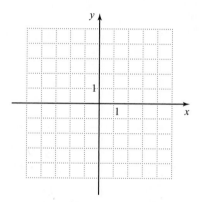

25. $f(x) = -x^2 + 6x - 9$
Vertex:
Axis of symmetry:
x-intercept(s):
y-intercept:

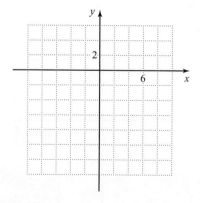

26. $f(x) = x^2 + 6x + 9$
Vertex:
Axis of symmetry:
x-intercept(s):
y-intercept:

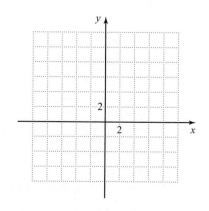

27. $f(x) = x^2 - 3x - 10$
Vertex:
Axis of symmetry:
x-intercept(s):
y-intercept:

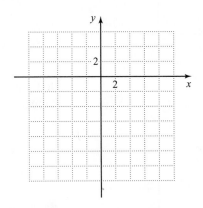

28. $f(x) = x^2 + x - 6$
Vertex:
Axis of symmetry:
x-intercept(s):
y-intercept:

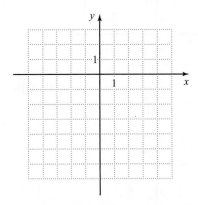

C *Complete the table.*

29.

Function	Opens Upward or Downward?	Maximum or Minimum Point?	Number of x-intercepts	Number of y-intercepts
$f(x) = x^2 + 5$				
$f(x) = 1 - 4x + 4x^2$				
$f(x) = 2 - 3x^2$				
$f(x) = -2x^2 - 5x - 8$				
$f(x) = 4x^2 - 4x - 1$				

30.

Function	Opens Upward or Downward?	Maximum or Minimum Point?	Number of x-intercepts	Number of y-intercepts
$f(x) = -6 - x^2$				
$f(x) = 2x^2$				
$f(x) = 3x^2 + 8x + 7$				
$f(x) = 5 + 2x - 3x^2$				
$f(x) = 9x^2 + 12x + 4$				

D *Graph each function. Then, determine the domain and range.*

31. $f(x) = 2x^2 - 1$

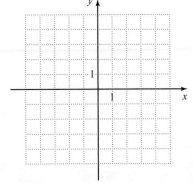

Domain:
Range:

32. $f(x) = 1 - 2x^2$

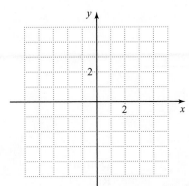

Domain:
Range:

33. $g(x) = -3x^2 + 6x - 2$

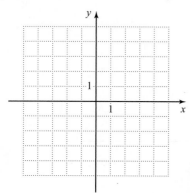

Domain:
Range:

34. $g(x) = 3x^2 - 12x + 6$

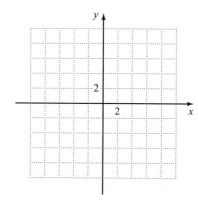

Domain:
Range:

35. $f(x) = 0.5x^2 + 2$

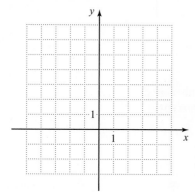

Domain:
Range:

36. $f(x) = -2 - 0.5x^2$

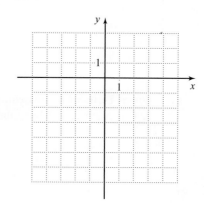

Domain:
Range:

37. $f(x) = x^2 - 3x - 4$

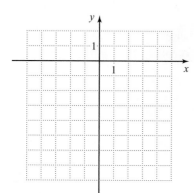

Domain:
Range:

38. $h(x) = x^2 + 5x + 1$

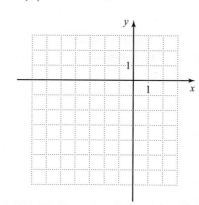

Domain:
Range:

39. $h(x) = -2x^2 + 2x - 3$

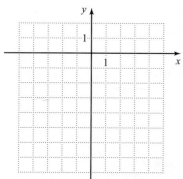

Domain:
Range:

40. $g(x) = 3x^2 - 3x + 1$

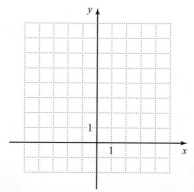

Domain:
Range:

▦ *Graph each function on a grapher. Then, use the grapher to find the vertex and*
x- and y-intercepts, rounding to the nearest hundredth.

41. $f(x) = x^2 + 0.2x - 1$ Vertex: *x*-intercept(s):   *y*-intercept:

42. $f(x) = x^2 - 1.1x - 3$ Vertex: *x*-intercept(s):   *y*-intercept:

43. $f(x) = -0.15x^2 - x + 0.5$ Vertex: *x*-intercept(s):   *y*-intercept:

44. $f(x) = -1.4x^2 + 2x + 7.1$ Vertex: *x*-intercept(s):   *y*-intercept:

45. $f(x) = 5x^2 + 3x + 7$ Vertex: *x*-intercept(s):   *y*-intercept:

46. $f(x) = -3x^2 + 2x - 8$ Vertex: *x*-intercept(s):   *y*-intercept:

Mixed Practice

Solve.

47. Complete the table. Then, plot the points, and sketch the graph of the parabola.

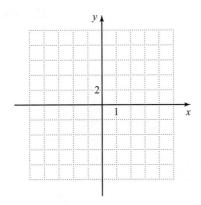

x	$y = f(x) = -3 + x^2$	(x, y)
-3		
-2		
-1		
0		
1		
2		
3		

48. Complete the table.

Function	Opens Upward or Downward?	Maximum or Minimum Point?	Number of *x*-intercepts	Number of *y*-intercepts
$f(x) = -x^2 - 5x + 5$				
$f(x) = 3x^2 + 8x + 7$				
$f(x) = x^2 - 6x + 9$				
$f(x) = x^2 - 2$				
$f(x) = 6x - 5 - 3x^2$				

Find the vertex, the axis of symmetry, and the x- and y-intercepts of the graph of the function. Then, sketch the graph.

49. $f(x) = -x^2 - 4x + 5$

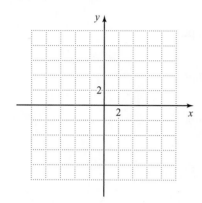

Vertex:
Axis of symmetry:
x-intercept(s):
y-intercept:

50. $f(x) = x^2 - 4x + 4$

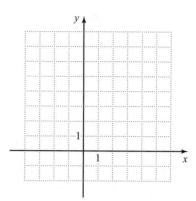

Vertex:
Axis of symmetry:
x-intercept(s):
y-intercept:

Graph each function. Then, determine the domain and range.

51. $g(x) = -x^2 + 2x + 1$

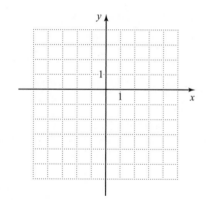

Domain: Range:

52. $h(x) = 2x^2 - 4x + 3$

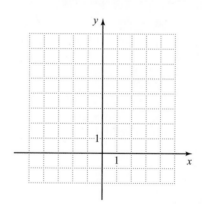

Domain: Range:

Applications

E *Solve.*

53. A stone is thrown straight upward with an initial velocity of 48 ft/sec from a bridge 280 ft above a river. The height of the stone above the river (in feet) t sec after it is thrown is given by the function $s(t) = -16t^2 + 48t + 280$.

   a. Find the vertex of the graph.
   b. Graph the function.
   c. When does the stone reach its maximum height? What is that height?

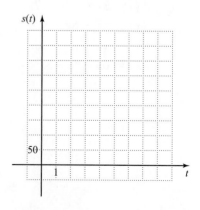

54. A toy manufacturer determines that the daily cost C for producing x units of a dump truck can be approximated by the function $C(x) = 0.005x^2 - x + 100$.

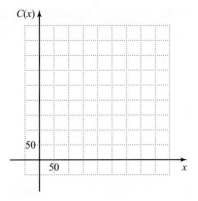

   a. How many toy dump trucks must the manufacturer produce per day in order to minimize the cost?

   b. Graph this function.

   c. What is the minimum daily cost?

55. A homeowner has 150 ft of fencing with which to build a rectangular exercise yard for her dog.

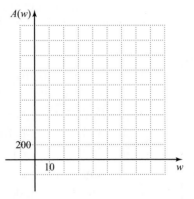

   a. Express the length l of the exercise yard in terms of the width w.

   b. Express in function notation the relationship between the area $A(w)$ and the width w of the exercise yard.

   c. Graph this function.

   d. What dimensions will maximize the area of the exercise yard? What is the maximum area?

56. A farmer plans to build a rectangular animal pen using the barn as one side, as shown in the figure to the right. The farmer has 300 ft of fencing with which to build the pen.

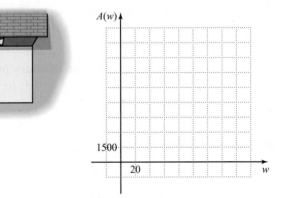

   a. Express the length l of the pen in terms of the width w.

   b. Express in function notation the relationship between the area $A(w)$ and the width w of the pen.

   c. Graph this function.

   d. What dimensions will maximize the area of the animal pen? What is the maximum area?

57. A company determines that its daily revenue R (in dollars) for selling x units of a product is modeled by the function $R(x) = x(120 - x)$.

   a. Find the vertex of the graph of this function.

   b. What number of units must be sold in order to maximize the revenue? What is the maximum revenue?

58. An architect is designing an office building. The building is to be rectangular in shape, with a perimeter of 360 ft and width w ft.

   a. Find a function $A(w)$ that represents the area of the building in terms of its width.

   b. What is the maximum possible area of the building?

• Check your answers on page A-45.

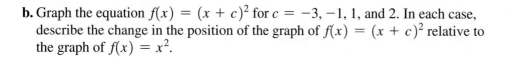

Investigation

1. Consider the graph of the equation $f(x) = x^2$.

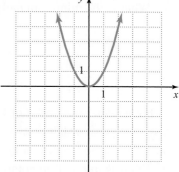

   a. Graph the equation $f(x) = x^2 + c$ for $c = -3, -1, 1$, and 2. In each case, describe the change in the position of the graph of $f(x) = x^2 + c$ relative to the graph of $f(x) = x^2$.

   b. Graph the equation $f(x) = (x + c)^2$ for $c = -3, -1, 1$, and 2. In each case, describe the change in the position of the graph of $f(x) = (x + c)^2$ relative to the graph of $f(x) = x^2$.

   c. Use your results from parts (a) and (b) to describe the change in the position of the graph relative to the graph of $f(x) = x^2$ of each of the following:

   i. $f(x) = x^2 + c$, for $c > 0$
   ii. $f(x) = x^2 + c$, for $c < 0$
   iii. $f(x) = (x + c)^2$, for $c > 0$
   iv. $f(x) = (x + c)^2$, for $c < 0$

   d. Without graphing, describe the position of the graph of $f(x) = (x - 2)^2 - 3$ relative to the graph of $f(x) = x^2$. Then, check your answer by graphing.

Critical Thinking

2. Suppose you are given three arbitrary points (x_1, y_1), (x_2, y_2), and (x_3, y_3) that lie on the graph of a parabola. The equation of the parabola, $y = ax^2 + bx + c$, that contains the points can be found by solving the following system of equations:

$$\textbf{(1)} \quad ax_1^2 + bx_1 + c = y_1$$
$$\textbf{(2)} \quad ax_2^2 + bx_2 + c = y_2$$
$$\textbf{(3)} \quad ax_3^2 + bx_3 + c = y_3$$

   Use this system to find the equation of the parabola that passes through the points $(-2, 0)$, $(-1, -4)$, and $(1, 6)$.

Mathematical Reasoning

3. Suppose the x-intercepts of the graph of a parabola are $(x_1, 0)$ and $(x_2, 0)$. What is the equation of the axis of symmetry of this graph? Give examples to justify your reasoning.

8.5 Solving Quadratic and Rational Inequalities

OBJECTIVES

A To solve a quadratic or a rational inequality

B To solve applied problems involving quadratic or rational inequalities

Recall that in Section 2.3 we discussed situations that could be described and solved by a linear inequality in one variable, such as $4x + 5 \leq 3x + 8$. There we saw that the solutions of an inequality are any values of the variable that make the inequality true. We expressed the solutions in one of three ways: as a simplified inequality, such as $x \leq 3$; in interval notation, such as $(-\infty, 3]$; or as a graph on the number line, such as:

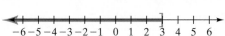

In this section, we consider situations that can be described and solved by a quadratic or a rational inequality in one variable. For instance, consider the following problem:

The owner of a bicycle shop determines that his monthly revenue for selling n bikes can be represented by $R = 430n - 6n^2$. He wants to find the number of bikes he must sell to generate revenue of at least \$7000 per month. To find the number of bikes he must sell, he can solve the following inequality:

$$430n - 6n^2 \geq 7000$$

In this section, we discuss techniques for solving such inequalities.

Quadratic Inequalities

A **quadratic inequality** in one variable is an inequality that contains a quadratic expression. Here are some examples of quadratic inequalities in one variable.

$$x^2 + 4x < -3 \qquad 2x^2 + 11x + 15 \leq 10$$
$$x^2 + x - 2 > 0 \qquad t^2 + 2t - 3 \geq 0$$

Consider, for instance, a quadratic inequality such as $x^2 + x - 2 > 0$. To solve, we find all the values of x for which the expression $x^2 + x - 2$ is greater than 0, that is, all values for which the expression is positive. Let's take a look at the graph of the function $f(x) = x^2 + x - 2$ to understand how to solve the related inequality.

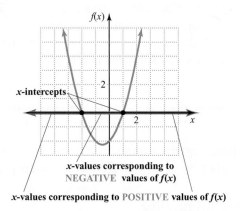

We observe from this graph that the solutions of $x^2 + x - 2 > 0$ are all the real numbers less than -2 or greater than 1. In interval notation, we can express the solutions as $(-\infty, -2) \cup (1, \infty)$. Note that the x-values for which $f(x)$ is positive are separated from those for which $f(x)$ is negative by the x-intercepts, that is, by the points where $f(x)$ is equal to 0.

It is usually easier to solve quadratic inequalities algebraically rather than graphically. The key to solving a quadratic inequality such as $x^2 + x - 2 > 0$ algebraically is to solve the related equation, $x^2 + x - 2 = 0$. Then, we draw a number line on which the solutions are plotted. The solutions serve as boundary points, separating the number line into intervals in which the expression $x^2 + x - 2$ is either greater than 0 or less than 0. To determine if the value of the expression is greater than or less than 0 in a particular interval, we use a *test value*, that is, an arbitrary point on the number line that lies within the interval.

Let's solve the inequality $x^2 + x - 2 > 0$ algebraically. To do this, we first solve the related equation $x^2 + x - 2 = 0$.

$$x^2 + x - 2 = 0$$
$$(x + 2)(x - 1) = 0 \qquad \text{Factor.}$$
$$x + 2 = 0 \quad \text{or} \quad x - 1 = 0 \qquad \text{Use the zero-product property.}$$
$$x = -2 \qquad\qquad x = 1$$

Next, we plot the two solutions, -2 and 1, on a number line. The solutions serve as *boundary points*, separating the line into three intervals.

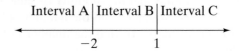

To determine if the quadratic expression is greater than 0 or less than 0 in a particular interval, we choose test values: say, -3 in Interval A, 0 in Interval B, and 2 in Interval C. Then, we evaluate the expression for each test value.

Interval	Test Value	Value of $x^2 + x - 2$	Conclusion
A	-3	$(-3)^2 + (-3) - 2 = 4$	$4 > 0$
B	0	$(0)^2 + (0) - 2 = -2$	$-2 < 0$
C	2	$(2)^2 + (2) - 2 = 4$	$4 > 0$

Value of $x^2 + x - 2$: $\quad > 0 \quad\quad < 0 \quad\quad > 0$

$$\xleftarrow{\qquad\qquad} \underset{-2}{\big|} \qquad \underset{1}{\big|} \xrightarrow{\qquad\qquad}$$

Because the inequality that we are considering is $x^2 + x - 2 > 0$, we are looking for intervals in which the value of the expression is positive. Since the original inequality involves the symbol $>$, we conclude that its solutions are all real numbers either less than -2 or greater than 1. We can express these solutions in interval notation as $(-\infty, -2) \cup (1, \infty)$, and graph them as follows:

$$\xleftarrow{\quad\;\;\;\;} \overset{)}{\underset{-6\;-5\;-4\;-3\;-2\;-1}{\;\;}} \;\; \overset{(}{\underset{0\;\;1\;\;2\;\;3\;\;4\;\;5\;\;6}{\;\;}} \xrightarrow{\quad}$$

Note that the algebraic solution agrees with the solutions we obtained earlier from graphing the corresponding function $f(x) = x^2 + x - 2$ on a coordinate plane.

To Solve a Quadratic Inequality

- Solve the related equation and plot the solutions on a number line, separating it into intervals.

- Choose a test value in each interval to check if it satisfies the original inequality.

- Identify the intervals that contain the solutions of the inequality. Check whether to include the boundary points.

EXAMPLE 1

Solve $x^2 - 2x - 3 \le 0$. Then, graph the solutions.

Solution We begin by solving the related equation $x^2 - 2x - 3 = 0$.

$$x^2 - 2x - 3 = 0$$
$$(x - 3)(x + 1) = 0 \qquad \text{Factor.}$$
$$x - 3 = 0 \quad \text{or} \quad x + 1 = 0 \qquad \text{Use the zero-product property.}$$
$$x = 3 \qquad\qquad x = -1$$

Next, we plot on a number line the two solutions, 3 and -1, which are the boundary points separating the number line into three intervals.

Interval A | Interval B | Interval C

$$-1 \qquad\qquad 3$$

Then, in each interval, we check to determine whether the expression is greater than 0 or less than 0. To do this, let's use the test values -2, 0, and 4.

Interval	Test Value	Value of $x^2 - 2x - 3$	Conclusion
A	-2	$(-2)^2 - 2(-2) - 3 = 5$	$5 > 0$
B	0	$(0)^2 - 2(0) - 3 = -3$	$-3 < 0$
C	4	$(4)^2 - 2(4) - 3 = 5$	$5 > 0$

Value of $x^2 - 2x - 3$: > 0 < 0 > 0

$$-1 \qquad 3$$

Because $x^2 - 2x - 3 \le 0$, we are looking for values of x for which the expression is either negative or 0. Since the original inequality involves the symbol $\le$, the solutions of the inequality are all real numbers between and including -1 and 3, or in interval notation, $[-1, 3]$. The corresponding graph is shown below:

$$-6\ -5\ -4\ -3\ -2\ -1\ \ 0\ \ 1\ \ 2\ \ 3\ \ 4\ \ 5\ \ 6$$

PRACTICE 1

Solve $x^2 - 3x - 4 \ge 0$. Then, graph the solutions.

$$-6\ -5\ -4\ -3\ -2\ -1\ \ 0\ \ 1\ \ 2\ \ 3\ \ 4\ \ 5\ \ 6$$

EXAMPLE 2

Solve $x^2 + 6x + 10 > 2$. Graph the solutions.

Solution First, we solve the related equation $x^2 + 6x + 10 = 2$.

$$x^2 + 6x + 10 = 2$$
$$x^2 + 6x + 8 = 0 \qquad \text{Write the equation in standard form.}$$
$$(x + 2)(x + 4) = 0 \qquad \text{Factor.}$$
$$x + 2 = 0 \quad \text{or} \quad x + 4 = 0 \qquad \text{Use the zero-product property.}$$
$$x = -2 \qquad\qquad x = -4$$

PRACTICE 2

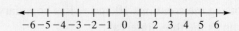

Solve $x^2 - 6x - 2 < -7$ and graph the solutions.

$$-6\ -5\ -4\ -3\ -2\ -1\ \ 0\ \ 1\ \ 2\ \ 3\ \ 4\ \ 5\ \ 6$$

EXAMPLE 2 (continued)

Plotting the solutions -2 and -4 as boundary points on a number line gives us the following intervals:

Interval A | Interval B | Interval C

-4 -2

Now, let's use the test values -5, -3, and 0 to determine whether the expression is greater than 2 or less than 2 in each interval.

Interval	Test Value	Value of $x^2 + 6x + 10$	Conclusion
A	-5	$(-5)^2 + 6(-5) + 10 = 5$	$5 > 2$
B	-3	$(-3)^2 + 6(-3) + 10 = 1$	$1 < 2$
C	0	$(0)^2 + 6(0) + 10 = 10$	$10 > 2$

Value of $x^2 + 6x + 10$: > 2 < 2 > 2

-4 -2

Because we are considering the inequality $x^2 + 6x + 10 > 2$, we are looking for values of x for which the expression is greater than 2. Since the original inequality involves the symbol $>$, it follows that the solutions of the inequality are all real numbers less than -4 or greater than -2, that is, $(-\infty, -4) \cup (-2, \infty)$. We can also represent the solutions graphically as follows:

$-6 -5 -4 -3 -2 -1 \; 0 \; 1 \; 2 \; 3 \; 4 \; 5 \; 6$

EXAMPLE 3

The monthly profit (in thousands of dollars) that a store makes on the sale of computers is modeled by the function $P(x) = x^2 - 6x - 30$, where x is the number of computers sold. How many computers must the store sell in order to make a profit of more than $10,000?

Solution To find the number of computers the store must sell, we need to solve the inequality $x^2 - 6x - 30 > 10$. To do this, we first solve the related equation.

$$x^2 - 6x - 30 = 10$$
$$x^2 - 6x - 40 = 0$$
$$(x - 10)(x + 4) = 0$$
$$x - 10 = 0 \quad \text{or} \quad x + 4 = 0$$
$$x = 10 \qquad\qquad x = -4$$

Plotting the solutions 10 and -4 as boundary points on a number line gives us the three intervals.

Interval A | Interval B | Interval C

-4 10

PRACTICE 3

The cost of operating a book-stand for x weeks can be approximated by the cost function $C(x) = 5x^2 + 25x + 70$. For how many weeks can the stand run if the cost must not exceed $400?

Let's use the test values -5, 0, and 11 for the three intervals.

Interval	Test Value	Value of $P(x) = x^2 - 6x - 30$	Conclusion
A	-5	$(-5)^2 - 6(-5) - 30 = 25$	$P(-5) > 10$
B	0	$(0)^2 - 6(0) - 30 = -30$	$P(0) < 10$
C	11	$(11)^2 - 6(11) - 30 = 25$	$P(11) > 10$

Value of $P(x)$: > 10 < 10 > 10

-4 10

We are looking for those values of x for which $P(x)$ is greater than 10. Since the number of computers sold cannot be negative, we do not include the solutions in Interval A. So for the store to make a profit greater than $10,000, each month it must sell more than 10 computers, that is, 11 or more computers.

Rational Inequalities

A **rational inequality** is an inequality that contains a rational expression. Some examples of rational inequalities are:

$$\frac{x-1}{x+3} < 0 \qquad \frac{2}{x} > -2 \qquad \frac{x}{2x-5} \leq 1 \qquad \frac{3x-1}{x+4} \geq 0$$

To solve rational inequalities, we use a procedure similar to that used for solving quadratic inequalities. However, when finding the boundary points that separate a number line into intervals, we must consider not only the solutions of the related equation but also the values for which the inequality is undefined.

To Solve a Rational Inequality

- Find all values for which the inequality is undefined.

- Solve the inequality's related equation.

- Plot the values found in the previous steps on a number line, separating it into intervals.

- Choose a test value in each interval to check if it satisfies the original inequality.

- Identify the intervals that contain the solutions of the inequality. Check whether to include the boundary points.

EXAMPLE 4

Solve $\dfrac{x-1}{x+3} < 0$. Graph the solutions.

Solution The inequality is undefined when its denominator, $x + 3$, is 0. So first, we set the denominator equal to 0, and solve for x.

$$x + 3 = 0$$
$$x = -3$$

PRACTICE 4

Solve $\dfrac{x-5}{x+5} < 0$. Graph the solutions.

EXAMPLE 4 (continued)

Next, we solve the related equation $\dfrac{x-1}{x+3} = 0$.

$$\frac{x-1}{x+3} \cdot (x+3) = 0 \cdot (x+3) \quad \text{Multiply each side of the equation by the LCD.}$$

$$x - 1 = 0 \quad \text{Simplify.}$$

$$x = 1$$

Then, we plot the boundary points -3 and 1, separating the number line into three intervals.

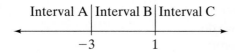

Now, we check a test value in each interval to determine whether the rational expression $\dfrac{x-1}{x+3}$ is greater than 0 or less than 0. Let's use -4, 0, and 2.

Interval	Test Value	Value of $\dfrac{x-1}{x+3}$	Conclusion
A	-4	$\dfrac{-4-1}{-4+3} = 5$	$5 > 0$
B	0	$\dfrac{0-1}{0+3} = -\dfrac{1}{3}$	$-\dfrac{1}{3} < 0$
C	2	$\dfrac{2-1}{2+3} = \dfrac{1}{5}$	$\dfrac{1}{5} > 0$

Value of $\dfrac{x-1}{x+3}$: $> 0 \quad < 0 \quad > 0$

$-3 \qquad 1$

Since we are solving the inequality $\dfrac{x-1}{x+3} < 0$, we are looking for values of x for which the expression is less than 0. The solutions are all real numbers between, but not including, -3 and 1, or in interval notation, $(-3, 1)$. The graph of the solutions is

$$\leftarrow\!+\!+\!+\!(\!+\!+\!+\!+\!+\!)\!+\!+\!+\!+\!+\!\rightarrow$$
$$-6\,-5\,-4\,-3\,-2\,-1\ \ 0\ \ 1\ \ 2\ \ 3\ \ 4\ \ 5\ \ 6$$

EXAMPLE 5

Solve $\dfrac{x+2}{x-4} \ge -2$. Then, graph the solutions.

Solution The inequality is undefined when the expression in the denominator, $x - 4$, is 0. Solving for x, we get:

$$x - 4 = 0$$

$$x = 4$$

PRACTICE 5

Solve $\dfrac{x-2}{x+3} \ge -1$. Then graph the solutions.

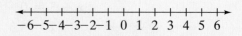

Now, we solve the related equation $\dfrac{x+2}{x-4} = -2$.

$$\frac{x+2}{x-4} \cdot (x-4) = -2 \cdot (x-4)$$
$$x + 2 = -2x + 8$$
$$3x = 6$$
$$x = 2$$

Plotting 2 and 4 as boundary points separates a number line into three intervals.

| Interval A | Interval B | Interval C |

$$\text{2} \qquad \text{4}$$

Next, we choose a test value in each interval: say, 0, 3, and 5.

Interval	Test Value	Value of $\dfrac{x+2}{x-4}$	Conclusion
A	0	$\dfrac{0+2}{0-4} = -\dfrac{1}{2}$	$-\dfrac{1}{2} > -2$
B	3	$\dfrac{3+2}{3-4} = -5$	$-5 < -2$
C	5	$\dfrac{5+2}{5-4} = 7$	$7 > -2$

Value of $\dfrac{x+2}{x-4}$: > -2 | < -2 | > -2

$$\text{2} \qquad \text{4}$$

Since the inequality under consideration is $\dfrac{x+2}{x-4} \geq -2$, we are looking for values of x for which the expression is greater than or equal to -2. So the solutions are all real numbers less than or equal to 2 or greater than 4, or in interval notation, $(-\infty, 2] \cup (4, \infty)$. Note that 4 is not a solution, since for this value of x the inequality is undefined. The graph of the solutions follows:

$$-6\ -5\ -4\ -3\ -2\ -1\ \ 0\ \ 1\ \ 2\ \ 3\ \ 4\ \ 5\ \ 6$$

Mathematically Speaking

Fill in each blank with the most appropriate term or phrase from the given list.

boundary point	quadratic	denominator
numerator	test value	linear
ratio	rational inequality	

1. To determine if the value of an expression is greater than or less than 0 in a particular interval, a(n) _____ is used.

2. A(n) _____ inequality in one variable is an inequality that contains a quadratic expression.

3. A(n) _____ is an inequality that contains a rational expression.

4. A rational inequality is undefined when the value of its _____ is 0.

A *Solve. Then, graph the solutions.*

5. $x^2 - x > 0$

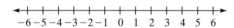

6. $x^2 + 4x > 0$

7. $x^2 < 4$

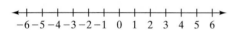

8. $9 > x^2$

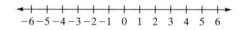

9. $x^2 - x - 2 \le 10$

10. $x^2 + 2x - 7 \le 8$

11. $x^2 + 6x + 9 \ge 0$

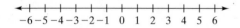

12. $x^2 - 4x + 4 \ge 0$

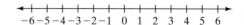

13. $6 + x - x^2 \le 0$

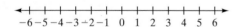

14. $20 - x - x^2 \le 0$

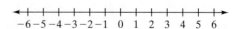

15. $x^2 + 5x + 4 < 0$

16. $x^2 - 4x + 3 \le 0$

17. $2x^2 - 3x + 1 \ge 0$

18. $2x^2 + 5x + 2 > 0$

19. $1 - 4x^2 \le 0$

20. $9x^2 - 4 < 0$

21. $3 > 6x^2 - 7x$

22. $4x^2 + 4x \ge 15$

23. $\dfrac{x}{x + 2} \le 0$

$\xleftarrow{\;+\!+\!+\!+\!+\!+\!+\!+\!+\!+\!+\!+\!+\;}\rightarrow$
$-6\,{-}5\,{-}4\,{-}3\,{-}2\,{-}1\;\;0\;\;1\;\;2\;\;3\;\;4\;\;5\;\;6$

24. $\dfrac{x - 1}{x} \le 0$

$\xleftarrow{\;+\!+\!+\!+\!+\!+\!+\!+\!+\!+\!+\!+\!+\;}\rightarrow$
$-6\,{-}5\,{-}4\,{-}3\,{-}2\,{-}1\;\;0\;\;1\;\;2\;\;3\;\;4\;\;5\;\;6$

25. $\dfrac{1}{6 - x} < 0$

$\xleftarrow{\;+\!+\!+\!+\!+\!+\!+\!+\!+\!+\!+\!+\!+\;}\rightarrow$
$0\;\;1\;\;2\;\;3\;\;4\;\;5\;\;6\;\;7\;\;8\;\;9\;\;10\;\;11\;\;12$

26. $\dfrac{2}{3 - x} < 0$

$\xleftarrow{\;+\!+\!+\!+\!+\!+\!+\!+\!+\!+\!+\!+\!+\;}\rightarrow$
$-6\,{-}5\,{-}4\,{-}3\,{-}2\,{-}1\;\;0\;\;1\;\;2\;\;3\;\;4\;\;5\;\;6$

27. $\dfrac{x + 3}{x - 3} > 2$

$\xleftarrow{\;+\!+\!+\!+\!+\!+\!+\!+\!+\!+\!+\!+\!+\;}\rightarrow$
$0\;\;1\;\;2\;\;3\;\;4\;\;5\;\;6\;\;7\;\;8\;\;9\;\;10\;\;11\;\;12$

28. $\dfrac{x - 6}{x + 6} > 4$

$\xleftarrow{\;+\!+\!+\!+\!+\!+\!+\!+\!+\!+\!+\!+\!+\;}\rightarrow$
$-12\,{-}11\,{-}10\,{-}9\,{-}8\,{-}7\,{-}6\,{-}5\,{-}4\,{-}3\,{-}2\,{-}1\;\;0$

29. $\dfrac{2x - 1}{x + 4} \ge 0$

$\xleftarrow{\;+\!+\!+\!+\!+\!+\!+\!+\!+\!+\!+\!+\!+\;}\rightarrow$
$-6\,{-}5\,{-}4\,{-}3\,{-}2\,{-}1\;\;0\;\;1\;\;2\;\;3\;\;4\;\;5\;\;6$

30. $\dfrac{x - 2}{3x + 2} \ge 0$

$\xleftarrow{\;+\!+\!+\!+\!+\!+\!+\!+\!+\!+\!+\!+\!+\;}\rightarrow$
$-6\,{-}5\,{-}4\,{-}3\,{-}2\,{-}1\;\;0\;\;1\;\;2\;\;3\;\;4\;\;5\;\;6$

31. $\dfrac{2x - 1}{2x + 5} \le 2$

$\xleftarrow{\;+\!+\!+\!+\!+\!+\!+\!+\!+\!+\!+\!+\!+\;}\rightarrow$
$-10\,{-}9\,{-}8\,{-}7\,{-}6\,{-}5\,{-}4\,{-}3\,{-}2\,{-}1\;\;0\;\;1\;\;2$

32. $\dfrac{3x + 9}{2x - 3} \le 3$

$\xleftarrow{\;+\!+\!+\!+\!+\!+\!+\!+\!+\!+\!+\!+\!+\;}\rightarrow$
$-2\,{-}1\;\;0\;\;1\;\;2\;\;3\;\;4\;\;5\;\;6\;\;7\;\;8\;\;9\;\;10$

Mixed Practice

Solve. Then, graph the solutions.

33. $x^2 + x - 5 < 7$

$\xleftarrow{\;+\!+\!+\!+\!+\!+\!+\!+\!+\!+\!+\!+\!+\;}\rightarrow$
$-6\,{-}5\,{-}4\,{-}3\,{-}2\,{-}1\;\;0\;\;1\;\;2\;\;3\;\;4\;\;5\;\;6$

34. $\dfrac{3}{5 - x} \ge 1$

$\xleftarrow{\;+\!+\!+\!+\!+\!+\!+\!+\!+\!+\!+\!+\!+\;}\rightarrow$
$-6\,{-}5\,{-}4\,{-}3\,{-}2\,{-}1\;\;0\;\;1\;\;2\;\;3\;\;4\;\;5\;\;6$

35. $\dfrac{x + 2}{x - 4} < 2$

$\xleftarrow{\;+\!+\!+\!+\!+\!+\!+\!+\!+\!+\!+\!+\!+\;}\rightarrow$
$0\;\;1\;\;2\;\;3\;\;4\;\;5\;\;6\;\;7\;\;8\;\;9\;\;10\;\;11\;\;12$

36. $-x^2 + 3x + 4 \le 0$

$\xleftarrow{\;+\!+\!+\!+\!+\!+\!+\!+\!+\!+\!+\!+\!+\;}\rightarrow$
$-6\,{-}5\,{-}4\,{-}3\,{-}2\,{-}1\;\;0\;\;1\;\;2\;\;3\;\;4\;\;5\;\;6$

37. $2x^2 + 5x - 3 \ge 0$

$\xleftarrow{\;+\!+\!+\!+\!+\!+\!+\!+\!+\!+\!+\!+\!+\;}\rightarrow$
$-6\,{-}5\,{-}4\,{-}3\,{-}2\,{-}1\;\;0\;\;1\;\;2\;\;3\;\;4\;\;5\;\;6$

38. $\dfrac{x - 1}{2x + 3} \ge 0$

$\xleftarrow{\;+\!+\!+\!+\!+\!+\!+\!+\!+\!+\!+\!+\!+\;}\rightarrow$
$-6\,{-}5\,{-}4\,{-}3\,{-}2\,{-}1\;\;0\;\;1\;\;2\;\;3\;\;4\;\;5\;\;6$

Applications

B *Solve.*

39. An outerwear manufacturer determines that its weekly revenue R (in dollars) for selling rain parkas at a price of p dollars each is modeled by the function $R(p) = 150p - p^2$. What range of prices for the parka will generate weekly revenue of at least $5000?

40. A penny is tossed upward with an initial velocity of 48 ft/sec. The height h of the penny (in feet) relative to the point of release t sec after it is tossed is modeled by the function $h(t) = -16t^2 + 48t$. For what interval of time is the penny above the point of release?

41. A furniture maker determines that the weekly cost C (in dollars) for producing x end tables is given by $C(x) = 2x^2 - 60x + 900$. How many end tables can be produced to keep the weekly cost under $500?

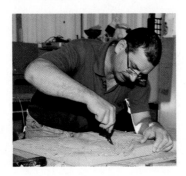

42. A ball is thrown straight upward with an initial velocity of 40 ft/sec from a height of 180 ft. The height h of the ball (in feet) above the ground t sec after it is thrown upward is given by $h(t) = -16t^2 + 40t + 180$. For what values of t is the ball at least 196 ft above the ground?

43. A gardener wants to enclose a rectangular flower bed using 90 ft of fencing. For what range of lengths will the area exceed 450 ft²?

44. The base of a box used for packaging has a perimeter of 60 in. For what range of widths will the area of the base be at least 200 in²?

45. A company determines that its average cost C (in dollars) for selling x units of a product is modeled by the function $C(x) = \dfrac{864 + 2x}{x}$. For what number of units will the average cost be less than $8?

46. A publishing company had revenue of $18 million last year. The company's financial analyst uses the formula $P(R) = \dfrac{100R - 1800}{R}$ to determine the percent growth P in the company's revenue this year over last year, where R is in millions of dollars. For what revenues R will the company's revenue grow by more than 10%?

• Check your answers on page A-46.

MINDStretchers

Writing

1. The inequalities $(x - 1)(x + 3) < 0$ and $\dfrac{x - 1}{x + 3} < 0$ have the same solutions. Explain why.

Mathematical Reasoning

2. Solve the compound inequality $4 < x^2 - 3x + 6 < 10$.

Groupwork

3. Working with a partner, describe a strategy for solving each of the following inequalities, and then solve the inequalities.

   a. $x^3 - 2x^2 - 8x \geq 0$

   b. $x^4 - 5x^2 + 4 < 0$

   c. $\dfrac{x - 2}{x^2 - 16} \leq 0$

   d. $\dfrac{x^2 - 4x}{x^2 - x - 6} > 0$

Key Concepts and Skills

Concept/Skill	Description	Example
[8.1] Quadratic equation	An equation that can be written in the form $ax^2 + bx + c = 0$, where a, b, and c are real numbers and $a \neq 0$.	$x^2 - 2x + 8 = 0$
[8.1] The square root property of equality	If b is a real number and $x^2 = b$, then $x = \pm\sqrt{b}$, that is, $x = \sqrt{b}$ or $x = -\sqrt{b}$.	$x^2 = 63$ $x = \pm\sqrt{63}$ $x = 3\sqrt{7}$ or $x = -3\sqrt{7}$
[8.1] To complete the square for the expression $x^2 + bx$	• Take one-half the coefficient of the linear term x, and then square it. Add this value to the expression. $$x^2 + bx + \left(\frac{1}{2}b\right)^2$$	Find the value of c that will make $x^2 + 6x + c$ a perfect square trinomial. $$\left(\frac{1}{2}\cdot 6\right)^2 = 3^2 = 9, \text{ so } c = 9$$ $x^2 + 6x + 9$
[8.1] To solve a quadratic equation $(ax^2 + bx + c = 0)$ by completing the square	• If $a \neq 1$, then divide each side of the equation by a, the coefficient of the quadratic (or second-degree) term. If $a = 1$, then proceed to the next step. • Move all terms with variables to one side of the equal sign and all constants to the other side. • Take one-half the coefficient of the linear term. Then, square it, and add this value to each side of the equation. • Factor the side of the equation containing the variables, writing it as the square of a binomial. • Use the square root property of equality, and solve the resulting equations.	$2x^2 + 8x - 6 = 0$ $x^2 + 4x - 3 = 0$ $x^2 + 4x = 3$ Since $\left(\frac{1}{2}\cdot 4\right)^2 = 2^2 = 4$, add 4 to both sides. $x^2 + 4x + 4 = 3 + 4$ $(x + 2)^2 = 7$ $x + 2 = \pm\sqrt{7}$ $x = -2 \pm \sqrt{7}$ Solutions: $-2 + \sqrt{7}$ and $-2 - \sqrt{7}$
[8.2] To solve a quadratic equation $(ax^2 + bx + c = 0)$ by using the quadratic formula	• Write the equation in standard form, if necessary. • Identify the coefficients a, b, and c. • Substitute the values for a, b, and c in the formula $$x = \frac{-b \pm \sqrt{b^2 - 4ac}}{2a}.$$ • Simplify.	$-2x^2 + 6x = 7$ $-2x^2 + 6x - 7 = 0$ $a = -2, b = 6,$ and $c = -7$ $$x = \frac{-6 \pm \sqrt{6^2 - 4(-2)(-7)}}{2(-2)}$$ $$= \frac{-6 \pm \sqrt{36 - 56}}{-4}$$ $$= \frac{-6 \pm \sqrt{-20}}{-4}$$ $$= \frac{-6 \pm 2i\sqrt{5}}{-4}$$ $$x = \frac{3 \pm i\sqrt{5}}{2}$$ Solutions: $\dfrac{3 + i\sqrt{5}}{2}$ and $\dfrac{3 - i\sqrt{5}}{2}$

continued

Concept/Skill	**Description**	**Example**
[8.2] Discriminant of a quadratic equation $(ax^2 + bx + c = 0)$	The expression in the radicand of the quadratic formula: $$b^2 - 4ac$$	For $-2x^2 + 6x - 7 = 0$, the value of the discriminant is: $$6^2 - 4(-2)(-7) = -20$$
[8.3] Equation quadratic in form	An equation that can be written in the form $au^2 + bu + c = 0$, where $a \neq 0$ and u represents an algebraic expression.	$x^4 - 3x^2 + 2 = 0$ is quadratic in form, since it can be written as $u^2 - 3u + 2 = 0$, where $u = x^2$.
[8.4] Quadratic function	A function of the form $f(x) = ax^2 + bx + c$.	$f(x) = 5x^2 - 3x + 9$
[8.4] Parabola	The graph of a quadratic function.	The graph of $f(x) = x^2 + 2x - 3$ is the following parabola:
[8.4] Vertex of a parabola	The highest point (with maximum y-value) or lowest point (with minimum y-value) of a parabola.	

Concept/Skill	Description	Example
[8.4] To find the vertex of a parabola	For the parabola given by the function $f(x) = ax^2 + bx + c$, • the x-coordinate of the vertex is $-\dfrac{b}{2a}$, and • the y-coordinate of the vertex is found by substituting $-\dfrac{b}{2a}$ for x in $f(x)$.	$f(x) = x^2 + 4x - 5$ The x-coordinate of the vertex is $-\dfrac{4}{2(1)} = -2$. The y-coordinate of the vertex is: $f(-2) = (-2)^2 + 4(-2) - 5$ $\qquad = 4 - 8 - 5$ $\qquad = -9$ The vertex is $(-2, -9)$.
[8.4] Equation of the axis of symmetry	For the graph of $f(x) = ax^2 + bx + c$, the equation of the axis of symmetry is: $$x = -\frac{b}{2a}$$	For $f(x) = x^2 + 4x - 5$, the equation of the axis of symmetry is: $$x = -\frac{4}{2(1)} = -2$$
[8.5] Quadratic inequality	An inequality that contains a quadratic expression.	$x^2 + 2x < -1$ and $t^2 + 3t - 5 \geq 0$
[8.5] To solve a quadratic inequality	• Solve the related equation, and plot the solutions on a number line separating it into intervals. • Choose a test value in each interval to check if it satisfies the original inequality. • Identify the intervals that contain the solutions of the inequality. Check whether to include the boundary points.	$x^2 + 3x - 10 < 0$ $x^2 + 3x - 10 = 0$ **Related equation** $(x - 2)(x + 5) = 0$ $x - 2 = 0$ or $x + 5 = 0$ $\quad x = 2 \qquad\qquad x = -5$ Let's use values -6, 0, and 3: $x^2 + 3x - 10$ $(-6)^2 + 3(-6) - 10 = 8 > 0$ $(0)^2 + 3(0) - 10 = -10 < 0$ $(3)^2 + 3(3) - 10 = 8 > 0$ Since the original inequality involves the symbol $<$, the solutions are all real numbers between, but not including, -5 and 2, or in interval notation, $(-5, 2)$.
[8.5] Rational inequality	An inequality that contains a rational expression.	$\dfrac{x + 1}{x - 3} \geq -5$ and $\dfrac{2x + 4}{2x + 3} < 0$

continued

Concept/Skill	Description	Example
[8.5] To solve a rational inequality	• Find all values for which the inequality is undefined. • Solve the inequality's related equation. • Plot the values found in the previous steps on a number line, separating it into intervals. • Choose a test value in each interval to check if it satisfies the original inequality. • Identify the intervals that contain the solutions of the inequality. Check whether to include the boundary points.	$\dfrac{3x}{x+1} \geq 2$ The rational expression is undefined when $x + 1 = 0$ or $x = -1$. $\dfrac{3x}{x+1} = 2$ $\dfrac{3x}{x+1} \cdot (x+1) = 2 \cdot (x+1)$ $\qquad\qquad 3x = 2x + 2$ $\qquad\qquad\ x = 2$ 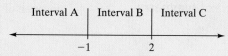 Let's use test values -2, 0, and 3: $\dfrac{3x}{x+1}$ $\dfrac{3(-2)}{-2+1} = 6 > 2$ $\dfrac{3(0)}{0+1} = 0 < 2$ $\dfrac{3(3)}{3+1} = \dfrac{9}{4} > 2$ Value of $\dfrac{3x}{x+1}$:  The solutions are all real numbers less than -1 or greater than or equal to 2, or in interval notation, $(-\infty, -1) \cup [2, \infty)$. 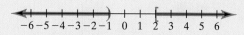

Value of the Discriminant ($b^2 - 4ac$)	Number and Type of Solutions
Zero	One real solution
Positive	Two real solutions
Negative	Two complex solutions (containing i)

Solving a Quadratic Equation ($ax^2 + bx + c = 0$)

Method	When to Use
Factoring	Use when the polynomial can be factored.
The Square Root Property	Use when $b = 0$.
The Quadratic Formula	Use when the first two methods do not apply.
Completing the Square	Use only when specified. This method is easier to use when $a = 1$ and b is even.

Say Why
Fill in each blank.

1. The equation $\sqrt{3}x^2 = 2$ _____ quadratic because
   is/is not
   _____
   _____
   _____ .

2. If $(x - 8)^2 = 24$, then $x - 8$ _____ equal to
   is/is not
   $\sqrt{24}$ or $-\sqrt{24}$ because _____
   _____ .

3. The equation $6x^2 - 9x + 4 = 0$ _____ have any
   does/does not
   real solutions because _____
   _____ .

4. The parabola given by $y = 2x^2 - 3x$ _____
   does/does not
   have a maximum because _____
   _____ .

5. The graph of $y = -x^2 - 1$ _____ lie in
   does/does not
   Quadrant I because _____
   _____ .

6. The equation $3x + \sqrt{3x} = 5$ _____ quadratic in
   is/is not
   form because _____
   _____ .

[8.1] *Solve by using the square root property.*

7. $x^2 - 81 = 0$

8. $3n^2 - 7 = 8$

9. $4(a - 5)^2 = 1$

10. $(2x + 1)^2 + 10 = 6$

11. $x^2 + 8x + 16 = 2$

12. $(n - 5)(n + 5) = -33$

Fill in the blank to make each trinomial a perfect square.

13. $x^2 + 10x + [\ \ ]$

14. $n^2 - 9n + [\ \ ]$

Solve by completing the square.

15. $x^2 - 6x + 2 = 0$

16. $a^2 + a - 3 = 0$

17. $2n^2 + 2n + 9 = 3 - 4n$

18. $3x^2 - 2x - 9 = 0$

[8.2] *Solve by using the quadratic formula.*

19. $x^2 + 7x + 6 = 0$

20. $x^2 - 4x + 5 = 0$

21. $3x^2 - 13x = 5 - 7x$

22. $4x^2 + 12x = -9$

23. $\frac{1}{3}x^2 + \frac{3}{2}x + 1 = 0$

24. $0.01x^2 + 0.1x + 0.34 = 0$

Use the discriminant to determine the number and type of solutions for each equation.

25. $2x^2 - x = 2x - 5$

26. $4x^2 + 9x - 3 = 0$

[8.3] *Solve.*

27. $\dfrac{2}{n-3} + \dfrac{1}{n-1} = -1$

28. $x^4 - 2x^2 - 24 = 0$

29. $x - 7\sqrt{x} + 12 = 0$

30. $(p-1)^2 + 3(p-1) + 2 = 0$

Find a quadratic equation that has the given solutions.

31. $x = -3, x = \frac{5}{2}$

32. $n = -7$

[8.4] *Find the vertex, axis of symmetry, and x- and y-intercepts of the graph of each function. Then, sketch the graph.*

33. $f(x) = x^2 - 6x + 5$
Vertex:
Axis of symmetry:
x-intercept(s):
y-intercept:

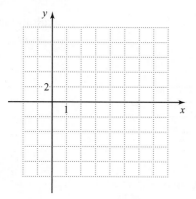

34. $f(x) = 3 + 2x - x^2$
Vertex:
Axis of symmetry:
x-intercept(s):
y-intercept:

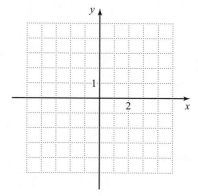

35. $f(x) = x^2 + 8x + 16$
Vertex:
Axis of symmetry:
x-intercept(s):
y-intercept:

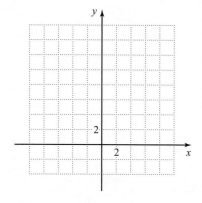

36. $f(x) = x^2 - 5x - 6$
Vertex:
Axis of symmetry:
x-intercept(s):
y-intercept:

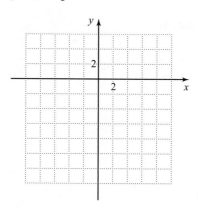

Complete the table.

Function	Opens Upward or Downward?	Maximum or Minimum Point?	Number of x-intercepts	Number of y-intercepts
37. $f(x) = x^2 + 9$				
38. $f(x) = 1 + 3x - 2x^2$				

Graph the function. Then, determine the domain and range.

39. $f(x) = 3x - 4x^2$

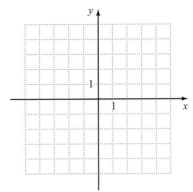

Domain: Range:

40. $f(x) = 2x^2 - x - 1$

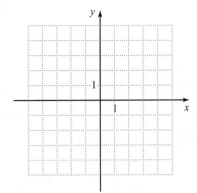

Domain: Range:

[8.5] *Solve. Then, graph the solutions.*

41. $x^2 - 4x < 12$

     $-6\ -5\ -4\ -3\ -2\ -1\ \ 0\ \ 1\ \ 2\ \ 3\ \ 4\ \ 5\ \ 6$

42. $10 + 3x - x^2 \le 0$

     $-6\ -5\ -4\ -3\ -2\ -1\ \ 0\ \ 1\ \ 2\ \ 3\ \ 4\ \ 5\ \ 6$

43. $2x^2 - 9x - 4 \ge -8$

     $-2\ -1\ \ 0\ \ 1\ \ 2\ \ 3\ \ 4\ \ 5\ \ 6\ \ 7\ \ 8\ \ 9\ \ 10$

44. $3 > 4x^2 - 4x$

     $-6\ -5\ -4\ -3\ -2\ -1\ \ 0\ \ 1\ \ 2\ \ 3\ \ 4\ \ 5\ \ 6$

45. $\dfrac{x + 3}{x - 5} > -3$

     $-1\ \ 0\ \ 1\ \ 2\ \ 3\ \ 4\ \ 5\ \ 6\ \ 7\ \ 8\ \ 9\ \ 10\ \ 11$

46. $\dfrac{4x - 12}{3x} \le 0$

     $-6\ -5\ -4\ -3\ -2\ -1\ \ 0\ \ 1\ \ 2\ \ 3\ \ 4\ \ 5\ \ 6$

Mixed Applications

Solve.

47. The distance s (in feet) that an object falls t sec after it is dropped is given by the equation $s = 16t^2$. How long will it take an object to fall 400 ft?

48. A cable television technician places a 10-foot ladder against a house. The ladder reaches a point on the house that is three times the distance d from the base of the house, as shown in the figure. To the nearest tenth of a foot, how far up the side of the house does the ladder reach?

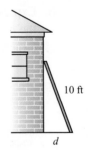

10 ft

d

49. Two trains leave the same station at the same time. The train going south travels at an average speed that is 8 mph faster than the train traveling west. Find the average speed, to the nearest mile per hour, of each train if they are 35 mi apart after 30 min.

50. A financial analyst advises a client to invest $5000 in a high-risk fund. To determine the average rate of return, r (in decimal form), on the investment after two years, the broker uses the equation $A = 5000\left[(1 + r)^2 - 1\right]$, where A is the amount earned on the investment. To the nearest tenth of a percent, what was the average rate of return if the investment earned $1100?

51. An open box is to be made from a 25-inch by 40-inch rectangular piece of cardboard by cutting squares of equal area from each corner and turning up the sides. If the specifications require the base of the box to have an area of 700 in^2, what size squares should be cut from each corner?

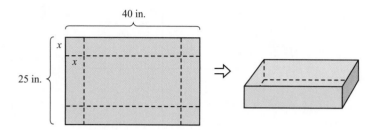

52. The owner of a small office-supply store uses the equation $R = 100x - 2x^2$ to determine the number of cases of paper x he needs to sell each week for revenue of R dollars. How many cases does he need to sell each week for revenue of $1250?

53. The number of Krispy Kreme stores S in operation for each year between 2004 and 2010 can be approximated by the equation $S = 6.69n^2 - 5.29n + 372.31$, where n represents the number of years after 2004. In what year were there about 582 stores? (*Source:* investor.krispykreme.com)

54. The number of Walmart stores S in operation for each year between 2006 and 2010 can be approximated by the equation $S = -16.9n^2 + 658.8n + 6042$, where n represents the number of years after 2006. In what year, during this time period, were there approximately 7863 stores? (*Source:* walmart.com)

55. A student bicycles to and from school each day. On a particular day, she bicycled to school at a rate that was 2 mph faster than her rate bicycling home. If she lives 8 mi from school and it took her 6 min longer to bicycle home, to the nearest tenth of a mile per hour, at what rate did she bicycle to school that day?

56. It takes an older printing press 5 hr longer to complete a print job than a newer printing press. If the printing presses work together, they can complete the job in 2 hr. To the nearest tenth of an hour, how long does it take each press to complete the job working alone?

57. A city parks department uses 300 ft of fencing to enclose a rectangular playground in one of its parks.
  a. Express the width w of the playground in terms of the length l.
  b. Express in function notation the relationship between the area $A(l)$ and the length of the playground.
  c. Graph the function.

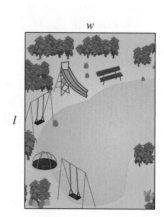

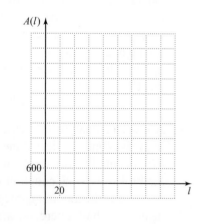

  d. What dimensions will maximize the area of the playground? What is the maximum area?

58. The owner of a factory determines that the daily cost (in dollars) of fabricating x units of a machine part is modeled by the function $C(x) = 0.05x^2 - 2x + 100$.

a. Graph the function.

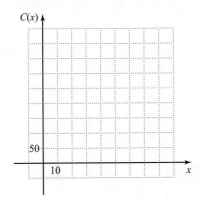

b. What number of units must be fabricated per day in order to minimize the cost?

c. What is the minimum cost?

59. A baseball is hit straight upward with an initial velocity of 80 ft/sec from a height of 4 ft. The height h (in feet) of the baseball above the ground t sec after it is hit is modeled by the function $h(t) = -16t^2 + 80t + 4$. For what values of t is the baseball 40 ft above the ground?

60. A homeowner wants to build a rectangular patio that has a perimeter of 44 ft. For what range of widths will the patio have an area of at least 120 ft^2?

• Check your answers on page A-47.

CHAPTER 8   Posttest

FOR
EXTRA
HELP

CHAPTER
Test Prep
VIDEOS

The Chapter Test Prep Videos with test solutions are available on DVD, in MyMathLab, and on You Tube ™ (search "AkstIntermediate Alg" and click on "Channels").

To see if you have mastered the topics in this chapter, take this test.

1. Solve by using the square root property:
$5n^2 - 11 = 29$

2. Solve by completing the square: $3p^2 - 6p = -24$

3. Solve by using the quadratic formula:
$4x^2 + 4x - 3 = 0$

4. Use the discriminant to determine the number and type of solutions of $2x^2 + 7x + 9 = 0$.

Solve.

5. $5(3n + 2)^2 - 90 = 0$

6. $x^2 + 8x = 6$

7. $x^2 - x - 2 = 4x - 13$

8. $4x^2 + 3x = 7 + 6x$

9. $2x^2 - 12x + 20 = 1$

10. $\frac{1}{2}x^2 + x - 2 = 0$

11. $x^{1/2} - 2x^{1/4} = -1$

12. Find a quadratic equation in n that has the solutions $\frac{1}{2}$ and $\frac{2}{3}$.

Identify the vertex, the equation of the axis of symmetry, and the x- and y-intercepts of the graph. Then, graph.

13. $f(x) = x^2 - 6x + 8$

14. $f(x) = -x^2 + 3x + 10$

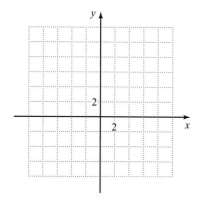

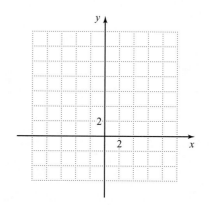

15. Graph the function $f(x) = 2x^2 - 4x - 1$. Then, identify the domain and range.

16. Solve the inequality $-x^2 - 3x + 18 < 0$. Then, graph the solutions.

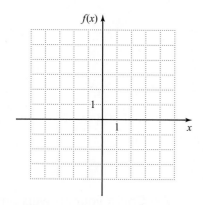

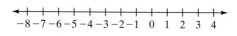

17. Two airplanes leave the same airport at the same time. The airplane flying west travels at an average speed that is 20 mph faster than the airplane flying south. If the two airplanes are 650 mi apart after 1 hr, find the average speed of each airplane to the nearest mile per hour.

18. A junior credit clerk takes 3 hr longer to process the same number of credit applications as a senior credit clerk. If they work together, they can process the applications in 2 hr. To the nearest tenth of an hour, how long does it take each clerk to process the applications working alone?

19. A toy rocket is launched straight upward with an initial velocity of 96 ft/sec from a 3-foot-high platform. The height h (in feet) of the rocket above the ground t sec after it is launched is modeled by the function $h(t) = -16t^2 + 96t + 3$. When will the rocket reach its maximum height? What is that height?

20. The owner of a factory determines that the daily cost C (in dollars) to produce x units of a product is modeled by the function $C(x) = 0.025x^2 - 8x + 800$. How many units can the factory produce each day in order to keep the daily cost below $320?

• Check your answers on page A-47.

Cumulative Review Exercises

To help you review, solve the following problems.

1. Simplify: $4n^5(2n^3)^{-1}$

2. Solve: $|2x - 3| + 1 = 7$

3. Graph: $x + 4y - 8 = 0$

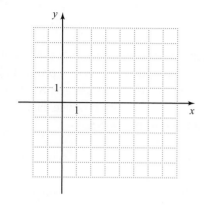

4. Does the relation $\{(-5, 2), (-3, -4), (0, 1), (3, 1), (4, 2)\}$ represent a function?

5. Solve the system:
$$\begin{array}{rcr} 2x - 3y + z &=& 5 \\ x + 3y + 2z &=& -2 \\ -2y + z &=& -3 \end{array}$$

6. Find the quotient: $(8 + 4a^3 - a) \div (2a + 3)$

7. Factor: $2x^2y^4 - 12x^3y^3 + 16x^4y^2$

8. Multiply: $(a - 2)(a^2 + 2a + 4)$

9. Simplify: $\dfrac{1 + \dfrac{7}{x} + \dfrac{12}{x^2}}{1 + \dfrac{2}{x} - \dfrac{8}{x^2}}$

10. Solve and check: $\dfrac{6b}{b - 6} - \dfrac{b^2}{b - 6} = 1$

11. Write using radical notation, simplifying if possible: $(-8x^9)^{1/3}$

12. Simplify: $\sqrt{27} - 3\sqrt{18} + 7\sqrt{3}$

13. Solve: $6x^2 - 2x = -1$

14. Find a quadratic equation that has the given solutions:
$$p = -3, p = \frac{1}{4}$$

15. Complete the table:

Function	Opens Upward or Downward?	Maximum or Minimum Point?	Number of x-intercepts	Number of y-intercepts
$f(x) = 4x - x^2 + 2$				

16. Typically, an audio file is 0.004 gigabytes (GB), and a TV episode video file is 0.35 GB. A certain portable multimedia player has 80 GB of storage. (*Source:* apple.com)

   a. Express as an inequality the possible number of average-sized audio *a* and video files *b* that can be stored on this player.

   b. Can the player store 1000 audio files and 200 video files?

17. In geometry, the formula $V = \pi r^2 h$ is used to find the capacity (volume) *V* of a cylinder in terms of the cylinder's radius *r* and height *h*.

   a. Find a formula for *r* in terms of *h* and *V*.

   b. A container of disinfecting wipes is cylindrical in shape. Use the formula in part (a) to find the radius of this container if its capacity is 1024π cm^3 and its height is 16 cm.

18. A farmer has less than 1000 ft of fencing to enclose a rectangular grazing area on his farm.

   a. Express this relationship as a linear inequality, letting *l* and *w* represent the length and the width of the fence, respectively.

   b. Graph the inequality.

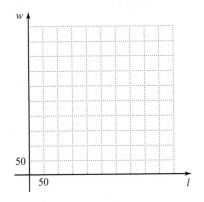

   c. What is one possible set of dimensions for the grazing area?

19. A long-distance phone service provider charges its customers a flat monthly fee and an additional fee for each minute of long-distance calls made during the month. A customer's long-distance phone bill is $13.50 when she makes 120 min of long-distance phone calls and $19.90 when she makes 248 min of long-distance phone calls.

   a. Write an equation in slope-intercept form that relates the monthly phone bill *B* (in dollars) and the number of minutes *m* of long-distance calls made that month.

   b. Identify the slope of the line and interpret its meaning in the context of the situation.

   c. Identify the *y*-intercept and interpret its meaning in the context of the situation.

20. The number of sales (in thousands) of Nissan Altima hybrid electric cars is approximated by the polynomial $53.5x^2 + 377.5x + 8388$, where *x* represents the number of years after 2007. (*Sources:* hybridcars.com: *Hybrid Vehicles Report*)

   a. In what year were the hybrid electric car sales of Nissan Altimas approximately 9300?

   b. If the trend described by the model continues, predict the year in which the total sales of hybrid Nissan Altimas will be approximately 11,000.

• Check your answers on page A-48.

CHAPTER 9

Exponential and Logarithmic Functions

Ecology and Exponential Functions

Ecologists study the effect on natural resources of growing industrialization. One of their concerns is the dwindling supply of fossil fuels, such as oil, natural gas, and coal. The known world coal reserves are estimated to be about 1×10^{12} short tons, with consumption currently running at approximately 7×10^9 short tons per year.

Suppose we assume that coal consumption increases by 2% per year, that is, each year it is 1.02 times that of the previous year. In this model, described by $C(t) = (7 \times 10^9)(1.02)^t$, consumption is an *exponential function* of time because consumption changes by a constant *factor* each year.

(*Sources:* eia.gov; Manuel C. Molles, *Ecology: Concepts and Applications*, McGraw-Hill Higher Education, 2009)

CHAPTER 9   PRETEST

To see if you have already mastered the topics in this chapter, take this test. Use a calculator, where necessary. Round answers to four decimal places.

1. Consider the functions $f(x) = 2x - 1$ and $g(x) = 2x^2 + 5x - 3$. Find:
   a. $(f + g)(x)$
   b. $(f - g)(x)$
   c. $(f \cdot g)(x)$
   d. $\left(\dfrac{f}{g}\right)(x)$

2. Given $f(x) = \dfrac{5}{x}, x \neq 0$, and $g(x) = 3x - 4$, find:
   a. $(f \circ g)(x)$
   b. $(g \circ f)(x)$
   c. $(f \circ g)(3)$
   d. $(g \circ f)(-5)$

3. Determine whether the function whose graph is shown on the coordinate plane below is one-to-one. If it is, sketch the graph of its inverse.

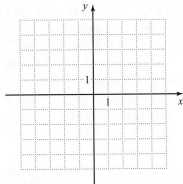

4. Find the inverse of the one-to-one function $f(x) = 3x - 7$.

5. Evaluate each function for the given value. If necessary, round the answer to the nearest thousandth.
   a. $f(x) = 2^{x-5}; f(3)$
   b. $f(x) = -e^{-x}; f(-2)$

6. Evaluate:
   a. $\log_7 1$
   b. $\log_9 \dfrac{1}{81}$

7. Graph:
   a. $f(x) = 2^x + 1$

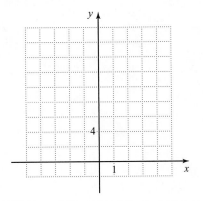

   b. $f(x) = -\log_4 x$

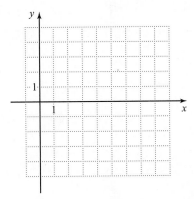

8. Evaluate:
   a. $\log_6 6^5$
   b. $\log_4 1$

9. Evaluate:
   a. $\log \dfrac{1}{10}$
   b. $\ln e^5$

10. Find and round to four decimal places: $\log_3 8$

11. Write each expression as a sum or difference of logarithms.

    a. $\log_5(5x^2)$

    b. $\log_8 \dfrac{3x^3}{y}$

12. Write each expression as a single logarithm.

    a. $3\log_7 2 + \log_7 5$

    b. $4\log_6 x - 2\log_6(x + 2)$

Solve.

13. $5^x = 25$

14. $4^{x+1} = 32^x$

15. $\log_x \dfrac{1}{2} = -1$

16. $\log_2(x - 4) = 3$

17. $\log_2 x + \log_2(x + 6) = 4$

18. Inflation is a sustained increase in the cost of goods and services. Suppose the annual rate of inflation will be 4% during the next 10 yr. Then, the cost C (in dollars) of goods and services in any year x during that decade will be modeled by the function $C(x) = P(1.04)^x$, where P is the present cost. If the present cost of a gallon of milk is \$3.89, what will it cost in the fifth year of the decade?

19. The effective annual interest rate r on an investment earning interest rate k, compounded continuously, can be determined from the logarithmic equation $\ln(r + 1) = k$, where r and k are expressed in decimal form. To the nearest tenth of a percent, find the effective annual interest rate on an investment earning 4.5% interest compounded continuously.

20. Gallium-67, a radioactive isotope used to detect tumors, has a half-life of about 3.3 days. The amount C of the isotope remaining after t days is given by the equation $C = C_0 e^{-0.21t}$, where C_0 is the initial amount present. Approximately how long will it take 100 mg of gallium-67 to decay to 12 mg?

• Check your answers on page A-48.

OBJECTIVES

In this section, we revisit to the discussion of functions that began in Section 3.6. There, we considered how to identify, evaluate, and graph a function, and also how to determine the domain and range of a function. Here, we deal with operations on functions, with particular attention to the operation of composition. We then move on to one-to-one and inverse functions. This final topic will be of particular importance in making the transition from Section 9.2 to Section 9.3, where we discuss exponential and logarithmic functions.

(A) To add, subtract, multiply, or divide functions

(B) To find the composition of two functions

(C) To determine if a function is one-to-one

(D) To find the inverse of a function

(E) To graph the inverse of a function

(F) To solve applied problems involving functions

Operations on Functions

Recall that the sum of two polynomials is a third polynomial. Similarly, the sum of two functions is a function. For example, suppose that $f(x) = 3x^2$ and $g(x) = 7x - 1$. We can define a new function (the "sum function") represented by $(f + g)(x)$ as follows:

$$(f + g)(x) = f(x) + g(x) = 3x^2 + 7x - 1$$

Similarly, we can subtract, multiply, or divide given functions to define other new functions.

There are many situations in which we define a new function in terms of old ones. For instance, suppose that a company is producing and selling a certain product. In this case, the profit P that the company makes can be computed by subtracting the cost C from the revenue R. If each of these three variables is a function of the number of products x produced and sold, then

$$P(x) = R(x) - C(x),$$

and we can write the profit function $P(x)$ as the difference function $(R - C)(x)$.

In general, we define the sum, difference, product, and quotient functions as follows:

DEFINITIONS

If f and g are functions, then

$(f + g)(x) = f(x) + g(x)$	**The sum of two functions**
$(f - g)(x) = f(x) - g(x)$	**The difference of two functions**
$(f \cdot g)(x) = f(x) \cdot g(x)$	**The product of two functions**
$\left(\dfrac{f}{g}\right)(x) = \dfrac{f(x)}{g(x)},$ for $g(x) \neq 0$	**The quotient of two functions**

EXAMPLE 1

Given that $f(x) = 3x - 5$ and $g(x) = 2x + 1$, find:

a. $(f + g)(x)$ **b.** $(f - g)(x)$

c. $(f \cdot g)(x)$ **d.** $\left(\dfrac{f}{g}\right)(x)$

Solution

a. $(f + g)(x) = f(x) + g(x)$
   $= (3x - 5) + (2x + 1)$ Substitute $(3x - 5)$ for $f(x)$ and $(2x + 1)$ for $g(x)$.
   $= 5x - 4$

PRACTICE 1

Given that $f(x) = x^2 - 1$ and $g(x) = 3x - 2$, find:

a. $(f + g)(x)$

b. $(f - g)(x)$

c. $(f \cdot g)(x)$

d. $\left(\dfrac{f}{g}\right)(x)$

EXAMPLE 1 (continued)

b. $(f - g)(x) = f(x) - g(x)$

$$= (3x - 5) - (2x + 1)$$
$$= 3x - 5 - 2x - 1$$
$$= x - 6$$

c. $(f \cdot g)(x) = f(x) \cdot g(x)$

$$= (3x - 5)(2x + 1)$$
$$= 6x^2 + 3x - 10x - 5$$
$$= 6x^2 - 7x - 5$$

d. $\left(\dfrac{f}{g}\right)(x) = \dfrac{f(x)}{g(x)}$

$$= \dfrac{3x - 5}{2x + 1}, \quad \text{where } x \neq -\dfrac{1}{2}$$

Can you explain in Example 1(d) why the value $-\dfrac{1}{2}$ is excluded?

Consider the functions $f(x) = 3x - 5$ and $g(x) = 2x + 1$ given in Example 1. Suppose we want to evaluate $(f + g)(x)$ for a particular value of x, for example, 3. Since $(f + g)(x) = f(x) + g(x)$, we get:

$$(f + g)(3) = f(3) + g(3)$$
$$= [3(3) - 5] + [2(3) + 1]$$
$$= (9 - 5) + (6 + 1)$$
$$= 4 + 7$$
$$= 11$$

So $(f + g)(3) = 11$.

Recall that in Example 1, we found the sum function $(f + g)(x) = 5x - 4$. Evaluating this function for $x = 3$ gives us:

$$(f + g)(3) = 5(3) - 4$$
$$= 15 - 4$$
$$= 11$$

The preceding examples suggest that to evaluate the sum of two functions for a particular value of x, we can *either evaluate the individual functions and then add the values, or find the sum function and then evaluate that function*. Similar statements can be made for the difference, product, and quotient of two functions.

EXAMPLE 2

Given that $f(x) = -x + 7$ and $g(x) = x^2 + 3x$, find:

a. $(f + g)(2)$ **b.** $(f - g)(-3)$ **c.** $(f \cdot g)(0)$ **d.** $\left(\dfrac{f}{g}\right)(1)$

Solution

a. $(f + g)(2) = f(2) + g(2)$ *Express as the sum of two functions.*

$$= (-2 + 7) + [2^2 + 3(2)] \quad \text{Evaluate } f(2) \text{ and } g(2).$$
$$= (5) + (4 + 6)$$
$$= 5 + 10$$
$$= 15$$

PRACTICE 2

Given that $f(x) = x - x^2$ and $g(x) = 2x + 5$, find:

a. $(f + g)(-1)$

b. $(f - g)(5)$

c. $(f \cdot g)(1)$

d. $\left(\dfrac{f}{g}\right)(2)$

b. $(f - g)(-3) = f(-3) - g(-3)$
$$= [-(-3) + 7] - [(-3)^2 + 3(-3)]$$
$$= (3 + 7) - (9 - 9)$$
$$= 10 - 0$$
$$= 10$$

c. $(f \cdot g)(0) = f(0) \cdot g(0)$
$$= (-0 + 7) \cdot [0^2 + 3(0)]$$
$$= 7 \cdot 0$$
$$= 0$$

d. $\left(\dfrac{f}{g}\right)(1) = \dfrac{f(1)}{g(1)}$
$$= \dfrac{-1 + 7}{1^2 + 3(1)}$$
$$= \dfrac{6}{4}$$
$$= \dfrac{3}{2}$$

The Composition of Functions

Another operation that can be performed on functions is *composition*, represented by the symbol ∘. When we take the composition of two functions, we find the value of one function, and then evaluate the other function at that value.

For instance, consider the following example. A size-4 dress in the United States is a size 36 in France. A function that converts dress sizes in the United States to those in France is $g(x) = x + 32$, where x is the U.S. size and $g(x)$ is the French size. Similarly, the function $f(x) = 2x - 40$ relates dress sizes in France to those in Italy, where x is the French size and $f(x)$ is the Italian size. Using these two functions, we can define a function—say, $h(x)$—that relates dress sizes in the United States to those in Italy.

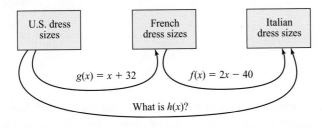

If $g(x) = x + 32$ and $f(x) = 2x - 40$, then:
$$h(x) = f(g(x)) = f(x + 32) = 2(x + 32) - 40 = 2x + 64 - 40 = 2x + 24$$

We call the function $h(x)$ the *composition of f and g.*

DEFINITION

If f and g are functions, then the **composition of f and g** is the function, represented by $f \circ g$, where

$$(f \circ g)(x) = f(g(x)).$$

Note that the *composite function* $f \circ g$ can be read either "f composed with g" or "f circle g." The composite function is defined for those values of x in the domain of g where $g(x)$ is also in the domain of f.

EXAMPLE 3

If $f(x) = 3x - 1$ and $g(x) = 2x + 1$, find:

a. $(f \circ g)(x)$ **b.** $(g \circ f)(x)$

c. $(f \circ g)(2)$ **d.** $(g \circ f)(2)$

Solution

a. $(f \circ g)(x) = f(g(x))$

$= f(2x + 1)$ Substitute $2x + 1$ for $g(x)$.

$= 3(2x + 1) - 1$ Substitute $2x + 1$ for x in $f(x)$.

$= 6x + 3 - 1$ Use the distributive property.

$= 6x + 2$

b. $(g \circ f)(x) = g(f(x))$

$= g(3x - 1)$ Substitute $3x - 1$ for $f(x)$.

$= 2(3x - 1) + 1$ Substitute $3x - 1$ for x in $g(x)$.

$= 6x - 2 + 1$ Use the distributive property.

$= 6x - 1$

c. $(f \circ g)(2) = f(g(2))$

$= f(2(2) + 1)$ Substitute 2 for x in $g(x)$.

$= f(5)$ Simplify.

$= 3(5) - 1$ Substitute 5 for x in $f(x)$.

$= 15 - 1$

$= 14$

d. $(g \circ f)(2) = g(f(2))$

$= g(3(2) - 1)$ Substitute 2 for x in $f(x)$.

$= g(5)$ Simplify.

$= 2(5) + 1$ Substitute 5 for x in $g(x)$.

$= 11$

PRACTICE 3

If $f(x) = 4x - 3$ and $g(x) = -x + 5$, find:

a. $(f \circ g)(x)$

b. $(g \circ f)(x)$

c. $(f \circ g)(3)$

d. $(g \circ f)(3)$

Note that in Example 3, the answers to parts (a) and (b) are different, as are the answers to parts (c) and (d), showing that the operation of composition is not commutative.

> **TIP**   Remember that the composite function $(f \circ g)(x) = f(g(x))$, whereas the product function $(f \cdot g)(x) = f(x) \cdot g(x)$.

One-to-One Functions

Recall that a function is a relation in which no two ordered pairs have the same *first* coordinates. Some functions satisfy the additional condition that no two ordered pairs have the same *second* coordinates. These functions are said to be *one-to-one*.

For instance, suppose that f is defined by the relation $\{(2, 5), (-4, 8), (1, 2)\}$. Note that every ordered pair has a different second coordinate, so that the function is one-to-one.

By contrast, the function g defined by $\{(-4, 1), (2, 3), (0, 1)\}$ is not one-to-one, since the second coordinate 1 corresponds to two first coordinates, -4 and 0.

One-to-One Function

A function f is said to be **one-to-one** if each y-value in the range corresponds to exactly one x-value in the domain.

In other words, a function is one-to-one if all the ordered pairs that define the function have different second coordinates.

It can be shown that any linear function $f(x) = mx + b$, where $m \neq 0$, is one-to-one. Consider for instance $f(x) = 3x - 1$. Assume that the y-values for two x-values, for example, a and b, are equal. In this case, we get:

$$f(a) = f(b)$$
$$3a - 1 = 3b - 1$$
$$a = b$$

So there can be only one x-value with this y-value. Therefore, the function is one-to-one.

By contrast, let's consider the function $g(x) = x^2$. This function has the same value, namely 4, for $x = 2$ and $x = -2$. Since there is more than one x-value for this y-value, g is not one-to-one.

Since the ordered pairs of a one-to-one function must have different second coordinates, it follows that the graph of a one-to-one function cannot have two points with different x-coordinates and the same y-coordinate. This suggests a test for determining if a graph represents a one-to-one function.

The Horizontal-Line Test

If every horizontal line intersects the graph of a function at most once, then the function is one-to-one.

Let's look again at the functions $f(x) = 3x - 1$ and $g(x) = x^2$. We see from the figure below that no horizontal line intersects the graph $f(x) = 3x - 1$ at more than one point. So f passes the horizontal-line test and is a one-to-one function.

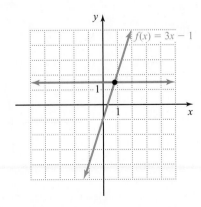

On the other hand, we see that a horizontal line intersects the graph of $g(x) = x^2$ more than once. So g fails the horizontal-line test, and is not a one-to-one function.

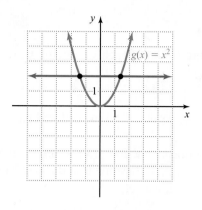

TIP To check if a graph defines a function, we use the *vertical-line* test. To check if a function is one-to-one, we use the *horizontal-line* test.

EXAMPLE 4

Consider the following graphs of functions. Determine whether each function is one-to-one.

a.

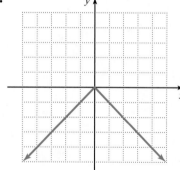

b.

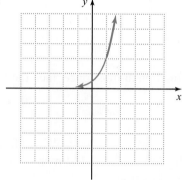

Solution

a. Since any horizontal line in Quadrants III and IV intersects the graph more than once, the graph fails the horizontal-line test. So the function is not one-to-one.

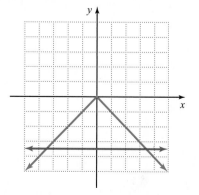

PRACTICE 4

Determine whether each function whose graph is shown below is a one-to-one function.

a.

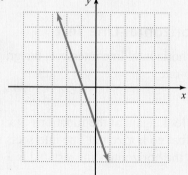

b.

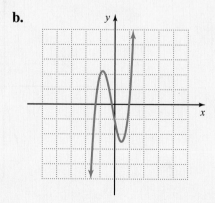

b. Since no horizontal line intersects the graph at more than one
point, the graph passes the horizontal-line test. So the function is
one-to-one.

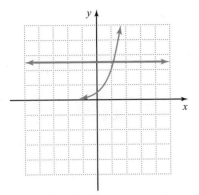

The Inverse of a Function

Every one-to-one function has an *inverse function* that is formed by interchanging the coordinates of the ordered pairs that define the original function.

For instance, let's consider a function defined by the following set of ordered pairs:
$\{(2, 5), (-4, 8), (1, 2)\}$. We have already noted that this function is one-to-one.
To find the inverse function, we interchange the first and second coordinates, getting
$\{(5, 2), (8, -4), (2, 1)\}$. This set of ordered pairs defines the function that is the inverse
of the original function.

DEFINITION

The **inverse of a one-to-one function f,** written f^{-1} and read "f inverse," is the set of
all ordered pairs (y, x) where f is the set of ordered pairs (x, y).

From this definition, it follows that if $f(x) = y$, then $f^{-1}(y) = x$. Since f^{-1} is found by interchanging x and y, the range of f^{-1} is the domain of f and the domain of f^{-1} is the range
of f.

Not every function has an inverse function. For instance, the function $f(x) = x^2$ does
not have an inverse function. Can you explain why?

EXAMPLE 5

Find the inverse of the function defined by
$\{(2, -5), (1, 0), (3, 3), (0, -1)\}$.

Solution   The inverse function is found by switching the coordinates of each ordered pair: $\{(-5, 2), (0, 1), (3, 3), (-1, 0)\}$

PRACTICE 5

If f is defined by $\{(-3, -1),$
$(-1, 0), (0, 2), (4, 3)\}$, find f^{-1}.

When a one-to-one function is defined by an equation rather than as a set of ordered
pairs, we can find its inverse function by interchanging the variables. The following rule gives
the procedure for finding the inverse of such a function:

To Find the Inverse of a Function

- Substitute y for $f(x)$.
- Interchange x and y.
- Solve the equation for y.
- Substitute $f^{-1}(x)$ for y.

EXAMPLE 6

Find the inverse function of $f(x) = 2x + 5$.

Solution

$$f(x) = 2x + 5$$
$$y = 2x + 5 \quad \text{Substitute } y \text{ for } f(x).$$
$$x = 2y + 5 \quad \text{Interchange } x \text{ and } y.$$
$$x - 5 = 2y \quad \text{Solve for } y.$$
$$\frac{x - 5}{2} = y$$
$$y = \frac{x - 5}{2}$$
$$f^{-1}(x) = \frac{x - 5}{2} \quad \text{Substitute } f^{-1}(x) \text{ for } y.$$

So the inverse function of $f(x)$ is $f^{-1}(x) = \dfrac{x - 5}{2}$.

PRACTICE 6

Find the inverse function of $g(x) = 3x - 1$.

Recall that if $f(x) = y$, then $f^{-1}(y) = x$. Therefore, $f^{-1}(f(x)) = f^{-1}(y) = x$. Similarly, $f(f^{-1}(y)) = f(x) = y$. These observations suggest the following property:

The Composition of a Function and Its Inverse

For any functions f and g, the following two conditions are equivalent:

- $f(x)$ and $g(x)$ are inverse functions of each other.
- $f(g(x)) = x$ and $g(f(x)) = x$.

We can use this property to check whether two functions are inverses of each other. For instance, consider the function in Example 6, $f(x) = 2x + 5$, where we found that $f^{-1}(x) = \dfrac{x - 5}{2}$. Now, let's evaluate the composition $(f^{-1} \circ f)(x)$.

$$(f^{-1} \circ f)(x) = f^{-1}(f(x))$$
$$= f^{-1}(2x + 5)$$
$$= \frac{(2x + 5) - 5}{2}$$
$$= \frac{2x}{2}$$
$$= x$$

So $(f^{-1} \circ f)(x) = x$, as the property predicts. Confirm that $(f \circ f^{-1})(x) = x$.

EXAMPLE 7

Is $g(x) = \dfrac{1}{4}x + 2$ the inverse of $f(x) = 4x - 8$?

Solution We can decide if g is the inverse of f by determining whether $f(g(x)) = x$ and $g(f(x)) = x$.

$$f(g(x)) = f\left(\frac{1}{4}x + 2\right)$$

$$= 4\left(\frac{1}{4}x + 2\right) - 8$$

$$= x + 8 - 8$$

$$= x$$

$$g(f(x)) = g(4x - 8)$$

$$= \frac{1}{4}(4x - 8) + 2$$

$$= x - 2 + 2$$

$$= x$$

Since $f(g(x)) = x$ and $g(f(x)) = x$, the functions are inverses of each another.

PRACTICE 7

Determine whether $p(x) = 5x - 2$ is the inverse of $q(x) = -5x + 2$.

The graph of a one-to-one function and its inverse function have a special relationship. If a point, say (a, b), lies on the graph of a one-to-one function, then the point (b, a), with the coordinates interchanged, lies on the graph of its inverse. Therefore, the graph of a function and the graph of its inverse function are mirror images of each other: They are reflected across the line $y = x$.

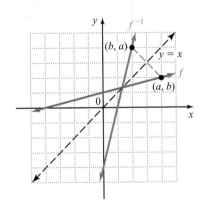

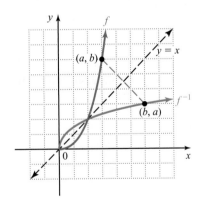

Consider, for instance, the one-to-one function $f(x) = 4x - 8$ and its inverse $f^{-1}(x) = \dfrac{1}{4}x + 2$. We graph both functions on the coordinate plane to the right.

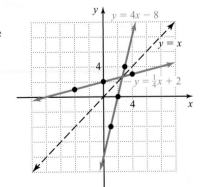

x	$y = f(x)$
1	-4
2	0
3	4

x	$y = f^{-1}(x)$
-4	1
0	2
4	3

Both functions are linear, so their graphs are straight lines. Note that, as predicted, each graph is the mirror image of the other, reflected across the line $y = x$. A point on either graph corresponds to a point on the other graph with the coordinates interchanged. For instance, $(2, 0)$ lies on the graph of $f(x)$, and $(0, 2)$ lies on the graph of $f^{-1}(x)$. We say that the graphs of f and f^{-1} are *symmetric* about the line $y = x$.

EXAMPLE 8

Given the graph of a function, graph its inverse.

a.

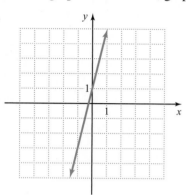

b.

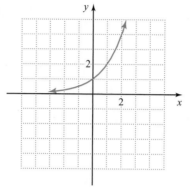

PRACTICE 8

Given the graph of a function, graph its inverse.

a.

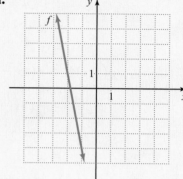

Solution

a.

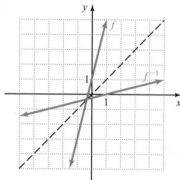

The graph of the inverse function is the mirror image of the original graph about the line $y = x$.

b.

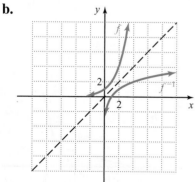

b.

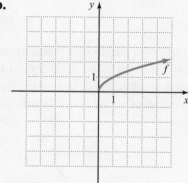

EXAMPLE 9

The function $f(x) = \dfrac{9}{5}x + 32$ can be used to convert a Celsius

temperature x to the equivalent Fahrenheit temperature $f(x)$.

a. Is f a one-to-one function? Explain how you know.

b. Identify the inverse function of f.

c. Explain the significance of the inverse function in this context.

Solution

a. To determine if f is one-to-one, let's graph it.

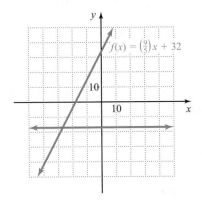

We see that the graph passes the horizontal-line test. So the function is one-to-one.

b. We need to identify the inverse function of f as follows:

$$f(x) = \dfrac{9}{5}x + 32$$

$$y = \dfrac{9}{5}x + 32$$

$$x = \dfrac{9}{5}y + 32$$

$$x - 32 = \dfrac{9}{5}y$$

$$\dfrac{5}{9}(x - 32) = y$$

$$y = \dfrac{5}{9}(x - 32)$$

$$f^{-1}(x) = \dfrac{5}{9}(x - 32)$$

c. Because f^{-1} is the inverse of f, it can be used to convert a Fahrenheit temperature x to the equivalent Celsius temperature.

PRACTICE 9

A rental agency charges $C(x)$ dollars to rent a six-passenger van for x hr, where $C(x) = 40x$.

a. Is C a one-to-one function? Explain how you know.

b. Identify C^{-1}.

c. Explain the significance of the inverse function in this context.

Mathematically Speaking

Fill in each blank with the most appropriate term or phrase from the given list.

product of $(f + g)$ and x	one-to-one	the y-axis
the line $y = x$	sum of f and g	product of f, g, and x
linear	product of f and g	composition of f and g
horizontal	difference of f and g	vertical
product of $(f - g)$ and x	composite functions	inverse functions of each other

1. The _____ is defined as
   $(f + g)(x) = f(x) + g(x)$.

2. The _____ is defined as
   $(f - g)(x) = f(x) - g(x)$.

3. The _____ is defined as
   $(f \cdot g)(x) = f(x) \cdot g(x)$.

4. The _____ is the function, represented by $f \circ g$, where $(f \circ g)(x) = f(g(x))$.

5. A function f is said to be _____ if each y-value in the range corresponds to exactly one x-value in the domain.

6. If every _____ line intersects the graph of a function at most once, then the function is one-to-one.

7. If $f(g(x)) = x$ and $g(f(x)) = x$, then $f(x)$ and $g(x)$ are _____.

8. The graphs of a function and its inverse are symmetric about _____.

A *Given $f(x)$ and $g(x)$, find $(f + g)(x)$, $(f - g)(x)$, $(f \cdot g)(x)$, and $\left(\dfrac{f}{g}\right)(x)$.*

9. $f(x) = 2x + 3$; $g(x) = 4x^2 - 9$

10. $f(x) = 3x^2 - 6x$; $g(x) = x - 2$

11. $f(x) = 3x^2 + 11x + 6$; $g(x) = 2x^2 + 5x - 3$

12. $f(x) = 1 - 4x - 5x^2$; $g(x) = x^2 + x$

13. $f(x) = \dfrac{1}{x + 3}$; $g(x) = x - 1$

14. $f(x) = \dfrac{-1}{x + 2}$; $g(x) = 2$

15. $f(x) = 2\sqrt{x}; g(x) = 4\sqrt{x} - 1$

16. $f(x) = \sqrt{x} + 3; g(x) = \sqrt{x} - 3$

Given $f(x) = x^2 - 3x - 4$ *and* $g(x) = 1 - x^2$, *find each of the following:*

17. $(f + g)(3)$ **18.** $(f + g)(-2)$ **19.** $(f - g)(-1)$ **20.** $(g - f)(0)$

21. $(f \cdot g)(2)$ **22.** $(f \cdot g)(-1)$ **23.** $\left(\dfrac{g}{f}\right)(5)$ **24.** $\left(\dfrac{f}{g}\right)(-4)$

B *Given* $f(x)$ *and* $g(x)$, *find each composition.*

25. $f(x) = 2x - 4$ and $g(x) = \dfrac{1}{2}x - 5$

    a. $(f \circ g)(x)$
    b. $(g \circ f)(x)$
    c. $(f \circ g)(3)$
    d. $(g \circ f)(-1)$

26. $f(x) = \dfrac{1}{3}x + 2$ and $g(x) = -3x + 6$

    a. $(f \circ g)(x)$
    b. $(g \circ f)(x)$
    c. $(f \circ g)(-2)$
    d. $(g \circ f)(4)$

27. $f(x) = x - 1$ and $g(x) = x^2 + 4x - 10$

    a. $(f \circ g)(x)$
    b. $(g \circ f)(x)$
    c. $(f \circ g)(1)$
    d. $(g \circ f)(-1)$

28. $f(x) = 4 + 5x - x^2$ and $g(x) = x + 3$

    a. $(f \circ g)(x)$
    b. $(g \circ f)(x)$
    c. $(f \circ g)(-3)$
    d. $(g \circ f)(3)$

29. $f(x) = \dfrac{3}{x}$ and $g(x) = 2x + 5$

    a. $(f \circ g)(x)$
    b. $(g \circ f)(x)$
    c. $(f \circ g)\left(\dfrac{1}{2}\right)$
    d. $(g \circ f)(-2)$

30. $f(x) = x - 6$ and $g(x) = \dfrac{2}{x + 4}$

    a. $(f \circ g)(x)$
    b. $(g \circ f)(x)$
    c. $(f \circ g)(-2)$
    d. $(g \circ f)(0)$

31. $f(x) = -\sqrt{x}$ and $g(x) = 1 - 4x$

    a. $(f \circ g)(x)$
    b. $(g \circ f)(x)$
    c. $(f \circ g)(-6)$
    d. $(g \circ f)(16)$

32. $f(x) = 3x + 5$ and $g(x) = 2\sqrt[3]{x}$

    a. $(f \circ g)(x)$
    b. $(g \circ f)(x)$
    c. $(f \circ g)(-8)$
    d. $(g \circ f)(1)$

C *Determine whether the function whose graph is shown is a one-to-one function.*

33.

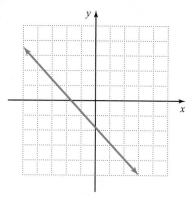

34.

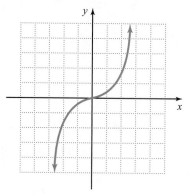

35.

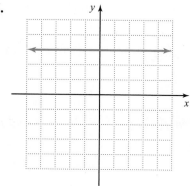

36.

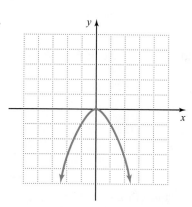

37.

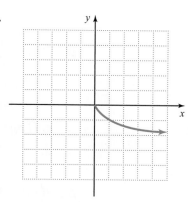

38.

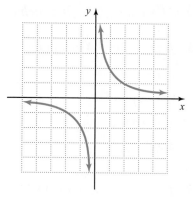

39.

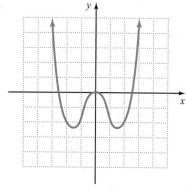

40.

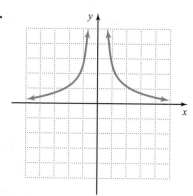

D *Match the graph of the function on the left with its inverse on the right.*

41.

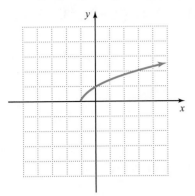

a.

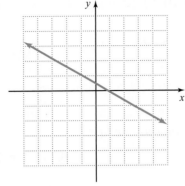

42.

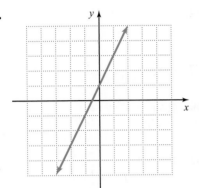

b.

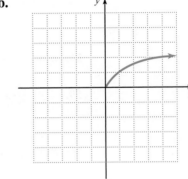

43.

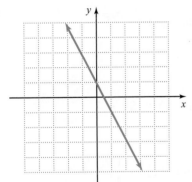

c.

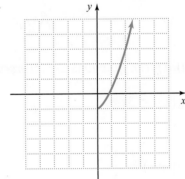

44.

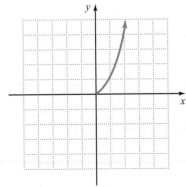

d.

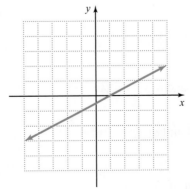

Find the inverse of the function represented by the set of ordered pairs.

45. $\{(-4, 5), (-2, 3), (0, 1), (2, -1), (4, -3)\}$

46. $\{(-9, 2), (-5, 3), (-2, 4), (2, -9), (3, -5), (4, -2)\}$

47. $\{(-27, -3), (-8, -2), (-1, -1), (0, 0), (1, 1),$ $(8, 2), (27, 3)\}$

48. $\{(36, -6), (25, -5), (16, -4), (9, -3), (4, -2),$ $(1, -1), (0, 0)\}$

Find the inverse of each one-to-one function.

49. $f(x) = 4x$

50. $f(x) = 7x$

51. $g(x) = \frac{1}{4}x$

52. $g(x) = \frac{1}{7}x$

53. $f(x) = -x - 5$

54. $f(x) = x + 10$

55. $f(x) = 5x + 2$

56. $h(x) = 4x - 9$

57. $g(x) = \frac{1}{2}x - 3$

58. $f(x) = \frac{1}{3}x + 1$

59. $f(x) = x^3 - 4$

60. $f(x) = x^3 + 2$

61. $h(x) = \frac{3}{x + 4}$

62. $g(x) = \frac{1}{x - 3}$

63. $f(x) = \frac{x}{2x - 1}$

64. $f(x) = \frac{2x}{3x + 2}$

65. $f(x) = \sqrt[3]{x + 4}$

66. $f(x) = \sqrt[3]{5 - x}$

Determine whether the two given functions are inverses of each other.

67. $f(x) = \frac{x - 1}{3}$ and $g(x) = 3x + 1$

68. $f(x) = \frac{x + 7}{2}$ and $g(x) = 2x - 7$

69. $p(x) = 5x - 10$ and $q(x) = \frac{1}{5}x + 2$

70. $f(x) = 3x + 12$ and $g(x) = \frac{1}{3}x + 4$

71. $f(x) = -4x - 1$ and $g(x) = \frac{x + 1}{4}$

72. $p(x) = -6x + 7$ and $q(x) = \frac{7 - x}{6}$

73. $g(x) = (x + 5)^3$ and $h(x) = \sqrt[3]{x} - 5$

74. $f(x) = x^3 - 10$ and $g(x) = \sqrt[3]{x + 10}$

75. $p(x) = \sqrt[3]{2x - 7}$ and $q(x) = 2x^3 + 7$

76. $p(x) = \sqrt[5]{x} - 1$ and $q(x) = (1 - x)^5$

77. $f(x) = \frac{2}{x} - 3$ and $g(x) = \frac{2}{x + 3}$

78. $f(x) = \frac{-5}{x - 8}$ and $g(x) = -\frac{5}{x} + 8$

E *Given the graph of a one-to-one function, sketch the graph of its inverse.*

79.

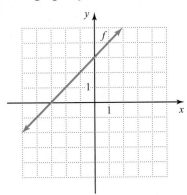

80.

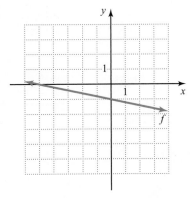

81.

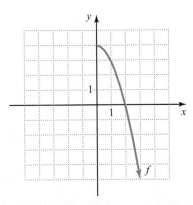

82.

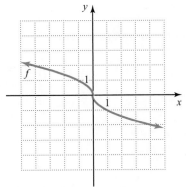

Mixed Practice

Determine whether the given functions are inverses of each other.

83. $g(x) = -4x + 9$ and $h(x) = \dfrac{9 - x}{4}$

84. $p(x) = \sqrt[4]{x} + 5$ and $q(x) = (x - 5)^4$

Solve.

85. Given $f(x) = 3x^2 - 5x + 2$ and $g(x) = x^2 - 8$, find $(g - f)(-3)$.

86. Determine whether the function shown by the graph is a one-to-one function.

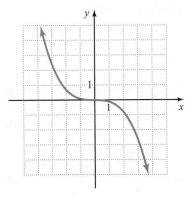

87. Given $f(x) = -2x + 7$ and $g(x) = \dfrac{4}{x + 2}$, find each composition:
   a. $(f \circ g)(x)$
   b. $(g \circ f)(x)$
   c. $(f \circ g)(6)$
   d. $(g \circ f)(0)$

88. Given $f(x) = 2x^2 - 5x + 7$ and $g(x) = -x + 2$, find $(f + g)(x)$, $(f - g)(x)$, $(f \cdot g)(x)$, and $\left(\dfrac{f}{g}\right)(x)$.

89. Given the following graph of a one-to-one function, sketch the graph of its inverse:

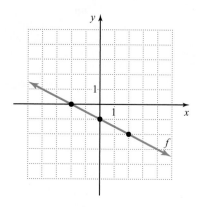

Find the inverse of each one-to-one function.

90. $f(x) = \dfrac{1}{4}x - 7$

91. $f(x) = \dfrac{3x}{2x + 3}$

92. $f(x) = x^3 - 6$

F Applications

Solve.

93. The accounting department of a manufacturing plant determines that the weekly cost C (in dollars), of producing x units of a product is approximated by the function $C(x) = 4.2x + 1000$. The number of units the plant produces in d days is modeled by the function $x(d) = 500d$.

a. Find $(C \circ x)(d)$, and interpret its meaning in this situation.

b. Determine the weekly cost if the manufacturing plant operates 5 days per week.

94. Each day, an employee at a toy factory receives a salary of $70 plus $0.50 for each item of doll clothing he produces. The number of items of doll clothing the employee can produce in h hr is modeled by the function $x(h) = 8h$.

a. Find the function $S(x)$ that expresses the employee's salary in terms of x, the number of items of doll clothing produced each day.

b. Find $(S \circ x)(h)$, and interpret its meaning in this situation.

c. Determine the employee's daily salary if he works 7.5 hours per day.

95. The sum of the measures of the interior angles of a polygon (in degrees) is modeled by the function $S(n) = 180n - 360$, where n is the number of sides of the polygon.

a. Identify the inverse function of S.

b. Explain the meaning of the inverse function in this situation.

c. Determine the number of sides of a polygon if the sum of the interior angles is 540 degrees.

96. A clinical dietitian can estimate the ideal body weight w (in pounds) for men using the function $w(h) = 6h - 254$, where h is height (in inches).

a. Identify the inverse function of w.

b. Explain the meaning of the inverse function in this situation.

c. Find the height of a man whose ideal body weight is 178 lb.

97. A long-distance telephone service provider charges a monthly service fee of $4.50 plus $0.05 for each minute (or part thereof) of long-distance calling.

  a. Find the function $B(t)$ that expresses the monthly bill in terms of t, the amount of time (in minutes) of long-distance calling.

  b. Find $B^{-1}(t)$.

  c. Explain the significance of the inverse within the context of the situation.

98. A young couple deposits $5000 into an account earning 3.5% simple interest. If no additional deposits are made, then the amount A in the account after t yr is modeled by the function $A(t) = 175t + 5000$.

  a. Find $A^{-1}(t)$.

  b. Explain the meaning of the inverse in the context of this situation.

• Check your answers on page A-48.

MIND*Stretchers*

Critical Thinking

1. Consider the graphs of the functions f and g shown on the coordinate plane below on the left.
   Use these graphs to graph $h(x) = (f + g)(x)$ and $k(x) = (f - g)(x)$ on the coordinate plane on the right.

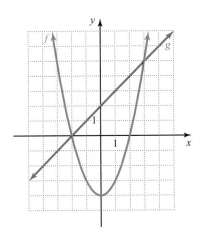
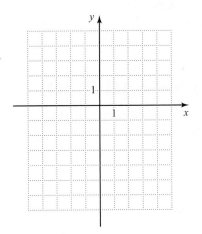

Mathematical Reasoning

2. Consider the function $h(x) = 6x - 1$. If $f(x) = 3x + 2$, find $g(x)$ such that $h(x) = (f \circ g)(x)$.

Groupwork

3. Working with a partner, give three examples of functions where each function is equal to its own inverse and explain why.

<table>
<tr><td>**9.2**</td><td># Exponential Functions</td></tr>
</table>

What Exponential Functions Are and Why They Are Important

In previous chapters, we discussed various types of functions (linear, quadratic, polynomial, and radical). In the remaining sections of this chapter, we focus on the *exponential function* and its inverse, the *logarithmic function*. We first consider exponential functions. These functions model data that increase or decrease rapidly.

As an example, suppose a bacteriologist is growing a certain culture of bacteria. The following table shows the number N of bacteria present in the culture after t hr:

Time t	0	1	2	3	4	5
Number N of Bacteria Present	1	2	4	8	16	32

From the table, we see that the number of bacteria in the culture doubles every hour. So after 24 hr, the number of bacteria in the culture will have grown from 1 bacterium to 2^{24}, or 16,777,216 bacteria. This type of growth can be modeled by the exponential function $N(t) = 2^t$.

Some other examples of real-life situations that are described by exponential functions include population growth, growth of an investment in which interest is compounded, and decay of radioactive substances.

We devote this section to a discussion of exponential functions (and equations) and how they are used as models in solving applied problems.

OBJECTIVES

A To identify various types of functions

B To evaluate an exponential function

C To graph an exponential function

D To solve an exponential equation of the form $b^x = b^n$

E To solve applied problems involving exponential functions

Evaluating Exponential Functions

In previous sections, we discussed polynomial functions, such as x^2 and $-5y^3$, in which a variable is raised to a constant power. In this section, we focus on exponential functions, such as 2^x and 5^{-n}, where a constant is raised to a power that is a variable expression.

> **DEFINITION**
>
> An **exponential function** is any function that can be written in the form
> $$f(x) = b^x,$$
> where x is a real number, $b > 0$, and $b \neq 1$.

Note that we call b the *base* of the exponential function. In more advanced courses, it is shown that b^x exists for all real values of x, both rational and irrational, as long as b is positive. Here are some other examples of exponential functions:

$$f(x) = 4^x \qquad g(x) = \left(\frac{1}{6}\right)^x \qquad h(x) = 10^{x+1}$$

Now, let's consider how to evaluate exponential functions. For some of these functions, a constant is raised to a variable power. For other functions, a constant is raised to a power that is a variable expression.

EXAMPLE 1

Evaluate:

a. $f(x) = 3^x$ for $x = 2$ **b.** $g(x) = \left(\frac{1}{2}\right)^x$ for $x = -4$

Solution

a. $f(x) = 3^x$

$f(2) = 3^2$ Substitute 2 for x.

    $= 9$

b. $g(x) = \left(\frac{1}{2}\right)^x$

$g(-4) = \left(\frac{1}{2}\right)^{-4}$ Substitute -4 for x.

    $= 2^4$

    $= 16$

PRACTICE 1

Evaluate:

a. $f(x) = 4^x$ for $x = 3$

b. $g(x) = \left(\frac{1}{3}\right)^x$ for $x = -2$

EXAMPLE 2

Given $f(x) = 3^{2x-1}$, find:

a. $f(1)$ **b.** $f(-1)$

Solution

a. $f(x) = 3^{2x-1}$

$f(1) = 3^{2(1)-1}$ Substitute 1 for x.

    $= 3^1 = 3$

b. $f(x) = 3^{2x-1}$

$f(-1) = 3^{2(-1)-1}$ Substitute -1 for x.

    $= 3^{-3}$

    $= \frac{1}{3^3} = \frac{1}{27}$

PRACTICE 2

Given $f(x) = 2^{3x-1}$, find:

a. $f(2)$

b. $f(-2)$

The exponential functions in previous examples have bases that are rational numbers. However, an important and frequently used base in applications of exponential functions is the irrational number e. Like the irrational number π, e has a nonterminating, nonrepeating decimal representation, which is approximately 2.7182818. The number e is studied in later mathematics courses, including calculus and statistics, and has important applications, as we shall see.

DEFINITION

The function defined by $f(x) = e^x$ is called the **natural exponential function.**

To evaluate the natural exponential function, we can use the $\boxed{e^x}$ feature on a calculator.

EXAMPLE 3

Evaluate the function for the given value. Round the answer to the nearest thousandth.

a. $f(x) = e^{3x}$ for $x = 2$ **b.** $g(x) = e^{x+3}$ for $x = -5$

Solution

a. $f(x) = e^{3x}$

$\quad f(2) = e^{3(2)}$

$\quad\quad = e^6 \approx 403.429$

b. $\quad g(x) = e^{x+3}$

$\quad\quad g(-5) = e^{-5+3}$

$\quad\quad\quad = e^{-2} \approx 0.135$

PRACTICE 3

Evaluate and round to the nearest thousandth.

a. $f(x) = e^{5x}$ for $x = -1$

b. $g(x) = e^{x-1}$ for $x = 3$

Graphing Exponential Functions

Now, let's consider some examples of graphing exponential functions.

EXAMPLE 4

Graph the exponential functions $f(x) = 2^x$ and $g(x) = 4^x$ on the same coordinate plane.

Solution   For each function, we find several values and list the results in a table. Then, we plot the ordered pairs and connect them with a smooth curve.

x	$f(x) = 2^x$	(x, y)	x	$g(x) = 4^x$	(x, y)
-2	$f(-2) = 2^{-2} = \frac{1}{4}$	$(-2, \frac{1}{4})$	-2	$g(-2) = 4^{-2} = \frac{1}{16}$	$(-2, \frac{1}{16})$
-1	$f(-1) = 2^{-1} = \frac{1}{2}$	$(-1, \frac{1}{2})$	-1	$g(-1) = 4^{-1} = \frac{1}{4}$	$(-1, \frac{1}{4})$
0	$f(0) = 2^0 = 1$	$(0, 1)$	0	$g(0) = 4^0 = 1$	$(0, 1)$
1	$f(1) = 2^1 = 2$	$(1, 2)$	1	$g(1) = 4^1 = 4$	$(1, 4)$
2	$f(2) = 2^2 = 4$	$(2, 4)$	2	$g(2) = 4^2 = 16$	$(2, 16)$
3	$f(3) = 2^3 = 8$	$(3, 8)$	3	$g(3) = 4^3 = 64$	$(3, 64)$

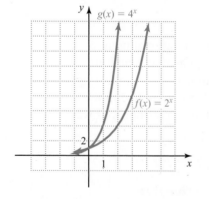

PRACTICE 4

Graph the exponential functions $f(x) = 3^x$ and $g(x) = 5^x$ on the same coordinate plane.

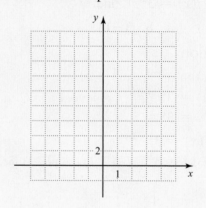

We observe from the graphs in Example 4 that each function is defined for any real number x and takes on any positive value. The y-intercept for both graphs is $(0, 1)$, and neither graph has an x-intercept. Also, each graph is increasing and passes both the vertical- and horizontal-line tests.

EXAMPLE 5

Graph the exponential functions $f(x) = \left(\dfrac{1}{2}\right)^x$ and $g(x) = \left(\dfrac{1}{4}\right)^x$ on the same set of axes.

Solution To graph each function, we find several values, listing the results in a table. Then, we plot the ordered pairs and connect them with a smooth curve.

x	$f(x) = \left(\tfrac{1}{2}\right)^x$	(x, y)
-3	$f(-3) = \left(\tfrac{1}{2}\right)^{-3} = 8$	$(-3, 8)$
-2	$f(-2) = \left(\tfrac{1}{2}\right)^{-2} = 4$	$(-2, 4)$
-1	$f(-1) = \left(\tfrac{1}{2}\right)^{-1} = 2$	$(-1, 2)$
0	$f(0) = \left(\tfrac{1}{2}\right)^{0} = 1$	$(0, 1)$
1	$f(1) = \left(\tfrac{1}{2}\right)^{1} = \tfrac{1}{2}$	$(1, \tfrac{1}{2})$
2	$f(2) = \left(\tfrac{1}{2}\right)^{2} = \tfrac{1}{4}$	$(2, \tfrac{1}{4})$

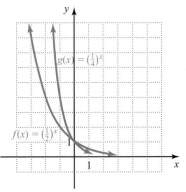

x	$g(x) = \left(\tfrac{1}{4}\right)^x$	(x, y)
-3	$g(-3) = \left(\tfrac{1}{4}\right)^{-3} = 64$	$(-3, 64)$
-2	$g(-2) = \left(\tfrac{1}{4}\right)^{-2} = 16$	$(-2, 16)$
-1	$g(-1) = \left(\tfrac{1}{4}\right)^{-1} = 4$	$(-1, 4)$
0	$g(0) = \left(\tfrac{1}{4}\right)^{0} = 1$	$(0, 1)$
1	$g(1) = \left(\tfrac{1}{4}\right)^{1} = \tfrac{1}{4}$	$(1, \tfrac{1}{4})$
2	$g(2) = \left(\tfrac{1}{4}\right)^{2} = \tfrac{1}{16}$	$(2, \tfrac{1}{16})$

PRACTICE 5

Graph the exponential functions

$$f(x) = \left(\dfrac{1}{3}\right)^x \text{ and } g(x) = \left(\dfrac{1}{5}\right)^x \text{ on}$$

the same coordinate plane.

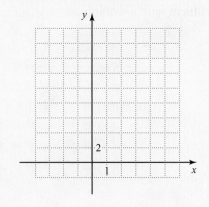

From the graphs in Example 5, we see that each function is defined for any real number x, and takes on any positive value. The y-intercept for both graphs is $(0, 1)$, and neither graph has an x-intercept. Also, each graph is decreasing and passes both the vertical- and horizontal-line tests.

Explain why the graphs in Example 4 are *increasing*, whereas the graphs in Example 5 are *decreasing*.

The following table summarizes some characteristics of exponential functions as shown in Examples 4 and 5:

Characteristics of the Exponential Function $f(x) = b^x$

- The domain of the function is the set of all real numbers, and the range is the set of all positive numbers.

- The vertical- and horizontal-line tests show that $f(x) = b^x$ is a one-to-one function.

- The function is increasing if the base is greater than 1. The function is decreasing if the base is between 0 and 1.

- The y-intercept of the graph of the function is $(0, 1)$, but the graph has no x-intercept. The graph approaches the x-axis, but never crosses it.

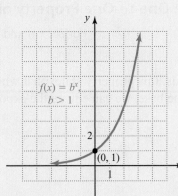

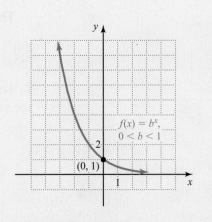

In the next example, we consider the graph of a constant raised to a variable expression.

EXAMPLE 6

Graph: $f(x) = 2^{x-1}$

Solution   First, we find and plot some ordered pairs. Then, we connect them with a smooth curve.

x	$f(x) = 2^{x-1}$	(x, y)
-2	$f(-2) = 2^{-2-1} = 2^{-3} = \frac{1}{8}$	$(-2, \frac{1}{8})$
-1	$f(-1) = 2^{-1-1} = 2^{-2} = \frac{1}{4}$	$(-1, \frac{1}{4})$
0	$f(0) = 2^{0-1} = 2^{-1} = \frac{1}{2}$	$(0, \frac{1}{2})$
1	$f(1) = 2^{1-1} = 2^{0} = 1$	$(1, 1)$
2	$f(2) = 2^{2-1} = 2^{1} = 2$	$(2, 2)$
3	$f(3) = 2^{3-1} = 2^{2} = 4$	$(3, 4)$

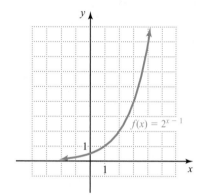

PRACTICE 6

Graph: $g(x) = 3^{x-1}$

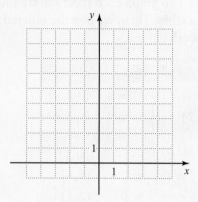

Solving Exponential Equations of the Form $b^x = b^n$

An equation that contains a variable as an exponent is called an *exponential equation*. Some examples of exponential equations are:

$$3^{x-1} = 3^{2x-1} \qquad 2^x = 8 \qquad 4^x = 9$$

We can use the fact that exponential functions are one-to-one to solve many exponential equations.

> ### The One-to-One Property of Exponential Functions
>
> For $b > 0$ and $b \neq 1$, if $b^x = b^n$, then $x = n$.

In words, this property states that when each side of an equation is expressed as a power of the same base, we can equate the exponents.

EXAMPLE 7

Solve:

a. $3^x = 81$ **b.** $4^x = 32$

Solution

a. The number 81 is a power of 3. So to solve this equation, we first write 81 as a power of 3:

$$3^x = 81$$
$$3^x = 3^4$$

Since the bases on each side of the equation are the same, we can use the one-to-one property of exponential functions and set the exponents equal. So $x = 4$. Substituting 4 for x in the original equation, we see that the solution checks.

b. Because both 4 and 32 are powers of 2, we get:

$$4^x = 32$$
$$(2^2)^x = 2^5$$
$$2^{2x} = 2^5$$

Since the bases are the same, we apply the one-to-one property of exponential functions.

$$2x = 5$$
$$x = \frac{5}{2}$$

So $4^x = 32$ when $x = \frac{5}{2}$. We can check the answer in the original equation.

PRACTICE 7

Solve:

a. $2^x = 64$

b. $9^x = 27$

Note that in some exponential equations, such as $4^x = 9$, the two sides cannot easily be expressed in terms of the same base. In a later section, we focus on solving equations like these using *logarithms*.

Exponential functions (and equations) are used to solve many real-life applications, such as those dealing with interest rates on investments or loans. Let's consider the following exponential equation:

$$A = P\left(1 + \frac{r}{n}\right)^{nt}.$$

This equation is used to compute the amount A accumulated (or owed) after P dollars are invested (or borrowed) at an annual interest rate r, compounded n times per year for t years. We call this equation the *compound interest formula*.

EXAMPLE 8

When a computer company downsized 3 yr ago, an employee received a lump-sum severance check for $10,000. At that time, she invested one-half of the money in a growth fund that yielded an annual rate of 9% in dividends, compounded quarterly. To the nearest cent, what is the value of her investment after 3 yr?

Solution   Use the formula $A = P\left(1 + \dfrac{r}{n}\right)^{nt}$, where

$P = \$5000$ (the amount invested that is one-half of $10,000)
$r = 9\% = 0.09$ (the annual interest rate)
$n = 4$ (the number of times interest is compounded each year: quarterly means 4 times)
$t = 3$ (the length of time of the investment, in years)

Substituting these values in the compound interest formula gives us:

$$A = P\left(1 + \frac{r}{n}\right)^{nt}$$
$$= 5000\left(1 + \frac{0.09}{4}\right)^{4\cdot3}$$
$$= 5000(1 + 0.0225)^{12}$$
$$= 5000(1.0225)^{12}$$

To find the approximate value of A, we use the $\boxed{\wedge}$ or $\boxed{y^x}$ key on a calculator.

$$A \approx 6530.25$$

So the amount that the fund is worth after 3 yr, to the nearest cent, is $6530.25.

PRACTICE 8

The half-life of a radioactive substance is the time it takes for half of the material to decay. Arsenic-74, with a half-life of about 18 days, is used to locate brain tumors. The amount A of a 90-milligram sample remaining after x days is modeled by the function $A(x) = 90\left(\dfrac{1}{2}\right)^{x/18}$.

Approximate, to the nearest milligram, the amount of the sample remaining after 3 days.

Mathematically Speaking

Fill in each blank with the most appropriate term or phrase from the given list.

imaginary	decreasing	exponential function
crosses	approaches	positive
natural exponential function	real	increasing
	irrational	

1. A(n) _____ is any function that can be written in the form $f(x) = b^x$, where x is a real number, $b > 0$, and $b \neq 1$.

2. The number e is _____.

3. The function defined by $f(x) = e^x$ is called the _____.

4. The range of an exponential function is the set of all _____ numbers.

5. An exponential function is _____ if the base is between 0 and 1.

6. The graph of an exponential function _____ the x-axis.

A *Determine whether each function is a polynomial function, a rational function, a radical function, or an exponential function.*

7. $f(x) = 2x^3 - 1$

8. $f(x) = -3x^2 + 2$

9. $g(x) = 8^{x+2}$

10. $g(x) = -3^x + 2$

11. $h(x) = x^{1/2}$

12. $f(x) = 3x^{-2} + 5$

13. $g(x) = \left(\dfrac{1}{2}\right)^{2x}$

14. $f(x) = e^{2x}$

B *Evaluate each function for the given values.*

15. $f(x) = 2^x$
  a. $f(-3)$
  b. $f(0)$
  c. $f(4)$

16. $f(x) = 6^x$
  a. $f(-1)$
  b. $f(2)$
  c. $f(-2)$

17. $g(x) = \left(\dfrac{1}{9}\right)^x$
  a. $g\left(-\dfrac{1}{2}\right)$
  b. $g\left(\dfrac{1}{2}\right)$
  c. $g(2)$

18. $h(x) = \left(\dfrac{1}{8}\right)^x$
  a. $h(-2)$
  b. $h\left(-\dfrac{1}{3}\right)$
  c. $h\left(\dfrac{1}{3}\right)$

19. $f(x) = 3^{x-4}$
  a. $f(1)$
  b. $f(3)$
  c. $f(6)$

20. $g(x) = 4^{x+1}$
  a. $g(-4)$
  b. $g(1)$
  c. $g(3)$

21. $h(x) = -2^x + 3$
  a. $h(-3)$
  b. $h(0)$
  c. $h(4)$

22. $f(x) = 10 - 3^x$
  a. $f(-2)$
  b. $f(0)$
  c. $f(2)$

Evaluate each function for the given values. Round to the nearest thousandth.

23. $f(x) = -e^x$
  a. $f(-5)$
  b. $f(1)$

24. $f(x) = e^{-x}$
  a. $f(-3)$
  b. $f(1)$

25. $g(x) = e^{-2x}$
  a. $g(2)$
  b. $g\left(-\dfrac{1}{2}\right)$

26. $g(x) = -e^{4x}$
  a. $g\left(-\dfrac{1}{4}\right)$
  b. $g(0)$

27. $f(x) = e^{3x-2}$
  a. $f\left(-\dfrac{1}{3}\right)$
  b. $f(0)$

28. $f(x) = e^{2x+1}$
  a. $f(2)$
  b. $f\left(\dfrac{1}{2}\right)$

C *Graph each function.*

29. $f(x) = 3^x$

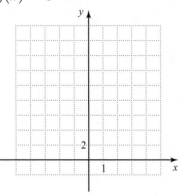

30. $f(x) = 4^x$

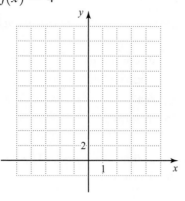

31. $f(x) = \left(\dfrac{1}{2}\right)^x$

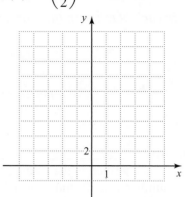

32. $f(x) = \left(\dfrac{1}{3}\right)^x$

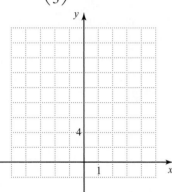

33. $f(x) = -2^x$

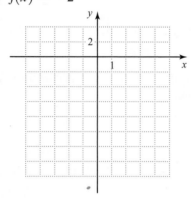

34. $f(x) = -3^x$

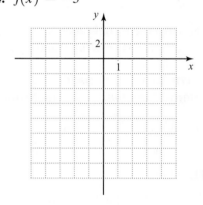

35. $f(x) = 2^{x-2}$

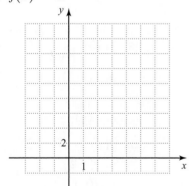

36. $f(x) = 3^{x+1}$

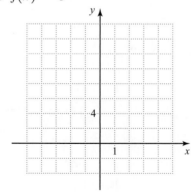

37. $f(x) = \left(\dfrac{1}{3}\right)^x + 2$

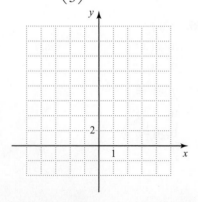

38. $f(x) = \left(\dfrac{1}{4}\right)^x - 3$

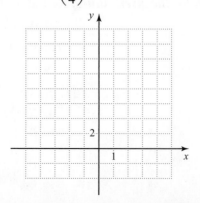

D *Solve.*

39. $5^x = 25$

40. $10^x = 10,000$

41. $2^x = \dfrac{1}{32}$

42. $3^x = \dfrac{1}{81}$

43. $36 = 6^{-x}$

44. $125 = 5^{-x}$

45. $8^x = 16$

46. $9^x = 243$

47. $4^{x+1} = 64$

48. $2^{x-3} = 32$

49. $7^{x-4} = 1$

50. $6^{x+2} = 1$

51. $16^{x+1} = 32$

52. $27^{x-1} = 81$

53. $9^{-x+3} = \dfrac{1}{27}$

54. $32^{-x+2} = \dfrac{1}{8}$

55. $100^{x-5} = 100,000^x$

56. $256^x = 32^{x+1}$

57. $64^{x-2} = 128^{x-3}$

58. $125^{x+1} = 625^{x-1}$

Mixed Practice

Solve.

59. $4^{-x+3} = \dfrac{1}{32}$

60. Graph the function $f(x) = -\left(\dfrac{1}{2}\right)^x$.

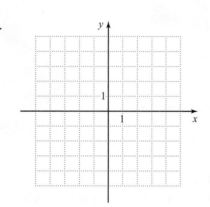

Determine whether each function is a polynomial or a rational, radical, or exponential function.

61. $g(x) = x^{3/2}$

62. $f(x) = 3^{2x}$

Evaluate each function for the given value.

63. $g(x) = -5^{x-2}$ for $x = 0$

64. $f(x) = -5^x + 1$ for $x = -1$

Evaluate each function for the given values. Round to the nearest thousandth.

65. $f(x) = e^{2x}$ for $x = -1$

66. $g(x) = e^{-3x}$ for $x = \dfrac{1}{4}$

Applications

Solve. Use a calculator, where appropriate.

E **67.** A new employee is given a starting salary of $28,000 and is guaranteed a 4% annual salary increase. The function $S(n) = 28,000(1.04)^n$ can be used to calculate the employee's salary S after n years of employment. What will the employee's salary be after 2 yr of employment?

68. The owner of a small business determines that a piece of office equipment purchased new depreciates at a rate of 15% per year. The value v of the equipment t yr after it is purchased is modeled by the function $v(t) = 1250(0.85)^t$.

a. To the nearest dollar, what will be the value of the equipment after 5 yr?

b. What was the value of the office equipment when it was purchased?

69. An environmental group estimates that the amount of pollution in a pond will decrease by 5% each month. To determine the concentration C of pollutants (in parts per million) after m mo, the group uses the function $C(m) = 40(0.95)^m$.

a. To the nearest tenth, what will the concentration of pollutants in the pond be after 1 yr?

b. What was the initial concentration of pollutants in the pond?

70. A botanist notes that the height of a certain tomato plant increases by 10% each week. Suppose the height of the plant is measured at 2 in. today. Then, its height h (in inches) x wk from today can be found using the function $h(x) = 2(1.1)^x$. To the nearest tenth of an inch, determine the height of the plant 8 wk from today.

71. An investor puts $8000 into an account with an annual interest rate of 6%, compounded monthly. To the nearest cent, calculate the amount in the account after 5 yr.

72. Strontium-85, a radioactive isotope used in bone imaging, has a half-life of about 65 days. The amount A of a 30-milligram sample remaining after t days is modeled by the function $A(t) = 30\left(\dfrac{1}{2}\right)^{t/65}$. Determine, to the nearest tenth of a milligram, the amount of the sample remaining after 130 days.

73. If the initial concentration of a certain drug in a patient's bloodstream is 50 mg/L, then the concentration C after t hr is modeled by the function $C(t) = 50e^{-0.125t}$. To the nearest milligram per liter, approximate the concentration in the bloodstream after 8 hr.

74. The population P of a town is modeled by the function $P(t) = 300e^{0.3t}$, where t is measured in years. What was the size of the population after 4 yr?

75. A bacteriologist found that the population of a certain bacterium doubles every 30 min. The number of bacteria N present after h hr is given by the equation $N(h) = N_0(2)^{2h}$, where N_0 is the number of bacteria present in the initial population. If 10 bacteria are present in the initial population, in how many hours will 1280 bacteria be present?

76. A professional organization determined that its membership tripled every 5 yr since its inception. If the organization had 500 members at its inception, then t yr after its inception the number of members m is given by the equation $m(t) = 500(3)^{t/5}$. How many years after its inception were there 13,500 members?

• Check your answers on page A-49.

MINDStretchers

Patterns

1. Consider the exponential expressions 2^x and 3^x.

   a. Complete the table.

x	-4	-3	-2	-1	0	1	2	3	4
2^x									
3^x									

   b. What do you notice about the value of 2^x relative to the value of 3^x when x is negative? What do you notice when x is positive?

   c. Use your results from parts (a) and (b) to solve the following inequalities:

   i. $2^x > 3^x$ **ii.** $2^x < 3^x$

   d. Consider the exponential expressions $\left(\dfrac{1}{2}\right)^x$ and $\left(\dfrac{1}{3}\right)^x$.

   Without making any calculations, solve the following inequalities:

   i. $\left(\dfrac{1}{2}\right)^x > \left(\dfrac{1}{3}\right)^x$ **ii.** $\left(\dfrac{1}{2}\right)^x < \left(\dfrac{1}{3}\right)^x$

Writing

2. In the definition of an exponential function, there are two restrictions on b, namely $b > 0$ and $b \neq 1$. Explain the importance of each restriction.

Mathematical Reasoning

3. The exponential function $f(x) = b^x$ is a one-to-one function. Use the graph of $f(x) = b^x$, where $b > 1$, to sketch a graph of its inverse. Then, identify the domain and range of the inverse function.

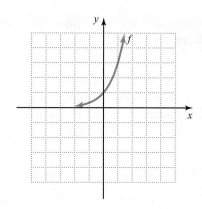

OBJECTIVES

Ⓐ To write equivalent exponential and logarithmic equations

Ⓑ To evaluate a logarithm

Ⓒ To solve a logarithmic equation

Ⓓ To graph a logarithmic function

Ⓔ To solve applied problems involving logarithms

9.3 Logarithmic Functions

What Logarithmic Functions Are and Why They Are Important

As with exponential functions, *logarithmic functions* help us to understand many natural phenomena. Consider, for instance, the magnitude of an earthquake. This magnitude is calculated from the amplitude of the largest seismic wave that the earthquake registers. The Richter scale is a measure of earthquake magnitude. From the Richter scale shown below, we see that an earthquake of magnitude 6 has a seismic wave amplitude 10 times as great as an earthquake of magnitude 5.

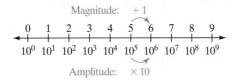

The exponents used on the Richter scale are examples of *logarithms*.

Writing Equivalent Exponential and Logarithmic Equations

From the previous section, we know that the exponential function $f(x) = b^x$ is one-to-one, and so has an inverse. Its inverse function $f^{-1}(x)$ is called a logarithmic function.

DEFINITION

The inverse of $f(x) = b^x$ is represented by the **logarithmic function**

$$f^{-1}(x) = \log_b x,$$

where $\log_b x$ is read either "the **logarithm** of x to *the base b*" or "the logarithm, base b, of x."

In general, we know that for a one-to-one function f, $f^{-1}(x) = y$ is equivalent to $f(y) = x$. Since the function $f^{-1}(x) = \log_b x$ is the inverse of the function $f(x) = b^x$, it follows that $y = \log_b x$ is equivalent to $b^y = x$.

Equivalent Logarithmic and Exponential Equations

If $b > 0$, $b \neq 1$, $x > 0$, and y is any real number,
then $y = \log_b x$ is equivalent to $b^y = x$.

In words, this property states that the logarithm of x to the base b is the exponent to which the base b must be raised to get x.

Logarithmic Equation	Exponential Equation
Exponent	Exponent
↓	↓
$y = \log_b x$	$b^y = x$
↑	↑
Base	Base

TIP Remember that a logarithm is an exponent.

The key to solving many logarithmic problems is to be able to write equivalent exponential and logarithmic equations. Examples of such equivalent equations are shown in the following table:

Logarithmic Equation	Equivalent Exponential Equation
$\log_2 16 = 4$	$2^4 = 16$
$\log_3 9 = 2$	$3^2 = 9$
$\log_8 1 = 0$	$8^0 = 1$
$\log_4 \dfrac{1}{64} = -3$	$4^{-3} = \dfrac{1}{64}$
$\log_9 3 = \dfrac{1}{2}$	$9^{1/2} = 3$

Let's consider some other examples of writing exponential equations in logarithmic notation and also logarithmic equations in exponential notation.

EXAMPLE 1

Write each logarithmic equation as an exponential equation.

a. $\log_{10} 1000 = 3$ **b.** $\log_7 7 = 1$ **c.** $\log_5 \sqrt{5} = \dfrac{1}{2}$

Solution We can write each logarithmic equation in its equivalent exponential form using the definition of a logarithm.

a. $\log_{10} 1000 = 3$ means $10^3 = 1000$

Logarithms are exponents.

b. $\log_7 7 = 1$ means $7^1 = 7$

c. $\log_5 \sqrt{5} = \dfrac{1}{2}$ means $5^{1/2} = \sqrt{5}$

PRACTICE 1

Write each logarithmic equation using exponential notation.

a. $\log_8 64 = 2$

b. $\log_5 \dfrac{1}{5} = -1$

c. $\log_2 \sqrt[3]{2} = \dfrac{1}{3}$

EXAMPLE 2

Write each exponential equation in its equivalent logarithmic form.

a. $5^4 = 625$ **b.** $8^{-1} = \dfrac{1}{8}$ **c.** $7^{1/3} = \sqrt[3]{7}$

Solution To find the equivalent logarithmic equation for each exponential equation, we use the definition of a logarithm.

a. $5^4 = 625$ is equivalent to $\log_5 625 = 4$.

b. $8^{-1} = \dfrac{1}{8}$ is equivalent to $\log_8 \dfrac{1}{8} = -1$.

c. $7^{1/3} = \sqrt[3]{7}$ is equivalent to $\log_7 \sqrt[3]{7} = \dfrac{1}{3}$.

PRACTICE 2

Write each exponential equation in logarithmic notation.

a. $2^5 = 32$

b. $7^0 = 1$

c. $10^{1/2} = \sqrt{10}$

Evaluating Logarithms

Because logarithms are exponents, some logarithms can be evaluated by inspection.

EXAMPLE 3

Evaluate:

a. $\log_2 8$ **b.** $\log_4 16$ **c.** $\log_{49} 7$

Solution   We know that the logarithm of x to the base b, written as $\log_b x$, is the exponent to which b must be raised to get x. We use this relationship to evaluate each of the following expressions:

a. $\log_2 8 = 3$ because $2^3 = 8$.

b. $\log_4 16 = 2$ because $4^2 = 16$.

c. $\log_{49} 7 = \dfrac{1}{2}$ because $49^{1/2} = \sqrt{49} = 7$.

PRACTICE 3

Evaluate:

a. $\log_{10} 10$

b. $\log_3 27$

c. $\log_8 2$

We can use properties of exponents to verify the special properties of logarithms shown below. Both of these properties can be useful in evaluating logarithms.

> ### Special Properties of Logarithms
> - $\log_b b = 1$ because $b^1 = b$; that is, 1 is the exponent to which b must be raised to get b.
> - $\log_b 1 = 0$ because $b^0 = 1$; that is, 0 is the exponent to which b must be raised to get 1.

EXAMPLE 4

Evaluate:

a. $\log_8 8$ **b.** $\log_4 1$

Solution

a. Because $\log_b b = 1$, we conclude that $\log_8 8 = 1$.

b. Because $\log_b 1 = 0$, we conclude that $\log_4 1 = 0$.

PRACTICE 4

Find the value of each logarithmic expression.

a. $\log_6 6$

b. $\log_3 1$

Solving Logarithmic Equations

Now, let's look at how to solve equations involving logarithms. The key is to write, and then solve, the equivalent exponential equations.

EXAMPLE 5

Solve:

a. $\log_3 x = 4$ **b.** $\log_5 x = -2$

PRACTICE 5

Solve:

a. $\log_4 x = 3$

b. $\log_2 x = -1$

Solution

a. $\log_3 x = 4$

$\quad 3^4 = x$ Write the equivalent exponential equation.

$\quad x = 81$

So the solution is 81.

b. $\log_5 x = -2$

$\quad 5^{-2} = x$ Write the equivalent exponential equation.

$\quad x = \dfrac{1}{25}$

So the solution is $\dfrac{1}{25}$.

EXAMPLE 6

Solve:

a. $\log_x 64 = 2$ **b.** $\log_x 5 = \dfrac{1}{2}$

Solution

a. $\log_x 64 = 2$

$\quad x^2 = 64$ Write the equivalent exponential equation.

$\quad x = 8$

Note that even though $(-8)^2$ is also equal to 64, the base of a logarithm must be positive. So the only solution is 8.

b. $\log_x 5 = \dfrac{1}{2}$

$\quad x^{1/2} = 5$ Write the equivalent exponential equation.

$\quad x = 25$

So the solution is 25.

PRACTICE 6

Solve:

a. $\log_x 125 = 3$

b. $\log_x 2 = \dfrac{1}{4}$

EXAMPLE 7

Solve: $\log_8 32 = x$

Solution

$\quad \log_8 32 = x$

$\quad\quad 8^x = 32$ Write the equivalent exponential equation.

$\quad (2^3)^x = 2^5$ Write each side of the equation as a power of 2.

$\quad\quad 2^{3x} = 2^5$

$\quad\quad 3x = 5$ Use the one-to-one property of exponential functions.

$\quad\quad x = \dfrac{5}{3}$

So the solution is $\dfrac{5}{3}$.

PRACTICE 7

Solve: $\log_9 27 = x$

Note that in Example 5, we solved the equation given its base and power. In Example 6, we solved for the base. And in Example 7, we found the exponent. Rewriting a logarithmic equation as the equivalent exponential equation and then solving for the unknown works in all three cases.

Graphing Logarithmic Functions

Recall that we defined a logarithmic function as the inverse of the corresponding exponential function. For example, if $f(x) = 3^x$, then its inverse is $f^{-1}(x) = \log_3 x$. Since a logarithmic function is the inverse of an exponential function, its graph is symmetric to the graph of the exponential function about the line $y = x$. So to graph $f^{-1}(x) = \log_3 x$, we can first graph $f(x) = 3^x$, and then use symmetry about the line $y = x$ to graph the inverse function. See the graph below:

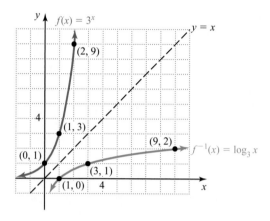

In graphing a logarithmic function, it is unnecessary to graph the corresponding exponential function. Instead, we can write the logarithmic function in exponential form and then graph, as shown in the following examples.

EXAMPLE 8

Graph: $f(x) = \log_2 x$

Solution First, we substitute y for $f(x)$ in the given function: $y = \log_2 x$. Next, we write $y = \log_2 x$ in exponential form: $x = 2^y$. Since $x = 2^y$ is already solved for x, let's choose some y-values and find the corresponding x values.

If $y = 0$, then $x = 2^0 = 1$.

If $y = 1$, then $x = 2^1 = 2$.

If $y = 2$, then $x = 2^2 = 4$.

If $y = -1$, then $x = 2^{-1} = \dfrac{1}{2}$.

Now, we plot the ordered pairs $(1, 0), (2, 1), (4, 2),$ and $\left(\frac{1}{2}, -1\right)$, then connect the points with a smooth curve.

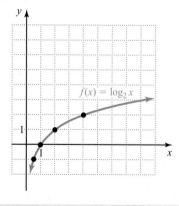

PRACTICE 8

Graph: $f(x) = \log_5 x$

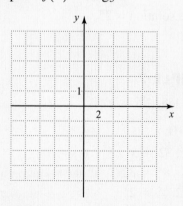

Note that the graph in Example 8 is defined for any positive value of x and takes on any real value of y. It is increasing and has x-intercept $(1, 0)$, but no y-intercept.

EXAMPLE 9

Graph: $f(x) = \log_{1/2} x$

Solution First, we substitute y for $f(x)$ in the given function:
$y = \log_{1/2} x$. Next, we write $y = \log_{1/2} x$ in exponential form:

$x = \left(\dfrac{1}{2}\right)^{y}$. Then, we find some ordered pair solutions of the equation.

Finally, we plot the ordered pairs, and connect the corresponding points with a smooth curve.

If $y = 0$, then $x = \left(\dfrac{1}{2}\right)^{0} = 1.$

If $y = 1$, then $x = \left(\dfrac{1}{2}\right)^{1} = \dfrac{1}{2}.$

If $y = 2$, then $x = \left(\dfrac{1}{2}\right)^{2} = \dfrac{1}{4}.$

If $y = -1$, then $x = \left(\dfrac{1}{2}\right)^{-1} = 2.$

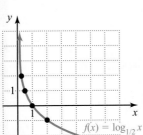

PRACTICE 9

Graph: $f(x) = \log_{1/5} x$

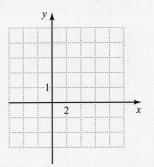

The graph in Example 9 is defined for any positive value x and takes on any real value of y. It is decreasing, has x-intercept $(1, 0)$, and has no y-intercept.

Some characteristics of logarithmic functions can be seen from the graphs in Examples 8 and 9.

Characteristics of the Logarithmic Function $f(x) = \log_b x$

- The domain of the function is the set of positive real numbers, and the range is the set of real numbers.

- The function is increasing if the base is greater than 1. The function is decreasing if the base is between 0 and 1.

- The x-intercept of the graph of the function is $(1, 0)$, but the graph has no y-intercept. The graph approaches the y-axis, but never intersects it.

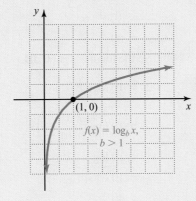

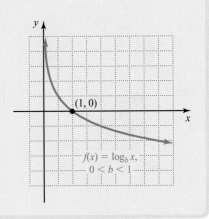

EXAMPLE 10

The loudness L of a sound, measured in decibels (dB), can be defined by the logarithmic equation $L = 10 \log_{10}\left(\dfrac{I}{I_0}\right)$, where I is the intensity of the sound (in watts per square meter) and I_0 is a constant equal to 10^{-12}. I_0 is called the zero decibel level, and is the intensity of a barely audible sound.

PRACTICE 10

In a typical hurricane, the barometric pressure is a function of the distance from the hurricane's eye. Researchers have found that x mi from the eye, the barometric pressure (in inches of

EXAMPLE 10 (continued)

Use the given logarithmic equation to complete the following table. Recall that the logarithm of a number is the power to which we raise the base in order to get that number.

Type of Sound	Intensity	Loudness
Busy street	10^{-5}	
Normal conversation	10^{-6}	
Quiet office	10^{-8}	

Solution Note that $\log_{10} 10^7 = 7$.

Type of Sound	Intensity	Loudness
Busy street	10^{-5}	$L = 10 \log_{10}\left(\dfrac{I}{I_0}\right)$ $= 10 \log_{10}\left(\dfrac{10^{-5}}{10^{-12}}\right)$ $= 10 \underbrace{\log_{10} 10^7}$ $= 10 \cdot 7$ $= 70$ So the loudness is 70 dB.
Normal conversation	10^{-6}	$L = 10 \log_{10}\left(\dfrac{I}{I_0}\right)$ $= 10 \log_{10}\left(\dfrac{10^{-6}}{10^{-12}}\right)$ $= 10 \underbrace{\log_{10} 10^6}$ $= 10 \cdot 6$ $= 60$ So the loudness is 60 dB.
Quiet office	10^{-8}	$L = 10 \log_{10}\left(\dfrac{I}{I_0}\right)$ $= 10 \log_{10}\left(\dfrac{10^{-8}}{10^{-12}}\right)$ $= 10 \underbrace{\log_{10} 10^4}$ $= 10 \cdot 4$ $= 40$ So the loudness is 40 dB.

mercury) can be approximated by the following function:

$$f(x) = 27 + 1.1 \log_{10}(x + 1)$$

Use this function to find the barometric pressure, to the nearest tenth of an inch, 9 mi from a hurricane's eye. (*Source:* A. Miller and R. Anthes, *Meteorology*)

Note in Example 10 that for each sound, after substituting the values for I and I_0 in the expression $\log_{10}\left(\dfrac{I}{I_0}\right)$ and then simplifying, we get an expression of the form $\log_{10} 10^n$. In each case, regardless of the value of n, we find that $\log_{10} 10^n = n$. That is, the logarithm, base 10, of 10 raised to a power equals that power.

Mathematically Speaking

Fill in each blank with the most appropriate term or phrase from the given list.

intersects	exponential function	increasing
decreasing		approaches
identical	logarithmic function	base
exponent		symmetric
b to the base x	x to the base b	

1. The inverse of $f(x) = b^x$ is represented by the _____ $f^{-1}(x) = \log_b x$.

2. The term $\log_b x$ may be read as "the logarithm of _____."

3. A logarithm is a(n) _____.

4. The graph of a logarithmic function is _____ to the graph of the corresponding exponential function.

5. The logarithmic function is _____ if the base is between 0 and 1.

6. The graph of a logarithmic function _____ the x-axis.

A *Write each logarithmic equation in its equivalent exponential form.*

7. $\log_3 81 = 4$

8. $\log_4 16 = 2$

9. $\log_{1/2} \dfrac{1}{32} = 5$

10. $\log_{1/3} \dfrac{1}{27} = 3$

11. $\log_5 \dfrac{1}{25} = -2$

12. $\log_{10} \dfrac{1}{10,000} = -4$

13. $\log_{1/4} 4 = -1$

14. $\log_{1/6} 36 = -2$

15. $\log_{16} 2 = \dfrac{1}{4}$

16. $\log_{64} 4 = \dfrac{1}{3}$

17. $\log_{10} \sqrt{10} = \dfrac{1}{2}$

18. $\log_6 \sqrt[5]{6} = \dfrac{1}{5}$

Write each exponential equation in its equivalent logarithmic form.

19. $3^5 = 243$

20. $10^6 = 1,000,000$

21. $\left(\dfrac{1}{4}\right)^1 = \dfrac{1}{4}$

22. $\left(\dfrac{1}{5}\right)^2 = \dfrac{1}{25}$

23. $2^{-4} = \dfrac{1}{16}$

24. $9^{-2} = \dfrac{1}{81}$

25. $\left(\dfrac{1}{3}\right)^{-4} = 81$

26. $\left(\dfrac{1}{6}\right)^{-1} = 6$

27. $49^{1/2} = 7$

28. $256^{1/4} = 4$

29. $11^{1/5} = \sqrt[5]{11}$

30. $15^{1/2} = \sqrt{15}$

B *Evaluate.*

31. $\log_5 125$

32. $\log_2 64$

33. $\log_3 9$

34. $\log_{10} 100$

35. $\log_6 6$

36. $\log_7 7$

37. $\log_{1/2} \dfrac{1}{16}$

38. $\log_{1/3} \dfrac{1}{9}$

39. $\log_4 \dfrac{1}{64}$

40. $\log_2 \dfrac{1}{2}$

41. $\log_{1/4} 16$

42. $\log_{1/10} 1000$

43. $\log_9 1$

44. $\log_4 1$

45. $\log_{27} 3$

46. $\log_{25} 5$

47. $\log_{1/16} \dfrac{1}{2}$

48. $\log_{1/64} \dfrac{1}{4}$

49. $\log_{36} \dfrac{1}{6}$

50. $\log_8 \dfrac{1}{2}$

C *Solve.*

51. $\log_3 x = 3$

52. $\log_2 x = 6$

53. $\log_6 x = -2$

54. $\log_3 x = -5$

55. $\log_{2/3} x = -1$

56. $\log_{1/6} x = -2$

57. $\log_4 x = \dfrac{1}{2}$

58. $\log_{64} x = \dfrac{1}{3}$

59. $\log_x 216 = 3$

60. $\log_x 81 = 2$

61. $\log_x 2 = \dfrac{1}{3}$

62. $\log_x 3 = \dfrac{1}{4}$

63. $\log_x \dfrac{9}{16} = 2$

64. $\log_x \dfrac{27}{8} = 3$

65. $\log_x 7 = -1$

66. $\log_x 16 = -4$

67. $\log_4 8 = x$

68. $\log_9 3 = x$

69. $\log_{27} 81 = x$

70. $\log_{16} 64 = x$

71. $\log_{1/3} 81 = x$

72. $\log_{1/2} 32 = x$

D *Graph each function.*

73. $f(x) = \log_3 x$

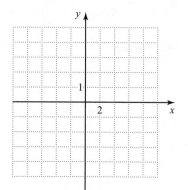

74. $f(x) = \log_4 x$

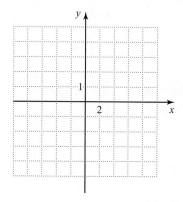

75. $f(x) = \log_{1/3} x$

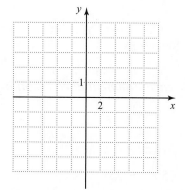

76. $f(x) = \log_{1/4} x$

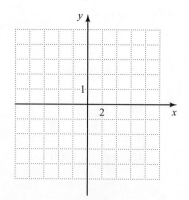

77. $f(x) = -\log_3 x$

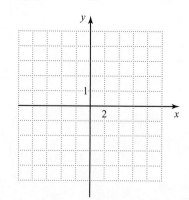

78. $f(x) = -\log_2 x$

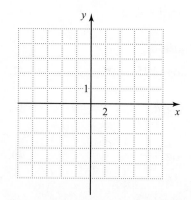

Mixed Practice

Solve.

79. $\log_x 5 = \dfrac{1}{3}$

80. $\log_x \dfrac{25}{64} = 2$

81. $\log_9 27 = x$

82. Graph the function $f(x) = -\log_4 x$.

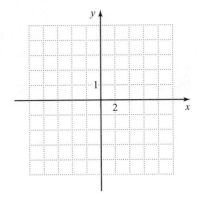

Evaluate.

83. $\log_2 \dfrac{1}{16}$

84. $\log_{81} 3$

Write each logarithmic equation in its equivalent exponential form.

85. $\log_{1/2} 16 = -4$

86. $\log_{81} 3 = \dfrac{1}{4}$

Write each exponential equation in its equivalent logarithmic form.

87. $\left(\dfrac{1}{7}\right)^{-2} = 49$

88. $32^{1/5} = 2$

Applications

E *Solve.*

89. The power gain P (in decibels) in an amplifier can be determined using the equation $P = 10 \log_{10}\left(\dfrac{W_2}{W_1}\right)$, where W_1 and W_2 represent the input and output power (in watts), respectively. What is the power gain in an amplifier if an input power of 10^{-2} watts gives an output power of 100 watts?

90. Scientists speak of *sound pressure* when referring to noise as perceived by a listener. The level of sound pressure S (in decibels) is defined by the logarithmic equation $S = 20 \log_{10}\left(\dfrac{P}{P_0}\right)$, where P is the actual sound pressure in newtons per square meter (N/m^2) and P_0 is the threshold of human hearing, which equals $2 \times 10^{-5}\,\text{N}/\text{m}^2$. A sound at the pain threshold has a sound pressure of about $20\,\text{N}/\text{m}^2$. Calculate the sound pressure level at the pain threshold.

91. The formula $\log_2 I = t$ relates amount of electrical current I (in amperes) flowing through a circuit to the length of time t (in seconds) that it takes for the current to pass through the circuit. Find the amount of current in a circuit when $t = 3$ sec.

92. The half-life of radium-226, a radioactive isotope used in the treatment of cervical cancer, is about 1600 yr. The equation $t = -1600 \log_2\left(\dfrac{b}{a}\right)$ can be used to determine the number of years it will take a mg of the radium to decay to b mg. How long will it take 40 mg of radium-226 to decay to 5 mg?

93. A picosecond is one trillionth (10^{-12}) of a second. Engineers found that the time t (in picoseconds) that a particular computer requires to carry out N computations is given by the equation $t = N + \log_2 N$. How long does it take this computer to carry out 64 computations?

94. In a certain chemical reaction, the time t (in seconds) required for 1 kg of a substance to be converted to N kg of a second substance is given by the equation $t = -10 \log_{10} N$. How long will it take for the reaction to produce $\dfrac{1}{10}$ kg of the second substance?

• Check your answers on page A-50.

MINDStretchers

Writing

1. Consider the logarithmic equation $\log_b y = x$. Explain why the value of y can never be equal to 0.

Research

2. Either in your college library or on the web, investigate the work of John Napier in the development of logarithms. Summarize your findings.

Critical Thinking

3. Suppose $f(x) = \log_2 x$ and $g(x) = 2^x$.
  a. Find $(f \circ g)(x)$.
  b. How could you have predicted the result you found in part (a)?

Cultural Note

Marie Curie, shown meeting with the physicist Albert Einstein, was a pioneer in the field of radioactivity. At the beginning of the twentieth century, she discovered radium and polonium—elements whose radioactive decay can be modeled by *exponential functions*. Madame Curie was the first person to win two Nobel prizes. Her work tremendously influenced basic science and ushered in a new era in medical research and treatment.

(*Source:* Julie Des Jardins, *The Madame Curie Complex: The Hidden History of Women in Science*, Feminist Press, 2010).

9.4 Properties of Logarithms

In the previous section, we introduced the concept of logarithmic functions. In this section, we consider *properties of logarithms,* which we can use to evaluate and to rewrite logarithmic expressions. The properties of logarithms are used in solving problems in many areas, such as chemistry, banking, and finance.

Using the Product Property

Since logarithms are exponents, they follow the laws of exponents. These laws lead to the properties of logarithms. To begin, we use the product rule of exponents, $b^x \cdot b^y = b^{x+y}$, to derive the *product property of logarithms*.

Suppose we let $x = \log_b M$ and $y = \log_b N$. If we write each equation in exponential form, we get $b^x = M$ and $b^y = N$. Forming the product of M and N, we have:

$$MN = b^x \cdot b^y = b^{x+y}$$

Writing $MN = b^{x+y}$ in logarithmic form gives us:

$$\log_b MN = x + y$$

Since $x = \log_b M$ and $y = \log_b N$, we can write:

$$\log_b MN = \log_b M + \log_b N,$$

which is called the *product property of logarithms*.

The Product Property of Logarithms

For any positive real numbers M, N, and b, $b \neq 1$,

$$\log_b MN = \log_b M + \log_b N.$$

In words, this property states that the logarithm of a product is the sum of the logarithms of the factors. Let's use this property to rewrite logarithmic expressions.

EXAMPLE 1

Write each expression using the product property.

a. $\log_4(6 \cdot 5)$ **b.** $\log_8 8x$

Solution We use the product property of logarithms as follows:

a. $\log_4(6 \cdot 5) = \log_4 6 + \log_4 5$

b. $\log_8 8x = \log_8 8 + \log_8 x$
$= 1 + \log_8 x$ Evaluate $\log_8 8$.

PRACTICE 1

Use the product property to rewrite each expression.

a. $\log_6(4 \cdot 9)$

b. $\log_2 2x$

EXAMPLE 2

Write as a single logarithm.

a. $\log_5 16 + \log_5 4$ **b.** $\log_3 x + \log_3(x + 1)$

Solution Using the product property of logarithms, we get:

a. $\log_5 16 + \log_5 4 = \log_5(16 \cdot 4)$

$\qquad\qquad\qquad\qquad = \log_5 64$

b. $\log_3 x + \log_3(x + 1) = \log_3[x(x + 1)]$

$\qquad\qquad\qquad\qquad\quad = \log_3(x^2 + x)$

PRACTICE 2

Write as a single logarithm.

a. $\log_2 9 + \log_2 15$

b. $\log_7(x + y) + \log_7 x$

Using the Quotient Property

Now, let's use the quotient rule of exponents, $\dfrac{b^x}{b^y} = b^{x-y}$, to derive the *quotient property of logarithms*. Again, we let $x = \log_b M$ and $y = \log_b N$. Next, we write each equation in exponential form: $M = b^x$ and $N = b^y$. Forming the quotient of M and N gives us:

$$\frac{M}{N} = \frac{b^x}{b^y} = b^{x-y}$$

We can write $\dfrac{M}{N} = b^{x-y}$ in logarithmic form as:

$$\log_b \frac{M}{N} = x - y$$

Substituting $\log_b M$ for x and $\log_b N$ for y, we have:

$$\log_b \frac{M}{N} = \log_b M - \log_b N,$$

which is called the *quotient property of logarithms*.

> ### The Quotient Property of Logarithms
>
> For any positive real numbers M, N, and b, $b \neq 1$,
>
> $$\log_b \frac{M}{N} = \log_b M - \log_b N.$$

In words, this property states that the logarithm of a quotient is the logarithm of the dividend minus the logarithm of the divisor.

EXAMPLE 3

Write each expression using the quotient property.

a. $\log_7 \dfrac{9}{5}$

b. $\log_a \dfrac{a}{x}$

PRACTICE 3

Write each expression using the quotient property.

a. $\log_4 \dfrac{3}{4}$

b. $\log_b \dfrac{u}{v}$

Solution We use the quotient property of logarithms.

a. $\log_7 \dfrac{9}{5} = \log_7 9 - \log_7 5$

b. $\log_a \dfrac{a}{x} = \log_a a - \log_a x$

$\phantom{\log_a \dfrac{a}{x}} = 1 - \log_a x$ Evaluate $\log_a a$.

EXAMPLE 4

Write as a single logarithm.

a. $\log_{10} 40 - \log_{10} 10$ **b.** $\log_2(x + 1) - \log_2(x - 3)$

Solution Using the quotient property of logarithms, we get:

a. $\log_{10} 40 - \log_{10} 10 = \log_{10} \dfrac{40}{10} = \log_{10} 4$

b. $\log_2(x + 1) - \log_2(x - 3) = \log_2 \dfrac{x + 1}{x - 3}$

PRACTICE 4

Write as a single logarithm.

a. $\log_7 49 - \log_7 98$

b. $\log_3(x - y) - \log_3(x + y)$

Using the Power Property

Finally, we consider the *power property of logarithms*. To derive this property, we use the power rule of exponents, $(b^x)^r = b^{xr}$. Again, if we let $x = \log_b M$ and then write this equation in exponential form, we get $b^x = M$. Raising both sides of this equation to the rth power gives:

$$(b^x)^r = M^r \quad \text{or} \quad b^{xr} = M^r$$

Writing $b^{xr} = M^r$ as a logarithmic equation, we have $\log_b M^r = xr$. Since $x = \log_b M$, we can now write

$$\log_b M^r = xr = (\log_b M)r = r \log_b M,$$

which is called the *power property of logarithms*.

> ### The Power Property of Logarithms
>
> For any positive real numbers M and b, $b \neq 1$, and any real number r,
>
> $$\log_b M^r = r \log_b M.$$

In words, this property states that the logarithm of a power of a number is that power times the logarithm of that number.

EXAMPLE 5

Write each expression using the power property.

a. $\log_8 n^2$ **b.** $\log_3 \sqrt{n}$

Solution

a. $\log_8 n^2 = 2 \log_8 n$

b. $\log_3 \sqrt{n} = \log_3 n^{1/2}$ Write the radical using a rational exponent.

$\phantom{\log_3 \sqrt{n}} = \dfrac{1}{2} \log_3 n$

PRACTICE 5

Use the power property to write each expression.

a. $\log_2 x^3$

b. $\log_5 \sqrt[3]{x}$

In the previous section, we noted that $\log_b b = 1$ and $\log_b 1 = 0$. Now, we consider two other special properties of logarithms. Both properties follow from the inverse relationship between exponential and logarithmic functions.

Special Properties of Logarithms

For $b > 0$ and $b \neq 1$,

- $\log_b b^x = x$
- $b^{\log_b x} = x$, where $x > 0$

In words, the first property states that the logarithm with base b of b raised to a power equals that power. The second property states that b raised to the logarithm of a number to the base b equals that number.

EXAMPLE 6

Find the value of each logarithm.

a. $\log_2 2^5$ **b.** $10^{\log_{10} 9}$

Solution

a. Since $\log_b b^x = x$, we conclude that

$$\log_2 2^5 = 5$$

b. Since $b^{\log_b x} = x$, we conclude that

$$10^{\log_{10} 9} = 9$$

PRACTICE 6

Evaluate:

a. $\log_{10} 10^7$

b. $3^{\log_3 29}$

The following table summarizes the properties of logarithms:

Properties of Logarithms

For any positive real numbers M, N, and b, where $b \neq 1$, and for any real numbers r and x,

The Product Property	$\log_b MN = \log_b M + \log_b N$
The Quotient Property	$\log_b \dfrac{M}{N} = \log_b M - \log_b N$
The Power Property	$\log_b M^r = r \log_b M$
Special Properties	$\log_b b = 1$
	$\log_b 1 = 0$
	$\log_b b^x = x$
	$b^{\log_b x} = x$, where $x > 0$

Using the Properties Together

Using the properties of logarithms together, we can write logarithmic expressions in alternative forms. These forms are important for solving equations with logarithms and in more advanced courses, as we will see in Section 9.6.

EXAMPLE 7

Use the properties of logarithms to write as a single logarithm.

a. $3 \log_4 2 + 2 \log_4 5$

b. $2(\log_5 x - \log_5 y)$

Solution

a. $3 \log_4 2 + 2 \log_4 5 = \log_4 2^3 + \log_4 5^2$ Use the power property.

$\qquad\qquad\qquad\qquad\quad = \log_4 8 + \log_4 25$

$\qquad\qquad\qquad\qquad\quad = \log_4(8 \cdot 25)$ Use the product property.

$\qquad\qquad\qquad\qquad\quad = \log_4 200$

b. $2(\log_5 x - \log_5 y) = 2 \log_5 x - 2 \log_5 y$ Use the distributive property.

$\qquad\qquad\qquad\qquad\quad = \log_5 x^2 - \log_5 y^2$ Use the power property.

$\qquad\qquad\qquad\qquad\quad = \log_5 \dfrac{x^2}{y^2}$ Use the quotient property.

PRACTICE 7

Write as a single logarithm using the properties of logarithms.

a. $2 \log_6 4 + 3 \log_6 2$

b. $3(\log_2 u - \log_2 v)$

EXAMPLE 8

Use the properties of logarithms to write each expression as a sum or difference of logarithms.

a. $\log_2 2x^5$ **b.** $\log_5 \sqrt[3]{\dfrac{x}{y}}$ **c.** $\log_n \dfrac{r^3 s}{t^2}$

Solution

a. $\log_2 2x^5 = \log_2 2 + \log_2 x^5$ Use the product property.

$\qquad\qquad\; = 1 + 5 \log_2 x$ Evaluate $\log_2 2$ and use the power property.

b. $\log_5 \sqrt[3]{\dfrac{x}{y}} = \log_5 \left(\dfrac{x}{y}\right)^{1/3}$ Write the radical using a rational exponent.

$\qquad\qquad\; = \dfrac{1}{3} \log_5 \dfrac{x}{y}$ Use the power property.

$\qquad\qquad\; = \dfrac{1}{3}(\log_5 x - \log_5 y)$ Use the quotient property.

$\qquad\qquad\; = \dfrac{1}{3} \log_5 x - \dfrac{1}{3} \log_5 y$ Use the distributive property.

c. $\log_n \dfrac{r^3 s}{t^2} = \log_n(r^3 s) - \log_n t^2$ Use the quotient property.

$\qquad\qquad\; = \log_n r^3 + \log_n s - \log_n t^2$ Use the product property.

$\qquad\qquad\; = 3 \log_n r + \log_n s - 2 \log_n t$ Use the power property.

PRACTICE 8

Use the properties of logarithms to write each expression as a sum or a difference of logarithms.

a. $\log_5 5x^2$

b. $\log_7 \sqrt{\dfrac{u}{v}}$

c. $\log_u \dfrac{b^3}{ac^5}$

EXAMPLE 9

Chemists use the *pH scale* for representing the acidity of a solution. The key formula is $pH = -\log_{10}[H^+]$, where $[H^+]$ stands for the concentration of hydrogen ions in the solution. A pH level of 7 is neutral. A level greater than 7 is basic, and a level less than 7 is acidic.

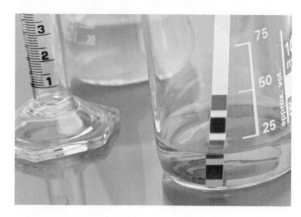

a. Find the pH of vinegar, which has a hydrogen ion concentration $[H^+]$ of 1×10^{-3}.

b. Determine whether vinegar is basic or acidic.

Solution

a. Using the formula $pH = -10 \log_{10}[H^+]$, we get:

$$
\begin{aligned}
pH &= -\log_{10}[H^+] \\
&= -\log_{10}(1 \times 10^{-3}) \\
&= -(\log_{10} 1 + \log_{10} 10^{-3}) \\
&= -\log_{10} 1 - \log_{10} 10^{-3} \\
&= 0 - (-3) \\
&= 3
\end{aligned}
$$

So the pH of vinegar is 3.

b. Since $3 < 7$, vinegar is acidic.

PRACTICE 9

The number N of bacteria present in a culture after x hr can be determined by using the formula $\log_{10} N = x(\log_{10} 6 - \log_{10} 2)$.

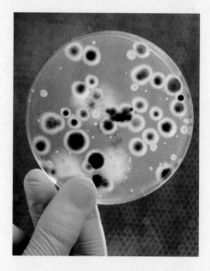

a. How many bacteria are present in the culture after 1 hr?

b. By what factor does the number of bacteria present in the culture increase each hour?

Mathematically Speaking

Fill in each blank with the most appropriate term or phrase from the given list.

base	divided by	sum
plus	product	times
number	minus	power

1. The logarithm of a product is the _____ of the logarithms of the factors.

2. The logarithm of a quotient is the logarithm of the numerator _____ the logarithm of the denominator.

3. The logarithm of a power of a number is the power _____ the logarithm of the number.

4. For any _____, the logarithm of 1 is 0.

5. The logarithm of b to the base b raised to a power equals that _____.

6. The number b raised to the logarithm of a number to the base b equals that _____.

A *Write each expression as the sum or difference of logarithms.*

7. $\log_2(16 \cdot 5)$

8. $\log_3(2 \cdot 27)$

9. $\log_6 \dfrac{7}{36}$

10. $\log_2 \dfrac{4}{9}$

11. $\log_3 8x$

12. $\log_5 15y$

13. $\log_n uv$

14. $\log_a pq$

15. $\log_b [a(b-1)]$

16. $\log_y [x(y-3)]$

17. $\log_4 \dfrac{64}{n}$

18. $\log_{10} \dfrac{p}{1000}$

19. $\log_7 \dfrac{v}{u}$

20. $\log_4 \dfrac{q}{p}$

21. $\log_a \dfrac{b-a}{b+a}$

22. $\log_k \dfrac{k+1}{k+5}$

Write each expression as a single logarithm.

23. $\log_2 5 + \log_2 7$

24. $\log_5 2 + \log_5 6$

25. $\log_7 10 - \log_7 2$

26. $\log_6 3 - \log_6 21$

27. $1 + \log_x 6$

28. $\log_a 18 + 1$

29. $1 - \log_b 16$

30. $\log_a 7 - 1$

31. $\log_5 2 + \log_5(x-5)$

32. $\log_3(y+2) + \log_3 6$

33. $\log_n x + \log_n z$

34. $\log_b c + \log_b a$

35. $\log_5 k - \log_5 n$

36. $\log_6 y - \log_6 x$

37. $\log_{10}(a+b) - \log_{10}(a-b)$

38. $\log_4 x - \log_4(x-y)$

B *Use the power property to rewrite each expression.*

39. $\log_6 8^2$

40. $\log_2 3^3$

41. $\log_3 x^4$

42. $\log_5 y^6$

43. $\log_a \sqrt[5]{b}$

44. $\log_x \sqrt[4]{y}$

45. $\log_4 a^{-1}$

46. $\log_7 y^{-2}$

Evaluate.

47. $\log_5 5^4$

48. $\log_3 3^7$

49. $\log_2 2^{1/2}$

50. $\log_6 6^{1/3}$

51. $\log_x x^{-3}$

52. $\log_a a^{-1}$

53. $8^{\log_8 15}$

54. $7^{\log_7 20}$

55. $9^{\log_9 x}$

56. $4^{\log_4 b}$

57. $x^{\log_x 10}$

58. $n^{\log_n 23}$

C *Write each expression as a single logarithm.*

59. $2 \log_2 6 + \log_2 3$

60. $3 \log_{10} 3 + \log_{10} 2$

61. $4 \log_7 2 - 3 \log_7 4$

62. $2 \log_3 8 - 3 \log_3 5$

63. $-\log_4 x + 6 \log_4 y$

64. $-3 \log_6 a + 5 \log_6 b$

65. $2 \log_b 3 + 3 \log_b 2 - \log_b 9$

66. $3 \log_x 4 - 2 \log_x 6 - 4 \log_x 2$

67. $\frac{1}{2} (\log_4 x - 2 \log_4 y)$

68. $\frac{1}{3}(-6 \log_7 u + \log_7 v)$

69. $2 \log_5 x + \log_5 (x - 1)$

70. $4 \log_2 n + \log_2 (n + 3)$

71. $\frac{1}{3}\left[\log_6(x^2 - y^2) - \log_6(x + y)\right]$

72. $\frac{1}{2}\left[\log_2(a^2 - b^2) - \log_2(a - b)\right]$

73. $-\log_2 z + 4 \log_2 x - 5 \log_2 y$

74. $-4 \log_8 b + 5 \log_8 a - 2 \log_8 c$

75. $\frac{1}{4} \log_5 a - 8 \log_5 b + \frac{3}{4} \log_5 c$

76. $\frac{2}{3} \log_7 u + \frac{1}{3} \log_7 v - 7 \log_7 w$

Write each expression as a sum or difference of logarithms.

77. $\log_6 3y^2$

78. $\log_2 5n^4$

79. $\log_x x^3 y^2$

80. $\log_b a^4 b^8$

81. $\log_2 4x^2 y^5$

82. $\log_3 3uv^3$

83. $\log_8 \sqrt{10x}$

84. $\log_5 \sqrt[3]{9v^2}$

85. $\log_4 \dfrac{x^2}{y}$

86. $\log_5 \dfrac{a^3}{b^4}$

87. $\log_7 \sqrt[4]{\dfrac{u^3}{v}}$

88. $\log_3 \sqrt[3]{\dfrac{x^2}{y^2}}$

89. $\log_c \dfrac{a^2 c^4}{b^3}$

90. $\log_z \dfrac{x^5 y^4}{z^2}$

91. $\log_6 \dfrac{x^3}{(x - y)^2}$

92. $\log_5 \dfrac{a + b}{a^4}$

93. $\log_a x^2 \sqrt[5]{y^3 z^2}$

94. $\log_n \dfrac{\sqrt[4]{a^3 c^2}}{b^5}$

Mixed Practice

Write each expression as a single logarithm.

95. $\log_b 8 - 1$

96. $\log_2 5 + \log_2(k + 3)$

97. $2 \log_a 6 - 4 \log_a 2 - 3 \log_a 3$

98. $\frac{1}{4}(-\log_5 a + 8 \log_5 b)$

Solve.

99. Use the power property to rewrite the expression $\log_p \sqrt[3]{q}$.

100. Evaluate: $4^{\log_4 25}$

Write each expression as a sum or difference.

101. $\log_2 8a^3 b^2$

102. $\log_5 \sqrt[4]{\dfrac{x^3}{y^2}}$

103. $\log_4 64a$

104. $\log_5 \dfrac{b}{25}$

Applications

D *Solve.*

105. The magnitude M of an earthquake on the Richter scale is given by the logarithmic equation $M = \log_{10} I$, where I is the intensity of the earthquake. The largest earthquake in a recent year, with an intensity of $10^{8.3}$, occurred in Hokkaido, Japan. Calculate the magnitude of that earthquake. (*Source:* U.S. Geological Survey, Earthquake Hazards Program)

106. In astronomy, the *apparent magnitude* of a star is a measure of its observed brightness as seen from Earth. The apparent magnitude m is given by the formula $m = -2.5 \log_{10} b$, where b is the observed brightness of the star. Vega, the fifth brightest star in the night sky, has brightness 1.0×10^{-8}. Find its apparent magnitude. (*Source:* wikipedia.org)

107. The sound pressure level S of acoustic sound pressure waves is defined by the logarithmic equation $S = 10 \log_{10} \left(\dfrac{P}{P_0}\right)^2$, where P is the actual pressure and P_0 is the pain threshold for human hearing. Use the properties of logarithms to write the expression on the right so that it does not contain the logarithm of a quotient.

108. The loudness of sound L (in decibels) can be defined by the logarithmic equation $L = 10 \log_{10} \dfrac{I}{10^{-16}}$, where I is the intensity of the sound (in watts per square centimeter). Use the properties of logarithms to rewrite the right side of the equation so that it does not contain the quotient of a logarithm.

109. The half-life of cobalt-60, a radioactive isotope used in industry to identify structural defects in metal parts, is approximately 5 yr. The amount C of cobalt remaining after t yr and the amount C_0 initially present are related by the equation

$$\log_{10} C = \log_{10} C_0 + 0.2t \log_{10} \frac{1}{2}.$$

a. Use the properties of logarithms to show that the equation can be written as
$$C = C_0 \left(\frac{1}{2}\right)^{0.2t}.$$

b. If 50 g of cobalt-60 are initially present, how many grams will remain after 10 yr?

110. The amount A in an account earning 5% interest, compounded annually, after t yr can be determined using the logarithmic equation $\log_{10} A = \log_{10} P + t \log_{10}(1.05)$, where P is the initial amount deposited into the account.

a. Use the properties of logarithms to show that the equation can be written as
$A = P(1.05)^t.$

b. Determine the amount in the account after 1 yr if $1000 is initially deposited.

• Check your answers on page A-50.

MIND Stretchers

Mathematical Reasoning

1. Show that the following properties of logarithms hold:

a. $\log_b b^x = x$, for any positive real number b, where $b \neq 1$, and for any real number x

b. $b^{\log_b x} = x$, for any positive real numbers b and x, where $b \neq 1$

Critical Thinking

2. Show that $3 + \log_3 2 = 1 + \log_3 18$.

Groupwork

3. Suppose $\log_b x = 3.5$, $\log_b y = 0.2$, and $\log_b z = 1.4$. Find the value of each of the following logarithms:

a. $\log_b(xy)$

b. $\log_b \frac{y}{x}$

c. $\log_b \frac{xy}{z}$

d. $\log_b z^5$

e. $\log_b(yz)^2$

OBJECTIVES

A To evaluate a common or natural logarithm

B To use the change-of-base formula

C To solve applied problems using common or natural logarithms

In the previous sections of this chapter, we discussed logarithms in which the base was any positive real number. In this section, we focus on two special logarithmic bases, 10 and e. A logarithm to the base 10 is called a *common logarithm*, and a logarithm to the base e is called a *natural logarithm*. Common logarithms and natural logarithms occur frequently in real-world situations.

Evaluating Common Logarithms

Recall the applications that we have considered involving logarithms to the base 10, such as the Richter scale used to measure the magnitude of an earthquake and the decibel scale used to find the loudness of sound. Base 10 is commonly used for logarithms, since this base makes it easy to find the logarithm of both powers of 10 and multiples of 10.

> **DEFINITION**
>
> A **common logarithm** is a logarithm to the base 10, where **log x** means $\log_{10} x$.

Note that when we write a common logarithm it is customary to omit the base. To evaluate common logarithms using a calculator, we use the $\boxed{\textbf{log}}$ key.

EXAMPLE 1

Use a calculator to approximate each logarithm to four decimal places.

a. log 9   **b.** log 35

Solution   Using the $\boxed{\textbf{log}}$ key on the calculator, we get:

a. $\log 9 \approx 0.9542$ **b.** $\log 35 \approx 1.5441$

PRACTICE 1

Using a calculator, approximate each logarithm to four decimal places.

a. log 6

b. log 27

We can evaluate common logarithms of powers of 10 without using a calculator. Recall from the properties of logarithms that:

$$\log_b b^x = x$$

So substituting 10 for b gives us:

$$\log_{10} 10^x = x \quad \text{or} \quad \log 10^x = x$$

EXAMPLE 2

Evaluate each logarithm.   **a.** log 10,000   **b.** $\log \dfrac{1}{100}$ **c.** $\log \sqrt{10}$

Solution   First, we write each logarithm as a power of 10. Then, we use the relationship $\log 10^x = x$ to simplify.

a. $\log 10{,}000 = \log 10^4 = 4$

b. $\log \dfrac{1}{100} = \log \dfrac{1}{10^2}$

$= \log 10^{-2}$

$= -2$

c. $\log \sqrt{10} = \log 10^{1/2} = \dfrac{1}{2}$

PRACTICE 2

Find the value of each logarithm.

a. log 1000

b. $\log \dfrac{1}{10}$

c. $\log \sqrt[3]{10}$

Evaluating Natural Logarithms

Now, let's focus our attention on another special logarithmic base, namely e. Recall from Section 9.2 that e is an irrational number that is approximately equal to 2.7182818. For the natural exponential function $f(x) = e^x$, the inverse is represented by:

$$f^{-1}(x) = \log_e x$$

Logarithms to the base e, or *natural logarithms*, model a wide range of real-life applications, such as continuous compound interest and population growth. The notation ln x (read "el en x" or "the natural logarithm of x") is used to represent a natural logarithm.

> **DEFINITION**
>
> A **natural logarithm** is a logarithm to the base e, where **ln x** means $\log_e x$.

Just as with common logarithms, natural logarithms can be evaluated using a calculator. To do this, we use the natural logarithm key $\boxed{\text{ln}}$ found on the calculator.

EXAMPLE 3

Use a calculator to approximate each logarithm to four decimal places.

a. ln 2   **b.** ln 30

Solution Using the $\boxed{\text{ln}}$ key, we get the following:

a. ln 2 $\approx$ 0.6931   **b.** ln 30 $\approx$ 3.4012

PRACTICE 3

Approximate each logarithm to four decimal places using a calculator.

a. ln 4

b. ln 25

We can evaluate natural logarithms of powers of e without using a calculator. From the property $\log_b b^x = x$, we know that $\log_e e^x = x$. So ln $e^x = x$, as in the next example.

EXAMPLE 4

Find the value of each logarithm.

a. ln e **b.** $\ln \dfrac{1}{e^4}$ **c.** $\ln \sqrt[3]{e}$

Solution Here, we use the relationship ln $e^x = x$ to evaluate each logarithm.

a. ln e = ln e^1 = 1

b. $\ln \dfrac{1}{e^4} = \ln e^{-4} = -4$

c. $\ln \sqrt[3]{e} = \ln e^{1/3} = \dfrac{1}{3}$

PRACTICE 4

Evaluate each logarithm.

a. ln e^2

b. $\ln \dfrac{1}{e^7}$

c. $\ln \sqrt[4]{e^3}$

TIP Remember that the base of common logarithms is understood to be 10, whereas the base of natural logarithms is understood to be e.

Using the Change-of-Base Formula

As we have just seen, calculators can be used to evaluate common and natural logarithms. However, in general, calculators do not directly evaluate logarithms with bases other than 10 or e. For these logarithms, we use the *change-of-base formula*.

The Change-of-Base Formula

For positive real numbers N, b, and c, where $b \neq 1$ and $c \neq 1$,

$$\log_b N = \frac{\log_c N}{\log_c b}.$$

In words, this formula states that if we know the logarithm of a number N to one base (the reference base c), we can use the formula to find the logarithm of N to some other base b simply by dividing by the logarithm of b to the base c. Since we can readily use a calculator to approximate common logarithms and natural logarithms, it is convenient to use either 10 or e as the reference base:

$$\log_b N = \frac{\log N}{\log b} \quad \text{and} \quad \log_b N = \frac{\ln N}{\ln b}$$

EXAMPLE 5

Use a calculator to approximate $\log_3 12$ to four decimal places.

Solution Using the change-of-base formula, we write the logarithm as the quotient of logarithms to base 10.

$$\log_b N = \frac{\log_c N}{\log_c b}$$

$$\log_3 12 = \frac{\log 12}{\log 3} \qquad \text{Substitute 12 for } N \text{ and 3 for } b.$$

$$\approx 2.2618595$$

So $\log_3 12 \approx 2.2619$.

PRACTICE 5

Using a calculator, approximate $\log_4 2.5$ to four decimal places.

In Example 5, we used 10 as the reference base to evaluate the given logarithm. Do you think that we would get the same answer if we used base e? Explain.

In previous sections, we saw several applications involving common logarithms. Now, let's consider some applications that use natural logarithms.

EXAMPLE 6

Some psychologists who study learning theory use the following formula:

$$t = \frac{1}{c} \ln \frac{A}{A - N}$$

Here, t represents the time (in weeks) that a subject needs to achieve the amount of learning N out of a maximum possible learning A, where c is a constant related to the subject's learning style. Using the above formula, determine how long, to the nearest tenth of a week, it would take a subject to learn 20 skills out of a maximum possible 32 skills. Assume that c is 0.22.

Solution Substituting 0.22 for c, 32 for A, and 20 for N, we get:

$$t = \frac{1}{c} \ln \frac{A}{A - N}$$

$$= \frac{1}{0.22} \ln \frac{32}{32 - 20}$$

$$= \frac{1}{0.22} \ln \frac{32}{12}$$

$$\approx (4.55)(0.9808)$$

$$\approx 4.5$$

So it will take the subject approximately 4.5 wk to learn 20 skills.

PRACTICE 6

The half-life of a radioactive substance is the time it takes for half of the material to decay. The amount A (in pounds) of substance remaining after t yr can be determined using the formula

$$\ln \frac{A}{C} = -\frac{t}{h} \ln 2,$$

where C is the initial amount (in pounds) and h is its half-life (in years). If the half-life of lead-210 is approximately 19 yr, how long, to the nearest year, will it take for 10 lb of this substance to decay to $\frac{1}{2}$ lb?

Mathematically Speaking

Fill in each blank with the most appropriate term or phrase from the given list.

change-of-base formula	common logarithm	quotient property
absolute value function	natural exponential function	natural logarithm

1. A(n) _____ is a logarithm to the base 10.

2. The inverse of the _____ $f(x) = e^x$ is $f^{-1}(x) = \log_e x$.

3. A(n) _____ is a logarithm to the base e.

4. According to the _____, $\log_2 5 = \dfrac{\log_6 5}{\log_6 2}$.

A ▦ *Use a calculator to find each logarithm, rounded to four decimal places.*

5. $\log 4$

6. $\log 7$

7. $\log 18$

8. $\log 22$

9. $\log \dfrac{2}{3}$

10. $\log \dfrac{1}{7}$

11. $\log 1.3$

12. $\log 0.9$

13. $\ln 3$

14. $\ln 6$

15. $\ln 17$

16. $\ln 13$

17. $\ln \dfrac{5}{8}$

18. $\ln \dfrac{3}{4}$

19. $\ln 1.2$

20. $\ln 0.24$

Evaluate.

21. $\log 1,000,000$

22. $\log 100,000$

23. $\log \dfrac{1}{1000}$

24. $\log \dfrac{1}{10,000}$

25. $\log 0.01$

26. $\log 0.000001$

27. $\log \sqrt[4]{1000}$

28. $\log \sqrt[3]{100}$

29. $\log 10^x$

30. $\log 10^a$

31. $\ln e^4$

32. $\ln e^{10}$

33. $\ln \dfrac{1}{e}$

34. $\ln \dfrac{1}{e^2}$

35. $\ln \sqrt{e}$

36. $\ln \sqrt[3]{e^2}$

37. $\ln e^b$

38. $\ln e^y$

39. $10^{\log 6}$

40. $10^{\log 2}$

41. $e^{\ln 3}$

42. $e^{\ln 5}$

B ▦ *Use a calculator to find each logarithm, rounded to four decimal places.*

43. $\log_2 7$

44. $\log_5 10$

45. $\log_6 21$

46. $\log_8 14$

47. $\log_{1/2} 11$

48. $\log_{1/3} 4$

49. $\log_3 \dfrac{1}{2}$

50. $\log_4 \dfrac{1}{3}$

51. $\log_5 3.6$

52. $\log_9 6.8$

53. $\log_7 0.023$

54. $\log_6 0.0044$

Mixed Practice

Evaluate.

55. $\log \sqrt{10,000}$

56. $\log 10^6$

57. $\ln \dfrac{1}{e^3}$

58. $e^{\ln 7}$

⊞ *Use a calculator to approximate each logarithm to four decimal places.*

59. $\log_{1/2} 7$ **60.** $\log_6 \dfrac{1}{4}$ **61.** $\log 2.7$ **62.** $\ln \dfrac{2}{3}$

Applications

C *Solve. Use a calculator, where appropriate.*

63. In chemistry, the pH of a substance can be determined using the formula pH $= -\log[\text{H}^+]$, where H^+ represents the hydrogen ion concentration. To the nearest tenth, determine the pH of household bleach, whose hydrogen ion concentration is 3.2×10^{-13}.

64. Based on a recent census, the population of a town is increasing according to the equation $t = 15.2 \ln\left(\dfrac{P}{P_0}\right)$, where P_0 is the initial population and P is the population after t yr. If the initial population of the town is 24,342 people, then to the nearest year, when will the population reach 50,000 people?

65. The annual revenue R (in millions of dollars) of a small computer company can be modeled by the logarithmic function $R(x) = 1.45 + 2 \log(x + 1)$, where x represents the number of years after 2000. To the nearest million dollars, calculate the company's revenue in 2005.

66. Students in a math class were given an exam at the end of the semester. To see how well the students remembered the material, they were given a similar exam each month for 1 yr after the semester ended. The average exam score s for the class on a test given x mo after the original exam is modeled by the equation $s = 80 - 8 \ln(x + 1)$. To the nearest whole number, what was the average exam score for the class after 6 mo?

67. The Beer-Lambert law relates the decay of the intensity of light to the properties of the material the light is passing through. The relationship is described by the equation $kd = -(\ln I - \ln I_0)$, where I is the intensity of light at depth d, I_0 is the initial intensity of the light, and k is a constant. Show that this equation can be written as $I = I_0 e^{-kd}$.

68. In the theory of rocket flight, the velocity v gained by a launch vehicle when its propellant is burned off is given by the equation $v = c \ln W - c \ln W_0$, where c is the exhaust velocity, W is the burnout weight, and W_0 is the takeoff weight. Write the equation so that the right side contains a single logarithm. (*Source:* NASA)

69. The population of a certain bacterium doubles every hour. If a culture has an initial population of 1 bacterium, then the number of hours it takes for the population to grow to 60 bacteria is given by the expression $\log_2 60$. Evaluate this expression, rounding the answer to the nearest tenth.

70. The half-life of carbon-14, a radioactive isotope used in carbon dating, is about 5730 yr. The time it takes for 35 g of carbon-14 to decay to 5 g is given by the expression $-5730 \log_2 \dfrac{1}{7}$. Evaluate this expression. Round the answer to the nearest year.

• Check your answers on page A-50.

MINDStretchers

Groupwork

1. Working with a partner, choose particular positive values of b, c, and N in the change-of-base formula $\log_b N = \dfrac{\log_c N}{\log_c b}$, where $b \neq 1$ and $c \neq 1$. Then, use a calculator to verify that the formula holds for these values. (*Hint:* Choose values of N and b so that N is a power of b.)

b	c	N	$\log_b N$	$\dfrac{\log_c N}{\log_c b}$

Mathematical Reasoning

2. Suppose $f(x) = x + 3$. Find all values of x such that $5^{2\log_5 f(x)} = f(x)$.

Technology

3. Consider the equation $y = \log_3 x$. Explain how you could graph this equation on a grapher that has the capability of evaluating only common and natural logarithms. Then, graph the equation on the grapher.

9.6 Exponential and Logarithmic Equations

In Section 9.2, we solved exponential equations of the form $b^x = b^n$. Similarly, in Section 9.3, we solved logarithmic equations of the form $y = \log_b x$. In this section, we solve a wider range of exponential and logarithmic equations by using the properties of logarithms. Then, we consider applied problems modeled by these types of equations.

A To solve an exponential equation

B To solve a logarithmic equation

C To solve applied problems using exponential or logarithmic equations

Solving Exponential Equations

Recall that we can solve exponential equations such as $3^x = 9$ by writing 9 as a power of 3 and using the one-to-one property of exponential functions, getting:

$$3^x = 9$$
$$3^x = 3^2$$
$$x = 2$$

By contrast, the exponential equation $3^x = 12$ cannot be solved this way because 12 is not a power of 3. Since $3^x = 12$, we can take the common logarithm of each side of the equation and set the two logarithms equal to each other, as follows:

$$\log 3^x = \log 12$$
$$x \log 3 = \log 12$$
$$x = \frac{\log 12}{\log 3}$$
$$x \approx 2.2619$$

So x is approximately 2.2619.

Consider the following example in which we solve an exponential equation by taking the logarithm of each side:

EXAMPLE 1

Solve $2^{x-1} = 7$. Approximate the answer to four decimal places.

Solution

$$2^{x-1} = 7$$

$\log 2^{x-1} = \log 7$ Take the common logarithm of each side of the equation.

$(x - 1)\log 2 = \log 7$ Use the power property of logarithms.

$x - 1 = \dfrac{\log 7}{\log 2}$ Divide each side of the equation by log 2.

$x = \dfrac{\log 7}{\log 2} + 1$ Add 1 to each side of the equation.

$x \approx 2.8074 + 1$

$x \approx 3.8074$

So x is approximately 3.8074.

PRACTICE 1

Solve $4^{x+3} = 15$. Approximate the answer to four decimal places.

Solving Logarithmic Equations

When solving a logarithmic equation, write the equivalent exponential equation, and solve it.

EXAMPLE 2

Solve: $\log_2(x - 4) = 4$

Solution

$\log_2(x - 4) = 4$

$2^4 = x - 4$ Write the equivalent exponential equation.

$16 = x - 4$

$x = 20$

Check

$\log_2(x - 4) = 4$

$\log_2(20 - 4) \stackrel{?}{=} 4$ Substitute 20 for x.

$\log_2 16 \stackrel{?}{=} 4$

$4 = 4$ True

So x is 20.

PRACTICE 2

Solve: $\log_4(3x + 1) = 2$

Use the properties of logarithms to try to write a logarithmic equation so that it contains a single logarithm on one side.

EXAMPLE 3

Solve: $\log_2 x + \log_2(x + 4) = 5$

Solution

$\log_2 x + \log_2(x + 4) = 5$

$\log_2 [x(x + 4)] = 5$ Use the product property.

$2^5 = x(x + 4)$ Write the equivalent exponential equation.

$32 = x^2 + 4x$

$0 = x^2 + 4x - 32$ Subtract 32 from each side of the equation.

$0 = (x + 8)(x - 4)$ Factor.

$x + 8 = 0$ or $x - 4 = 0$

$x = -8$ $x = 4$

PRACTICE 3

Solve: $\log_3 x + \log_3(x - 6) = 3$

EXAMPLE 3 (continued)

Check

Substitute -8 for x.

$$\log_2 x + \log_2(x + 4) = 5$$
$$\log_2(-8) + \log_2(-8 + 4) \overset{?}{=} 5$$

Logarithms of negative numbers are undefined.

Substitute 4 for x.

$$\log_2 x + \log_2(x + 4) = 5$$
$$\log_2 4 + \log_2(4 + 4) \overset{?}{=} 5$$
$$\log_2 4 + \log_2 8 \overset{?}{=} 5$$
$$2 + 3 \overset{?}{=} 5$$
$$5 = 5 \qquad \text{True}$$

Since -8 is an extraneous solution, the only solution is 4.

EXAMPLE 4

Solve: $\log_5(2x) - \log_5(x - 1) = 1$

Solution

$$\log_5(2x) - \log_5(x - 1) = 1$$

$$\log_5 \frac{2x}{x - 1} = 1 \qquad \text{Use the quotient property.}$$

$$5^1 = \frac{2x}{x - 1} \qquad \begin{array}{l}\text{Write the equivalent} \\ \text{exponential equation.}\end{array}$$

$$5 = \frac{2x}{x - 1}$$

$$5(x - 1) = 2x$$

$$5x - 5 = 2x$$

$$3x = 5$$

$$x = \frac{5}{3}$$

So the solution is $\frac{5}{3}$. We can check the solution in the original equation.

PRACTICE 4

Solve: $\log_2(2x + 3) - \log_2 x = 3$

Recall that in Section 9.2 we discussed the compound interest formula

$$A = P\left(1 + \frac{r}{n}\right)^{nt},$$

which is used to calculate the number of dollars A accrued (or owed) after P dollars are invested (or loaned) at an annual interest rate r (written as a decimal), compounded n times per year for t yr.

When the number of times per year n that interest is compounded is high (for instance, when the compounding is daily instead of monthly), the amount of money A can be approximated using the following exponential equation:

$$A = Pe^{rt},$$

This equation is called the *continuous compound interest formula*. Many investment firms and banks use this formula to compound interest. In this method, which is known as *continuous compounding*, interest is said to be compounded every instant.

EXAMPLE 5

An account pays 10% interest compounded continuously. To the nearest year, how long does it take a deposit to triple at this rate?

Solution   If P dollars are deposited in the account, the amount of money after t yr is given by $A = Pe^{0.1t}$. Since we want to know how long it takes for this amount to grow to $3P$, we solve the exponential equation $3P = Pe^{0.1t}$ for t.

$$3P = Pe^{0.1t}$$
$$3 = e^{0.1t}$$
$$\ln 3 = \ln e^{0.1t}$$
$$\ln 3 = 0.1t \ln e$$
$$\ln 3 = 0.1t$$
$$\frac{\ln 3}{0.1} = t$$
$$t \approx 10.9861$$

So it took approximately 11 yr for a deposit in this account to triple.

PRACTICE 5

In predicting population growth, scientists use the formula $P = P_0 e^{kt}$. This formula models the size P of a population that has an annual growth rate k (in decimal form), where t is time (in years) and P_0 is the initial population at time 0. In 2000, the population of a town was 36,695. By 2010, the population had grown to 61,102. To the nearest tenth of a percent, find the annual growth rate over this period.

In Example 5, note that since the base of the exponential function is e, we take the natural logarithm of each side of the equation.

A *Solve. Round each answer to four decimal places.*

1. $3^x = 18$

2. $6^x = 56$

3. $\left(\dfrac{1}{2}\right)^x = 9$

4. $\left(\dfrac{1}{3}\right)^x = 15$

5. $4^x = \dfrac{3}{4}$

6. $7^x = \dfrac{6}{7}$

7. $5.4^x = 0.0034$

8. $8.1^x = 0.097$

9. $2^{3x} = 4.6$

10. $5^{2x} = 7.5$

11. $100^x = 55$

12. $1000^{-x} = 95$

13. $3^{x+4} = 38$

14. $2^{x+1} = 68$

15. $5^{2x-3} = 43$

16. $6^{3x-5} = 324$

17. $e^{0.25x} = 3.1$

18. $e^{-0.012x} = 7.5$

B *Solve.*

19. $\log_3(x + 10) = 4$

20. $\log_2(x + 7) = 5$

21. $\log_5(4x - 3) = 1$

22. $\log_4(3x - 5) = 3$

23. $\log_2(x + 1) = -3$

24. $\log_3(x - 1) = -2$

25. $\log(x^2 - 21) = 2$

26. $\log_3(x^2 - 19) = 4$

27. $\log_4 x + \log_4 6 = 2$

28. $\log_2 x + \log_2 12 = 6$

29. $\log_3 x - \log_3 7 = 2$

30. $\log_5 x - \log_5 2 = 1$

31. $\log_2 x + \log_2(x - 3) = 2$

32. $\log_3(x + 6) + \log_3 x = 3$

33. $\log_2(x + 2) + \log_2(x - 5) = 3$

34. $\log_7(x - 3) + \log_7(x + 3) = 1$

35. $\log_4(2x - 1) + \log_4(6x - 1) = 2$

36. $\log(3x + 4) + \log(5x - 9) = 1$

37. $\log_3(7x) - \log_3(x - 1) = 2$

38. $\log_2(10x + 4) - \log_2 x = 4$

39. $\log_2(3x + 7) - \log_2(x - 1) = 3$

40. $\log_5(2x - 1) - \log_5(x - 8) = 1$

41. $\log_8 x + \log_8(x + 1) = \dfrac{1}{3}$

42. $\log_9(x - 6) + \log_9 x = \dfrac{3}{2}$

43. $\log_2(3x - 8) - \log_2(x + 4) = -1$

44. $\log_3(x - 9) - \log_3(2x + 3) = -2$

Mixed Practice

Solve.

45. $\log_3(x^2 - 16) = 2$

46. $\log_4 x - \log_4 3 = 2$

47. $\log_2 x - \log_2(x - 4) = 2$

48. $\log_4(x - 2) + \log_4 x = \dfrac{3}{2}$

Use a calculator to solve. Round each answer to four decimal places.

49. $\left(\dfrac{1}{4}\right)^x = 11$

50. $4^{2x+5} = 75$

Applications

C *Solve. Use a calculator, where appropriate.*

51. Between the years 2000 and 2050, the population P (in millions) of the United States is projected to grow according to the model $P = 284(1.007)^t$, where t represents the number of years after 2000. To the nearest year, when is the population of the United States expected to be 350 million people? (*Source:* U.S. Bureau of the Census)

52. In the United States, total private health care expenditures E (in billions of dollars) is projected to grow according to the model $E = 1334(1.073)^t$, where t represents the number of years after 2000. To the nearest year, when did the private expenditures reach $2500 billion? (*Source:* U.S. Centers for Medicare and Medicaid Services)

53. The compound interest formula, $A = P\left(1 + \frac{r}{n}\right)^{nt}$, is used to calculate the amount A (in dollars) in an account t yr after P dollars are invested. The interest rate r (in decimal form) is compounded n times per year. Suppose a student invests $1000 in an account at 5% interest, compounded quarterly. Approximately how long will it take for the amount in the account to double?

54. The value of a car V is declining exponentially, according to the formula $V = V_0 e^{kt}$. Here, V represents the value of the car t yr after its purchase, V_0 represents the original selling price, and k is a constant. The car is currently 2 yr old and has a value of $16,000. Brand new, it sold for $31,000. One year from now, how much, to the nearest thousand dollars, will the car be worth?

55. The pH of a substance is given by the formula $\text{pH} = -\log[\text{H}^+]$, where $[\text{H}^+]$ is the concentration of hydrogen ions. Find the approximate concentration of hydrogen ions in blood, which has a pH of 7.4.

56. The loudness L of a sound in decibels (dB) is defined by the equation $L = 10 \log\left(\dfrac{I}{I_0}\right)$, where I is the intensity of the sound (in watts per square meter) and I_0 equals 10^{-12} W/m^2. An airplane taking off has a loudness of about 140 dB. What is the intensity of the sound?

57. A magazine publisher determines that the circulation C (in thousands) of one of her magazines has grown according to the equation $C = 300e^{0.2x}$, where x represents the number of years after the magazine was launched.
   a. Solve the equation for x.
   b. Approximately how many years after it was launched did the magazine circulation reach 1,000,000?

58. The effective annual interest rate R on an investment with an annual interest rate r (expressed as a decimal), compounded continuously, is given by the equation $R = e^r - 1$.
   a. Solve the equation for r.
   b. To the nearest tenth of a percent, calculate the annual interest rate on an investment if the effective annual interest rate is 5.65%.

59. The amount A (in dollars) in an account with an annual interest rate r, compounded continuously, is given by the formula $A = 12{,}000e^{rt}$, where t is time in years.
   a. What was the initial amount invested in the account? Explain how you know.
   b. What is the annual interest rate, to the nearest tenth of a percent, if the amount in the account after 15 yr will be $31,814?

60. The number N of wild rabbits in a forest after y yr is modeled by the equation $N = 7(2^y)$.

   a. How many wild rabbits were initially present in the forest? Explain how you know.
   b. How long will it take the population to triple?

61. A company determines that its revenue R (in millions of dollars) on the sales of a new line of products is given by the equation $R = 1.7 + 2.3 \ln(x + 1)$, where x is the number of months the products have been on the market. After how many months, to the nearest month, did the company's revenue reach $7 million?

62. The time t (in hours) it takes for the initial concentration C_0 (in milligrams per liter) of a certain medication in the bloodstream to decrease to a concentration of C mg/L is modeled by the equation $t = \dfrac{1}{k}(\ln C - \ln C_0)$. Suppose that half of the initial concentration of 4 milligrams per liter remains in the bloodstream 3 hr after it is administered to a patient intravenously. Calculate the value of k to the nearest thousandth and interpret its meaning in the situation.

• Check your answers on page A-50.

MINDStretchers

Technology

1. Consider the exponential equation $2^{3x} = 6$.

a. Explain how you can use a grapher to solve this equation. Then, solve with a grapher, rounding the answer to four decimal places.

b. Verify your solution algebraically.

Mathematical Reasoning

2. Financial analysts use the *Rule of 72* to estimate how long it takes for an investment generating interest to double in value. The rule states that an investment with annual interest rate r (expressed as a percent) will double in approximately $\dfrac{72}{r}$ yr. According to this rule, an investment paying 8% annual interest, for example, will be worth twice its original value in about 9 yr, since $\dfrac{72}{8}$ equals 9. Consider the continuous interest formula $A = Pe^{rt}$, which gives the current value (in dollars) of an investment after t yr on an initial investment of P dollars with interest rate r (in decimal form). Use this formula to show why the Rule of 72 works for this case.

Groupwork

3. Working with a partner, solve the following inequalities:

a. $3^x \geq 9$

b. $\log_2(x - 1) > 1$

c. $\log(x^2 + 1) \leq 1$

d. $\log_2(x - 3) < 2$

Key Concepts and Skills

Concept/Skill	Description	Example
[9.1] Operations on functions	If f and g are functions, then **Sum:** $(f + g)(x) = f(x) + g(x)$ **Difference:** $(f - g)(x) = f(x) - g(x)$ **Product:** $(f \cdot g)(x) = f(x) \cdot g(x)$ **Quotient:** $\left(\dfrac{f}{g}\right)(x) = \dfrac{f(x)}{g(x)}$, for $g(x) \neq 0$	$f(x) = x + 2$ and $g(x) = x^2 - 4$ $\begin{aligned}(f + g)(x) &= f(x) + g(x)\\ &= (x + 2) + (x^2 - 4)\\ &= x^2 + x - 2\end{aligned}$ $\begin{aligned}(f - g)(x) &= f(x) - g(x)\\ &= (x + 2) - (x^2 - 4)\\ &= -x^2 + x + 6\end{aligned}$ $\begin{aligned}(f \cdot g)(x) &= f(x) \cdot g(x)\\ &= (x + 2)(x^2 - 4)\\ &= x^3 + 2x^2 - 4x - 8\end{aligned}$ $\begin{aligned}\left(\dfrac{f}{g}\right)(x) &= \dfrac{f(x)}{g(x)}\\[4pt] &= \dfrac{x + 2}{x^2 - 4}\\[4pt] &= \dfrac{1}{x - 2}, x \neq -2, 2\end{aligned}$
[9.1] Composition of two functions	The function represented by $f \circ g$, where $$(f \circ g)(x) = f(g(x)).$$	$f(x) = 3x + 5$ and $g(x) = 2x - 1$ $\begin{aligned}(f \circ g)(x) &= f(2x - 1)\\ &= 3(2x - 1) + 5\\ &= 6x - 3 + 5\\ &= 6x + 2\end{aligned}$
[9.1] One-to-one function	A function f in which each y-value in the range corresponds to exactly one x-value in the domain.	$f(x) = 4x + 5$ $g(x) = \sqrt{x}$
[9.1] Horizontal-line test	If every horizontal line intersects the graph of a function at most once, then the function is one-to-one.	One-to-one function: $f(x) = 3x - 3$

continued

Concept/Skill	Description	Example
		Not a one-to-one function:
[9.1] **Inverse of a one-to-one function**	The set of all ordered pairs (y, x), where f is the set of ordered pairs (x, y); written f^{-1} and read "f inverse."	If f is defined by $\{(1, 4), (3, -1), (2, 5)\}$, then f^{-1} is defined by $\{(4, 1), (-1, 3), (5, 2)\}$.
[9.1] **To find the inverse of a function**	• Substitute y for $f(x)$. • Interchange x and y. • Solve the equation for y. • Substitute $f^{-1}(x)$ for y.	$f(x) = 2x - 1$ $y = 2x - 1$ $x = 2y - 1$ $y = \dfrac{x + 1}{2}$ $f^{-1}(x) = \dfrac{x + 1}{2}$
[9.1] **Composition of a function and its inverse**	For any functions f and g, the following two conditions are equivalent: • $f(x)$ and $g(x)$ are inverse functions of one another. • $f(g(x)) = x$ and $g(f(x)) = x$.	$f(x) = \dfrac{1}{4}x + 3$ and $g(x) = 4x - 12$ $f(g(x)) = f(4x - 12)$ $\qquad = \dfrac{1}{4}(4x - 12) + 3$ $\qquad = x - 3 + 3$ $\qquad = x$ $g(f(x)) = g\left(\dfrac{1}{4}x + 3\right)$ $\qquad = 4\left(\dfrac{1}{4}x + 3\right) - 12$ $\qquad = x + 12 - 12$ $\qquad = x$
[9.2] **Exponential function**	Any function that can be written in the form $$f(x) = b^x,$$ where x is a real number, $b > 0$, and $b \neq 1$.	$f(x) = 4^x$ $g(x) = \left(\dfrac{1}{6}\right)^x$
[9.2] **Natural exponential function**	The function defined by $f(x) = e^x$.	$f(t) = e^t$
[9.2] **The one-to-one property of exponential functions**	For $b > 0$ and $b \neq 1$, if $b^x = b^n$, then $x = n$.	$2^x = 64$ $2^x = 2^6$ $x = 6$

CONCEPT	SKILL

Concept/Skill	Description	Example
[9.3] Logarithmic function	The inverse of $f(x) = b^x$ is represented by the logarithmic function $f^{-1}(x) = \log_b x$, where $\log_b x$ is read either "the logarithm of x to the base b" or "the logarithm, base b, of x."	If $f(x) = 3^x$, then $f^{-1}(x) = \log_3 x$ is the logarithmic function associated with f. $g(x) = \log_5 x$ is another logarithmic function.
[9.3] Equivalent logarithmic and exponential equations	If $b > 0$, $b \neq 1$, $x > 0$, and y is any real number, then $$y = \log_b x$$ is equivalent to $b^y = x$.	Logarithmic equation: $y = \log_4 x$ Exponential equation: $4^y = x$
[9.3] Special properties of logarithms	• $\log_b b = 1$ because $b^1 = b$, that is, 1 is the exponent to which b must be raised to get b. • $\log_b 1 = 0$ because $b^0 = 1$, that is, 0 is the exponent to which b must be raised to get 1.	$\log_5 5 = 1$ because $5^1 = 5$ $\log_5 1 = 0$ because $5^0 = 1$
[9.4] The product property of logarithms	For any positive real numbers M, N, and b, $b \neq 1$, $$\log_b MN = \log_b M + \log_b N.$$	$\log_n 6x = \log_n 6 + \log_n x$
[9.4] The quotient property of logarithms	For any positive real numbers M, N, and b, $b \neq 1$, $$\log_b \frac{M}{N} = \log_b M - \log_b N.$$	$\log_5 \frac{a}{2} = \log_5 a - \log_5 2$
[9.4] The power property of logarithms	For any positive real numbers M and b, $b \neq 1$, and any real number r, $$\log_b M^r = r \log_b M.$$	$\log_3 x^5 = 5 \log_3 x$
[9.4] Special properties of logarithms	For $b > 0$ and $b \neq 1$, • $\log_b b^x = x$ • $b^{\log_b x} = x$, where $x > 0$.	$\log_n n^3 = 3$ $n^{\log_n 7} = 7$
[9.5] Common logarithm	A logarithm to the base 10, where **log** x means $\log_{10} x$.	$\log 15 = \log_{10} 15 \approx 1.1761$ $\log \frac{1}{2} = \log_{10} \frac{1}{2} \approx -0.3010$
[9.5] Natural logarithm	A logarithm to the base e, where **ln** x means $\log_e x$.	$\ln 9 = \log_e 9 \approx 2.1972$ $\ln \frac{2}{3} = \log_e \frac{2}{3} \approx -0.4055$
[9.5] The change-of-base formula	For positive real numbers N, b, and c, where $b \neq 1$ and $c \neq 1$, $$\log_b N = \frac{\log_c N}{\log_c b}.$$	$\log_6 43 = \frac{\log 43}{\log 6} \approx 2.0992$

continued

Characteristics of the Exponential Function $f(x) = b^x$

- The domain of the function is the set of all real numbers, and the range is the set of all positive real numbers.

- The vertical- and horizontal-line tests show that $f(x) = b^x$ is a one-to-one function.

- The function is increasing if the base is greater than 1. The function is decreasing if the base is between 0 and 1.

- The y-intercept of the graph of the function is $(0, 1)$, but the graph has no x-intercept. The graph approaches the x-axis, but never intersects it.

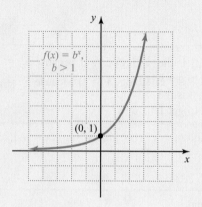 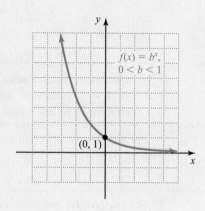

Characteristics of the Logarithmic Function $f(x) = \log_b x$

- The domain of the function is the set of positive real numbers, and the range is the set of real numbers.

- The function is increasing if the base is greater than 1. The function is decreasing if the base is between 0 and 1.

- The x-intercept of the graph of the function is $(1, 0)$, but the graph has no y-intercept. The graph approaches the y-axis, but never intersects it.

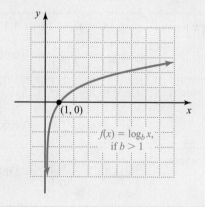 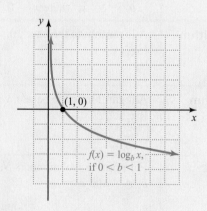

CHAPTER 9   Review Exercises

Say Why
Fill in each blank.

1. $f(x) = 4x - 9$ and $g(x) = \dfrac{x}{4} + 9$ _____ inverse
   (are/are not)
   functions because _____
   _____ .

2. A function that fails the horizontal-line test
   _____ have an inverse because _____
   (does/does not)
   _____ .

3. The function $f(x) = x^e$ _____ an exponential
   (is/is not)
   function because _____
   _____ .

4. If $5^8 = 5^x$, then x _____ equal to 8 because
   (is/is not)
   _____
   _____ .

5. The graph of $f(x) = \log_3 x$ _____ pass through
   (does/does not)
   the point $(1, 0)$ because _____
   _____ .

6. The expressions $\log_5(3 \cdot 4)$ and $(\log_5 3 + \log_5 4)$
   _____ equal because _____
   (are/are not)
   _____ .

[9.1] *Given $f(x)$ and $g(x)$, find $(f + g)(x)$, $(f - g)(x)$, $(f \cdot g)(x)$, and $\left(\dfrac{f}{g}\right)(x)$.*

7. $f(x) = 5 - 6x$; $g(x) = x - 3$

8. $f(x) = 3x^2 + 1$; $g(x) = 2x^2 - 4x$

9. $f(x) = \dfrac{2}{x - 3}$; $g(x) = \dfrac{3}{x^2 - 9}$

10. $f(x) = 7\sqrt{x} + 2$; $g(x) = \sqrt{x} - 2$

Given $f(x) = x^2 - 6x + 5$ and $g(x) = x^2 - x$, find:

11. $(f + g)(-2)$

12. $(f - g)(3)$

13. $(f \cdot g)(1)$

14. $\left(\dfrac{f}{g}\right)(-4)$

Given $f(x)$ and $g(x)$, find each composition.

15. $f(x) = x + 2$ and $g(x) = x^2 - 5x + 9$
   a. $(f \circ g)(x)$
   b. $(g \circ f)(x)$
   c. $(f \circ g)(0)$
   d. $(g \circ f)(-2)$

16. $f(x) = \sqrt{x}$ and $g(x) = 2x - 9$
   a. $(f \circ g)(x)$
   b. $(g \circ f)(x)$
   c. $(f \circ g)(9)$
   d. $(g \circ f)(5)$

Determine whether the function whose graph is shown is one-to-one.

17.

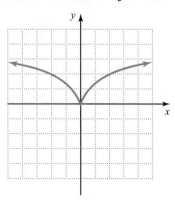

18.

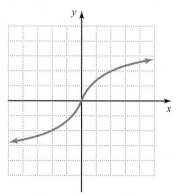

Find the inverse of each function.

19. $f(x) = 8x + 3$

20. $f(x) = \dfrac{1}{6}x - 1$

21. $f(x) = x^3 - 5$

22. $f(x) = \sqrt[3]{x + 10}$

23. $f(x) = \dfrac{2}{x + 3}$

24. $f(x) = \dfrac{x}{3x + 1}$

Determine whether the two given functions are inverses of each other.

25. $f(x) = \dfrac{2}{3}x + 2$ and $g(x) = \dfrac{3}{2}x - 3$

26. $f(x) = \dfrac{6}{x} - 4$ and $g(x) = \dfrac{6}{x} + 4$

Given the graph of a one-to-one function, sketch the graph of its inverse.

27.

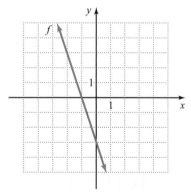

28.

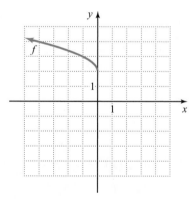

[9.2] *Evaluate each function for the given value. Round each answer to four decimal places, if necessary.*

29. $f(x) = -3^x; f(-3)$

30. $f(x) = \left(\dfrac{1}{4}\right)^{-x}; f(2)$

31. $g(x) = 2^{x+3}; g(-1)$

32. $g(x) = 9^x - 5; g\left(\dfrac{3}{2}\right)$

33. $f(x) = e^{2x}; f\left(\dfrac{1}{2}\right)$

34. $h(x) = e^{-x+1}; h(3)$

Graph each function.

35. $f(x) = 3^{-x}$

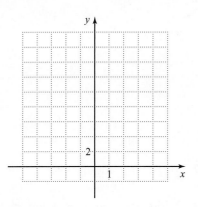

36. $f(x) = 2^x + 1$

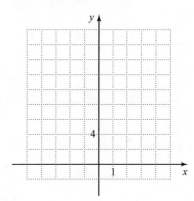

Solve.

37. $4^x = 32$

38. $27^x = \dfrac{1}{9}$

39. $2^{x-3} = 16$

40. $25^{x+2} = 125^{3-x}$

[9.3] *Write each logarithmic equation as an exponential equation.*

41. $\log_6 216 = 3$

42. $\log_{1/3} 9 = -2$

Write each exponential equation as a logarithmic equation.

43. $5^{-2} = \dfrac{1}{25}$

44. $81^{1/4} = 3$

Evaluate.

45. $\log_8 8$

46. $\log_{1/2} \dfrac{1}{32}$

47. $\log_6 1$

48. $\log_{16} 2$

Solve.

49. $\log_4 x = 3$

50. $\log_2 x = -6$

51. $\log_x 7 = -1$

52. $\log_x \dfrac{4}{9} = 2$

53. $\log_4 64 = x$

54. $\log_{1/2} 4 = x$

Graph each function.

55. $f(x) = -\log_4 x$

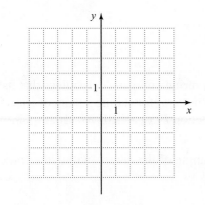

56. $f(x) = \log_2 x$

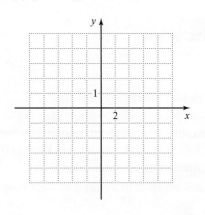

[9.4] *Use the properties of logarithms to write each expression as the sum or difference of logarithms.*

57. $\log_3 27x$

58. $\log_6 \dfrac{6}{a}$

59. $\log_n \dfrac{x}{x-5}$

60. $\log_9\big[u(u+1)\big]$

61. $\log_5 x^3 y$

62. $\log_2 \sqrt[5]{x^2 y^3}$

63. $\log_3 \dfrac{x^2 z^3}{y^2}$

64. $\log_b \dfrac{b^2}{(a-b)^4}$

Write each expression as a single logarithm.

65. $\log_4 10 + \log_4 n$

66. $\log_9 x - \log_9 5$

67. $\log_3 a + 2 \log_3 b$

68. $5 \log_b a - 4 \log_b c$

69. $\dfrac{1}{4}(3 \log_6 x - 8 \log_6 y)$

70. $2 \log_a x - \log_a y - 4 \log_a z$

Evaluate.

71. $\log_8 8^3$

72. $\log_x x^{-1/3}$

73. $5^{\log_5 y}$

74. $a^{\log_a 14}$

[9.5] *Use a calculator to approximate each logarithm to four decimal places.*

75. $\log 23$

76. $\log 9.4$

77. $\ln 48$

78. $\ln \dfrac{2}{3}$

Evaluate.

79. $\log 0.1$

80. $\log \sqrt[4]{1000}$

81. $\ln e^{100}$

82. $\ln \dfrac{1}{e^y}$

Use a calculator to approximate each logarithm to four decimal places.

83. $\log_5 121$

84. $\log_9 6.1$

[9.6] *Solve. Round each answer to four decimal places.*

85. $7^x = 72$

86. $6^{2x} = 0.58$

87. $2^{x-5} = 20$

88. $3^{2x+3} = 63$

Solve.

89. $\log_7(x+8) = 0$

90. $\log_2 12 - \log_2 x = -3$

91. $\log_6 x + \log_6(x-5) = 2$

92. $\log_5(x+2) + \log_5(x+6) = 1$

93. $\log_4(x-3) - \log_4 x = -1$

94. $\log_3(2x-1) - \log_3(x-4) = 2$

Mixed Applications

Solve.

95. A real estate agent makes a monthly base salary of $2500 plus a commission c on his total sales for the month. His total monthly salary can be modeled by the function $f(c) = 2500 + c$. The function $c(s) = 0.008s$ represents his commission on his total sales s.

   a. Find $(f \circ c)(s)$, and interpret its meaning in this situation.

   b. Determine the agent's total monthly salary if his monthly sales totaled $400,000.

96. A clothing manufacturing company determines that its weekly profit P (in dollars) for selling x pairs of jeans is modeled by the function $P(x) = 10x - 150$.

   a. Find $P^{-1}(x)$.

   b. Explain the meaning of the inverse within the context of this situation.

   c. How many pairs of jeans must the company sell in order to make a weekly profit of $1200?

97. The compound interest formula $A = P\left(1 + \dfrac{r}{n}\right)^{nt}$

is used to compute amount A (in dollars) in an account after P dollars are invested. The annual interest rate r (expressed as a decimal) is compounded n times per year for t yr.

   a. If an investor puts $6500 in an account earning 5.4% interest, compounded twice a year, how much will be in the account after 5 yr?

   b. To the nearest year, how long will it take for this investment to double?

98. A stamp collector determines that the value of a particular stamp quadruples every 15 yr. The value of the stamp v (in dollars) in t yr is given by the equation

$t = 15 \log_4 \dfrac{v}{v_0}$, where v_0 is the initial value of the stamp.

If the initial value of the stamp is $3, to the nearest tenth of a year, when will it have a value of $24?

99. Students in a physics class were given a final exam at the end of the course. As part of a study on memory, the students were given an equivalent exam each month for 1 yr. The average exam score for the class on a test given x months after the course ended is modeled by the equation $s = \log \dfrac{10^{90}}{(x + 1)^{16}}$. Use the properties of logarithms to show that the equation can be written as $s = 90 - 16 \log(x + 1)$.

100. To determine the approximate time of death of a person, a medical examiner uses the equation $\ln(T - A) = \ln(98.6 - A) - kt \ln e$, where T is the temperature t hr after death, A is the temperature of the surrounding environment, and k is a constant. Show that this equation can be written as $T - A = (98.6 - A)e^{-kt}$.

101. In chemistry, the pH of a substance can be determined using the formula $\text{pH} = -\log[\text{H}^+]$, where $[\text{H}^+]$ represents the hydrogen ion concentration. To the nearest tenth, determine the pH of milk, which has a hydrogen ion concentration of 2.5×10^{-7}.

102. The time t (in years) it takes for a continuously compounded investment of P dollars at an annual interest rate r (in decimal form) to grow to x times the original amount is modeled by the equation $t = \dfrac{\ln x}{r}$. How long, to the nearest year, will it take an investment earning 4.4% interest to triple?

103. A car purchased new for A dollars depreciates according to the equation $V = A(0.85)^t$, where V is the value of the car t yr after it is purchased. To the nearest year, how many years after a car is purchased new for $28,500 will its value depreciate to $15,000?

104. The amount A (in dollars) in an account at time t (in years), with an annual interest rate r compounded continuously, is given by the formula $A = Pe^{rt}$, where P is the initial amount invested. If $8000 is invested in an account at 4.8% interest compounded continuously, then to the nearest year, how long will it take the amount in the account to grow to $12,300?

• Check your answers on page A-50.

CHAPTER 9   Posttest

FOR EXTRA HELP

CHAPTER Test Prep VIDEOS

The Chapter Test Prep Videos with test solutions are available on DVD, in MyMathLab, and on YouTube™ (search "AkstIntermediate Alg" and click on "Channels").

To see if you have mastered the topics in this chapter, take this test. Use a calculator, where necessary.

1. Consider the functions $f(x) = 3x^2 - 4x - 4$ and $g(x) = 3x^2 + 2x$. Find:

   a. $(f + g)(x)$ **b.** $(f - g)(x)$

   c. $(f \cdot g)(x)$ **d.** $\left(\dfrac{f}{g}\right)(x)$

2. Given $f(x) = \sqrt{x}$ and $g(x) = 2x + 5$, find:

   a. $(f \circ g)(x)$ **b.** $(g \circ f)(x)$

   c. $(f \circ g)(2)$ **d.** $(g \circ f)(16)$

3. Determine whether the following function whose graph is shown is one-to-one. If it is, sketch the graph of its inverse.

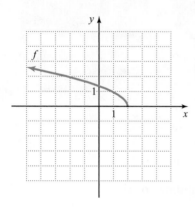

4. Find the inverse of the one-to-one function
$$f(x) = \frac{1}{x - 7}.$$

5. Evaluate each function for the given value. Round to four decimal places.

   a. $f(x) = 3^x - 2; f(-2)$
   b. $f(x) = e^{x+2}; f(5)$

6. Evaluate:

   a. $\log_9 1$ **b.** $\log_4 \dfrac{1}{2}$

7. Graph:

   a. $f(x) = -\left(\dfrac{1}{2}\right)^{-x}$

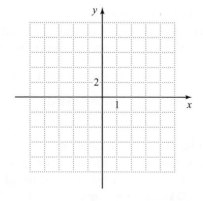

   b. $f(x) = -\log_3 x$

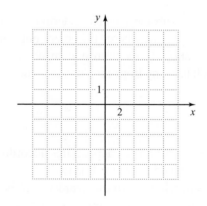

8. Write each expression as a sum or difference of logarithms.

a. $\log_3(9x^4)$

b. $\log_6 \dfrac{x^5}{6y^2}$

9. Write each expression as a single logarithm.

a. $2 \log_5 n + 5 \log_5 2$

b. $3 \log_b(a - 2) - 4 \log_b(a + 2)$

10. Evaluate:

a. $\log_9 9^7$ **b.** $6^{\log_6 x}$

11. Evaluate:

a. $\log \sqrt{1000}$ **b.** $\ln \dfrac{1}{e^5}$

12. Evaluate, rounding to four decimal places: $\log_5 42$

Solve.

13. $7^x = 49$

14. $27^{2x-3} = \left(\dfrac{1}{3}\right)^{x-5}$

15. $\log_x 4 = -2$

16. $\log_5(2x + 3) = 2$

17. $\log_2(x + 3) + \log_2(x - 4) = 3$

18. The population P (in millions) of India is modeled by the function $P(t) = 1108(1.014)^t$, where t represents the number of years after 2006. Calculate the predicted population of India in 2014. (*Source:* U.S. Bureau of the Census, International Database)

19. The loudness L (in decibels) of a sound can be determined by the equation $L = 10(\log I + 12)$, where I is the intensity of the sound (in watts per square meter). Normal conversation has a loudness of 60 dB. What is the intensity of the sound?

20. If the initial concentration of a certain drug in a patient's bloodstream is 20 mg/L, then the concentration C after t hr is given by the equation $C = 20e^{-0.125t}$. To the nearest hour, approximate the time it takes for the concentration to decrease to 5 mg/L.

• Check your answers on page A-51.

Cumulative Review Exercises

To help you review, solve the following:

1. Solve: $3(2x - 7) - 4x = -\dfrac{1}{3}(9x + 12) - 8$

2. Solve: $|5x - 1| > 9$

3. Find the point-slope form and the slope-intercept form of the equation of the line that contains the points $(4, 1)$ and $(-2, 7)$.

4. Solve by graphing: $4y - x = -8$
$4y + 2x = 4$

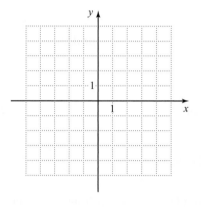

5. Factor: $81x^4y^4 - 16$

6. Solve by factoring: $5t^2 + 7t - 6 = 0$

7. Find the constant of variation and the variation equation, if y varies inversely as the square of x and directly as z

and if $y = 1$ when $x = 4$ and $z = 8$.

8. Simplify, if possible: $\dfrac{1}{3}\sqrt{162x} + \sqrt{50x^5}$

9. Multiply: $(4 - \sqrt{-16})(7 - \sqrt{-1})$

10. Solve: $\sqrt{x^2 - 16} = \sqrt{3x + 12}$

11. Solve: $2x^2 + 4x - 5 = 0$

12. Solve $3x^2 - 7x + 2 \geq 0$ and graph the solutions.

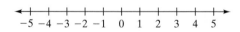

13. Solve: $\log_{\frac{1}{4}} x = -2$

14. Write as a single logarithm:
$2\log_3(a + 2) - \log_3(a^2 - 4)$

15. Solve: $\log_4(x + 3) - \log_4 x = 2$

16. Meteorologists use the formula

$$W = 35.74 + 0.6215T - 35.75V^{0.16} + 0.4275TV^{0.16},$$

where T is the air temperature (in degrees Fahrenheit) and V is the speed of the wind (in miles per hour), to calculate the windchill temperature W (in degrees Fahrenheit). What is the windchill temperature, to the nearest degree Fahrenheit, if the air temperature is 40°F and the speed of the wind is 18 mph?

17. A manufacturer of ice cream cones plans to double the capacity of his cones by increasing the radius of the cone without changing its height.

a. If the radii of the new and old cones are R and r, respectively, express R in terms of r. (Reminder: The capacity (volume) of a cone is given by the formula $V = 1/3\ \pi r^2 h$, where r is the radius and h is the height.)

b. If $r = 30$ mm, find R to the nearest millimeter.

18. A boat in the Bering Sea is sinking. From a platform 30 ft above the water, the captain launches a distress flare straight upward, with an initial velocity of 175 ft/sec. The height of the flare (in feet) is modeled by the function $f(t) = -16t^2 + 175t + 30$.

a. To the nearest second, how long will it take the flare to reach its maximum height?

b. To the nearest hundred feet, determine the maximum height of the flare.

19. Three sixth-grade classes and adult chaperones, to include a total of 67 people, want to take a field trip to the top of the Prudential Center observation deck in Boston. The school has budgeted $592 for the tickets, which cost $12 for an adult and $8 for a child. Determine how many adults and how many children can go on the trip for the school to be exactly on budget.

20. Determine each of the following functions:
a. $d(x)$, which represents the selling price of an item after applying a 20% discount to the regular price x

b. $s(x)$, which represents the selling price, including a 6% sales tax, of an item that sells for x prior to applying the tax

c. $(s \circ d)(x)$, which represents the selling price of an item after both the discount and the sales tax are applied

• Check your answers on page A-51.

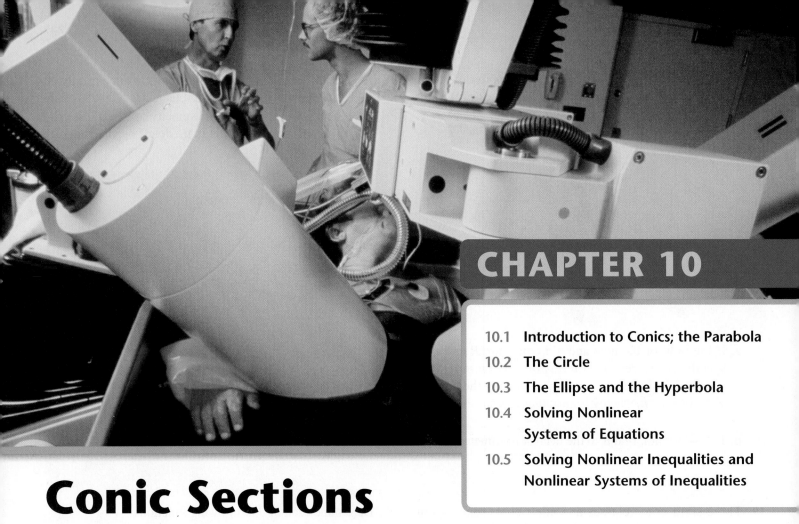

CHAPTER 10

Conic Sections

Medicine and Conic Sections

The ellipse, a kind of conic section, has a surprising property that underlies many applications: Any light, sound, or energy that emanates from one of the foci and reflects off the mirrored surface of the ellipse will be reflected to the other focus.

As shown in the diagram, this property is used in the medical procedure known as *lithotripsy*, in which a kidney stone is pulverized by placing the patient in an elliptical tank of water, with the kidney stone at one focus. At the other focus, high-energy sound waves, which are concentrated on the kidney stone, are generated.

(*Sources:* mwstone.com; Davi-Ellen Chabner, *The Language of Medicine*, W. B. Saunders, 1996)

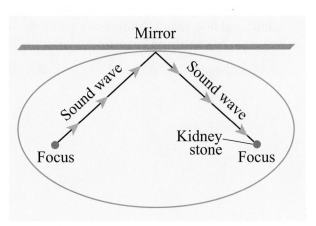

1. Write the equation of the parabola $y = 3x^2 - 12x + 23$ in standard form. Then, identify the vertex.

2. Find the distance between each pair of points.
   a. $(-5, 9)$ and $(-7, 13)$
   b. $(1.2, -1)$ and $(0.8, -0.7)$

3. Find the midpoint of the line segment joining each pair of points.
   a. $(11, -6)$ and $(-3, 10)$
   b. $(1, -4)$ and $(0, 1)$

4. Find the equation of a circle with center $(7, -3)$ and radius $\sqrt{10}$.

5. Find the center and radius of the circle $x^2 + y^2 - 2x - 5 = 0$.

Graph.

6. $x = -2(y + 1)^2 + 3$

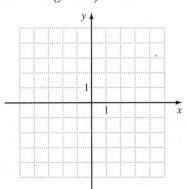

7. $x^2 + y^2 = 81$

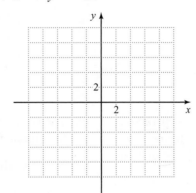

8. $(x + 2)^2 + (y - 1)^2 = 16$

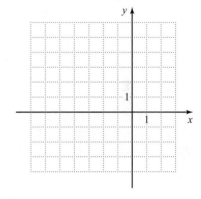

9. $\dfrac{x^2}{9} + y^2 = 1$

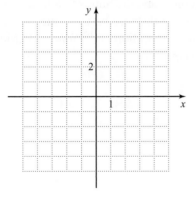

10. $12x^2 + 3y^2 = 48$

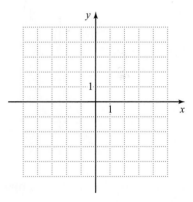

11. $\dfrac{y^2}{25} - \dfrac{x^2}{9} = 1$

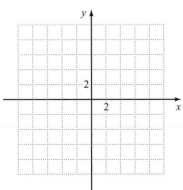

12. $x^2 - 4y^2 = 16$

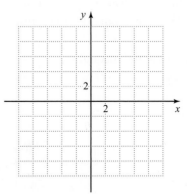

Solve.

13. $x^2 + y^2 = 9$
$\quad x = y - 3$

14. $y = x^2 - 2$
$\quad y = -x^2 - 6$

15. $4x^2 + 2y^2 = 26$
$\quad 5x^2 - 2y^2 = 28$

Solve by graphing.

16. $x^2 - 9y^2 < 36$

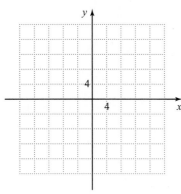

17. $x^2 + (y + 3)^2 \le 4$
$\quad 8x^2 + 32y^2 \ge 128$

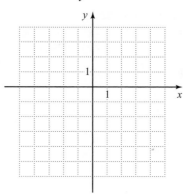

18. The cross section of a television satellite dish is parabolic. Find an equation that represents the parabola formed by the cross section of the satellite dish shown in the diagram below.

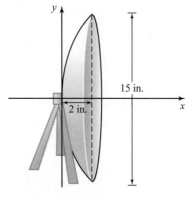

19. The comet Encke travels in an elliptical orbit with the Sun at one focus. An equation that approximates the comet's orbit is $\dfrac{x^2}{354^2} + \dfrac{y^2}{181^2} = 1$, where the units are in millions of kilometers. The maximum distance between the comet and the Sun is 658 million km. What is the minimum distance between the comet and the Sun, to the nearest 10 million km? (*Source:* Patrick Moore, *Guide to Comets*)

20. The security officers on a campus communicate using walkie-talkies. The walkie-talkie has a maximum range of 1000 ft.

a. Write an inequality to represent the situation.

b. Graph the inequality.

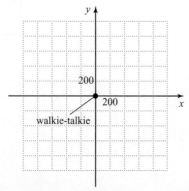

• Check your answers on page A-52.

10.1 Introduction to Conics; the Parabola

What Conics Are and Why They Are Important

The *conic sections* are among the earliest mathematical objects to be systematically studied. Originally viewed in geometric terms, each **conic section** was defined as the intersection of a plane and a (double) cone. Changing the angle at which the plane slices the cone results in a different curve. Depending on the angle, the curve could be a *parabola*, a *circle*, an *ellipse,* or a *hyperbola*.

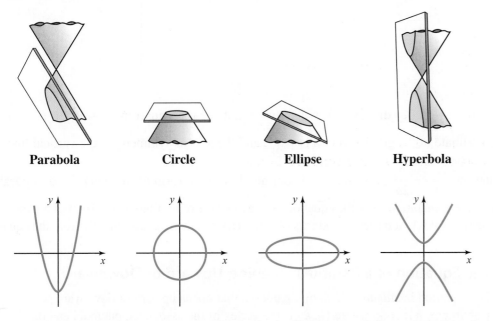

| Parabola | Circle | Ellipse | Hyperbola |

Conic sections can also be represented algebraically by second-degree equations in two variables. For instance, the following equations represent particular conic sections:

$$y = x^2 \qquad x^2 + y^2 = 1 \qquad \frac{x^2}{4} + \frac{y^2}{9} = 1 \qquad \frac{x^2}{9} - \frac{y^2}{16} = 1$$

Parabola   Circle   Ellipse   Hyperbola

Much of this chapter is devoted to investigating equations of conic sections and finding their graphs on a coordinate plane.

The first few sections focus on the conic sections the parabola, the circle, the ellipse, and the hyperbola, in that order. In the rest of the chapter, we consider how to solve *nonlinear* systems of equations, that is, systems containing equations that are not linear, and how to solve nonlinear inequalities and nonlinear systems of inequalities.

The range of applications of conic sections is wide, from the path of thrown objects to medical procedures, from the shape of objects in nature to that of the most modern architectural structures.

The Parabola

The first of the conic sections that we will discuss is one that we have considered previously, namely the *parabola*. Let's review the earlier discussion before extending it.

Recall from Section 8.4 that the graph of a quadratic equation such as $y = x^2 + 3x - 1$ is a U-shaped curve called a parabola. Every parabola has a *vertex* and an *axis of symmetry,* features of the curve that are useful in graphing the corresponding equation. In general, the graph of $y = ax^2 + bx + c$ opens upward if a is positive and downward if a is negative. In either case, the axis of symmetry is parallel to the y-axis.

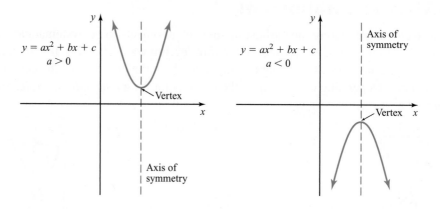

In our earlier discussion, we showed that for the equation $y = ax^2 + bx + c$, the x-coordinate of the graph's vertex is $-\dfrac{b}{2a}$, and the axis of symmetry is the vertical line with equation $x = -\dfrac{b}{2a}$. Now, let's consider an alternative form of the equation of a parabola called the *standard form* of the equation. The advantage of writing the equation of a parabola in standard form is that its vertex and axis of symmetry can then be read directly from the equation.

The Equation of a Parabola Opening Upward or Downward

The equation in *standard form* of a parabola that opens upward or downward is $y = a(x - h)^2 + k$, where (h, k) is the vertex of the associated parabola and the equation of the axis of symmetry is $x = h$.

EXAMPLE 1

Graph: $y = -2(x + 3)^2 + 8$

Solution   The given equation can be written in standard form as $y = -2[x - (-3)]^2 + 8$, where $a = -2$, $h = -3$, and $k = 8$. So the vertex is the point $(-3, 8)$, and the axis of symmetry is the line with equation $x = -3$.

To graph, let's find the intercepts. To identify any y-intercepts, we substitute 0 for x.

$$y = -2(x + 3)^2 + 8$$
$$y = -2(0 + 3)^2 + 8$$
$$y = -10$$

So the y-intercept is the point $(0, -10)$.

PRACTICE 1

Graph: $y = 5(x + 1)^2 - 3$

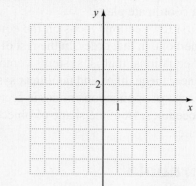

To find any x-intercepts, we substitute 0 for y.

$$y = -2(x + 3)^2 + 8$$
$$0 = -2(x + 3)^2 + 8$$
$$-8 = -2(x + 3)^2$$
$$4 = (x + 3)^2$$
$$\pm 2 = x + 3$$

$$2 = x + 3 \quad \text{or} \quad -2 = x + 3$$
$$-1 = x \qquad\qquad -5 = x$$
$$x = -1 \qquad\qquad x = -5$$

So the x-intercepts are $(-1, 0)$ and $(-5, 0)$.

We know that the graph of our equation is a parabola that opens downward, since a is negative. To sketch the graph, first plot the vertex and the x- and y-intercepts. Then, find a few additional points on each side of the axis of symmetry, and plot them.

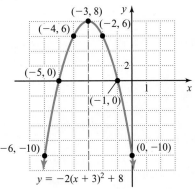

When the equation of a parabola is given in the form $y = ax^2 + bx + c$, we can rewrite the equation in standard form by completing the square. Consider, for instance, the equation $y = 4x^2 - 24x - 13$.

To write this equation in standard form, we first group the x-terms and factor out the coefficient of x^2.

$$y = (4x^2 - 24x) - 13$$
$$y = 4(x^2 - 6x) - 13$$

Next, we complete the square for $x^2 - 6x$ by taking one-half of the coefficient of the x-term and squaring it. That is, $\left[\dfrac{1}{2}(-6)\right]^2 = (-3)^2 = 9$. Now, we add this value to $x^2 - 6x$ in the parentheses, which gives us $x^2 - 6x + 9$. Since each term of the trinomial is multiplied by 4, when we add 9 within the parentheses we are actually *adding* 4(9), or 36. In order to keep the value of the right side of the equation unchanged, we must also *subtract* the same amount 36 outside the parentheses since the terms are on the same side of the equal sign, as shown below:

Add 4(9) = 36.

$$y = 4(x^2 - 6x + 9) - 13 - 36$$

Subtract 4(9) = 36.

Finally, we write the perfect square trinomial $x^2 - 6x + 9$ as the square of a binomial and simplify, getting:

$$y = 4(x - 3)^2 - 49$$

So the equation $y = 4x^2 - 24x - 13$ written in standard form is $y = 4(x - 3)^2 - 49$.

EXAMPLE 2

Consider the equation $y = 3x^2 - 6x + 8$, whose graph is a parabola.

a. Write this equation in standard form.

b. Identify the vertex and the axis of symmetry of the parabola.

c. Graph this parabola.

Solution

a. To write an equation of the form $y = ax^2 + bx + c$ in standard form, we complete the square as follows:

$$y = 3x^2 - 6x + 8$$
$$y = (3x^2 - 6x) + 8 \qquad \text{Group the } x\text{-terms.}$$
$$y = 3(x^2 - 2x) + 8 \qquad \text{Factor out the coefficient of } x^2.$$

Add $3(1) = 3$.

$$y = 3(x^2 - 2x + 1) + 8 - 3 \qquad \text{Complete the square of the polynomial in parentheses, and subtract } 3(1), \text{ or } 3,$$

Subtract $3(1) = 3$. $\qquad$ to keep the value of the right side of the equation unchanged.

$$y = 3(x - 1)^2 + 5 \qquad \text{Write } x^2 - 2x + 1 \text{ as the square of a binomial and simplify.}$$

So the equation $y = 3x^2 - 6x + 8$ written in standard form is $y = 3(x - 1)^2 + 5$.

b. From the standard form of the equation, we see that $h = 1$, and $k = 5$. So the vertex of the parabola is the point $(1, 5)$ and the axis of symmetry is $x = 1$.

c. To graph the parabola, let's find the intercepts. To find any y-intercepts, we substitute 0 for x.

$$y = 3x^2 - 6x + 8$$
$$y = 3(0)^2 - 6(0) + 8$$
$$y = 8$$

So the y-intercept is $(0, 8)$.
To find any x-intercepts, we substitute 0 for y.

$$y = 3x^2 - 6x + 8$$
$$0 = 3x^2 - 6x + 8$$

Recall from Section 8.2 that we can use the discriminant of a quadratic equation to determine if the solutions are real numbers or complex numbers (containing i). Evaluating the discriminant $b^2 - 4ac$ of the equation $0 = 3x^2 - 6x + 8$ yields:

$$(-6)^2 - 4(3)(8) = -60$$

Since the discriminant is negative, the equation has no real solution. So the parabola has no x-intercept.

PRACTICE 2

The graph of the equation $y = 4x^2 - 8x + 7$ is a parabola.

a. Write this equation in standard form.

b. Identify the vertex and the axis of symmetry of the parabola.

c. Graph this parabola.

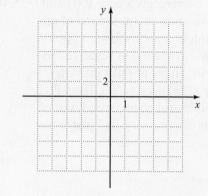

We know that the graph of our equation is a parabola that opens upward, since a is positive. To sketch the graph, let's first plot the vertex and the y-intercept. Then, we plot a few additional points on each side of the axis of symmetry.

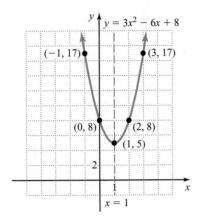

So far in our discussion of parabolas we have only considered parabolas that open either upward or downward. However, some parabolas open to the left or to the right. Equations of the form $x = ay^2 + by + c$ have as their graphs parabolas that open to the right if $a > 0$ and to the left if $a < 0$. In either case, the axis of symmetry is parallel to the x-axis.

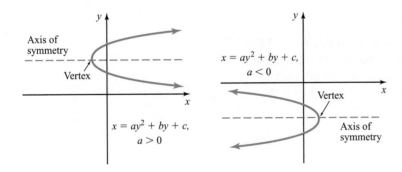

Just as we can write in standard form the equation of a parabola that opens upward or downward, we can also write in standard form the equation of a parabola that opens either to the left or to the right.

The Equation of a Parabola Opening to the Left or Right

The equation in *standard form* of a parabola that opens to the left or right is $x = a(y - k)^2 + h$, where (h, k) is the vertex of the associated parabola and the equation of the axis of symmetry is $y = k$.

EXAMPLE 3

Graph: $x = 2(y - 2)^2 - 2$

Solution The given equation is of the form $x = a(y - k)^2 + h$, where $h = -2$ and $k = 2$. So the vertex is the point $(-2, 2)$, and the axis of symmetry is the horizontal line with equation $y = 2$.

To graph, let's find the intercepts. To identify any y-intercepts, we substitute 0 for x.

$$x = 2(y - 2)^2 - 2$$
$$0 = 2(y - 2)^2 - 2$$
$$2 = 2(y - 2)^2$$
$$1 = (y - 2)^2$$
$$\pm 1 = y - 2$$
$$1 = y - 2 \quad \text{or} \quad -1 = y - 2$$
$$y = 3 \quad | \quad y = 1$$

So there are two y-intercepts, $(0, 3)$ and $(0, 1)$.

To find any x-intercepts, we substitute 0 for y.

$$x = 2(y - 2)^2 - 2$$
$$x = 2(0 - 2)^2 - 2$$
$$x = 2(-2)^2 - 2$$
$$x = 6$$

So there is only one x-intercept, $(6, 0)$.

Since $a = 2$ is positive, the graph opens to the right. We plot the vertex, the x- and y-intercepts, and a few additional points on each side of the axis of symmetry, $y = 2$. Then, we sketch the smooth curve passing through them, obtaining the graph of $x = 2(y - 2)^2 - 2$.

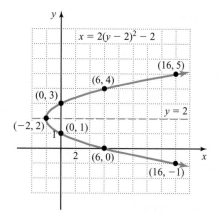

PRACTICE 3

Graph: $x = -(y - 3)^2 + 4$

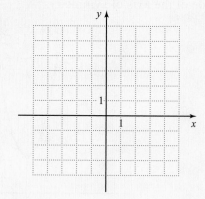

Note that the parabola in Example 3 is not the graph of a function. Can you explain why?

EXAMPLE 4

Write the equation of the parabola $x = -4y^2 + 16y - 3$ in standard form. Then, identify its vertex.

Solution To write the given equation in standard form, we complete the square.

$$x = -4y^2 + 16y - 3$$
$$x = (-4y^2 + 16y) - 3 \qquad \text{Group the } y\text{-terms.}$$
$$x = -4(y^2 - 4y) - 3 \qquad \text{Factor out the coefficient of } y^2.$$

Add $-4(4) = -16$.

$$x = -4(y^2 - 4y + 4) - 3 - (-16) \qquad \text{Complete the square of the polynomial in parentheses, and}$$

Subtract $-4(4) = -16$.

subtract $-4(4)$, or -16, to keep the value of the right side of the equation unchanged.

$$x = -4(y - 2)^2 + 13 \qquad \text{Write } y^2 - 4y + 4 \text{ as the square of a binomial, and simplify.}$$

So the equation of the parabola $x = -4y^2 + 16y - 3$ in standard form is $x = -4(y - 2)^2 + 13$. The vertex of the graph of this equation is the point $(13, 2)$.

EXAMPLE 5

The Gateway Arch, seen in the photo below, was built in 1965 and has become a symbol of the city of St. Louis. Approximately parabolic in shape, the arch is 630 ft high and 630 ft wide at the base. (See diagram.)

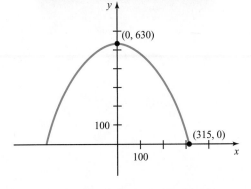

Express in standard form the equation of the parabola $x = -5y^2 - 20y - 9$. Find its vertex.

A toy rocket is launched straight upward from the ground with an initial velocity of 64 ft/sec. Two seconds after it is launched, the rocket reaches its maximum height of 64 ft, as shown.

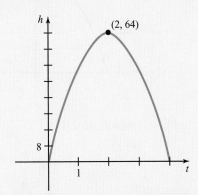

a. Find an equation in standard form for the height h of the rocket t sec after it is launched.

b. Find the height of the rocket 1 sec after it is launched.

EXAMPLE 5 (continued)

a. Find the equation of the arch.

b. Find the height of the arch 105 ft from either base.

Solution

a. Because the arch is approximately parabolic, we can write its equation in standard form: $y = a(x - h)^2 + k$. We know that the vertex of the parabola is the point $(0, 630)$, so $h = 0$ and $k = 630$. Therefore, the equation is $y = a(x - 0)^2 + 630$, or $y = ax^2 + 630$. Since the width of the parabola at its base is 630 ft, the point $(315, 0)$ lies on the graph. We can substitute the coordinates of the point $(315, 0)$ into the equation and solve for a.

$$y = ax^2 + 630$$
$$0 = a(315)^2 + 630$$
$$0 = 99{,}225a + 630$$
$$-630 = 99{,}225a$$
$$\frac{-630}{99{,}225} = a$$
$$a = -\frac{2}{315}$$

So the equation of the arch is $y = -\dfrac{2}{315}x^2 + 630$.

b. We need to find the height of the arch 105 ft from either base, in particular, above the point on the x-axis with x-coordinate $315 - 105$, or 210.

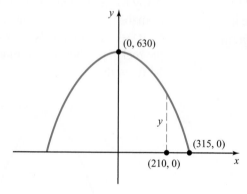

To find y, we substitute 210 for x:

$$y = -\frac{2}{315}x^2 + 630$$
$$y = -\frac{2}{315}(210)^2 + 630$$
$$y = -280 + 630$$
$$y = 350$$

So the height of the arch 105 ft from either base is 350 ft.

Mathematically Speaking

Fill in each blank with the most appropriate term or phrase from the given list.

| solving for *x* | function | upward or downward |
| to the left or to the right | completing the square | equation |

1. The equation of a parabola that opens _____ is in standard form if it is written as $y = a(x - h)^2 + k$, where (h, k) is the parabola's vertex and $x = h$ is the equation of the axis of symmetry.

2. When the equation of a parabola is given in the form $y = ax^2 + bx + c$, it can be written in standard form by _____.

3. The equation of a parabola that opens _____ is in standard form if it is written as $x = a(y - k)^2 + h$, where (h, k) is the parabola's vertex and $y = k$ is the equation of the axis of symmetry.

4. A parabola that opens to the left or to the right is not the graph of a(n) _____.

A *Identify the vertex and axis of symmetry of each parabola. Then, graph.*

5. $y = (x - 3)^2 - 4$

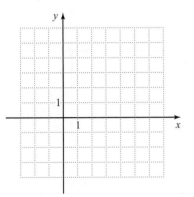

6. $y = (x - 1)^2 + 1$

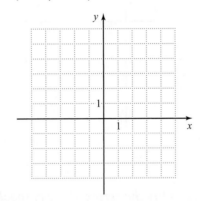

7. $y = -\dfrac{1}{2}(x + 2)^2 - 5$

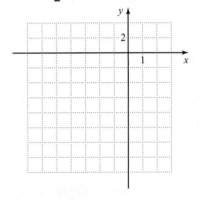

8. $y = -\dfrac{1}{3}(x + 3)^2 - 1$

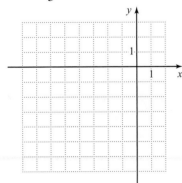

9. $x = 4(y - 1)^2$

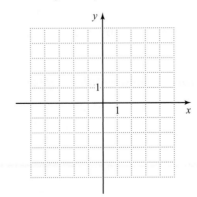

10. $x = 2y^2 - 8$

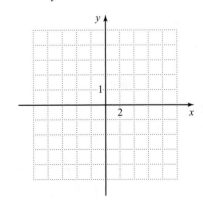

11. $x = -y^2 + 2y + 2$

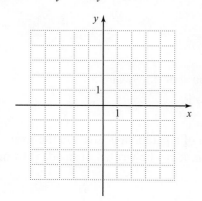

12. $x = -y^2 - 4y + 1$

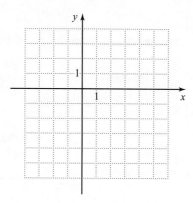

13. $y = -2x^2 + 12x - 10$

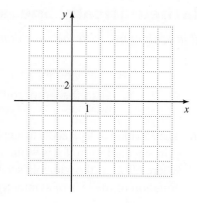

14. $y = 3x^2 - 6x - 9$

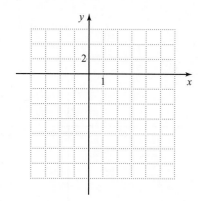

15. $x = 3y^2 + 12y + 10$

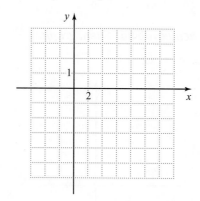

16. $x = -4y^2 + 8y - 1$

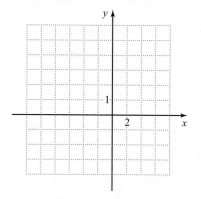

B *Write each equation in standard form. Then, identify the vertex of the parabola.*

17. $y = x^2 - x + 1$

18. $y = x^2 + 3x - 5$

19. $y = 5x^2 - 50x + 57$

20. $y = 4x^2 + 56x + 207$

21. $x = -2y^2 - 32y - 95$

22. $x = -3y^2 + 36y - 136$

23. $x = 3y^2 + 9y + 11$

24. $x = 6y^2 - 6y + 5$

Find the equation in standard form of the parabola whose graph is shown.

25.

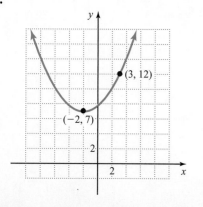

26.

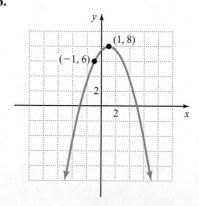

27.

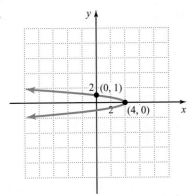

28.

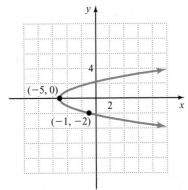

Mixed Practice

Write each equation in standard form. Then, identify the vertex of the parabola.

29. $y = x^2 + 5x + 2$

30. $x = -4y^2 + 24y - 40$

31. Find the equation in standard form of the parabola whose graph is shown below.

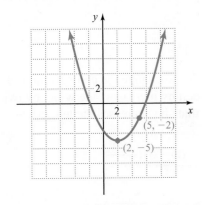

Identify the vertex and axis of symmetry of each parabola. Then, graph.

32. $y = \dfrac{1}{2}(x + 3)^2 + 2$

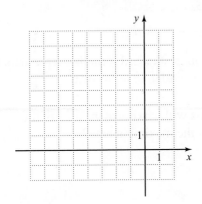

33. $x = -2y^2 - 4y + 1$

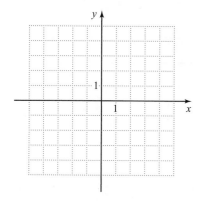

34. $y = -2x^2 + 4x - 6$

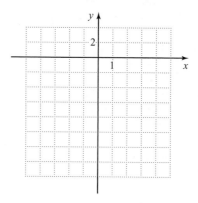

Applications

C *Solve.*

35. The owner of a cosmetics store determines that her weekly revenue R (in dollars) from selling x bottles of perfume is given by the equation of the parabola $R = -(x - 40)^2 + 1600$. Identify the vertex of the parabola, and explain its significance in this situation.

36. The height h (in feet) of an object relative to the point of release t sec after it is thrown straight upward with an initial velocity of 32 ft/sec is given by the equation of the parabola $h = -16(t - 1)^2 + 16$. Identify the vertex of the parabola and explain its significance in this situation.

37. A farmer has 500 ft of fencing to enclose a rectangular grazing field.

  a. Write an equation in standard form that represents the area A of the enclosed field in terms of the width, w, of the field.

  b. What dimensions will produce a field with a maximum area? What is the maximum area?

38. A homeowner decides to change the configuration of her rectangular garden by increasing the width by twice the number of feet that she decreases the length. (See figure.)

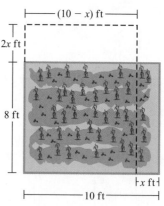

  a. Write an equation in standard form that represents the area A of the new garden in terms of x.

  b. What dimensions will maximize the area of the garden? What is the maximum area?

39. The cross section of a radio telescope is parabolic. Find an equation that represents the parabola formed by the cross section of the radio telescope shown in the diagram.

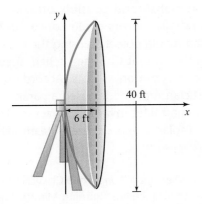

40. The cables suspended between two towers of a suspension bridge form a curve that can be approximated by a parabola. Suppose that plans for a new suspension bridge require the distance between two towers that rise 116 m above the road surface to be 1120 m. If the center of the cable is 4 m above the road surface, find an equation in standard form that represents the parabola formed by the cable, as shown in the diagram.

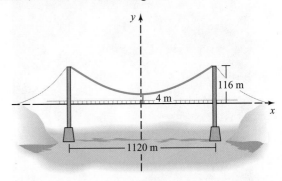

• Check your answers on page A-52.

MIND*Stretchers*

Research

1. Either in your college library or on the web, investigate the contribution to the study of conic sections by the Greek mathematician Apollonius of Perga, who lived more than 2000 years ago.

Writing

2. In addition to the four conic sections shown at the beginning of the section, there are three "degenerate" conic sections (extreme cases)—a point, two intersecting straight lines, and a single straight line. Explain how a plane could intersect a double cone to produce each of these.

Mathematical Reasoning

3. Consider a fixed line l and a fixed point F not on the line. It can be shown that the set of all points on a coordinate plane that are the same distance from F and from l form a parabola.

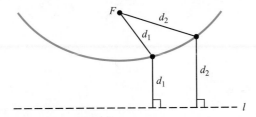

Determine the coordinates of the vertex of a parabola formed by the set of all points that are the same distance from the line $y = 4$ and the fixed point $(3, 10)$.

Cultural Note

Thousands of years ago, the ancient Chinese, Greeks, Incas, Romans—and possibly others—discovered that parabolic mirrors could concentrate the Sun's rays onto a flammable object, causing the object to ignite in seconds. In the ancient Chinese kitchen, these "burning mirrors" were as common as pots. According to legend, the mathematician Archimedes used a parabolic mirror to destroy the Roman fleet invading Syracuse. Today, parabolic mirrors in solar towers generate steam, which in turn drives turbines.

(*Sources: Greek Mathematical Works*, Loeb Classical Library, Harvard University Press, 1941; J. Perlin, "Burning Mirrors: Snagging Pure Fire from the Rays of the Sun," *Whole Earth*, Winter 1999; G. J. Toomer, *Diocles on Burning Mirrors*, Springer Verlag, 1976)

10.2 The Circle

OBJECTIVES

In this section, we discuss the most familiar of all conic sections, the *circle*. There are many applications of circles found in nature, from the ripples on the surface of a lake to the cross section of a planet. Some applications of circles involve traveling a fixed distance from a given point.

First, we consider two formulas about pairs of points on the coordinate plane—the *distance formula* and the *midpoint formula*.

A To find the distance between two points by using the distance formula

B To find the midpoint of a line segment by using the midpoint formula

C To graph a circle

D To find the equation of a circle in standard form

E To solve applied problems involving circles

The Distance Formula and the Midpoint Formula

Recall from Section 7.2 that for any two points on a coordinate plane, (x_1, y_1) and (x_2, y_2), we can determine the distance d between them by finding the length of the line segment connecting them. To do this, we draw a vertical and a horizontal line through the points to form a right triangle, as shown in the diagram below. Note that the lengths of the two legs of the right triangle are $x_2 - x_1$ and $y_2 - y_1$. Then, we apply the Pythagorean theorem to find the length of the hypotenuse.

$$d^2 = (x_2 - x_1)^2 + (y_2 - y_1)^2$$
$$d = \sqrt{(x_2 - x_1)^2 + (y_2 - y_1)^2},$$

which is the distance formula.

Distance Formula

The distance between any two points (x_1, y_1) and (x_2, y_2) on a coordinate plane is given by the formula $d = \sqrt{(x_2 - x_1)^2 + (y_2 - y_1)^2}$.

EXAMPLE 1

Find the distance between the points $(3, 0)$ and $(-1, 4)$, rounded to the nearest tenth.

Solution   Let $(3, 0)$ stand for (x_1, y_1) and $(-1, 4)$ stand for (x_2, y_2). Substituting into the distance formula, we get:

$$\begin{aligned} d &= \sqrt{(x_2 - x_1)^2 + (y_2 - y_1)^2} \\ &= \sqrt{(-1 - 3)^2 + (4 - 0)^2} \\ &= \sqrt{(-4)^2 + (4)^2} \\ &= \sqrt{16 + 16} \\ &= \sqrt{32} \\ &= 4\sqrt{2} \\ &\approx 5.7 \end{aligned}$$

Using a calculator, we find that the distance between the two points is approximately 5.7 units.

PRACTICE 1

Find the distance between the points $(1, -2)$ and $(0, -5)$, rounded to the nearest tenth.

Our second formula concerns the *midpoint* of the line segment, which connects two given points called the *endpoints* of the segment. By midpoint, we mean the point on the segment located halfway between the two endpoints. The following formula gives the coordinates of the midpoint in terms of the coordinates of the endpoints.

> ## Midpoint Formula
>
> The line segment with endpoints (x_1, y_1) and (x_2, y_2) has as its midpoint
>
> $$\left(\frac{x_1 + x_2}{2}, \frac{y_1 + y_2}{2} \right).$$

In words, this formula states that the coordinates of the midpoint are the average of the *x*-coordinates and the average of the *y*-coordinates of the endpoints.

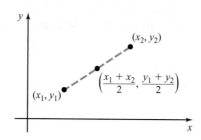

Can you show that the point $\left(\dfrac{x_1 + x_2}{2}, \dfrac{y_1 + y_2}{2} \right)$ lies on this segment?

EXAMPLE 2

Find the midpoint of the line segment joining the points $(-3, 2)$ and $(1, 5)$.

Solution   Let $(-3, 2)$ stand for (x_1, y_1) and $(1, 5)$ stand for (x_2, y_2). Substituting into the midpoint formula, we get:

$$\frac{x_1 + x_2}{2} = \frac{-3 + 1}{2} = \frac{-2}{2} = -1$$

$$\frac{y_1 + y_2}{2} = \frac{2 + 5}{2} = \frac{7}{2}$$

So the midpoint is $\left(-1, \dfrac{7}{2} \right)$.

PRACTICE 2

Find the midpoint of the line segment joining the points $(3, 1)$ and $(7, 4)$.

The Circle

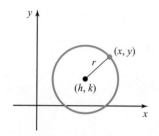

A **circle** is the set of all points on a coordinate plane that are a fixed distance from a given point. The given point is called the **center** of the circle, and is represented by (h, k). The fixed distance is called the **radius** of the circle, and is represented by r.

Let's consider the simplest case—a circle centered at the origin. Suppose, for instance, we want to find the equation of a circle with radius 2 and with center $(0, 0)$, and then to graph it. If (x, y) is an arbitrary point on the circle, we can apply the distance formula as follows:

$$d = \sqrt{(x_2 - x_1)^2 + (y_2 - y_1)^2}$$
$$2 = \sqrt{(x - 0)^2 + (y - 0)^2}$$
$$4 = x^2 + y^2$$
$$x^2 + y^2 = 4$$

This last equation is the equation of the circle with radius 2 centered at the origin. To graph $x^2 + y^2 = 4$, we start at the center $(0, 0)$, and then plot points 2 units to the left, to the right, up, and down from the center. Finally, we sketch a smooth curve through the four points $(-2, 0)$, $(2, 0)$, $(0, 2)$, and $(0, -2)$, as shown in the figure to the right.

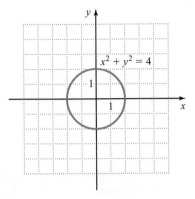

The Equation of a Circle Centered at the Origin

The equation of a circle with center $(0, 0)$ and with radius r is

$$x^2 + y^2 = r^2.$$

EXAMPLE 3

Graph: $x^2 + y^2 = 25$

Solution To get the equation in the form $x^2 + y^2 = r^2$, we can write it as $x^2 + y^2 = 5^2$. From the standard form we see that the center is the point $(0, 0)$, and the radius is 5. The graph of the equation is as follows:

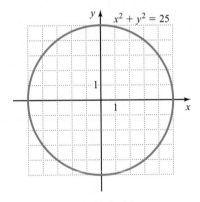

PRACTICE 3

Graph: $x^2 + y^2 = 9$

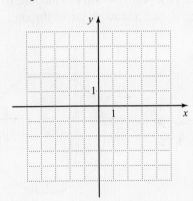

Note from Example 3 that in the equation $x^2 + y^2 = 25$, y is not a function of x. Can you explain why?

More generally, suppose that we wish to find the equation of a circle with given radius r and with arbitrary center (h, k). If (x, y) is any point on the circle, the distance formula gives us:

$$r = \sqrt{(x - h)^2 + (y - k)^2}$$
$$r^2 = (x - h)^2 + (y - k)^2$$
$$(x - h)^2 + (y - k)^2 = r^2,$$

which is the equation of the circle in *standard form*.

> ## The Equation of a Circle Centered at (h, k)
>
> The equation in standard form of the circle with center (h, k) and radius r is
> $$(x - h)^2 + (y - k)^2 = r^2.$$

Note that we can determine the center and radius of a circle just by looking at the equation in standard form.

EXAMPLE 4

Consider the circle whose equation is $(x - 2)^2 + (y + 1)^2 = 16$.

a. Identify the center and radius of the circle.

b. Graph the circle.

Solution

a. To get the equation in the form $(x - h)^2 + (y - k)^2 = r^2$, we write it as $(x - 2)^2 + [y - (-1)]^2 = 4^2$. From the standard form we see that $h = 2, k = -1$, and $r = 4$. So the center of the circle is the point $(2, -1)$, and the radius is 4.

b. To graph the circle, we first plot the center $(2, -1)$. From the center point we plot points 4 units to the left, to the right, up, and down, and then sketch a smooth curve through these four points.

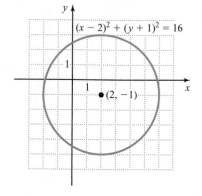

PRACTICE 4

The graph of the equation
$$(x + 6)^2 + (y - 2)^2 = 100$$
is a circle.

a. Identify the center and radius of the circle.

b. Graph the circle.

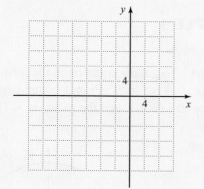

EXAMPLE 5

Find the equation of the circle with center $(-2, 5)$ and with radius 6.

Solution   For this circle, $h = -2$ and $k = 5$, and $r = 6$. So the equation of the circle is $[x - (-2)]^2 + (y - 5)^2 = 6^2$, or $(x + 2)^2 + (y - 5)^2 = 36$.

PRACTICE 5

Determine the equation of the circle centered at $(7, -1)$ with radius 8.

Sometimes, the equation of a circle is not given in standard form. We can write the equation in standard form by *completing the square*.

EXAMPLE 6

Find the center and the radius of the circle with equation
$x^2 + y^2 + 2x - 8y + 15 = 0$.

Solution To find the center and radius of the circle, first let's write the given equation in standard form by completing the squares for both x-terms and y-terms.

$$x^2 + y^2 + 2x - 8y + 15 = 0$$

$$x^2 + y^2 + 2x - 8y = -15 \qquad \text{Subtract 15 from each side of the equation.}$$

$$(x^2 + 2x) + (y^2 - 8y) = -15 \qquad \text{Group the } x\text{- and } y\text{-terms.}$$

$$(x^2 + 2x + 1) + (y^2 - 8y + 16) = -15 + 1 + 16$$

Complete the squares, adding both 1 and 16 to each side of the equation.

$$(x + 1)^2 + (y - 4)^2 = 2 \qquad \text{Express both the polynomial in } x \text{ and the polynomial in } y \text{ as the square of a binomial.}$$

Since $(x + 1)^2 + (y - 4)^2 = 2$ is in the form $(x - h)^2 + (y - k)^2 = r^2$, we see that $h = -1$ and $k = 4$. Note that since $r^2 = 2$, $r = \sqrt{2}$. So the center of the circle is the point $(-1, 4)$, and its radius is $\sqrt{2}$.

PRACTICE 6

Find the center and radius of the circle with equation
$x^2 + y^2 - 4x - 2y - 19 = 0$.

EXAMPLE 7

A family is on a sailboat that is sinking at sea. A Coast Guard cutter, located 7 mi east and 5 mi north of a harbor, can travel 2 mi in the time that it takes the sailboat to sink.

a. Find the equation of the circle that consists of the farthest points that the cutter can reach before the sailboat sinks.

b. If the sailboat is 8 mi east and 6 mi north of the harbor, can the cutter reach the sailboat before it sinks?

PRACTICE 7

A computer hub in a public park provides free wi-fi access up to 10 yd away.

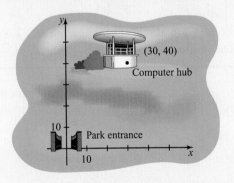

a. Find the equation of the circle that represents the boundary for surfing the web in the park. Let units be in yards.

b. Can a visitor sitting 35 yd east and 18 yd north of the entrance in the park surf the web?

EXAMPLE 7 (continued)

Solution

a. Let's use the harbor as the origin of the coordinate system. The farthest points that the cutter can reach are all 2 mi from the point $(7, 5)$. So the equation of this circle is
$(x - 7)^2 + (y - 5)^2 = 2^2$, or $(x - 7)^2 + (y - 5)^2 = 4$,
where units are in miles.

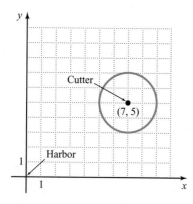

b. One way to see if the cutter can reach the sailboat in time is to calculate the distance between the cutter and the sailboat. The cutter starts at the point $(7, 5)$, and the sailboat is at $(8, 6)$. We can find out how far apart they are by using the distance formula.

$$d = \sqrt{(x_2 - x_1)^2 + (y_2 - y_1)^2}$$
$$= \sqrt{(8 - 7)^2 + (6 - 5)^2}$$
$$= \sqrt{1 + 1}$$
$$= \sqrt{2}$$
$$\approx 1.4$$

So they are about 1.4 mi apart. Since $1.4 < 2$, the cutter can reach the sailboat before it sinks.

What is another way to solve part (b)?

Mathematically Speaking

Fill in each blank with the most appropriate term or phrase from the given list.

solving	the difference between	the sum of
length	completing the squares	radius
centered at	midpoint	at the origin
circle	at any point	passing through

1. The distance between any two points (x_1, y_1) and (x_2, y_2) on a coordinate plane is equal to the square root of _____ $(x_2 - x_1)^2$ and $(y_2 - y_1)^2$.

2. The _____ of a line segment with end-points (x_1, y_1) and (x_2, y_2) is $\left(\dfrac{x_1 + x_2}{2}, \dfrac{y_1 + y_2}{2} \right)$.

3. A(n) _____ is the set of all points on a co-ordinate plane that are a fixed distance from a given point.

4. The equation of a circle centered _____ with radius r is $x^2 + y^2 = r^2$.

5. The equation of a circle _____ (h, k) with radius r is $(x - h)^2 + (y - k)^2 = r^2$.

6. The equation $x^2 + y^2 + 3x - 2y = 0$ can be put in standard form by _____ for both x-terms and y-terms.

A *Find the distance between the two points on the coordinate plane. If necessary, round the answer to the nearest tenth.*

7. $(3, 2)$ and $(9, 10)$

8. $(2, 8)$ and $(7, -4)$

9. $(1, 6)$ and $(-1, 2)$

10. $(3, 12)$ and $(4, 5)$

11. $(8, -4)$ and $(5, -6)$

12. $(-7, 10)$ and $(-9, 8)$

13. $(-3.2, 1.7)$ and $(-1.2, -5.3)$

14. $(8.1, -2.6)$ and $(9.1, -14.6)$

15. $\left(\dfrac{1}{4}, \dfrac{1}{3} \right)$ and $\left(\dfrac{1}{2}, \dfrac{2}{3} \right)$

16. $\left(\dfrac{1}{6}, \dfrac{3}{4} \right)$ and $\left(\dfrac{5}{6}, \dfrac{1}{4} \right)$

17. $(-4\sqrt{3}, -2\sqrt{2})$ and $(\sqrt{3}, -3\sqrt{2})$

18. $(\sqrt{5}, -\sqrt{6})$ and $(2\sqrt{5}, 3\sqrt{6})$

B *Find the coordinates of the midpoint of the line segment joining the points.*

19. $(5, 9)$ and $(7, 1)$

20. $(2, 8)$ and $(4, -4)$

21. $(3, 11)$ and $(-12, -1)$

22. $(6, -7)$ and $(-5, 3)$

23. $(-8, -11)$ and $(-13, -2)$

24. $(-12, -9)$ and $(-5, 10)$

25. $(3.4, -1.1)$ and $(0.6, -3.9)$

26. $(-2.7, 0.3)$ and $(-6.3, 5.7)$

27. $\left(\dfrac{3}{4}, -\dfrac{2}{5} \right)$ and $\left(-\dfrac{1}{3}, \dfrac{1}{2} \right)$

28. $\left(\dfrac{1}{9}, \dfrac{1}{6} \right)$ and $\left(-\dfrac{2}{3}, -\dfrac{3}{2} \right)$

29. $(-7\sqrt{3}, 5\sqrt{6})$ and $(7\sqrt{3}, 3\sqrt{6})$

30. $(-9\sqrt{2}, 5\sqrt{5})$ and $(-3\sqrt{2}, -5\sqrt{5})$

 *Find the center and radius of each circle. Then, graph.*

31. $x^2 + y^2 = 36$

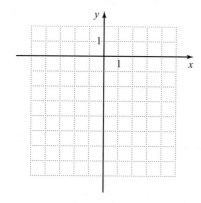

32. $x^2 + y^2 = 16$

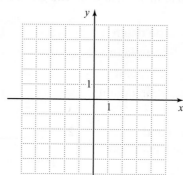

33. $(x - 3)^2 + (y - 2)^2 = 1$

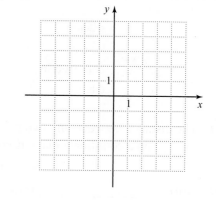

34. $(x + 1)^2 + (y + 4)^2 = 9$

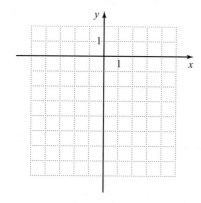

35. $(x + 2)^2 + (y - 4)^2 = 25$

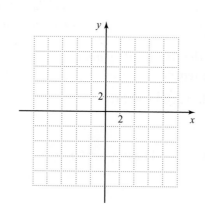

36. $(x - 3)^2 + (y + 3)^2 = 4$

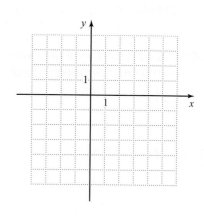

37. $x^2 + y^2 - 2y = 8$

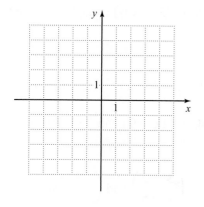

38. $x^2 + y^2 - 4x = 12$

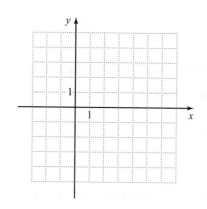

39. $x^2 + y^2 + 2x + 2y - 23 = 0$

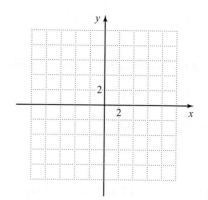

40. $x^2 + y^2 + 4x + 6y - 3 = 0$ **41.** $x^2 + y^2 - 8x - 6y + 16 = 0$ **42.** $x^2 + y^2 + 4x + 2y + 1 = 0$

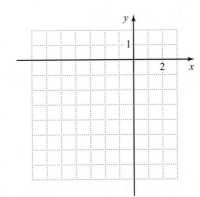   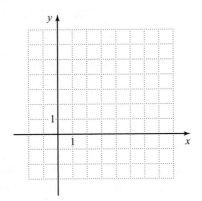   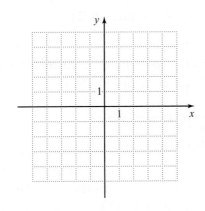

D *Use the given information to find the equation of each circle.*

43. Center: $(-7, 2)$; radius: 9

44. Center: $(6, -10)$; radius: 8

45. Center: $(0, 4)$; radius: $\sqrt{5}$

46. Center: $(-1, 0)$; radius: $\sqrt{2}$

47. Center: $(3, 5)$; passes through the point $(-1, 9)$

48. Center: $(-2, 8)$; passes through the point $(-3, 5)$

49. Endpoints of a diameter: $(-6, 1)$ and $(2, 11)$

50. Endpoints of a diameter: $(7, 7)$ and $(3, -5)$

Find the center and radius of each circle.

51. $x^2 + y^2 + 3x + 4y = 0$

52. $x^2 + y^2 - 8x - y + 4 = 0$

53. $x^2 + y^2 - 10x + 6y - 4 = 0$

54. $x^2 + y^2 + 14x - 2y = 0$

55. $3x^2 + 3y^2 - 12x - 24 = 0$

56. $2x^2 + 2y^2 + 20y + 10 = 0$

Mixed Practice

Find the center and radius of each circle. Then, graph.

57. $(x - 4)^2 + (y + 1)^2 = 9$

58. $x^2 + y^2 + 2x - 4y - 11 = 0$

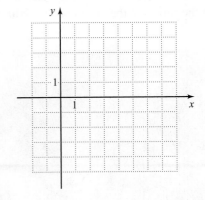   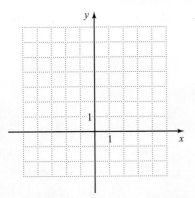

Solve.

59. Find the equation of the circle with center $(2, -4)$ that passes through the point $(-1, -2)$.

60. Find the center and radius of the circle:
$x^2 + y^2 - 3x + 6y + 9 = 0$

Find the distance between the two points on the coordinate plane. If necessary, round the answer to the nearest tenth.

61. $(-4, 7)$ and $(-6, 2)$

62. $(-6.8, -2.9)$ and $(-3.8, 3.1)$

Find the coordinates of the midpoint of the line segment joining the points.

63. $(8, 5)$ and $(-3, -11)$

64. $(-4.8, 0.7)$ and $(-9.4, 5.3)$

Applications

E *Solve.*

65. A lawyer drives from her home, located 11 mi east and 8 mi north of the town courthouse, to her office, located 1 mi west and 1 mi south of the courthouse. Find the distance between the lawyer's home and her office.

66. On a particular sales route, a sales representative first drives to a community college located 20 mi east and 17 mi south from his home. From there, he drives to a state college located 25 mi east and 7 mi south of the community college. Then, he drives home. Calculate the distance between the state college and the sales representative's home.

67. In geometry, an *altitude* of a triangle is a line segment through a vertex that is perpendicular to the opposite side of the triangle. Consider an isosceles triangle formed by the points $A(2, 1)$, $B(1, 6)$, and $C(6, 5)$.

   a. Find the coordinates of the midpoint D of the base $\overline{AC}$.

   b. By considering its slope, show that $\overline{BD}$ is an altitude of triangle ABC.

   c. Calculate the area of triangle ABC.

68. The *perpendicular bisector* of a segment is the line perpendicular to the segment that intersects the segment at its midpoint. Consider the line segment with endpoints $(3, 5)$ and $(9, 7)$.

   a. Find the equation of the perpendicular bisector of the line segment in slope-intercept form.

   b. Find the coordinates of the x-intercept of the perpendicular bisector.

   c. Show that the x-intercept of the perpendicular bisector is equidistant from the endpoints of the line segment.

69. An oil barge runs aground, producing a circular oil slick. If the slick is 6 mi across, find an equation that represents the boundary of the slick. Assume the center of the oil slick is at the origin of a coordinate plane.

70. Circus acts are performed in a circular area called a *ring*. If a ring has a diameter of 42 ft, find an equation that represents the boundary of the ring. Assume that the center of the ring is at the origin of a coordinate plane.

71. A radio station broadcasts a signal that can be received by a radio within a 36-mile radius of the station.

   a. If the radio station is located 8 mi west and 15 mi north of the center of City A, find an equation of the circle that represents the boundary of the radio signal.

   b. Can a student whose apartment is located 15 mi east and 3 mi south of the center of City A listen to the station's broadcast on his radio?

72. A power outage affected all homes and businesses within a 5-mile radius of the power station.

   a. If the power station is located 2 mile east and 6 mi south of the center of town, find an equation of the circle that represents the boundary of the power outage.

   b. Will a mall located 4 mi east and 7 mi north of the power station be affected by the outage?

• Check your answers on page A-53.

MINDStretchers

Mathematical Reasoning

1. Use the distance formula to show that $\left(\dfrac{x_1 + x_2}{2}, \dfrac{y_1 + y_2}{2}\right)$ is equidistant from (x_1, y_1) and (x_2, y_2).

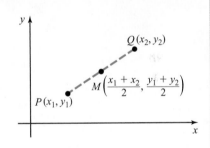

Critical Thinking

2. The equation of the smallest circle shown is $x^2 + y^2 = r^2$. Find the equation of the largest circle.

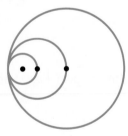

Technology

3. Consider the equation $(x - 5)^2 + (y - 2)^2 = 4$.

   a. Is y a function of x? Explain how you know.

   b. Solve the equation for y.

   c. Explain how you can use a graphing calculator or computer software to graph this equation.

10.3 The Ellipse and the Hyperbola

In this section, we discuss two additional conic sections, namely, the *ellipse* and the *hyperbola*. There are many applications of ellipses and hyperbolas. For instance, ellipses are found in orbits of planets, supporting arches of bridges, and medical procedures. Applications of hyperbolas include the paths of certain comets and the mirrors of some telescopes.

In considering these conics, we restrict our discussion to ellipses and hyperbolas that are centered at the origin.

The Ellipse

An **ellipse** is the set of all points on a coordinate plane the *sum* of whose distances from two fixed points is constant. The fixed points are called **foci** (the plural of **focus**). The point halfway between the two foci is called the **center** of the ellipse. In the ellipse shown, the center is the origin.

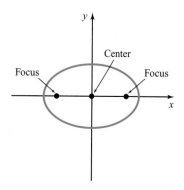

Note that an ellipse is not the graph of a function. Can you explain why?

An ellipse centered at the origin has two *x*-intercepts and two *y*-intercepts. The equation of such an ellipse has the following *standard form:*

The Equation of an Ellipse Centered at the Origin

The equation in standard form of an ellipse with center $(0, 0)$, *x*-intercepts $(-a, 0)$ and $(a, 0)$, and *y*-intercepts $(0, -b)$ and $(0, b)$ is

$$\frac{x^2}{a^2} + \frac{y^2}{b^2} = 1.$$

To graph an ellipse of the form $\dfrac{x^2}{a^2} + \dfrac{y^2}{b^2} = 1$, we begin by plotting the *x*- and *y*-intercepts, which we can read from its equation. Then, we sketch an oval-shaped curve passing through these four points, as shown below.

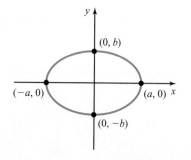

EXAMPLE 1

Graph: $\dfrac{x^2}{25} + \dfrac{y^2}{16} = 1$

Solution We recognize the equation as that of an ellipse centered at the origin in standard form: $\dfrac{x^2}{a^2} + \dfrac{y^2}{b^2} = 1$. Since $a^2 = 25$ and $b^2 = 16$, it follows that $a = 5$ and $b = 4$. So the x-intercepts are $(-5, 0)$ and $(5, 0)$, and the y-intercepts are $(0, -4)$ and $(0, 4)$. Finally, we sketch the smooth curve passing through these intercepts.

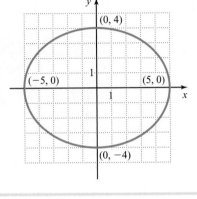

PRACTICE 1

Graph: $\dfrac{x^2}{100} + \dfrac{y^2}{49} = 1$

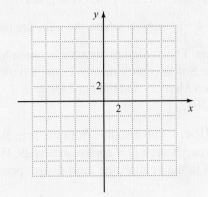

EXAMPLE 2

Graph: $9x^2 + 4y^2 = 36$

Solution Let's divide both sides of the equation by 36, so as to get 1 on the right side.

$$9x^2 + 4y^2 = 36$$

$$\frac{x^2}{4} + \frac{y^2}{9} = 1$$

$$\frac{x^2}{2^2} + \frac{y^2}{3^2} = 1$$

The result is the equation of an ellipse centered at the origin in standard form, where $a = 2$ and $b = 3$. The x-intercepts of the graph are $(-2, 0)$ and $(2, 0)$, and the y-intercepts are $(0, -3)$ and $(0, 3)$. We can sketch the smooth curve passing through these points.

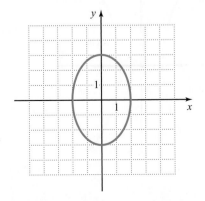

PRACTICE 2

Graph: $25x^2 + 4y^2 = 100$

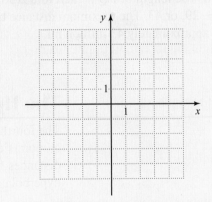

EXAMPLE 3

The planet Mercury travels in an elliptical orbit with the Sun at one focus. An equation that approximates Mercury's orbit is given by $\frac{x^2}{1296} + \frac{y^2}{1247} = 1$, where the units are millions of miles. The minimum distance between Mercury and the Sun is approximately 29 million mi. What is the maximum distance between Mercury and the Sun? (*Source: The World Almanac and Book of Facts 2012*)

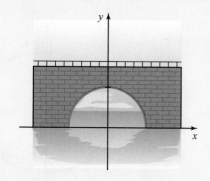

Solution The distance from the Sun, at point S, to the nearest point in Mercury's orbit Q is approximately 29. We are looking for the distance from S to the point P, the most distant point from the Sun.

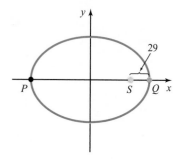

From the equation $\frac{x^2}{1296} + \frac{y^2}{1247} = 1$, we see that $a^2 = 1296$, so $a = 36$. The length of $\overline{PQ}$ is therefore 2(36), or 72. So the length of $\overline{PS}$ is $72 - 29$, or 43. The maximum distance between Mercury and the Sun is approximately 43 million mi.

PRACTICE 3

The arch supporting the bridge shown below is in the shape of half an ellipse. The equation of the ellipse is $\frac{x^2}{100} + \frac{y^2}{144} = 1$, where x and y are in feet. Determine the maximum width and maximum height of the arch.

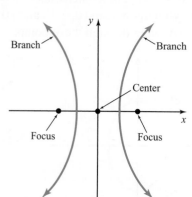

The Hyperbola

The fourth and last conic section we consider is the **hyperbola.** A hyperbola is the set of all points on a coordinate plane that satisfy the following property: The *difference* of the distances from two fixed points is constant. Each fixed point is said to be a **focus** of the hyperbola. The point halfway between the two foci is called the **center** of the hyperbola.

Note that each **branch** of a hyperbola extends indefinitely on both sides. Although a hyperbola looks somewhat like two parabolas, in fact the shape of a parabola and the shape of the branch of a hyperbola differ.

Some hyperbolas centered at the origin have two x-intercepts but no y-intercepts, as shown to the right. The form of the equations of these hyperbolas is as follows:

$$\frac{x^2}{a^2} - \frac{y^2}{b^2} = 1$$

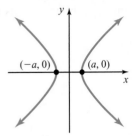

The x-intercepts are $(-a, 0)$ and $(a, 0)$.

Other hyperbolas centered at the origin have two y-intercepts but no x-intercepts, as shown to the right. These hyperbolas have equations of the following form:

$$\frac{y^2}{b^2} - \frac{x^2}{a^2} = 1$$

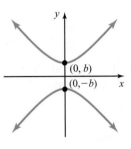

The y-intercepts are $(0, -b)$ and $(0, b)$.

The equation of a hyperbola in the form of either $\dfrac{x^2}{a^2} - \dfrac{y^2}{b^2} = 1$ or $\dfrac{y^2}{b^2} - \dfrac{x^2}{a^2} = 1$ is said to be in *standard form*.

The Equation of a Hyperbola Centered at the Origin

The equation in standard form of a hyperbola with center $(0, 0)$ and x-intercepts $(-a, 0)$ and $(a, 0)$ is

$$\frac{x^2}{a^2} - \frac{y^2}{b^2} = 1.$$

The equation in standard form of a hyperbola with center $(0, 0)$ and y-intercepts $(0, -b)$ and $(0, b)$ is

$$\frac{y^2}{b^2} - \frac{x^2}{a^2} = 1.$$

As we move farther from the origin, each branch of a hyperbola gets closer and closer to a straight line called an **asymptote**. Sketching the two asymptotes of a hyperbola allows us to graph the hyperbola more accurately. The hyperbolas that we are considering have as their asymptotes the lines $y = \dfrac{b}{a}x$ and $y = -\dfrac{b}{a}x$, as shown on the following coordinate planes.

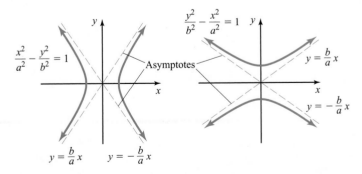

Each branch of a hyperbola approaches, but never intersects, an asymptote. Note that the asymptotes themselves are not part of the hyperbola, as indicated by the broken lines.

To graph a hyperbola centered around the origin, given its equation, we first find the x- or y-intercepts and the equations of the asymptotes. Then, we plot the intercepts, and sketch the asymptotes using broken lines. Finally, we draw the branches of the hyperbola passing through the intercepts and approaching, but not touching, the asymptotes.

EXAMPLE 4

Graph: $\dfrac{y^2}{25} - \dfrac{x^2}{4} = 1$

Solution The equation is of the form $\dfrac{y^2}{b^2} - \dfrac{x^2}{a^2} = 1$. Let's find the intercepts of this hyperbola.

Writing the equation in the form $\dfrac{y^2}{5^2} - \dfrac{x^2}{2^2} = 1$ with $a = 2$ and $b = 5$,

we see that the y-intercepts are $(0, -5)$ and $(0, 5)$, and that there are no x-intercepts. Next, we find the equations of the asymptotes of the hyperbola, getting:

$$y = \frac{b}{a}x \quad \text{and} \quad y = -\frac{b}{a}x$$
$$y = \frac{5}{2}x \qquad \quad y = -\frac{5}{2}x$$

Then, we plot the intercepts, and sketch the asymptotes using broken lines.

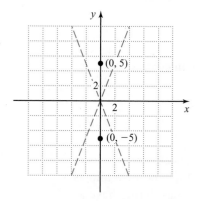

Finally, we draw the graph, sketching each branch of the hyperbola through a y-intercept and approaching, but not touching, the asymptotes.

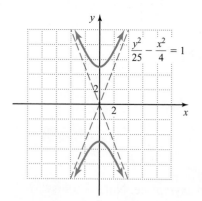

PRACTICE 4

Graph: $\dfrac{y^2}{16} - \dfrac{x^2}{9} = 1$

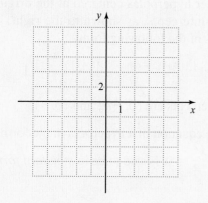

EXAMPLE 5

Graph: $64x^2 - 4y^2 = 64$

Solution To write the equation in standard form, we divide each side of the equation by 64 to get 1 on the right side.

$$64x^2 - 4y^2 = 64$$

$$\frac{x^2}{1} - \frac{y^2}{16} = 1$$

$$\frac{x^2}{1^2} - \frac{y^2}{4^2} = 1$$

This equation is of the form $\frac{x^2}{a^2} - \frac{y^2}{b^2} = 1$, with $a = 1$ and $b = 4$. So the x-intercepts are $(-1, 0)$ and $(1, 0)$, and there are no y-intercepts. The equations of the asymptotes are $y = \frac{4}{1}x$ and $y = -\frac{4}{1}x$, or $y = 4x$ and $y = -4x$. To graph, plot the intercepts and sketch the asymptotes. Then, draw a smooth curve through each intercept approaching, but not touching, the asymptotes.

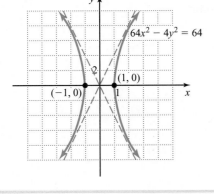

PRACTICE 5

Graph: $4x^2 - 16y^2 = 144$

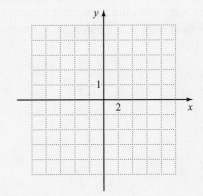

EXAMPLE 6

When a jet breaks the sound barrier, the region on the ground bounded by the hyperbola $\frac{x^2}{9} - \frac{y^2}{25} = 1$, where the units are in miles, experiences a sonic boom shock wave. In the following diagram, the coordinate axes are positioned so that the origin is at the center of the hyperbola. Find, to the nearest mile, the width of the region 3 mi from the hyperbola's x-intercept.

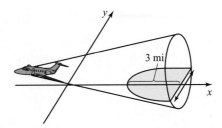

PRACTICE 6

Many comets have a hyperbolic path, with the Sun as one of the foci of the hyperbola. A comet has as its path the graph of the equation

$$\frac{x^2}{8100} - \frac{y^2}{2700} = 1,$$

where the units are in millions of miles. If Earth is the center of the hyperbola, how close does the comet come to Earth?

(continued)

EXAMPLE 6 (continued)

Solution   Since the equation is of the form $\dfrac{x^2}{a^2} - \dfrac{y^2}{b^2} = 1$ with

$a = 3$ and $b = 5$, the x-intercept is $(3, 0)$. The point $(6, 0)$ on the x-axis is 3 mi from the hyperbola's x-intercept. To find the width of the hyperbola at $x = 6$, we calculate the value of y when x is 6, and then double it.

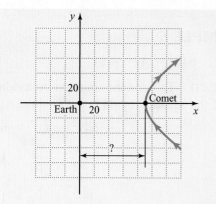

$$\frac{x^2}{9} - \frac{y^2}{25} = 1$$

$$\frac{6^2}{9} - \frac{y^2}{25} = 1$$

$$-\frac{y^2}{25} = 1 - \frac{36}{9}$$

$$-\frac{y^2}{25} = -3$$

$$y^2 = 75$$

$$y = \pm\sqrt{75}, \text{ or } \pm 5\sqrt{3}$$

Since y cannot be negative, we conclude that y is $5\sqrt{3}$. Doubling this value, we find that the width of the region experiencing the sonic boom shock wave is $10\sqrt{3}$ mi, or approximately 17 mi.

Mathematically Speaking

Fill in each blank with the most appropriate term or phrase from the given list.

circle	ellipse	hyperbola
sum of	difference between	asymptotes

1. A(n) _____ is the set of all points on a coordinate plane the sum of whose distances from two fixed points is constant.

2. If an ellipse has center $(0, 0)$ and passes through $(-a, 0)$, $(a, 0)$, $(0, -b)$, and $(0, b)$, then the _____ $\dfrac{x^2}{a^2}$ and $\dfrac{y^2}{b^2}$ is 1.

3. A(n) _____ is the set of all points on a coordinate plane the difference of whose distances from two fixed points is constant.

4. If a hyperbola has center $(0, 0)$ and x-intercepts $(-a, 0)$ and $(a, 0)$, then the _____ $\dfrac{x^2}{a^2}$ and $\dfrac{y^2}{b^2}$ is 1.

A *Graph each ellipse.*

5. $\dfrac{x^2}{16} + \dfrac{y^2}{4} = 1$

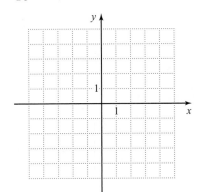

6. $\dfrac{x^2}{25} + \dfrac{y^2}{9} = 1$

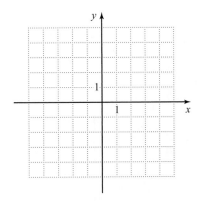

⊙ 7. $\dfrac{x^2}{36} + \dfrac{y^2}{81} = 1$

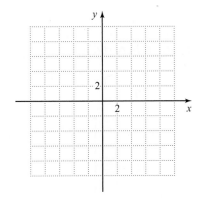

8. $\dfrac{x^2}{16} + \dfrac{y^2}{49} = 1$

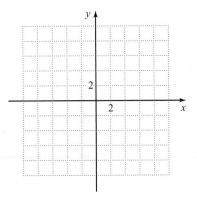

9. $64x^2 + 16y^2 = 64$

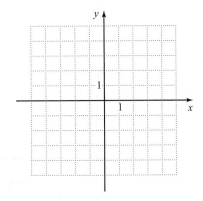

10. $4x^2 + 36y^2 = 36$

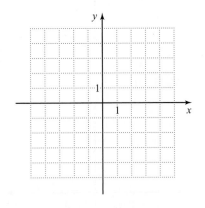

11. $x^2 + 4y^2 = 100$

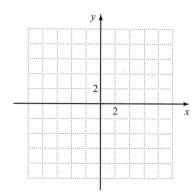

12. $2x^2 + 8y^2 = 128$

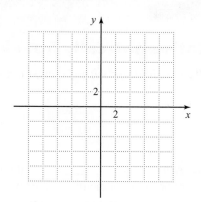

13. $36x^2 + 9y^2 = 144$

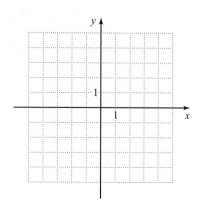

14. $18x^2 + 2y^2 = 162$

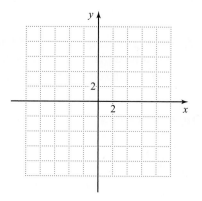

B *Graph each hyperbola.*

15. $\dfrac{x^2}{64} - \dfrac{y^2}{9} = 1$

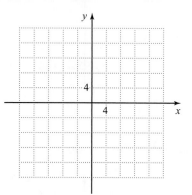

16. $\dfrac{x^2}{25} - \dfrac{y^2}{4} = 1$

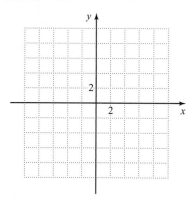

17. $\dfrac{y^2}{81} - \dfrac{x^2}{36} = 1$

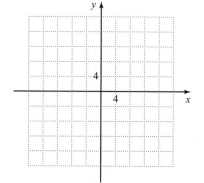

18. $\dfrac{y^2}{49} - \dfrac{x^2}{9} = 1$

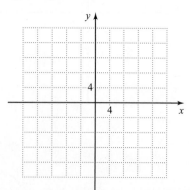

19. $x^2 - 9y^2 = 225$

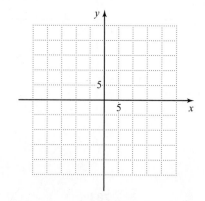

20. $4x^2 - y^2 = 64$

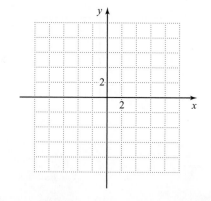

21. $3y^2 - 12x^2 = 108$

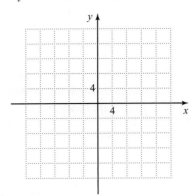

22. $2y^2 - 8x^2 = 200$

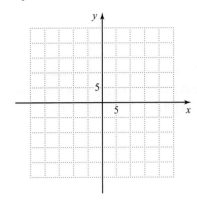

C *Identify whether the conic section whose equation is given is a parabola, a circle, an ellipse, or a hyperbola.*

23. $8x^2 + 8y^2 = 32$

24. $6x^2 + 6y^2 = 18$

25. $x^2 - 2y = 10$

26. $9y^2 + x = 27$

27. $3x^2 + 5y^2 = 45$

28. $10y^2 - 12x^2 = 120$

D *Use the given information to find the equation in standard form of each conic section.*

29. An ellipse centered at the origin that passes through the points $(-11, 0), (11, 0), (0, -3),$ and $(0, 3)$.

30. An ellipse centered at the origin that passes through the points $(-6, 0), (6, 0), (0, -13),$ and $(0, 13)$.

31. An ellipse centered at the origin that passes through the points $(-\sqrt{7}, 0), (\sqrt{7}, 0), (0, -2\sqrt{2}),$ and $(0, 2\sqrt{2})$.

32. An ellipse centered at the origin that passes through the points $(-4\sqrt{3}, 0), (4\sqrt{3}, 0), (0, -\sqrt{5}),$ and $(0, \sqrt{5})$.

33. A hyperbola centered at the origin that passes through the points $(-4, 0)$ and $(4, 0)$ and whose graph approaches the asymptotes $y = -\dfrac{1}{4}x$ and $y = \dfrac{1}{4}x$.

34. A hyperbola centered at the origin that passes through the points $(0, -9)$ and $(0, 9)$ and whose graph approaches the asymptotes $y = -\dfrac{9}{7}x$ and $y = \dfrac{9}{7}x$.

Mixed Practice

Graph each hyperbola.

35. $\dfrac{y^2}{4} - \dfrac{x^2}{16} = 1$

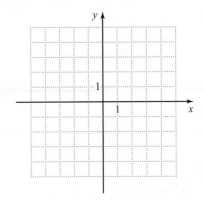

36. $2x^2 - 8y^2 = 72$

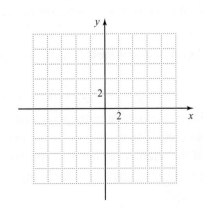

Identify whether the conic section whose equation is given is a parabola, a circle, an ellipse, or a hyperbola.

37. $3x^2 + 6y^2 = 24$

38. $2x^2 - 3y = 9$

Solve.

39. Find the equation in standard form of an ellipse centered at the origin that passes through the points $(-3\sqrt{5}, 0), (3\sqrt{5}, 0), (0, -\sqrt{3}), (0, \sqrt{3})$.

40. Find the equation in standard form of a hyperbola centered at the origin that passes through the points $(0, -3), (0, 3)$ and whose graph approaches the asymptotes $y = -\dfrac{3}{4}x$ and $y = \dfrac{3}{4}x$.

Graph each ellipse.

41. $\dfrac{x^2}{49} + \dfrac{y^2}{25} = 1$

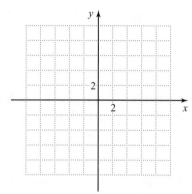

42. $12x^2 + 3y^2 = 108$

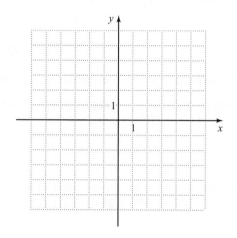

Applications

E *Solve.*

43. According to the rules of the game, the playing field in Australian soccer is elliptical, and must have a length between 135 m and 185 m and a width between 110 m and 155 m. An equation of the ellipse that represents a team's playing field is $\dfrac{x^2}{8100} + \dfrac{y^2}{5625} = 1$, where units are in meters. (*Source:* Australian Football League)

   a. What are the length and the width of this playing field?

   b. Does the team's field meet the requirements stated?

44. An elliptical bicycle path is constructed in a local park. City planners required the length of the path to be between 20 and 40 yd longer than the width. The equation of the ellipse that represents a proposed path is given by $\dfrac{x^2}{4225} + \dfrac{y^2}{2500} = 1$, where units are in yards.

   a. What are the length and the width of the bicycle path?

   b. Does the proposed path meet the city planners' requirement?

45. The amount of rubber A in a 25-foot-long cylindrical rubber hose is given by the equation

$$A = 25\pi(R^2 - r^2),$$

where R is the radius of the outside of the hose, and r is the inside radius. R and r are in feet, and A is in cubic feet. The amount of rubber in the hose is 100π.

a. Sketch a graph of the relationship between the two radii on a coordinate plane.

b. Explain why the part of the graph that lies in the first quadrant is of particular interest in this situation.

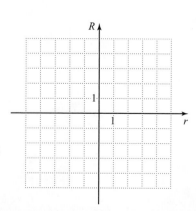

46. A square parcel of land has sides y ft in length. At each of the four corners of the land, a square house is built with side x ft long. The area of the land on which houses are not built is given by the expression $y^2 - 4x^2$, where units are in feet. This area is measured to be 6400 ft^2.

a. Sketch a graph of the relationship between x and y on the coordinate plane.

b. Describe the portion of the graph in Quadrant I.

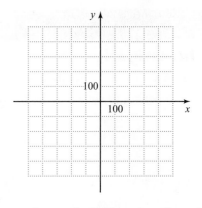

47. The cross section of an ellipsoidal passenger capsule on an observation wheel is an ellipse. (See figure.) The cross section is 6 m long and 4 m wide. Find an equation of the ellipse that represents the cross section of the capsule.

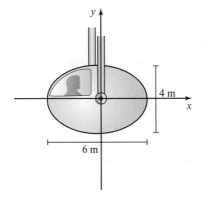

48. Sound coming from one focus of an ellipse is reflected to the other focus. In an elliptically shaped *whispering gallery*, a person speaking near one focus can be heard at the other focus but not at many places in between. (*Source:* stpauls.co.uk)

a. Suppose the length of a whispering gallery is 150 ft and its width is 70 ft. On a coordinate plane where the origin represents the center of the gallery and the length of the gallery is vertical, write an equation that models this ellipse.

b. On the coordinate plane below, sketch the ellipse.

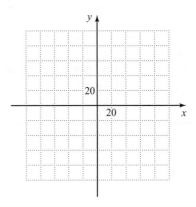

• Check your answers on page A-54.

MINDStretchers

Investigation

1. An ellipse can be drawn by using a string attached to two fixed tacks and a pencil, as shown in the figure.

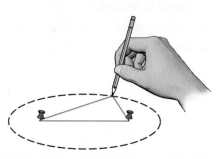

Varying the length of the string and the distance between the tacks, draw several ellipses. What do you notice about the shape of the ellipse as the tacks are moved closer together? What do you notice about the shape as the tacks are moved farther apart?

Mathematical Reasoning

2. Consider a hyperbola with equation either $\dfrac{x^2}{a^2} - \dfrac{y^2}{b^2} = 1$ or $\dfrac{y^2}{b^2} - \dfrac{x^2}{a^2} = 1$. The points

   $(a, b), (-a, b), (-a, -b)$, and $(a, -b)$ can be used to form the *fundamental rectangle* of the hyperbola.

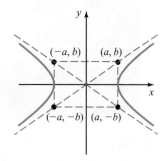 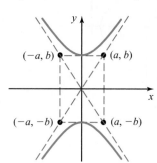

Show that the lines containing the diagonals of this rectangle are the asymptotes of the hyperbola.

Groupwork

3. Working with a partner, use a flashlight to project a cone of light on a wall. Create a parabola, a circle, an ellipse, and a hyperbola, and describe your results.

10.4 Solving Nonlinear Systems of Equations

In the previous sections of this chapter, we discussed nonlinear equations, such as $y = 2x^2$, $x^2 + y^2 = 25$, $x^2 + 3y^2 = 10$, and $4x^2 - y^2 = 16$, and their graphs. Here, we focus on solving **nonlinear systems of equations.** These are systems in which at least one equation is nonlinear. Many applications in astronomy, astrophysics, and communications involve such systems.

Recall that in Section 4.2, we solved systems of linear equations in two variables by using the substitution and elimination methods. We now apply these two methods in solving nonlinear systems of equations in two variables. As with systems of linear equations, a **solution of a nonlinear system of equations** in two variables is an ordered pair of numbers that makes each equation in the system true.

First, we consider the substitution method, and then the elimination method.

Solving Nonlinear Systems by Substitution

From the following example, we see that the substitution method particularly lends itself to solving nonlinear systems in which one of the equations contains linear terms.

EXAMPLE 1

Solve by substitution.

$$\textbf{(1)} \quad x^2 + y^2 = 25$$
$$\textbf{(2)} \quad y - x = 1$$

Solution Since equation (2) is linear, we first solve it for x or y. Solving for y, we get:

$$\textbf{(2)} \quad y - x = 1$$
$$y = x + 1$$

Next, we substitute the expression $x + 1$ for y in equation (1), and solve the resulting equation for x.

$$
\begin{aligned}
\textbf{(1)} \quad x^2 + y^2 &= 25 \\
x^2 + (x + 1)^2 &= 25 \quad \text{Substitute } x + 1 \text{ for } y. \\
x^2 + x^2 + 2x + 1 &= 25 \\
2x^2 + 2x - 24 &= 0 \\
2(x^2 + x - 12) &= 0 \\
x^2 + x - 12 &= 0 \\
(x + 4)(x - 3) &= 0 \\
x + 4 = 0 \quad \text{or} \quad x - 3 &= 0 \\
x = -4 \qquad\qquad x &= 3
\end{aligned}
$$

To find the corresponding values of y, we substitute the values of x in either equation of the original system. Substituting in equation (2) gives us:

$$
\begin{aligned}
\textbf{(2)} \quad y - x &= 1 & y - x &= 1 \\
y - (-4) &= 1 \quad \text{Substitute } -4 \text{ for } x. & y - 3 &= 1 \quad \text{Substitute 3} \\
y &= -3 & y &= 4 \quad \text{for } x.
\end{aligned}
$$

We see that when $x = -4$, $y = -3$, and when $x = 3$, $y = 4$.

PRACTICE 1

Solve by substitution.

$$\textbf{(1)} \quad x^2 + y^2 = 100$$
$$\textbf{(2)} \quad y - x = 2$$

EXAMPLE 1 (continued)

When we check $(-4, -3)$ and $(3, 4)$ in the equations of the system, we find that both ordered pairs make the equations true. So the solutions are $(-4, -3)$ and $(3, 4)$.

The graph of each equation in this system is shown to the right. Note that the line intersects the circle at two points, $(-4, -3)$ and $(3, 4)$, which also confirms the solutions of the system.

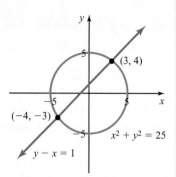

EXAMPLE 2

Solve by substitution.

(1) $\quad x^2 + y^2 = 4$

(2) $\qquad y = x^2 - 2$

Solution   Since both equations in the system have an x^2-term, we can solve equation (2) for x^2. Then, we substitute for x^2 in equation (1).

(2) $\qquad\qquad\qquad y = x^2 - 2$

$\qquad\qquad\qquad y + 2 = x^2$

(1) $\qquad\qquad\qquad x^2 + y^2 = 4$

$\qquad\qquad y + 2 + y^2 = 4 \qquad$ Substitute $y + 2$ for x^2.

$\qquad\qquad y + 2 + y^2 = 4$

$\qquad\qquad y^2 + y + 2 = 4$

$\qquad\qquad y^2 + y - 2 = 0$

$\qquad\quad (y + 2)(y - 1) = 0$

$\qquad y + 2 = 0 \quad$ or $\quad y - 1 = 0$

$\qquad\qquad y = -2 \qquad\qquad y = 1$

Next, we find the corresponding values of x by substituting into equation (2) the values found for y.

(2) $\quad y = x^2 - 2 \qquad\qquad\qquad y = x^2 - 2$

$\quad -2 = x^2 - 2 \quad$ Substitute -2 for y. $\quad 1 = x^2 - 2 \quad$ Substitute 1 for y.

$\qquad 0 = x^2 \qquad\qquad\qquad\qquad\quad 3 = x^2$

$\qquad x = 0 \qquad\qquad\qquad\qquad\quad x = \pm\sqrt{3}$

We see that when $y = -2, x = 0$, and when $y = 1, x = \pm\sqrt{3}$. So the solutions to the system are $(0, -2)$, $(\sqrt{3}, 1)$, and $(-\sqrt{3}, 1)$.

The graph of each equation in this system is shown to the right. Note the parabola intersects the circle at three points, namely $(0, -2), (\sqrt{3}, 1)$, and $(-\sqrt{3}, 1)$, which confirms the solutions of the system.

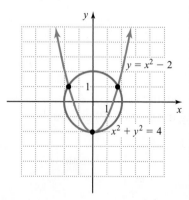

PRACTICE 2

Use the substitution method to solve.

(1) $\qquad\qquad y = x^2 - 4$

(2) $\quad x^2 + y^2 = 4$

Solving Nonlinear Systems by Elimination

When the variable terms in each equation of a nonlinear system are of second degree, we can use the elimination method to solve the system. In this case, we can eliminate an x^2- or y^2-term using a procedure similar to the one shown in Section 4.2.

EXAMPLE 3

Solve by elimination.

$$(1) \quad x^2 + y^2 = 10$$
$$(2) \quad 9x^2 + y^2 = 18$$

Solution   Note that if we multiply equation (1) by -1 and then add the equations, we eliminate the y^2-terms.

$(1) \quad x^2 + y^2 = 10$ $\xrightarrow{\text{Multiply by } -1.}$

$$\begin{aligned}
-x^2 - y^2 &= -10 \\
9x^2 + y^2 &= 18 \\
\hline
8x^2 &= 8 \\
x^2 &= 1 \\
x &= \pm 1
\end{aligned}$$

$(2) \quad 9x^2 + y^2 = 18$

When $x = 1$ or $x = -1$, $x^2 = 1$. So to find the corresponding values of y, we can substitute 1 for x^2 in equation (2).

$$\begin{aligned}
(2) \quad 9x^2 + y^2 &= 18 \\
9(1) + y^2 &= 18 \\
9 + y^2 &= 18 \\
y^2 &= 9 \\
y &= \pm 3
\end{aligned}$$

We see that when $x = 1$, $y = 3$ or -3, and when $x = -1$, $y = 3$ or -3. So the solutions are $(1, 3)$, $(1, -3)$, $(-1, 3)$, and $(-1, -3)$.

The graph of each equation in this system is shown below. Note that the circle intersects the ellipse at $(1, 3)$, $(1, -3)$, $(-1, 3)$, and $(-1, -3)$, confirming the solutions of the system.

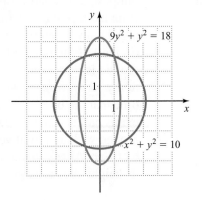

PRACTICE 3

Use the elimination method to solve.

$$(1) \quad x^2 + y^2 = 9$$
$$(2) \quad 2x^2 - y^2 = -6$$

EXAMPLE 4

Use the elimination method to solve.

$$\textbf{(1)} \quad x^2 - 4y^2 = 16$$
$$\textbf{(2)} \quad x^2 + y^2 = 1$$

Solution Here, the goal is to eliminate the x^2-term, and then solve for y.

$\textbf{(1)} \quad x^2 - 4y^2 = 16 \xrightarrow{\text{Multiply by} -1.} \quad -x^2 + 4y^2 = -16$
$\textbf{(2)} \quad x^2 + y^2 = 1 \underline{x^2 + y^2 = 1}$
$$ 5y^2 = -15 \quad \text{Add the}$$
$$ y^2 = -3 \quad \text{equations.}$$
$$ y = \pm i\sqrt{3}$$

When $y = \pm i\sqrt{3}$, $y^2 = -3$. So we can substitute -3 for y^2 in equation (2) to find the corresponding x-values.

$$\textbf{(2)} \quad x^2 + y^2 = 1$$
$$x^2 - 3 = 1$$
$$x^2 = 4$$
$$x = \pm 2$$

We see that when $y = \pm i\sqrt{3}$, $x = \pm 2$. So the solutions are $(2, i\sqrt{3})$, $(-2, i\sqrt{3})$, $(2, -i\sqrt{3})$, and $(-2, -i\sqrt{3})$.

The graph of each equation in this system is shown to the right. Note that the hyperbola does not intersect the circle, which confirms that there are no real solutions.

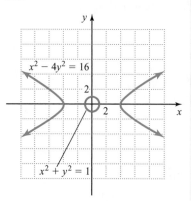

PRACTICE 4

Solve by elimination.

$$\textbf{(1)} \quad x^2 - y^2 = -9$$
$$\textbf{(2)} \quad 4x^2 + y^2 = 4$$

EXAMPLE 5

Suppose the equation $\dfrac{x^2}{4} + \dfrac{y^2}{16} = 1$ describes the elliptical orbit of a planet and the equation $y = x^2 - 4$ describes the parabolic path of a comet. Where will the path of the comet intersect the orbit of the planet?

Solution The nonlinear system that we consider is:

$$\textbf{(1)} \quad \frac{x^2}{4} + \frac{y^2}{16} = 1$$
$$\textbf{(2)} \quad y = x^2 - 4$$

To eliminate the fractions in equation (1), we multiply each side of the equation by 16, getting:

$$\textbf{(1)} \quad 4x^2 + y^2 = 16$$
$$\textbf{(2)} \quad y = x^2 - 4$$

PRACTICE 5

Suppose a baseball is dropped from the top of a 144-foot building. At any given time t, the height (in feet) of the baseball above the ground is represented by the equation $y = -16t^2 + 144$. At the same time that the baseball is dropped, a tennis ball is thrown upward from the ground with an initial velocity of 64 ft/sec. At any given time t, its height above the ground is

Solving equation (2) for x^2, we have:

$$(2) \qquad y = x^2 - 4$$
$$y + 4 = x^2$$

Now, we substitute $(y + 4)$ for x^2 in equation (1), which gives us:

$$(1) \qquad 4x^2 + y^2 = 16$$
$$4(y + 4) + y^2 = 16$$
$$4y + 16 + y^2 = 16$$
$$y^2 + 4y = 0$$
$$y(y + 4) = 0$$
$$y = 0 \quad \text{or} \quad y + 4 = 0$$
$$y = -4$$

Finally, substituting the y-values in equation (2), we find the corresponding x-values:

$$(2) \qquad y = x^2 - 4 \qquad\qquad (2) \qquad y = x^2 - 4$$
$$0 = x^2 - 4 \qquad\qquad\qquad -4 = x^2 - 4$$
$$0 = (x + 2)(x - 2) \qquad\qquad 0 = x^2$$
$$x + 2 = 0 \quad \text{or} \quad x - 2 = 0 \qquad 0 = x$$
$$x = -2 \qquad\qquad x = 2 \qquad\qquad x = 0$$

The points of intersection are the ordered pairs $(-2, 0)$, $(2, 0)$, and $(0, -4)$. So the path of the comet intersects the orbit of the planet at $(-2, 0)$, $(2, 0)$, and $(0, -4)$.

represented by the equation $y = -16t^2 + 64t$. At what time will the two balls be the same height above the ground?

A *Solve by substitution or elimination.*

1. $y = x^2 - 2$
$y = 2x + 1$

2. $y = x^2 + 7$
$y = 3x + 5$

3. $x^2 + y^2 = 12$
$x = y^2 - 6$

4. $x^2 + y^2 = 16$
$x = y^2 - 4$

5. $y = x^2 - 5$
$y = -x^2 + 11$

6. $x = y^2 - 8$
$x = -y^2 + 4$

7. $x^2 + y^2 = 16$
$x - y = 4$

8. $x^2 + y^2 = 25$
$x + y = 5$

9. $x = 2y^2 - y - 3$
$y = \dfrac{1}{4}x$

10. $y = 3x^2 - 2x - 4$
$x = \dfrac{1}{2}y$

11. $x^2 + y^2 = 32$
$y = x^2 - 2$

12. $x^2 + y^2 = 84$
$x = y^2 + 6$

13. $-x^2 + 2y^2 = -8$
$x^2 + 3y^2 = 18$

14. $4x^2 + y^2 = 30$
$5x^2 - y^2 = 15$

15. $3x^2 + y^2 = 24$
$x^2 + y^2 = 16$

16. $x^2 + y^2 = 20$
$x^2 + 4y^2 = 32$

17. $5x^2 + 6y^2 = 24$
$5x^2 + 5y^2 = 25$

18. $2x^2 + 8y^2 = 40$
$5x^2 + 8y^2 = 16$

19. $5x^2 - 3y^2 = 35$
$3x^2 - 3y^2 = 3$

20. $4x^2 - 6y^2 = 16$
$4x^2 - 2y^2 = 32$

21. $6x^2 + 6y^2 = 96$
$x^2 + 9y^2 = 144$

22. $3x^2 - 3y^2 = 108$
$2x^2 + y^2 = 72$

23. $\dfrac{x^2}{4} + \dfrac{y^2}{16} = 1$
$\dfrac{x^2}{2} + \dfrac{y^2}{24} = 1$

24. $\dfrac{x^2}{3} + \dfrac{y^2}{9} = 1$
$\dfrac{x^2}{4} + \dfrac{y^2}{6} = 1$

25. $\dfrac{y^2}{8} - \dfrac{x^2}{4} = 1$
$\dfrac{y^2}{11} + \dfrac{x^2}{11} = 1$

26. $-\dfrac{x^2}{9} + y^2 = 1$
$\dfrac{x^2}{16} - \dfrac{y^2}{4} = 1$

27. $(x - 4)^2 + y^2 = 4$
$(x + 2)^2 + y^2 = 16$

28. $x^2 + (y + 1)^2 = 9$
$y = x^2 + 2$

Mixed Practice

Solve by substitution or elimination.

29. $x^2 + y^2 = 27$
$\quad\quad y = x^2 + 3$

30. $x^2 + y^2 = 13$
$\quad\quad 3x^2 + y^2 = 21$

31. $x^2 + y^2 = 25$
$\quad\quad x = y^2 - 5$

32. $y = x^2 - 7$
$\quad\quad y = -x^2 + 3$

33. $3x^2 + 6y^2 = 24$
$\quad\quad 3x^2 + 5y^2 = 19$

34. $\dfrac{x^2}{18} + \dfrac{y^2}{3} = 1$
$\quad\quad \dfrac{x^2}{9} + \dfrac{y^2}{6} = 1$

Applications

B *Solve.*

35. The revenue R of an electronics manufacturer from selling x handheld organizers is given by $R = 0.25x^2 - 100x$. The revenue of a competing manufacturer from selling similar products is $R = 75x$. Under what circumstances will both manufacturers have the same revenue?

36. A company's annual operating cost C (in millions of dollars) is given by $C = t^2 - t + 4$, where t is the number of years after 2000. The corresponding equation for the company's annual revenue R (in millions of dollars) is $R = 0.5t^2 + 1.5t + 2$. In what year(s) did the company break even?

37. Suppose the equation $x^2 + 2y^2 = 7$ represents the elliptical path traveled by one comet and the equation $x^2 - 3y^2 = 2$, where $x \geq 0$, represents the hyperbolic path traveled by another comet. At what point(s) will the paths of the two comets intersect?

38. Two elliptical pathways in a park intersect as shown below. If the equations of the ellipses that represent the outermost perimeter of the pathways are $x^2 + 4y^2 = 10{,}000$ and $4x^2 + y^2 = 10{,}000$, at what points do the ellipses intersect?

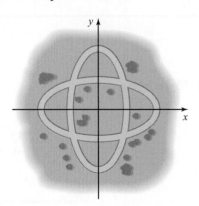

39. One square has a side length 3 in. greater than that of a second square. The difference between the areas of the two squares is 75 in². Find the dimensions of the two squares.

40. Points *A* and *B* lie on a coordinate plane, as shown below. A third point *C* also lies on the plane. If the length of $\overline{AC}$ is $\sqrt{41}$ and the length of $\overline{BC}$ is $\sqrt{13}$, find the possible locations of *C*.

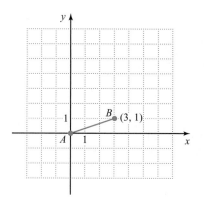

• Check your answers on page A-54.

MINDStretchers

Technology

1. Use a graphing calculator or a computer with graphing software to solve the following nonlinear system, rounding to the nearest hundredth.

$$x^2 + y^2 = 9$$
$$y = e^x$$

Mathematical Reasoning

2. Show algebraically that the graph of the hyperbola whose equation is $4x^2 - y^2 = 16$ does not intersect its asymptotes.

Critical Thinking

3. Solve:

$$(1) \quad x^2 + y^2 = 15$$
$$(2) \quad 2x^2 + y^2 = 22$$
$$(3) \quad 3x^2 - y^2 = 13$$

OBJECTIVES

In this section, we solve nonlinear inequalities and nonlinear systems of inequalities by using the same general methods for graphing linear inequalities and systems of linear inequalities that we discussed in Chapters 3 and 4.

Ⓐ To graph a nonlinear inequality in two variables

Ⓑ To solve a nonlinear system of inequalities by graphing

Solving Nonlinear Inequalities by Graphing

Let's look at how to graph a nonlinear inequality in two variables. Recall from Section 3.5 that to graph a linear inequality such as $x + 2y < 2$, we begin by graphing the corresponding linear equation. This graph is called the boundary line. Then, we choose a test point not on the boundary to determine the region that satisfies the inequality. This method can also be used to graph a nonlinear inequality such as $x^2 + y^2 < 49$.

Ⓒ To solve applied problems involving nonlinear inequalities or nonlinear systems of inequalities

EXAMPLE 1

Graph the solutions of $x^2 + y^2 < 49$.

Solution The corresponding equation is $x^2 + y^2 = 49$, which is a circle with center $(0, 0)$ and radius 7. Since the original inequality involves the symbol $<$, we draw a *broken curve* to indicate that the boundary $x^2 + y^2 = 49$ is not part of the graph of $x^2 + y^2 < 49$. The following graph is a circle that divides the plane into two regions—the "inner" and "outer" regions of the circle:

PRACTICE 1

Graph: $4x^2 + y^2 > 4$

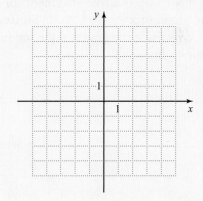

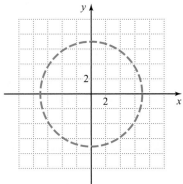

To determine in which region the original inequality is satisfied, we choose a test point in either region, and substitute the coordinates of this point in the inequality. Let's choose the test point $(0, 0)$. Substituting in the inequality, we get:

$x^2 + y^2 < 49$

$0^2 + 0^2 < 49$ Substitute 0 for x and 0 for y.

$0 + 0 < 49$

$0 < 49$ True

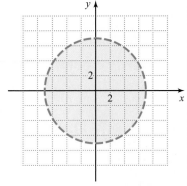

Since the inequality is a true statement, the region containing the test point is the graph $x^2 + y^2 < 49$. So the inner region of the circle, shown to the right, is the graph of the solutions of the inequality.

EXAMPLE 2

Graph: $x^2 - 9y^2 \geq 36$

Solution The corresponding equation is $x^2 - 9y^2 = 36$, or $\dfrac{x^2}{36} - \dfrac{y^2}{4} = 1$.

Since the original inequality involves the symbol $\geq$, we draw a *solid curve* to indicate that the boundary $x^2 - 9y^2 = 36$ is part of the graph of $x^2 - 9y^2 \geq 36$. The graph to the right is a hyperbola that divides the plane into three regions.

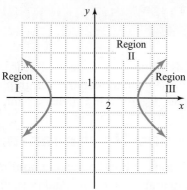

To determine in which region the original inequality is satisfied, we choose a test point in each region, but not on the boundary.

Region I

Test point $(-7, 0)$

$x^2 - 9y^2 \geq 36$

$(-7)^2 - 9(0)^2 \geq 36$

$49 - 0 \geq 36$

$49 \geq 36$ True

Region II

Test point $(0, 0)$

$x^2 - 9y^2 \geq 36$

$(0)^2 - 9(0)^2 \geq 36$

$0 - 0 \geq 36$

$0 \geq 36$ False

Region III

Test point $(7, 0)$

$x^2 - 9y^2 \geq 36$

$(7)^2 - 9(0)^2 \geq 36$

$49 - 0 \geq 36$

$49 \geq 36$ True

Since the inequality is a true statement for Regions I and III, those regions are part of the graph of $x^2 - 9y^2 \geq 36$. So Region I, Region III, and the boundary hyperbola constitute the graph of the inequality, which is shown on the coordinate plane to the right.

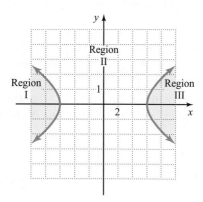

PRACTICE 2

Graph the solutions of $4x^2 - 25y^2 \leq 100$.

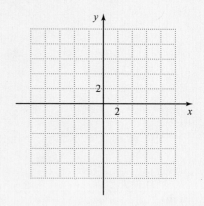

> **TIP** When graphing nonlinear inequalities:
> - draw a solid curve for an inequality that involves either the symbol $\leq$ or $\geq$.
> - draw a broken curve for an inequality that involves either the symbol $<$ or $>$.

Solving Nonlinear Systems of Inequalities

In Section 4.5, we solved systems of linear inequalities in two variables by graphing each inequality on the same coordinate plane. In such systems, every point in the region of overlap is a solution of the system. Here, we use the same general methods for graphing nonlinear systems of inequalities in two variables. As with systems of linear inequalities, **a solution of a system of nonlinear inequalities** is an ordered pair of real numbers that makes each inequality in the system true.

EXAMPLE 3

Graph the solutions of the following system:

$$x^2 + y^2 \leq 16$$
$$y \leq x - 4$$

Solution   Let's begin by graphing each inequality. First, we graph $x^2 + y^2 \leq 16$. The boundary is a circle, the graph of the equation $x^2 + y^2 = 16$, and is a solid curve because the original inequality involves the symbol $\leq$. The test point $(0, 0)$ gives us a true statement for the inequality $x^2 + y^2 \leq 16$. So the inner region of the circle is part of our graph, which we shade in as follows:

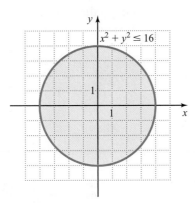

Next, we graph $y \leq x - 4$. The boundary line is the graph of $y = x - 4$, drawn as a solid line since the original inequality involves the symbol $\leq$. The test point $(0, 0)$ gives us a false statement for the inequality $y \leq x - 4$. So we shade the region below the line, as shown to the right.

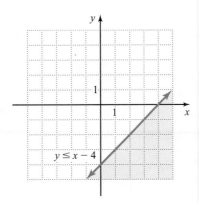

PRACTICE 3

Solve by graphing:

$$x + y \geq 3$$
$$x^2 + y^2 \leq 25$$

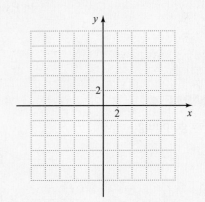

EXAMPLE 3 (continued)

Finally, we draw each inequality on the same coordinate plane. A solution of the nonlinear system of inequalities must satisfy each inequality. So the solutions are all the points that lie in *both* shaded regions, that is, in the purple overlapping region of the two graphs, including the boundary of the region.

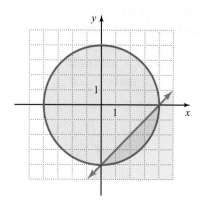

In Example 3, can you describe the overlapping region of the two graphs? Explain.

EXAMPLE 4

Solve by graphing:

$$\frac{x^2}{16} + \frac{y^2}{25} < 1$$

$$\frac{x^2}{25} - \frac{y^2}{4} > 1$$

Solution We graph each inequality separately on the same coordinate plane. First, we draw the ellipse $\frac{x^2}{16} + \frac{y^2}{25} = 1$ as a broken curve. The test point $(0, 0)$ gives us a true statement in the inequality $\frac{x^2}{16} + \frac{y^2}{25} < 1$, so we shade the region containing this point. Next, we graph the hyperbola $\frac{x^2}{25} - \frac{y^2}{4} = 1$ as a broken curve.

In the inequality $\frac{x^2}{25} - \frac{y^2}{4} > 1$, the test points $(-6, 0)$ and $(6, 0)$ give us true statements and the test point $(0, 0)$ gives us a false statement. So we shade the regions containing the points $(-6, 0)$ and $(6, 0)$. The graphs have no overlapping regions. So there are no real solutions.

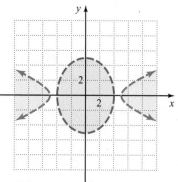

PRACTICE 4

Graph the solutions of the following system:

$$x^2 + y^2 \leq 1$$

$$\frac{x^2}{9} + \frac{y^2}{4} \geq 1$$

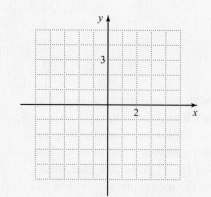

Now, let's look at how nonlinear systems of inequalities are applied in real-world situations.

EXAMPLE 5

In business, a company breaks even when its cost C equals its revenue R. When its cost is greater than its revenue, the company has a loss. When its revenue exceeds its cost, the company makes a profit. The profit region of a company that produces novelty items can be found by solving the following system:

$$C \geq 40x + 500$$
$$R \leq 100x - 0.2x^2$$

where x is the number of novelty items produced and sold each month. Find the profit region of this company.

Solution To solve the system, we graph each inequality on the same coordinate plane. First, we graph $C \geq 40x + 500$ as a solid line. Choose a test point, say $(0, 0)$. Since this test point does not satisfy the inequality, we shade the region above the line.

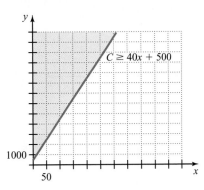

Next, we draw the parabola $R = 100x - 0.2x^2$ as a solid curve. Then, we choose a test point, say $(50, 1000)$. Since the test point $(50, 1000)$ satisfies the inequality, we shade the region below the parabola.

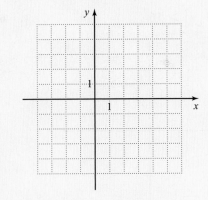

The profit region of the company is the overlapping region.

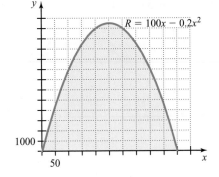

PRACTICE 5

Two underground water sprinklers are placed 2 ft apart. Each sprinkler waters a circular area of the lawn within a 3-foot radius. The region of lawn that the two sprinklers both water can be found by solving the following system:

$$x^2 + y^2 \leq 9$$
$$(x - 2)^2 + y^2 \leq 9$$

Graph this region.

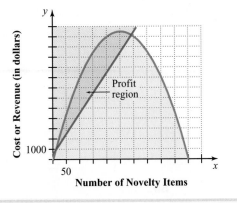

Mathematically Speaking

Fill in each blank with the most appropriate term or phrase from the given list.

solid	each	broken
below	at least one	above

1. In graphing a nonlinear inequality that involves either the $<$ or the $>$ symbol, a(n) _____ curve is drawn.

2. In graphing a nonlinear inequality that involves either the $\leq$ or the $\geq$ symbol, a(n) _____ curve is drawn.

3. A solution of a nonlinear system of inequalities must satisfy _____ inequality in the system.

4. The graph of $y < 3x^2 - 7x$ is the region _____ the parabola $y = 3x^2 - 7x$.

A *Graph the solutions of each nonlinear inequality.*

5. $x^2 + y^2 > 16$

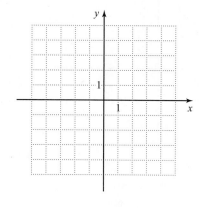

6. $x^2 + y^2 \geq 4$

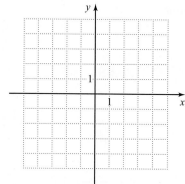

7. $(x - 1)^2 + (y + 1)^2 \leq 25$

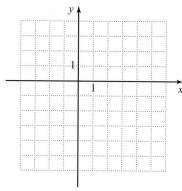

8. $(x + 3)^2 + (y - 2)^2 < 1$

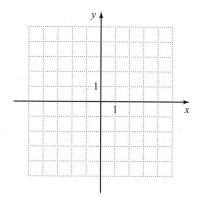

9. $y > -x^2 + 5x - 4$

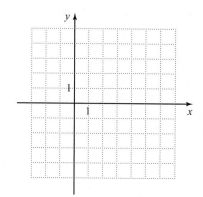

10. $y < 2x^2 - 8x$

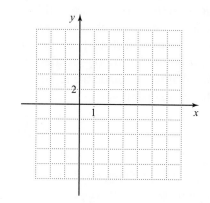

11. $9x^2 + y^2 \geq 36$

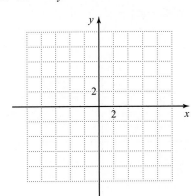

12. $x^2 + 16y^2 > 64$

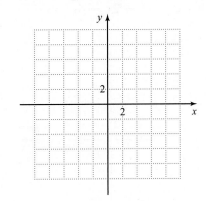

13. $16x^2 + 36y^2 < 144$

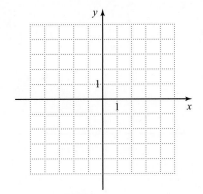

14. $25x^2 + 9y^2 \leq 225$

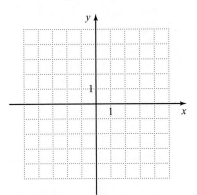

15. $4x^2 - y^2 > 100$

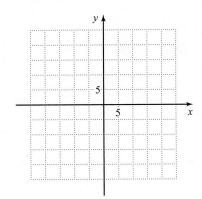

16. $2y^2 - 8x^2 < 32$

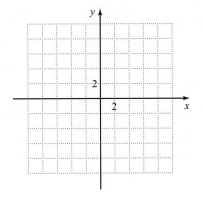

B *Graph the solutions of each nonlinear system of inequalities.*

17. $y > x^2 - 3$
      $y \leq -x + 1$

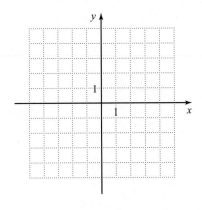

18. $y \leq -x^2 + 5$
      $y \leq \dfrac{1}{2}x - 1$

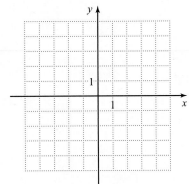

19. $x^2 + y^2 \geq 9$
      $x^2 + y^2 < 25$

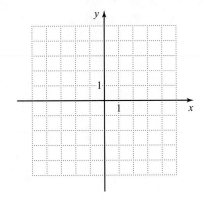

20. $x^2 + y^2 > 16$
$x^2 + y^2 < 4$

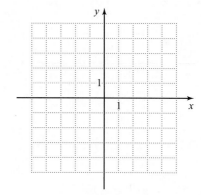

21. $y > x^2 - 3x + 2$
$y \le -x^2 - 2x - 4$

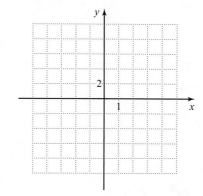

22. $y < x^2 + 2x + 1$
$y < -x^2 - 2x - 4$

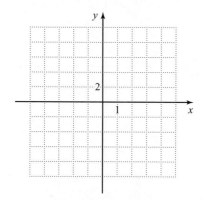

23. $8x^2 + 2y^2 \le 72$
$x^2 + 4y^2 \le 36$

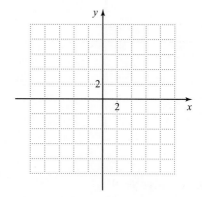

24. $12x^2 + 27y^2 > 108$
$25x^2 + \quad y^2 < \quad 25$

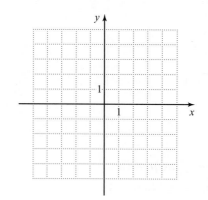

25. $x^2 - 4y^2 < 16$
$x^2 + \quad y^2 \ge \quad 1$

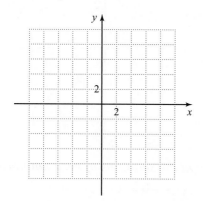

26. $16x^2 + 4y^2 > 64$
$y^2 - 9x^2 \le 36$

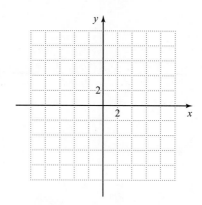

Mixed Practice

Graph the solutions of each nonlinear system of inequalities.

27. $x^2 + y^2 < 4$
$\quad\ y \geq x^2 + 3$

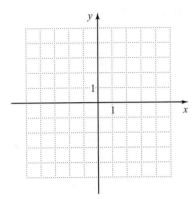

28. $x > y^2 - 2y + 3$
$\quad\ x \geq -y^2 + 2y - 3$

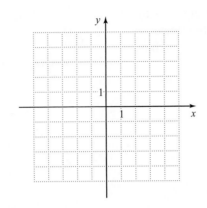

29. $x^2 + 16y^2 > 16$
$\quad\ 4x^2 + 9y^2 < 36$

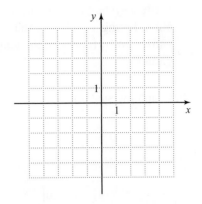

Graph the solutions of each nonlinear inequality.

30. $y \leq -x^2 + 2x + 3$

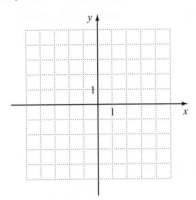

31. $4x^2 + y^2 > 16$

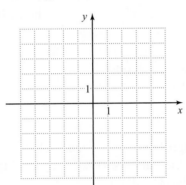

32. $9x^2 - y^2 \leq 36$

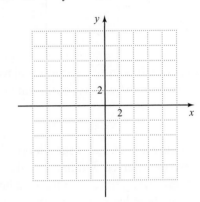

Applications

C *Solve.*

33. In a search-and-rescue mission, a team maps out an elliptical search region from the last known location of a pair of hikers. The search region can be represented by the inequality $4x^2 + 9y^2 \leq 144$. Graph this region.

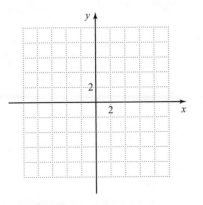

34. As a safety precaution, emergency workers evacuated residents within a 2-mile radius of a chemical spill. The evacuated region can be represented by the inequality $x^2 + y^2 \leq 4$. Graph this region.

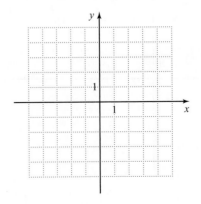

35. A homeowner wants to plant a flower garden and a vegetable garden on two separate square plots of land. She determines that the length of the vegetable garden is to be at least 2 ft longer than the length of the flower garden and that the combined area of the two gardens is not to exceed 2500 ft².

    a. Write a system of inequalities to represent the situation, letting x represent the length of the flower garden and y represent the length of the vegetable garden.

    b. Graph the system.

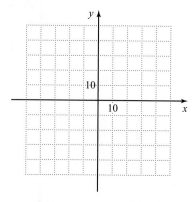

    c. Are all points in the solution region possible solutions within the context of the situation? Explain.

36. A company determines that its cost (in dollars) of producing x machine parts is given by $C = 0.3x^2 + 9$ and that its revenue (in dollars) for selling x machine parts is given by $R = 12x - 0.2x^2$. The profit region for producing and selling machine parts can be found by solving the following system:

$$C \geq 0.3x^2 + 9$$
$$R \leq 12x - 0.2x^2$$

    a. Graph the system.

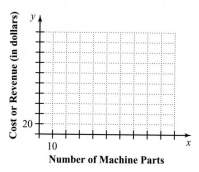

    b. Is every point in the solution region a possible solution within the context of the situation? Explain.

37. A classic rock radio station broadcasts a signal that can be received within a 30-mile radius of the station's transmitting tower. An R&B radio station located 50 mi away broadcasts a signal that can be received within a 40-mile radius of its tower. The system of inequalities

$$x^2 + y^2 \leq 900$$
$$(x - 50)^2 + y^2 \leq 1600$$

can be used to determine the region in which both radio station signals can be received. Find this region.

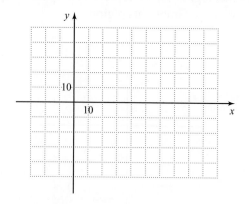

38. A hotel builds an elliptical swimming pool surrounded by an elliptical walkway. The system of inequalities

$$\frac{x^2}{400} + \frac{y^2}{225} \geq 1$$
$$\frac{x^2}{625} + \frac{y^2}{400} \leq 1$$

represents the region used for the walkway. Show this region.

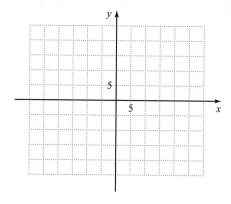

• Check your answers on page A-55.

MINDStretchers

Mathematical Reasoning

1. Is it possible for a single point to be the only solution of a system of nonlinear inequalities? If so, give an example of such a system. If not, explain why.

Critical Thinking

2. Graph the solution of the following system:

$$x^2 + 9y^2 \geq 36$$
$$x^2 - 4y^2 < 16$$
$$x - 2y > 4$$

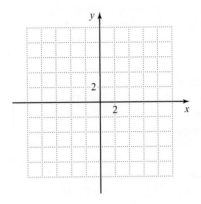

Groupwork

3. Working with a partner, write a system of inequalities with the solutions shown on the following graph:

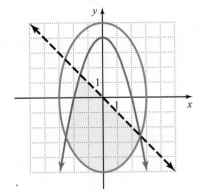

Key Concepts and Skills

CONCEPT   SKILL

Concept/Skill	Description	Example
[10.1] **Conic section**	A curve formed by the intersection of a plane and a (double) cone.	Circle, Ellipse, Parabola, Hyperbola
[10.1] **Equation of a parabola opening upward or downward**	The equation $y = a(x - h)^2 + k$ is written in standard form, where (h, k) is the vertex of the associated parabola and the equation of the axis of symmetry is $x = h$.	The equation $y = 2(x - 4)^2 + 5$ is in standard form with vertex $(4, 5)$ and axis of symmetry $x = 4$.
[10.1] **Equation of a parabola opening to the left or right**	The equation $x = a(y - k)^2 + h$ is written in standard form, where (h, k) is the vertex of the associated parabola and the equation of the axis of symmetry is $y = k$.	The equation $x = -2(y - 5)^2 + 4$ is in standard form with vertex $(4, 5)$ and axis of symmetry $y = 5$.
[10.2] **Distance formula**	The distance between any two points (x_1, y_1) and (x_2, y_2) on a coordinate plane is given by the formula $$d = \sqrt{(x_2 - x_1)^2 + (y_2 - y_1)^2}.$$	The distance between $(2, 6)$ and $(-3, 1)$ is $$\begin{aligned} d &= \sqrt{(-3 - 2)^2 + (1 - 6)^2} \\ &= \sqrt{(-5)^2 + (-5)^2} \\ &= \sqrt{25 + 25} \\ &= \sqrt{50} \\ &= 5\sqrt{2} \quad \text{units} \end{aligned}$$
[10.2] **Midpoint formula**	The line segment with endpoints (x_1, y_1) and (x_2, y_2) has as its midpoint $$\left(\frac{x_1 + x_2}{2}, \frac{y_1 + y_2}{2} \right).$$	The midpoint of the line segment joining $(6, 7)$ and $(-3, 5)$ is $$\left(\frac{6 + (-3)}{2}, \frac{7 + 5}{2} \right) = \left(\frac{3}{2}, 6 \right).$$
[10.2] **Circle**	The set of all points on a coordinate plane that are a fixed distance from a given point. The given point is called the **center** and is represented by (h, k). The fixed distance is called the **radius** of the circle and is represented by r.	
[10.2] **Equation of a circle centered at the origin**	The equation of a circle with center $(0, 0)$ and radius r is $x^2 + y^2 = r^2$.	$x^2 + y^2 = 9^2$ is the equation of a circle with center $(0, 0)$ and radius 9.
[10.2] **Equation of a circle centered at (h, k)**	The equation $(x - h)^2 + (y - k)^2 = r^2$, written in standard form, where (h, k) is the center of the associated circle and r is the radius.	$(x - 3)^2 + (y - 2)^2 = 7^2$ is the equation of a circle with center $(3, 2)$ and radius 7.

CONCEPT	SKILL

Concept/Skill	Description	Example
[10.3] **Ellipse**	The set of all points on a coordinate plane the *sum* of whose distances from two fixed points is constant. The fixed points of this graph are called **foci** (the plural of **focus**). The point halfway between the two foci is called the **center** of the ellipse.	
[10.3] **Equation of an ellipse centered around the origin**	The equation $\dfrac{x^2}{a^2} + \dfrac{y^2}{b^2} = 1$, written in standard form, where $(0,0)$ is the center of the associated ellipse, $(-a, 0)$ and $(a, 0)$ are the x-intercepts, and $(0, -b)$ and $(0, b)$ are the y-intercepts.	$\dfrac{x^2}{3^2} + \dfrac{y^2}{4^2} = 1$ is the equation of an ellipse with center $(0,0)$, x-intercepts $(-3, 0)$ and $(3, 0)$, and y-intercepts $(0, -4)$ and $(0, 4)$.
[10.3] **Hyperbola**	The set of all points on a coordinate plane that satisfy the following property: The *difference* of the distances from two fixed points is constant. Each fixed point is said to be a **focus** of the hyperbola. The point halfway between the two foci is called the **center** of the hyperbola. Each **branch** of a hyperbola extends indefinitely, getting closer and closer to two straight lines called **asymptotes**.	
[10.3] **Equation of a hyperbola centered at the origin**	The equation $\dfrac{x^2}{a^2} - \dfrac{y^2}{b^2} = 1$, written in standard form, where $(0,0)$ is the center of the associated hyperbola and $(-a, 0)$ and $(a, 0)$ are the x-intercepts. The equation $\dfrac{y^2}{b^2} - \dfrac{x^2}{a^2} = 1$, written in standard form, where $(0,0)$ is the center of the associated hyperbola and $(0, -b)$ and $(0, b)$ are the y-intercepts.	$\dfrac{x^2}{5^2} - \dfrac{y^2}{6^2} = 1$ is the equation of a hyperbola with center $(0,0)$ and x-intercepts $(-5, 0)$ and $(5, 0)$. $\dfrac{y^2}{6^2} - \dfrac{x^2}{5^2} = 1$ is the equation of a hyperbola with center $(0,0)$ and y-intercepts $(0, -6)$ and $(0, 6)$.
[10.4] **Nonlinear system of equations**	A system of equations in which at least one equation is nonlinear.	$x^2 + y^2 = 25$ $x - y = 1.$
[10.4] **Solution of a system of nonlinear equations**	An ordered pair of numbers that makes each equation in the system true.	The ordered pairs $(4, 3)$ and $(-3, -4)$ are solutions of the system $x^2 + y^2 = 25$ $x - y = 1.$
[10.5] **Solution of a system of nonlinear inequalities**	An ordered pair of real numbers that makes each inequality in the system true.	The ordered pair $(1, 2)$ is a solution of the system of inequalities $x^2 + y^2 \le 16$ $y \le x + 4.$

continued

Summary of Conic Sections

Conic Section	Graph	Equation	Comment on Equation
Parabola		$y = ax^2 + bx + c$ $y = a(x - h)^2 + k$ for $a > 0$	Contains an x^2-term, but no y^2-term.
		$y = ax^2 + bx + c$ $y = a(x - h)^2 + k$ for $a < 0$	Contains an x^2-term, but no y^2-term.
		$x = ay^2 + by + c$ $x = a(y - k)^2 + h$ for $a > 0$	Contains a y^2-term, but no x^2-term.
		$x = ay^2 + by + c$ $x = a(y - k)^2 + h$ for $a < 0$	Contains a y^2-term, but no x^2-term.
Circle centered at the origin		$x^2 + y^2 = r^2$	Contains an x^2-term and a y^2-term with the same positive coefficient.
Circle centered at (h, k)		$(x - h)^2 + (y - k)^2 = r^2$	Contains an x^2-term and a y^2-term with the same positive coefficient.
Ellipse centered at the origin		$\dfrac{x^2}{a^2} + \dfrac{y^2}{b^2} = 1$	Contains an x^2-term and a y^2-term with different positive coefficients.

Conic Section	Graph	Equation	Comment on Equation
Hyperbola centered at the origin		$$\frac{x^2}{a^2} - \frac{y^2}{b^2} = 1$$	Contains an x^2-term with a positive coefficient and a y^2-term with a negative coefficient.
		$$\frac{y^2}{b^2} - \frac{x^2}{a^2} = 1$$	Contains a y^2-term with a positive coefficient and an x^2-term with a negative coefficient.

Say Why

Fill in each blank.

1. The graph of the equation $y = (x - 8)^2 + 4$ _____ a parabola because _____
   is/is not
   _____.

2. The graph of the equation $x^2 + 6x + y^2 - 2y = 15$ _____ a circle because _____
   is/is not
   _____.

3. The graph of the equation $2x^2 + 5y^2 = 10$ _____ an ellipse because _____
   is/is not
   _____.

4. The parabola with equation $y = -4x^2 + 2x - 2$ _____ open upward because _____
   does/does not
   _____.

5. The graph of the equation $3x^2 - 5y^2 = 15$ _____ have two branches because _____
   does/does not
   _____.

6. The distance between the origin and the point $(-1, 7)$ _____ less than 7 because _____
   is/is not
   _____.

[10.1] *Identify the vertex and axis of symmetry of each parabola. Then, graph.*

7. $y = -2(x + 3)^2 + 5$

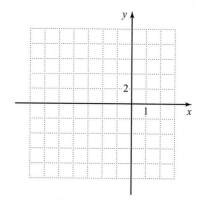

8. $x = 3(y - 2)^2 - 10$

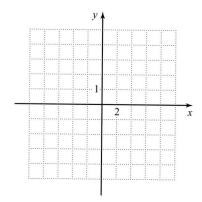

9. $x = 4y^2 + 32y + 64$

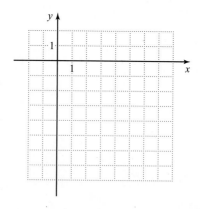

10. $y = -x^2 - 6x - 7$

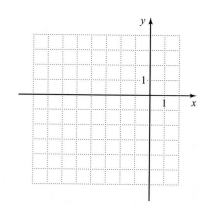

Write each equation in standard form. Then, identify the vertex of each parabola.

11. $y = 8x^2 - 56x + 74$

12. $x = 5y^2 + 15y + 7$

[10.2] *Find the distance between the two points on the coordinate plane. If necessary, round the answer to the nearest tenth.*

13. $(9, -5)$ and $(6, 1)$

14. $(-8, 10)$ and $(-2, 11)$

15. $(-1.3, -0.7)$ and $(-1.8, 0.5)$

16. $(\sqrt{7}, -\sqrt{2})$ and $(-2\sqrt{7}, -5\sqrt{2})$

Find the coordinates of the midpoint of the line segment joining each pair of points.

17. $(-16, 9)$ and $(24, 17)$

18. $(-7, -13)$ and $(10, -5)$

19. $\left(\dfrac{5}{2}, \dfrac{1}{5}\right)$ and $\left(-\dfrac{3}{4}, \dfrac{3}{10}\right)$

20. $(-0.35, -2.8)$ and $(-1.05, -4.5)$

Find the center and radius of each circle. Then, graph.

21. $x^2 + y^2 = 64$

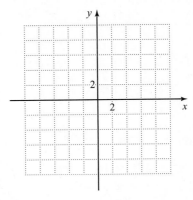

22. $(x + 2)^2 + (y - 4)^2 = 4$

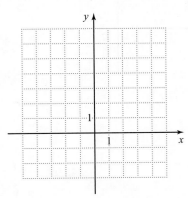

23. $x^2 + y^2 - 4x + 6y - 12 = 0$

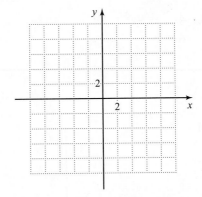

24. $x^2 + y^2 + 8x - 2y + 8 = 0$

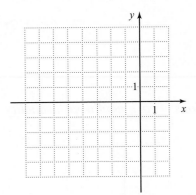

Use the given information to find the equation of each circle.

25. Center: $(9, 0)$; radius: 13

26. Center: $(-6, 10)$; radius: $2\sqrt{5}$

Find the center and radius of each circle.

27. $x^2 + y^2 - 12x - 14y - 35 = 0$

28. $x^2 + y^2 + 16x + 10y + 9 = 0$

[10.3] Graph.

29. $\dfrac{x^2}{64} + \dfrac{y^2}{9} = 1$

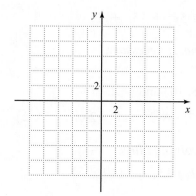

30. $\dfrac{x^2}{16} - \dfrac{y^2}{36} = 1$

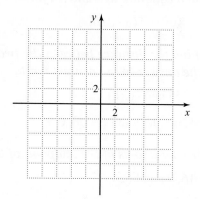

31. $y^2 - 4x^2 = 4$

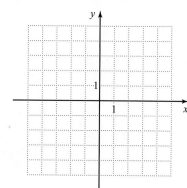

32. $4x^2 + y^2 = 64$

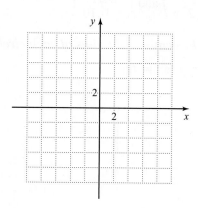

[10.4] Solve.

33. $y = 2x^2 + 1$
    $y = 2x + 5$

34. $x^2 + y^2 = 10$
    $y = 3x$

35. $x^2 + 4y^2 = 16$
    $x = y^2 - 4$

36. $x^2 + y^2 = 8$
    $3x^2 + y^2 = 12$

37. $6x^2 + 2y^2 = 16$
    $2x^2 - y^2 = 17$

38. $4x^2 - 9y^2 = 36$
    $6x^2 + 6y^2 = 54$

[10.5] Graph the solutions of each nonlinear inequality.

39. $y < -x^2 + 1$

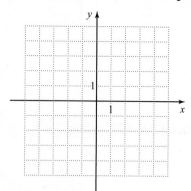

40. $(x + 3)^2 + y^2 \geq 4$

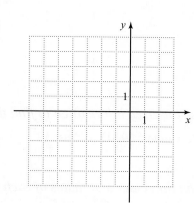

41. $8x^2 + 2y^2 \leq 32$

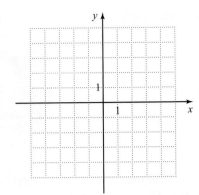

42. $9y^2 - 36x^2 > 144$

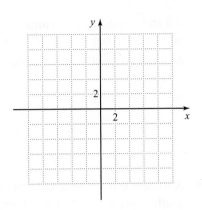

Solve each nonlinear system of inequalities by graphing.

43. $x^2 + y^2 < 64$

$y \geq \dfrac{1}{2}x + 3$

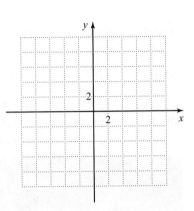

44. $x \geq y^2 - 4$

$x \leq -y^2 + 4$

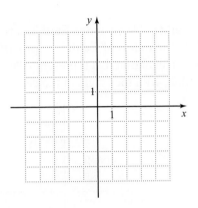

45. $x^2 + y^2 < 16$
$x^2 + y^2 > 49$

46. $4x^2 + 25y^2 \leq 100$
$9x^2 - \quad y^2 \geq 9$

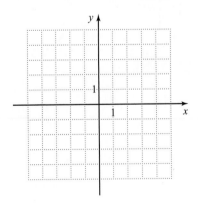

Mixed Applications

Solve.

47. A small manufacturing plant determines that its revenue R (in dollars) from selling x units of a product is given by $R = 20x - 0.5x^2$.

   a. Write the equation in standard form.

   b. Identify the vertex and explain its significance within the context of the situation.

48. An arch supporting a bridge is in the shape of a parabola. The maximum height of the arch is 50 ft and it spans 80 ft across a river.

   a. Find an equation that represents the parabolic arch of the bridge.

   b. What is the height of the arch 24 ft from the center of the arch?

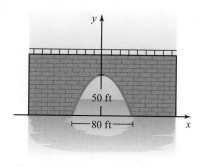

49. A student leaving his apartment drives 8 mi east and 15 mi south to get to school. From there, he drives 4 mi west and 7 mi north to get to work. Find the distance between the student's apartment and work.

50. An amusement park Ferris wheel has a diameter of 120 ft. Find an equation that represents the wheel. Assume the center of the wheel is at the origin of a coordinate plane.

51. A satellite 500 mi above Earth's surface has a circular orbit. The radius of Earth is approximately 4000 mi. Find an equation that represents the circular orbit of the satellite, with the origin at the center of Earth.

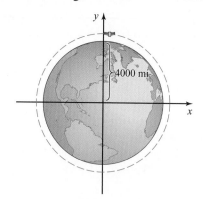

52. The dwarf planet Pluto travels in an elliptical orbit with the Sun at one focus. An equation that approximates Pluto's orbit is $\dfrac{x^2}{16} + \dfrac{y^2}{15} = 1$, where units are in billions of kilometers. The minimum distance between Pluto and the Sun is approximately 3 billion km. What is the maximum distance between Pluto and the Sun?

53. Suppose the playing surface of an elliptical pool table is 8 ft long and 4 ft wide. Find an equation of the ellipse that represents the playing surface of the pool table. Assume that the center of the pool table is at the origin of a coordinate plane.

54. As part of a physics experiment, a student throws a tennis ball straight downward with an initial velocity of 32 ft/sec from a height of 120 ft. The height h (in feet) of the ball above the ground t sec after it is thrown is given by $h = -16t^2 - 32t + 120$. At the same time, another student drops a golf ball from a height of 80 ft. The height of the golf ball above the ground t sec after it is dropped is given by $h = -16t^2 + 80$. How long after the balls are released will they be at the same height?

55. In a particular school district, a student living within a 1-mile radius of the school is not eligible for bus service.

   a. Write an inequality that represents the eligibility region. Assume that the school is at the origin of a coordinate plane.

   b. Graph this region.

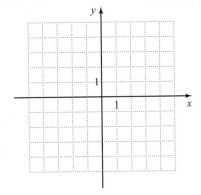

56. The owner of a daycare center wants to build two square sandboxes—one in the toddler playground and one in the preschool playground. The combined areas of the two sandboxes must not exceed 225 ft². The length of the sandbox in the preschool playground is to be greater than the length of the sandbox in the toddler playground. The system of inequalities

$$x^2 + y^2 \le 225$$
$$x < y,$$

where x represents the length of the sandbox in the toddler playground and y represents the length of the sandbox in the preschool playground, can be used to determine the possible dimensions of the sandboxes.

   a. Graph this system.

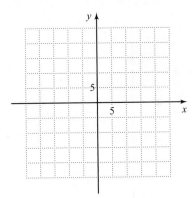

   b. Are all points in the solution region possible solutions within the context of the situation? Explain.

• Check your answers on page A-55.

CHAPTER 10 Posttest

FOR EXTRA HELP

CHAPTER **Test Prep** VIDEOS

The Chapter Test Prep Videos with test solutions are available on DVD, in MyMathLab, and on You**Tube** ™ (search "AkstIntermediate Alg" and click on "Channels").

To see if you have mastered the topics in this chapter, take this test.

1. Write in standard form the equation of the parabola $x = -2y^2 + 16y - 91$. Then, identify the vertex.

2. Find the distance between each pair of points.
   a. $(-2, -1)$ and $(-4, 1)$
   b. $(7.9, -2.4)$ and $(-4.1, 2.6)$

3. Find the midpoint of the line segment joining each pair of points.
   a. $(-8, -15)$ and $(-4, 9)$
   b. $(-3, 7)$ and $(-6, 3)$

4. Find the equation of a circle with center $(-2, 8)$ and radius $3\sqrt{2}$.

5. Find the center and radius of the circle $x^2 + y^2 + 10x - 2y + 14 = 0$.

Graph.

6. $x = 3(y - 1)^2 - 5$

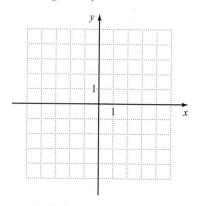

7. $x^2 + y^2 = 144$

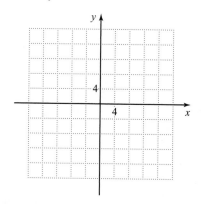

8. $(x - 4)^2 + (y + 5)^2 = 1$

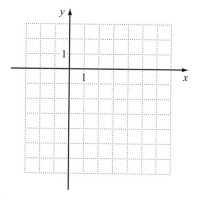

9. $\dfrac{x^2}{100} + \dfrac{y^2}{64} = 1$

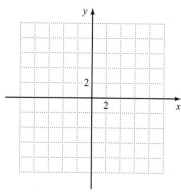

10. $64x^2 + 4y^2 = 256$

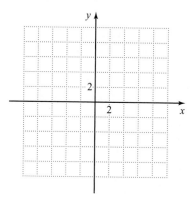

11. $\dfrac{y^2}{49} - \dfrac{x^2}{25} = 1$

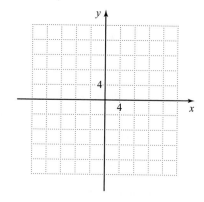

12. $6x^2 - 54y^2 = 216$

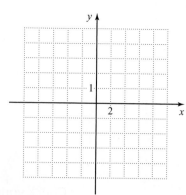

Solve.

13. $x^2 + y^2 = 25$
$\qquad y = x + 5$

14. $y = x^2 + 4$
$\qquad y = 2x^2 - 6$

15. $3x^2 + y^2 = 9$
$\qquad 9x^2 + 2y^2 = 15$

Solve by graphing.

16. $x^2 + 9y^2 > 36$

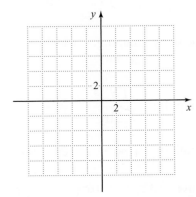

17. $4x^2 + 16y^2 \leq 64$
$\qquad 4x^2 - y^2 \leq 100$

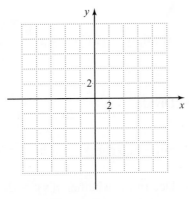

18. The curves of the cables suspended between two towers of a suspension bridge can be approximated by a parabola. Suppose that the distance between two towers that rise 110 m above the road is 900 m. If the center of the cable is 2 m above the road, find an equation in standard form that represents the parabola formed by the cable. (See diagram.)

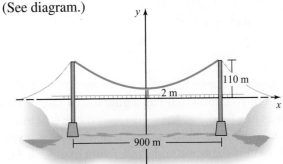

19. Cricket, a team sport similar to baseball, is played on an elliptical field. Find an equation of the ellipse shown.

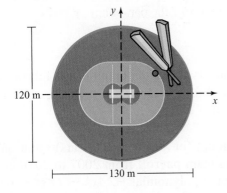

20. The owner of a factory determines that his cost C (in dollars) for producing x units of machine parts is $C = 0.5x^2 + 500$ and that his revenue R (in dollars) from selling x units is $R = 50x$. The profit region for producing and selling machine parts can be found by solving the following system:

$$C \geq 0.5x^2 + 500$$
$$R \leq 50x$$

a. Graph the system.

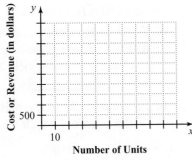

b. Is every point in the solution region a possible solution within the context of the situation? Explain.

● Check your answers on page A-57.

Cumulative Review Exercises

To help you review, solve the following problems.

1. Simplify: $-2[1 - (3 - 9)^2]$

2. Find the slope-intercept form of the equation of a line perpendicular to the line $y = \dfrac{3}{4}x - 1$ that contains the point $(9, 4)$.

3. Multiply: $(2x + y)(x - 3y)^2$

4. Factor by grouping: $18m^3 + 12m^2n + 3mn^2 + 2n^3$

5. Divide: $\dfrac{x + 3}{x^3 + 125} \div \dfrac{x^2 - 2x - 15}{x^2 - 25}$

6. Simplify: $\sqrt[3]{\dfrac{7w^3}{216z^9}}$

7. Solve and check: $\sqrt{2n + 5} - 2 = \sqrt{4n - 7}$

8. Multiply: $(-4 - 2i)(2 - 4i)$

9. Determine whether $f(x) = 2x - 3$ and $g(x) = \dfrac{x + 3}{2}$ are inverses of one another.

10. Solve: $32^{x-4} = 16^{2x}$

11. Write the following equation in standard form, and identify the vertex of the parabola: $y = 2x^2 + 12x + 23$

12. Find the radius and center of the circle $x^2 + y^2 - 10x - 4y + 25 = 0$.

13. Determine whether the graph of the following equation is a parabola, a circle, an ellipse, or a hyperbola: $25x^2 - 9y^2 = 225$

14. Solve: $\begin{aligned} x^2 + y^2 &= 21 \\ 4x^2 - y^2 &= -6 \end{aligned}$

15. The amount (in millions of dollars) spent on the Texas Rangers team payroll from 2007 to 2010 can be modeled by the polynomial $-2.4x^3 + 7.8x^2 - 6.0x + 68.3$, where x represents the number of years after 2007. The corresponding polynomial for the San Francisco Giants is $-1.6x^3 + 14.6x^2 - 26.6x + 90.2$. Express as a polynomial in x the difference between the San Francisco Giants payroll and the Texas Rangers payroll. (*Source:* usatoday.com)

16. An investor wants to invest twice the amount of money in a high-risk fund that has a guaranteed rate of return of $5\frac{1}{2}\%$ as in a fund that has a guaranteed rate of return of 4%. What is the minimum amount that should be invested in each fund if she wants to have a return of at least \$1500?

17. On a road trip, a group of students traveled 192 mi before stopping at a rest area. After leaving the rest area, they traveled an additional 224 mi at an average speed that was 8 mph slower than their average speed before stopping. If the entire trip took 7 hr of driving time, what was the average speed during each part of the trip?

18. A football punter kicked the football from ground level. The ball reached a maximum height of 35 ft, and hit the ground 150 ft from the kick.

    a. Write an equation that models the parabolic path of the football, where y is the height of the ball, x is the horizontal distance from the kick, and the origin is the spot from which the ball was kicked.

    b. Find the height of the ball when its horizontal distance from the kick was 45 ft.

19. The amount A in an account at an annual interest rate r compounded continuously after t yr is given by the formula $A = Pe^{rt}$, where P is the initial amount invested. If \$6800 is invested in an account at 3.85% interest compounded continuously, approximately how long will it take the amount in the account to grow to \$10,000?

20. A department store gift wrapper is required to wrap each box with at most 64 in^2 of wrapping paper. She needs to wrap a box with a square base and a height of 2 in.

   a. Find a formula for the surface area S of this box in terms of the side x of the square.

   b. What should the side of the base be to ensure that she can wrap the box with the given amount of wrapping paper?

• Check your answers on page A-57.

Appendix A.1

Table of Symbols

$+$	add
$-$	subtract
$\times, \cdot, (a)(b), 2y$	multiply
$\div, \dfrac{a}{b}, x + 1\overline{)x^2 - 1}$	divide
$=$	is equal to
$\approx$	is approximately equal to
$\neq$	is not equal to
$<$	is less than
$\leq$	is less than or equal to
$>$	is greater than
$\geq$	is greater than or equal to
$\cup$	the union of two sets
$\cap$	the intersection of two sets
$(\ )$	parentheses (a grouping symbol)
$[\ ]$	brackets (a grouping symbol)
$\{\ \}$	braces (a grouping symbol)
(a, b)	the set of real numbers between a and b, excluding a and b
$[a, b]$	the set of real numbers between a and b, including a and b
$[a, b)$	the set of real numbers between a and b, including a but excluding b
$(a, b]$	the set of real numbers between a and b, excluding a but including b
$\{a, b, c\}$	the set with elements a, b, and c
∞	infinity
π	pi, an irrational number approximately equal to 3.14
e	an irrational number approximately equal to 2.718
$-a$	the opposite, or additive inverse, of a
$\dfrac{1}{a}$	the reciprocal, or multiplicative inverse, of a
$|n|$	the absolute value of n
x^n	x raised to the power n
$\sqrt{a}$	the principal square root of a
$\sqrt[n]{a}$	the nth root of a
(x, y)	an ordered pair whose first coordinate is x and whose second coordinate is y
$f(x)$	the function f of x

$f^{-1}(x)$	the inverse function of $f(x)$
$f \circ g$	the composition of functions f and g
i	the imaginary number $\sqrt{-1}$
$a + bi$	a complex number
log	the common logarithm
ln	the natural logarithm
°	degree (for angles)
$\overleftrightarrow{AB}$	line AB
$\overline{AB}$	line segment AB
$\angle A$	angle A

Appendix A.2

Introduction to Graphing Calculators

This appendix covers the basic graphing features of a graphing calculator (or graphing software) used in this text. Note that the keystrokes and screens presented here and in the calculator inserts found throughout the text may be different from those on your graphing calculator. Refer to your user's manual for specific information and instructions on accessing and using the features of your particular model.

Graphing Equations

To graph an equation, it must be entered into the graphing calculator's *equation editor* in "$y =$" form. On many graphing calculators, the equation editor can be displayed by pressing the $\boxed{Y =}$ key. Note that you may enter more than one equation in the equation editor.

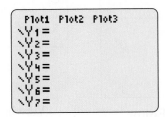

Equation Editor

For example, to graph the equation $4x + 2y = 6$, you must first solve the equation for y, getting $y = -2x + 3$. Then, enter the expression $-2x + 3$ to the right of $\backslash Y1 =$ in the equation editor. Finally, pressing the $\boxed{GRAPH}$ key displays the graph on a coordinate plane.

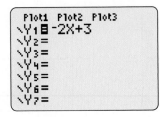
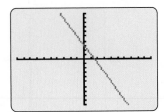

Note that most graphing calculators have two separate keys to distinguish between subtraction and negation. The $\boxed{-}$ key is used for subtraction and the $\boxed{(-)}$ key is used for negation.

You must be careful when entering equations containing fractions. To ensure that the graphing calculator interprets the input correctly, you may need to enclose all or part of the fraction in parentheses. For instance, to graph the equation $y = \dfrac{1}{2}x$, enter the expression $(1/2)x$ to the right of $\backslash Y1 =$ in the equation editor. Then, press $\boxed{GRAPH}$ to display the graph of the equation.

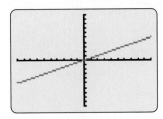

Viewing Windows

The screen on which a graph is displayed is called the *viewing window*, and it represents a portion of a coordinate plane. The viewing window is defined by the following values:

Xmin: the minimum x value displayed on the x-axis

Xmax: the maximum x value displayed on the x-axis

Xscl: the distance between adjacent tick marks on the x-axis

Ymin: the minimum y value displayed on the y-axis

Ymax: the maximum y value displayed on the y-axis

Yscl: the distance between adjacent tick marks on the y-axis

You can set the viewing window by entering the minimum and maximum values and the scales for the axes in the *window editor*. The window editor can be displayed by pressing the WINDOW key. The screen on the left shows the window editor, and displays the settings for the corresponding *standard viewing window* shown on the right.

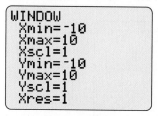

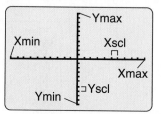

Window Editor   **Viewing Window**

The graph of an equation can be easily misinterpreted if an inappropriate viewing window is selected. For instance, compare the graph of the quadratic equation $y = x^2 - 15x + 14$, as shown in the following three viewing windows:

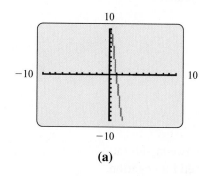

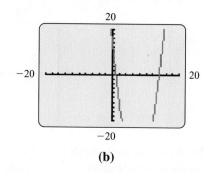

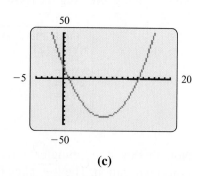

(a)   **(b)**   **(c)**

Although each of these viewing windows displays the graph of the equation, the viewing window in (c) is best because it shows all the key features of the graph, that is, the x- and y-intercepts and the vertex. Selecting an appropriate viewing window may require some practice, but familiarity with key features of the graphs of the equations discussed in the text will facilitate this process.

Other graphing calculator features, such as TRACE, ZOOM, and INTERSECT, are discussed in calculator inserts throughout the text.

Appendix A.3

Determinants and Cramer's Rule

In Chapter 4, we discussed several methods of solving systems of two or three linear equations. Here we consider another method, based on the concept of *determinants*.

To begin, recall from Section 4.4 that a *matrix* is a rectangular array of numbers. The numbers of the matrix are called *elements* or *entries*. The *rows* of the matrix are horizontal, and the *columns* are vertical.

The following are examples of **square matrices**, that is, matrices that have the same number of rows as columns:

$$\begin{bmatrix} -2 & 0 \\ 0 & 3 \end{bmatrix} \longleftarrow \text{2} \times \text{2 (read ``two by two'') matrix}$$

$$\begin{bmatrix} 5 & 1 & 2 \\ 0 & 3 & -1 \\ 4 & -2 & 1 \end{bmatrix} \longleftarrow \text{3} \times \text{3 matrix}$$

Determinants of 2 × 2 Matrices and Cramer's Rule

A determinant is a real number associated with a square matrix. Let's consider determinants of 2 × 2 matrices.

> **DEFINITION**
> The **determinant of the 2 × 2 matrix** $\begin{bmatrix} a & b \\ c & d \end{bmatrix}$, written $\begin{vmatrix} a & b \\ c & d \end{vmatrix}$, is the real number $ad - bc$.

EXAMPLE 1

Evaluate $\begin{vmatrix} 4 & -1 \\ 3 & 7 \end{vmatrix}$.

Solution Here $a = 4, b = -1, c = 3$, and $d = 7$.

$$\begin{vmatrix} 4 & -1 \\ 3 & 7 \end{vmatrix} = ad - bc$$
$$= (4)(7) - (-1)(3)$$
$$= 28 + 3$$
$$= 31$$

PRACTICE 1

Find the value of the following determinant:

$$\begin{vmatrix} 5 & -2 \\ 8 & -3 \end{vmatrix}$$

We can apply the concept of determinant in solving a system of linear equations. Consider the general system of two linear equations in two unknowns, x and y.

$$a_1 x + b_1 y = k_1$$
$$a_2 x + b_2 y = k_2$$

If we solve this system by elimination, the result is:

$$x = \frac{k_1 b_2 - k_2 b_1}{a_1 b_2 - a_2 b_1} \quad \text{and} \quad y = \frac{a_1 k_2 - a_2 k_1}{a_1 b_2 - a_2 b_1},$$

where $a_1 b_2 - a_2 b_1 \neq 0$.

We can write the values of x and y in terms of the following determinants:

$$x = \frac{\begin{vmatrix} k_1 & b_1 \\ k_2 & b_2 \end{vmatrix}}{\begin{vmatrix} a_1 & b_1 \\ a_2 & b_2 \end{vmatrix}} \quad \text{and} \quad y = \frac{\begin{vmatrix} a_1 & k_1 \\ a_2 & k_2 \end{vmatrix}}{\begin{vmatrix} a_1 & b_1 \\ a_2 & b_2 \end{vmatrix}}$$

It is common practice to represent the numerator for x as D_x, the numerator for y as D_y, and the denominator of both variables as D. Note the relationship between these determinants and the numbers in the original system:

$$D_x = \begin{vmatrix} k_1 & b_1 \\ k_2 & b_2 \end{vmatrix} \qquad D_y = \begin{vmatrix} a_1 & k_1 \\ a_2 & k_2 \end{vmatrix} \qquad D = \begin{vmatrix} a_1 & b_1 \\ a_2 & b_2 \end{vmatrix}$$

Substituting the names of these determinants in the equations for x and y, we get:

$$x = \frac{D_x}{D} \quad \text{and} \quad y = \frac{D_y}{D}$$

We can now state the determinant formula for solving a system of two linear equations in two variables that is known as *Cramer's Rule*.

Cramer's Rule for Solving a System of Two Linear Equations in Two Unknowns

The solution to the system

$$a_1 x + b_1 y = k_1$$
$$a_2 x + b_2 y = k_2$$

is $x = \dfrac{D_x}{D}$ and $y = \dfrac{D_y}{D}$, where $D_x = \begin{vmatrix} k_1 & b_1 \\ k_2 & b_2 \end{vmatrix}$, $D_y = \begin{vmatrix} a_1 & k_1 \\ a_2 & k_2 \end{vmatrix}$, and $D = \begin{vmatrix} a_1 & b_1 \\ a_2 & b_2 \end{vmatrix}$,

where $D \neq 0$.

Note that for a given system of linear equations, if the value of D is equal to 0, then Cramer's Rule does not apply. In this case, there is not a unique solution, since either the system is inconsistent or the equations are dependent.

EXAMPLE 2

Use Cramer's Rule to solve the following system:

$$x + 3y = 2$$
$$x - 4y = 9$$

Solution   To begin, let's evaluate D_x, D_y, and D.

Constants

$$D_x = \begin{vmatrix} k_1 & b_1 \\ k_2 & b_2 \end{vmatrix} = \begin{vmatrix} 2 & 3 \\ 9 & -4 \end{vmatrix} = (2)(-4) - (3)(9) = -35$$

y-coefficients

Constants

$$D_y = \begin{vmatrix} a_1 & k_1 \\ a_2 & k_2 \end{vmatrix} = \begin{vmatrix} 1 & 2 \\ 1 & 9 \end{vmatrix} = (1)(9) - (2)(1) = 7$$

x-coefficients

x-coefficients

$$D = \begin{vmatrix} a_1 & b_1 \\ a_2 & b_2 \end{vmatrix} = \begin{vmatrix} 1 & 3 \\ 1 & -4 \end{vmatrix} = (1)(-4) - (3)(1) = -7$$

y-coefficients

Cramer's Rule gives us:

$$x = \frac{D_x}{D} = \frac{-35}{-7} = 5 \quad \text{and} \quad y = \frac{D_y}{D} = \frac{7}{-7} = -1$$

So the solution is $(5, -1)$. We leave the check as an exercise.

PRACTICE 2

Use Cramer's Rule to solve:

$$2x - y = 5$$
$$x + 5y = -14$$

Determinants of 3×3 Matrices and Cramer's Rule

Determinants of 3×3 matrices can be used to solve systems of three linear equations in three variables. Such determinants are more complex to evaluate than determinants of 2×2 matrices.

DEFINITION

The **3×3 determinant** $\begin{vmatrix} a_1 & b_1 & c_1 \\ a_2 & b_2 & c_2 \\ a_3 & b_3 & c_3 \end{vmatrix}$ is equal to

$$a_1 \cdot \begin{vmatrix} b_2 & c_2 \\ b_3 & c_3 \end{vmatrix} - a_2 \cdot \begin{vmatrix} b_1 & c_1 \\ b_3 & c_3 \end{vmatrix} + a_3 \cdot \begin{vmatrix} b_1 & c_1 \\ b_2 & c_2 \end{vmatrix}.$$

Note that the determinant of a 3 × 3 matrix is defined in terms of three related 2 × 2 matrices. Each of these smaller matrices is called a **minor** and can be found by considering the larger matrix and then crossing out the appropriate row and column in which each value of a appears, as shown below.

$$\begin{array}{ccc} a_1 & b_1 & c_1 \\ a_2 & b_2 & c_2 \\ a_3 & b_3 & c_3 \end{array} \quad \text{The minor of } a_1 \text{ is } \begin{vmatrix} b_2 & c_2 \\ b_3 & c_3 \end{vmatrix}.$$

$$\begin{array}{ccc} a_1 & b_1 & c_1 \\ a_2 & b_2 & c_2 \\ a_3 & b_3 & c_3 \end{array} \quad \text{The minor of } a_2 \text{ is } \begin{vmatrix} b_1 & c_1 \\ b_3 & c_3 \end{vmatrix}.$$

$$\begin{array}{ccc} a_1 & b_1 & c_1 \\ a_2 & b_2 & c_2 \\ a_3 & b_3 & c_3 \end{array} \quad \text{The minor of } a_3 \text{ is } \begin{vmatrix} b_1 & c_1 \\ b_2 & c_2 \end{vmatrix}.$$

So we conclude that:

$$\begin{vmatrix} a_1 & b_1 & c_1 \\ a_2 & b_2 & c_2 \\ a_3 & b_3 & c_3 \end{vmatrix} = a_1(\text{the minor of } a_1) - a_2(\text{the minor of } a_2) + a_3(\text{the minor of } a_3)$$

Evaluating the determinant of a 3 × 3 matrix in this way is called *expanding by the minors of the first column*. Such a determinant can, in fact, be evaluated by considering minors of any row or column as long as they are preceded by the appropriate sign, a subject pursued in more advanced texts.

EXAMPLE 3

Evaluate $\begin{vmatrix} 5 & -2 & 1 \\ 1 & -1 & 0 \\ -4 & 3 & 2 \end{vmatrix}$.

Solution   We evaluate the determinant of this 3 × 3 matrix expanding by the minors of the first column.

$$\begin{vmatrix} 5 & -2 & 1 \\ 1 & -1 & 0 \\ -4 & 3 & 2 \end{vmatrix} = 5 \cdot \begin{vmatrix} -1 & 0 \\ 3 & 2 \end{vmatrix} - 1 \cdot \begin{vmatrix} -2 & 1 \\ 3 & 2 \end{vmatrix} + (-4) \cdot \begin{vmatrix} -2 & 1 \\ -1 & 0 \end{vmatrix}$$

Alternate the signs.

$$= 5[(-1)(2) - (0)(3)] - 1[(-2)(2) - (1)(3)] + (-4)[(-2)(0) - (1)(-1)]$$
$$= (5)(-2) - (1)(-7) + (-4)(1)$$
$$= -7$$

PRACTICE 3

Evaluate $\begin{vmatrix} 3 & -1 & 5 \\ -2 & 2 & -3 \\ 1 & 0 & 4 \end{vmatrix}$.

Just as with a system of two linear equations in two unknowns, we can also use Cramer's Rule to solve a system of three linear equations in three unknowns.

Cramer's Rule for Solving a System of Three Linear Equations in Three Unknowns

The solution to the system

$$a_1x + b_1y + c_1z = k_1$$
$$a_2x + b_2y + c_2z = k_2$$
$$a_3x + b_3y + c_3z = k_3$$

is $x = \dfrac{D_x}{D}$, $y = \dfrac{D_y}{D}$, and $z = \dfrac{D_z}{D}$, where $D_x = \begin{vmatrix} k_1 & b_1 & c_1 \\ k_2 & b_2 & c_2 \\ k_3 & b_3 & c_3 \end{vmatrix}$, $D_y = \begin{vmatrix} a_1 & k_1 & c_1 \\ a_2 & k_2 & c_2 \\ a_3 & k_3 & c_3 \end{vmatrix}$,

$D_z = \begin{vmatrix} a_1 & b_1 & k_1 \\ a_2 & b_2 & k_2 \\ a_3 & b_3 & k_3 \end{vmatrix}$, and $D = \begin{vmatrix} a_1 & b_1 & c_1 \\ a_2 & b_2 & c_2 \\ a_3 & b_3 & c_3 \end{vmatrix}$, where $D \neq 0$.

EXAMPLE 4

Solve using Cramer's Rule:

$$3x - 4y + 2z = 9$$
$$2x - 3y + z = 5$$
$$x + 2y + 3z = 8$$

Solution

$$D_x = \begin{vmatrix} k_1 & b_1 & c_1 \\ k_2 & b_2 & c_2 \\ k_3 & b_3 & c_3 \end{vmatrix} = \begin{vmatrix} 9 & -4 & 2 \\ 5 & -3 & 1 \\ 8 & 2 & 3 \end{vmatrix}$$

$$= 9\begin{vmatrix} -3 & 1 \\ 2 & 3 \end{vmatrix} - 5\begin{vmatrix} -4 & 2 \\ 2 & 3 \end{vmatrix} + 8\begin{vmatrix} -4 & 2 \\ -3 & 1 \end{vmatrix}$$

$$= 9(-11) - 5(-16) + 8(2)$$

$$= -3$$

$$D_y = \begin{vmatrix} a_1 & k_1 & c_1 \\ a_2 & k_2 & c_2 \\ a_3 & k_3 & c_3 \end{vmatrix} = \begin{vmatrix} 3 & 9 & 2 \\ 2 & 5 & 1 \\ 1 & 8 & 3 \end{vmatrix}$$

$$= 3\begin{vmatrix} 5 & 1 \\ 8 & 3 \end{vmatrix} - 2\begin{vmatrix} 9 & 2 \\ 8 & 3 \end{vmatrix} + 1\begin{vmatrix} 9 & 2 \\ 5 & 1 \end{vmatrix}$$

$$= 3(7) - 2(11) + 1(-1)$$

$$= -2$$

$$D_z = \begin{vmatrix} a_1 & b_1 & k_1 \\ a_2 & b_2 & k_2 \\ a_3 & b_3 & k_3 \end{vmatrix} = \begin{vmatrix} 3 & -4 & 9 \\ 2 & -3 & 5 \\ 1 & 2 & 8 \end{vmatrix}$$

$$= 3\begin{vmatrix} -3 & 5 \\ 2 & 8 \end{vmatrix} - 2\begin{vmatrix} -4 & 9 \\ 2 & 8 \end{vmatrix} + 1\begin{vmatrix} -4 & 9 \\ -3 & 5 \end{vmatrix}$$

$$= 3(-34) - 2(-50) + 1(7)$$

$$= 5$$

PRACTICE 4

Use Cramer's Rule to solve:

$$4x + y - z = 5$$
$$x - y + 2z = 8$$
$$x + y + z = 5$$

EXAMPLE 4 (continued)

$$D = \begin{vmatrix} a_1 & b_1 & c_1 \\ a_2 & b_2 & c_2 \\ a_3 & b_3 & c_3 \end{vmatrix} = \begin{vmatrix} 3 & -4 & 2 \\ 2 & -3 & 1 \\ 1 & 2 & 3 \end{vmatrix}$$

$$= 3\begin{vmatrix} -3 & 1 \\ 2 & 3 \end{vmatrix} - 2\begin{vmatrix} -4 & 2 \\ 2 & 3 \end{vmatrix} + 1\begin{vmatrix} -4 & 2 \\ -3 & 1 \end{vmatrix}$$

$$= 3(-11) - 2(-16) + 1(2)$$

$$= 1$$

By Cramer's Rule, we get:

$$x = \frac{D_x}{D} = \frac{-3}{1} = -3 \qquad y = \frac{D_y}{D} = \frac{-2}{1} = -2 \qquad z = \frac{D_z}{D} = \frac{5}{1} = 5$$

So the solution is $(-3, -2, 5)$. We leave the check as an exercise.

Evaluate each determinant.

1. $\begin{vmatrix} 2 & 5 \\ 1 & -1 \end{vmatrix}$

2. $\begin{vmatrix} 8 & -5 \\ 1 & 2 \end{vmatrix}$

3. $\begin{vmatrix} 7 & 2 \\ 9 & 4 \end{vmatrix}$

4. $\begin{vmatrix} 11 & 3 \\ 3 & 0 \end{vmatrix}$

5. $\begin{vmatrix} -6 & 4 \\ -12 & 8 \end{vmatrix}$

6. $\begin{vmatrix} 3 & 1 \\ 6 & 2 \end{vmatrix}$

Solve using Cramer's Rule.

7. $\begin{aligned} 4x - 5y &= 13 \\ 3x - y &= 7 \end{aligned}$

8. $\begin{aligned} x + 2y &= 4 \\ -x + y &= 5 \end{aligned}$

9. $\begin{aligned} 3x + 2y &= 11 \\ x - 2y &= 1 \end{aligned}$

10. $\begin{aligned} -2x + y &= -10 \\ 2x + 6y &= 10 \end{aligned}$

11. $\begin{aligned} 3x + 2y &= 1 \\ -7x + 10y &= -6 \end{aligned}$

12. $\begin{aligned} 2x - 6y &= 6 \\ -x + 9y &= 1 \end{aligned}$

Evaluate each determinant.

13. $\begin{vmatrix} 2 & 1 & 5 \\ 0 & 3 & 1 \\ 4 & 2 & 4 \end{vmatrix}$

14. $\begin{vmatrix} 0 & 2 & 1 \\ 5 & 4 & 3 \\ 1 & -1 & 3 \end{vmatrix}$

15. $\begin{vmatrix} 3 & 3 & 1 \\ 2 & 2 & -6 \\ 2 & 3 & -4 \end{vmatrix}$

16. $\begin{vmatrix} 6 & -3 & 5 \\ 1 & -2 & 4 \\ 2 & -4 & 7 \end{vmatrix}$

Solve using Cramer's Rule.

17. $\begin{aligned} 2x - 3y + z &= 8 \\ 5x + 4y + 2z &= -1 \\ 7x + 2y + 3z &= 5 \end{aligned}$

18. $\begin{aligned} 2x - y + z &= 2 \\ 3x - 2y - z &= -1 \\ x - 3y + 6z &= -1 \end{aligned}$

19. $\begin{aligned} x + 3y - 4z &= -12 \\ 3x + z &= 2 \\ 5x - y + z &= -3 \end{aligned}$

20. $\begin{aligned} 2x + 3y &= 4 \\ 3x + 7y - 4z &= -3 \\ x - y + 2z &= 9 \end{aligned}$

• Check your answers on page A-57.

Appendix A.4

Synthetic Division

In Section 5.3, we discussed the division of polynomials. Here we consider a shortcut for dividing polynomials, known as **synthetic division**, which eliminates the repetition in the division process. This procedure is only used in a particular situation, namely, when we are dividing a polynomial by a binomial of the form $x - k$.

In Example 2(b) of Section 5.3, we carried out the following division:

$$
\begin{array}{r}
3x +7 \\
x - 5\overline{)\smash{3x^2} - 8x - 40} \\
\underline{3x^2 - 15x} \\
7x - 40 \\
\underline{7x - 35} \\
-5 \leftarrow \text{Remainder}
\end{array}
$$

To see how this shortcut was developed, let's first rewrite the example above by omitting the variables as shown to the right:

$$
\begin{array}{r}
3 7 \\
1 - 5\overline{)\smash{3} - 8 - 40} \\
\underline{3 - 15} \\
7 - 40 \\
\underline{7 - 35} \\
-5 \leftarrow \text{Remainder}
\end{array}
$$

Note that several of the numbers in this division are exactly the same as numbers written directly above. So we can simplify the division process further by omitting these repeated numbers. We can also omit the 1 in the divisor, since we are only considering divisors with leading coefficient 1:

$$
\begin{array}{r}
3 7 \\
1 - 5\overline{)\smash{3} - 8 - 40} \\
\underline{3 - 15} \\
7 - 40 \\
\underline{7 - 35} \\
-5 \leftarrow \text{Remainder}
\end{array}
\qquad
\begin{array}{r}
3 7 \\
- 5\overline{)\smash{3} - 8 - 40} \\
\underline{- 15} \\
7 \\
\underline{- 35} \\
-5 \leftarrow \text{Remainder}
\end{array}
$$

Finally, we can simplify the arithmetic by changing the -5 to its additive inverse 5 so that we can add rather than subtract in each column. Using a compressed format, we have:

$$
\begin{array}{r}
\text{Coefficients of the dividend} \\
\overbrace{} \\
5\rfloor \quad 3 \qquad -8 \qquad -40 \\
\underline{ \qquad 15 \qquad 35} \\
3 \qquad 7 \qquad -5 \\
\underbrace{} \qquad \uparrow
\end{array}
$$

k in the divisor $x - k$ — Coefficients of the quotient — Remainder

We can use the division method modeled above to see how synthetic division works, step by step.

EXAMPLE 1

Use synthetic division to divide $2x^2 - x + 1$ by $x - 4$.

Solution   Since the divisor is $x - 4$, k is 4. So we write 4 as $\underline{4|}$. Turning to the dividend, we write the coefficients and constant term without the variable: 2, -1, and 1.

$$
\begin{array}{r|rrr}
\underline{4|} & 2 & -1 & 1 \\
& \downarrow & & \\
\hline
& 2 & &
\end{array}
$$

Draw a line, and bring down the leading coefficient of the dividend 2.

$$
\begin{array}{r|rrr}
\underline{4|} & 2 & -1 & 1 \\
& & 8 & \\
\hline
& 2 & 7 &
\end{array}
$$

Multiply the 2 by 4, getting 8. Write the 8 under the -1, adding to get 7.

$$
\begin{array}{r|rrr}
\underline{4|} & 2 & -1 & 1 \\
& & 8 & 28 \\
\hline
& 2 & 7 & 29
\end{array}
$$

Multiply the 7 by 4, getting 28. Write the 28 under the 1, adding to get 29.

Since the dividend is a second-degree polynomial, the quotient is first degree, that is, one degree less. From the bottom row, we see that the solution is $2x + 7$, with remainder 29. That is,

$$\frac{2x^2 - x + 1}{x - 4} = 2x + 7 + \frac{29}{x - 4}.$$

EXAMPLE 2

Find the quotient using synthetic division: $\dfrac{x^3 + 6x - 1}{x + 2}$

Solution   The divisor is $x + 2$, which in the form $x - k$ is $x - (-2)$. So k is -2, which we write as $\underline{-2|}$. Since there is no x^2-term in the dividend, we insert a coefficient of 0 for the missing term. So the dividend coefficients and constant term are: 1, 0, 6, and -1.

$$
\begin{array}{r|rrrr}
\underline{-2|} & 1 & 0 & 6 & -1 \\
& & -2 & 4 & -20 \\
\hline
& 1 & -2 & 10 & -21
\end{array}
$$

Since the dividend is a third-degree polynomial, the quotient is second degree. The quotient is $x^2 - 2x + 10$, with remainder -21.

So $\dfrac{x^3 + 6x - 1}{x + 2} = x^2 - 2x + 10 - \dfrac{21}{x + 2}.$

PRACTICE 1

Use synthetic division to find the quotient: $(5x^2 - 3x - 1) \div (x - 3)$

PRACTICE 2

Divide using synthetic division:
$$\frac{x^3 + 4x^2 + 7}{x + 5}$$

Find the quotient, using synthetic division.

1. $\dfrac{x^2 + x - 1}{x - 2}$

2. $\dfrac{x^2 - 4x - 2}{x - 4}$

3. $(x^2 - 5x - 7) \div (x + 3)$

4. $(x^2 + 2x - 4) \div (x + 5)$

5. $\dfrac{5x^2 - 3x + 2}{x + 1}$

6. $\dfrac{2x^2 - 4x - 21}{x + 3}$

7. $(3x^2 - 8x - 3) \div (x - 3)$

8. $(9x^2 - 4x - 5) \div (x - 1)$

9. $(3x^2 - 8x) \div (x - 4)$

10. $(2x^2 - 11) \div (x - 2)$

11. $\dfrac{4x^3 + x^2 - 3x + 5}{x - 1}$

12. $\dfrac{2x^3 + x^2 - 6x - 24}{x - 2}$

13. $(-y^3 + 17y - 40) \div (y + 5)$

14. $(7t^3 + 8t^2 - 1) \div (t + 1)$

15. $(x^3 - 5x) \div (x - 2)$

16. $(x^3 + 3) \div (x - 1)$

17. $\dfrac{x^4 + 3x^3 + 5x}{x + 3}$

18. $\dfrac{x^4 - 4x^2 - 8}{x + 2}$

19. $(2x^3 + 7x^2 - 2x + 3) \div (x - 0.5)$ 20. $(6x^3 + 5x^2 - 9x + 2) \div (x + 0.5)$

• Check your answers on page A-57.

Answers

Chapter 1   Pretest, p. 2

1. [number line: point at -2.5... actually a dot]
$$-5\ -4\ -3\ -2\ -1\ \ 0\ \ 1\ \ 2\ \ 3\ \ 4\ \ 5$$
2. 8   **3.** $<$
4. The commutative property of addition   **5.** -13 **6.** -5 **7.** 40   **8.** 36
9. 1   **10.** 3   **11.** 6   **12.** 0   **13.** y^2 **14.** $-27a^{12}b^6$ **15.** $-\frac{9}{x^7}$
16. $2ab + 10$ **17.** $(0.9)^3$ **18.** $400c$ dollars   **19.** Nitrogen melts at a higher temperature, since $-210 > -259.34$.   **20.** About 15,000 bachelor's degrees were awarded in mathematics or statistics in 2009.

Section 1.1   Practices: Section 1.1, pp. 5–10

1, p. 5:
[number line: -2.5 and $\frac{7}{4}$ marked]
$$-5\ -4\ -3\ -2\ -1\ \ 0\ \ 1\ \ 2\ \ 3\ \ 4\ \ 5$$
2, p. 6: a. 41   **b.** $8\frac{1}{2}$
c. -1.7 **d.** $\frac{2}{5}$ **3, p. 7: a.** $\frac{2}{5}$ **b.** 0   **c.** 2.9   **d.** -9 **4, p. 8: a.** False
b. True   **c.** True   **d.** False
5, p. 8:
[number line: -2.6, -1.4, $-\frac{1}{2}$, and 3 marked] $3, -\frac{1}{2}, -1.4, -2.6$
$$-5\ -4\ -3\ -2\ -1\ \ 0\ \ 1\ \ 2\ \ 3\ \ 4\ \ 5$$
6, p. 9: Kennedy Airport had the best visibility, since $\frac{3}{8} < \frac{1}{2} < \frac{3}{4}$.
7, p. 10: a.
[number line] $[2, \infty)$
$$-5\ -4\ -3\ -2\ -1\ \ 0\ \ 1\ \ 2\ \ 3\ \ 4\ \ 5$$
b.
[number line] $(-\infty, 0)$
$$-5\ -4\ -3\ -2\ -1\ \ 0\ \ 1\ \ 2\ \ 3\ \ 4\ \ 5$$
c.
[number line] $[-2, 3)$
$$-5\ -4\ -3\ -2\ -1\ \ 0\ \ 1\ \ 2\ \ 3\ \ 4\ \ 5$$

Exercises 1.1, pp. 11–15

1. whole numbers   **3.** rational numbers   **5.** additive inverses   **7.** infinity

	Whole Numbers	Integers	Rational Numbers	Irrational Numbers	Real Numbers
9. -5		✓	✓		✓
11. -3.9			✓		✓
13. 7	✓	✓	✓		✓
15. $\sqrt{3}$				✓	✓

17.
[number line]
$$-5\ -4\ -3\ -2\ -1\ \ 0\ \ 1\ \ 2\ \ 3\ \ 4\ \ 5$$
19.
[number line]
$$-5\ -4\ -3\ -2\ -1\ \ 0\ \ 1\ \ 2\ \ 3\ \ 4\ \ 5$$
21.
[number line]
$$-5\ -4\ -3\ -2\ -1\ \ 0\ \ 1\ \ 2\ \ 3\ \ 4\ \ 5$$
23. 5   **25.** $-\frac{1}{4}$ **27.** 12   **29.** -2.7 **31.** $\frac{1}{4}$ **33.** 9   **35.** -2
37. -5.3 **39.** 6 and -6 **41.** Impossible; absolute value is always positive or zero.   **43.** False   **45.** False   **47.** True   **49.** True   **51.** False
53. $>$ **55.** $>$ **57.** $<$ **59.** $=$ **61.** $<$
63.
[number line: -4, $-\frac{2}{5}$, $\frac{3}{4}$, 5 marked] $-4, -\frac{2}{5}, \frac{3}{4}, 5$
$$-5\ -4\ -3\ -2\ -1\ \ 0\ \ 1\ \ 2\ \ 3\ \ 4\ \ 5$$
65.
[number line: -3.5, -1.8, 1, 4.3 marked] $-3.5, -1.8, 1, 4.3$
$$-5\ -4\ -3\ -2\ -1\ \ 0\ \ 1\ \ 2\ \ 3\ \ 4\ \ 5$$
67.
[number line] $[-4, \infty)$
$$-5\ -4\ -3\ -2\ -1\ \ 0\ \ 1\ \ 2\ \ 3\ \ 4\ \ 5$$
69.
[number line] $(-\infty, 3)$
$$-5\ -4\ -3\ -2\ -1\ \ 0\ \ 1\ \ 2\ \ 3\ \ 4\ \ 5$$

71.
[number line] $[-4, 3]$
$$-5\ -4\ -3\ -2\ -1\ \ 0\ \ 1\ \ 2\ \ 3\ \ 4\ \ 5$$
73. Impossible; absolute value is always positive or zero.
75. Irrational number, real number.
77.
[number line] $[-2, 1)$
$$-5\ -4\ -3\ -2\ -1\ \ 0\ \ 1\ \ 2\ \ 3\ \ 4\ \ 5$$
79. It is warmer in January at the South Pole; $-16°F > -74°F$.
81. -27 m > -400 m   **83. a.** Arizona, Florida, and Texas *gained* representatives as a result of the 2010 census. Illinois, New York, Ohio, and Pennsylvania lost representatives as a result of the 2010 census.
b. The population of Florida was greater than the population of Ohio in 2010, and so Florida had more representatives than Ohio. **c.** Illinois and Pennsylvania had the same number of representatives. This would indicate that the total populations of Illinois and Pennsylvania were about the same.
85. $x < y$ (or $y > x$)   **87.** $x > y$ (or $y < x$)
89. a.
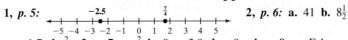
Greg Mahoney   Ian Hoffman   Brian Richey
$$-14\ -12\ -10\ -8\ -6\ -4\ -2\ \ 0\ \ 2\ \ 4\ \ 6\ \ 8\ \ 10\ \ 12\ \ 14$$
Nick Jones   Tim Lamb
b. A score of par   **c.** Nick Jones

Section 1.2   Practices, pp. 16–23

1, p. 16: -29 **2, p. 16:** 1.8   **3, p. 17:** 9   **4, p. 17:** -14.9
5, p. 18: 100   **6, p. 18:** -3.1 **7, p. 19: a.** 49   **b.** -49
8, p. 19: a. -7 **b.** 0.7   **9, p. 20: a.** -2 **b.** $\frac{1}{32}$ **c.** -50
10, p. 21: -2 **11, p. 21:** 0   **12, p. 22: a.** -6 **b.** -3 **c.** 26
13, p. 22: a. 3   **b.** -1 **c.** $\sqrt{3}$ **14, p. 23:** A: **80** B: **90** C: **25**

Exercises 1.2, pp. 24–28

1. negative   **3.** product   **5.** positive   **7.** 2   **9.** 4   **11.** -11 **13.** -1
15. -8 **17.** 0   **19.** -3.6 **21.** 0.3   **23.** $-\frac{4}{4} = -1$ **25.** $-\frac{2}{10} = -\frac{1}{5}$
27. $-\frac{5}{5} = -1$ **29.** 4.4   **31.** 14.46   **33.** 40   **35.** -40 **37.** 7
39. -5 **41.** -5 **43.** -11.5 **45.** 8.75   **47.** -2.2 **49.** 30   **51.** 6
53. -24 **55.** $-\frac{1}{3}$ **57.** $\frac{2}{3}$ **59.** 64   **61.** 9   **63.** -4 **65.** 12   **67.** 39
69. 8   **71.** -27 **73.** 15   **75.** -22 **77.** 6   **79.** 11   **81.** -1 **83.** 0
85. 5   **87.** 7   **89.** 9   **91.** -24 **93.** $\sqrt{5}$ **95.** $\sqrt{6}$ **97.** -1.9 **99.** 3
101. -7 **103.** 5000 ft   **105.** A loss of 5 yd   **107.** The object is 16 ft above the ground.   **109. a.** 2005: $-\$0.32$, 2006: $-\$0.25$, 2007: $-\$0.16$, 2008: $-\$0.46$, 2009: $-\$1.42$, 2010: $-\$1.55$ **b.** The budget showed an average annual deficit of about \$0.69 trillion ($-\$0.69$ trillion).   **111.** The average daily low temperature was 2°F.   **113.** The patient will take 45 mg of the medication in one day.   **115. a.** 2008: $+2.6$, 2009: -1.7, 2010: -1.5, 2011: $+1.1$ **b.** The average annual change in revenue was \$125,000. **c.** The average annual revenue over the five-year period was \$17,980,000.

Section 1.3   Practices, pp. 29–36

1, p. 29: a. $5 + (-3)$ **b.** $3a + b$ **2, p. 30: a.** $(2)(-8)$ **b.** $n(-4)$
3, p. 30: a. $8 + [(-1) + 2]$ **b.** $x + (3y + z)$
4, p. 31: a. $(-3)[(5)(-2)]$ **b.** $[3(-6)]n$ **5, p. 31:** -17
6, p. 31: -120 **7, p. 32: a.** -5 **b.** $6y$ **c.** -2 **d.** $-5x$
8, p. 32: a. 2   **b.** $\frac{2}{3}$ **c.** $-y$ **9, p. 33: a.** 5   **b.** $\frac{1}{2}$ **c.** $-\frac{1}{5}$ **d.** $-\frac{7}{2}$
10, p. 34: a. 0   **b.** 0   **c.** 0   **11, p. 34: a.** $(-2)(9) + (-2)(4.3)$

b. $0.2a + 0.2b$ c. $2q - pq$ **12, p. 35: a.** 0 **b.** Not possible, because $-2n \div 0$ is undefined. **c.** 0 **13, p. 35:** (a) The associative property of multiplication; (b) The multiplicative inverse property; (c) The multiplicative identity property **14, p. 36:** $1n = n$ dollars by the multiplicative identity property.

Exercises 1.3, pp. 37–40

1. commutative property of multiplication **3.** additive identity property **5.** additive inverse property **7.** multiplication property of zero **9.** $2 + 3.7$ **11.** $(-1) + [(-6) + 7]$ **13.** -3 **15.** $3 \cdot 1 + 3 \cdot 9$ **17.** $5(2 + 7)$ **19.** $2(a + b)$ **21.** The commutative property of multiplication **23.** The commutative property of addition **25.** The distributive property **27.** The multiplication property of zero **29.** The additive inverse property **31.** The associative property of addition **33.** The multiplicative inverse property **35.** 0 **37.** 0 **39.** 0 **41.** 6.33 **43.** 42 **45.** -8 **47.** -700 **49.** 0 **51.** Not possible because dividing by 0 is undefined. **53.** -2 **55.** 7 **57.** $\frac{1}{7}$ **59.** -1 **61.** $(-4)(2) + (-4)(5)$ **63.** $3 \cdot x + 3 \cdot 10 = 3x + 30$ **65.** $(-1)(a) + (-1)(6b) = -a + (-6b)$, or $-a - 6b$ **67.** $n \cdot n - n \cdot 2 = n^2 - 2n$ **69.** (a) The commutative property of multiplication (b) The associative property of multiplication (c) Multiplication of real numbers **71.** (a) The associative property of multiplication (b) The multiplicative inverse property (c) The multiplicative identity property **73.** 0.2 **75.** The associative property of multiplication **77.** The multiplicative inverse property **79.** The distance traveled to work is the same as the distance traveled going home. By the commutative property of addition: (distance traveled on bus) + (distance walked) = (distance walked) + (distance traveled on bus). **81.** By the distributive property, $r(p + q) = rp + rq$, so the shopper will pay the same amount regardless of which expression is used. **83.** Using the additive identity property, $p + 0 = p$, so his weight at the end of the week is p lb. **85.** According to the commutative property of multiplication, the calculations give the same result.

Section 1.4   Practices, pp. 43–55

1, p. 43: a. $-3y$ **b.** -1 **c.** 5 **d.** $-7a + b$ **2, p. 43: a.** n^{10} **b.** $(-4x)^4$ **c.** $(y + 3)^5$ **d.** Cannot be simplified because the bases are not the same **3, p. 44: a.** s^6 **b.** $(-3r)^1 = -3r$ **c.** $(t - 2)^2$ **d.** Cannot apply the quotient rule because the bases are not the same **4, p. 45: a.** x^{13} **b.** a^7b^2 **c.** y^3 **5, p. 46: a.** $\frac{1}{a^9}$ **b.** $-\frac{1}{4y}$ **c.** $\frac{1}{(-5 + x)^3}$ **d.** p^4 **e.** $\frac{1}{x^7}$ **f.** $\frac{1}{m^3}$ **6, p. 46: a.** y^3 **b.** $5x^2$ **c.** mn^2 **d.** $4r^3s^3$ **7, p. 47: a.** y^{14} **b.** $-n^{-5} = -\frac{1}{n^5}$ **c.** p^9 **8, p. 48: a.** $16x^2$ **b.** $25a^{18}$ **c.** $q^{14}r^{16}$ **d.** $-6a^{15}b^9$ **e.** $\frac{x^8y^2}{49}$ **9, p. 49: a.** $\frac{y^5}{32}$ **b.** $\frac{u^8}{v^{12}}$ **c.** $\frac{-125a^{15}}{8b^6}$ or $-\frac{125a^{15}}{8b^6}$ **d.** $\frac{u^8}{v^{12}}$ **10, p. 50:** $\frac{16y^8}{81x^{12}}$ **11, p. 50: a.** The pollution level will be $30 \times (0.95)^{11}$ ppm in the twelfth month. **b.** $(0.95)^5$ times as great **12, p. 57: a.** 5193 **b.** 0.0000037 **13, p. 52: a.** 4.0×10^{12} or 4×10^{12} **b.** 6.7×10^{-11} **14, p. 53: a.** 1.024×10^3 **b.** 2.2×10^7 **15, p. 54: a.** 234 **b.** 0.036 **c.** Gold has a higher coefficient. **d.** 18.0 **16, p. 55:** Answers may vary. Possible answer: 6.4E$-$17 **17, p. 55:** Answers may vary. Possible answer: 3.36E11

Exercises 1.4, pp. 56–60

1. base **3.** added **5.** multiplied **7.** 7^4 **9.** a^5 **11.** $(-5x)^4$ **13.** $(a + b)^3$ **15.** 16 **17.** -27 **19.** -27 **21.** 100 **23.** 1 **25.** -4 **27.** 1 **29.** $3n$ **31.** 1 **33.** 3^4 **35.** n^{13} **37.** Cannot be simplified **39.** x^5 **41.** $(x + 2)^4$ **43.** 6^4 **45.** x^5 **47.** Cannot be simplified **49.** t^7 **51.** $(2n + 1)^3$ **53.** y^{14} **55.** p^6q^{10} **57.** $-7x^3y^3z^6$ **59.** n^4 **61.** r^2 **63.** $\frac{1}{5^2}$ **65.** $\frac{1}{t^8}$ **67.** $\frac{1}{n^{10}}$ **69.** $-\frac{1}{3y}$ **71.** $\frac{1}{(a + 6)^5}$ **73.** x^6 **75.** $-\frac{1}{n^8}$ **77.** $\frac{y^2}{x^4}$ **79.** $\frac{y^2}{2}$ **81.** $-\frac{3}{t^2}$ **83.** $2x^4$ **85.** $-\frac{p^3q}{5}$ **87.** $\frac{3}{a^2b}$ **89.** n^3 **91.** $\frac{1}{x^7}$ **93.** 5^8 **95.** $-x^{15}$ **97.** $\frac{1}{n^{12}}$ **99.** $-a^5$ **101.** $-16p^4$

103. $-32n^5$ **105.** $a^{10}b^4$ **107.** $-27n^{15}$ **109.** $-2a^8b^{12}$ **111.** $\frac{y^{12}}{x^{27}}$ **113.** $\frac{64b^{16}}{a^{16}}$ **115.** $\frac{x^3}{8}$ **117.** $\frac{p^{20}}{q^{12}}$ **119.** $-\frac{4n^6}{25m^4}$ **121.** $-\frac{64r^6}{s^3t^{12}}$ **123.** $\frac{9a^6}{b^2}$ **125.** $\frac{16v^6}{9u^8}$ **127.** $-\frac{p^4r^4t^2}{q^4}$ **129.** 4×10^8 **131.** 9.26×10^{11} **133.** 4.2×10^{-6} **135.** 7.4×10^{-10} **137.** 2,430,000 **139.** 0.003027 **141.** 0.000001 **143.** 1×10^7 **145.** 8.1×10^{-5} **147.** 2.3×10^3 **149.** 6×10^5 **151.** 1.4086683×10^{-2} **153.** 2.631564845×10^8 **155.** x^6 **157.** -5 **159.** $\frac{1}{(n + 1)^3}$ **161.** $(2x)^3$ **163.** 8.4×10^6 **165.** No; the volume of the large storage locker is x^3 and the volume of the small storage locker is $(\frac{1}{2}x)^3 = \frac{1}{8}x^3$, which is not equal to $\frac{1}{2}x^3$. **167.** $2^6n = 64n$ bacteria **169.** 6×10^7 m **171.** 0.000000001 sec **173. a.** $6000 **b.** $6714.56 **c.** The total interest earned in the five-year period was $714.56. **175.** In 2010, there were about 335 people per sq km in Japan. **177.** Income taxes: 8.97×10^{11}; Social Security: 8.65×10^{11}; Corporate taxes: 1.92×10^{11}; Other: 2.08×10^{11} **179.** 1.44×10^{13} kg of matter are converted in an hour.

Section 1.5   Practices, pp. 62–68

1, p. 62: a. $\frac{1}{3}$ of p **b.** 6 minus x **c.** The quotient of s and -4 **d.** The sum of n and -10 **e.** $\frac{5}{8}$ times m **2, p. 62: a.** 7 less than the product of 6 and x **b.** The sum of 3 and twice m **c.** 10 times the sum of a and b **d.** 3 divided by the difference of p and q **3, p. 63: a.** $\frac{1}{2}n$, where n represents the number **b.** $x + (-1)$, where x represents the number **c.** $x - (-2)$ **d.** $\frac{4}{n}$ **e.** $-3d$ **4, p. 63: a.** $u + (-v)$ **b.** $3y - 12$ **c.** $\frac{a + b}{a - b}$ **d.** $-2(x - y)$ **5, p. 64: a.** 19 **b.** -16 **c.** 16 **d.** -9 **6, p. 64: a.** -1 **b.** 11 **c.** -20 **d.** 32 **7, p. 65: a.** The runners are $5L$ m apart. **b.** The distance between the runners after 3 laps is 15 m. **8, p. 65: a.** r and $-9r$; like **b.** $-x$ and 5; unlike **c.** $2x^2y$ and $-xy^2$; unlike **d.** $m, 5m$, and $-m$; like **9, p. 66: a.** $-8y$ **b.** $-3a + 8b$ **10, p. 66: a.** Cannot be simplified **b.** $2n^2 + 2n^3$ **c.** $8xy^2$ **11, p. 66:** $5y + 6$ **12, p. 67:** $-4a + 9b$ **13, p. 67:** $y + 12$ **14, p. 67:** $4y + 10$ **15, p. 68:** $6y + 21$ **16, p. 68:** $(400 - y)$ rooms

Exercises 1.5, pp. 69–74

1. variable **3.** term **5.** expression **7.** like **9.** $\frac{1}{2}$ of n **11.** 8 less than x **13.** The quotient of y and 5 **15.** a plus -2 **17.** The product of 3 and x **19.** Twice a decreased by b **21.** 9 more than 6 times n **23.** Negative 3 times the sum of x and y **25.** The product of 4 and the sum of u and v **27.** The sum of r and s, all divided by 9 **29.** Negative 3 divided by the sum of x and y **31.** The difference between 4 times p and 5 times q **33.** Twice x times y, divided by the quantity x minus y **35.** $\frac{1}{4}n$ **37.** $n + (-3)$ **39.** $x - (-2)$ **41.** $\frac{b}{a}$ **43.** $-4y$ **45.** $x + (-y)$ **47.** $3h - 7$ **49.** $2c + 4d$ **51.** $\frac{p - q}{p + q}$ **53.** $-2(a + 5)$

55.

x	-2	-1	0	1	2
$3x + 2$	-4	-1	2	5	8

57.

y	0	1	2	3	4
$2y - 1.5$	-1.5	0.5	2.5	4.5	6.5

59.

x	-4	-2	0	2	4
$-0.5x$	2	1	0	-1	-2

61.

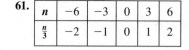

n	-6	-3	0	3	6
$\frac{n}{3}$	-2	-1	0	1	2

63.

g	-2	-1	0	1	2
$-2g^2$	-8	-2	0	-2	-8

65.

a	-2	-1	0	1	2
$a^2 - 3a + 2$	12	6	2	0	0

67. a. 16 **b.** 24 **c.** -30 **d.** 19 **69. a.** -3 **b.** $\frac{1}{-3}$, or $-\frac{1}{3}$
c. -16 **d.** 9 **71.** $3a$ and $5b$; unlike **73.** x and $-7x$; like **75.** $2x^2$
and $-6x$; unlike **77.** $8a^2b$ and $-5a^2b$; like **79.** $6x$ **81.** $10a^2$
83. $7t + 7$ **85.** $6x - 9$ **87.** $-3x^2 + 9x$ **89.** Cannot be simplified
91. $-3a^2b$ **93.** $\frac{1}{2}x - 7$ **95.** $-3y + 1$ **97.** $3x - 13$ **99.** -22
101. $20x - 12y + 21$ **103.** 49 **105.** $-62n + 54$ **107.** $n - (-4)$
109. The quotient of the difference between a and b and 2 **111.** $11 - 24y$
113. 2.1 **115.** The female was 63 in. tall. **117.** The pressure at the
bottom of the swimming pool is 2612.8 pounds per sq ft.
119. $2(x + 4) + 2(3x + 4)$; $(8x + 16)$ units **121. a.** $\frac{h}{b}$
b. Ichiro Suzuki: .315, Ryan Braun: .304, James Loney: .267, Prince
Fielder: .261 **123. a.** Q1: 592, Q2: 558, Q3: 524, Q4: 490
b. Each quarter, revenue is decreasing by about \$34 million.
c. About \$388 million. **125.** The area of the ledge is 138 ft^2.
127. She needs to buy $(4w + 10)$ ft of fencing.
129. a. $10.5h$

Number of Hours, h	30	32.5	35	38	40
Weekly Pay (in dollars)	315	341.25	367.50	399	420

b. $(15.75h - 210)$ dollars **c.** \$640.50 **131.** The cineplex has
$(200x + 3000)$ seats. **133.** $-50 + \frac{3c}{2} + \frac{b}{2}$

Chapter 1 Review Exercises, pp. 80–83

1. is not; Possible answer: it can be written as the ratio of two integers,
namely, -17 and 100. **2.** is; Possible answer: $|-4|$ equals 4, so its op-
posite, or additive inverse, is -4. **3.** can; Possible answer: of the distributive property. **4.** is not; Possible answer: 11.852 is not between 1 and
10. **5.** have; Possible answer: any nonzero number raised to the power 0
is equal to 1. **6.** does; Possible answer: 1 and $5x^2$ are the terms.

7. False **8.** True **9.** ⟨number line with point at -2, from -4 to 4⟩ **10.** 8 **11.** 6

12. False **13.** True **14. a.** ⟨number line with bracket at 0, from -5 to 5⟩
b. $(-\infty, 0]$ **15.** -2.5 **16.** 7 **17.** 13.5 **18.** $\frac{1}{16}$ **19.** 2 **20.** 18
21. 36 **22.** -2 **23.** 16 **24.** 6 **25.** -7 **26.** $\sqrt{13}$ **27.** $9 + 3$
28. $(-3)(1) + (-3)(9)$ **29.** associative property of addition
30. additive inverse property **31.** 7 **32.** -160 **33.** -4 **34.** $\frac{3}{2}$
35. $3a - 12b$ **36.** $-x + 5$ **37.** $(-5x)^3$ **38.** -64 **39.** 1 **40.** $-3xy$
41. $\frac{1}{7^2} = \frac{1}{49}$ **42.** $\frac{1}{2y}$ **43.** n^{12} **44.** a^3b^3 **45.** Cannot be simplified
46. $(2n + 1)^3$ **47.** $-y^{15}$ **48.** p^4q^2 **49.** $\frac{u^8}{v^8}$ **50.** $-\frac{4}{t^2}$ **51.** $2p^3q$
52. $x^1 = x$ **53.** $\frac{q^{10}}{4p^4}$ **54.** $\frac{9y^2}{x^8}$ **55.** 2×10^8 **56.** 3.1×10^{-4}
57. 286,000 **58.** 0.00502 **59.** 1.5×10^7 **60.** 4×10^2 **61.** 4 less
than x **62.** 3 divided by the sum of p and q **63.** $x - (-5)$ **64.** $\frac{p + q}{q - p}$
65.

x	-2	-1	0	1	2
$2x - 5$	-9	-7	-5	-3	-1

66. -39 **67.** $2a$ and $5b$; unlike **68.** $3a^2b$ and $-a^2b$; like **69.** $-4x$
70. $3ab^2$ **71.** $3t + 2$ **72.** $14x^2 - 6x$ **73.** $-5y + 1$ **74.** $6x + 42$
75. a. Wednesday **b.** $-16 < -15$ **c.** The average daily low temperature was about $-13°$F for the week.

76. a.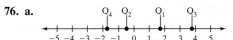

b. The company showed losses in quarters 2 and 4. **c.** The average quarterly profit was \$812,500 (\$0.8125 million). **77.** The new balance is
$-\$6.63$. **78.** An 8-oz glass of milk contains about 300 mg of calcium.
79. 6280 m **80. a.** Christie Kerr: -2, Se Ri Pak: -3, Angela Stanford:
$+1$, Karrie Webb: $+2$, Michelle Wie: 0 **b.** The golfers scored below par.
81. 8.9×10^9 **82.** 30,500,000 adults were living with HIV/AIDS in
2009. **83.** The object is 74.4 m above the ground. **84.** The ideal body
weight of this man is 172 lb. **85. a.** 25 drops per min **b.** 125 mL per
hour **c.** About 31 drops per min are required to fill the order.

86. a.

Time, t (in sec)	0	1	2	3	4	5
Height (in ft)	300	304	276	216	124	0

b. A height of 0 ft represents ground level. **87.** $(2700 - 0.01x)$
dollars
88. $(1.5h + 10.5)$ dollars

Chapter 1 Posttest, p. 84

1. ⟨number line with point at -3, from -5 to 5⟩ **2.** 9 **3.** False
4. $(-4 + x) + y$ **5.** 9 **6.** 13 **7.** -30 **8.** -1 **9.** $-\frac{1}{4}$ **10.** 46
11. -27 **12.** 5 **13.** $2p^3q^2$ **14.** a^9b^{13} **15.** $\frac{1}{x^8y^4z^2}$ **16.** 4 times
the difference between p and q **17.** -32 **18.** $-12x + 28$
19. $(5000 + 80x)$ dollars **20.** The distance from the Sun to the planet
Venus is approximately 6.5×10^7 mi.

CHAPTER 2

Chapter 2 Pretest, p. 86

1. No, -2 is not a solution. **2.** $\frac{5}{2}$ **3.** 2 **4.** 1
5. $y = \frac{4}{3}x - 2$ **6.** Yes, 5 is a solution.
7. $x \geq -6$ ⟨number line, bracket at -6, from -12 to 0⟩
8. $-3 \leq x < 1$ ⟨number line from -6 to 6⟩
9. $x \leq -6$ or $x \geq -1$ ⟨number line from -10 to 2⟩
10. $-5 < x < 2$ **11.** 5.9 **12.** $-14, -4$ **13.** $-1, 1$ **14.** $-4, \frac{6}{5}$
15. $x \leq 0$ or $x \geq 1$ ⟨number line from -6 to 6⟩
16. The copier will reach its salvage value in five years. **17.** The saline
concentration of the new solution is 39%. **18.** He can drive at most 500
mi in a five-day workweek. **19.** Between 5000 and 7500 bottles of juice
should be produced per month. **20.** The homeowner could pay a minimum of \$9500 or a maximum of \$14,500 for the kitchen remodeling.

Section 2.1 Practices, pp. 88–100

1, p. 88: Yes, 6 is a solution. **2, p. 89: a.** 2 **b.** -25.3 **3, p. 90:**
a. -3 **b.** -25 **c.** 10.6 **4, p. 91: a.** -5 **b.** 39 **5, p. 92:**
a. -2 **b.** $\frac{4}{5}$ **6, p. 93: a.** -2 **b.** -4 **7, p. 95:** There will be 80 radio
commercials in the advertising campaign. **8, p. 95:** He must negotiate
a starting salary of \$48,500. **9, p. 96:** The perimeters of the rectangle
and of the triangle are 28 in. **10, p. 96:** The average speed of the bus is
about 48 mph. **11, p. 97:** The second car will overtake the first car in
3 hr. **12, p. 98:** They will have traveled 1.5 hr by the time they are 96
km apart. **13, p. 99:** She invested \$80,000 in the investment that gained
12% and \$120,000 in the investment that lost 8%. **14, p. 99:** The percent
of antifreeze in the new solution is 55%.

Exercises 2.1, pp. 101–105

1. equation **3.** linear equation **5.** Not a solution **7.** Not a solution
9. A solution **11.** Not a solution **13.** A solution **15.** -2 **17.** 1.7
19. $-\frac{11}{8}$ **21.** -5 **23.** 4 **25.** $\frac{2}{3}$ **27.** -9 **29.** -1 **31.** 22 **33.** -4
35. 0 **37.** 21 **39.** -15 **41.** -6 **43.** 0 **45.** $-\frac{4}{3}$ **47.** 0 **49.** 6
51. -1 **53.** 3 **55.** $\frac{13}{15}$ **57.** -11 **59.** $\frac{9}{2}$ **61.** 0.8 **63.** 4 **65.** 7
67. 6 **69.** -17 **71.** $\frac{2}{3}$ **73.** $-\frac{2}{3}$ **75.** -6 **77.** $-\frac{2}{3}$ **79.** 1 **81.** $-\frac{10}{3}$
83. -16 **85.** No, not a solution **87.** 9.2 **89.** 5.25 **91.** 1
93. a. $37.00 **b.** $7.40 was saved purchasing the book online.
95. The distance between each hurdle is 35 m. **97.** The monthly cost of
Plan B will be the same as the monthly cost of Plan A for 300 min.
99. 12 small pillows and 6 large pillows are made for the family
room. **101.** 326 nonmembers registered for the conference. **103.** The
third number assigned was 1205. **105.** The measures of the angles are
$40°$ and $77°$. **107.** It will take her $2\frac{1}{2}$ hr to finish the walkathon. **109.** It
took his mother 5 min to catch up with him. **111.** The second group
catches up to the first group at 9:00 A.M. **113.** One person walks at a
rate of 3 mph and the other walks at a rate of 2.5 mph. **115.** He invested
$8000 in the fund that made a 7% profit and $4000 in the fund that lost
9%. **117.** The broker invested $70,000. **119.** The student must add
2 mL of pure alcohol to make the 16% alcohol solution. **121. a.** She
needs 4 gal of 87 octane and 6 gal of 92 octane to make the required
amount. **b.** She paid $39.80 to fill her gas tank.

Section 2.2 Practices, pp. 107–109

1, p. 107: $x = \frac{y - b}{m}$ **2, p. 107:** $x = \frac{y - b + ma}{m}$ **3, p. 108:**
a. $b_2 = \frac{2A - b_1 h}{h}$ **b.** 50 ft **4, p. 109: a.** $S = 2a^2 + 4ah$
b. $h = \frac{S - 2a^2}{4a}$ **c.** 18 in.

Exercises 2.2, pp. 110–112

1. $n = -7m + 2$ **3.** $a = \frac{b}{3}$ **5.** $x = \frac{3z}{2y}$ **7.** $a = \frac{c - b}{5}$,
or $a = \frac{1}{5}(c - b)$ **9.** $y = -\frac{c - ax}{b}$, or $y = \frac{ax - c}{b}$ **11.** $r = -\frac{t}{qs}$
13. $x = y(v + z)$ **15.** $b = cd + a$ **17.** $s = \frac{t(v + w)}{r}$
19. $w = \frac{u + s + y}{x}$ **21.** $c = \frac{d - g - ab}{a}$ **23.** $R = \frac{V}{I}$ **25.** $m = \frac{Fr}{v^2}$
27. $w = \frac{P - 2l}{2}$ **29.** $r_1 = \frac{A - \pi s r_2}{\pi s}$ **31.** $a_n = \frac{2S - a_1 n}{n}$
33. $F = \frac{b - c - dG}{d}$ **35.** $h = \frac{A - 2\pi r^2}{2\pi r}$ **37. a.** $a = \frac{V^2 - v^2}{2s}$
b. $a = -1.6\ \text{m/sec}^2$ **39. a.** $h = \frac{l + 254}{6}$ **b.** The man is
74 in. tall (or 6 ft 2 in. tall). **41. a.** $y = -\frac{E - 9.6w - 1.7h - 655}{4.7}$,
or $y = \frac{-E + 9.6w + 1.7h + 655}{4.7}$ **b.** The woman is approximately 32 yr
old. **43. a.** $E = 9\left(\frac{R}{I}\right)$ **b.** $R = \frac{1}{9}EI$ **c.** Felix Hernandez, 63; Josh Johnson, 47; Clay Bucholz, 45; Adam Wainwright, 62. **45. a.** $A = \frac{100}{d}$
b. The accommodation is 8 diopters.

Section 2.3 Practices, pp. 114–121

1, p. 114: -4 is a solution. **2, p. 115:** 10 is not a solution.

3, p. 115: a.
b. **4, p. 116:** $x \leq -1\frac{1}{2}$, or
$\left(-\infty, -1\frac{1}{2}\right]$
5, p. 117: $y \geq -4$, or $[-4, \infty)$
6, p. 118: $n > \frac{5}{2}$, or $\left(\frac{5}{2}, \infty\right)$
7, p. 119: $x > 9$, or $(9, \infty)$
8, p. 119: $x \leq -10$, or $(-\infty, -10]$
9, p. 120: $y \geq -\frac{2}{3}$, or $\left[-\frac{2}{3}, \infty\right)$

10, p. 121: The lowest score she can get is 98. **11, p. 121:** She needs to
buy at least $30,000 worth of bonds.

Exercises 2.3, pp. 122–126

1. inequality **3.** solve **5.** includes **7.** addition property
9. A solution **11.** A solution **13.** Not a solution
15. Not a solution **17.** A solution

19. $(-\infty, -1)$
21. $[2, \infty)$
23. $\left(-3\frac{1}{2}, \infty\right)$
25. $(-\infty, 7.5]$
27. $n \geq -2; [-2, \infty)$
29. $y < -4; (-\infty, -4)$
31. $x > -2; (-2, \infty)$
33. $-2 \geq y$, or $y \leq -2; (-\infty, -2]$
35. $x < 3; (-\infty, 3)$
37. $x \geq 0; [0, \infty)$
39. $x \leq 1.5; (-\infty, 1.5]$
41. $x \geq -10; [-10, \infty)$
43. $n > -5; (-5, \infty)$
45. $6 \geq y$, or $y \leq 6; (-\infty, 6]$
47. $x < -\frac{3}{2}; \left(-\infty, -\frac{3}{2}\right)$
49. c **51.** d **53.** $x \leq 4; (-\infty, 4]$ **55.** $x > 42; (42, \infty)$
57. $y \geq -3; [-3, \infty)$ **59.** $y < 0; (-\infty, 0)$ **61.** $y > 7; (7, \infty)$
63. $x \leq -\frac{45}{2}; \left(-\infty, -\frac{45}{2}\right]$ **65.** $z > 0.25; (0.25, \infty)$
67. $n < -7; (-\infty, -7)$ **69.** $n \geq \frac{9}{2}; \left[\frac{9}{2}, \infty\right)$ **71.** $x \leq 5; (-\infty, 5]$
73. $n < \frac{5}{3}; \left(-\infty, \frac{5}{3}\right)$ **75.** $x < \frac{33}{10}; \left(-\infty, \frac{33}{10}\right)$ **77.** $x \leq \frac{7}{2}; \left(-\infty, \frac{7}{2}\right]$
79. A solution **81.** $(-\infty, -1]$
83. $n > 0$; interval notation: $(0, \infty)$ **85.** The daily rate is better if you
park in the garage fewer than 12 days per month. **87.** You can make at
most 25 min of calls outside the network. **89.** She can spend at most
$170 per day on remaining expenses. **91.** The company's profit must be
at least $97,995 in the fourth quarter. **93.** She must bike at a minimum
rate of 10.5 mph. **95.** She invested less than $19,000 in CDs. **97.** The
courier service must make at least 292 deliveries per month. **99.** The
first option should be selected on a car to be leased more than 25 months.

Section 2.4 Practices, pp. 128–132

1, p. 128: $0 \leq x \leq 4$, interval notation: $[0, 4]$
2, p. 129: $-1 < x < 1$, interval notation: $(-1, 1)$
3, p. 129: $-3 < t < 5$, interval notation: $(-3, 5)$
4, p. 130: No solution **5, p. 130:** $x < -2$ or $x > 3$; interval notation:
$(-\infty, -2) \cup (3, \infty)$
6, p. 131: All real numbers; interval notation: $(-\infty, \infty)$
7, p. 132: Any value between 6.2 and 8,
including 6.2 and 8, will make the average pH normal.

Exercises 2.4, pp. 133–136

1. inequalities   **3.** or   **5.** $(2, 5)$

7. $[-3, 0)$

9. $(-\infty, -4) \cup (3, \infty)$

11. $(-\infty, -2) \cup [0.5, \infty)$

13. $-5 < x < 3$; interval notation: $(-5, 3)$

15. $-6 < t \le 5$; interval notation: $(-6, 5]$

17. No solution

19. $-4 < a \le 2$; interval notation: $(-4, 2]$

21. $-5 < x < \frac{5}{2}$; interval notation: $(-5, \frac{5}{2})$

23. $h < -\frac{3}{2}$; interval notation: $(-\infty, -\frac{3}{2})$

25. $-\frac{11}{2} \le x \le -\frac{5}{2}$; interval notation: $[-\frac{11}{2}, -\frac{5}{2}]$

27. $x < -1$ or $x > 2$; interval notation: $(-\infty, -1) \cup (2, \infty)$

29. $r < -8$ or $r \ge -1$; interval notation: $(-\infty, -8) \cup [-1, \infty)$

31. All real numbers; interval notation: $(-\infty, \infty)$

33. $a \le 3$; interval notation: $(-\infty, 3]$

35. $n \le \frac{9}{2}$ or $n > 6$; interval notation: $(-\infty, \frac{9}{2}] \cup (6, \infty)$

37. All real numbers; interval notation: $(-\infty, \infty)$

39. $-3 < x < -1$ **41.** $x < 2$ or $x \ge 12$ **43.** No solution   **45.** All real numbers   **47.** $x \ge 1$ **49.** $4 \le x < 11$ **51.** $-6 < x \le 6$
53. $-7 \le x \le -4$ **55.** $-0.5 < x < 3$
57. $(-\infty, -2] \cup (1, \infty)$

59. $3 < x \le 4$; $(3, 4]$

61. $\frac{5}{2} < x \le \frac{11}{2}$; $(\frac{5}{2}, \frac{11}{2}]$

63. She can score between 83 and 99 to get a B average for the semester.
65. A person 5 ft 8 in. tall is considered to have an unhealthy weight if s/he weighs less than 122 lb or more than 164 lb.   **67.** Aircrafts with speeds between 740 mph and 1480 mph have Mach numbers between 1.0 and 2.0.   **69.** The couple can invite between 100 and 150 guests.
71. The owner must sell between 30 and 55 gift baskets per month.

Section 2.5   Practices, pp. 137–144

1, p. 137: $-5, 5$ **2, p. 138: a.** 0
b. No solution   **3, p. 138:** $5, -5$ **4, p. 139:** $-2, -10$
5, p. 139: $3, -\frac{7}{3}$ **6, p. 139:** No solution   **7, p. 140:** $-1, -\frac{1}{3}$
8, p. 140: $-\frac{1}{2}$ **9, p. 141:** The maximum and minimum acceptable weights are 7.2515 oz and 7.2485 oz, respectively.   **10, p. 141:** $-5 < x < 5$;

interval notation: $(-5, 5)$

11, p. 142: $x < -5$ or $x > 5$; interval notation: $(-\infty, -5) \cup (5, \infty)$

12, p. 142: $-7 < x < 3$; interval notation: $(-7, 3)$

13, p. 143: $x \le -2$ or $x \ge 6$; interval notation: $(-\infty, -2] \cup [6, \infty)$

14, p. 143: $-1 \le x \le \frac{5}{3}$; interval notation: $[-1, \frac{5}{3}]$

15, p. 143: $x < \frac{1}{2}$ or $x > 1$; interval notation: $(-\infty, \frac{1}{2}) \cup (1, \infty)$

16, p. 144:   a. $|x - 454| \le 5$ **b.** The range of acceptable weights is between, and including, 449 g and 459 g.

Exercises 2.5, pp. 145–148

1. the opposite of that number   **3.** two solutions   **5.** 6   **7.** 7   **9.** 7.5
11. $-7, 7$ **13.** $-\frac{2}{3}, \frac{2}{3}$ **15.** $-4, 4$ **17.** No solution   **19.** 0   **21.** $-2, 2$
23. No solution   **25.** $-7, 7$ **27.** $-\frac{3}{2}, \frac{3}{2}$ **29.** $-6, 6$ **31.** $-3, -11$
33. No solution   **35.** $-17, 4$ **37.** $-\frac{3}{2}, 7$ **39.** $-\frac{7}{3}, 3$ **41.** No solution
43. $-4, \frac{44}{7}$ **45.** $2, 12$ **47.** $-12, -4$ **49.** $-2, 12$ **51.** $0, 2$ **53.** $-3, 1$
55. -4 **57.** 3   **59.** $-5, -\frac{3}{5}$ **61.** No solution   **63.** $x < -3$ or $x > 3$;
interval notation: $(-\infty, -3) \cup (3, \infty)$

65. $-\frac{1}{2} \le x \le \frac{1}{2}$; interval notation: $[-\frac{1}{2}, \frac{1}{2}]$

67. $x \le -3$ or $x \ge 3$; interval notation: $(-\infty, -3] \cup [3, \infty)$

69. $-8 < x < 0$; interval notation: $(-8, 0)$

71. $x \le 2$ or $x \ge 8$; interval notation: $(-\infty, 2] \cup [8, \infty)$

73. $x \le -1$ or $x \ge 1$; interval notation: $(-\infty, -1] \cup [1, \infty)$

75. $-1 < x < 1$; interval notation: $(-1, 1)$

77. $x \le -7$ or $x \ge 4$; interval notation: $(-\infty, -7] \cup [4, \infty)$

79. $-1 < x < 5$; interval notation: $(-1, 5)$

81. All real numbers; interval notation: $(-\infty, \infty)$

83. $x < \frac{1}{2}$ or $x > \frac{13}{2}$; interval notation: $(-\infty, \frac{1}{2}) \cup (\frac{13}{2}, \infty)$

85. $-2 \le x \le \frac{7}{2}$; interval notation: $[-2, \frac{7}{2}]$

87. 5   **89.** $-11, 3$
91. $x < -\frac{5}{2}$ or $x > 3$; interval notation: $(-\infty, -\frac{5}{2}) \cup (3, \infty)$

93. The maximum possible temperature is 37.2°C and the minimum possible temperature is 36.4°C.   **95.** The maximum possible speed was 38 mph and the minimum possible speed was 36 mph.

97. a. $|x - 68| \leq 4$ **b.** Opposing the construction of the runway are 64% to 72%, inclusive, of the residents.   **99. a.** $|I - 39{,}000| > 5000$ **b.** The resident makes either more than \$44,000 or less than \$34,000.   **101.** The speed must satisfy the inequality $|s - 45| \leq 10$.

Chapter 2   Review Exercises, pp. 152–156

1. is not; possible answer: substituting 5 for x does not make the equation a true statement   **2.** is; possible answer: it can be written as $8c + (-1) = 0$ **3.** are; possible answer: they have the same solution   **4.** is; possible answer: it is an equation that relates two variables   **5.** does not; possible answer: a solution for x would mean that $|x|$ equals -1, but no absolute value can be negative   **6.** is; possible answer: this interval contains all real numbers   **7.** A solution   **8.** Not a solution   **9.** Not a solution   **10.** A solution   **11.** -6 **12.** -4 **13.** -8 **14.** $-\frac{3}{4}$ **15.** 3 **16.** 0.7 **17.** $-\frac{14}{3}$ **18.** $\frac{13}{6}$ **19.** -5 **20.** -1 **21.** $y = -5x - 1$ **22.** $x = \frac{1}{4}y - 2$ **23.** $c = -\frac{ab}{d}$ **24.** $x = \frac{z + v}{6y}$ **25.** $b = \frac{d + ac}{a}$ **26.** $t = -\frac{p - q + rs}{r}$ **27.** $h = \frac{A - 2\pi r^2}{2\pi r}$ **28.** $A = \frac{150D}{y}$ **29.** $T = \frac{PV}{nR}$ **30.** $w = \frac{V}{lh}$ **31.** Not a solution   **32.** A solution   **33.** b   **34.** a   **35.** d   **36.** c

37. $x > -6$

38. $n \leq 8$

39. $y \geq 0$

40. $x < -\frac{1}{4}$

41. $y \leq 3; (-\infty, 3]$ **42.** $x \geq 8; [8, \infty)$ **43.** $x < -2; (-\infty, -2)$ **44.** $n > 4; (4, \infty)$

45. $(-4, 1]$

46. $[-2, 3)$

47. $(-\infty, 0] \cup [2.5, \infty)$

48. $(-\infty, -3) \cup (4, \infty)$

49. $2 < x < 5$; interval notation: $(2, 5)$

50. $-4 \leq x < 3$; interval notation: $[-4, 3)$

51. $-3 < x \leq 2$; interval notation: $(-3, 2]$

52. $-4 \leq x \leq \frac{1}{2}$; interval notation: $[-4, \frac{1}{2}]$

53. $x < -1$ or $x \geq \frac{7}{2}$; interval notation: $(-\infty, -1) \cup [\frac{7}{2}, \infty)$

54. All real numbers; $(-\infty, \infty)$

55. $x \leq -2$ or $x > 0$ **56.** No solution   **57.** 11   **58.** 0.7   **59.** $-15, 15$ **60.** $-6, 6$ **61.** $-9, 9$ **62.** $-1, 1$ **63.** No solution   **64.** $\frac{2}{3}, 1$ **65.** $-24, -\frac{6}{7}$ **66.** 0   **67.** $-1.5 < x < 1.5$; interval notation: $(-1.5, 1.5)$

68. $1 \leq x \leq 7$; interval notation: $[1, 7]$

69. All real numbers; interval notation: $(-\infty, \infty)$

70. $x < \frac{1}{2}$ or $x > \frac{9}{2}$; interval notation: $(-\infty, \frac{1}{2}) \cup (\frac{9}{2}, \infty)$

71. $x \leq 2$ or $x \geq \frac{5}{2}$; interval notation: $(-\infty, 2] \cup [\frac{5}{2}, \infty)$

72. $1 < x < 6$; interval notation: $(1, 6)$

73. 30 in. are required for the width.   **74.** The height can be a maximum of 12 in.   **75.** The person was driving at a speed of 72 mph.   **76.** The cost will be equal to the revenue if 35 quilts are sold.   **77.** The last four digits are 3456.   **78.** The area of the poster is 32 ft^2.   **79.** It took the car $\frac{5}{6}$ hr, or 50 min, to catch up to the bus.   **80.** One friend is driving at a rate of 36 mph and the other is driving at a rate of 30 mph.   **81.** She needs 4.8 L of the 15% solution and 3.2 L of the 75% solution.   **82.** They invested \$5000 in the high-risk fund and \$15,000 in the low-risk fund.   **83. a.** $r = \frac{I}{Pt}$ **b.** The interest rate is $5\frac{1}{2}\%$, or 5.5%.   **84. a.** $D = \frac{Ay}{y + 12}$ **b.** 30 mg   **85.** At least 60 seat fillers and ushers   **86.** The jet was traveling at 1000 mph.   **87.** Moving company B is better if more than 200 mi are driven.   **88.** It can have at most 37 words in the ad.   **89.** The range in distances is 135 mi to 195 mi in 3 hr.   **90.** The target heart rate is between and includes 92.5 and 129.5 beats per minute.   **91.** The maximum length is $30\frac{1}{2}$ in. and the minimum length is $29\frac{1}{2}$ in.   **92.** The possible approval ratings are between and include 47% and 51%.

Chapter 2   Posttest, p. 157

1. Yes, -3 is a solution.   **2.** $-\frac{1}{2}$ **3.** -7 **4.** 1   **5.** $y = \frac{c + b - ax}{a}$ **6.** No, -1 is not a solution.   **7.** $x > 9$

8. $\frac{1}{2} \leq x < \frac{9}{2}$

9. $n < -4$ or $n > 1$

10. $\frac{1}{2} < x \leq 2$ **11.** 3.8   **12.** $\frac{1}{3}, 3$ **13.** $-6, 5$ **14.** $\frac{13}{2}$

15. $\frac{1}{3} < x < 3$

16. The spa provided 42 relaxation massages and 11 deep-tissue massages.   **17.** He invested \$27,500 in the fund that lost 8% and \$22,500 in the fund that gained 6%.   **18.** Option A is a better deal if the gym is used more than 20 hr per month.   **19.** She can have between 100 and 1100 flyers printed.   **20.** The areas range from 2303.904 in.^{2} to 2304.096 in.^{2}.

Chapter 2   Cumulative Review, p. 158

1. True   **2.** 6   **3.** $2 \cdot (5 + 3)$ **4.** $a^6 b^2$ **5.** $\frac{6}{n}$ **6.** $\frac{1}{3}(2n + 9)$ **7.** 8   **8.** $x - 8$ **9.** -4 **10.** $c = \dfrac{d(a - e)}{b}$ **11.** No

12. $x \geq -3$

13. $x < 1$ or $x \geq 3$ **14.** No solution   **15.** $-\$0.04$ **16.** 30.0 astronomical units   **17.** $(2^3 \times 3^1) \times (2^2 \times 5^2) = 2^5 \times 3^1 \times 5^2$ **18. a.** $m = \frac{s}{5}$ **b.** 12.5 sec   **19.** About 75% of the U.S. population   **20.** $|x - 69.65| > 3.85$

CHAPTER 3

Chapter 3   Pretest, pp. 160–162

1.

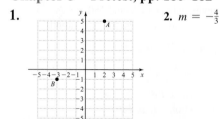

2. $m = -\frac{4}{3}$

3.

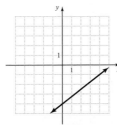

4. The slope of $\overleftrightarrow{PQ}$ is -1 and the slope of $\overleftrightarrow{RS}$ is -1. Since the slopes are the same, the lines are parallel.   **5.** Yes, $(3, -2)$ is a solution.
6. x-intercept: $(-2, 0)$, y-intercept: $(0, 3)$
7.

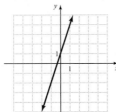

8.

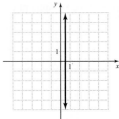

9.

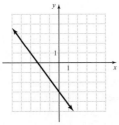

10. Point-slope form: $y - 5 = 2(x + 1)$; slope-intercept form: $y = 2x + 7$ **11.** Point-slope form: $y + 1 = -\frac{3}{2}(x - 0)$; slope-intercept form: $y = -\frac{3}{2}x - 1$ **12.** Not a solution
13.

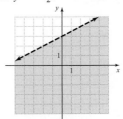

14.

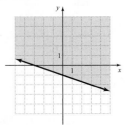

15. 20   **16.** Domain: $[-2, \infty)$; range: $(-\infty, 4]$

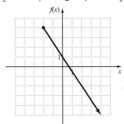

17. a.

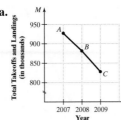

b. Slope of $\overline{AB}$: $-45,407$; Slope of $\overline{BC}$: $-53,667$;   **c.** Since the slopes of the lines are not equal, the number of take-offs and landings did not change at the same rate over the three years.

18. a. $B = 0.15k + 20$ **b.**

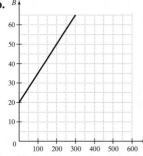

c. The monthly bill was about \$65.   **19. a.** $s = -8t + 40$
b. -8; the slope represents the rate of decrease of the car's velocity, that is, its deceleration.   **c.** $(0, 40)$; the s-intercept of the graph represents the speed of the car, 40 mph, when the brakes were first applied.   **d.** The car comes to a complete stop 5 sec after the brakes are applied.
20. a. $1.5x + y < 20$ **b.**

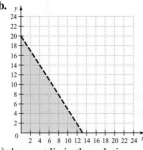

c. Since the point $(10, 8)$ does not lie in the solution region, the customer cannot make 10 point-of-sale transactions.

Section 3.1   Practices, pp. 165–166

1, p. 165: a. Ohio has two senators.   **b.** The number of senators from each of the three states is the same, two, regardless of the size of the state's population.
2, p. 165: a.

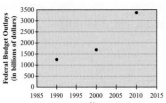

b. The points indicate an upward trend

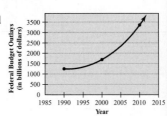

Exercises 3.1, pp. 167–170

1. x-axis   **3.** ordered pair   **5.** vertical   **7.** Quadrant IV
9.

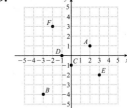

11.

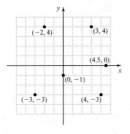

13. II 15. I 17. III 19. IV 21. II 23. I
25.

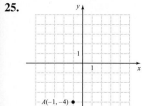

27. I

29. **a.** $A(400, 340)$; $B(450, 400)$; $C(470, 590)$; $D(520, 720)$; $E(600, 500)$ **b.** Students C and D **c.** Student D

31.

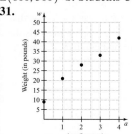

Age (in years)

33. a

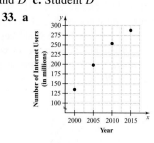

Year

b. The points indicate an upward trend; that is, each year the number of Internet users increased.   **35. a.** $3 **b.** The average price was below $2 in December. **c.** The greatest decrease in price occurred between October and November. The decrease was about $1 per gallon.   **37.** Graph II best illustrates the trip. As the friends drive to the beach, the distance from their apartment increases (line segment slants up to the right). When they get to the beach, the distance from their apartment does not change (horizontal line segment). As they head home, the distance from the apartment decreases (line segment slants down to the right). When they stop for dinner, the distance from the apartment does not change (horizontal line segment). Finally, as they complete the drive home, their distance from the apartment decreases and eventually returns to 0 (line segment slants down to the right).

Section 3.2   Practices, pp. 174–182

1, *p. 174:* $m = \frac{5}{3}$

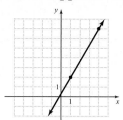

2, *p. 175:* $m = -\frac{3}{4}$

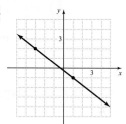

3, *p. 175:* $m = 0$

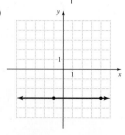

4, *p. 176:* The slope is undefined.

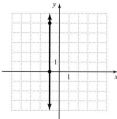

5, *p. 177:* Slope of $\overleftrightarrow{PQ}$: $-\frac{4}{5}$; slope of $\overleftrightarrow{RS}$: $-\frac{1}{4}$

6, *p. 177:* a. The slope of this line is negative, since it slants downward from left to right. **b.** As the number of years since purchasing the equipment increases, the value of the equipment decreases.

7, *p. 178:*

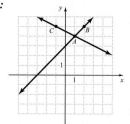

8, *p. 179:* a. The line that passes through $(1, 100)$ and $(3, 350)$ has slope 125. So the rate of infection between the first and third weeks is 125 residents per week. **b.** The line that passes through $(3, 350)$ and $(4, 450)$ has slope 100. So the rate of infection between the third and fourth weeks is 100 residents per week. **c.** Since $125 > 100$, the rate of infection was greater between the first and third weeks than between the third and fourth weeks. **9, *p. 180:*** Since the slope of $\overleftrightarrow{EF}$, $-\frac{3}{2}$, is not equal to the slope of $\overleftrightarrow{GH}$, -2, the lines are not parallel. **10, *p. 180:*** The brother's weight increased at a rate of 8 lb/yr and the sister's weight increased at a rate of $8\frac{1}{2}$ lb/yr. Since the rates of increase are not equal, their weights did not increase at the same rate.

11, *p. 181:*

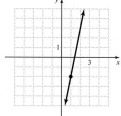

The slope of $\overleftrightarrow{AB}$ is 1, and the slope of $\overleftrightarrow{AC}$ is $-\frac{1}{2}$. Since the product of the slopes is not equal to -1, the lines are not perpendicular. **12, *p. 182:*** The slope of $\overline{AB}$ is -3, the slope of $\overline{BC}$ is $-\frac{1}{2}$, and the slope of $\overline{AC}$ is $\frac{1}{3}$. Since the slopes of $\overline{AB}$ and $\overline{AC}$ are negative reciprocals, the sides are perpendicular and the triangle is a right triangle. So the shelf will fit.

Exercises 3.2, pp. 183–191

1. y-values to the change in x-values   **3.** decreasing
5. horizontal **7.** parallel
9. -1

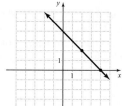

11. 2

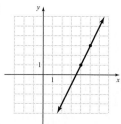

13. $\frac{4}{5}$ 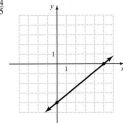   **15.** 0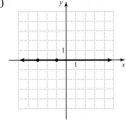

61.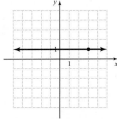

17. Undefined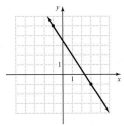

63. The slope of the ramp is $\frac{1}{21}$ and $\frac{1}{21} < \frac{1}{20}$; the curb ramp does meet the guideline.   **65. a.** The slope is positive.   **b.** The slope would indicate that as the amount of a purchase increases, the amount of sales tax charged also increases.   **c.** The slope is $\frac{1}{25} = 0.04$; it represents the sales tax rate, which is 4%, or $4 for every $100 purchased.   **67. a.** The slope represents the cost of the coffee per pound.   **b.** Since the slope of the line for Coffee Shop A, 5, is greater than the slope of the line for Coffee Shop B, 4.5, Coffee Shop A charges more for Kona coffee.

69. a.   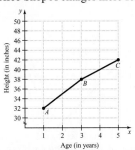   **b.** Slope of $\overline{AB}$: 3; slope of $\overline{BC}$: 2

19. $-\frac{3}{2}$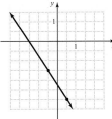

c. Since the slopes are not the same, the child's rate of growth was not constant. It was 3 inches per year from age 1 to age 3, and 2 inches per year from age 3 to age 5.   **71.** The slope of the line for provider A is $\frac{7}{100} = 0.07$, and the slope of the line for provider B is $\frac{7}{100} = 0.07$. Since the slopes are equal, the providers charge the same per-minute fee, namely 7 cents. But the flat monthly fee for provider B is lower.   **73.** The slope of $\overline{AB}$ is -1, the slope of $\overline{BC}$ is 1, the slope of $\overline{CD}$ is -1, and the slope of $\overline{AD}$ is 1. Since the slopes of each adjacent pair of line segments—$\overline{AB}$ and $\overline{BC}$, $\overline{BC}$ and $\overline{CD}$, $\overline{CD}$ and $\overline{AD}$, and $\overline{AB}$ and $\overline{AD}$—are negative reciprocals, the sides are perpendicular and the quadrilateral is a rectangle. So the insert will fit.   **75.** Plan A charges a monthly flat fee up to a certain number of call minutes (horizontal line segment) and a per-minute fee for any additional number of call minutes (line segment slants up to the right). Plan B charges a flat monthly fee plus a per-minute fee for each minute of calling.

21. -1.5 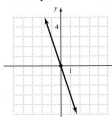   **23.** Slope of $\overleftrightarrow{AB}$: $\frac{7}{6}$; slope of $\overleftrightarrow{CD}$: $-\frac{2}{3}$

Section 3.3   Practices, pp. 192–198

1, *p. 192:*   **a.** Not a solution   **b.** A solution

2, *p. 193:*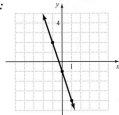

25. Negative slope   **27.** Undefined slope   **29.** Positive slope
31. Zero slope
33.   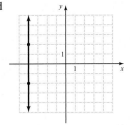   **35.**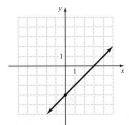

37.   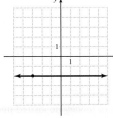   For Exercises 39–49, answers may vary.

3, *p. 194:*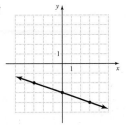

39. $(1, 6)$ and $(2, 8)$ **41.** $(-4, -2)$ and $(-3, -6)$ **43.** $(-2, 2)$ and $(0, 3)$ **45.** $(-4, -4)$ and $(0, -7)$ **47.** $(0, -7)$ and $(-2, -7)$
49. $(-2.4, 5)$ and $(-2.4, 0)$ **51.** Perpendicular   **53.** Parallel
55. Neither   **57.** Negative   **59.** perpendicular

4, p. 196:

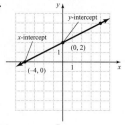

5, p. 197:

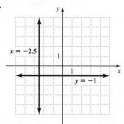

a. The slope is 0.
b. The slope is undefined.

6, p. 197: a. $0.1d + 0.05n = 2$ **b.**

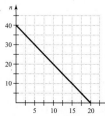

c. The d-intercept represents the number of coins in the cash register if it contained only dimes. The n-intercept represents the number of coins in the register if it contained only nickels. **d.** The slope of the line is -2. Negative; the given conditions imply that as the number of dimes in the register increases, the number of nickels decreases, indicating a negative slope.

Exercises 3.3, pp. 199–206

1. ordered pair   **3.** three points   **5.** y-intercept   **7.** Not a solution
9. A solution   **11.** A solution   **13.** Not a solution   **15.** A solution
17. b   **19.** d   **21.** x-intercept: $(4, 0)$; y-intercept: $(0, -4)$
23. x-intercept: $(-9, 0)$; y-intercept: $(0, 3)$ **25.** x-intercept: $(-2, 0)$;
y-intercept: $(0, -4)$ **27.** x-intercept: $(2, 0)$; y-intercept: $(0, -5)$
29. x-intercept: $(\frac{1}{3}, 0)$; y-intercept: $(0, -\frac{1}{4})$ **31.** No x-intercept; y-intercept: $(0, -\frac{3}{2})$ **33.** x-intercept: $(6, 0)$; no y-intercept

35.

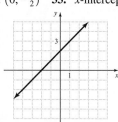

37.

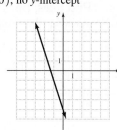

39.

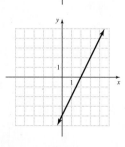

41.

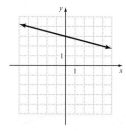

43.

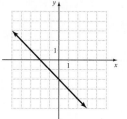

45.

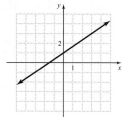

47.

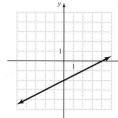

49.

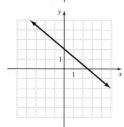

51.

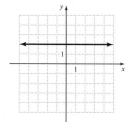

53.

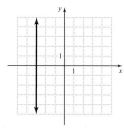

55.

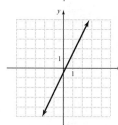

57.

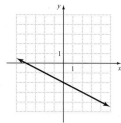

59. Not a solution   **61.**

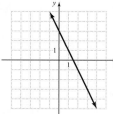

63. a. 1000; 1080; 1160; **b.** 1240

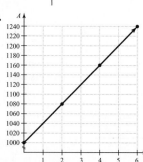

c. The A-intercept represents the initial amount of money deposited into the savings account. **d.** From the graph it appears that it will take five years for the account to grow to $1200.   **65. a.** $A = 200w + 50$

b.

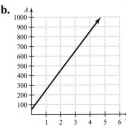

c. Since both quantities in this situation are positive, only points whose co-ordinates are positive need to be considered. The only points in the coordinate plane for which both coordinates are positive lie in Quadrant I.

67.

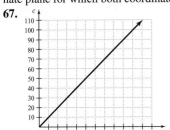

$c = 10m$

69. a. $c = 45n + 99$ **b.**

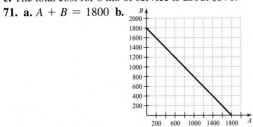

c. The total cost for 6 mo of service is about $370.

71. a. $A + B = 1800$ **b.**

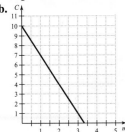

c. The A-intercept is $(1800, 0)$, and the B-intercept is $(0, 1800)$. The A-intercept represents the number of associate's degrees awarded if no bachelor's degrees were awarded. The B-intercept represents the number of bachelor's degrees awarded if no associate's degrees were awarded.
d. Only positive integer values make sense in this situation, since the college cannot award fractions of degrees. **73. a.** $300m + 100c = 1000$
b.

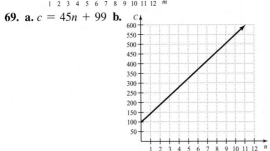

c. The m-intercept represents the number of servings she would need to meet the daily minimum requirement if she uses only milk as her source of calcium. The c-intercept represents the number of servings she would need to meet the daily minimum requirement if she uses only cottage cheese as her source of calcium.

Section 3.4 Practices, pp. 209–215

1, p. 209: $y = \frac{3}{2}x - \frac{1}{2}$ **2, p. 210:**

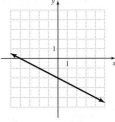

3, p. 210: $y = -2x + 3$ **4, p. 211:** $y = -6x - 2$
5, p. 211: $y = -\frac{1}{3}x$ **6, p. 211: a.** $P = 1800 - 100m$
b.

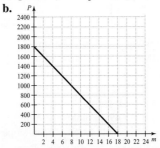

c. It will take about 18 months to pay off the debt. **7, p. 213:**
$y - 5 = 4(x - 2)$ **8, p. 213:** $y - 0 = \frac{1}{3}(x - 0)$, or $y = \frac{1}{3}x$ **9, p. 214:**
Point-slope form: $l - 10 = \frac{1}{250}(T - 0)$, or $l - 10 = \frac{1}{250}T$; slope-intercept
form: $l = \frac{1}{250}T + 10$ **10, p. 215:**

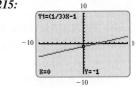

The displayed coordinates of the y-intercept are $x = 0$ and $y = -1$.

Exercises 3.4, pp. 216–222

1. slope-intercept form **3.** y-intercept **5.** one point on the line
7. $y = 2x$; slope: 2; y-intercept: $(0, 0)$ **9.** $y = -4x - 6$; slope: -4;
y-intercept: $(0, -6)$ **11.** $y = \frac{1}{2}x - 1$; slope: $\frac{1}{2}$; y-intercept: $(0, -1)$
13. $y = \frac{4}{5}x + 10$; slope: $\frac{4}{5}$; y-intercept: $(0, 10)$ **15.** $y = -\frac{3}{2}x + \frac{1}{2}$;
slope: $-\frac{3}{2}$; y-intercept: $(0, \frac{1}{2})$ **17.** $y = -3x + 29$; slope: -3; y-intercept:
$(0, 29)$ **19.** b **21.** a
23.

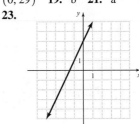

25.

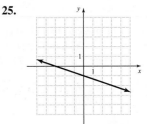

27.

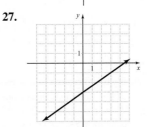

29.

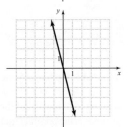

31.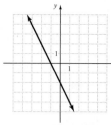

33.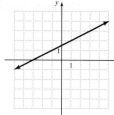

35.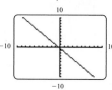

37.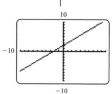

39. Point-slope form: $y - 8 = -\frac{3}{5}(x - 0)$, or $y - 8 = -\frac{3}{5}x$; slope-intercept form: $y = -\frac{3}{5}x + 8$ **41.** Point-slope form: $y + 4 = -\frac{1}{3}(x - 0)$, or $y + 4 = -\frac{1}{3}x$; slope-intercept form: $y = -\frac{1}{3}x - 4$ **43.** Point-slope form: $y - 0 = -\frac{1}{2}(x + 5)$, or $y = -\frac{1}{2}(x + 5)$; slope-intercept form: $y = -\frac{1}{2}x - \frac{5}{2}$ **45.** $x = -2.3$ **47.** Point-slope form: $y - 2 = -(x - 3)$; slope-intercept form: $y = -x + 5$ **49.** Point-slope form: $y - 5 = \frac{3}{2}(x + 6)$; slope-intercept form: $y = \frac{3}{2}x + 14$ **51.** Point-slope form: $y + 1 = -\frac{1}{2}(x - 2)$; slope-intercept form: $y = -\frac{1}{2}x$ **53.** Point-slope form: $y + 6 = -\frac{4}{5}(x + 10)$; slope-intercept form: $y = -\frac{4}{5}x - 14$ **55.** Point-slope form: $y - 7 = (x - 3)$, or $y - 9 = (x - 5)$; slope-intercept form: $y = x + 4$ **57.** Point-slope form: $y - 1 = -\frac{5}{3}(x - 0)$, or $y - 1 = -\frac{5}{3}x$, or $y + 4 = -\frac{5}{3}(x - 3)$; slope-intercept form: $y = -\frac{5}{3}x + 1$ **59.** Point-slope form: $y - 1 = \frac{3}{10}(x - 6)$, or $y + 2 = \frac{3}{10}(x + 4)$; slope-intercept form: $y = \frac{3}{10}x - \frac{4}{5}$ **61.** Point-slope form: $y - 4 = \frac{1}{2}(x - 2)$; slope-intercept form: $y = \frac{1}{2}x + 3$ **63.** $y = -\frac{1}{4}x + 3$; slope: $-\frac{1}{4}$; y-intercept: $(0, 3)$ **65.** $y = -\frac{2}{3}x - 1$ **67. a.** $A = 40x + 100$

b.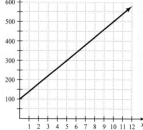

c. The A-intercept represents the initial consultation fee.
69. a. $C - 0 = \frac{5}{9}(F - 32)$, or $C = \frac{5}{9}(F - 32)$ **b.** The equivalent Fahrenheit temperature is $-121°$F. **71. a.** $v = -32t + 84$

b.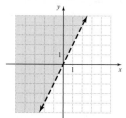

c. The slope represents the rate at which the velocity changes, that is, the deceleration. **d.** The t-intercept is the point at which the motion changes from an upward direction to a downward direction. **73. a.** $f = 1.5m + 2.5$
b. The slope represents the per-mile meter rate. **c.** 13 mi **75.** $d = 56t$
77. Point-slope form: $n - 26 = 19(t - 0)$, or $n - 83 = 19(t - 3)$; slope-intercept form: $n = 19t + 26$

Section 3.5 Practices, pp. 224–227

1, *p. 224:* Yes, $(3, -1)$ is a solution.
2, *p. 225:*

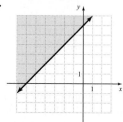

3, *p. 226:*

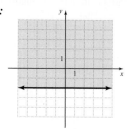

4, *p. 226:*

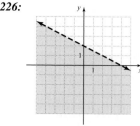

5, *p. 227:* **a.** $8f + 10s \geq 200$
b.

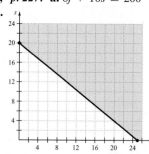

c. Answers may vary. Possible answers: 15 hr at the first job and 12 hr at the second job; 5 hr at the first job and 20 hr at the second job

Exercises 3.5, pp. 228–233

1. linear inequality **3.** solid **5.** is **7.** A solution **9.** A solution
11. Not a solution **13.** Not a solution **15.** c **17.** a
19. 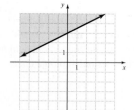 **21.**

23. 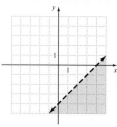 **25.**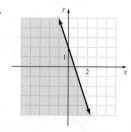

27.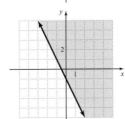

29.

31.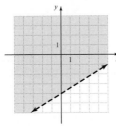

33.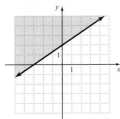

35.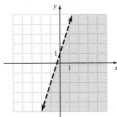

37.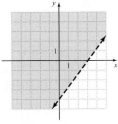

39.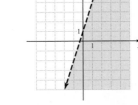

41. A solution

43. a. $30x + 21y \le 5000$ **b.**
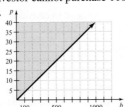

c. Since the point $(110, 110)$ does not lie in the solution region, the investor cannot purchase 110 shares of each stock.   **45. a.** $p \ge 0.04b$

b. 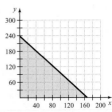   **c.** Answers may vary. Possible answer: $30

47. a. $160a + 60c \le 2000$ **b.**
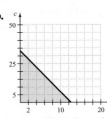

c. Answers may vary. Possible answer: 10 adults and 5 children

49. a. $12c + 8r \le 360$ **b.**
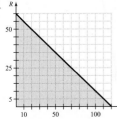

c. Yes, since the solution point $(20, 15)$ lies on the boundary line and all points on the boundary line are solutions of the inequality
51. a. $\frac{1}{2}T + R \le 60$ **b.**
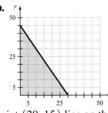

c. No, since the point $(50, 40)$ does not lie in the solution region

Section 3.6   Practices, pp. 236–243

1, p. 236: a. Not a function **b.** A function   **2, p. 237: a.** Domain: $\{2, 5, 8\}$; range: $\{-6, 0, 3\}$ **b.** Domain: $\{1, 2, 5\}$; range: $\{-1, 2, 4\}$
c. Domain: $\{1, 2, 3, 4, 5\}$; range: $\{6.5, 7, 7.5, 8, 9\}$ **3, p. 238: a.** 5
b. -4 **c.** $3n - 1$ **d.** $3n + 2$

4, p. 240: 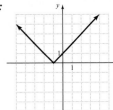   Domain: $(-\infty, \infty)$; range: $[0, \infty)$

5, p. 241: 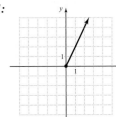   Domain: $[0, \infty)$; range: $[0, \infty)$

6, p. 242: a. Not a function **b.** A function
7, p. 243: a. $C(m) = 0.1m + 40$ **b.** $C(200); C(200) = \$60$
c.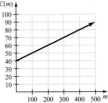

d. Domain: $[0, \infty)$; range: $[40, \infty)$ **e.** The graph in part (c) passes the vertical-line test, showing that the relation is a function.

Exercises 3.6, pp. 244–251

1. relation **3.** dependent **5.** range **7.** horizontal line
9. A function **11.** Not a function **13.** Not a function **15.** A function **17.** Domain: $\{-2, -1, 0, 1, 2\}$; range: $\{6, 8, 10, 12, 14\}$
19. Domain: $\{-3, -1, 0, 1, 3\}$; range: $\{-27, -1, 0, 1, 27\}$

21. Domain: $\{-4, -3, -2, -1, 1, 2, 3, 4\}$; range: $\{-8, -5, 0, 7\}$
23. Domain: $\{-4, -2, 0, 1, 2, 3.5, 5\}$; range: $\{1, 3, 5\}$
25. Domain: $\{-7, -3, -1, 2, 4\}$; range: $\{-4, -2, 1, 3, 7\}$
27. Domain: $\{2007, 2008, 2009, 2010, 2011\}$; range: $\{17, 29, 31\}$
29. a. -2 **b.** 13 **c.** 5 **d.** -1 **31. a.** 5 **b.** -11.8 **c.** $2.4a - 7$
d. $2.4a^2 - 7$ **33. a.** 3 **b.** 1 **c.** $|-2t + 3|$ **d.** $\left|\frac{1}{2}t\right|$ **35. a.** -9
b. 0 **c.** $3n^2 + 6n - 9$ **d.** $12n^2 - 12n - 9$ **37. a.** 10 **b.** 10 **c.** 10
d. 10 **39.**

Domain: $(-\infty, \infty)$;
range: $(-\infty, \infty)$

41.

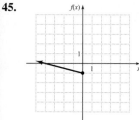

Domain: $(-\infty, \infty)$; range: $[2, \infty)$

43.

Domain: $(-\infty, \infty)$; range: $\{-5\}$

45.

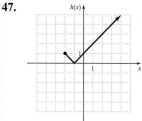

Domain: $(-\infty, 0]$; range: $[-1, \infty)$

47.

Domain: $[-2, \infty)$; range: $[0, \infty)$

49.

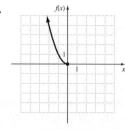

Domain: $(-\infty, 0]$; range: $[0, \infty)$

51. A function **53.** Not a function **55.** A function **57.** A function
59. A function **61.** A function **63.** Domain: $\{0, 2, 4, 6, 8, 9\}$;
range: $\{-3, -1, 0, 1, 5, 6\}$ **65.**

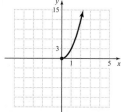

67. a. $d(a) = 0.2a$ **b.** $d(150)$; $d(150) = 30$, so $30 was saved.

c.

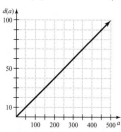

d. Domain: $[0, \infty)$; range: $[0, \infty)$ **69. a.** $V(t) = 22{,}500 - 1875t$
b. $V(6)$ represents the value of the car six years after it is purchased.
The value of the car after six years is $11,250.

c.

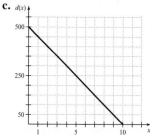

71. a. $d(x) = 500 - 50x$ **b.** $d(2) = 400$; after two weeks the patient's
dosage is 400 mg.

c.

73. a. $h(0.5) = 180, h(3) = 160$ **b.** No; there is no need to consider
any values of $t > 5$, since the object reaches the ground in 5 sec.

Chapter 3   Review Exercises, pp. 257–266

1. are; possible answer: their slopes are equal **2.** are not; possible answer: the product of their slopes is not -1 **3.** is not; possible answer:
the power of x is not 1 **4.** cannot; possible answer: the y-value of this
point is not 0 **5.** is; possible answer: substituting $(2, 5)$ into the inequality gives a true statement **6.** does not; possible answer: the 1 is paired
with two different second coordinates
7.

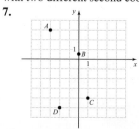

8. $A(5,4), B(-2,0), C(-5,-3), D(4,-5)$ **9.** IV **10.** III
11. I or III **12.** II or IV
13. 2

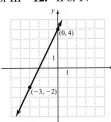

14. $\frac{1}{3}$

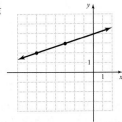

15. Undefined

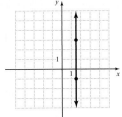

16. $-\frac{5}{3}$ **17.** Positive slope

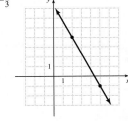

18. Undefined slope **19.** Zero slope **20.** Negative slope
21.

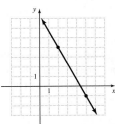

22.

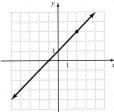

23.

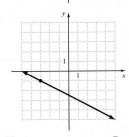

24.

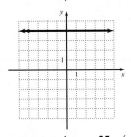

25–28: Answers may vary. Possible answers are given. **25.** $(-4,-1)$ and $(-3,-5)$ **26.** $(7,-6)$ and $(8,-3)$ **27.** $(4,-1)$ and $(-4,-1)$
28. $(-9,-4)$ and $(1,-10)$ **29.** Perpendicular **30.** Parallel
31. A solution **32.** Not a solution **33.** x-intercept: $(3,0)$; y-intercept: $(0,-9)$ **34.** x-intercept: $(-4,0)$; no y-intercept **35.** No x-intercept; y-intercept: $(0,2)$ **36.** x-intercept: $\left(-\frac{2}{3},0\right)$; y-intercept: $\left(0,\frac{3}{2}\right)$
37.

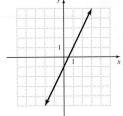

38.

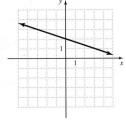

39.

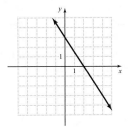

40.

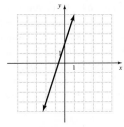

41.

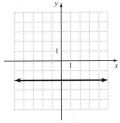

42.

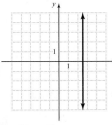

43. $y = 5x - 1$; slope: 5, y-intercept: $(0,-1)$
44. $y = \frac{2}{3}x - 3$; slope: $\frac{2}{3}$, y-intercept: $(0,-3)$
45. $y = -\frac{2}{7}x + 2$; slope: $-\frac{2}{7}$, y-intercept: $(0,2)$
46. $y = -\frac{3}{2}x + 11$; slope: $-\frac{3}{2}$, y-intercept: $(0,11)$
47.

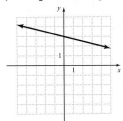

48.

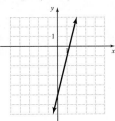

49.

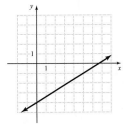

50.

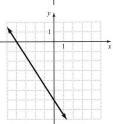

51. Point-slope form: $y - 3 = -6(x + 1)$; slope-intercept form: $y = -6x - 3$ **52.** Point-slope form: $y - 7 = \frac{2}{5}(x - 5)$; slope-intercept form: $y = \frac{2}{5}x + 5$ **53.** $y = -2.4$ **54.** Point-slope form: $y - 1 = \frac{1}{4}(x + 8)$; slope-intercept form: $y = \frac{1}{4}x + 3$ **55.** Point-slope form: $y - 7 = \frac{1}{5}(x - 0)$, or $y - 7 = \frac{1}{5}x$; slope-intercept form: $y = \frac{1}{5}x + 7$
56. Point-slope form: $y + 5 = -3(x + 2)$, or $y + 8 = -3(x + 1)$; slope-intercept form: $y = -3x - 11$ **57.** Not a solution **58.** A solution
59.

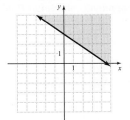

60.

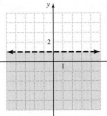

61.

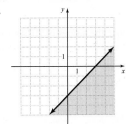

62.

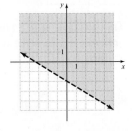

63. A function **64.** Not a function **65.** Domain: $\{-7, -5, -3, -1\}$; range: $\{3\}$ **66.** Domain: $\{-6, -4, -3, -2, 0, 2, 3, 4, 6\}$; range: $\{-6, -3, 0, 1, 2, 5, 8\}$ **67.** Domain: $\{-27, -8, -1, 0, 1, 8, 27\}$; range: $\{-3, -2, -1, 0, 1, 2, 3\}$ **68.** Domain: $\{20, 24, 26, 30, 32, 38\}$; range: $\{190, 228, 247, 285, 304, 361\}$ **69. a.** 3 **b.** 7.2 **c.** $a + 6$ **d.** $2a + 2$ **70. a.** 5 **b.** 10 **c.** $|8n - 7|$ **d.** $|n - 3|$
71. Domain: $(-\infty, \infty)$; range: $(-\infty, \infty)$

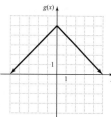

72. Domain: $(-\infty, \infty)$; range: $(-\infty, 0]$

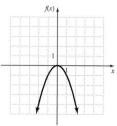

73. Domain: $(-\infty, \infty)$; range: $(-\infty, 5]$

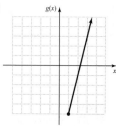

74. Domain: $[1, \infty)$; range: $[-5, \infty)$

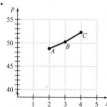

75. Not a function **76.** A function **77. a.** The slope represents the employee's hourly pay rate. **b.** Employee B has a higher hourly pay, since the slope of the line for Employee A, 7.5, is less than the slope of the line for Employee B, $13\frac{1}{3}$. **78.** The childcare center charges a flat fee for a certain number of minutes (horizontal line segment). Then an additional per-minute fee is charged for any number of minutes thereafter (line segment slants up).
79. a.

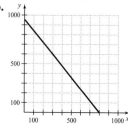

b. Slope of $\overline{AB}$: 1.44; slope of $\overline{BC}$: 2.08 **c.** Since the slopes of $\overline{AB}$ and $\overline{BC}$ are not the same, the increase in the average ticket price was not constant over the three years.

80. a.

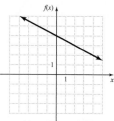

b. The slope is 50; it represents the monthly fee the garage charges.
c. The A-intercept is $(0, 20)$; it represents the activation fee.
d. A new customer pays about \$370 for 7 mo of parking.
81. a. $1.5x + 1.25y = 1200$
b.

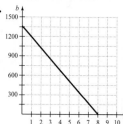

c. The x-intercept is $(800, 0)$, and the y-intercept is $(0, 960)$. The x-intercept represents the number of cups of coffee sold at Friday night's game if no hot chocolate was sold. The y-intercept represents the number of cups of hot chocolate sold if no coffee was sold. **d.** If 300 cups of hot chocolate were sold, then about 550 cups of coffee were sold.
82. a. $b = -170n + 1360$ **b.**

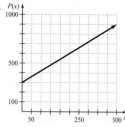

c. The n-intercept represents the number of months a customer will pay in order to pay off the balance. The b-intercept represents the annual premium amount. **83. a.** $b - 165 = 55(m - 3)$ or $b - 330 = 55(m - 6)$
b. The slope represents the number of bottles the machine can fill per minute.
c. The machine can fill 3300 bottles in 1 hr. **84. a.** $P(x) = 1.2x + 300$
b. $P(200) = 540$; the company makes a monthly profit of \$540 for selling 200 bottles of nail polish. **c.**

d. Domain: $[0, \infty)$; range: $[300, \infty)$
85. a. $5c + 10d \geq 500$ **b.**

c. No, since the point $(30, 25)$ does not lie in the solution region

86. a. $0.59x + 1.49y \le 100$ **b.**

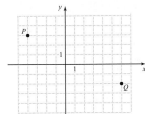

c. Answers may vary. Possible answer: 100 min of calling within the network and 20 min of calls outside the network.

Chapter 3   Posttest, pp. 267–269

1.

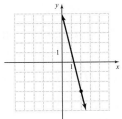

2. $m = -2$

3.

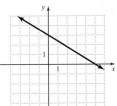

4. Perpendicular

5. Not a solution   **6.** x-intercept: $\left(-\frac{1}{2}, 0\right)$; y-intercept: $\left(0, \frac{1}{3}\right)$

7.

8.

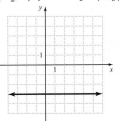

9.

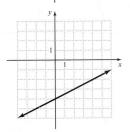

10. Point-slope form: $y + 1 = 4(x - 0)$, or $y + 1 = 4x$; slope-intercept form: $y = 4x - 1$ **11.** Point-slope form: $y - 5 = \frac{7}{2}(x - 2)$ or $y + 2 = \frac{7}{2}(x - 0)$, or $y + 2 = \frac{7}{2}x$; slope-intercept form: $y = \frac{7}{2}x - 2$

12. $-2a + 5$

13. Domain: $(-\infty, 5]$; range: $(-\infty, 1]$

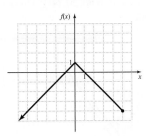

14. No, $(-10, -20)$ is not a solution.

15.

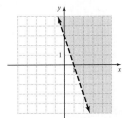

16.

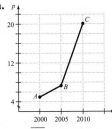

17. a.

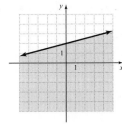

b. Slope of $\overline{AB} \approx 0.47$; slope of $\overline{BC} \approx 2.58$ **c.** No; since the slopes of $\overline{AB}$ and $\overline{BC}$ are not the same, the price of silver did not increase at the same rate over the ten years.   **18. a.** $4r + 8l = 1200$

b.

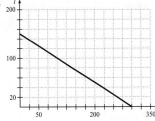

c. The r-intercept is $(300, 0)$ and the l-intercept is $(0, 150)$. The r-intercept represents the number of roses used in a floral arrangement with no lilies. The l-intercept represents the number of lilies used in the floral arrangements with no roses. **d.** 75 lilies are used with 150 roses.

19. a. $d = \frac{1}{8}t$ **b.** The slope of the line is $\frac{1}{8}$; it represents the jogger's running speed in miles per minute. **c.** It would take her 80 min to run 10 mi.

20. a. $8x + 14y \ge 4000$

b.

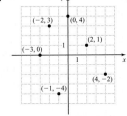

c. Yes, since the point $(200, 250)$ lies in the solution region

Chapter 3   Cumulative Review, pp. 270–271

1. $=$ **2.** -10 **3.** Distributive property   **4.** $\frac{1}{8x^6y^3}$ **5.** $-3n + 5$ **6.** $\frac{5}{3}$

7. $x \le -6$ **8.** $7 \le x < 10$

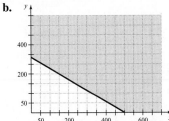

9. $-\frac{2}{3}, -\frac{14}{3}$ **10.** x-intercept: $(-2, 0)$; y-intercept: $\left(0, \frac{1}{2}\right)$

11.

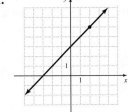

12.

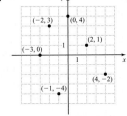

13.

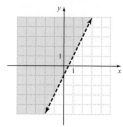

14.

15.

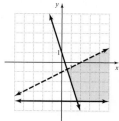

16. Both taxi companies will charge the same amount for a 2.5-mi ride. **17.** The buses will be 171 mi apart after $1\frac{1}{2}$ hr.   **18.** She can combine 8 L of the 25% alcohol solution with 12 L of the 50% alcohol solution to obtain the required amount.   **19.** 545 small, 84 medium, and 178 large bags of popcorn were sold on Wednesday.

15. a. 4.4 **b.** -11.2 **c.** $1.2a - 4$ **d.** $3.6a - 4$ **16.**   10 to 1   **17.** $175
18.   She needs 6 L of the 80% solution and 14 L of the 30% solution to make the required solution.   **19. a.** $15x + 5y \leq 50$, or $3x + y \leq 10$

20. a. $4.5t + 6.5r \leq 400$ $t \geq 50$ $r \geq 20$
b.

b.

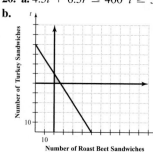

c. Possible purchase: 53 turkey sandwiches and 23 roast beef sandwiches

c. Answers may vary. Possible solution: 3 new movies and 1 old movie
20. a. $V(w) = 48w$ **b.** $V(5.1) = 244.8 \approx 245$; the value of the gold in a 24 karat ring that weighed 5.1 g was approximately \$245.

c.

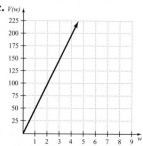

Section 4.1   Practices, pp. 276–282

1, p. 276: a. A solution **b.** Not a solution
2, p. 277: The solution is $(3, -3)$. **3, p. 278:** The system has no solution.

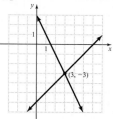

4, p. 278: The system has infinitely many solutions, namely all points on the line.

CHAPTER 4

Chapter 4   Pretest, pp. 273–274

1. Yes, it is a solution.
2. The solution is $(1, -1)$.   **3.** The solution is $(0, 2)$.

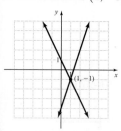

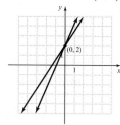

4. The solution is $(-4, 4)$.

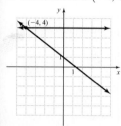

5. $\left(-\frac{1}{2}, 1\right)$ **6.** No solution
7. $u = 1, v = 1$ **8.** $(4, 6)$ **9.** $a = \frac{1}{4}$, $b = \frac{3}{4}$ **10.** Infinitely many solutions, namely all ordered pairs that satisfy both equations   **11.** No, it is not a solution.   **12.** $\left(-\frac{3}{5}, -\frac{3}{5}, \frac{6}{5}\right)$
13. $\left(-4, \frac{1}{2}, -3\right)$ **14.** $(14, -23, -55)$

5, p. 279: a. $C = 2.5v + 10$
(Uptown Video) $C = 2v + 20$
(Midtown Video)

b.

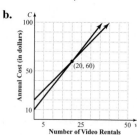

The solution is $(20, 60)$. This means that the memberships cost the same, \$60, when 20 videos are rented annually.   **6, p. 280: a.** $R(x) = 7500x$
b. $C(x) = 5900x + 80,000$

c.

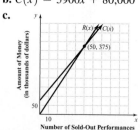

The solution is $(50, 375,000)$. The production will need to have 50 sold-out performances in order to break even.

7, p. 281: 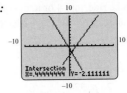 The approximate solution is $(0.444, -2.111)$.

Exercises 4.1, pp. 283–289

1. system **3.** one **5.** infinitely many **7.** A solution
9. Not a solution **11.** A solution **13.** d **15.** c
17. The solution is $(0, 0)$. **19.** The solution is $(4, -1)$.

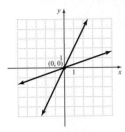 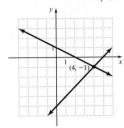

21. The solution is $(-2, -3)$. **23.** The solution is $(1, 4)$.

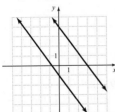 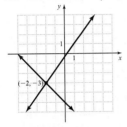

25. No solution **27.** The solution is $(2, 3)$.

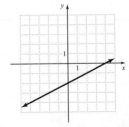 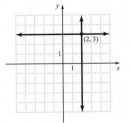

29. **31.** The solution is $(-2, 1)$.

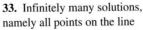 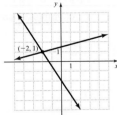

Infinitely many solutions, namely
all points on the line

33. Infinitely many solutions, namely all points on the line **35.** The approximate solution is $(0.111, -1.667)$.

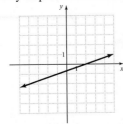 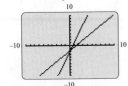

37. The approximate solution is $(1.023, -3.707)$. **39.** b

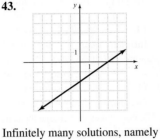

41.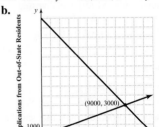

The solution is $(3, -2)$.

43.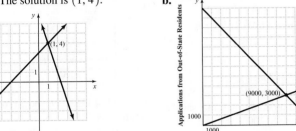

Infinitely many solutions, namely all points on the line

45. a. $x + y = 12,000; x = 3y$
b.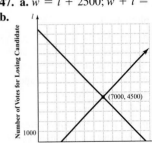
c. 9000 applications were from in-state residents, and 3000 were from out-of-state residents.

47. a. $w = l + 2500; w + l = 11,500$
b.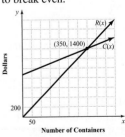
c. The winning candidate received 7000 votes, and the losing candidate received 4500 votes.

49. The company must sell 350 containers in order to break even.

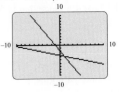

51. She works 6 hr at the coffee shop and 14 hr at the office.

Section 4.2 Practices, pp. 291–298

1, p. 291: $(0, 2)$ **2, p. 292:** $\left(\frac{3}{4}, \frac{1}{2}\right)$ **3, p. 293:** No solution
4, p. 293: Infinitely many solutions, namely all ordered pairs that satisfy both equations **5, p. 294:** $(2, -2)$ **6, p. 294:** No solution **7, p. 295:** $(8, -2)$ **8, p. 295:** $(-4, 1)$ **9, p. 296:** 25 pairs of $20 shoes were sold.

10, p. 297: a. $d = 20(t + 3)$; $d = 32t$ **b.** The Coast Guard cutter will reach the boat 5 hr after it leaves the harbor. **11, p. 298:** She should use 3 L of the sauce that is 70% tomato paste and 1 L of the sauce that is 40% tomato paste.

Exercises 4.2, pp. 299–302

1. addition **3.** remaining variable **5.** no solution **7.** $(-4, -7)$
9. $(-18, -6)$ **11.** $(6, -13)$ **13.** Infinitely many solutions, namely all ordered pairs that satisfy both equations **15.** $(-\frac{3}{2}, 6)$ **17.** $a = \frac{1}{5}$, $b = -3$ **19.** $(\frac{3}{2}, \frac{1}{2})$ **21.** No solution **23.** $s = 5, t = 1$ **25.** $(-\frac{7}{2}, -\frac{16}{3})$ **27.** $(4, -1)$ **29.** $(\frac{4}{3}, 1)$ **31.** $(-\frac{5}{2}, -\frac{12}{5})$ **33.** $c = 0, d = 0$
35. Infinitely many solutions, namely, all ordered pairs that satisfy both equations **37.** $(-8, 10)$ **39.** $(-\frac{5}{4}, -\frac{3}{2})$ **41.** No solution **43.** $(3, -2)$ **45.** $(3, -2)$ **47.** No solution **49.** $a = 0.4, b = 4$ **51.** $(-6, 3)$ **53.** $(1, -1)$ **55.** No solution **57.** $(25, 4)$ **59. a.** $C = 25d + 60$ (Uptown Towing Company); $C = 20d + 75$ (Downtown Towing Company) **b.** The companies will charge the same for three days of storage. **61.** The length of the garden is 15 ft and the width is 11 ft. **63. a.** $x - y = 18$; $x + y = 22$ **b.** The speed of the boat in still water is 20 mph and the speed of the current is 2 mph. **65.** 12-oz cups: 127, 20-oz cups: 381 **67.** The facility has 20 large storage lockers. **69.** $1750 was invested in the low-risk fund and $3500 was invested in the high-risk fund. **71.** She must combine 3.75 L of the 65% solution with 6.25 L of the 25% solution.

Section 4.3   Practices, pp. 305–312

1, p. 305: a. Not a solution **b.** A solution **2, p. 306:** $(1, 2, -1)$ **3, p. 308:** $(\frac{1}{2}, 1, -\frac{1}{2})$ **4, p. 309:** $(-1, -2, 2)$ **5, p. 310:** No solution **6, p. 311:** Infinitely many solutions, namely all ordered triples that satisfy the three equations **7, p. 311:** She should invest $1300 in the growth fund, $1250 in the income fund, and $650 in the money market fund.

Exercises 4.3, pp. 313–316

1. A solution **3.** Not a solution **5.** A solution **7.** $(4, 5, -2)$
9. $(-2, 3, -2)$ **11.** $(\frac{1}{2}, \frac{3}{2}, -4)$ **13.** $(3, 1, 0)$ **15.** $(\frac{1}{5}, \frac{1}{4}, \frac{1}{3})$
17. $(-\frac{3}{2}, -4, 9)$ **19.** No solution **21.** $(1, -1, 4)$ **23.** Infinitely many solutions, namely, all ordered triples that satisfy the three equations **25.** $(4, 0, -5)$ **27.** A solution **29.** Infinitely many solutions, namely, all ordered triples that satisfy the three equations **31.** No solution **33.** 84 beef dinners, 66 chicken dinners, and 42 fish dinners were ordered. **35.** The package does meet the parcel post restrictions, since length + girth = 30 + 2(20 + 15) = 100 in., which is less than 130 in. **37.** The thickness of a nickel is 1.95 mm, of a dime is 1.35 mm, and of a quarter is 1.75 mm. **39.** There are four par-3 holes, ten par-4 holes, and four par-5 holes on the course. **41.** 8 $10 traveler's checks, 5 $20 traveler's checks, and 2 $50 traveler's checks

Section 4.4   Practices, pp. 318–323

1, p. 318: $(3, -3)$ **2, p. 319:** $(2, 5, \frac{1}{2})$ **3, p. 321:** $(-1, 4, -6)$
4, p. 322: The top speed of the boat in still water is 22.5 mph and the speed of the current is 7.5 mph.

Exercises 4.4, pp. 324–326

1. matrix **3.** rows **5.** coefficients **7.** adding the products to
9. $(-2, -2)$ **11.** $(1, -1)$ **13.** Infinitely many solutions, namely all ordered pairs that satisfy both equations **15.** $(-8, 9)$ **17.** $(3, 1, -2)$
19. $(\frac{1}{2}, \frac{1}{4}, 2)$ **21.** $(-5, -2, -3)$ **23.** No solution **25.** $(0, 5, -4)$
27. $(\frac{25}{8}, \frac{1}{4}, \frac{9}{16})$ **29.** $(-4, 3)$ **31.** $(-8, 3)$ **33.** Infinitely many solutions, namely, all ordered triples that satisfy the three equations **35.** She needs 3 L of the 15% solution and 2 L of the 40% solution to get the required solution. **37.** It took him 1 hr 20 min (or $\frac{4}{3}$ hr) to complete the swim,

6 hr 40 min (or $\frac{20}{3}$ hr) to complete the bike ride, and 4 hr to complete the run. **39.** 3 cheeseburgers, 1 medium french fries, and 2 baked apple pies

Section 4.5   Practices, pp. 328–331

1, p. 328:

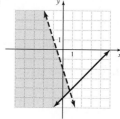

2, p. 329:

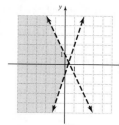

3, p. 330:

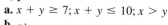

4, p. 331:
a. $x + y \geq 7$; $x + y \leq 10$; $x > y$
b.

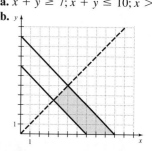

c. $(4, 3)$: 4 freshmen and 3 returning students; $(5, 2)$: 5 freshmen and 2 returning students; $(5, 3)$: 5 freshmen and 3 returning students; $(5, 4)$: 5 freshmen and 4 returning students; $(6, 1)$: 6 freshmen and 1 returning student; $(6, 2)$: 6 freshmen and 2 returning students; $(6, 3)$: 6 freshmen and 3 returning students; $(6, 4)$: 6 freshmen and 4 returning students; $(7, 0)$: 7 freshmen and 0 returning students; $(7, 1)$: 7 freshmen and 1 returning student; $(7, 2)$: 7 freshmen and 2 returning students; $(7, 3)$: 7 freshmen and 3 returning students; $(8, 0)$: 8 freshmen and 0 returning students; $(8, 1)$: 8 freshmen and 1 returning student; $(8, 2)$: 8 freshmen and 2 returning students; $(9, 0)$: 9 freshmen and 0 returning students; $(9, 1)$: 9 freshmen and 1 returning student; $(10, 0)$: 10 freshmen and 0 returning students. In this situation, only nonnegative integer solutions can be considered. The number of solutions in the overlapping region is limited to those listed here.

Exercises 4.5, pp. 332–337

1. a system of **3.** coordinate plane
5.

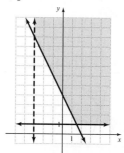

7.

9.

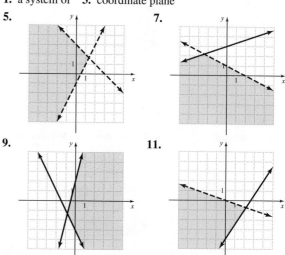

11.

13. 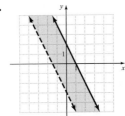   **15.**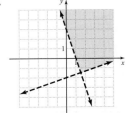

17. 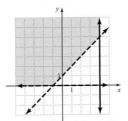   **19.**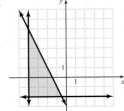

21. 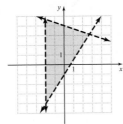   **23.**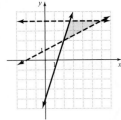

25. 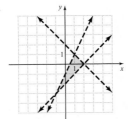   **27.**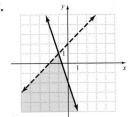

29. 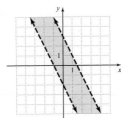   **31.**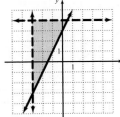

33. a. $2x + 1.5y \leq 360; 3x + y \leq 400$
b.

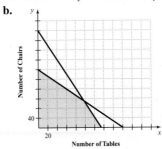

35. a. $150x + 225y < 15,000; x > y; x > 20$
b.

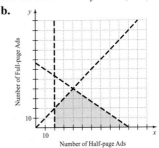

c. Possible answer: 30 half-page ads and 10 full-page ads

37. a. $x + y \geq 35; x \leq 20; y \leq 30$
b.

c. She must work between 20 and 30 hr, inclusive, at her office job.

39. a. $2l + 2w \leq 400; l \geq w + 25; w \geq 25$
b.

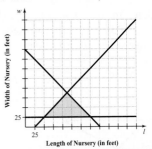

c. The solution region represents all possible dimensions for the nursery.

Chapter 4   Review Exercises, pp. 343–346

1. has; possible answer: the graphs of the equations are the same line
2. is not; possible answer: it is not a solution to the second equation
3. would; possible answer: the coefficient of y in one of the equations is 1
4. would not; possible answer: then neither variable would drop out
5. are; possible answer: both methods involve changing a given system of two equations in two variables to one equation in one variable   **6.** is not; possible answer: the first number represents the number of rows and the second represents the number of columns   **7.** A solution   **8.** A solution
9. $(-1, 3)$

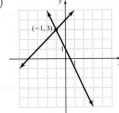

10. $(0, 3)$

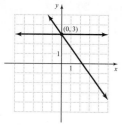

11. No solution

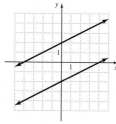

12. Infinitely many solutions, namely all points on the line

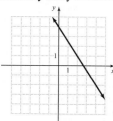

13. $(4, 19)$ **14.** $(2, 6)$ **15.** No solution **16.** $a = 4, b = 12$
17. $(7, -1)$ **18.** $p = 1, q = -2$ **19.** $s = 5, t = 3$ **20.** Infinitely
many solutions, namely all ordered pairs that satisfy both equations
21. A solution **22.** Not a solution **23.** $(0, 1, 2)$ **24.** $\left(\frac{1}{2}, -\frac{1}{3}, \frac{5}{6}\right)$
25. No solution **26.** $(-4, 3, 4)$ **27.** $(-3, -3)$ **28.** $\left(\frac{1}{2}, 1, \frac{1}{3}\right)$
29. **30.**

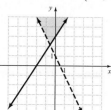 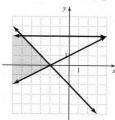

31. The family's monthly income is $2875 and the mortgage is $690.
32. 4123 male students and 5377 female students attend the college.
33. The area of the soccer field is 9600 sq yd. **34.** The athlete swims at
a rate of 1.6 mph in still water and the rate of the current is 0.5 mph.
35. Both plumbing services will charge the same amount in $2\frac{1}{2}$ hr.
36. The manufacturer breaks even when 250 helmets are produced and sold.
37. There were 120 employees in one department and 60 employees in the
other department before the layoffs. **38.** It will take the boy 10 min to catch
up to his sister. **39.** The movie theater sold 846 tickets before 5:00 P.M.
and 3384 tickets after 5:00 P.M. **40.** The original price of the boots was
$150 and the original price of the outerwear was $300. **41.** 1-point baskets: 14; 2-point baskets: 25; 3-point baskets: 16 **42.** 1-credit courses:
0, 2-credit courses: 4, 4-credit courses: 6 **43.** 16-oz drinks; 106, 24-oz
drinks; 167, 32-oz drinks:192 **44.** The median salary of a registered
nurse is $62,450, of a licensed practical nurse is $39,030, and of a medical assistant is $28,300. **45. a.** First plan: $t = 0.2e - 2$; second plan:
$t = 0.3e - 4.5$

b.

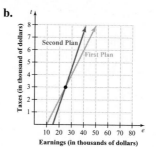

c. $25 thousand **d.** Under the second plan **46. a.** $x + y \leq 12,000$;
$0.08 y > 0.06x; y \geq 4500$ **b.**

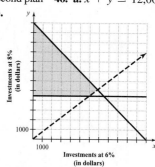

Chapter 4 Posttest, pp. 347–348

1. No, it is not a solution.
2. $(2, -1)$

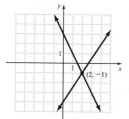

3. $(4, -4)$

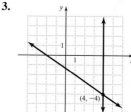

4. No solution

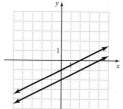

5. $\left(-7, \frac{2}{3}\right)$ **6.** $p = 1, q = 4$ **7.** Infinitely many solutions, namely
all ordered pairs that satisfy both equations **8.** $\left(2, -\frac{1}{2}\right)$ **9.** $(0, 4)$
10. No solution **11.** Yes, it is a solution. **12.** $(3, -5, 2)$
13. $(-1, 1, 3)$ **14.** $(-3, -2, 1)$
15.

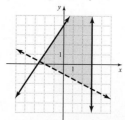

16. The length is 115 yd and the width is 60 yd.   **17.** The rate of the airplane in still air was 477 mph.   **18.** 8 adults and 11 children
19. 2 servings of yogurt, 1 plain bagel, and 3 servings of orange juice
20. a. $64x + 60y \geq 300; x + y \leq 12; x > y$
b.

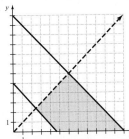

c. Answers may vary. Possible answers: $(6, 4)$: The faster student drives 6 hr and the slower student drives 4 hr. $(5, 3)$: The faster student drives 5 hr and the slower student drives 3 hr.

Chapter 4   Cumulative Review, pp. 349–351

1. 14   **2.**

x	-1	-0.5	0	0.5	1
$0.3x$	-0.3	-0.15	0	0.15	0.3

3. 5×10^8 **4.** 4   **5.** 10, 0
6. $-1 \leq x < 3$

7.

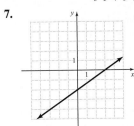

8. $y = \frac{1}{4}x - 2$ **9.** 7   **10.** No   **11.**

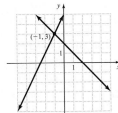

12. No solution   **13.** $(2, -2)$ **14.**

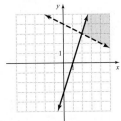

15. $2420   **16.** The customer was charged for 6 hr of labor.
17. The customer could be charged between $3.55 and $3.70 for the purchase.   **18. a.** $v = -32t + 120$
b.

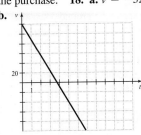

c. t-intercept: $\left(\frac{15}{4}, 0\right)$; v-intercept: $(0, 120)$. The t-intercept represents the time at which the object changes from an upward direction to a downward direction. The v-intercept represents the initial velocity of the object.

19. a. $7v + 5c \geq 29; v \leq 5; c \leq 5$
b.

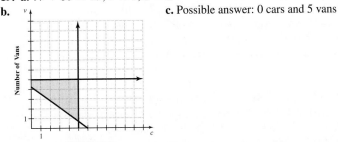

c. Possible answer: 0 cars and 5 vans

20. She scored 102 one-point free throws, 142 two-point field goals, and 40 three-point field goals.

CHAPTER 5

Chapter 5   Pretest, p. 353

1. a. $6x^4, -2x, \frac{x^3}{4}, -x^6, 3x^2$ **b.** $6, -2, \frac{1}{4}, -1, 3$
c. $-x^6 + 6x^4 + \frac{x^3}{4} + 3x^2 - 2x$ **d.** 6 **e.** Leading term: $-x^6$; leading coefficient: -1 **2.** $-2n^3 + 3n^2 + 8n - 3$
3. $18x^4 - 2x^2 - 7x - 1$ **4.** $2p^3q^2 - p^2q^3 + 4pq^4$
5. $6x^2 + 13xy - 5y^2$ **6.** $36n^2 - 12n + 1$ **7.** $64a^2 - 9b^2$
8. $u^3 + 8$ **9.** $3x^3 - 2x^2 + x + \frac{-9}{2x + 1}$ **10.** $(x + 4)(x - 8)$
11. $(3x - 4)(x + 6)$ **12.** $2q(5p - 2q)^2$ **13.** $(6ab + 11)(6ab - 11)$
14. $(3x + 2)(3x - 2)(x - y)$ **15.** $0, \frac{3}{2}$ **16.** 3, 7 **17.** $375
18. $(2x^2 + 240x + 7000) \, \text{m}^2$ **19.** $a(T_1^2 + T_2^2)(T_1 + T_2)(T_1 - T_2)$
20. The average rate of return on the investment was 12.5%.

Section 5.1   Practices, pp. 355–360

1, p. 355: **a.** Yes **b.** No **2, p. 355:** **a.** $-x^3, 9x^2, -3x$, and 13
b. $-1, 9, -3$, and 13 **3, p. 356:** **a.** Binomial of degree 1
b. Monomial of degree 0 **c.** Polynomial of degree 5 **d.** Trinomial of degree 5 **4, p. 356:** **a.** $-5x^4 + 7x^2 + 4x + 11$; leading term: $-5x^4$, leading coefficient: -5 **b.** $9x^5 - 3x^3 + 12x$; leading term: $9x^5$, leading coefficient: 9 **5, p. 357:** $4q^2 - 8pq + 4p^2$ **6, p. 357:**
$-2x^3 + 8x^2 + 12$ **7, p. 357:** **a.** 21 **b.** 45 **8, p. 358:** $14 trillion
9, p. 358: $9x^2 + 3x - 43$ **10, p. 359:** $5n^2 + 8n - 2$
11, p. 359: $-3x^2 - 6x$ **12, p. 359:** $2p^2 + 16pq - 15q^2$ **13, p. 360:**
$2x^2 + 18xy + 2y^2$ **14, p. 360:** **a.** $f(x) + g(x) = 2x^2 + 7x + 4$
b. $g(x) - f(x) = 4x^2 + 11x - 4$ **15, p. 360:**
$(178x^3 - 906x^2 - 307x + 144{,}257)$ thousand

Exercises 5.1, pp. 361–365

1. monomial **3.** variables **5.** highest **7.** constant term **9.** Polynomial; terms: $5x^3, -x^2, -6x, 7$; coefficients: $5, -1, -6, 7$ **11.** Not a polynomial
13. Polynomial; terms: $2x, \frac{x^2}{7}, -x^3, 5x^4$; coefficients: $2, \frac{1}{7}, -1, 5$
15. Polynomial terms: $3a^2, -8ab, 5b^2$; coefficients: $3, -8, 5$
17. Polynomial; terms: $\frac{p^2}{10}, -\frac{pq}{10}$; coefficients: $\frac{1}{10}, -\frac{1}{10}$
19. Polynomial of degree 3 **21.** Binomial of degree 2 **23.** Monomial of degree 1 **25.** Trinomial of degree 5 **27.** $-3x^2 + 9x + 2$; leading term: $-3x^2$, leading coefficient: -3 **29.** $5x^5 + 8x^4 - 2x^3 - x + 10$; leading term: $5x^5$, leading coefficient: 5 **31.** $-x^6 + 8x^4 - x^3 + 4x^2 + 5x - 11$; leading term: $-x^6$, leading coefficient: -1
33. $-x^3 - 4x^2y + 5xy^2 + y^3$ **35.** $p^5q^4 + 4p^4 + 2p^3q + 3p^2q^2 - pq^3$
37. $-2x^2 + 10x - 9$ **39.** $3n^3 + n^2 + 2n$ **41.** $-6x^4 - \frac{1}{6}x^2 + \frac{9}{8}$
43. a. 0 **b.** 45 **45. a.** 22 **b.** 6 **47.** $2x^2 - x + 10$
49. $a^3 - 10a^2 - a + 6$ **51.** $4x^4 + x^3 - 6x^2 - x - 1$
53. $-p^5 - p^4 - 4p^3 + 7p^2 + 1$

A-38   Answers

55. $2x^2 + 11x^2y + 2xy - 14xy^2 + 5y^2$
57. $5y^3 + 5y^2 + y + 2$ **59.** $3p^4 - 3p^2 + 10p - 26$
61. -1 **63.** $-3x^3 + 10x^2y - 3xy^2 + 7y^3$ **65.** $2x^2 - 6x + 7$
67. $n^3 + 9n^2 + 7n + 21$ **69.** $f(x) + g(x) = 4x^2 + 7x - 3;$
$f(x) - g(x) = -7x - 13$ **71.** $f(x) + g(x) = -x^4 - 3x^3 +$
$8x^2 - x - 3; f(x) - g(x) = -9x^4 + 3x^3 + 4x^2 + x - 3$
73. $-2a^4 - 2a^3 + 3a^2 - a + 1$ **75.** $6n^2 + 10$
77. $5x^3 + 5x^2 - 8x + 7$ **79.** Polynomial   **81.** $-5a^3 + 2a^2 - 9a + 11$
83. $520, 504, 456, 376$ **85.** The factory's revenue in May was $31,800.
87. $(-56x^3 + 341x^2 - 752x + 49{,}770)$ thousand
89. $(7x^3 - 23x^2 + 40x + 357)$ thousand
91. $P(x) = -0.005x^2 + 28x - 10{,}000$

Section 5.2   Practices, pp. 367–374

1, p. 367: a. $24x^5$ **b.** $-42a^4b^4$ **c.** $100x^6y^4$ **2, p. 368:**
a. $-8x^3 + 18x$ **b.** $18a^6b^7 + 3a^4b^4 - 30a^3b^3$
c. $-12y^5 + 18y^4 + 12y^3 + 2y^2$ **3, p. 368:** $8n^2 + 22n - 21$
4, p. 369: a. $5x^2 + 3x - 2$ **b.** $9x^2 - 24x + 16$ **c.** $6m^2 - mn - n^2$
5, p. 370: a. $5x^4 + 41x^3 + 8x^2 - 2x - 16$
b. $3x^4 + 8x^3 - 8x^2 + 5x - 2$ **6, p. 371:** $p^3 + 6p^2q + 12pq^2 + 8q^3$
7, p. 372: a. $16p^2 + 56p + 49$ **b.** $100 - 40x + 4x^2$, or
$4x^2 - 40x + 100$ **c.** $25x^2 + 30xy + 9y^2$ **d.** $36m^6 - 12m^3n^4 + n^8$
8, p. 372: a. $4s^2 - 81$ **b.** $16a^2 - 49b^2$ **c.** $64x^8 - y^4$ **9, p. 373:**
$-5x^3 + 21x^2 - 14x + 2$ **10, p. 373: a.** $4x^2 + 16x + 15$
b. $16x + 16$ **c.** $8xh + 4h^2$ **11, p. 374:** $R = -\frac{1}{4}p^2 + 100p$

Exercises 5.2, pp. 375–379

1. product rule of exponents   **3.** positive   **5.** square of a binomial
7. $-30n^6$ **9.** $6r^4t^3$ **11.** $20x^{12}$ **13.** $-6p^4q^3r$ **15.** $-64a^3b^{15}$
17. $-10x + 8x^2$, or $8x^2 - 10x$ **19.** $24n^3 - 4n^2$
21. $-3x^3 + 12x^2 - 15x$ **23.** $6m^2n^3 + 4n^4$
25. $28x^7y^4 - 16x^5y^4 + 4x^2y^4$ **27.** $16a^5b^3 - 6a^4b^4 + 8a^3b^5 - 2a^2b^6$
29. $-5y^3 + 2y^2 + 2y$ **31.** $10x^5 - 3x^4 - 4x^3$
33. $2p^4 - 3p^3q + 5pq^3 - 3q^4$ **35.** $x^2 + 6x + 8$ **37.** $n^2 - 9n + 18$
39. $-a^2 - 2a + 35$ **41.** $y^2 + 14y + 49$ **43.** $2x^2 + 7x - 4$
45. $-6x^2 + 5x + 6$ **47.** $30x^2 + 43x + 15$ **49.** $16x^2 - 72x + 81$
51. $2a^2 + ab - 3b^2$ **53.** $18p^2 + 11pq - 10q^2$
55. $56x^2 - 137xy + 77y^2$ **57.** $\frac{1}{3}x^2 + 8xy + 48y^2$
59. $n^3 + 2n^2 - n - 2$ **61.** $3x^3 - 13x^2 + 8x + 12$
63. $x^4 - 2x^3 - 5x^2 + 20x - 50$
65. $-14x^5 - 7x^4 + 42x^2 + 2x^3y + x^2y - 6y$
67. $x^4 - 9x^2 - 24x - 16$ **69.** $3x^5 + x^3y + 12x^2y^4 + 4y^5$
71. $12p^4 + 7p^2q^2 - 12q^4$ **73.** $t^2 - 100$ **75.** $x^2 + 16x + 64$
77. $16n^2 - 9$ **79.** $4n^2 - 20n + 25$ **81.** $b^2 - \frac{2}{3}ab + \frac{1}{9}a^2$, or
$\frac{1}{9}a^2 - \frac{2}{3}ab + b^2$ **83.** $25x^2 - y^2$ **85.** $121y^2 - 81x^2$
87. $16p^2 + 56pq + 49q^2$ **89.** $4x^4 - 36x^2y^4 + 81y^8$ **91.** $p^8 - 64q^4$
93. $x^2 + 8x + 16 - y^2$ **95.** $a^2 + b^2 - 2a + 2ab - 2b + 1$
97. $16x^3 - 48x^2 - x + 3$ **99.** $8n^3 - 12n^2 + 6n - 1$
101. $6a^3 - 29a^2b - 41ab^2 - 6b^3$ **103.** $x^3 + 6x^2y + 12xy^2 + 8y^3$
105. $16x^4 - 81y^4$ **107.** $f(x) \cdot g(x) = 15x^3 + 6x^2 - 5x - 2$
109. $f(x) \cdot g(x) = x^4 + x^3 - 45x^2 + 18x$ **111. a.** $n^2 - 4n + 8$
b. $4n^2 + 8n + 8$ **113. a.** $x^2 - 5x - 3$ **b.** $-4x + 6$ **c.** $h^2 + 2xh - h$
115. $15a^4 + 7a^3 - 2a^2$ **117.** $8x^3 - 18x^2 - 5x$ **119.** $2x^3 - x^2 + 4x$
121. $x^3 - 5x^2 + 8x - 4$ **123.** $16x^2 - 40xy + 25y^2$ **125.** $pV_2 - pV_1$
127. a. $x^2 + 18x + 72$ **b.** The size of her patio will increase by
$(x^2 + 18x)$ ft^2.   **129. a.** Longest panel: $(x + 0.1)$ in. by $(x + 0.1)$ in.
Shortest panel: $(x - 0.1)$ in. by $(x - 0.1)$ in. **b.** The difference in area
between the largest and smallest possible panel is $0.4x$ in^2.
131. $(4x^3 - 60x^2 + 216x)$ in^3

Section 5.3   Practices, pp. 380–384

1, p. 380: a. $4x^2$ **b.** $-x^2 + 2x$ **c.** $3x^5 + 5x^2 - 2$
d. $-a^6b^3 + 2ab - \frac{1}{5b^2}$ **2, p. 382: a.** $x - 7$ **b.** $2x + 1 + \frac{11}{x - 6}$

c. $-2n - 3$ **3, p. 383:** $2n + \frac{11n - 1}{n^2 - 3}$ **4, p. 383:** $2x + 1 + \frac{6}{2x - 1}$
5, p. 384:

$$
\begin{array}{r}
5r^2 + 5r + 5 \\
-r + 1{\overline{\smash{\big)}\,-5r^3 + 0r^2 + 0r + 5}} \\
\underline{-5r^3 + 5r^2} \\
-5r^2 + 0r \\
\underline{-5r^2 + 5r} \\
-5r + 5 \\
\underline{-5r + 5} \\
0
\end{array}
$$

Exercises 5.3, pp. 385–387

1. dividend   **3.** missing terms   **5.** $3x$ **7.** $-5y^2$ **9.** $3x^4y$
11. $2x^2 + 3x$ **13.** $-6n^3 + \frac{5}{2}n^2 - n$ **15.** $9a^2 - a + 6$
17. $-8t^2 - 5t + 9 + \frac{4}{t}$ **19.** $\frac{2p^2}{q^2} - \frac{4p}{q} + 7$ **21.** $-3x^2 - 2xy + 4y^2$
23. $2a^2b - a + \frac{1}{4b} - \frac{3}{a}$ **25.** $x - 2$ **27.** $n^2 + 4n - 5$
29. $-3x^2 - 7x + 4 + \frac{-9}{5 - x}$ **31.** $x - 3$ **33.** $-n^2 + 3n - 4 + \frac{28}{3n + 4}$
35. $2x^3 - 3x^2 + x - 1$ **37.** $-2a^2 - 5a + \frac{-7}{1 - 3a}$
39. $y^4 - y^3 + y^2 - y + 1$ **41.** $4x^2 - 2x + 1$ **43.** $x^2 - x + 2$
45. $2x - 1 + \frac{8x + 1}{x^2 - 4}$ **47.** $p^2 - 4$ **49.** $n + 2 + \frac{n - 4}{n^2 - 2n + 3}$
51. $\frac{f(x)}{g(x)} = x + 3$ **53.** $\frac{f(x)}{g(x)} = 2x^2 - 3x + \frac{11}{2x + 3}$ **55.** $x - 7$
57. $-5y + 15 - \frac{9}{y}$ **59.** $-2x^2 - 4x + 3 - \frac{4}{3 - x}$ **61.** $5n^2 - 2n + 3$
63. a. $\frac{7.95}{x} + 0.05$ **b.** The average cost for 5 hr of long-distance calls is
$0.08 per minute.   **65. a.** $30{,}000r^2 + 30{,}000r + 30{,}000$
b. The employee will earn a total of $93,648 in 3 yr.

Section 5.4   Practices, pp. 390–392

1, p. 390: a. $4y(3y + 2)$ **b.** $3(3x^2 + 4x - 2)$ **c.** Not factorable
2, p. 390: a. $7a(-2ab - 3)$ **b.** $6ab(a^3 - 6a^2b + 2b^2)$
3, p. 390: $l = \frac{S - 2wh}{2w + 2h}$ **4, p. 391:** $(a - 2)(6b + 7)$ **5, p. 391:**
$(n - 1)(4p - 3)$ **6, p. 391: a.** $(a + 3)(5a - 2b)$
b. $(y - z)(10 - y^4)$ **7, p. 392:** $\frac{1}{2}n(n - 3)$

Exercises 5.4, pp. 393–395

1. product   **3.** common factors   **5.** opposite binomial   **7.** $2a(a^2 + 4)$
9. $8(4 - 5a + 3a^2)$ **11.** $x^2(3x^2 + 7x - 9)$ **13.** Not factorable
15. $3a^3(-5a^2 + 3a - 6)$, or $-3a^3(5a^2 - 3a + 6)$
17. $6y^2(x^2 + 3xy + 6y^2)$ **19.** $9p^3q^2(-3p^3 - 5p^2q^3 - 4pq + 8q^2)$,
or $-9p^3q^2(3p^3 + 5p^2q^3 + 4pq - 8q^2)$ **21.** $ab(-a^2b - 3ab^3c - c^2)$,
or $-ab(a^2b + 3ab^3c + c^2)$ **23.** $(x - 5)(3x + 2)$
25. $(2x - 3)(3x - 2y)$ **27.** $(x + y)^2(x - y)$
29. $(x + 2y^2)(x^2 - 3y + 1)$ **31.** $S = \frac{A}{1 + t}$ **33.** $h = \frac{2A}{b_1 + b_2}$
35. $(y - 7)(x - 10)$ **37.** $(z - y)(4x + 1)$ **39.** $(2x - y)(1 - 2x)$,
or $(2x - y)(-2x + 1)$ **41.** $(b + 5)(a + c)$ **43.** $(x - 9)(x + 2)$
45. $(3x - 1)(3y + 1)$ **47.** $(5t - 3)(2s - 5)$
49. $(14u + 15v)(w - 2x)$ **51.** $(a - 2b)(2a + b)$
53. $(2x + y)(3x - z)$ **55.** $(x^2 - y)(x - y)$
57. $(3c^2 + d^2)(4c + 3d)$ **59.** $(b - 2)(3a + 4c)$
61. $(4 - r)(p^2 + q^3)$ **63.** $p = \frac{2M}{q + r}$ **65.** $(y - z)(3x + 1)$
67. $(3r - 7)(2s + 5)$ **69.** $3xyz(-y - 3xz + 2)$
71. a. $16(45 - t^2)$ **b.** The stone is 320 ft above the ground 5 sec after
it is dropped.   **73. a.** $l(l - 25)$ **b.** The length is 50 m. In the expression
$l(l - 25), l - 25$ represents the width of the pool. So $l - 25 = 25$.
Solving the equation for l yields $l = 50$, or 50 m.
75. $(P + Pr) + (P + Pr)r = P(1 + r) + Pr(1 + r) =$
$(1 + r)(P + Pr) = (1 + r)[P(1 + r)] = P(1 + r)^2$
77. a. $P = \dfrac{A}{1 + rt}$ **b.** The initial investment was $2000.

Section 5.5 Practices, pp. 397–402

1, *p. 397:* $(x + 1)(x + 2)$ **2,** *p. 398:* $(x - 1)(x - 9)$
3, *p. 398:* $(p + 4)(p - 7)$ **4,** *p. 399:* $(p - 5q)(p + 8q)$
5, *p. 399:* **a.** $y(y + 1)(y - 8)$ **b.** $-(n + 2)(n - 5)$
6, *p. 400:* $(n^2 + 3)(n^2 + 5)$ **7,** *p. 400:* $(2x + 1)(2x + 7)$
8, *p. 401:* $8(2c - 3d)(c + d)$ **9,** *p. 402:* $(3x + 4)(x - 2)$
10, *p. 402:* $5x(x + 9)$

Exercises 5.5, pp. 403–406

1. trial-and-error   **3.** factors of c **5.** number ac whose sum is b
7. $x - 3$ **9.** $x + 7$ **11.** $x + 4y$ **13.** $(x + 3)(x + 4)$
15. $(n - 5)(n - 7)$ **17.** $(t - 6)(t + 8)$ **19.** $(x + 6)(x - 9)$
21. $(-y + 2)(y + 9)$, or $(2 - y)(9 + y)$, or $-(y - 2)(y + 9)$
23. $(x + 6)(x + 6) = (x + 6)^2$ **25.** Prime polynomial
27. $(x - 2y)(x - 3y)$ **29.** $(a + 3b)(a - 5b)$ **31.** Prime polynomial
33. $(c + 5d)(c - 6d)$ **35.** $5(p + 2)(p + 2) = 5(p + 2)^2$
37. $-(y + 3)(y - 7)$ **39.** $2c(c + 6)(c - 8)$
41. $-q^2(p - 2)(p + 10)$ **43.** $-2m^2n^2(n - 1)(n - 5)$
45. $(ab - 8)(ab + 9)$ **47.** $(x^2 + 3)(x^2 - 10)$
49. $(y^3 - 4)(y^3 + 13)$ **51.** $n(-n^4 + 4)(n^4 + 16)$, or
$-n(n^4 - 4)(n^4 + 16)$ **53.** $(p^3q^2 + 7)(p^3q^2 - 9)$ **55.** $2x + 1$
57. $2x + 5$ **59.** $4x - 9y$ **61.** $(3x + 2)(x + 3)$ **63.** $(2a - 9)(a - 2)$
65. $(2t - 3)(2t + 5)$ **67.** $(3n + 4)(2n - 3)$
69. $(3m + 10)(-m + 2)$, or $-(3m + 10)(m - 2)$ **71.** Prime
polynomial   **73.** $(2p + 3)(2p + 3) = (2p + 3)^2$
75. $(2x - 5y)(x - 6y)$ **77.** $(4p - 7q)(2p + 3q)$
79. $(3c - 2d)(2c + 3d)$ **81.** Prime polynomial
83. $5(3n + 2)(n - 4)$ **85.** $4y^2(2x + 1)(2x + 1) = 4y^2(2x + 1)^2$
87. $-2(4a - 9b)(a + 4b)$ **89.** $3ab^2(5b - 3a)(2b + a)$
91. $(6xy + 5)(2xy + 5)$ **93.** $(5n^2 - 4)(n^2 + 6)$
95. $(4a^2 - b^3)(a^2 - 3b^3)$ **97.** $x^2y(3x + 4y^2)(2x + 5y^2)$
99. $a - 8$ **101.** $(m^2 - 7)(m^2 - 8)$ **103.** $5a^2b^2(2a + b)(a - b)$
105. $3b(b - 4)(b + 3)$ **107.** $(2p + 7q)(2p + 7q) = (2p + 7q)^2$
109. **a.** $(x + 40)(x + 100)$ **b.** The dimensions of the original ice-
skating rink were 100 ft by 40 ft.   **111.** $-16(t - 2)(t + 6)$
113. $4n^2 - 4n - 15 = (2n + 3)(2n - 5)$; the difference of the factors
is $(2n + 3) - (2n - 5) = 2n + 3 - 2n + 5 = 8.$

Section 5.6 Practices, pp. 408–413

1, *p. 408:* **a.** Perfect square trinomial **b.** Not a perfect square trinomial
c. Perfect square trinomial   **2,** *p. 409:* **a.** $(n + 4)^2$ **b.** $(t - 3)^2$
c. $(4c - d)^2$ **3,** *p. 410:* **a.** Difference of squares **b.** Not a difference
of squares **c.** Difference of squares   **4,** *p. 410:* **a.** $(y + 9)(y - 9)$
b. $(x + 10y)(x - 10y)$ **c.** $2(2x + 5)(2x - 5)$
5, *p. 411:* **a.** $(4 + 2x - y)(4 - 2x + y)$ **b.** $(8x^4 + 5y)(8x^4 - 5y)$
6, *p. 412:* **a.** Difference of cubes **b.** Sum of cubes **c.** Neither
7, *p. 412:* **a.** $(5 - y)(25 + 5y + y^2)$
b. $(3mn^2 + 1)(9m^2n^4 - 3mn^2 + 1)$ **c.** $2(3x^2 + 1)$
8, *p. 413:* $4(5 + 2t)(5 - 2t)$, or $-4(2t + 5)(2t - 5)$

Exercises 5.6, pp. 414–417

1. sum of the squares   **3.** difference of the squares   **5.** Perfect square
trinomial   **7.** Neither   **9.** Perfect square trinomial   **11.** Difference of
squares   **13.** Neither   **15.** Difference of squares   **17.** Neither
19. $(x - 6)^2$ **21.** $(n + 10)^2$ **23.** $(2a - 5)^2$ **25.** Prime polynomial
27. $(x + 4)(x - 4)$ **29.** $(11 + y)(11 - y)$ **31.** $(3x + 2)(3x - 2)$
33. $(0.4r + 0.9)(0.4r - 0.9)$ **35.** Prime polynomial
37. $2(6u - 7v)(6u + 7v)$ **39.** $(\frac{1}{3}a + 2b)(\frac{1}{3}a - 2b)$ **41.** $(\frac{1}{2}u - v)^2$
43. $(7 - 4ab)^2$, or $(4ab - 7)^2$ **45.** $(2u - v + 8)(2u - v - 8)$
47. $(7 + 2x + 2y)(7 - 2x - 2y)$ **49.** $(p^3 - 11)^2$ **51.** $(3a^4 + 8b)^2$
53. $(2a^2 + 15)(2a^2 - 15)$ **55.** $(7x^3 + 12y^2)(7x^3 - 12y^2)$

57. $(10p^2q + 3r)(10p^2q - 3r)$ **59.** $5p(p + 2q)$ **61.** Difference of
cubes   **63.** Neither   **65.** Sum of cubes   **67.** $(x + 1)(x^2 - x + 1)$
69. $(p - 2)(p^2 + 2p + 4)$ **71.** Prime polynomial
73. $(\frac{1}{2} - a)(\frac{1}{4} + \frac{1}{2}a + a^2)$ **75.** $(5x + y)(25x^2 - 5xy + y^2)$
77. $(0.4b - 0.3a)(0.16b^2 + 0.12ba + 0.09a^2)$ **79.** Prime polynomial
81. $(a^2 - 2)(a^4 + 2a^2 + 4)$ **83.** $(4x^3 + 3y)(16x^6 - 12x^3y + 9y^2)$
85. $(2p^4 - q^3)(4p^8 + 2p^4q^3 + q^6)$
87. $-(ab + 4c^2)(a^2b^2 - 4abc^2 + 16c^4)$ **89.** $(2 - a)(a^2 + 5a + 13)$
91. $2x(x^2 + 12)$ **93.** $3x^2(3x - 1)^2$ **95.** $4xy(2x + 5y)(2x - 5y)$
97. $2uv^3(3u + 4v)(9u^2 - 12uv + 16v^2)$ **99.** $2x(2x + y)^2(2x - y)^2$
101. $-y^4(x^4 + 16)(x^2 + 4)(x + 2)(x - 2)$
103. $3(9p^2 + 4q^2)(3p + 2q)(3p - 2q)$ **105.** $(y - z)(x + 1)(x - 1)$
107. $(v + w)(u + 2)(u^2 - 2u + 4)$
109. $(2x + 3)(2x - 3)(y^2 + z^2)(y + z)(y - z)$
111. $(p - 2q + 6)(p - 2q - 6)$ **113.** $(2c + a + 5b)(2c - a - 5b)$
115. $(x^{2n} + 4)(x^n + 2)(x^n - 2)$ **117.** $(t^{2a} + s)(t^{2a} - s)$
119. $(5a^n + 2b^m)(5a^n - 2b^m)$ **121.** Difference of cubes
123. $(10a - 5b + 3)(10a - 5b - 3)$ **125.** $4a^2(a - 3)^2$
127. $(8a + \frac{1}{6}b)(8a - \frac{1}{6}b)$ **129.** $2xy(4x + 3y)(4x - 3y)$
131. $(x + 8)(x - 8)$ in^2 **133.** $4(9 + 2t)(9 - 2t)$
135. $10{,}000(1 + r + \frac{1}{4}r^2) = 10{,}000(1 + \frac{1}{2}r)^2 = 10{,}000(1 + \frac{r}{2})^2$
137. $(x + y)(x^2 - xy + y^2)$

Section 5.7 Practices, pp. 419–423

1, *p. 419:* $-\frac{1}{3}, 7$ **2,** *p. 420:* $-4, -5$ **3,** *p. 421:* **a.** $-\frac{2}{3}, 2$ **b.** $-3, 5$
4, *p. 422:* 25%   **5,** *p. 423:* The slower car traveled 3 mi and the faster
car traveled 4 mi.

Exercises 5.7, pp. 424–426

1. quadratic equation   **3.** square of the length of the hypotenuse
5. $-3, 4$ **7.** $\frac{3}{4}, 2$ **9.** $-\frac{1}{2}, -\frac{3}{2}$ **11.** 3   **13.** $0, -5$ **15.** $0, \frac{1}{2}$
17. $-6, -2$ **19.** $1, 2$ **21.** $-4, 9$ **23.** $-\frac{7}{2}, 1$ **25.** $\frac{5}{2}$ **27.** $-\frac{3}{2}, \frac{3}{2}$
29. $-\frac{5}{3}, -3$ **31.** $-7, 7$ **33.** $0, \frac{3}{5}$ **35.** $-8, 4$ **37.** $-\frac{4}{3}, \frac{9}{2}$ **39.** $\frac{1}{3}, 3$
41. $-2, 4$ **43.** $-4, -\frac{5}{3}$ **45.** $\frac{1}{2}, \frac{4}{3}$ **47.** $-3, 5$ **49.** $-\frac{1}{4}, 1$ **51.** $\frac{1}{2}, 4$
53. $3, 4$ **55.** $-4, \frac{3}{2}$ **57.** $-\frac{9}{2}, 2$ **59.** $4, 6$ **61.** $-\frac{1}{3}, 2$ **63.** $-\frac{3}{4}, -1$
65. $2, 4$ **67.** $\frac{1}{2}, -5$ **69.** $-5, 8$ **71.** 8 teams   **73.** The soccer field is
50 yd by 100 yd.   **75.** The average annual rate of return on the invest-
ment was 10%.   **77.** The ladder should be placed 9 ft from the house.
79. The screen is 12 in. by 9 in.

Chapter 5 Review Exercises, pp. 432–436

1. is not; possible answer: the variable in the expression is raised to a
negative power   **2.** is; possible answer: it can be expressed as the sum
or difference of monomials   **3.** is not; possible answer: 10 is the product and not the sum of the exponents of the variables   **4.** is; possible
answer: 2 is the coefficient of the term with the highest degree   **5.** is;
possible answer: 4 is the GCF of the coefficients, x^2 is the highest power
of x common to both monomials, and y is the highest power of y common to both monomials   **6.** is; possible answer: it can be written in
the form $ax^2 + bx + c = 0$, where a, b, and c are real numbers and
$a \neq 0$ **7.** Polynomial   **8.** Not a polynomial   **9.** Terms: $3n$ and -1;
coefficients: 3 and -1 **10.** Terms: x^5, $-4x^3$, $-x^2$, and $\frac{x}{2}$; coefficients:
$1, -4, -1,$ and $\frac{1}{2}$ **11.** Binomial of degree 3   **12.** Monomial of degree 2
13. Polynomial of degree 4   **14.** Trinomial of degree 5
15. $-y^6 - 4y^4 + y^3 + 5y^2 - 3y + 10$; leading term: $-y^6$, leading
coefficient: -1 **16.** $-\frac{x^3}{5} + 9x^2 - 2x + 7$; leading term: $-\frac{x^3}{5}$, leading
coefficient: $-\frac{1}{5}$ **17.** $2n^3 - 7n^2 + 11n + 8$ **18.** $\frac{7}{4}x^4 + 4x^2 - \frac{3}{2}x$
19. a. 21 **b.** $\frac{27}{2}$ **20. a.** 17 **b.** 13   **21.** $-3y^3 - y^2 + 3y + 4$
22. $7n^5 + 5n^2 - 4n - 1$ **23.** $-a^3 - 5a^2 - a + 12$
24. $2x^3 + 3x^2 + 10x - 7$ **25.** $4x - 3$ **26.** $5n^3 + 6n^2 - 4n + 5$

27. $12x^3 - 6x^2$ **28.** $3a^3b^3 - 6a^2b^4 - ab^5$ **29.** $x^3 + 5x$
30. $16a^3b - 10a^2b^3 + 5b^2$ **31.** $n^2 + 4n - 60$ **32.** $3t^2 + 7t + 2$
33. $4x^2 - 24x + 35$ **34.** $12x^2 - 7xy - 12y^2$
35. $3p^4 - p^3q + 3p^2q - 2p^2q^2 - pq^2 - 2q^3$ **36.** $u^6 - 64v^6$
37. $2x^4 - 5x^3 + 6x^2 + 15x - 18$
38. $6a^5 - 11a^4 - 14a^3 + 10a^2 + 25a - 4$ **39.** $25x^2 - 9$
40. $q^2 - 9p^4$ **41.** $16n^2 - 56n + 49$ **42.** $81a^2 + 36ab + 4b^2$
43. $8a^3 + 36a^2 + 54a + 27$ **44.** $16x^3 + 16x^2y - xy^2 - y^3$
45. $4a^3 - 2a + 3 - \frac{1}{2a}$ **46.** $-3pq^3 + q^2 - \frac{2q}{p} + \frac{1}{3p^2}$
47. $4y^2 - 2yz + 3xy$ **48.** $x + 7$ **49.** $5x - 1 + \frac{4}{x^2 + 2}$
50. $25x^2 + 10x + 4$ **51.** $4x^3 - 2x^2 + x + 4 + \frac{6}{x - 4}$
52. $3x^2 + x + 1$ **53.** $9x^2y(4x^3y^2 - 3x^2 - xy + 2y^3)$
54. $(p - q)(2p + 1)$ **55.** $(x + 7)(x + 2y)$ **56.** $(a - 2)(2a - 3b)$
57. $(b - 5)(a - 8)$ **58.** $(1 + 3v)(3u^2 - 2w^2)$ **59.** $n = \frac{2S}{a + b}$
60. $k = \frac{T}{x_1 + x_2}$ **61.** $(x - 4)(x - 5)$ **62.** Prime polynomial
63. $(a + 3b)(a + 7b)$ **64.** $-4p^2q^2(p + 3q)(p - 4q)$
65. $n(n^2 - 2)(n^2 + 16)$ **66.** $(2a - 7)(2a - 7) = (2a - 7)^2$
67. Prime polynomial **68.** $2(3u + 2v)(2u - 3v)$
69. $-6xy(5x + 2)(x + 1)$ **70.** $(4p^3 + q^2)(2p^3 - q^2)$
71. $(2n - 11)^2$ **72.** $(3a + 5b)^2$ **73.** $(10y + 7x)(10y - 7x)$
74. $3(u + 3)(u^2 - 3u + 9)$ **75.** $(4c - 3d)(16c^2 + 12cd + 9d^2)$
76. $3(2x^3y^2 - 1)^2$ **77.** $(x + y + z)(x + y - z)$
78. $(3a + 2)(9a^2 + 3a + 1)$ **79.** $2(4u^n + v^m)(4u^n - v^m)$
80. $(x - 1)(x + 3)(x - 3)$ **81.** $-1, \frac{4}{5}$ **82.** $-8, 8$ **83.** $0, 3$
84. $-2, \frac{4}{3}$ **85.** $\frac{5}{2}$ **86.** $\frac{3}{2}, 2$ **87.** $-\frac{2}{3}$ **88.** $-\frac{7}{4}, 1$ **89.** $R = 250x - \frac{1}{4}x^2$
90. $P(x) = (0.15x - 1500)$ dollars **91.** $4\pi(r_2 + r_1)(r_2 - r_1)$
92. $2000r + 2000$, or $2000(1 + r)$ **93.** 50 DVDs must be sold per
day to generate revenue of $5000. **94.** The slower airplane flew 450 mi
and the faster airplane flew 600 mi. **95. a.** $(x + 18)(x + 40)$ **b.** The
original dog run had an area of 720 ft² **96.** An 8-inch by 8-inch square
should be cut from each corner. **97.** The technician used 17 ft of wire.
98. $(-x^2 + 8x + 84)$ ft² **99.** $(x + 2y)^2$ **100. a.** $0.85(p + 2q)$
b. $102 **c.** The customer saved $18 with the discount.

Chapter 5   Posttest, p. 437

1. a. $1, 5n^3, -4n^5, -n^4, -\frac{n}{3}$ **b.** $1, 5, -4, -1, -\frac{1}{3}$
c. $-4n^5 - n^4 + 5n^3 - \frac{n}{3} + 1$ **d.** 5 **e.** Leading term: $-4n^5$; leading
coefficient: -4 **2.** $5x^4 + x^3 - 7x + 8$ **3.** $-y^3 + y^2 - 3y - 3$
4. $-8x^5y^4 - 6x^3y^4 + 4x^2y^4 + 10x^2y^5$ **5.** $14a^2 - 45ab - 14b^2$
6. $9p^2 - 30p + 25$ **7.** $16u^2 - 81v^2$
8. $-2x^4 - 11x^3 - 11x^2 + 11x - 2$ **9.** $3x^3 + x^2 - 3x - 1$
10. $(4x + 7)(2x - 5)$ **11.** $-6ab(ab + 2)(ab + 4)$
12. $(9p + 10q)(9p - 10q)$ **13.** $(4 - n)(16 + 4n + n^2)$
14. $(x + y)(x + 2z)(x - 2z)$ **15.** $-2, 2$ **16.** $-\frac{7}{2}, 3$ **17.** 439 acres
18. a. After two years: $(1000r^2 + 2000r + 1000)$ dollars; after three
years: $(1000r^3 + 3000r^2 + 3000r + 1000)$ dollars
b. $(1000r^3 + 2000r^2 + 1000r)$ dollars **19.** 4 sec **20.** The ladder
should be placed 12 ft from the building.

Chapter 5   Cumulative Review, pp. 438–439

1. $-x + 3.4y$ **2.** $\frac{3b^3}{4a^2c^2}$ **3.** $\frac{5}{2}$ **4.** $y_1 = y_2 - m(x_2 - x_1)$
5. $-6 \le x \le 3$
6.   **7.**

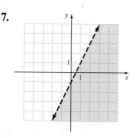

8. $(1, 2, -1)$ **9.** $(3, 0, -4)$
10.

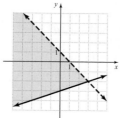

11. $x^2 + \frac{4}{3}xy + \frac{4}{9}y^2$ **12.** $x^2 + x - 4$ **13.** $b^2(a - 3)(a + 6)$
14. $(2xy + 3)(4x^2y^2 - 6xy + 9)$ **15.** $b = 0.08h + 4$ **16.** They
should invest at most $2500 in the account earning 5% interest
and at least $12,500 in the account earning 6% interest.
17. a. $y = 2000x + 110{,}000$ **b.** The population last year was about
108,000. **18.** The store sold 366 paperback books and 182 hardcover books.
19. a. $H(1)$ **b.** $78.95 **20.** The temperature was 55°F at 4 P.M.

CHAPTER 6

Chapter 6   Pretest, p. 441

1. a. $0, 3$ **b.** $-7, -2$ **2.** $\frac{6a^2 - 9b + ab^2}{3ab}$ **3.** $-\frac{1}{2p + 1}$ **4.** $\frac{y - 7}{2y + 3}$
5. $-\frac{2}{3x - y}$ **6.** $\frac{t + 1}{t}$ **7.** $\frac{1}{q(p + 8q)}$ **8.** $\frac{4n + 3}{(4n - 3)(2n + 3)}$ **9.** $\frac{5y - 4}{9y^2(y - 2)}$
10. $\frac{r^2 + 3r - 2}{r^2 - 1}$ **11.** $\frac{x^2 + x - 10}{(x - 4)(x + 3)(x - 2)}$ **12.** $\frac{8p + 3}{(2p - 3)^2(p + 1)}$
13. a. $\frac{3y - x}{3xy}$ **b.** $-(x + 5)(x - 4)$ **14.** -5 **15.** $-7, -2$ **16.** 2
17. $k = 60; y = \frac{60}{x}$ **18.** The average speed of the car was 60 mph, and
the average speed of the train was 48 mph. **19.** Working together, it
would take the technicians $3\frac{3}{5}$ hr (or 3 hr 36 min) to fill the prescriptions.
20. The spring will stretch $7\frac{1}{2}$ in.

Section 6.1   Practices, pp. 443–451

1, p. 443: a. -4 **b.** $-6, 6$ **2, p. 444: a.** $\frac{6}{mn}$ **b.** $\frac{a^2}{5 - 3a}$ **3, p. 445: a.** $\frac{5}{3}$
b. -1 **c.** Cannot be simplified **4, p. 446: a.** $\frac{n + 9}{n + 5}$ **b.** $\frac{1}{y + 1}$ **5, p. 446:** $\frac{\pi}{6}$
6, p. 447: a. $-\frac{14}{3x^2}$ **b.** $\frac{x(x - y)}{x + y}$ **c.** $-\frac{(a + 1)(a + 3)}{4(a + 5)(a - 1)}$ **7, p. 449: a.** $\frac{2}{15c}$
b. $x - 3$ **c.** $\frac{1}{4(t - 4)}$ **8, p. 450:** $\frac{r + s}{2r - 3s}$ **9, p. 450: a.** $g(t) \cdot h(t) = \frac{(t + 3)^2}{t(t - 3)}$
b. $g(t) \div h(t) = \frac{t}{t - 3}$ **10, p. 451:** $\frac{(b - a)(d - c)}{bd}$

Exercises 6.1, pp. 452–456

1. rational expression **3.** has an asymptote **5.** equivalent rational
expression **7.** opposites **9.** 0 **11.** The expression is defined for all
real numbers. **13.** $\frac{1}{2}$ **15.** $2, 6$ **17.** $-3, 3$ **19.** $-\frac{3}{2}, 1$ **21.** Equivalent
23. Not equivalent **25.** Not equivalent **27.** $\frac{8a^2}{b^2}$ **29.** $\frac{3x^2 - 2}{x}$
31. $-\frac{3p^2q}{2p + q}$ **33.** 1 **35.** Cannot be simplified **37.** $-\frac{3}{4}$ **39.** $\frac{x - 4}{4}$
41. $\frac{n - 3}{n + 4}$ **43.** $-\frac{r - 3}{3(r + 4)}$ **45.** $\frac{2x + 3y}{3x - 5y}$ **47.** $\frac{a + 7}{a^2b}$ **49.** $\frac{y^2 + 4y + 16}{y + 4}$
51. xy **53.** $\frac{2}{y - 2}$ **55.** $\frac{2(p - 5)}{p(p + 6)}$ **57.** $-\frac{x(x - y)}{x^2 + y^2}$ **59.** $-\frac{r + 5}{r + 8}$
61. $\frac{n - 8}{3n - 2}$ **63.** $-\frac{1}{x^2 - 2x + 4}$ **65.** $\frac{6y(2x - 7y)}{(2x + 7y)^2}$ **67.** $\frac{20x}{9y^3}$ **69.** $\frac{2t + 3}{2t}$
71. $\frac{1}{3a}$ **73.** $-\frac{(p + 4q)^2}{2q(p - 8q)}$ **75.** $\frac{2x - 3}{2x + 3}$ **77.** $6(t + 4)$ **79.** $-\frac{4p + 9q}{q^2 + 2pq + 4p^2}$
81. $\frac{x - 6}{x + 5}$ **83.** $f(x) \cdot g(x) = \frac{2(x - 4)}{(x + 1)^2}; f(x) \div g(x) = \frac{x - 4}{2x^2}$
85. $f(x) \cdot g(x) = -(x - 3)^2; f(x) \div g(x) = -\frac{(x - 4)}{(x + 3)^2}$ **87.** They are
equivalent. **89.** $-10, 4$ **91.** $-\frac{b - 5}{b - 3}$ **93.** $\frac{5a^2 - 2}{a}$ **95.** The expression
is undefined for cleaning 100% of the pollutants. **97.** $a(r^2 + r + 1)$
99. $\frac{V^2}{R}$ **101.** $\frac{i}{100 + i}$

Section 6.2   Practices, pp. 457–464

1, p. 457: a. $\frac{5y}{x+y}$ **b.** 3   **2, p. 458: a.** $\frac{-16s}{s+4r}$ **b.** $\frac{4n-3}{n^2+5}$

3, p. 459:   a. LCD $= 12x^2y^3$ **b.** LCD $= (m-n)(m+n)$

c. LCD $= (x-y)^2(x+y)^2$ **4, p. 460: a.** $\frac{3n(n-4)}{2(3n+4)(n-4)}$

and $\frac{10(3n+4)}{2(3n+4)(n-4)}$ **b.** $\frac{7(x+4)}{(3x-1)(x+4)}$ and $\frac{2x+3}{(3x-1)(x+4)}$

5, p. 461: a. $\frac{x^2+4x-3}{(x-1)(x+1)}$ **b.** $\frac{7t-25}{(t-5)(t-4)}$ **6, p. 462: a.** $\frac{x+1}{x-3}$

b. $\frac{-11y+7}{(y+3)(y+1)^2}$ **7, p. 463:** $\frac{4x^2-11x+8}{x(2x-3)(x+2)(x-2)}$ **8, p. 463:**

a. $f(x)+g(x)=\frac{5x-15}{(x+5)(x-5)}$ **b.** $f(x)-g(x)=\frac{-x+15}{(x+5)(x-5)}$

9, p. 464: a. The cost per check is $\frac{4}{n(n+1)}$ dollars less. **b.** \$0.04

Exercises 6.2, pp. 465–468

1. same denominator   **3.** different denominators   **5.** $\frac{n-2}{n^2}$ **7.** 2

9. $\frac{1}{x+y}$ **11.** $\frac{y+4}{(y+3)(y-4)}$ **13.** LCD $= 24x^3y^3$; $\frac{9x}{24x^3y^3}$ and $\frac{4y}{24x^3y^3}$

15. LCD $= (n-2)(n-1)$; $\frac{n(n-1)}{(n-2)(n-1)}$ and $\frac{7(n-2)}{(n-2)(n-1)}$

17. LCD $= 6t^2(t+6)$; $\frac{9t}{6t^2(t+6)}$ and $\frac{8}{6t^2(t+6)}$

19. LCD $= (p+3q)(p-3q)$; $\frac{2p-q}{(p+3q)(p-3q)}$ and $\frac{3(p+3q)}{(p+3q)(p-3q)}$

21. LCD $= (n-2)(n-4)(n+6)$; $\frac{2n(n+6)}{(n-2)(n-4)(n+6)}$ and

$\frac{3n(n-4)}{(n-2)(n-4)(n+6)}$ **23.** LCD $= (3n-1)(n+2)(n-3)$;

$\frac{n-3}{(3n-1)(n+2)(n-3)}$, $\frac{n(3n-1)}{(3n-1)(n+2)(n-3)}$, and $\frac{(n-4)(n+2)}{(3n-1)(n+2)(n-3)}$

25. $\frac{3y+2x}{24x^2y^2}$ **27.** $\frac{3-4p^2}{p}$ **29.** $\frac{n^2+3n+8}{(n+4)(n+1)}$ **31.** $\frac{12x-9y}{(x-2y)(2x-y)}=$

$\frac{3(4x-3y)}{(x-2y)(2x-y)}$ **33.** 2 **35.** $\frac{-3r+14}{6r(r-2)}$ **37.** $\frac{2t-9}{(t+4)(t-4)}$ **39.** $\frac{n^2-4n+8}{(n-3)(n-7)}$

41. $\frac{9u-9v}{(u+5v)(u+6v)(u-4v)}=\frac{9(u-v)}{(u+5v)(u+6v)(u-4v)}$ **43.** $\frac{8-x}{(x-3)(x+2)}$

45. $-\frac{11x}{(2x-3)(x+4)(3x+1)}$ **47.** $\frac{a^2-15a+2}{(a+4)(a-2)^2}$

49. $\frac{4x^2-11x-2}{6x^2(2x-5)(x+1)(x-1)}$ **51.** $\frac{-v^2+v+4}{(v+3)(v-3)(v+1)}$ **53.** $\frac{8x^2-18x-11}{2(x+4)(2x-1)^2}$

55. $f(x)+g(x)=\frac{10x-60}{(x+4)(x+6)(x-9)}=\frac{10(x-6)}{(x+4)(x+6)(x-9)}$;

$f(x)-g(x)=\frac{6x-84}{(x+4)(x+6)(x-9)}=\frac{6(x-14)}{(x+4)(x+6)(x-9)}$

57. $f(x)+g(x)=\frac{x^2-2x+24}{3x^2(x-4)}$; $f(x)-g(x)=\frac{x^2+10x-24}{3x^2(x-4)}$

$=\frac{(x+12)(x-2)}{3x^2(x-4)}$ **59.** $\frac{x^2-5x-18}{2x^2(x+3)}$, $\frac{x^2+7x+18}{2x^2(x+3)}$

61. $\frac{-y^3+2y^2+20y-12}{3(y-1)(y+2)}$ **63.** $\frac{8(2x-3y)}{(x-3y)(3x-y)}$ **65.** $\frac{1}{a-b}$ **67.** $\frac{V_1-V_2}{\pi r^2}$

69. $\frac{100(S_1-S_0)}{S_0}$ **71.** $\frac{C_2+C_1}{C_1C_2}$ **73.** $\frac{2ms}{(s+w)(s-w)}$ **75.** The whole

trip took $\frac{6}{r}$ hr.

Section 6.3   Practices, pp. 470–474

1, p. 470: n **2, p. 471:** $\frac{y+3}{2y^3}$ **3, p. 471:** $\frac{4mn}{n-m}$ **4, p. 473:** $\frac{6y(y-3)}{5}$

5, p. 473: $\frac{n}{3n+2}$ **6, p. 474:** $\frac{q^2-qp+p^2}{p^2q^2}$

7, p. 474: $\frac{R_1R_2R_3R_4}{R_2R_3R_4+R_1R_3R_4+R_1R_2R_4+R_1R_2R_3}$

Exercises 6.3, pp. 475–477

1. rational   **3.** division method   **5.** $20n$ **7.** $\frac{3}{2x^3}$ **9.** $\frac{2}{a}$ **11.** $\frac{2}{5r}$

13. $-\frac{3(s+4)}{4s^2}$ **15.** $\frac{2(n-5)}{5(n+3)}$ **17.** $\frac{3}{n-1}$ **19.** $\frac{t}{t-2}$ **21.** $\frac{p-3}{p}$

23. $-\frac{2x+y}{x}$ **25.** $\frac{3(3-2n)}{2n(8+3n)}$ **27.** $\frac{2x-3}{x+2}$ **29.** $-\frac{ab}{3}$ **31.** $\frac{5x-y}{x+5y}$

33. $\frac{x(x+2)}{2(x-3)}$ **35.** $-\frac{n+4}{n-4}$ **37.** $\frac{x+1}{3x}$ **39.** $\frac{b}{a^2(a+b)}$ **41.** $\frac{6}{2r-3}$

43. $\frac{2(t+3)}{3t^2}$ **45.** $-\frac{6(x+5)}{x+4}$ **47.** $\frac{gF}{w}$ **49.** $\frac{Sf}{S+v}$ **51. a.** $\frac{2rR}{R+r}$

b. Her average speed was 48 mph.   **53.** $\frac{VR(R-2)}{2R-2}$

Section 6.4   Practices, pp. 480–488

1, p. 480: $-\frac{1}{6}$ **2, p. 481:** No solution   **3, p. 482:** 0 **4, p. 483:** $-6,4$

5, p. 484: -3 **6, p. 484:** $-3,2$ **7, p. 484:** 9 mo **8, p. 485:** 45 mph

9, p. 486: 2 km/hr **10, p. 487:** $7\frac{1}{5}$ hr (or 7 hr 12 min) **11, p. 488:**

a. $x=\frac{2ab}{a+b}$ **b.** $\frac{24}{5}$ in., or 4.8 in.

Exercises 6.4, pp. 489–492

1. LCD **3.** ratio **5.** cross-product **7.** -12 **9.** 2 **11.** -15 **13.** 8

15. -7 **17.** 2 **19.** No solution **21.** 1, 2 **23.** $-6,6$ **25.** -5

27. $-3,\frac{5}{4}$ **29.** $-4,-2$ **31.** $-2,5$ **33.** -8 **35.** $\frac{4}{3}$ **37.** 11

39. $\frac{3}{2}$ **41.** $-2,9$ **43.** $-2,0$ **45.** $-2,8$ **47.** No solution

49. 5 **51.** $-\frac{7}{3}$ **53.** $t=\frac{W}{P}$ **55.** $I_2=\frac{a^2I_1}{b^2}$ **57.** $R_1=\frac{RR_2}{R_2-R}$

59. $m_1=\frac{m_2v_2-Vm_2}{V-v_1}=\frac{m_2(v_2-V)}{V-v_1}$ **61.** 21 **63.** $-1,-2$ **65.** $-2,9$

67. $\frac{4(x-3)}{x(x+2)}$ **69.** 5 **71.** $-6,3$ **73.** -2 **75.** $-2,6$ **77.** Approximately

932 employees are satisfied with their benefits package.   **79.** The speed
of the wind was 30 mph.   **81.** She ran at a rate of 6 mph during her
cardio workout.   **83.** The team rows at a rate of 19 mph in still water.
85. Working together, it will take $3\frac{3}{5}$ hr (or 3 hr 36 min) to prepare the field.
87. It takes the other employee $2\frac{2}{5}$ hr (or 2 hr 24 min) to load the truck
working alone. **89. a.** $n=\frac{r_1r_2-r_1f+r_2f}{f(r_2-r_1)}$ **b.** The index of refraction is 1.5.

Section 6.5   Practices, pp. 494–499

1, p. 494: $k=\frac{4}{5}$; $y=\frac{4}{5}x$ **2, p. 494:** 220 mg of the drug should be
administered. **3, p. 496:** $k=48$; $y=\frac{48}{x}$ **4, p. 496:** The f-stop of the
lens is 4. **5, p. 497:** $k=\frac{1}{2}$; $y=\frac{1}{2}xz$ **6, p. 497:** The other employee
must invest \$2250. **7, p. 498: a.** $w=\frac{kxy}{z^2}$ **b.** $k=24$ **c.** 9
8, p. 499: The patient's BMI is approximately 19.

Exercises 6.5, pp. 500–502

1. directly **3.** direct variation **5.** inverse variation **7.** Decreases;
inverse variation **9.** Increases; direct variation **11.** Increases; direct
variation **13.** Decreases; inverse variation **15.** $k=3$; $y=3x$
17. $k=\frac{1}{6}$; $y=\frac{1}{6}x$ **19.** $k=9$; $y=9x$ **21.** $k=\frac{3}{2}$; $y=\frac{3}{2}x$
23. $k=39$; $y=\frac{39}{x}$ **25.** $k=27$; $y=\frac{27}{x}$ **27.** $k=\frac{7}{25}$; $y=\frac{7}{25x}$
29. $k=18$; $y=\frac{18}{x}$ **31.** $k=4$; $y=4xz$ **33.** $k=\frac{6}{5}$; $y=\frac{6}{5}xz$
35. $k=3$; $y=3xz$ **37.** $k=25$; $y=25xz$ **39.** $k=125$; $y=\frac{125x}{z^2}$
41. $k=500$; $y=\frac{500}{xz^2}$ **43.** $k=\frac{1}{10}$; $y=\frac{xw}{10z^2}$ **45.** $k=30$; $y=\frac{30xw}{z^3}$
47. $k=\frac{2}{3}$; $y=\frac{2}{3}x$ **49.** $k=12$; $y=12x$ **51.** Decreases; inverse variation **53.** Increases; direct variation **55. a.** $A=\frac{3}{100}i$, or $A=0.03i$,
the constant of variation represents the income tax rate, which is 3%.
b. A person will pay \$795 in income tax. **57.** 3.06 m **59.** 28.4 J
61. 5.4 lm/m²

Chapter 6   Review Exercises, pp. 508–510

1. is; possible answer: the denominator is equal to 0 for this value

2. is; possible answer: it can be written as $\frac{(3x-5)(x+1)}{3x-5}$

3. is; possible answer: this rational expression has a rational expression in
its numerator **4.** will not; possible answer: it is not a solution
5. does; possible answer: $C=(2\pi)r$ **6.** does not; possible answer:
their product is constant **7.** $\frac{1}{2}$ **8.** 2, 4 **9.** Equivalent **10.** Not equivalent

11. $\frac{2x^2y}{3x-4y}$ **12.** $-\frac{1}{2}$ **13.** $\frac{a-2b}{5}$ **14.** Cannot be simplified **15.** $\frac{x+8}{x-7}$

16. $-\frac{2t-1}{3t+8}$ **17.** $6u^3v^2$ **18.** $\frac{1}{10pq}$ **19.** $\frac{3(7a-1)}{2b}$ **20.** $-\frac{1}{3y}$

21. $\frac{2(x+5)}{x^2}$ **22.** $-\frac{r+10}{r^2+r+1}$ **23.** $\frac{1}{(x-4)(2x+3)}$ **24.** $\frac{n+9}{3n+4}$

25. LCD $= x^2(x+1)^2$; $\frac{(x-1)(x+1)}{x^2(x+1)^2}$ and $\frac{2x^2}{x^2(x+1)^2}$

26. LCD $= (x-4)(x-12)(x+2)$; $\frac{(x+5)(x+2)}{(x-4)(x-12)(x+2)}$ and

$\dfrac{(2x-3)(x-4)}{(x-4)(x-12)(x+2)}$ **27.** $\dfrac{4}{x^2y}$ **28.** 1 **29.** $\dfrac{-2p+16}{(p-5)(p-4)}$ **30.** $\dfrac{a-2}{8a^2}$

31. $\dfrac{t-2}{t+6}$ **32.** $\dfrac{3y-5}{(y+2)(y-2)}$ **33.** $\dfrac{2x^2-x+3}{(x+3)(x-1)^2}$ **34.** $\dfrac{3u^2-7u-12}{(2u+3)(u+4)(u-9)}$

35. $4x^2$ **36.** $\dfrac{3x+1}{x}$ **37.** $\dfrac{p-5}{p+1}$ **38.** $\dfrac{3x+5}{-x+5}$ **39.** $\dfrac{3r+4}{r-2}$ **40.** $2x+x^2$

41. -3 **42.** No solution **43.** $-\dfrac{1}{9}$ **44.** 22 **45.** 1 **46.** 5 **47.** $-1, 6$

48. $-3, -4$ **49.** $m=\dfrac{Fr}{v^2}$ **50.** $R_2=\dfrac{RR_1}{R_1-R}$ **51.** $k=0.4; y=0.4x$

52. $k=\dfrac{3}{2}; y=\dfrac{3}{2x}$ **53.** $k=6; y=6xz$ **54.** $k=8; y=\dfrac{8x^2}{z^2}$ **55.** 0

56. $\dfrac{T_2-T_1}{T_2}$ **57.** $\dfrac{V(R_2+R_1)}{R_1R_2}$ **58.** $\dfrac{pq}{q+p}$ **59.** Her average speed was 63 mph on the drive home and 42 mph on the way back to school.
60. Working together, it will take them $6\frac{6}{7}$ hr to complete the job.
61. Its velocity after 5 sec is 49 m/sec. **62.** The accommodation is 10 diopters.

Chapter 6   Posttest, p. 511

1. a. $0, 2$ **b.** 4 **2.** $\dfrac{4y-2xy^2+3x^2}{6xy}$ **3.** $-\dfrac{1}{3a+2}$ **4.** $\dfrac{3n-2}{n+8}$

5. $\dfrac{p+6q}{3pq-1}$ **6.** $\dfrac{y}{y+2}$ **7.** $a(a-5b)$ **8.** $\dfrac{1}{3x-y}$ **9.** $\dfrac{5r-6}{12r^3(r-3)}$

10. $\dfrac{5t-10}{(2t+1)(2t-1)}=\dfrac{5(t-2)}{(2t+1)(2t-1)}$ **11.** $\dfrac{n+6}{(n-4)(n+1)}$

12. $\dfrac{-13y+6}{(3y+2)^2(y-3)}$ **13. a.** $\dfrac{1-2x}{3x}$ **b.** $\dfrac{a-2}{a-1}$ **14.** $3, 1$ **15.** $\dfrac{11}{5}$

16. $-8, -3$ **17.** $k=0.2; y=0.2xz$ **18.** The speed of the river's current is 3.5 km/h. **19.** It would take the second gardener $4\frac{4}{5}$ hr (or 4 hr 48 min) to mow the lawn working alone. **20.** $30°$

Chapter 6   Cumulative Review, pp. 512–513

1. $\dfrac{y^2}{4x^4}$ **2.** $\dfrac{7}{16}$ **3.** $m=-\dfrac{1}{3}$

4.
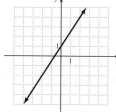

5. Domain: $(-\infty, \infty)$; range: $[-2, \infty)$
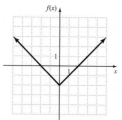

6. No solution
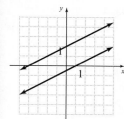

7. $(-1, -2, 3)$
8. $x^4-x^3-4x^2+5x-5$
9. $(3x-5)(2y+3)$
10. $(2x-7y)(x-4y)$
11. $(3x-5y^3)^2$ **12.** $\dfrac{4}{3(x+3)^2(3-x)}$
13. $-\dfrac{y}{3y+x}$ **14.** $-5, \dfrac{2}{3}$
15. $(27-4t)(27+4t)$
16. The speed of the westbound train is 56 mph. **17.** The maximum amount of cola in each case is 288.5 oz and the minimum amount is 287.5 oz. **18.** She invested $12,000 in the high-risk fund and $36,000 in the low-risk fund.

19. a. $\dfrac{-27x^3+245x^2+4380x+81,500}{3x+293}$ dollars **b.** $300

20. $\dfrac{9(P+A)}{D}$, or $\dfrac{9P+9A}{D}$

CHAPTER 7

Chapter 7   Pretest, p. 515

1. a. 12 **b.** -4 **2.** $10uv^2$ **3.** $(\sqrt[4]{81x^4})^3; 27x^3$ **4. a.** $\dfrac{4}{3}$ **b.** $\sqrt[3]{xy^2}$
5. $\sqrt{42ab}$ **6.** $\sqrt[3]{6r}$ **7.** $7x^2y^2\sqrt{3x}$ **8.** $\dfrac{\sqrt{2p}}{5q^4}$ **9.** $7\sqrt{3}$

10. $2x-15\sqrt{x}-8$ **11.** $\dfrac{\sqrt{6xy}}{9y}$ **12.** $\dfrac{n-\sqrt{3n}}{n-3}$ **13.** 32 **14.** 8 **15.** 2
16. $5-37i$ **17.** $-\dfrac{3}{5}+\dfrac{4}{5}i$ **18.** $\dfrac{\sqrt{h-d}}{4}$ **19.** The diagonal of the court is $2\sqrt{2834}$ ft, or about 106.5 ft. **20.** $p=\dfrac{f^2}{14,400}$

Section 7.1   Practices, pp. 517–524

1, *p. 517:* **a.** 5 **b.** $\dfrac{3}{10}$ **c.** -36 **d.** Not a real number **2,** *p. 517:* 2.449
3, *p. 518:* **a.** $-2y$ **b.** $6x^3y^3$ **4,** *p. 519:* **a.** 6 **b.** -3 **c.** $\dfrac{1}{5}$ **d.** $-4x^2$
5, *p. 520:* **a.** 2 **b.** -2 **c.** $4y^2$ **6,** *p. 521:* **a.** $\sqrt{36}$; 6 **b.** $\sqrt[3]{27}$; 3
c. $-\sqrt[4]{n}$ **d.** $\sqrt[3]{-125y^9}$; $-5y^3$ **7,** *p. 522:* **a.** $(\sqrt[4]{81})^3$; 27
b. $(\sqrt[3]{-64})^2$; 16 **c.** $(\sqrt{\frac{4}{9}y^4})^5$; $\dfrac{32}{243}y^{10}$ **8,** *p. 523:* **a.** $-\dfrac{1}{3}$ **b.** $\dfrac{1}{81a^8}$
9, *p. 523:* **a.** 27 **b.** $\sqrt[5]{n}$ **c.** $\sqrt{r}$ **d.** $\dfrac{x^2}{2y}$ **10,** *p. 524:* $\dfrac{2\pi d^{3/2}}{(Gm)^{1/2}}$, or $\dfrac{2\pi d^{3/2}}{G^{1/2}m^{1/2}}$

Exercises 7.1, pp. 525–527

1. square root **3.** radicand **5.** irrational **7.** index **9.** 8 **11.** -10
13. Not a real number **15.** 8 **17.** 3 **19.** -10 **21.** 4 **23.** -8
25. $\dfrac{3}{4}$ **27.** $-\dfrac{2}{5}$ **29.** 0.2 **31.** 4.583 **33.** 6.782 **35.** 3.775 **37.** 4.820
39. 2.724 **41.** x^4 **43.** $4a^3$ **45.** $9p^4q^2$ **47.** $2x^5y$ **49.** $-5u^3$
51. $12uv^4$ **53.** $2t^3$ **55.** pq^3 **57.** $\sqrt{16}$; 4 **59.** $-\sqrt{16}$; -4
61. $\sqrt[3]{-64}$; -4 **63.** $6\sqrt[4]{x}$ **65.** $\sqrt{36a^2}$; $6a$ **67.** $\sqrt[3]{-216u^6}$; $-6u^2$
69. $(\sqrt[3]{27})^4$; 81 **71.** $-(\sqrt{16})^3$; -64 **73.** $(\sqrt[3]{-27y^3})^2$; $9y^2$
75. $-\dfrac{1}{(\sqrt[4]{81})^3}$; $-\dfrac{1}{27}$ **77.** $\sqrt{\frac{4}{x^{10}}}$; $\dfrac{2}{x^5}$ **79.** 64 **81.** $\sqrt[6]{6}$ **83.** $\dfrac{1}{2}$ **85.** 3
87. $4\sqrt[5]{n^3}$ **89.** $\sqrt{y}$ **91.** $\dfrac{1}{2x}$ **93.** $5\sqrt[4]{r}$ **95.** $\dfrac{\sqrt[3]{a^2}}{b^2}$ **97.** $12x^5$ **99.** $\dfrac{4}{p^2}$
101. $2a^3b^2$ **103.** $3\sqrt{n}$ **105.** 0.095 **107.** $\sqrt{81x^6}$; $9x^3$ **109.** $\dfrac{5}{11}$
111. It takes the stone $\dfrac{5}{2}$, or 2.5, sec to reach the ground. **113. a.** The length of the side of the frame is 5 in. **b.** A 3-inch by 3-inch photograph fits in the frame. **115. a.** $8000\sqrt[3]{0.5^t}$ **b.** The value of the equipment 6 yr after it was purchased is $2000.

Section 7.2   Practices, pp. 529–533

1, *p. 529:* **a.** $\sqrt[4]{y}$ **b.** $\sqrt[3]{5}$ **c.** $\sqrt[10]{n^7}$ **d.** $\sqrt[4]{t}$ **2,** *p. 529:* **a.** $\sqrt[12]{n}$
b. $\sqrt[4]{ab^3}$ **c.** Cannot be simplified **3,** *p. 530:* **a.** $\sqrt{21}$ **b.** $\sqrt[5]{2y^3}$
c. $\sqrt[3]{28p^2}$ **d.** $\sqrt{\frac{15}{uv}}$ **4,** *p. 530:* **a.** $6\sqrt{2}$ **b.** $-21\sqrt{2}$ **c.** $2\sqrt[3]{7}$
d. $2\sqrt[4]{3}$ **5,** *p. 531:* **a.** $y^3\sqrt{y}$ **b.** $3x^2\sqrt{2x}$ **c.** $3ab^2\sqrt[3]{3a}$ **6,** *p. 532:*
a. $\sqrt{7n}$ **b.** 6 **c.** $\sqrt[3]{2y}$ **7,** *p. 532:* **a.** $\dfrac{4}{5}$ **b.** $\dfrac{\sqrt{n}}{8}$ **c.** $\dfrac{4\sqrt[3]{x}}{5y^3}$ **d.** $\dfrac{p^3q\sqrt{7}}{6r^2}$
8, *p. 533:* $2\sqrt{10}$ units **9,** *p. 533:* **a.** $(-20, 50)$ and $(10, 60)$
b. The drivers were $10\sqrt{10}$ mi (or about 31.6 mi) apart.

Exercises 7.2, pp. 534–536

1. the same index **3.** radicands **5.** $\sqrt[3]{n}$ **7.** $\sqrt{2}$ **9.** $\sqrt{x}$
11. Cannot be simplified **13.** $\sqrt[12]{x}$ **15.** $\sqrt[4]{y}$ **17.** $x^2\sqrt{y}$ **19.** x
21. $\sqrt[3]{y}$ **23.** $\sqrt{pq}$ **25.** $\sqrt{30}$ **27.** $\sqrt{6xy}$ **29.** $\sqrt[3]{36a^2b}$
31. $\sqrt[4]{21n^3}$ **33.** $\sqrt{\frac{3x}{y}}$ **35.** $2\sqrt{6}$ **37.** $-12\sqrt{5}$ **39.** $3\sqrt[3]{3}$ **41.** $2\sqrt[4]{6}$
43. $2\sqrt[5]{2}$ **45.** $6x^3\sqrt{x}$ **47.** $2\sqrt{5y}$ **49.** $10r^2\sqrt{2}$ **51.** $3x^2y^3\sqrt{6xy}$
53. $2n\sqrt[3]{4n^2}$ **55.** $2n\sqrt[5]{2n^3}$ **57.** $2x^2y^3\sqrt[3]{9x}$ **59.** $2xy^2\sqrt[4]{4xy^2}$
61. 3 **63.** $\sqrt[3]{5n}$ **65.** $2x\sqrt{y}$ **67.** $\sqrt[3]{4u}$ **69.** $\sqrt[4]{6ab}$ **71.** $\dfrac{5}{4}$ **73.** $\dfrac{\sqrt{7}}{9}$
75. $\dfrac{\sqrt{2}}{n^3}$ **77.** $\dfrac{\sqrt{7a}}{11}$ **79.** $\dfrac{\sqrt{a}}{3b^2}$ **81.** $\dfrac{3\sqrt{u}}{5v}$ **83.** $\dfrac{3\sqrt[3]{a^2}}{4}$ **85.** $\dfrac{\sqrt[3]{9a}}{2b^2c^3}$
87. $\dfrac{a\sqrt[4]{2b^3}}{3c^2}$ **89.** $\dfrac{a^2\sqrt{13b}}{3c^3d}$ **91.** $6\sqrt{2}$ units **93.** $3\sqrt{5}$ units **95.** 5 units
97. $\sqrt[4]{30x^3y^3}$ **99.** x^2 **101.** $5r^3s^3\sqrt{2s}$ **103.** $\dfrac{\sqrt[3]{25x}}{4yz^3}$ **105.** $4uv^2\sqrt[3]{2u}$
107. $8\sqrt{d}$ **109.** The height of the screen is $\sqrt{245}$ m, or 16 m, to the nearest meter. **111.** The length of string let out is $20\sqrt{13}$ ft, or about 72.1 ft. **113. a.** The second college is located at $(48, 20)$. **b.** The second college is 52 mi from her home.

Section 7.3   Practices, pp. 537–540

1, *p. 537:* **a.** $2\sqrt[4]{9}$ **b.** $5\sqrt{p}$ **c.** $-4\sqrt{t^2-4}$ **d.** Cannot be combined
2, *p. 538:* **a.** $8\sqrt{3}$ **b.** $-19\sqrt{6}$ **c.** $-5\sqrt[4]{2}$ **d.** $8+\sqrt{7}$

3, *p. 539:* **a.** $21\sqrt{n}$ **b.** $(3b-11)\sqrt[3]{a}$ **c.** $(5-4x)\sqrt{2x}$
4, *p. 539:* **a.** $p(x)+q(x)=5x\sqrt[3]{3x}$ **b.** $p(x)-q(x)=x\sqrt[3]{3x}$
5, *p. 540:* **a.** $\sqrt{337}$ in. and 20 in. **b.** $(20-\sqrt{337})$ in., or approximately 1.6 in.

Exercises 7.3, pp. 541–543

1. like **3.** simplified **5.** $5\sqrt{3}$ **7.** $-3\sqrt[3]{2}$ **9.** Cannot be combined
11. $-20\sqrt{y}$ **13.** Cannot be combined **15.** 0 **17.** $8\sqrt[4]{p}$
19. $13y\sqrt{3x}$ **21.** $-4\sqrt{2r-1}$ **23.** $2\sqrt{x^2-9}$ **25.** $3\sqrt{6}$
27. $4\sqrt{2}$ **29.** $10\sqrt{2}$ **31.** $-4\sqrt[3]{3}$ **33.** $-2x\sqrt{6x}$
35. $(2n^2+11)\sqrt{2n}$ **37.** $(-4y^2+7x^2)\sqrt{xy}$ **39.** $-7p^3$
41. $(10-2a)\sqrt[4]{3ab^3}$ **43.** $7\sqrt{5}+2$ **45.** $\sqrt[3]{6}-3$
47. $f(x)+g(x)=13x\sqrt{5x}; f(x)-g(x)=7x\sqrt{5x}$
49. $f(x)+g(x)=3x\sqrt[4]{4x}; f(x)-g(x)=5x\sqrt[4]{4x}$
51. $-9\sqrt{6}$ **53.** $-5\sqrt{3b}$ **55.** $3(k-2)\sqrt{5k}$ **57.** The difference in velocity is $16\sqrt{5}$ ft/sec, or 35.8 ft/sec. **59.** The side of the larger tile is $2\sqrt{2}$ in., or about 2.8 in. longer than the smaller tile.
61. a. The length of the ramp is $2\sqrt{5}$ ft. **b.** The length of the ramp increases by approximately 1.9 ft.

Section 7.4 Practices, pp. 544–551

1, *p. 544:* **a.** $6y^4\sqrt{2}$ **b.** $-14\sqrt[3]{15}$ **2, *p. 544:*** **a.** $10-4\sqrt{5}$
b. $\sqrt[4]{35}+6\sqrt[4]{10}$ **c.** $\sqrt{5}-7$ **d.** $6x+\sqrt{x}-1$ **3, *p. 545:*** **a.** -3
b. $2x-9$ **c.** $p-16q$ **4, *p. 546:*** **a.** $3+4\sqrt{3b}+4b^2$, or $4b^2+4\sqrt{3b}+3$ **b.** $n-6\sqrt{n+1}+10$ **5, *p. 546:*** $2\sqrt{30}$ cm^2
6, *p. 547:* **a.** 10 **b.** $-\frac{\sqrt[3]{9p^2}}{5}$ **7, *p. 547:*** **a.** $\frac{y}{3x^4}$ **b.** $\frac{n\sqrt[3]{2n^2}}{3}$ **8, *p. 548:***
a. $\frac{4\sqrt{7}}{7}$ **b.** $\frac{\sqrt{2y}}{8}$ **c.** $\frac{p^2\sqrt{2}}{3}$ **d.** $\frac{\sqrt[3]{20pqr}}{2r}$ **9, *p. 548:*** **a.** $\frac{\sqrt{3xy}}{3y}$ **b.** $\frac{\sqrt[3]{5a^2}}{a}$
10, *p. 549:* **a.** $\sqrt{2}-3\sqrt{3}$ **b.** $\frac{5\sqrt{b}+b}{b}$ **11, *p. 550:*** **a.** $4-2\sqrt{2}$
b. $\frac{a\sqrt{b}+a\sqrt{c}}{b-c}$ **12, *p. 551:*** **a.** $6x^2\sqrt[3]{9x^2}$ **b.** $\frac{3}{2}$ **13, *p. 551:*** **a.** $\frac{\sqrt[3]{6\pi^2 V}}{2\pi}$
b. No; $\frac{\sqrt[3]{6\pi^2(2V)}}{2\pi}=\frac{\sqrt[3]{12\pi^2 V}}{2\pi}\neq 2\left(\frac{\sqrt[3]{6\pi^2 V}}{2\pi}\right)$

Exercises 7.4, pp. 552–555

1. distributive **3.** quotient **5.** perfect power **7.** $4\sqrt{6}$ **9.** $-4\sqrt{21}$
11. $45\sqrt[3]{2}$ **13.** $70x^3\sqrt{2}$ **15.** $-16a^2b^2$ **17.** $-4xy\sqrt[3]{3y^2}$ **19.** $4-4\sqrt{2}$
21. $12\sqrt{2}$ **23.** $-4\sqrt{15}+36$ **25.** $10\sqrt[3]{6}+2\sqrt[3]{12}$ **27.** $x^2+x\sqrt{2}$
29. $3a\sqrt[4]{2a}-a\sqrt[4]{10}$ **31.** $-10+\sqrt{2}$ **33.** $-28-10\sqrt{3}$
35. $10+5\sqrt{6}$ **37.** $16r-20\sqrt{r}-24$ **39.** 4 **41.** $x-64$
43. $5x-y$ **45.** $x-26$ **47.** $9x-12\sqrt{x}+4$ **49.** $6a+8\sqrt{6a}+16$
51. $n-2\sqrt{n+7}+8$ **53.** -1 **55.** $2x\sqrt{3}$ **57.** $-n^3\sqrt{3}$ **59.** $\frac{\sqrt[4]{2r^3}}{3}$
61. $\frac{\sqrt{5a}}{4}$ **63.** $\frac{2\sqrt{3x}}{y^3}$ **65.** $\frac{3\sqrt[3]{2}}{n^3}$ **67.** $\frac{4a^2\sqrt{2a}}{3b^4}$ **69.** $\frac{4\sqrt{5}}{5}$ **71.** $\frac{\sqrt{10y}}{20}$
73. $\frac{\sqrt{6a}}{6}$ **75.** $\frac{\sqrt{uv}}{2v}$ **77.** $\frac{5x^2\sqrt{3y}}{3y}$ **79.** $\frac{\sqrt{10xyz}}{3z}$ **81.** $\frac{\sqrt[3]{36xy}}{2y}$ **83.** $\frac{\sqrt{11x}}{x}$
85. $\frac{\sqrt{21xy}}{3y}$ **87.** $\frac{x\sqrt{15xy}}{12y^2}$ **89.** $\frac{\sqrt{2a^2bc^2}}{4c^2}$ **91.** $\frac{2\sqrt{6}-3\sqrt{2}}{6}$ **93.** $\frac{\sqrt{ab}-b}{b}$
95. $\frac{\sqrt{3t}+2t\sqrt{15}}{3t}$ **97.** $\frac{\sqrt[3]{x^2}-\sqrt[4]{x}}{x}$ **99.** $\frac{2-\sqrt{2}}{2}$ **101.** $-2\sqrt{2}-2\sqrt{5}$
103. $\frac{8-4\sqrt{2x}}{2-x}$ **105.** $\frac{x-y\sqrt{x}}{x-y^2}$ **107.** $\frac{a+\sqrt{2a}+3\sqrt{a}+3\sqrt{2}}{a-2}$
109. $\frac{x-2\sqrt{xy}+y}{x-y}$ **111.** $\frac{6a+13\sqrt{ab}+6b}{9a-4b}$ **113.** $f(x)\cdot g(x)=$
$2x^2\sqrt{3}; \frac{f(x)}{g(x)}=3x\sqrt{3}$ **115.** $f(x)\cdot g(x)=x-1; \frac{f(x)}{g(x)}=\frac{x+2\sqrt{x}+1}{x-1}$
117. $9+9\sqrt{2}$ **119.** $-56a^3b^2$ **121.** $\frac{5\sqrt{3y}}{z^2}$ **123.** $\frac{\sqrt{35xy}}{7y}$
125. $\frac{p+\sqrt{3p}-4\sqrt{p}-4\sqrt{3}}{p-3}$ **127. a.** $A=\frac{\sqrt{3}}{4}a^2$ **b.** $16\sqrt{3}$ in^2
129. $\frac{\pi\sqrt{2L}}{4}$ **131.** $\frac{\sqrt{\pi A}}{2\pi}$

Section 7.5 Practices, pp. 557–562

1, *p. 557:* 64 **2, *p. 557:*** 10 **3, *p. 558:*** No solution **4, *p. 558:*** 3
5, *p. 559:* 3, 11 **6, *p. 560:*** 3, 4 **7, *p. 561:*** 0 **8, *p. 561:*** $P=I^2R$
9, *p. 562:* **a.** $A=P(r+1)^3$ **b.** The value after 3 yr is about \$10,927.27.

Exercises 7.5, pp. 563–564

1. 12 **3.** 3 **5.** 2 **7.** -6 **9.** No solution **11.** 121 **13.** 27
15. 24 **17.** 85 **19.** -2 **21.** 6 **23.** No solution **25.** -15
27. 2, 3 **29.** 5, -1 **31.** -2 **33.** $\frac{3}{4}$ **35.** 3, 5 **37.** $-\frac{1}{4}$, 0 **39.** 3
41. No solution **43.** $\frac{5}{4}$ **45.** -1, 3 **47.** -2, 0 **49.** No solution
51. 8 **53.** -6, -4 **55.** The satellite is 100 mi above Earth's surface.
57. a. $L=\frac{S^2}{30f}$ **b.** 60 ft **59.** The painter must place the ladder 9 ft from the side of the house.

Section 7.6 Practices, pp. 567–572

1, *p. 567:* **a.** $6i$ **b.** $i\sqrt{2}$ **c.** $-5i$ **2, *p. 568:*** **a.** $4-i$ **b.** $-12+4i$
c. $5+19i$ **3, *p. 568:*** **a.** -12 **b.** $-\sqrt{15}$ **c.** 40 **d.** $12-10i$
4, *p. 569:* **a.** $1+18i$ **b.** $-3-11i$ **5, *p. 569:*** **a.** $-1+7i$
b. $8-9i$ **c.** $3i$ **6, *p. 570:*** 29 **7, *p. 570:*** **a.** $\frac{7}{5}+\frac{14}{5}i$ **b.** $\frac{2}{5}i$
8, *p. 571:* **a.** $-\frac{1}{10}+\frac{7}{10}i$ **b.** $\frac{1}{4}$ **9, *p. 572:*** -1 **10, *p. 572:*** **a.** $I=\frac{V}{Z}$
b. $(-1+4i)$ amps **c.** $V=(-1+4i)(1-i)=3+5i$

Exercises 7.6, pp. 573–575

1. imaginary number **3.** real part **5.** complex conjugates **7.** $2i$
9. $\frac{1}{4}i$ **11.** $i\sqrt{3}$ **13.** $3i\sqrt{2}$ **15.** $10i\sqrt{5}$ **17.** $-3i$ **19.** $\frac{3i\sqrt{5}}{2}$
21. $-\frac{i\sqrt{3}}{2}$ **23.** $1+20i$ **25.** $5-6i$ **27.** $-6i$ **29.** $-7-3i$
31. $-2-2i$ **33.** $12-17i$ **35.** -10 **37.** 9 **39.** -63 **41.** 28
43. $3+3i$ **45.** $7-12i$ **47.** $-12+21i$ **49.** $-4+6i$
51. $2+16i$ **53.** $46+56i$ **55.** $14+56i$ **57.** $-16+12i$ **59.** 61
61. $5+12i$ **63.** $-5-12i$ **65.** $-7+i\sqrt{2}$ **67.** $1-10i$; 101
69. $7+i$; 50 **71.** $-9-6i$; 117 **73.** $-8i$; 64 **75.** $11i$; 121
77. $\frac{28}{17}-\frac{7}{17}i$ **79.** $-\frac{3}{26}-\frac{15}{26}i$ **81.** $-\frac{5}{4}i$ **83.** $\frac{2}{7}i$ **85.** $-3-4i$
87. $4+i$ **89.** $\frac{3}{4}+\frac{9}{4}i$ **91.** $\frac{2}{5}$ **93.** $-i$ **95.** -1 **97.** $-i$ **99.** i
101. $-i$ **103.** $-8-5i$; 89 **105.** $\frac{4i}{5}\sqrt{3}$ **107.** $\frac{9}{2}-\frac{5i}{2}$
109. $-8+10i$ **111.** $-2-4i$ **113.** $(8+i)$ ohms **115.** Yes, because $(2+3i)^2-4(2+3i)=-13$.

Chapter 7 Review Exercises, pp. 580–582

1. is; possible answer: cubing and taking a cube root undo each other
2. is; possible answer: it is the cube of the rational number $\frac{1}{4}$
3. is; possible answer: to add like radicals, combine the coefficients and keep the radical **4.** are; possible answer: their indexes are equal and their radicands are equal **5.** is not; possible answer: the radicand is a fraction
6. is not; possible answer: the denominator is a radical **7.** is; possible answer: of the product rule of radicals **8.** is; possible answer: their imaginary parts have opposite coefficients **9.** -66 **10.** -10 **11.** $\frac{1}{3}$
12. 0.6 **13.** $9y^4$ **14.** $-7a^3b$ **15.** $-6x^3$ **16.** $3p^3$ **17.** $-\sqrt{64}$; -8
18. $7\sqrt[3]{x}$ **19.** $-(\sqrt[4]{16n^4})^3$; $-8n^3$ **20.** $\frac{1}{(\sqrt[3]{8})^2}$; $\frac{1}{4}$ **21.** $\sqrt[4]{x^3}$ **22.** $\frac{\sqrt{r}}{6}$
23. $\frac{1}{5y}$ **24.** $\frac{\sqrt[3]{a}}{12}$ **25.** $\sqrt[4]{x}$ **26.** n **27.** $\sqrt[3]{y}$ **28.** $q\sqrt{p}$ **29.** $\sqrt[6]{a}$
30. $\sqrt{xy}$ **31.** $\sqrt{30rs}$ **32.** $\sqrt[3]{28p^2q^2}$ **33.** $10n\sqrt{3n}$ **34.** $3x^2y^2\sqrt{5x}$
35. $4t^2\sqrt[3]{2t}$ **36.** $2ab^2\sqrt[4]{6ab^2}$ **37.** $\sqrt{5a}$ **38.** $\sqrt[3]{2p}$ **39.** $\frac{\sqrt{n}}{5}$
40. $\frac{\sqrt{6}}{7y^2}$ **41.** $\frac{4\sqrt[3]{u^2}}{5v^3}$ **42.** $\frac{p\sqrt[4]{4q^3}}{3rs^2}$ **43.** $4\sqrt{x}$ **44.** $11\sqrt[3]{q^2}$ **45.** $7\sqrt{3}$
46. 0 **47.** $(12a-3)\sqrt[3]{7a}$ **48.** $3p\sqrt[3]{p^2q}$ **49.** $-18a\sqrt{2}$
50. $20\sqrt{2}-10$ **51.** $n\sqrt{2}-2\sqrt[3]{n}$ **52.** $4t-17\sqrt{t}+15$
53. $6-x$ **54.** $2y-2\sqrt{2y}+1$ **55.** $\frac{n\sqrt{3n}}{2}$ **56.** $\frac{4\sqrt{2a}}{3b^2}$ **57.** $\frac{\sqrt{2}}{4}$
58. $\frac{2\sqrt{xy}}{y}$ **59.** $\frac{p\sqrt{42q}}{3q}$ **60.** $\frac{\sqrt[3]{20uv}}{6u^2}$ **61.** $\frac{5\sqrt{2}-3\sqrt{5}}{5}$ **62.** $\frac{2a\sqrt{3}+\sqrt{a}}{a}$
63. $2\sqrt{3}+2$ **64.** $\frac{x+\sqrt{5x}+2\sqrt{x}+2\sqrt{5}}{x-5}$ **65.** 8 **66.** 25 **67.** No solution **68.** 7 **69.** $-\frac{5}{2}$ **70.** 4 **71.** $6i$ **72.** $5i\sqrt{5}$ **73.** $8+5i$
74. $4-2i$ **75.** -9 **76.** $10-2i$ **77.** $33-15i$ **78.** $24-10i$
79. $-\frac{1}{8}-\frac{1}{8}i$ **80.** $\frac{13}{20}-\frac{9}{20}i$ **81.** -1 **82.** i **83. a.** $N\sqrt[20]{2^t}$
b. 80 bacteria are present after 1 hr. **84.** $8\sqrt{R}$ **85. a.** The velocity of the car is $10\sqrt{2}$ m/sec (or about 14.1 m/sec). **b.** The velocity is

$(20 - 10\sqrt{2})$ m/sec (or about 5.9 m/sec) greater. **86.** $\frac{\sqrt{kl}}{l}$
87. $\frac{2\pi\sqrt{mk}}{k}$ **88.** The office building is $4\sqrt{5}$ mi (or approximately 8.9 mi) from her home. **89.** The wire needs to be anchored to the ground $4\sqrt{11}$ ft (or approximately 13.3 ft) from the pole. **90.** 85 bottles per week are demanded.

Chapter 7 Posttest, p. 583

1. a. -27 **b.** -6 **2.** $12a^3b$ **3.** $\sqrt[5]{(32x^{10})^2}; 4x^4$ **4. a.** $8\sqrt[6]{p}$
b. $\sqrt[4]{x^3y}$ **5.** $\sqrt[3]{20p^2q}$ **6.** $2\sqrt{2}$ **7.** $3xy^3\sqrt{13xy}$ **8.** $\frac{\sqrt{6u}}{7v^3}$
9. $-9\sqrt{6}$ **10.** $1 - 14\sqrt{2}$ **11.** $\frac{\sqrt{6ab}}{10b}$ **12.** $\frac{x + \sqrt{xy}}{x - y}$ **13.** No solution
14. 1, 2 **15.** -2 **16.** $31 - 5i$ **17.** $-\frac{9}{13} - \frac{19}{13}i$ **18. a.** $\frac{\sqrt{\pi S}}{2\pi}$
b. The radius of the beach ball is $\frac{8\sqrt{2\pi}}{\pi}$ in. **19.** The distance to the horizon is $200\sqrt{41}$ mi (or approximately 1280.6 mi). **20.** The daily demand is 149 units.

Chapter 7 Cumulative Review, pp. 584–585

1. $x < -5$

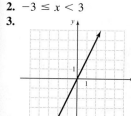

2. $-3 \le x < 3$
3.

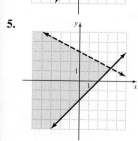

4.

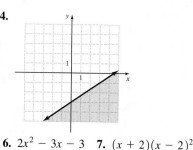

5.

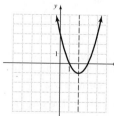

6. $2x^2 - 3x - 3$ **7.** $(x + 2)(x - 2)^2$
8. $(3x^2 - \frac{1}{2}y)(9x^4 + \frac{3}{2}x^2y + \frac{1}{4}y^2)$
9. $-1, 3$ **10.** $\frac{x + 14}{(x - 1)^2(x + 4)}$ **11.** $\frac{1 + 2x}{3x}$
12. $-12, 4$ **13.** $3x^2y\sqrt[3]{2y}$ **14.** $\frac{1}{2} - \frac{7}{2}i$
15. The least acceptable width is 2.998 in., and the greatest is 3.002 in.
16. a. $p = 11h$ **b.** The slope represents the employee's hourly rate, which is $11/hr.

17. $(0.2a^2 + 40.2a + 2020)$ dollars **18.** The object will be 115 ft above the ground $\frac{15}{4}$ sec, or 3.75 sec, after it is thrown upward. **19.** It will take 8 painters 12.5 hr to paint the house. **20.** A basketball's radius is $\sqrt[3]{35}$ (or between 3 and 4) times a baseball's radius.

CHAPTER 8

Chapter 8 Pretest, pp. 587–588

1. $-2i, 2i$ **2.** $-\frac{7}{2}, \frac{1}{2}$ **3.** $\frac{3 + \sqrt{3}}{2}, \frac{3 - \sqrt{3}}{2}$ **4.** Two real solutions
5. $\frac{3 + 2\sqrt{2}}{2}, \frac{3 - 2\sqrt{2}}{2}$ **6.** $2 + 2i\sqrt{2}, 2 - 2i\sqrt{2}$ **7.** $-2, -\frac{8}{3}$
8. $\frac{-4 + 3\sqrt{2}}{2}, \frac{-4 - 3\sqrt{2}}{2}$ **9.** $\frac{5 + i\sqrt{5}}{5}, \frac{5 - i\sqrt{5}}{5}$ **10.** $\frac{3}{2}$ **11.** $\pm 3, \pm 2i\sqrt{2}$
12. $2x^2 - 5x - 12 = 0$

13.

14.

$(2, -1); x = 2; (1, 0)$ and $(3, 0); (0, 3)$

$(-\frac{1}{2}, \frac{49}{4}); x = -\frac{1}{2}; (-4, 0)$ and $(3, 0); (0, 12)$

15.

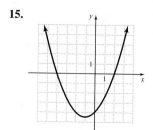

Domain: $(-\infty, \infty)$;
range: $\left[-\frac{9}{2}, \infty\right)$

16.

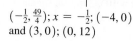

$(-\infty, 2)\cup(4, \infty)$

17. The object is 50 ft below the point of release $\frac{5}{2}$ sec, or 2.5 sec, after it is tossed upward. **18.** The speed of the current is 3 mph.
19. The radius is approximately 6.1 ft.
20. The company must sell between and including 30 and 80 bottles for revenue of at least $1200.

Section 8.1 Practices, pp. 589–596

1, p. 589: a. $-7, 7$ **b.** $-3i, 3i$ **2, p. 590: a.** $\pm 2\sqrt{2}$ **b.** $\pm 3i\sqrt{2}$
3, p. 591: a. $\frac{-1 + 2\sqrt{3}}{2}, \frac{-1 - 2\sqrt{3}}{2}$ **b.** $-3 + i\sqrt{2}, -3 - i\sqrt{2}$ **c.** $-\frac{4}{3}, \frac{10}{3}$
4, p. 592: $3 + 2\sqrt{6}, 3 - 2\sqrt{6}$ **5, p. 593: a.** 9 **b.** $\frac{1}{4}$
6, p. 594: $\frac{-3 + \sqrt{13}}{2}, \frac{-3 - \sqrt{13}}{2}$ **7, p. 595:** $\frac{1 + i\sqrt{7}}{4}, \frac{1 - i\sqrt{7}}{4}$
8, p. 595: $-3 + \sqrt{11}, -3 - \sqrt{11}$ **9, p. 596:** 100

Exercises 8.1, pp. 597–599

1. $-4, 4$ **3.** $-2\sqrt{6}, 2\sqrt{6}$ **5.** $-5i, 5i$ **7.** $-\sqrt{2}, \sqrt{2}$
9. $-6\sqrt{2}, 6\sqrt{2}$ **11.** $-\sqrt{5}, \sqrt{5}$ **13.** $-\frac{i\sqrt{6}}{3}, \frac{i\sqrt{6}}{3}$
15. $1 + 4\sqrt{3}, 1 - 4\sqrt{3}$ **17.** $\frac{-5 + 5\sqrt{3}}{2}, \frac{-5 - 5\sqrt{3}}{2}$ **19.** $\frac{3 + 2i\sqrt{2}}{4}, \frac{3 - 2i\sqrt{2}}{4}$
21. $-\frac{25}{4}, -\frac{7}{4}$ **23.** $3 + 4\sqrt{5}, 3 - 4\sqrt{5}$ **25.** $\frac{-3 + 4\sqrt{2}}{2}, \frac{-3 - 4\sqrt{2}}{2}$
27. $-\frac{4i\sqrt{3}}{3}, \frac{4i\sqrt{3}}{3}$ **29.** $\frac{1 + 3\sqrt{2}}{2}, \frac{1 - 3\sqrt{2}}{2}$ **31.** $r = \frac{\sqrt{\pi Vh}}{\pi h}$
33. $v = \pm\frac{\sqrt{Frm}}{m}$ **35.** 36 **37.** $\frac{49}{4}$ **39.** $\frac{4}{9}$ **41.** 0, 8 **43.** $-1, 4$
45. $-2 + \sqrt{6}, -2 - \sqrt{6}$ **47.** -2 **49.** $5 + 2i, 5 - 2i$
51. $2 + 2i\sqrt{2}, 2 - 2i\sqrt{2}$ **53.** $2 + 4\sqrt{2}, 2 - 4\sqrt{2}$
55. $\frac{-5 + \sqrt{37}}{2}, \frac{-5 - \sqrt{37}}{2}$ **57.** $\frac{3 + i\sqrt{11}}{2}, \frac{3 - i\sqrt{11}}{2}$ **59.** $\frac{2 + 2\sqrt{10}}{3}, \frac{2 - 2\sqrt{10}}{3}$
61. $-2, -\frac{3}{4}$ **63.** $\frac{-1 + 3i\sqrt{7}}{4}, \frac{-1 - 3i\sqrt{7}}{4}$ **65.** $2 + \sqrt{7}, 2 - \sqrt{7}$
67. $-1 + \sqrt{3}, -1 - \sqrt{3}$ **69.** $r = \frac{\sqrt{3V\pi h}}{\pi h}$ **71.** $3 + \sqrt{7}, 3 - \sqrt{7}$
73. $-4 + 3\sqrt{2}, -4 - 3\sqrt{2}$ **75.** $5 + 3\sqrt{5}, 5 - 3\sqrt{5}$ **77.** The team is searching 5 mi from the last known location of the hikers. **79.** The interest rate is 5%. **81. a.** It will return to the original height 2 sec after it is thrown upward. **b.** The object will be 12 ft above the point of release 0.5 sec and 1.5 sec after it is thrown upward. **83. a.** The patio is 18 ft by 8 ft. **b.** It will cost $508.30 to enclose the patio. **85.** The average speed of the truck driving north is 48 mph, and the average speed of the truck driving east is 64 mph.

Section 8.2 Practices, pp. 603–607

1, p. 603: $\frac{-3 + \sqrt{29}}{2}, \frac{-3 - \sqrt{29}}{2}$ **2, p. 603:** $\frac{1 + i\sqrt{5}}{6}, \frac{1 - i\sqrt{5}}{6}$
3, p. 604: $-2.820, 5.320$ **4, p. 604:** -9 **5, p. 606: a.** 1, 6, 9, 0, 1, Real number **b.** 1, 6, -9, 72, 2, Real numbers **c.** 4, 2, 3, -44, 2, Complex numbers (containing i) **6, p. 607:** In 2006

Exercises 8.2, pp. 608–610

1. quadratic formula **3.** one **5.** $-2, -1$ **7.** $3 + \sqrt{10}, 3 - \sqrt{10}$
9. $\frac{1 + 3\sqrt{5}}{2}, \frac{1 - 3\sqrt{5}}{2}$ **11.** $2 + i, 2 - i$ **13.** $\frac{-3 + \sqrt{30}}{3}, \frac{-3 - \sqrt{30}}{3}$
15. $\frac{1}{2}, \frac{2}{3}$ **17.** $\frac{-4 + i\sqrt{2}}{2}, \frac{-4 - i\sqrt{2}}{2}$ **19.** $\frac{-1 + \sqrt{2}}{3}, \frac{-1 - \sqrt{2}}{3}$ **21.** $\frac{3}{2}$
23. $1 + i\sqrt{11}, 1 - i\sqrt{11}$ **25.** $\frac{-2 + \sqrt{19}}{3}, \frac{-2 - \sqrt{19}}{3}$ **27.** $-4, -1$
29. $2 + 2i, 2 - 2i$ **31.** $-4 + 2\sqrt{3}, -4 - 2\sqrt{3}$ **33.** $\frac{5 + 2\sqrt{6}}{2}, \frac{5 - 2\sqrt{6}}{2}$
35. $-0.036, 1.965$ **37.** $-71.421, -5.246$ **39.** Two complex solutions (containing i) **41.** One real solution **43.** Two real solutions
45. Two real solutions **47.** Two complex solutions (containing i) **49.** Two complex solutions (containing i)

51. $\dfrac{2+i\sqrt6}{2}, \dfrac{2-i\sqrt6}{2}$ **53.** $\dfrac{4+\sqrt{22}}{3}, \dfrac{4-\sqrt{22}}{3}$ **55.** The stone is 300 ft above the ground 5 sec after it is thrown downward.   **57.** Squares measuring 2.5 in. by 2.5 in. should be cut from each corner.   **59.** The company must sell 16 tickets to generate revenue of $3520.   **61.** In 2008

Section 8.3   Practices, pp. 611–616

1, p. 611: $-1, 0$ **2, p. 612:** $\pm1, \pm2\sqrt2$ **3, p. 613:** 25
4, p. 614: $-64, 125$ **5, p. 614:** Working alone, it would take the slower painter about 10.5 hr and his partner about 9.5 hr to paint the apartment.
6, p. 616: $y^2 + 3y = 0$ **7, p. 616:** $24n^2 - 22n + 3 = 0$

Exercises 8.3, pp. 617–618

1. quadratic   **3.** in terms of a new variable   **5.** $\frac12, 1$ **7.** 2, 10
9. $\pm\sqrt2, \pm\sqrt6$ **11.** $\pm\frac12, \pm i\sqrt3$ **13.** 4   **15.** 256   **17.** $-8, -1$
19. 5, 6   **21.** 6   **23.** $x^2 - 9x + 14 = 0$ **25.** $y^2 - 5y - 36 = 0$
27. $3t^2 + 4t - 4 = 0$ **29.** $6n^2 + 17n + 12 = 0$ **31.** $x^2 - 8x + 16 = 0$
33. $t^2 - 2 = 0$ **35.** $x^2 + 9 = 0$ **37.** $-1, 8$ **39.** $\pm3i, \pm\sqrt2$
41. $5a^2 - 13a - 6 = 0$ **43.** The person's height is approximately 1.6 m.
45. She drives to work at an average speed of about 35 mph and she drives home at an average speed of about 27 mph.   **47.** If only one drain is open, it takes the large drain 3.4 hr and the small drain 4.9 hr to empty the pool.

Section 8.4   Practices, pp. 621–632

1, p. 621:

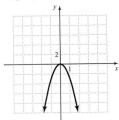

2, p. 623:

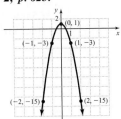

3, p. 624: **a.** $(2, 9); x = 2$
b. Opens downward **c.** $(-1, 0)$ and $(5, 0); (0, 5)$
d.

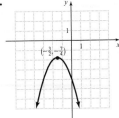

4, p. 626: **a.** Opens upward and has a minimum point; two x-intercepts and one y-intercept **b.** Opens downward and has a maximum point; no x-intercepts and one y-intercept **c.** Opens downward and has a maximum point; one x-intercept and one y-intercept

5, p. 627: **a.** $\left(-\frac32, -\frac74\right)$ **b.** Opens downward
c.

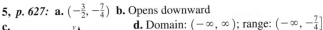

d. Domain: $(-\infty, \infty)$; range: $\left(-\infty, -\frac74\right]$

6. p. 628: a. $l = 100 - w$ **b.** $A(w) = 100w - w^2$
c.

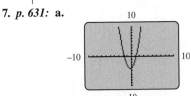

d. A garden measuring 50 ft by 50 ft will maximize the area of the garden. The maximum area is 2500 ft^2.

7. p. 631: a.

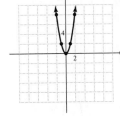

b. $(-0.25, -4.13)$
c. x-intercepts: $(-1.69, 0)$ and $(1.19, 0)$; y-intercept: $(0, -4)$

Exercises 8.4, pp. 633–640

1. quadratic   **3.** vertex   **5.** upward   **7.** two
9. $8, (-2, 8); 2, (-1, 2); 0, (0, 0);$ $2, (1, 2); 8, (2, 8)$
11. $8, (-4, 8); 2, (-2, 2);$ $0, (0, 0); 2, (2, 2); 8, (4, 8)$

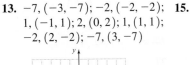

13. $-7, (-3, -7); -2, (-2, -2);$ $1, (-1, 1); 2, (0, 2); 1, (1, 1);$ $-2, (2, -2); -7, (3, -7)$ **15.**

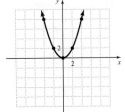

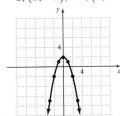

$(4, -16); x = 4; (0, 0)$ and $(8, 0); (0, 0)$

17.

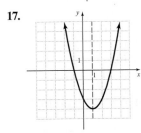

$(1, -4); x = 1; (-1, 0)$ and $(3, 0); (0, -3)$

19.

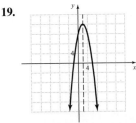

$(2, 16); x = 2; (-2, 0)$ and $(6, 0); (0, 12)$

ANSWERS

21.

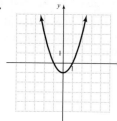

$(0, -1); x = 0; (-1, 0)$ and $(1, 0); (0, -1)$

23.

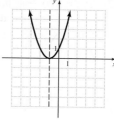

$(-1, 0); x = -1; (-1, 0); (0, 1)$

25.

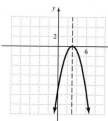

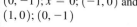

$(3, 0); x = 3; (3, 0); (0, -9)$

27.

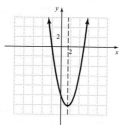

$\left(\frac{3}{2}, -\frac{49}{4}\right); x = \frac{3}{2}; (-2, 0)$ and $(5, 0); (0, -10)$

29. Upward, Minimum, 0, 1; Upward, Minimum, 1, 1; Downward, Maximum, 2, 1; Downward, Maximum, 0, 1; Upward, Minimum, 2, 1

31.

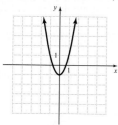

$(-\infty, \infty), [-1, \infty)$

33.

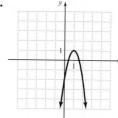

$(-\infty, \infty), (-\infty, 1]$

35.

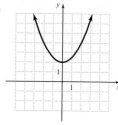

$(-\infty, \infty), [2, \infty)$

37.

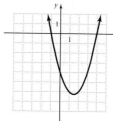

$(-\infty, \infty), \left[-\frac{25}{4}, \infty\right)$

39.

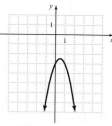

$(-\infty, \infty), \left(-\infty, -\frac{5}{2}\right]$

41. $(-0.10, -1.01); (-1.10, 0),$ $(0.90, 0); (0, -1)$
43. $(-3.33, 2.17); (-7.13, 0),$ $(0.47, 0); (0, 0.5)$
45. $(-0.30, 6.55);$ None; $(0, 7)$

47. $6, (-3, 6); 1, (-2, 1);$ $-2, (-1, -2); -3, (0, -3);$ $-2, (1, -2); 1, (2, 1); 6, (3, 6)$

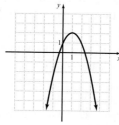

49.

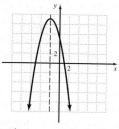

$(-2, 9); x = -2; (-5, 0)$ and $(1, 0); (0, 5)$

51.

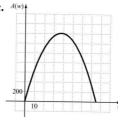

$(-\infty, \infty), (-\infty, 2]$

53. **a.** $\left(\frac{3}{2}, 316\right)$

b.

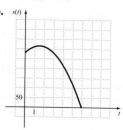

c. At $\frac{3}{2}$ sec (or 1.5 sec) after it is thrown, the stone reaches its maximum height of 316 ft. **55.** **a.** $l = 75 - w$ **b.** $A(w) = 75w - w^2$
c.

d. An exercise yard measuring 37.5 ft by 37.5 ft will produce a maximum area of 1406.25 ft^2.

57. **a.** $(60, 3600)$ **b.** 60 units must be sold to generate the maximum revenue of $3600.

Section 8.5 Practices, pp. 645–649

1, p. 645: $(-\infty, -1] \cup [4, \infty)$

2, p. 645: $(1, 5)$

3, p. 646: The stand can run for six weeks or less.
4, p. 647: $(-5, 5)$

5, p. 648: $(-\infty, -3) \cup \left[-\frac{1}{2}, \infty\right)$

Exercises 8.5, pp. 650–652

1. test value **3.** rational inequality
5. $(-\infty, 0) \cup (1, \infty)$

7. $(-2, 2)$

9. $[-3, 4]$

11. $(-\infty, \infty)$

13. $(-\infty, -2] \cup [3, \infty)$

15. $(-4, -1)$

17. $\left(-\infty, \frac{1}{2}\right] \cup [1, \infty)$

19. $\left(-\infty, -\frac{1}{2}\right] \cup \left[\frac{1}{2}, \infty\right)$

21. $\left(-\frac{1}{3}, \frac{3}{2}\right)$

23. $(-2, 0]$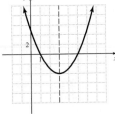

25. $(6, \infty)$

27. $(3, 9)$

29. $(-\infty, -4) \cup [\frac{1}{2}, \infty)$

31. $(-\infty, -\frac{11}{2}] \cup (-\frac{5}{2}, \infty)$

33. $(-4, 3)$

35. $(-\infty, 4) \cup (10, \infty)$

37. $(-\infty, -3] \cup [\frac{1}{2}, \infty)$

39. Any price between, and including, \$50 and \$100 will generate revenue of at least \$5000.   **41.** Producing between, but not including, 10 and 20 end tables per week will keep the cost under \$500.   **43.** The area will exceed 450 ft^2 for any length between, but not including, 15 ft and 30 ft.   **45.** The average cost will be less than \$8 if more than 144 units are sold.

Chapter 8   Review Exercises, pp. 657–661

1. is; possible answer: it can be written as $\sqrt{3}x^2 + 0x - 2 = 0$, which is in the form $ax^2 + bx + c = 0$, where a, b, and c are real numbers and $a \neq 0$ **2.** is; possible answer: of the square root property of equality   **3.** does not; possible answer: its discriminant is negative   **4.** does not; possible answer: the coefficient of the x^2-term is positive   **5.** does not; possible answer: all points on the graph have a negative y-value   **6.** is; possible answer: it can be written as $(\sqrt{3x})^2 + \sqrt{3x} - 5 = 0$, or as $u^2 + u - 5 = 0$, where $u = \sqrt{3x}$ **7.** $-9, 9$ **8.** $-\sqrt{5}, \sqrt{5}$ **9.** $\frac{9}{2}, \frac{11}{2}$ **10.** $\frac{-1 + 2i}{2}, \frac{-1 - 2i}{2}$ **11.** $-4 + \sqrt{2}, -4 - \sqrt{2}$ **12.** $2i\sqrt{2}, -2i\sqrt{2}$ **13.** 25 **14.** $\frac{81}{4}$ **15.** $3 + \sqrt{7}, 3 - \sqrt{7}$ **16.** $\frac{-1 + \sqrt{13}}{2}, \frac{-1 - \sqrt{13}}{2}$ **17.** $\frac{-3 + i\sqrt{3}}{2}, \frac{-3 - i\sqrt{3}}{2}$ **18.** $\frac{1 + 2\sqrt{7}}{3}, \frac{1 - 2\sqrt{7}}{3}$ **19.** $-1, -6$ **20.** $2 + i, 2 - i$ **21.** $\frac{3 + 2\sqrt{6}}{3}, \frac{3 - 2\sqrt{6}}{3}$ **22.** $-\frac{3}{2}$ **23.** $\frac{-9 + \sqrt{33}}{4}, \frac{-9 - \sqrt{33}}{4}$ **24.** $-5 + 3i, -5 - 3i$ **25.** Two complex solutions (containing i)   **26.** Two real solutions   **27.** $-1, 2$ **28.** $\pm 2i, \pm \sqrt{6}$ **29.** $9, 16$ **30.** $-1, 0$ **31.** $2x^2 + x - 15 = 0$ **32.** $n^2 + 14n + 49 = 0$

33.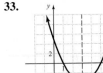

$(3, -4); x = 3; (1, 0)$ and $(5, 0); (0, 5)$

34.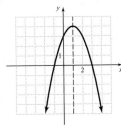

$(1, 4); x = 1; (-1, 0)$ and $(3, 0); (0, 3)$

35.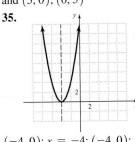

$(-4, 0); x = -4; (-4, 0);$ $(0, 16)$

36.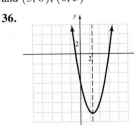

$(\frac{5}{2}, -\frac{49}{4}); x = \frac{5}{2}; (-1, 0)$ and $(6, 0); (0, -6)$

37. Upward, Minimum, None, One   **38.** Downward, Maximum, Two, One

39.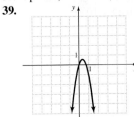

$(-\infty, \infty), (-\infty, \frac{9}{16}]$.

40.

$(-\infty, \infty), [-\frac{9}{8}, \infty)$

41. $(-2, 6)$

42. $(-\infty, -2] \cup [5, \infty)$

43. $(-\infty, \frac{1}{2}] \cup [4, \infty)$

44. $(-\frac{1}{2}, \frac{3}{2})$

45. $(-\infty, 3) \cup (5, \infty)$

46. $(0, 3]$

47. It will take 5 sec for the object to fall 400 ft.   **48.** The ladder reaches approximately 9.5 ft up the side of the house.   **49.** The speed of the train traveling south is approximately 53 mph and the speed of the train traveling west is approximately 45 mph.   **50.** The average rate of return was about 10.5%.   **51.** $\frac{5}{2}$-inch by $\frac{5}{2}$-inch squares should be cut from each corner.   **52.** He needs to sell 25 cases per week.   **53.** In 2010   **54.** In 2009   **55.** She bicycled to school at a rate of 13.7 mph that day.   **56.** Working alone, it takes the newer printing press about 2.7 hr and the older printing press about 7.7 hr to complete the job.   **57. a.** $w = 150 - l$ **b.** $A(l) = 150l - l^2$

c.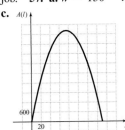

d. A park measuring 75 ft by 75 ft will produce a maximum area of 5625 ft^2.

58. a.

b. 20 units must be fabricated per day in order to minimize the cost. **c.** The minimum cost is \$80.

59. The baseball is 40 ft above the ground at $\frac{1}{2}$ sec and $\frac{9}{2}$ sec.   **60.** Any width between, and including, 10 ft and 12 ft will produce an area of at least 120 ft^2.

Chapter 8   Posttest, pp. 662–663

1. $-2\sqrt{2}, 2\sqrt{2}$ **2.** $1 + i\sqrt{7}, 1 - i\sqrt{7}$ **3.** $-\frac{3}{2}, \frac{1}{2}$ **4.** Two complex solutions (containing i)   **5.** $\frac{-2 + 3\sqrt{2}}{3}, \frac{-2 - 3\sqrt{2}}{3}$ **6.** $-4 + \sqrt{22}, -4 - \sqrt{22}$ **7.** $\frac{5 + i\sqrt{19}}{2}, \frac{5 - i\sqrt{19}}{2}$ **8.** $-1, \frac{7}{4}$ **9.** $\frac{6 + i\sqrt{2}}{2}, \frac{6 - i\sqrt{2}}{2}$ **10.** $-1 + \sqrt{5}, -1 - \sqrt{5}$ **11.** 1 **12.** $6n^2 - 7n + 2 = 0$

13. $(3, -1)$; $x = 3$; $(2, 0)$ and $(4, 0)$; $(0, 8)$

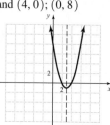

14. $(\frac{3}{2}, \frac{49}{4})$; $x = \frac{3}{2}$; $(-2, 0)$ and $(5, 0)$; $(0, 10)$

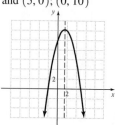

15.

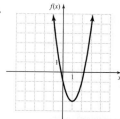

Domain: $(-\infty, \infty)$; Range: $[-3, \infty)$

16. $(-\infty, -6) \cup (3, \infty)$

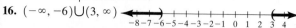

$-8\,-7\,-6\,-5\,-4\,-3\,-2\,-1\ \ 0\ \ 1\ \ 2\ \ 3\ \ 4$

17. The average speed of the airplane flying west is approximately 470 mph and the average speed of the airplane flying south is approximately 450 mph. **18.** Working alone, the senior clerk can process the applications in 3 hr and the junior clerk can process the applications in 6 hr. **19.** The rocket will reach its maximum height of 147 ft in 3 sec. **20.** The factory can produce between, but not including, 80 and 240 units each day.

Chapter 8   Cumulative Review, pp. 664–665

1. $2n^2$ **2.** $-\frac{3}{2}, \frac{9}{2}$ **3.**

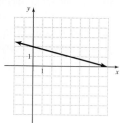

4. Yes, the relation represents a function.   **5.** $(4, 0, -3)$

6. $2a^2 - 3a + 4 - \frac{4}{2a + 3}$ **7.** $2x^2y^2(y - 2x)(y - 4x)$ **8.** $a^3 - 8$

9. $\frac{x + 3}{x - 2}$ **10.** -1 **11.** $-2x^3$ **12.** $10\sqrt{3} - 9\sqrt{2}$ **13.** $\frac{1 + i\sqrt{5}}{6}, \frac{1 - i\sqrt{5}}{6}$

14. $4p^2 + 11p - 3 = 0$ **15.** Downward, Maximum, 2, 1

16. a. $0.004a + 0.35b \leq 80$ **b.** Yes   **17. a.** $r = \sqrt{\frac{V}{\pi h}}$, or

$r = \frac{\sqrt{V\pi h}}{\pi h}$ **b.** 8 cm   **18. a.** $2l + 2w < 1000$

b.

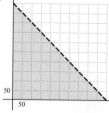

c. 300 ft by 150 ft

19. a. $B = 0.05m + 7.5$ **b.** The slope is 0.05. It represents the per-minute fee for long-distance calls. **c.** The y-intercept is $(0, 7.5)$. It represents the flat monthly fee that the phone service charges. **20. a.** In 2009 **b.** In 2011

CHAPTER 9

Chapter 9   Pretest, pp. 667–668

1. a. $2x^2 + 7x - 4$ **b.** $-2x^2 - 3x + 2$ **c.** $4x^3 + 8x^2 - 11x + 3$ **d.** $\frac{1}{x + 3}, x \neq -3, \frac{1}{2}$ **2. a.** $\frac{5}{3x - 4}, x \neq \frac{4}{3}$ **b.** $\frac{15}{x} - 4, x \neq 0$ **c.** 1 **d.** -7 **3.** One-to-one

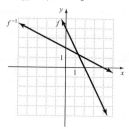

4. $f^{-1}(x) = \frac{x + 7}{3}$ **5. a.** $\frac{1}{4}$ **b.** -7.389 **6. a.** 0 **b.** -2

7. a.

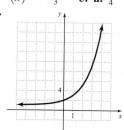 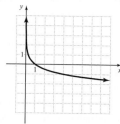

b.

8. a. 5 **b.** 0 **9. a.** -1 **b.** 5 **10.** 1.8928 **11. a.** $1 + 2\log_5 x$ **b.** $\log_8 3 + 3\log_8 x - \log_8 y$ **12. a.** $\log_7 40$ **b.** $\log_6 \frac{x^4}{(x + 2)^2}$ **13.** 2 **14.** $\frac{2}{3}$ **15.** 2 **16.** 12 **17.** 2 **18.** A gallon of milk will cost about $4.73 in the fifth year. **19.** The effective annual interest rate is approximately 4.6%. **20.** It will take about 10.1 days for 100 mg of gallium-67 to decay to 12 mg.

Section 9.1   Practices, pp. 669–679

1, p. 669: a. $(f + g)(x) = x^2 + 3x - 3$ **b.** $(f - g)(x) = x^2 - 3x + 1$ **c.** $(f \cdot g)(x) = 3x^3 - 2x^2 - 3x + 2$ **d.** $\left(\frac{f}{g}\right)(x) = \frac{x^2 - 1}{3x - 2}, x \neq \frac{2}{3}$

2, p. 670: a. 1 **b.** -35 **c.** 0 **d.** $-\frac{2}{9}$ **3, p. 672: a.** $-4x + 17$ **b.** $-4x + 8$ **c.** 5 **d.** -4 **4, p. 674: a.** One-to-one function **b.** Not a one-to-one function **5, p. 675:** $\{(-1, -3), (0, -1), (2, 0), (3, 4)\}$

6, p. 676: $g^{-1}(x) = \frac{x + 1}{3}$ **7, p. 677:** p is not the inverse of q.

8, p. 678: a.

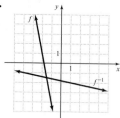

b.

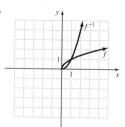

9, p. 679: a. C is a one-to-one function since its graph is linear and passes the horizontal-line test. **b.** $C^{-1}(x) = \frac{1}{40}x$ **c.** The inverse can be used to determine the number of hours that the van was rented based on the amount of money charged.

Exercises 9.1, pp. 680–687

1. sum of f and g **3.** product of f and g **5.** one-to-one **7.** inverse functions of each other

9. $(f + g)(x) = 4x^2 + 2x - 6$;
$(f - g)(x) = -4x^2 + 2x + 12$;
$(f \cdot g)(x) = 8x^3 + 12x^2 - 18x - 27$;
$\left(\frac{f}{g}\right)(x) = \frac{1}{2x - 3}, x \neq -\frac{3}{2}, \frac{3}{2}$

11. $(f + g)(x) = 5x^2 + 16x + 3$;

$(f - g)(x) = x^2 + 6x + 9$;

$(f \cdot g)(x) = 6x^4 + 37x^3 + 58x^2 - 3x - 18$;

$\left(\frac{f}{g}\right)(x) = \frac{3x + 2}{2x - 1}, x \neq -3, \frac{1}{2}$

13. $(f + g)(x) = \frac{x^2 + 2x - 2}{x + 3}, x \neq -3$;

$(f - g)(x) = \frac{-x^2 - 2x + 4}{x + 3}, x \neq -3$;

$(f \cdot g)(x) = \frac{x - 1}{x + 3}, x \neq -3$;

$\left(\frac{f}{g}\right)(x) = \frac{1}{(x + 3)(x - 1)}, x \neq -3, 1$

15. $(f + g)(x) = 6\sqrt{x} - 1; (f - g)(x) = -2\sqrt{x} + 1$;

$(f \cdot g)(x) = 8x - 2\sqrt{x}; \left(\frac{f}{g}\right)(x) = \frac{8x + 2\sqrt{x}}{16x - 1}, x \neq \frac{1}{16}$

17. -12 **19.** 0 **21.** 18 **23.** -4 **25. a.** $x - 14$ **b.** $x - 7$ **c.** -11
d. -8 **27. a.** $x^2 + 4x - 11$ **b.** $x^2 + 2x - 13$ **c.** -6 **d.** -14
29. a. $\frac{3}{2x + 5}, x \neq -\frac{5}{2}$ **b.** $\frac{6}{x} + 5, x \neq 0$ **c.** $\frac{1}{2}$ **d.** 2
31. a. $-\sqrt{1 - 4x}, x \leq \frac{1}{4}$ **b.** $1 + 4\sqrt{x}, x \geq 0$ **c.** -5 **d.** 17
33. One-to-one **35.** Not one-to-one **37.** One-to-one
39. Not one-to-one **41.** c **43.** a **45.** $\{(5, -4), (3, -2), (1, 0),$
$(-1, 2), (-3, 4)\}$ **47.** $\{(-3, -27), (-2, -8), (-1, -1), (0, 0),$
$(1, 1), (2, 8), (3, 27)\}$ **49.** $f^{-1}(x) = \frac{1}{4}x$ **51.** $g^{-1}(x) = 4x$
53. $f^{-1}(x) = -x - 5$ **55.** $f^{-1}(x) = \frac{x - 2}{5}$ **57.** $g^{-1}(x) = 2x + 6$
59. $f^{-1}(x) = \sqrt[3]{x + 4}$ **61.** $h^{-1}(x) = \frac{3 - 4x}{x}$ **63.** $f^{-1}(x) = -\frac{x}{1 - 2x}$,
or $\frac{x}{2x - 1}$ **65.** $f^{-1}(x) = x^3 - 4$ **67.** Inverses **69.** Inverses
71. Not inverses **73.** Inverses **75.** Not inverses **77.** Inverses
79.

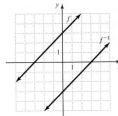

81.

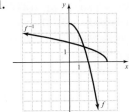

83. Inverses **85.** -43 **87. a.** $-\frac{8}{x + 2} + 7, x \neq -2$ **b.** $\frac{4}{-2x + 9}, x \neq \frac{9}{2}$
c. 6 **d.** $\frac{4}{9}$ **89.**

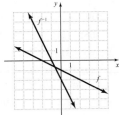

91. $f^{-1}(x) = \frac{3x}{3 - 2x}$

93. a. $(C \circ x)(d) = 2100d + 1000$; it represents the plant's weekly cost
for d days of operation. **b.** The weekly cost is $11,500.
95. a. $S^{-1}(n) = \frac{n + 360}{180}$ **b.** The inverse can be used to calculate the
number of sides of a polygon if the sum of the interior angles is known.
c. The polygon has five sides. **97. a.** $B(t) = 0.05t + 4.5$
b. $B^{-1}(t) = 20t - 90$ **c.** The inverse can be used to determine the
number of minutes of long-distance calling time a customer was billed for.

Section 9.2 Practices, pp. 689–694

1, p. 689: a. 64 **b.** 9 **2, p. 689: a.** 32 **b.** $\frac{1}{128}$ **3, p. 690: a.** 0.007
b. 7.389 **4, p. 690:**

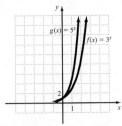

5, p. 691:

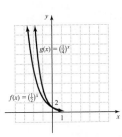

6, p. 692:

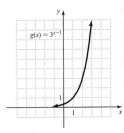

7, p. 693: a. 6 **b.** $\frac{3}{2}$ **8, p. 694:** Approximately 80 mg will remain after
3 days.

Exercises 9.2, pp. 695–698

1. exponential function **3.** natural exponential function **5.** decreasing
7. Polynomial function **9.** Exponential function **11.** Radical function
13. Exponential function **15. a.** $\frac{1}{8}$ **b.** 1 **c.** 16 **17. a.** 3 **b.** $\frac{1}{3}$
c. $\frac{1}{81}$ **19. a.** $\frac{1}{27}$ **b.** $\frac{1}{3}$ **c.** 9 **21. a.** $\frac{23}{8}$ **b.** 2 **c.** -13
23. a. -0.007 **b.** -2.718 **25. a.** 0.018 **b.** 2.718 **27. a.** 0.050
b. 0.135 **29.**

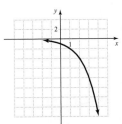

31.

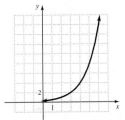

33.

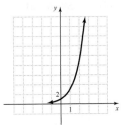

35.

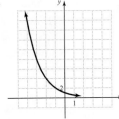

37.

39. 2 **41.** -5 **43.** -2 **45.** $\frac{4}{3}$ **47.** 2

49. 4 **51.** $\frac{1}{4}$ **53.** $\frac{9}{2}$ **55.** $-\frac{10}{3}$ **57.** 9 **59.** $x = \frac{11}{2}$ **61.** radical
function **63.** $-\frac{1}{25}$ **65.** 0.135 **67.** The salary will be $30,284.80
69. a. The concentration will be approximately 21.6 parts per million.
b. The initial concentration of pollutants was 40 parts per million.
71. The amount in the account will be $10,790.80. **73.** The concentra-
tion after 8 hr is approximately 18 mg/L. **75.** 1280 bacteria will be
present in 3.5 hr.

Section 9.3 Practices, pp. 701–706

1, p. 701: a. $8^2 = 64$ **b.** $5^{-1} = \frac{1}{5}$ **c.** $2^{1/3} = \sqrt[3]{2}$
2, p. 701: a. $\log_2 32 = 5$ **b.** $\log_7 1 = 0$ **c.** $\log_{10} \sqrt{10} = \frac{1}{2}$
3, p. 702: a. 1 **b.** 3 **c.** $\frac{1}{3}$ **4, p. 702: a.** 1 **b.** 0 **5, p. 702: a.** 64
b. $\frac{1}{2}$ **6, p. 703: a.** 5 **b.** 16 **7, p. 703:** $\frac{3}{2}$

8, p. 704:   **9. p. 705:**

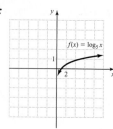
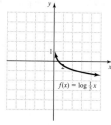

10, p. 705: The barometric pressure is about 28.1 in. of mercury.

Exercises 9.3, pp. 707–710

1. logarithmic function   **3.** exponent   **5.** decreasing   **7.** $3^4 = 81$
9. $\left(\frac{1}{2}\right)^5 = \frac{1}{32}$ **11.** $5^{-2} = \frac{1}{25}$ **13.** $\left(\frac{1}{4}\right)^{-1} = 4$ **15.** $16^{1/4} = 2$
17. $10^{1/2} = \sqrt{10}$ **19.** $\log_3 243 = 5$ **21.** $\log_{1/4} \frac{1}{4} = 1$ **23.** $\log_2 \frac{1}{16} = -4$
25. $\log_{1/3} 81 = -4$ **27.** $\log_{49} 7 = \frac{1}{2}$ **29.** $\log_{11} \sqrt[5]{11} = \frac{1}{5}$ **31.** 3
33. 2   **35.** 1   **37.** 4   **39.** -3 **41.** -2 **43.** 0   **45.** $\frac{1}{3}$ **47.** $\frac{1}{4}$
49. $-\frac{1}{2}$ **51.** 27   **53.** $\frac{1}{36}$ **55.** $\frac{3}{2}$ **57.** 2   **59.** 6   **61.** 8   **63.** $\frac{3}{4}$ **65.** $\frac{1}{7}$
67. $\frac{3}{2}$ **69.** $\frac{4}{3}$ **71.** -4 **73.**

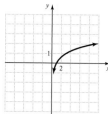

75.   **77.**

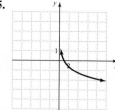
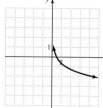

79. 125   **81.** $\frac{3}{2}$ **83.** -4 **85.** $\left(\frac{1}{2}\right)^{-4} = 16$ **87.** $\log_{1/7} 49 = -2$
89. The power gain is 40 dB.   **91.** The amount of current is 8 amp.
93. It takes the computer 70 picoseconds.

Section 9.4   Practices, pp. 711–716

1, p. 711: a. $\log_6 4 + \log_6 9$ **b.** $1 + \log_2 x$ **2, p. 712: a.** $\log_2 135$
b. $\log_7(x^2 + xy)$ **3, p. 712: a.** $\log_4 3 - 1$ **b.** $\log_b u - \log_b v$
4, p. 713: a. $\log_7 \frac{1}{2}$ **b.** $\log_3 \frac{x-y}{x+y}$ **5, p. 713: a.** $3 \log_2 x$ **b.** $\frac{1}{3} \log_5 x$
6, p. 714: a. 7   **b.** 29   **7, p. 715: a.** $\log_6 128$ **b.** $\log_2 \frac{u^3}{v^3}$
8, p. 715: a. $1 + 2 \log_5 x$ **b.** $\frac{1}{2} \log_7 u - \frac{1}{2} \log_7 v$
c. $3 \log_u b - \log_u a - 5 \log_u c$ **9, p. 716: a.** 3 bacteria are present
after 1 hr.   **b.** The number of bacteria in the culture increases by a factor
of 3 each hour.

Exercises 9.4, pp. 717–720

1. sum   **3.** times   **5.** power   **7.** $4 + \log_2 5$
9. $\log_6 7 - 2$ **11.** $\log_3 8 + \log_3 x$ **13.** $\log_n u + \log_n v$
15. $\log_b a + \log_b(b - 1)$ **17.** $3 - \log_4 n$ **19.** $\log_7 v - \log_7 u$
21. $\log_a(b - a) - \log_a(b + a)$ **23.** $\log_2 35$ **25.** $\log_7 5$
27. $\log_x 6x$ **29.** $\log_b \frac{b}{16}$ **31.** $\log_5(2x - 10)$ **33.** $\log_n xz$
35. $\log_5 \frac{k}{n}$ **37.** $\log_{10} \frac{a+b}{a-b}$ **39.** $2 \log_6 8$ **41.** $4 \log_3 x$ **43.** $\frac{1}{5} \log_a b$
45. $-\log_4 a$ **47.** 4   **49.** $\frac{1}{2}$ **51.** -3 **53.** 15   **55.** x **57.** 10
59. $\log_2 108$ **61.** $\log_7 \frac{1}{4}$ **63.** $\log_4 \frac{x}{x}$ **65.** $\log_b 8$ **67.** $\log_4 \frac{\sqrt{x}}{y}$
69. $\log_5(x^3 - x^2)$ **71.** $\log_6 \sqrt[3]{x - y}$ **73.** $\log_2 \frac{x^4}{y^5 z}$ **75.** $\log_5 \frac{\sqrt[4]{ac^3}}{b^8}$

77. $\log_6 3 + 2 \log_6 y$ **79.** $3 + 2 \log_x y$ **81.** $2 + 2 \log_2 x + 5 \log_2 y$
83. $\frac{1}{2} \log_8 10 + \frac{1}{2} \log_8 x$ **85.** $2 \log_4 x - \log_4 y$ **87.** $\frac{3}{4} \log_7 u - \frac{1}{4} \log_7 v$
89. $2 \log_c a + 4 - 3 \log_c b$ **91.** $3 \log_6 x - 2 \log_6(x - y)$
93. $2 \log_a x + \frac{3}{5} \log_a y + \frac{2}{5} \log_a z$ **95.** $\log_b \frac{8}{b}$ **97.** $\log_a \frac{1}{12}$
99. $\frac{1}{3} \log_p q$ **101.** $3 + 3 \log_2 a + 2 \log_2 b$ **103.** $3 + \log_4 a$
105. The magnitude of the earthquake was 8.3.
107. $S = 20 \log_{10} P - 20 \log_{10} P_0$
109. a. $\log_{10} C = \log_{10} C_0 + \log_{10}\left(\frac{1}{2}\right)^{0.2t} \Rightarrow$
$$\log_{10} C = \log_{10}\left[C_0\left(\tfrac{1}{2}\right)^{0.2t}\right] \Rightarrow$$
$$C = 10^{\log_{10}\left[C_0(1/2)^{0.2t}\right]} \Rightarrow$$
$$C = C_0\left(\tfrac{1}{2}\right)^{0.2t}$$
b. 12.5 g will remain after 10 yr.

Section 9.5   Practices, pp. 721–724

1, p. 721: a. 0.7782   **b.** 1.4314   **2, p. 721: a.** 3   **b.** -1 **c.** $\frac{1}{3}$
3, p. 722: a. 1.3863   **b.** 3.2189   **4, p. 722: a.** 2   **b.** -7 **c.** $\frac{3}{4}$
5, p. 723: 0.6610   **6, p. 724:** It takes approximately 82 yr.

Exercises 9.5, pp. 725–726

1. common logarithm   **3.** natural logarithm   **5.** 0.6021   **7.** 1.2553
9. -0.1761 **11.** 0.1139   **13.** 1.0986   **15.** 2.8332   **17.** -0.4700
19. 0.1823   **21.** 6   **23.** -3 **25.** -2 **27.** $\frac{3}{4}$ **29.** x **31.** 4
33. -1 **35.** $\frac{1}{2}$ **37.** b **39.** 6   **41.** 3   **43.** 2.8074   **45.** 1.6992
47. -3.4594 **49.** -0.6309 **51.** 0.7959   **53.** -1.9386 **55.** 2
57. -3 **59.** -2.8074 **61.** 0.4314   **63.** The pH of household bleach is
approximately 12.5.   **65.** The company's revenue was approximately
\$3 million in the year 2005.   **67.** $kd = -\ln \frac{I}{I_0} \Rightarrow kd = \ln \frac{I_0}{I} \Rightarrow$
$e^{kd} = \frac{I_0}{I} \Rightarrow Ie^{kd} = I_0 \Rightarrow I = \frac{I_0}{e^{kd}} \Rightarrow I = I_0 e^{-kd}$ **69.** The population will
grow to 60 bacteria in approximately 5.9 hr.

Section 9.6   Practices, pp. 728–731

1, p. 728: -1.0466 **2, p. 729:** 5   **3, p. 729:** 9   **4, p. 730:** $\frac{1}{2}$
5, p. 731: The annual rate of population growth was approximately 5.1%.

Exercises 9.6, pp. 732–734

1. 2.6309   **3.** -3.1699 **5.** -0.2075 **7.** -3.3705 **9.** 0.7339
11. 0.8702   **13.** -0.6889 **15.** 2.6685   **17.** 4.5256   **19.** 71   **21.** 2
23. $-\frac{7}{8}$ **25.** $-11, 11$ **27.** $\frac{8}{3}$ **29.** 63   **31.** 4   **33.** 6   **35.** $\frac{3}{2}$ **37.** $\frac{9}{2}$
39. 3   **41.** 1   **43.** 4   **45.** $-5, 5$ **47.** $\frac{16}{3}$ **49.** -1.7297 **51.** The
population will be 350 million people in the year 2030.   **53.** It will take
approximately 14 yr for the amount in the account to double.   **55.** The
concentration of hydrogen ions in blood is about 4.0×10^{-8}.
57. a. $x = 5 \ln \frac{C}{300}$ **b.** The circulation reached 1,000,000 about 6 yr after
it was launched.   **59. a.** The initial investment was \$12,000; the initial
investment was the amount in the account when $t = 0$.   **b.** The annual interest rate is about 6.5%.   **61.** The company's revenue reached \$7 million
9 months after the products were on the market.

Chapter 9   Review Exercises, pp. 739–743

1. are not; possible answer: their composition is not x **2.** does not; possible answer: it is not one-to-one   **3.** is not; possible answer: the variable
is the base and not the exponent   **4.** is; possible answer: of the one-to-one
property of exponential functions   **5.** does; possible answer: $3^0 = 1$
6. are; possible answer: of the product property of logarithms

7. $(f + g)(x) = -5x + 2, (f - g)(x) = -7x + 8, (f \cdot g)(x) = $
$-6x^2 + 23x - 15, \frac{f}{g}(x) = \frac{5 - 6x}{x - 3}, x \neq 3$
8. $(f + g)(x) = 5x^2 - 4x + 1, (f - g)(x) = x^2 + 4x + 1,$
$(f \cdot g)(x) = 6x^4 - 12x^3 + 2x^2 - 4x, \frac{f}{g}(x) = \frac{3x^2 + 1}{2x^2 - 4x}; x \neq 0, 2$

9. $(f + g)(x) = \frac{2x + 9}{(x + 3)(x - 3)}, x \neq -3, 3;$

$(f - g)(x) = \frac{2x + 3}{(x + 3)(x - 3)}, x \neq -3, 3;$

$(f \cdot g)(x) = \frac{6}{(x - 3)(x^2 - 9)}, x \neq -3, 3;$

$\frac{f}{g}(x) = \frac{2(x + 3)}{3}, x \neq -3, 3$

10. $(f + g)(x) = 8\sqrt{x}, (f - g)(x) = 6\sqrt{x} + 4,$

$(f \cdot g)(x) = 7x - 12\sqrt{x} - 4, \frac{f}{g}(x) = \frac{7x + 16\sqrt{x} + 4}{x - 4}, x \neq 4$

11. 27 **12.** -10 **13.** 0 **14.** $\frac{9}{4}$ **15. a.** $x^2 - 5x + 11$
b. $x^2 - x + 3$ **c.** 11 **d.** 9 **16. a.** $\sqrt{2x - 9}$ **b.** $2\sqrt{x} - 9$ **c.** 3
d. $2\sqrt{5} - 9$ **17.** Not one-to-one **18.** One-to-one
19. $f^{-1}(x) = \frac{x - 3}{8}$ **20.** $f^{-1}(x) = 6x + 6$ **21.** $f^{-1}(x) = \sqrt[3]{x + 5}$
22. $f^{-1}(x) = x^3 - 10$ **23.** $f^{-1}(x) = \frac{2 - 3x}{x}$ **24.** $f^{-1}(x) = \frac{x}{1 - 3x}$
25. Inverses **26.** Not inverses

27. 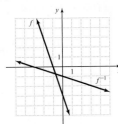 **28.**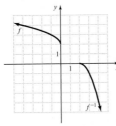

29. $-\frac{1}{27}$ **30.** 16 **31.** 4 **32.** 22 **33.** 2.7183 **34.** 0.1353

35. 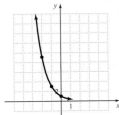 **36.**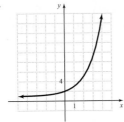

37. $\frac{5}{2}$ **38.** $-\frac{2}{3}$ **39.** 7 **40.** 1 **41.** $6^3 = 216$ **42.** $(\frac{1}{3})^{-2} = 9$
43. $\log_5 \frac{1}{25} = -2$ **44.** $\log_{81} 3 = \frac{1}{4}$ **45.** 1 **46.** 5 **47.** 0 **48.** $\frac{1}{4}$
49. 64 **50.** $\frac{1}{64}$ **51.** $\frac{1}{7}$ **52.** $\frac{2}{3}$ **53.** 3 **54.** -2

55. 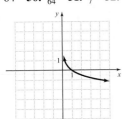 **56.**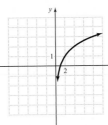

57. $3 + \log_3 x$ **58.** $1 - \log_6 a$ **59.** $\log_n x - \log_n(x - 5)$
60. $\log_9 u + \log_9(u + 1)$ **61.** $3 \log_5 x + \log_5 y$ **62.** $\frac{2}{5} \log_2 x + \frac{3}{5} \log_2 y$
63. $2 \log_3 x + 3 \log_3 z - 2 \log_3 y$ **64.** $2 - 4 \log_b (a - b)$
65. $\log_4(10n)$ **66.** $\log_9 \frac{x}{5}$ **67.** $\log_3(ab^2)$ **68.** $\log_b \frac{a^5}{c^4}$ **69.** $\log_6 \frac{\sqrt[4]{x^3}}{y^2}$
70. $\log_a \frac{x^2}{yz^4}$ **71.** 3 **72.** $-\frac{1}{3}$ **73.** y **74.** 14 **75.** 1.3617 **76.** 0.9731
77. 3.8712 **78.** -0.4055 **79.** -1 **80.** $\frac{3}{4}$ **81.** 100 **82.** $-y$
83. 2.9798 **84.** 0.8230 **85.** 2.1978 **86.** -0.1520 **87.** 9.3219
88. 0.3856 **89.** -7 **90.** 96 **91.** 9 **92.** -1 **93.** 4 **94.** 5
95. a. $(f \circ c)(s) = 0.008s + 2500$; it represents the agent's total
monthly salary if his monthly sales totaled s dollars. **b.** $5700
96. a. $P^{-1}(x) = \frac{1}{10} x + 15$ **b.** The inverse can be used to determine the
number of jeans the company must sell in order to make a certain profit.
c. 135 pairs of jeans **97. a.** $8484.33 **b.** It will take about 13 yr for this
investment to double. **98.** The stamp will have a value of $24 in 22.5 yr.
99. $s = \log 10^{90} - \log(x + 1)^{16} \Rightarrow$

$s = 90 \log 10 - 16 \log(x + 1) \Rightarrow$

$s = 90 - 16 \log(x + 1)$

100. $\ln(T - A) - \ln(98.6 - A) = -kt \ln e \Rightarrow$

$\ln \frac{T - A}{98.6 - A} = -kt \Rightarrow$

$e^{-kt} = \frac{T - A}{98.6 - A} \Rightarrow$

$(98.6 - A)e^{-kt} = T - A \Rightarrow$

$T - A = (98.6 - A)e^{-kt}$

101. The pH of milk is about 6.6. **102.** It will take approximately 25
yr for the investment to triple. **103.** Its value will depreciate to $15,000
about 4 yr after it is purchased. **104.** It will take approximately 9 yr.

Chapter 9 Posttest, pp. 744–745

1. a. $6x^2 - 2x - 4$ **b.** $-6x - 4$ **c.** $9x^4 - 6x^3 - 20x^2 - 8x$
d. $\frac{x - 2}{x}, x \neq -\frac{2}{3}, 0$ **2. a.** $\sqrt{2x + 5}$ **b.** $2\sqrt{x} + 5$ **c.** 3 **d.** 13

3. One-to-one

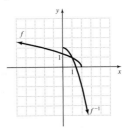

4. $f^{-1}(x) = \frac{7x + 1}{x}$ **5. a.** $-\frac{17}{9}$, or -1.8889 **b.** 1096.6332
6. a. 0 **b.** $-\frac{1}{2}$

7. a. 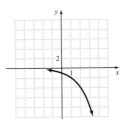 **b.**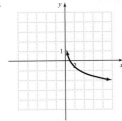

8. a. $2 + 4 \log_3 x$ **b.** $5 \log_6 x - 2 \log_6 y - 1$ **9. a.** $\log_5 32n^2$
b. $\log_b \frac{(a - 2)^3}{(a + 2)^4}$ **10. a.** 7 **b.** x **11. a.** $\frac{3}{2}$ **b.** -5 **12.** 2.3223 **13.** 2
14. 2 **15.** $\frac{1}{2}$ **16.** 11 **17.** 5 **18.** Approximately 1238 million (or
1.238 billion) people **19.** The intensity is 10^{-6} watts per square meter.
20. It takes about 11 hr.

Chapter 9 Cumulative Review, pp. 746–747

1. $\frac{9}{5}$ **2.** $(-\infty, -\frac{8}{5}) \cup (2, \infty)$ **3.** Point–slope form:
$y - 1 = -(x - 4)$; slope-intercept form: $y = -x + 5$
4.

5. $(3xy + 2)(3xy - 2)(9x^2y^2 + 4)$
6. $-2, \frac{3}{5}$ **7.** $k = 2; y = \frac{2z}{x^2}$
8. $(5x^2 + 3)\sqrt{2x}$ **9.** $24 - 32i$
10. $-4, 7$ **11.** $\frac{-2 + \sqrt{14}}{2}, \frac{-2 - \sqrt{14}}{2}$

12. $(-\infty, \frac{1}{3}] \cup [2, \infty)$ **13.** 16

14. $\log_3 \left(\frac{a + 2}{a - 2} \right)$ **15.** $\frac{1}{5}$ **16.** The windchill temperature is
approximately 31°F. **17. a.** $R = r\sqrt{2}$ **b.** 42 mm **18. a.** The flare
will take approximately 5 sec to reach its maximum height. **b.** The maximum height of the flare is 500 ft. **19.** 53 children and 14 adult chaperones **20. a.** $d(x) = 0.8x$ **b.** $s(x) = 1.06x$ **c.** $(s \circ d)(x) = 1.06(0.8x)$,
or $(s \circ d)(x) = 8.48x$

ANSWERS

CHAPTER 10

Chapter 10   Pretest, pp. 749–750

1. $y = 3(x - 2)^2 + 11; (2, 11)$ **2. a.** $2\sqrt{5}$ units **b.** 0.5 unit
3. a. $(4, 2)$ **b.** $(\frac{1}{2}, -\frac{3}{2})$ **4.** $(x - 7)^2 + (y + 3)^2 = 10$
5. $(1, 0); \sqrt{6}$

6.

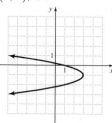

7.

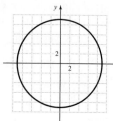

8.

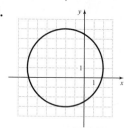

9.

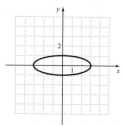

10.

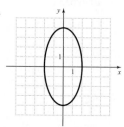

11.

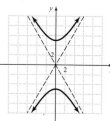

12.

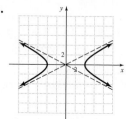

13. $(-3, 0), (0, 3)$
14. $(-i\sqrt{2}, -4), (i\sqrt{2}, -4)$
15. $(-\sqrt{6}, -1), (-\sqrt{6}, 1),$
$(\sqrt{6}, -1), (\sqrt{6}, 1)$

16.

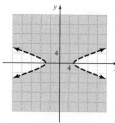

17.

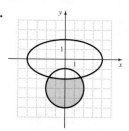

18. $x = \frac{8}{225}y^2$ **19.** The minimum distance is about 50 million km.
20. a. $x^2 + y^2 \le 1,000,000$
b.

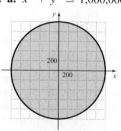

Section 10.1   Practices, pp. 752–757

1. p. 752:

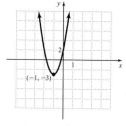

2. p. 754:   a. $y = 4(x - 1)^2 + 3$
b. Vertex: $(1, 3)$; axis of symmetry:
$x = 1$
c.

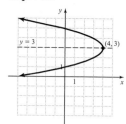

3. p. 756:

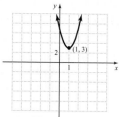

4. p. 757: $x = -5(y + 2)^2 + 11; (11, -2)$
5. p. 757:   a. $h = -16(t - 2)^2 + 64$
b. 48 ft

Exercises 10.1, pp. 759–763

1. Upward or downward   **3.** To the left or to the right
5. Vertex: $(3, -4)$; axis of symmetry: $x = 3$

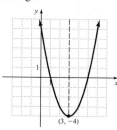

7. Vertex: $(-2, -5)$;
axis of symmetry: $x = -2$

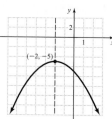

9. Vertex: $(0, 1)$; axis of
symmetry: $y = 1$

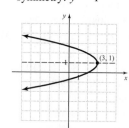

11. Vertex: $(3, 1)$; axis of
symmetry: $y = 1$

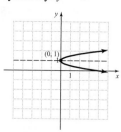

13. Vertex: $(3, 8)$; axis of
symmetry: $x = 3$

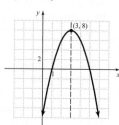

15. Vertex: $(-2, -2)$; axis of symmetry: $y = -2$

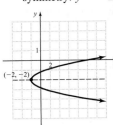

17. $y = (x - \frac{1}{2})^2 + \frac{3}{4}$; $(\frac{1}{2}, \frac{3}{4})$

19. $y = 5(x - 5)^2 - 68$; $(5, -68)$

21. $x = -2(y + 8)^2 + 33$; $(33, -8)$

23. $x = 3(y + \frac{3}{2})^2 + \frac{17}{4}$; $(\frac{17}{4}, -\frac{3}{2})$

25. $y = \frac{1}{5}(x + 2)^2 + 7$

27. $x = -4y^2 + 4$

29. $y = (x + \frac{5}{2})^2 - \frac{17}{4}$; $(-\frac{5}{2}, -\frac{17}{4})$ **31.** $y = \frac{1}{3}(x - 2)^2 - 5$

33. Vertex: $(3, -1)$; axis of symmetry: $y = -1$

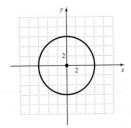

35. $(40, 1600)$; it shows that the company will have maximum revenue of $1600 when 40 bottles of perfume are sold.
37. a. $A = -(w - 125)^2 + 15{,}625$
b. Dimensions of 125 ft by 125 ft will produce a field with a maximum area of 15,625 ft². **39.** $x = \frac{3}{200}y^2$

Section 10.2   Practices, pp. 765–770

1. p. 765: $\sqrt{10} \approx 3.2$ units   **2. p. 766:** $(5, \frac{5}{2})$

3. p. 767:

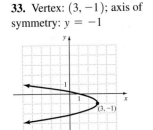

4. p. 768:   a. Center: $(-6, 2)$; radius: 10
b.

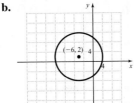

5. p. 768: $(x - 7)^2 + (y + 1)^2 = 8^2$, or $(x - 7)^2 + (y + 1)^2 = 64$
6. p. 769: Center: $(2, 1)$; radius: $2\sqrt{6}$ **7. p. 769:**
a. $(x - 30)^2 + (y - 40)^2 = 10^2$, or $(x - 30)^2 + (y - 40)^2 = 100$
b. No, since $\sqrt{509} \approx 22.6 > 10$

Exercises 10.2, pp. 771–775

1. the sum of   **3.** circle   **5.** centered at   **7.** 10 units   **9.** $2\sqrt{5} \approx 4.5$ units   **11.** $\sqrt{13} \approx 3.6$ units   **13.** $\sqrt{53} \approx 7.3$ units   **15.** $\frac{5}{12}$ unit
17. $\sqrt{77} \approx 8.8$ units   **19.** $(6, 5)$ **21.** $(-\frac{9}{2}, 5)$ **23.** $(-\frac{21}{2}, -\frac{13}{2})$
25. $(2, -2.5)$ **27.** $(\frac{5}{24}, \frac{1}{20})$ **29.** $(0, 4\sqrt{6})$
31. Center: $(0, 0)$; radius: 6   **33.** Center: $(3, 2)$; radius: 1

35. Center: $(-2, 4)$; radius: 5

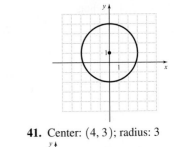

37. Center: $(0, 1)$; radius: 3

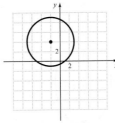

39. Center: $(-1, -1)$; radius: 5

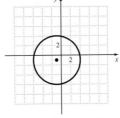

41. Center: $(4, 3)$; radius: 3

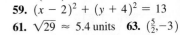

43. $(x + 7)^2 + (y - 2)^2 = 81$ **45.** $x^2 + (y - 4)^2 = 5$
47. $(x - 3)^2 + (y - 5)^2 = 32$ **49.** $(x + 2)^2 + (y - 6)^2 = 41$
51. $(-\frac{3}{2}, -2)$; $\frac{5}{2}$ **53.** $(5, -3)$; $\sqrt{38}$ **55.** $(2, 0)$; $2\sqrt{3}$
57. Center: $(4, -1)$; radius: 3

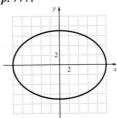

59. $(x - 2)^2 + (y + 4)^2 = 13$
61. $\sqrt{29} \approx 5.4$ units   **63.** $(\frac{5}{2}, -3)$
65. The distance between her home and her office is 15 mi.
67. a. $D(4, 3)$ **b.** The slope of $\overline{BD}$ is -1 and the slope of $\overline{AC}$ is 1. Since the slopes are negative reciprocals, $\overline{BD}$ is perpendicular to the base $\overline{AC}$ and is an altitude of the triangle.   **c.** 12 sq units
69. $x^2 + y^2 = 9$ **71. a.** $(x + 8)^2 + (y - 15)^2 = 36^2$, or $(x + 8)^2 + (y - 15)^2 = 1296$ **b.** Yes, since $\sqrt{853} \approx 29.2 < 36$

Section 10.3   Practices, pp. 777–782

1. p. 777:

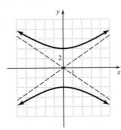

2. p. 777:

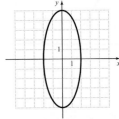

3. p. 778: The maximum width is 20 ft, and the maximum height of the arch is 12 ft.

4. p. 780:

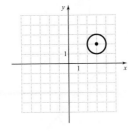

5. p. 781:

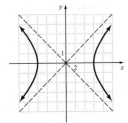

6. p. 781: The closest the comet comes to Earth is 90 million mi.

Exercises 10.3, pp. 783–787

1. ellipse **3.** hyperbola

5.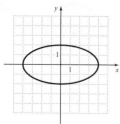

7.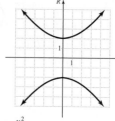

9.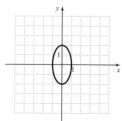

11.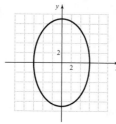

13.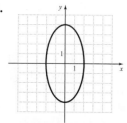

15.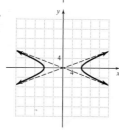

17.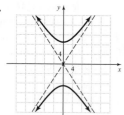

19.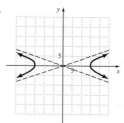

21.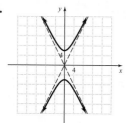

23. Circle **25.** Parabola **27.** Ellipse **29.** $\frac{x^2}{121} + \frac{y^2}{9} = 1$

31. $\frac{x^2}{7} + \frac{y^2}{8} = 1$ **33.** $\frac{x^2}{16} - \frac{y^2}{1} = 1$, or $\frac{x^2}{16} - y^2 = 1$

35.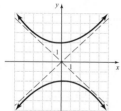

37. Ellipse **39.** $\frac{x^2}{45} + \frac{y^2}{3} = 1$

41.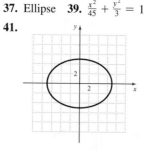

43. a. The length is 180 m and the width is 150 m. **b.** Yes

45. a. 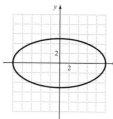 **b.** In Quadrant I, both the r- and R-coordinate of each point on the graph are nonnegative. Neither radius can be negative.

47. $\frac{x^2}{9} + \frac{y^2}{4} = 1$

Section 10.4 Practices, pp. 789–793

1. *p. 789:* $(-8, -6), (6, 8)$
2. *p. 790:* $(-2, 0), (2, 0), (-\sqrt{3}, -1), (\sqrt{3}, -1)$
3. *p. 791:* $(-1, -2\sqrt{2}), (-1, 2\sqrt{2}), (1, -2\sqrt{2}), (1, 2\sqrt{2})$
4. *p. 792:* $(-i, -2\sqrt{2}), (-i, 2\sqrt{2}), (i, -2\sqrt{2}), (i, 2\sqrt{2})$
5. *p. 792:* The two balls will be the same height above the ground $\frac{9}{4}$ sec, or 2.25 sec, after they are released.

Exercises 10.4, pp. 794–796

1. $(-1, -1), (3, 7)$ **3.** $(-3, -\sqrt{3}), (-3, \sqrt{3}), (2, -2\sqrt{2}), (2, 2\sqrt{2})$
5. $(-2\sqrt{2}, 3), (2\sqrt{2}, 3)$ **7.** $(0, -4), (4, 0)$ **9.** $(-2, -\frac{1}{2}), (12, 3)$
11. $(-2i, -6), (2i, -6), (-\sqrt{7}, 5), (\sqrt{7}, 5)$
13. $(-2\sqrt{3}, -\sqrt{2}), (-2\sqrt{3}, \sqrt{2}), (2\sqrt{3}, -\sqrt{2}), (2\sqrt{3}, \sqrt{2})$
15. $(-2, -2\sqrt{3}), (-2, 2\sqrt{3}), (2, -2\sqrt{3}), (2, 2\sqrt{3})$
17. $(-\sqrt{6}, -i), (-\sqrt{6}, i), (\sqrt{6}, -i), (\sqrt{6}, i)$
19. $(-4, -\sqrt{15}), (-4, \sqrt{15}), (4, -\sqrt{15}), (4, \sqrt{15})$
21. $(0, -4), (0, 4)$ **23.** $(-1, -2\sqrt{3}), (-1, 2\sqrt{3}), (1, -2\sqrt{3}), (1, 2\sqrt{3})$
25. $(-1, -\sqrt{10}), (-1, \sqrt{10}), (1, -\sqrt{10}), (1, \sqrt{10})$ **27.** $(2, 0)$
29. $(-3i, -6), (3i, -6), (-\sqrt{2}, 5), (\sqrt{2}, 5)$
31. $(-5, 0), (4, -3), (4, 3)$
33. $(-i\sqrt{2}, -\sqrt{5}), (i\sqrt{2}, -\sqrt{5}), (-i\sqrt{2}, \sqrt{5}), (i\sqrt{2}, \sqrt{5})$
35. They will have the same revenue if they sell 0 or 700 organizers.
37. The paths will intersect at $(\sqrt{5}, 1)$ and $(\sqrt{5}, -1)$. **39.** The smaller square has dimensions 11 in. by 11 in., in contrast to 14 in. by 14 in. for the larger square.

Section 10.5 Practices, pp. 797–801

1. *p. 797:*

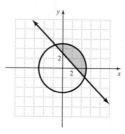

2. *p. 798:*

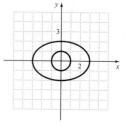

3. *p. 799:*

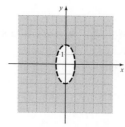

4. *p. 800:* No real solutions

5. *p. 801:*

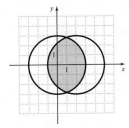

Exercises 10.5, pp. 802–806

1. broken   **3.** each

5.

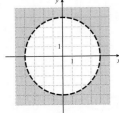

7.

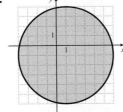

9.

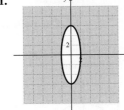

11.

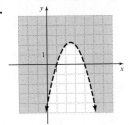

13.

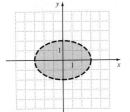

15.

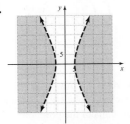

17.

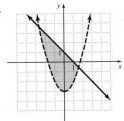

19.

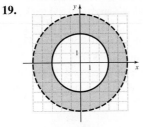

21.

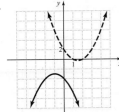

23.

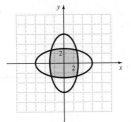

25.

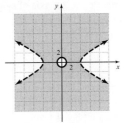

27. No real solutions

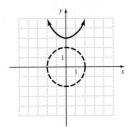

29.

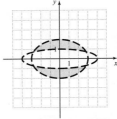

31.

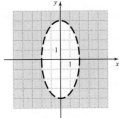

33.

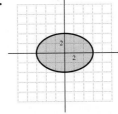

35. a. $x^2 + y^2 \le 2500$; $y \ge x + 2$ **37.**

b.

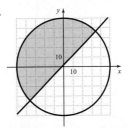

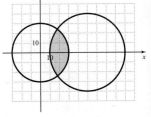

c. No, only points in Quadrant I are possible solutions since the lengths of the gardens cannot be negative.

Chapter 10   Review Exercises, pp. 812–817

1. is; possible answer: it is in the form $y = (x - h)^2 + k$ **2.** is; possible answer: it can be written in the form $(x - h)^2 + (y - k)^2 = r^2$

3. is; possible answer: it can be written in the form $\dfrac{x^2}{a^2} + \dfrac{y^2}{b^2} = 1$

4. does not; possible answer: the coefficient of the x^2-term is negative

5. does; possible answer: it is a hyperbola   **6.** is not; possible answer: according to the distance formula, it is $\sqrt{50} \approx 7.07$

7. Vertex: $(-3, 5)$; axis of symmetry: $x = -3$

8. Vertex: $(-10, 2)$; axis of symmetry: $y = 2$

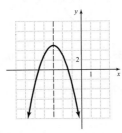

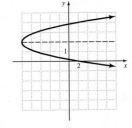

9. Vertex: $(0, -4)$; axis of symmetry: $y = -4$

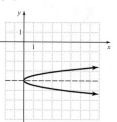

10. Vertex: $(-3, 2)$; axis of symmetry: $x = -3$

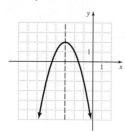

11. $y = 8\left(x - \frac{7}{2}\right)^2 - 24$; $\left(\frac{7}{2}, -24\right)$ **12.** $x = 5\left(y + \frac{3}{2}\right)^2 - \frac{17}{4}$; $\left(-\frac{17}{4}, -\frac{3}{2}\right)$

13. $3\sqrt{5} \approx 6.7$ units   **14.** $\sqrt{37} \approx 6.1$ units   **15.** 1.3 units

16. $\sqrt{95} \approx 9.7$ units   **17.** $(4, 13)$ **18.** $\left(\frac{3}{2}, -9\right)$ **19.** $\left(\frac{7}{8}, \frac{1}{4}\right)$

20. $(-0.7, -3.65)$

21. Center: $(0, 0)$; radius: 8

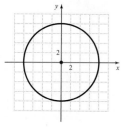

22. Center: $(-2, 4)$; radius: 2

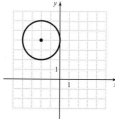

23. Center: $(2, -3)$; radius: 5

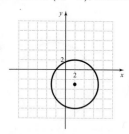

24. Center: $(-4, 1)$; radius: 3

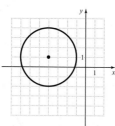

25. $(x - 9)^2 + y^2 = 169$ **26.** $(x + 6)^2 + (y - 10)^2 = 20$

27. Center: $(6, 7)$; radius: $2\sqrt{30}$ **28.** Center: $(-8, -5)$; radius: $4\sqrt{5}$

29.

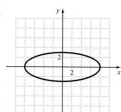

30.

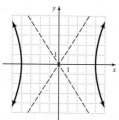

31.

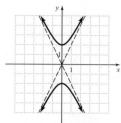

32.

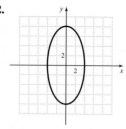

33. $(-1, 3), (2, 9)$ **34.** $(-1, -3), (1, 3)$ **35.** $(-4, 0), (0, -2), (0, 2)$

36. $\left(-\sqrt{2}, -\sqrt{6}\right), \left(-\sqrt{2}, \sqrt{6}\right), \left(\sqrt{2}, -\sqrt{6}\right), \left(\sqrt{2}, \sqrt{6}\right)$

37. $\left(-\sqrt{5}, -i\sqrt{7}\right), \left(-\sqrt{5}, i\sqrt{7}\right), \left(\sqrt{5}, -i\sqrt{7}\right), \left(\sqrt{5}, i\sqrt{7}\right)$

38. $(-3, 0), (3, 0)$

39.

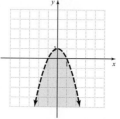

40.

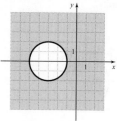

41.

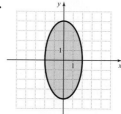

42.

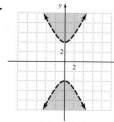

43.

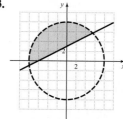

44.

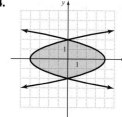

45. No real solutions

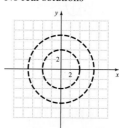

46.

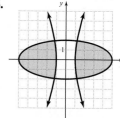

47. a. $R = -0.5(x - 20)^2 + 200$ **b.** $(20, 200)$; it shows that the manufacturing plant will make a maximum revenue of $200 when it sells 20 units of the product.   **48. a.** $y = -\frac{1}{32}x^2 + 50$ **b.** The height of the arch 24 ft from the center is 32 ft.   **49.** The distance between the student's apartment and work is $4\sqrt{5}$ mi, or approximately 8.9 mi.

50. $x^2 + y^2 = 60^2$, or $x^2 + y^2 = 3600$ **51.** $x^2 + y^2 = 4500^2$, or $x^2 + y^2 = 20,250,000$ **52.** The maximum distance is approximately 5 billion km.   **53.** $\frac{x^2}{4^2} + \frac{y^2}{2^2} = 1$, or $\frac{x^2}{16} + \frac{y^2}{4} = 1$ **54.** They will be at the same height $\frac{5}{4}$ sec, or 1.25 sec, after they are released.

55. a. $x^2 + y^2 > 1$
b.

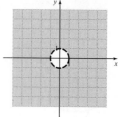

56. a.

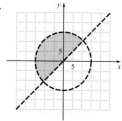

b. No, only points in Quadrant I are possible solutions since the lengths of the sandboxes must be nonnegative quantities.

Chapter 10 Posttest, pp. 818–819

1. $x = -2(y - 4)^2 - 59$; $(-59, 4)$ **2. a.** $2\sqrt{2}$ units **b.** 13 units **3. a.** $(-6, -3)$ **b.** $(-\frac{9}{2}, 5)$ **4.** $(x + 2)^2 + (y - 8)^2 = 18$ **5.** $(-5, 1)$; $2\sqrt{3}$

6.

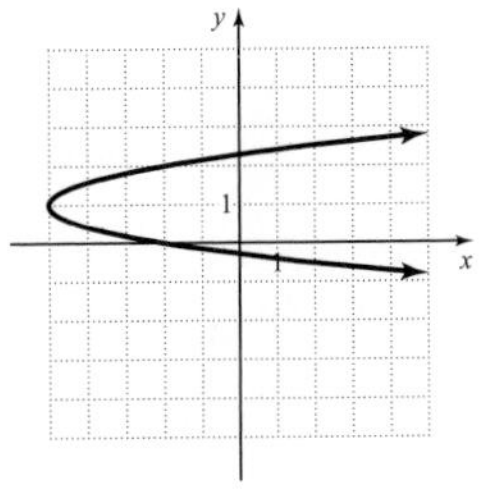

7.

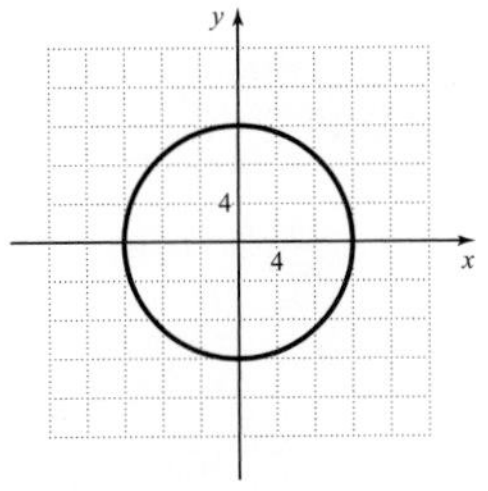

8.

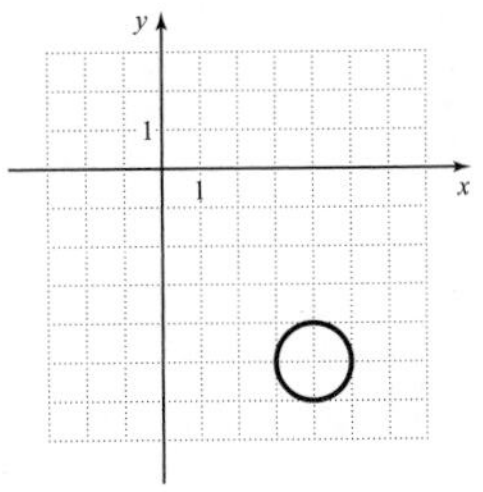

9.

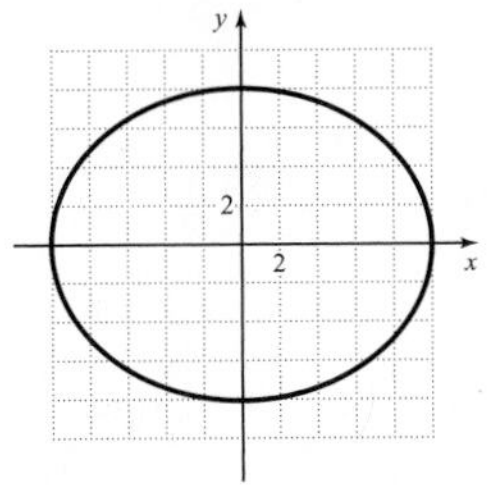

10.

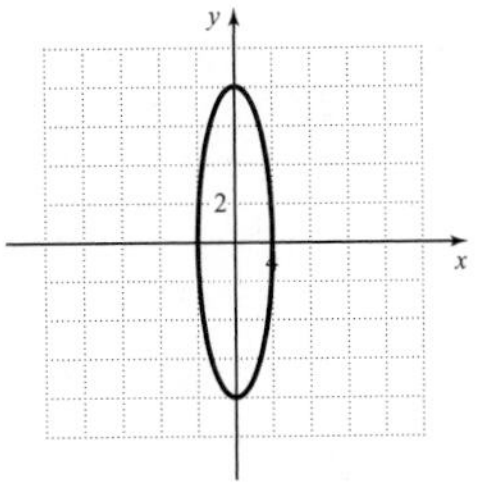

11.

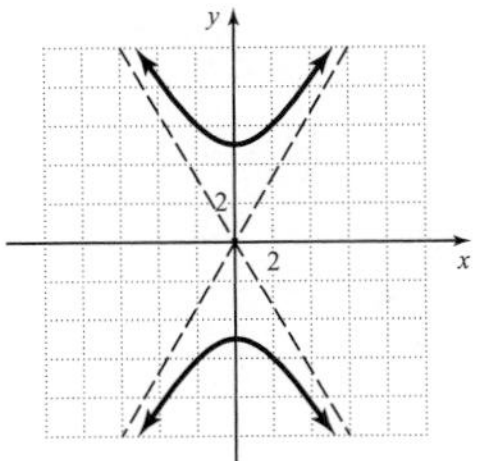

12.

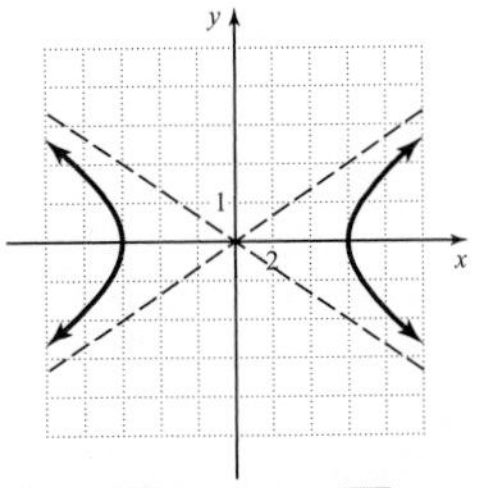

13. $(-5, 0)$, $(0, 5)$

14. $(-\sqrt{10}, 14)$, $(\sqrt{10}, 14)$

15. $(-i, -2\sqrt{3})$, $(-i, 2\sqrt{3})$, $(i, -2\sqrt{3})$, $(i, 2\sqrt{3})$

16.

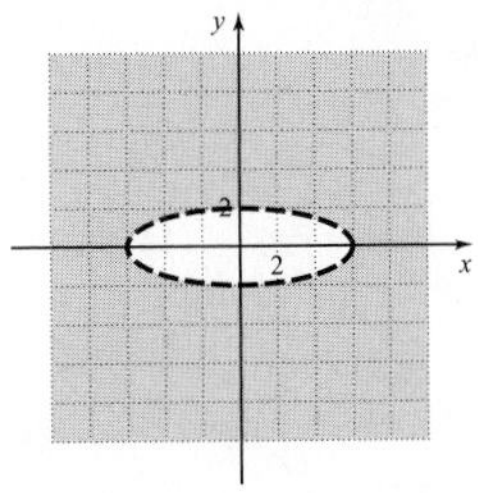

17.

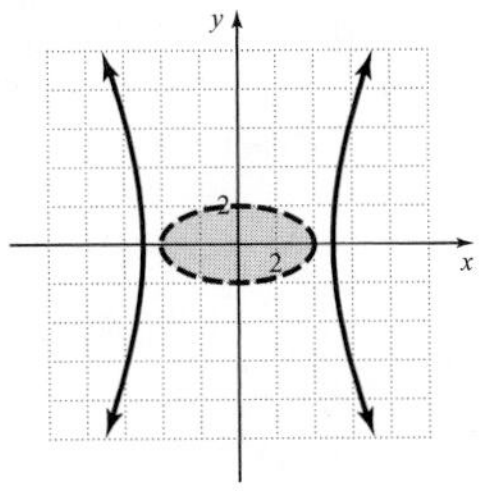

18. $y = \frac{6}{11{,}250}x^2 + 2$ **19.** $\frac{x^2}{65^2} + \frac{y^2}{60^2} = 1$, or $\frac{x^2}{4225} + \frac{y^2}{3600} = 1$

20. a.

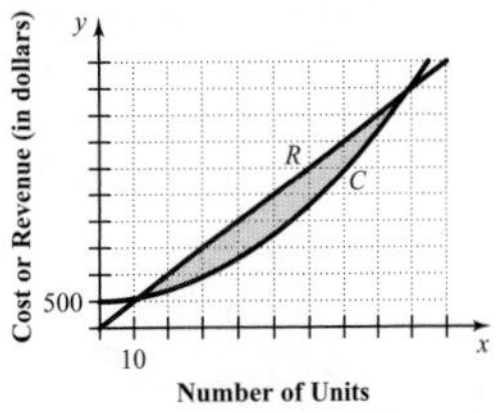

b. No, only points with a positive-integer x-coordinate are possible solutions since it does not make sense to produce and sell a fraction of a machine part.

Chapter 10 Cumulative Review, pp. 820–821

1. 70 **2.** $y = -\frac{4}{3}x + 16$ **3.** $2x^3 - 11x^2y + 12xy^2 + 9y^3$ **4.** $(3m + 2n)(6m^2 + n^2)$ **5.** $\dfrac{1}{x^2 - 5x + 25}$ **6.** $\dfrac{w}{6z^3}\sqrt[3]{7}$ **7.** 2 **8.** $-16 + 12i$ **9.** Inverses **10.** $-\frac{20}{3}$ **11.** $y = 2(x + 3)^2 + 5$; $(-3, 5)$ **12.** Center: (5, 2); radius: 2 **13.** A hyperbola **14.** $(\sqrt{3}, 3\sqrt{2})$, $(\sqrt{3}, -3\sqrt{2})$, $(-\sqrt{3}, 3\sqrt{2})$, $(-\sqrt{3}, -3\sqrt{2})$ **15.** $0.8x^3 + 6.8x^2 - 20.6x + 21.9$ **16.** She must invest at least \$10,000 in the fund with a guaranteed rate of return of 4% and \$20,000 in the fund that has a guaranteed rate of return of $5\frac{1}{2}$%. **17.** The average speed during the first part of the trip was 64 mph, and the average speed during the second part of the trip was 56 mph. **18. a.** $y = -\dfrac{7}{1125}(x - 75)^2 + 35$ **b.** 29.4 ft **19.** It would take approximately 10 yr. **20. a.** $S = 2x^2 + 8x$ **b.** The side of the base must be less than or equal to 4 in.

APPENDIX

Appendix A.3 Practices, pp. A-5–A-10

1, ***p. A-5:*** 1 **2,** ***p. A-7:*** $(1, -3)$ **3,** ***p. A-8:*** 9 **4,** ***p. A-9:*** $(2, 0, 3)$

Exercises A.3, p. A-11

1. -7 **3.** 10 **5.** 0 **7.** $(2, -1)$ **9.** $(3, 1)$ **11.** $(\frac{1}{2}, -\frac{1}{4})$ **13.** -36 **15.** 20 **17.** $(3, -2, -4)$ **19.** $(-1, 3, 5)$

Appendix A.4 Practices, p. A-13

1, ***p. A-13:*** $5x + 12 + \frac{35}{x - 3}$ **2,** ***p. A-13:*** $x^2 - x + 5 - \frac{18}{x + 5}$

Exercises A.4, p. A-14

1. $x + 3 + \frac{5}{x - 2}$ **3.** $x - 8 + \frac{17}{x + 3}$ **5.** $5x - 8 + \frac{10}{x + 1}$ **7.** $3x + 1$ **9.** $3x + 4 + \frac{16}{x - 4}$ **11.** $4x^2 + 5x + 2 + \frac{7}{x - 1}$ **13.** $-y^2 + 5y - 8$ **15.** $x^2 + 2x - 1 - \frac{2}{x - 2}$ **17.** $x^3 + 5 - \frac{15}{x + 3}$ **19.** $2x^2 + 8x + 2 + \frac{4}{x - 0.5}$

Glossary

The numbers in brackets following each glossary term represent the section in which that term is discussed.

absolute value [1.1] The absolute value of a number is its distance from 0 on a number line. The absolute value of the number n is written $|n|$.

absolute value function [3.6] A function in the form $f(x) = |x|$.

ac method [5.5] A method for factoring a trinomial of the form $ax^2 + bx + c$, which is based on grouping.

addition property of equality [2.1] This property states that when adding any real number to each side of an equation, the result is an equivalent equation.

additive identity property [1.3] The additive identity property states that the sum of any number and zero is the original number. For any real number a, $a + 0 = a$ and $0 + a = a$.

additive inverse property [1.3] The additive inverse property states that the sum of any number and its opposite is zero. For any real number a, there is exactly one number, $-a$, such that $a + (-a) = 0$ and $-a + a = 0$.

additive inverses [1.1] Two real numbers that are the same distance from 0, but on opposite sides of 0, on a number line are called additive inverses. For any real number a, the additive inverse of a is written $-a$.

algebraic expression [1.5] An expression in which constants and variables are combined using standard arithmetic operations.

altitude of a triangle [10.2] A line segment though a vertex that is perpendicular to the opposite side of the triangle.

associative property of addition [1.3] The associative property of addition states that when adding three numbers, their sum is the same regardless of how they are grouped. For any real numbers a, b and c, $(a + b) + c = a + (b + c)$.

associative property of multiplication [1.3] The associative property of multiplication states that when multiplying three numbers, their product is the same regardless of how they are grouped. For any real numbers a, b and c, $(a \cdot b) \cdot c = a \cdot (b \cdot c)$.

axis of symmetry [8.4] The vertical line that passes through the vertex of a parabola.

base [1.4] The base is the number that is a repeated factor when written with an exponent.

binomial [5.1] A binomial is a polynomial with two terms.

break-even point [4.1] In a business, the point at which revenue (income) equals cost (expenses).

center of a circle [10.2] The given point that is a fixed distance from all the points on a circle.

center of a hyperbola [10.3] The point halfway between the two foci.

center of an ellipse [10.3] The point halfway between the two foci.

change-of-base formula [9.5] The change-of-base formula states that for positive real numbers N, b, and c, where $b \neq 1$ and $c \neq 1$, $\log_b N = (\log_c N)/(\log_c b)$.

circle [10.2] A circle is the set of all points on a coordinate plane that are a fixed distance from a given point.

coefficient [1.5, 5.1] The numerical factor of a variable term; for example, in the expression $5x$, 5 is called the coefficient.

columns of a matrix [4.4] The vertical elements in a matrix.

common logarithm [9.5] A logarithm to the base 10, where $\log x$ means $\log_{10} x$.

commutative property of addition [1.3] The commutative property of addition states that the sum of two numbers is the same, regardless of order. For any real numbers a and b, $a + b = b + a$.

commutative property of multiplication [1.3] The commutative property of multiplication states that the product of two numbers is the same, regardless of order. For any real numbers a and b, $a \cdot b = b \cdot a$.

completing the square [8.1] A method for solving quadratic equations in which we transform one side of an equation to a perfect square trinomial.

complex conjugates [7.6] The complex numbers $a + bi$ and $a - bi$.

complex number [7.6] A number that can be written in the form $a + bi$, where a and b are real numbers and $i = \sqrt{-1}$.

complex rational expression (complex algebraic fraction) [6.3] A rational expression whose numerator, denominator, or both contain one or more rational expressions.

composition of f and g [9.1] If f and g are functions, then the composition of f and g is the function, represented by $f \circ g$, where $(f \circ g)(x) = f(g(x))$.

compound inequality [2.4] Two inequalities that are joined by the word *and* or the word *or* form a compound inequality.

conic section [10.1] A curve formed by the intersection of a plane and a (double) cone.

conjugate [7.4] The conjugate of $a + b$ is $a - b$.

consistent system [4.1] A system of linear equations with at least one solution.

constant function [3.6] A linear function defined by $f(x) = k$, where k is a real number.

constant of variation (constant of proportionality) [6.5] In the equations, $y = kx$, $y = k/x$ and $y = kxz$, k is called the constant of variation or the constant of proportionality.

constant term [5.1] In a polynomial, the constant term is the term of degree 0.

coordinate plane [3.1] The flat surface on which we draw graphs.

cross-product method [6.4] A method for solving a proportion in which the cross products are set equal to each other.

cube root [7.1] The number b is the cube root of a if $b^3 = a$, for any real numbers a and b.

degree of a monomial [5.1] The degree of a monomial in one variable is the power of the variable in the monomial.

degree of a polynomial [5.1] The highest degree of any of its terms.

dependent equations [4.1] Equations in a linear system that have infinitely many solutions.

difference of squares [5.6] An expression in the form $a^2 - b^2$, which can be factored as $(a + b)(a - b)$.

difference of two functions [9.1] If f and g are functions, then the difference of two functions is $(f - g)(x) = f(x) - g(x)$.

difference of two perfect cubes [5.6] An expression in the form $a^3 - b^3$, which can be factored as $(a - b)(a^2 + ab + b^2)$.

direct variation [6.5] Direct variation occurs if a relationship between two variables is described by an equation of the form $y = kx$, where k is a positive constant.

direct variation equation [6.5] The equation $y = kx$.

discriminant [8.2] The expression $b^2 - 4ac$ in the quadratic formula that allows us to determine the number and type of solutions to a quadratic equation without actually solving the equation.

distance formula [7.2, 10.2] The formula, $d = \sqrt{(x_2 - x_1)^2 + (y_2 - y_1)^2}$. In words, this formula states the distance between two points on the coordinate plane is the square root of the sum of the squares of the difference of the x-values and the difference of the y-values.

distributive property [1.3] The distributive property states that a number times the sum of two quantities is equal to the number times one quantity plus the number times the other quantity. For any real numbers a, b, and c, $a \cdot (b + c) = a \cdot b + a \cdot c$.

domain [3.6, 8.4] The domain of a function is the set of all values of the independent variable.

element [1.1] Each object in a set.

elements (entries) of a matrix [4.4] The numbers of a matrix.

elimination (or addition) method [4.2] An algebraic method used to solve a system of linear equations in two variables that is based on the following property of equality: If $a = b$ and $c = d$, then $a + c = b + d$.

ellipse [10.3] The set of all points on a coordinate plane the *sum* of whose distances from two fixed points is constant.

empty set [1.1] A set with no elements.

equation [2.1] A mathematical statement that two expressions are equal.

equation of a circle centered at (h, k) [10.2] The equation in standard form of a circle with center (h, k) and radius r is $(x - h)^2 + (y - k)^2 = r^2$.

equation of a circle centered at the origin [10.2] The equation of a circle with center $(0, 0)$ and with radius r is $x^2 + y^2 = r^2$.

equation of a hyperbola centered at the origin [10.3] The equation in standard form of a hyperbola with center $(0, 0)$ and x-intercepts $(-a, 0)$ and $(a, 0)$, is $\dfrac{x^2}{a^2} - \dfrac{y^2}{b^2} = 1$. The equation in standard form of a hyperbola with center $(0, 0)$ and y-intercepts $(0, -b)$ and $(0, b)$, is $\dfrac{y^2}{b^2} - \dfrac{x^2}{a^2} = 1$.

equation of an ellipse centered at the origin [10.3] The equation in standard form of an ellipse with center $(0, 0)$, x-intercepts $(-a, 0)$ and $(a, 0)$, and y-intercepts $(0, -b)$ and $(0, b)$ is $\dfrac{x^2}{a^2} + \dfrac{y^2}{b^2} = 1$.

equation of a parabola [10.1] The equation in standard form of a parabola that opens upward or downward is $y = a(x - h)^2 + k$, where (h, k) is the vertex of the associated parabola and the equation of the axis of symmetry is $x = h$.

(continued)

The equation in standard form of a parabola that opens to the left or right is $x = a(y - k)^2 + h$, where (h, k) is the vertex of the associated parabola and the equation of the axis of symmetry is $y = k$.

equivalent equations [2.1] Equations that have the same solution.

equivalent matrix [4.4] A matrix corresponding to a system of equations equivalent to the original system.

exponent (or power) [1.4] A number that indicates how many times another number is multiplied by itself.

exponential function [9.2] Any function that can be written in the form $f(x) = b^x$, where x is a real number, $b > 0$, and $b \neq 1$.

exponential notation [1.4] A shorthand method for representing repeated multiplication of the same factor.

extraneous solutions [6.4] Extraneous solutions are *not* solutions of the original equation.

factoring by grouping [5.4] When trying to factor a polynomial that has four terms, it may be possible to group pairs of terms in such a way that a common binomial factor can be found. This method is called factoring by grouping.

finite set [1.1] A set whose elements can be counted.

focus (plural, foci) [10.3] The fixed point or points on the graph of an ellipse or a hyperbola.

FOIL method [5.2, 7.6] A method for multiplying two binomials: Multiply First terms, Outer terms, Inner terms, and Last terms, then combine like terms.

formula [2.2] A special type of literal equation that provides a symbolic description of some real-world object or action.

function [3.6] A relation in which no two ordered pairs have the same first coordinates.

graph [3.3] The graph of a linear equation in two variables consists of all points whose coordinates make the equation true.

greatest common factor (GCF) [5.4] The greatest common factor (GCF) of two or more integers is the largest integer that is a factor of each integer.

greatest common factor (GCF) of two or more monomials [5.4] The product of the greatest common factor of the coefficients and the highest powers of the variables common to all of the monomials.

horizontal-line test [9.1] The horizontal-line test states that if every horizontal line intersects the graph of a function at most once, then the function is one-to-one.

hyperbola [10.3] The set of all points on a coordinate plane that satisfy the following property: The *difference* of the distances from two fixed points is constant.

hypotenuse [5.7] In a right triangle, the side opposite the 90° angle.

imaginary number [7.6] The imaginary number i is $\sqrt{-1}$; that is, $i^2 = -1$.

imaginary part [7.6] In the complex number $a + bi$, b is called the imaginary part.

inconsistent system [4.1] A system of linear equations that has no solution.

independent equations [4.1] Equations in a linear system that have one or no solutions.

index [7.1] In the radical $\sqrt[n]{a}$, n is called the index of the radical.

inequality [2.3] A mathematical statement containing $<, \leq, >, \geq,$ or $\neq$.

inequality symbols [1.1] The symbols $\neq, <, \leq, >,$ and $\geq$, which are used to compare numbers.

infinite set [1.1] A set whose elements go on without end.

integers [1.1] The integers are the numbers . . . . $-4, -3, -3, -1, 0, +1, +2, +3, +4, \ldots$ continuing infinitely in both directions.

intersection [1.1] The intersection of sets A and B, written $A \cap B$, is the set of all elements that are in *both* A and B.

inverse of a one-to-one function f [9.1] The inverse of a one-to-one function f, written f^{-1} and read "f inverse," is the set of all ordered pairs (y, x) where f is the set of ordered pairs (x, y).

inverse variation [6.5] Inverse variation occurs if a relationship between two variables is described by an equation of the form $y = k/x$, where k is a positive constant.

inverse variation equation [6.5] The equation $y = k/x$.

irrational numbers [1.1, 7.1] Real numbers that cannot be written as the quotient (or ratio) of two integers.

joint variation [6.5] Joint variation occurs if a relationship among three variables is described by an equation of the form $y = kxz$, where k is a positive constant.

joint variation equation [6.5] The equation $y = kxz$.

leading coefficient [5.1] The coefficient of the leading term in a polynomial.

leading term [5.1] The leading term of a polynomial is the term in the polynomial with the highest degree.

least common denominator (LCD) of rational expressions [6.2] The LCD of rational expressions is the product of the different factors in the denominators, where the power of each factor is the greatest number of times that it occurs in any single denominator.

legs [5.7] In a right triangle, the two sides that form the 90° angle.

like radicals [7.3] Radicals that have the same index and also the same radicand.

like terms [1.5] Terms that have the same variables and the same exponents.

linear equation in one variable [2.1] An equation that can be written in the form $ax + b = c$, where a, b, and c are real numbers and $a \neq 0$.

linear equation in two variables [3.3] An equation that can be written in the *general form* $Ax + By = C$, where A, B, and C are real numbers and A and B are not both 0.

linear inequality in two variables [3.5] An inequality that can be written in the form $Ax + By < C$, where A, B, and C are real numbers and A and B are not both 0. The inequality symbol can be $<$, $>$, $\leq$, or $\geq$.

literal equation [2.2] An equation describing the relationship between two or more variables.

linear function [3.6] The function $f(x) = mx + b$, where m and b are real numbers.

logarithmic function [9.3] The inverse of $f(x) = b^x$ is represented by the logarithmic function $f^{-1}(x) = \log_b x$.

matrix [4.4] A rectangular array of numbers.

maximum [8.4] A parabola that opens downward has a highest point with a maximum y-value occurring at the vertex.

member [1.1] Each object in a set.

midpoint formula [10.2] The line segment with endpoints (x_1, y_1) and (x_2, y_2) has as its midpoint, $\left(\dfrac{x_1 + x_2}{2}, \dfrac{y_1 + y_2}{2} \right)$.

minimum [8.4] A parabola that opens upward has a lowest point with a minimum y-value occurring at the vertex.

monomial [5.1] An expression that is the product of a real number and variables raised to nonnegative integer powers.

multiplication property of equality [2.1] This property states that when multiplying each side of an equation by any real number, the result is an equivalent equation.

multiplicative identity property [1.3] The multiplicative identity property states that the product of any number and one is the original number. For any real number a, $a \cdot 1 = a$ and $1 \cdot a = a$.

multiplicative inverse property [1.3] The multiplicative inverse property states that the product of a number and its multiplicative inverse is one. For any nonzero real number a, there is exactly one number $1/a$ such that $a \cdot 1/a = 1$ and $1/a \cdot a = 1$.

multiplicative inverses [1.2] Two real numbers whose product is 1. For any nonzero real number a, the multiplicative inverse of a is written $1/a$.

multiplication property of zero [1.3] The multiplication property of zero states that the product of any number and zero is zero. For any real number a, $a \cdot 0 = 0$ and $0 \cdot a = 0$.

natural exponential function [9.2] The function defined by $f(x) = e^x$ is called the natural exponential function.

natural logarithm [9.5] A logarithm to the base e, where $\ln x$ means $\log_e x$.

negative number [1.1] A number to the left of 0 on the number line.

negative slope [3.2] On a graph, the slope of a line falling from left to right.

nonlinear system of equations [10.4] A system of equations in which at least one equation is nonlinear.

nth root [7.1] The number b is the nth root of a if $b^n = a$, for any real number a and for any positive integer n greater than 1.

one-to-one function [9.1] A function f is said to be one-to-one if each y-value in the range corresponds to exactly one x-value in the domain.

one-to-one property of exponential functions [9.2] The one-to-one property of exponential functions states that for $b > 0$ and $b \neq 1$, if $b^x = b^n$, then $x = n$.

opposites [1.1] Two real numbers that are the same distance from 0, but on opposite sides of 0, on a number line are called opposites.

ordered pair [3.1] A pair of coordinates that represent a point in the coordinate plane.

ordered triple [4.3] A set of three numbers that represents a solution of a system of linear equations in three variables.

order of operations rule [1.2] The order of operations rule tells us in which order to carry out the operations in the expression so that its value is unambiguous.

origin [1.1, 3.1] On the number line, the point 0; in the coordinate plane, the point where the axes intersect, $(0, 0)$.

parabola [8.4] The graph of a quadratic function.

parallel lines [3.2] Two nonvertical lines are parallel if and only if their slopes are equal.

perfect cube [7.1] A rational number is said to be a perfect cube if it is the cube of another rational number.

perfect square [7.1] A nonnegative rational number is said to be a perfect square if it is the square of another rational number.

perfect square trinomial [5.6] A trinomial that can be factored as the square of a binomial, for example,

$$a^2 + 2ab + b^2 = (a + b)^2 \quad \text{and}$$
$$a^2 - 2ab + b^2 = (a - b)^2.$$

perpendicular bisector [10.2] The perpendicular bisector of a segment is the line perpendicular to the segment that intersects the segment at its midpoint.

perpendicular lines [3.2] Two nonvertical lines are perpendicular if and only if the product of their slopes is -1.

point-slope form [3.4] The point-slope form of a linear equation in two variables is written as $y - y_1 = m(x - x_1)$, where x_1, y_1 and m are real numbers. In this form, m is the slope, and (x_1, y_1) is a point that lies on the graph of the equation.

polynomial [5.1] An algebraic expression with one or more monomials added or subtracted.

polynomial function [5.1] A function in which the rule that defines the function is a polynomial.

positive number [1.1] A number to the right of 0 on the number line.

positive slope [3.2] On a graph, the slope of a line rising from left to right.

power property of logarithms [9.4] The power property of logarithms states that for any positive real numbers M and $b, b \neq 1$, and any real number r, $\log_b M^r = r \log_b M$.

prime polynomials [5.5] Polynomials that are not factorable.

principle square root [7.1] The square root of a number that is positive.

product of two functions [9.1] If f and g are functions, then the product of two functions is $(f \cdot g)(x) = f(x) \cdot g(x)$.

product property of logarithms [9.4] The product property of logarithms states that for any positive real numbers M, N, and $b, b \neq 1$, $\log_b MN = \log_b M + \log_b N$.

proportion [6.4] A statement that two ratios are equal.

Pythagorean theorem [5.7] The Pythagorean theorem states that for every right triangle, the sum of the squares of the lengths of the legs (a and b) equals the square of the length of the hypotenuse (c): $a^2 + b^2 = c^2$.

quadrant [3.1] One of four regions of a coordinate plane separated by the x- and y-axes.

quadratic equation (second-degree equation) [5.7, 8.1] An equation that can be written in the form $ax^2 + bx + c = 0$, where a, b, and c are real numbers and $a \neq 0$.

quadratic formula [8.2] The quadratic formula states that if $ax^2 + bx + c = 0$, where a, b and c are real numbers and $a \neq 0$, then $x = (-b \pm \sqrt{b^2 - 4ac})/2a$.

quadratic function [3.6, 8.4] A function of the form $f(x) = ax^2 + bx + c$.

quadratic inequality [8.5] A quadratic inequality in one variable is an inequality that contains a quadratic expression.

quadratic in form [8.3] An equation is quadratic in form if it can be rewritten in the form $au^2 + bu + c = 0$, where $a \neq 0$ and u represents an algebraic expression.

quotient of two functions [9.1] If f and g are functions, then the quotient of two functions is $(f/g)(x) = f(x)/g(x)$, for $g(x) \neq 0$.

quotient property of logarithms [9.4] A property that states for any positive real numbers M, N, and $b, b \neq 1$, $\log_b (M/N) = \log_b M - \log_b N$.

radical equation [7.5] An equation in which a variable appears in one or more radicands.

radical expression [7.1] An algebraic expression that contains a *radical*.

radical sign [7.1] The symbol $\sqrt{}$.

radicand [7.1] The number under a radical sign.

radius [10.2] The fixed distance from the center of a circle to any point on the circle.

range [3.6, 8.4] The range of a function is the set of all values of the dependent variable.

rate of change [3.2] Slope can be interpreted as a rate of change. It indicates how fast the graph of a line is changing and if it increases or decreases.

ratio [6.4] A comparison of two numbers or two quantities expressed as a quotient.

rational equation (fractional equation) [6.4] An equation that contains one or more rational expressions.

rational exponents [7.1] Exponents that are rational numbers.

rational expression [6.1] A rational expression, P/Q, is an algebraic expression that can be written as the quotient of two polynomials P and Q where $Q \neq 0$.

rational function [3.6, 6.1] A function in which the rule that defines the function is a rational expression. A rational function is of the form $f(x) = P(x)/Q(x)$, where $P(x)$ and $Q(x)$ are polynomials and $Q(x) \neq 0$.

rational inequality [8.5] An inequality that contains a rational expression.

rational number [1.1] A real number that can be written in the form a/b, where a and b are integers and $b \neq 0$.

rationalize the denominator [7.4] To rewrite the expression in an equivalent form that contains no radical in its denominator.

real numbers [1.1] Numbers that can be represented as points on a number line.

real part [7.6] In the complex number $a + bi$, a is called the real part.

reciprocals [1.2] Two real numbers whose product is 1. For any nonzero real number a, the reciprocal of a is written $1/a$.

relation [3.6] A set of ordered pairs.

rows of a matrix [4.4] The horizontal elements in a matrix.

scientific notation [1.4] A number is expressed in scientific notation if it is written in the form $a \times 10^n$, where n is an integer and a is greater than or equal to 1 but less than 10 ($1 \leq a < 10$).

second-degree equation (quadratic equation) [5.7] An equation that can be written in the form $ax^2 + bx + c = 0$, where a, b, and c are real numbers and $a \neq 0$.

set [1.1] A collection of objects.

slope [3.2] The ratio of the change in y-values to the change in x-values along a line. The slope m of a line passing through the points (x_1, y_1) and (x_2, y_2) is defined to be $m = (y_2 - y_1)/(x_2 - x_1)$, where $x_1 \neq x_2$.

slope-intercept form [3.4] A linear equation in two variables is in slope-intercept form if it is written as $y = mx + b$, where m and b are real numbers. In this form, m is the slope of the line and $(0, b)$ is the y-intercept of the graph of the equation.

solution of a linear inequality in two variables [3.5] An ordered pair of real numbers that when substituted for the variables makes the inequality a true statement.

solution of an equation [2.1] A value of the variable that makes the equation a true statement.

solution of an equation in two variables [3.3] An ordered pair of numbers that makes the equation true.

solution of an inequality [2.3] A solution of an inequality is any value of the variable that makes the inequality true.

solution of a nonlinear system of equations [10.4] A solution of a nonlinear system of equations in two variables is an ordered pair of numbers that makes each equation in the system true.

solution of a system of equations [4.1, 4.3] A solution of a system of equations in two variables is an ordered pair of numbers that makes both equations in the system true. A solution of a system of linear equations in three variables is an ordered triple of numbers that makes all three equations in the system true.

solution of a system of inequalities [4.5] A solution of a system of inequalities in two variables is an ordered pair of numbers that makes both inequalities in the system true.

solution of a system of nonlinear inequalities [10.5] A solution of a system of nonlinear inequalities is an ordered pair of real numbers that makes each inequality in the system true.

solve an inequality [2.3] To solve an inequality is to find all of its solutions.

square matrix [4.4] A matrix that has the same number of rows as columns.

square root [7.1] The number b is a square root of a if $b^2 = a$, for any real numbers a and b and for a nonnegative.

square root property of equality [8.1] If b is a real number and $x^2 = b$, then $x = \pm \sqrt{b}$, that is $x = \sqrt{b}$ or $x = -\sqrt{b}$.

substitution method [4.2] An algebraic method for solving a system of equations in which one linear equation is solved for one of the variables and then the result is substituted into the other equation.

sum of two functions [9.1] If f and g are functions, then the sum of two functions is $(f + g)(x) = f(x) + g(x)$.

sum of two perfect cubes [5.6] An expression in the form $a^3 + b^3$ that can be factored as $(a + b)(a^2 - ab + b^2)$.

system of equations [4.1] Two or more equations considered simultaneously, that is, together.

system of inequalities [4.5] Two or more inequalities considered simultaneously, that is, together.

term [1.5] A number, a variable, or the product or quotient of numbers and variables.

trial-and-error method [5.5] A method of factoring a trinomial that involves listing all the possible combinations of factors to find the correct factorization.

trinomial [5.1] A polynomial with three terms.

union [1.1] The union of sets A and B, written $A \cup B$, is the set of all elements that are in either A or B or both.

unlike terms [1.5] Terms that do not have the same variable or the variables are not raised to the same exponents.

variable [1.3] A quantity that is unknown, that is, one that can change or vary in value.

varies directly (directly proportional) [6.5] In a relationship between two variables, y varies directly as x if there is a positive constant k such that $y = kx$.

varies inversely (inversely proportional) [6.5] In a relationship between two variables, y varies inversely as x if there is a positive constant k such that $y = k/x$.

varies jointly (jointly proportional) [6.5] In a relationship among three variables, y varies jointly as x and z, if there is a positive constant k such that $y = kxz$.

vertex [8.4] The highest or lowest point of a parabola.

vertical line test [3.6] If any vertical line intersects a graph at more than one point, the graph does not represent a function. If no such line exists, then the graph represents a function.

whole numbers [1.1] The whole numbers consist of 0, 1, 2, 3, 4, 5, ...

x-axis [3.1] The horizontal number line in the coordinate plane.

x-coordinate [3.1] The first coordinate in an ordered pair that represents the horizontal distance in a coordinate plane.

x-intercept [3.3, 8.4] The x-intercept of a graph is the point where the graph intersects the x-axis.

y-axis [3.1] The vertical number line in the coordinate plane.

y-coordinate [3.1] The second coordinate in an ordered pair that represents the vertical distance in a coordinate plane.

y-intercept [3.3, 8.4] The y-intercept of a graph is the point where the graph intersects the y-axis.

zero-product property [5.7] A property that states if the product of two factors is zero, then either one or both of the factors must be zero.

Index

U.S. Customary Units

Length	Weight	Capacity	Time
12 in. = 1 ft	16 oz = 1 lb	16 fl oz = 1 pt	60 sec = 1 min
3 ft = 1 yd	2,000 lb = 1 ton	2 pt = 1 qt	60 min = 1 hr
5,280 ft = 1 mi		4 qt = 1 gal	24 hr = 1 day
			7 days = 1 wk
			52 wk = 1 yr
			12 mo = 1 yr
			365 days = 1 yr

Metric Units

Length	Weight	Capacity
1,000 mm = 1 m	1,000 mg = 1 g	1,000 mL = 1 L
100 cm = 1 m	1,000 g = 1 kg	1,000 L = 1 kL
1,000 m = 1 km		

Key U.S./Metric Conversions

Length	Weight	Capacity
1 in. ≈ 2.5 cm	1 oz ≈ 28 g	1 pt ≈ 470 mL
1 ft ≈ 30 cm	1 lb ≈ 450 g	1.1 qt ≈ 1 L
39 in. ≈ 1 m	2.2 lb ≈ 1 kg	1 gal ≈ 3.8 L
3.3 ft ≈ 1 m	1 ton ≈ 910 kg	
1 mi ≈ 1,600 m		
1 mi ≈ 1.6 km		